CALCULUS APPLIED TO THE REAL WORLD

CALCULUS
APPLIED
TO THE
REAL
WORLD

STEFAN WANER
Hofstra University

STEVEN R. COSTENOBLE
Hofstra University

HarperCollins*CollegePublishers*

Sponsoring Editor: Kevin M. Connors
Developmental Editors: Louise Howe and Lynn Mooney
Design Administrator: Jess Schaal
Text and Cover Design: Lesiak/Crampton Designs: Cindy Crampton
Cover Photo: Russell Phillips
Production Administrator: Randee Wire
Project Coordination: Elm Street Publishing Services, Inc.
Compositor: Interactive Composition Corporation
Printer and Binder: R. R. Donnelley & Sons Company
Cover Printer: Phoenix Color Corporation

Calculus Applied to the Real World

Library of Congress Cataloging-in-Publication Data

Waner, Stefan, 1949–
    Calculus applied to the real world / Stefan Waner, Steven R.
Costenoble.
        p.   cm.
    Includes index.
    ISBN 0-06-501824-9
    1. Calculus.   I. Costenoble, Steven R., 1961– . II. Title.
QA303.W358   1996
515—dc20   95–11117
                                                     CIP

95 96 97 98 9 8 7 6 5 4 3 2 1

To my parents, Ben and Mary Waner—Stefan Waner

To my parents, Earl and Kayla Costenoble, and to my wife, Nancy and my son, Alexander—Steven R. Costenoble

Contents ◄

Preface

This book is intended for a course in basic calculus for students majoring in business, the social sciences, or the liberal arts. It is designed to address the considerable challenge of generating enthusiasm and developing mathematical sophistication in an audience that is often ill-prepared for and disaffected by the traditional basic calculus courses offered on many college campuses.

This text is ambitious; we take the positive view that interested and motivated students can overcome any deficiencies in their mathematical background and attain a surprising degree of mathematical sophistication. The strong emphasis in this book on developing mathematical concepts and the abundance of relevant applications are dictated by this view. No less importantly, this book is one that a student whose primary interest is not mathematics can relate to and enjoy reading.

OUR APPROACH AND PHILOSOPHY

Our approach has been influenced by the current calculus reform movement. Within a framework of fairly traditional topics, we incorporate important features of the various calculus reform projects, including a thorough integration of graphing technology, a focus on real applications, and an emphasis on mathematical concepts through the extensive use of conceptual exercises and pedagogical techniques such as the Rule of Three (numerical, geometric, and algebraic approach to concepts). In fact, we include a fourth rule: verbal communication of mathematical concepts. We implement this element through verbalizing mathematical concepts, rephrasing sentences into forms that translate to mathematical statements, and writing exercises.

At the same time, we retain the strong features of more traditional texts, including an abundance of practice and drill exercises where appropriate, large numbers of applications to choose from, and inclusion of the standard topics in applied calculus.

In addition to combining what we view as the strongest features of reform and traditional texts, we have worked to create a unique and fresh approach to pedagogy and style.

EMPHASIS ON CONCEPTS

As we develop each mathematical concept, we steer the student directly toward the most important ideas with as few obstacles as possible. We do this by equipping the student with relevant skills and a working knowledge of a topic before considering its abstract foundations in detail. We avoid prematurely side-tracking the student with rigor and abstract subtleties; the appropriate time for abstraction and rigor is later, once the student has a firm grasp of the underlying concepts. Our approach enables students to learn mathematics in the most natural way—by example. Just as one learns to speak before mastering grammar or to play a musical instrument before studying harmony, so the understanding of new mathematical concepts is best established on a solid foundation of relevant skills rather than upon abstraction devoid of context.

So that learning of concepts by example will be effective, discussion and explanation of key concepts within each example must not obscure the mathematical simplicity of the solution. We address this issue in many examples with a separate *Before we go on* discussion following the solution. Here we explain, discuss, and explore the example's solution. We sometimes remind the student that an example is not always finished when a solution is found— we should check the solution and examine its implications.

More globally, the organization of material within each chapter and section has also been planned with conceptual development in mind. We present a new concept or an application of an old one as directly as possible, with many worked examples and references to actual data, even if it means postponement of some of the underlying theory. Once the concept has progressed to the point where the student is sufficiently comfortable with the material and can relate it to the real world, we return for a retroactive in-depth look at the foundations of what we have been doing and then develop the concept further, if need be. In this way, the theory reinforces the student's knowledge and view of the world and is not seen as meaningless abstraction. For example, we introduce the student to the derivative as the rate of change, along with formulas and real applications, before we formally discuss limits.

THOROUGH INTEGRATION OF GRAPHING TECHNOLOGY

The use of graphing calculators and computer software has been thoroughly integrated throughout the discussion, examples, and exercise sets, beginning with the first example of the graph of an equation in Chapter 1. In many examples we discuss how to use either graphing technology or computer spreadsheet software to solve the example; the sections are marked with the ▧ symbol. Groups of exercises for which the use of graphing calculators or computers is suggested or required are also fully integrated into the exercise sets and carry the ▧ symbol.

This focus on graphing technology plays an important conceptual and pedagogical role in the presentation of many topics. For example, Chapter 1 includes a section on the numerical solution of equations using a graphing calculator. Chapter 5's treatment of curve sketching was written with the graphing calculator in mind. It also follows the increasingly popular approach of using graphing calculators to draw the graphs and then using calculus to explain the results. As a result, our approach to curve sketching is more concise, less rigid, and also less long-winded than the standard treatments.

Appendix B: Using a Graphing Calculator introduces the student to a graphing calculator and provides programs and interesting applications. Appendix D: Using a Computer Spreadsheet shows how to use spreadsheet software. Discussion and exercises on proper entry of functions into graphing calculators and computers are included in the first section of Appendix A: Algebra Review.

Although we emphasize graphing technology throughout the text, we are mindful of the varying degree of emphasis on and use of graphing calculators in college courses. The text is not dependent on this technology. Students who are equipped with nothing more than a scientific calculator will not find themselves at a disadvantage.

FOCUS ON REAL APPLICATIONS

We are particularly proud of the diversity, breadth, and sheer abundance of examples and exercises that are based on real, referenced data from business, economics, the life sciences, and the social sciences. This focus on real data has helped create a text that students in diverse fields can relate to and that instructors can use to demonstrate the importance and relevance of calculus in the real world.

Our coverage of real applications begins with the very first example of Chapter 1, where the Dow Jones Average is used to introduce the discussion on coordinates and graphs. It continues uniformly throughout the text, which includes innumerable examples and exercises based on real data.

At the same time, we have been careful to strike a pedagogically sound balance between applications based on real data and more traditional generic applications. The density and selection of real data-based applications have been tailored to the pedagogical goals and appropriate difficulty level for each section.

STYLE

It is a common complaint that many students do not actually read mathematics texts but simply search through them for examples that match the assigned exercises. We would like students to read this book—we would like students to *enjoy* reading this book. Thus we have written this book in a conversational and student-oriented style. We make frequent use of a question-and-answer dialogue format (indicated with the Q/A symbol) in order to encourage the development of the student's mathematical curiosity and intuition. We hope that this text will give the student insight into how a mathematician develops and thinks about mathematical ideas and their applications.

EXERCISE SETS

The strength of our exercise sets is one of the best features of this text. Our collection of almost 4,000 exercises provides a wealth of material that can be used to challenge students at almost every level of preparation. The exercises include everything from straightforward drills to interesting and rather challenging applications. We also include, in virtually every section of every chapter, applications based on real data (including data from approximately 100 corporations and government agencies); conceptual and discussion exercises useful for writing assignments; graphing calculator exercises; and what we hope are amusing exercises. In addition, every chapter contains a collection of chapter review exercises. Communication and Reasoning exercises appear at appropriate places in the text.

Many of the scenarios used in application examples and exercises are revisited several times throughout the book. Thus, for instance, students will find themselves using a variety of techniques, from graphing through the use of derivatives to elasticity of demand, to maximize revenue in the same application. The Cobb-Douglas production function is used in several different contexts throughout the text, including applications of derivatives, implicit differentiation, related rates, maxima and minima, and, in the chapter on calculus of several variables, linear regression, where we show the student how to obtain a best-fit Cobb-Douglas function. Our treatment of the logistic function is similar. Reusing scenarios and important functions this way provides unifying threads and shows students the complex texture of real-life problems.

PEDAGOGICAL FEATURES

- **The Rule of Four** All of the central concepts, such as functions, limits, derivatives, and integrals, are discussed *numerically, graphically,* and *algebraically*. We have gone to some lengths to draw the student's attention to these distinctions. (See, for instance, the section headings in Chapter 6 in the Table of Contents.) In addition, we *verbally* communicate mathematical concepts as a fourth element. We do so through conceptual exercises at the end of each section, our emphasis on verbalizing mathematical concepts, and our discussions of rephrasing sentences into forms that easily go over to mathematical statements.
- ***Q/A*** An important pedagogical tool in this text is the frequent use of informal question-and-answer dialogues. These often anticipate the kind of questions that may occur to the student and also guide the student through the development of new concepts.

- **You're the Expert** Each chapter begins with an application—an interesting problem—and returns to it at the end of that chapter in a section titled *You're the Expert*. This extended application uses and illustrates the central ideas of the chapter. The themes of these applications are varied, and they are designed to be as unintimidating as possible. For example, we do not pull complicated formulas out of thin air but focus instead on the development of mathematical models appropriate to the topics. Among the more notable of the *You're the Expert* applications are an early example of modeling poultry demand based on actual data and an example using marginal analysis to design a strategy for regulating sulfur emissions. These applications are also ideal for assignment as projects, and it is to this end that we have included groups of exercises at the end of each.

- **Before We Go On** Most examples are followed by supplementary discussions under the heading *Before we go on*. These discussions may include a check on the answer to a problem, a discussion of the feasibility and significance of a solution, or an in-depth look at what the solution means.

- **Communication and Reasoning Exercises** These exercises are designed to broaden the student's grasp of the mathematical concepts. The student might be asked to provide examples, to illustrate a point, to fill in the blank, or to discuss and debate. These exercises often have no single correct answer.

- **Conceptual and Computational Devices** The text features a wide variety of novel devices to assist the student in overcoming hurdles. These include a calculation thought experiment for the analysis and differentiation of complicated functions, the use of the remarkably quick and efficient *column integration* method for integration by parts, and a *template* method for evaluating definite integrals that avoids the common errors in signs.

- **Cautions, Hints, and Notes** Most sections include suggestions to assist students in avoiding common errors and in tracking down their source when they do occur in a calculation.

- **Footnotes** Footnotes throughout the text provide sources, interesting background, extended discussion, and various asides.

MORE DISTINGUISHING FEATURES

- **Mathematical Rigor** Rather than present a sequence of formal theorems without proof, we have taken considerable care to motivate and present informal proofs of virtually *all* stated results.

▪ **Algebra Review Appendix** The algebra review in Appendix A is comprehensive—not a rushed afterthought—and has been written with the same care and thoroughness as the main body of the text. It is designed to serve both as a self-contained refresher course in basic algebra techniques and also as a reference. Appendix A covers basic concepts in considerable detail and deals at length with the often neglected art of solving equations. The first section also includes important drill exercises in the proper entry of functions into graphing calculators or computer programs.

▪ **Graphing Calculator Appendix** Appendix B introduces the student to the use of a graphing calculator in calculus. This appendix includes several programs referred to in the text. Although written specifically for the TI-82, the appendix should be useful for students using other brands and models.

▪ **Spreadsheet Appendix** Appendix C shows the student how to use a computer spreadsheet program to help with numerical calculations. Written using Lotus 1-2-3 as its example, this appendix should be helpful with any computer spreadsheet.

RELATED BOOKS IN THE SERIES

This is one of three books by the authors on mathematics for business, the social sciences, and the liberal arts. *Finite Mathematics Applied to the Real World* covers the topics of linear mathematics, including linear programming, set theory, probability, statistics and the mathematics of finance. *Finite Mathematics and Calculus Applied to the Real World* combines most of *Finite Mathematics Applied to the Real World* with this book to provide a text suitable for use in a two-semester sequence in finite mathematics and calculus.

ORGANIZATION AND COURSE OPTIONS

Care has been taken in the format of this text to provide flexibility in course design. For example, the chapter on logarithmic and exponential functions presented early in the text can be regarded as a "floating chapter" (see the chart on the next page) and can be studied at any time in the course prior to Section 3 in Chapter 4 (on derivatives of logarithmic and exponential functions).

The following chart shows the logical dependence of the chapters. Notice again that Chapter 2 (Logarithmic and Exponential Functions) can be covered at any time prior to Section 3 in Chapter 4. Following are suggested chapter orders for possible one-semester courses in applied calculus. Coverage of specific sections within each chapter could vary with the emphasis of the particular course.

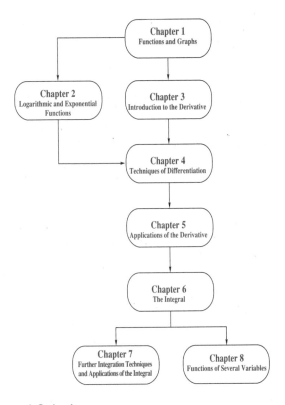

Precalculus and Calculus

Algebra Review (Appendix A)

Chapter 1: Functions and Graphs

Chapter 2: Logarithmic and Exponential Functions

Chapter 3: Introduction to the Derivative

Chapter 4: Techniques of Differentiation

Chapter 5: Applications of the Derivative

Chapter 6: The Integral

Basic Applied Calculus

Chapter 1: Functions and Graphs (last section optional)

Chapter 2: Logarithmic and Exponential Functions (here or just before Chapter 4, section 3)

Chapter 3: Introduction to the Derivative

Chapter 4: Techniques of Differentiation

Chapter 5: Applications of the Derivative

Chapter 6: The Integral

One or both of the following

Chapter 7 Further Integration Techniques and Applications of the Integral

Chapter 8 Functions of Several Variables

Applied Calculus with a Focus on Graphing Technology

Algebra Review, Section 1 (Appendix A)

Chapter 1: Functions and Graphs
Self-study or lab: Using a Graphing Calculator, Sections 1, 2 (Appendix B)

Chapter 2: Logarithmic and Exponential Functions

Chapter 3: Introduction to the Derivative
Self-study or lab: Appendix B, Section 3

Chapter 4: Techniques of Differentiation

Chapter 5: Applications of the Derivative

Chapter 6: The Integral
Self-study or lab: Appendix B, Section 4

Chapter 7: Further Integration Techniques and Applications of the Integral
Self-study or lab: Appendix B, Section 5

SUPPLEMENTS

For the Instructor

HarperCollins Test Generator/Editor for Mathematics with QuizMaster is fully networkable and is available in both IBM and Macintosh versions. The system features printed graphics and accurate mathematical symbols. The program allows the instructor to choose problems either randomly from a section or problem type or manually while viewing them on the screen, with the options to regenerate variables or scramble the order of questions while printing if desired. The editing feature allows instructors to customize the chapter data disks by adding their own problems. The Test Generator comes free to adopters.

Instructor's Resource Manual provides detailed discussion of the material in each section, complete solutions to the *Communications and Reasoning Exercises* and to the *You're the Expert* exercise sets, teaching tips, and a large collection of sample test questions.

Instructor's Complete Solution Manual contains solutions to every exercise in the text.

Graf Software is a graphing tool for the Macintosh that can be used either by students for independent exploration or by instructors as demonstration software.

For the Student

Interactive Tutorial Software with Management System is available in IBM and Macintosh versions and is fully networkable. As with the Test Generator/Editor, this innovative software is algorithm-driven, automatically regenerating constants so that a student will not see values repeated if he or she revisits any particular problem for additional practice. The tutorial is self-paced and provides unlimited opportunities to review lessons and to practice problem solving. If a student gives a wrong answer, he or she can request to see the problem worked out and get a textbook page reference. The program is menu-driven for ease of use, and on-screen help can be obtained at any time with a single keystroke. Students' scores are automatically recorded and can be printed for a permanent record. The optional **Management System** lets instructors record student scores on disk and print diagnostic reports for individual students or classes. This software is free to adopters, but may also be purchased by students for home use. (Macintosh version ISBN 0-06-502593-8; IBM version 0-06-502592-X)

GraphExplorer provides students and instructors with a comprehensive graphing utility able to graph rectangular, conic, polar, and parametric equations; zoom; transform functions; and experiment with families of equations quickly and easily. It is available in IBM and Macintosh formats. Printing capabilities further set this apart from other programs.

Graf Software, described above, is a Macintosh program.

Student's Resource Manual (ISBN 0-06-501825-7) provides complete, worked-out solutions to the odd-numbered exercises in the text. The manual also includes comprehensive chapter summaries and true/false quizzes for each chapter that help students both review and test their understanding.

The Electronic Spreadsheet and Elementary Calculus by Sam Spero, Cuyahoga Community College (ISBN 0-673-46595-0) helps students get started with graphing and problem solving by means of the spreadsheet. Knowledge of spreadsheets is not assumed, and the approach is adaptable to all spreadsheet programs.

CLASS TESTED

This book was used in manuscript form by many of our colleagues at Hofstra University. Their reactions and input were crucial to the development of this book. We thank them for all of their help; their names appear in the Acknowledgments section on the following page.

ACCURACY

Accuracy checking was carried out at every stage of the production process. In particular, each chapter was checked by at least two different mathematicians. We gratefully acknowledge their help in this critical aspect of the project: Susan Boyer, University of Maryland–College Park; Pamela G. Coxson, Lawrence Berkeley National Laboratory; Carol DeVille; Louisiana Technical University; Richard Leedy, Polk Community College; Ron Netzel; Richard Porter, Northeastern University; Jane Rood, Eastern Illinois University.

ACKNOWLEDGMENTS

This project would not have been possible without the contributions and suggestions of numerous colleagues, students, and friends. We are particularly grateful to our many colleagues who class tested the various preliminary editions of this book, and to our editors at HarperCollins for their encouragement and guidance throughout the project. Specifically, we would like to thank: George Duda, sponsoring editor, for his enthusiasm and for believing this would work; Kevin Connors, sponsoring editor, for vision in the final phases of the project; Louise Howe, developmental editor, for pushing us in the right directions; Lynn Mooney, developmental editor, for keeping us in check and seeing it through to the end; David Knee for his detailed critiques of the text; Safwan Akbik and Michael Steiner for numerous helpful suggestions; Daniel Rosen for his careful analysis of the material and numerous suggestions, especially for his version of tabular integration by parts; Edward Ostling for his encouragement and for class testing several versions of the calculus portion.

We would also like to thank the numerous reviewers who read carefully and commented on successive drafts, providing many helpful suggestions that have shaped the development of this book:

William L. Armacost, California State University—Dominquez Hills

Stephen A. Bacon, Central Connecticut State University

Louise B. Bernauer, County College of Morris

Chris Boldt, Eastfield College

Barbara M. Brook, Camden City College

Charles E. Cleaver, The Citadel

Richard L. Conlon, University of Wisconsin—Stevens Point

B. Jan Davis, University of Southern Missouri

Kenneth A. Dodaro, Florida State University

William L. Etheridge, University of North Carolina—Wilmington

Elise Fischer, Johnson County Community College

Carol J. Flakus, Lower Columbia College

Donald R. Goral, Northern Virginia Community College

John Gregory, Southern Illinois University

Joan F. Guetti, Seton Hall University

Dianne Hendrickson, Becker College

Martin Kotler, Pace University

Richard Leedy, Polk Community College

James T. Loats, Metropolitan State College of Denver

Vicky Lymbery, Stephen F. Austin State University

Randall Maddox, Pepperdine University

Steven E. Martin, Richard Bland College

Gertrude Okhuysen, Mississippi State University

Kevin O'Neil, Kankakee Community College

Kathleen R. Pirtle, Franklin University

Georgia B. Pyrros, University of Delaware

Ken Reeves, San Antonio College

Mohamad Riaza, Fort Hays State University

Judith F. Ross, San Diego Mesa College

Daniel E. Scanlon, Orange Coast College

Richard H. Schroeder, Ball State University

Richard Semmler, Northern Virginia Community College

Sally Sestini, Cerritos College

Hari Shankar, Ohio University

Robert E. Sharpton, Miami-Dade Community College

Mahendra P. Singhal, University of Wisconsin—Parkside

Clifford W. Sloyer, University of Delaware

Stanley L. Stephens, Anderson University

James Wooland, Florida State University

Anne L. Young, Loyola College

Fredric Zerla, University of South Florida

Stefan Waner
Steven R. Costenoble

November 1995

CALCULUS APPLIED TO THE REAL WORLD

Boston Chicken Wields Midas Touch

By FLOYD NORRIS

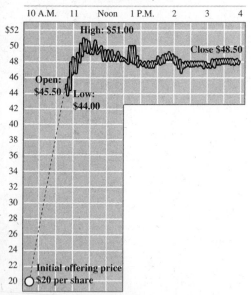

Nasdaq trading in Boston Chicken stock November 9, 1993

High: $51.00

Close $48.50

Open: $45.50

Low: $44.00

Initial offering price $20 per share

Source: Bloomberg Financial Markets
The New York Times, Nov. 10, 1993 p. D1

In the most successful initial public offering of stock in years, Boston Chicken, a fast-food chain that has yet to earn its first annual profit, went public and promptly more than doubled in price.

At yesterday's closing price of $48.50 a share, Boston Chicken had a market value of $839 million, or almost 27 times revenues over the last 12 months. While the offering had been tipped for weeks as this year's hottest, the amount of the increase stunned both company officials and Wall Street traders.

The appetite for the stock reflected a strong growth story for the chain. Based in the Chicago suburb of Naperville, Ill., Boston Chicken claims to be the largest franchise operation specializing in rotisserie chicken, a segment that many think will grow rapidly as consumers become more health-conscious.

The top officers of Boston Chicken include several early investors in Blockbuster Video, including Scott A. Beck, the 35-year-old chairman and chief executive of Boston Chicken who "retired" last year as Blockbuster vice chairman. The connection gave Boston Chicken a following on Wall Street. In a highly unusual move, those officers yesterday bought stock in a companion offering, a fact that helped to persuade investors to buy.

Boston Chicken sold 2.06 million shares to the public at $20, in an offering whose lead underwriters were Merrill Lynch and Alex Brown & Sons.

The shares began trading in the public market at $45.50—a more than twofold gain—and traded as high as $51 before closing at $48.50. Volume was 8.4 million shares, meaning the average share in the offering was traded more than four times during the day.

The price of $20, while well above the range of $15 to $17 the company said it would seek when it filed to go public, proved to be far less than the demand. Leaving all that money on the table meant the company had less money than it might have had to finance its ambitious expansion plans, but Mark W. Stephens, the company's chief financial officer and a former investment banker, said the company was happy with the price.

Source: From Floyd Norris, "Boston Chicken Wields Midas Touch," *The New York Times,* November 10, 1993, p. D1.

Functions and Graphs

APPLICATION ▶ A government agency plans to regulate the price of poultry with the goal of increasing revenue to poultry producers. Economists have determined that the amount of poultry people will buy depends on both the price of poultry and the price of beef. They have devised an equation relating the demand for poultry (measured by annual consumption) to the price of poultry and the price of beef. The agency has hired you to develop a pricing policy and to analyze the effect of fluctuations in the beef price on revenue to poultry producers under your proposed policy. How do you go about developing the policy?

INTRODUCTION   ▶ To advise the government agency, we need to identify, given the price of beef, the price of poultry that will maximize revenue to the poultry industry. This will tell us how the price of poultry should depend on the price of beef. By the time we get to the end of this chapter we will be able to give the government agency some good advice.

But where do we start? In order to model real-life situations, we need to build on a solid understanding of basic concepts and simple examples. Most problems that we try to solve using mathematics involve a relationship between two or more quantities, such as the price and the annual sales of poultry or the length and width of a rectangle. When we look at two related quantities, the relationship usually can be expressed as an equation in which the two quantities are the only unknown numbers. Such an equation is called an "equation in two unknowns." A very useful way of picturing such a relationship is to draw the *graph* of the equation. The simplest relationships, with the simplest graphs, are the linear and quadratic equations in two unknowns, and we shall look at these in some detail (in Sections 3 through 5). In our discussion of linear equations, we shall meet the concepts of *slope* and *rate of change,* which are the starting point of calculus.

Mathematics once revolved around the concept of an equation, but a more sophisticated notion evolved during the seventeenth and eighteenth centuries: a *function.* Briefly, if one quantity (the demand for poultry, for instance) depends on another (the price of poultry), then we say that the first quantity is a function of the second. In Section 2, we shall discuss the concept of a function and see again that graphs are a very useful tool.

You are urged to spend as much time as possible visualizing the functions you come across by means of their graphs—especially if you are using a graphing calculator. In particular, you should learn the graphs of the standard functions discussed in Section 2. Much of the intuition behind calculus comes from such pictures. You are also encouraged to glance through the Algebra Review in Appendix A, particularly Sections A.6 through A.9, before you begin your study of this chapter, and to refer to the Review as needed.

The material in the first part of this chapter is too simple to require the use of a graphing calculator or computer program. If you plan on using one of these devices later in the course, however, now is a good time to start practicing, so we have put comments throughout the chapter pointing out places where you can do so. A graphing calculator is handy for Section 2, and Section 6 discusses one particularly good use for these devices.

1.1 COORDINATES AND GRAPHS

PLOTTING DATA IN THE COORDINATE PLANE

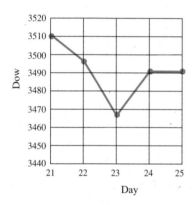

FIGURE 1

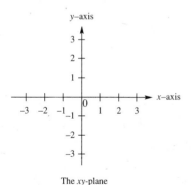

The *xy*-plane

FIGURE 2

The following table lists the daily closing value of the Dow Jones Industrial Average over the course of the week from June 21–25, 1993.*

Day	June 21	22	23	24	25
Dow	3511	3498	3467	3491	3491

Although this table tells you everything you need to know about the behavior of the closing values that week, it is much more striking to see the *graph* of the Dow, as in Figure 1. This is a graph you might see published in the newspaper.

The movements up and down, and their relative sizes, are easier to see in the graph than in the tabulated data. This is the purpose of a graph: it is a way of *visualizing information.* Throughout this book we will be using graphs to visualize the relationships between quantities (here, time and the Dow), and so we need to recall some basic facts about the Cartesian plane, or "*xy*-plane." Although you probably think you know what the *xy*-plane is, you have probably never actually tried to *define* what it is. Thus, we pose the following question.

Q Just what is the *xy*-plane?

A The *xy*-plane is an infinite flat surface with two perpendicular lines, usually labeled the *x*-**axis** and *y*-**axis.** These axes are calibrated as shown in Figure 2.

Thus the *xy*-plane is nothing more than a very large—in fact, infinitely large—flat surface. The purpose of the axes is to allow us to locate specific positions, or **points,** on the plane, with the use of **coordinates.** (If Captain Picard wants to have himself beamed to a specific location, he must supply its coordinates, or he's in trouble.)

▼ *Source: Newsday,* issues of June 22–26, 1993.

Q So how do we use coordinates to locate points?

A The rule is simple. Each point in the plane has two coordinates, an **x-coordinate** and a **y-coordinate.** These can be determined in two ways:

1. The x-coordinate measures a point's distance to the right or left of the y-axis. It is positive if the point is to the right of the axis, negative if it is to the left of the axis, and 0 if it is on the axis. The y-coordinate measures a point's distance above or below the x-axis. It is positive if the point is above the axis, negative if it is below the axis, and 0 if it is on the axis. Briefly, the x-coordinate tells us the *horizontal* position (distance left or right), and the y-coordinate tells us the *vertical* position (height).

2. Given a point P, we get its x-coordinate by drawing a vertical line from P and seeing where it intersects the x-axis. Similarly, we get the y-coordinate by extending a horizontal line from P and seeing where it intersects the y-axis.

Here are a few examples to help you review coordinates.

▼ **EXAMPLE 1**

Find the coordinates of the indicated points. (See Figure 3. The grid lines are placed at intervals of one unit.)

SOLUTION   Taking them in alphabetical order, we start with the origin O. This point has height zero and is also zero units to the right of the y-axis, so its coordinates are $(0, 0)$. Turning to P, dropping a vertical line gives $x = 2$ and extending a horizontal line gives $y = 5$. Thus, P has coordinates $(2, 5)$. For practice, determine the coordinates of the remaining points, and check your work against the list that follows.

$$Q(-1, 3), R(-4, -3), S(3, -3), T(1, 0),$$
$$U(-\tfrac{9}{2}, 0), V(\tfrac{7}{2}, \tfrac{5}{2})$$

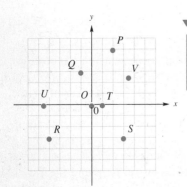

FIGURE 3

▼ **EXAMPLE 2**

Locate the following points in the xy-plane.

$$P(2, 3), Q(-4, 2), R(3, -\tfrac{5}{2}), S(0, -3), T(3, 0),$$
$$U(\tfrac{5}{2}, \tfrac{2}{3}).$$

SOLUTION   In order to locate each of these points, we start at the origin $(0, 0)$, and proceed as follows (see Figure 4):

To locate P, we move 2 units to the right and 3 up, as shown.

To locate Q, we move -4 units to the right (i.e., 4 to the *left*) and 2 up, as shown.

To locate R, we move 3 units right and $\tfrac{5}{2}$ down, and so on.

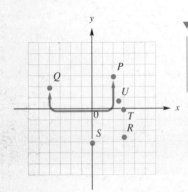

FIGURE 4

▶ NOTES

1. The correspondence between a *point* on the xy-plane and the *pair of numbers* representing its coordinates is a one-to-one correspondence. That is,

(i) to every point in the xy-plane, there corresponds one and only one pair of numbers (as in Example 1);

(ii) to every pair of numbers, there corresponds one and only one point (as in Example 2).

As a result, we usually think of a point as "being" a pair of numbers and vice-versa. Analogously, points in *three*-dimensional space may be thought of as *triples of numbers.*

2. One drawback of graphs is that it is difficult to plot points with fractional coordinates (such as R and U) with perfect accuracy. For instance, imagine trying to graph the point with coordinates $(1.0000000001, 3.999998889997659)$.

3. This way of assigning coordinates to points in the plane is often called the system of **Cartesian** coordinates, in honor of the mathematician and philosopher René Descartes (1596–1650), who was the first to use them extensively. There are other useful coordinate systems that we shall not discuss. ◀

It is sometimes convenient to think of the xy-plane as being divided by the axes into four **quadrants** (first, second, third, and fourth). These consist of the following sets of points. (See Figure 5.)

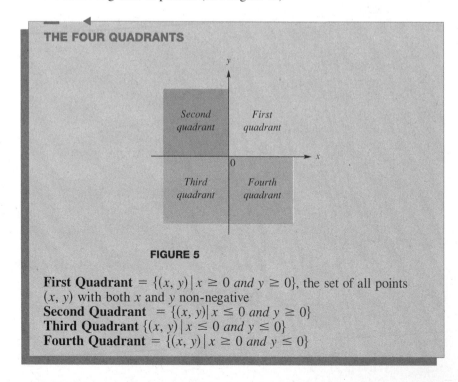

THE FOUR QUADRANTS

FIGURE 5

First Quadrant = $\{(x, y) \mid x \geq 0 \text{ and } y \geq 0\}$, the set of all points (x, y) with both x and y non-negative

Second Quadrant = $\{(x, y) \mid x \leq 0 \text{ and } y \geq 0\}$

Third Quadrant $\{(x, y) \mid x \leq 0 \text{ and } y \leq 0\}$

Fourth Quadrant = $\{(x, y) \mid x \geq 0 \text{ and } y \leq 0\}$

THE GRAPH OF AN EQUATION

One of the more surprising developments of mathematics was the realization that equations, which are algebraic objects, can be represented by graphs, which are geometric objects. The kinds of equations that we have in mind are equations in two variables, such as

$$y = 4x - 1,$$
$$2x^2 - y = 0,$$
$$q = 3p^2 + 1,$$
$$y = \sqrt{x - 1}.$$

In order to describe the graph of an equation, we must first say something about the *solutions* of an equation.

SOLUTION TO AN EQUATION

A **solution** to an equation in one or more unknowns is an assignment of numerical values to each of the unknowns so that when these values are substituted for the unknowns, the equation becomes a *true statement about numbers.* We say that a solution **satisfies** the equation.

Thus, for example, one solution to the equation

$$x + y^2 = 5$$

is $x = 1$, $y = -2$, since substituting these values for the unknowns gives

$$1 + (-2)^2 = 5,$$

which is a true statement about numbers. There are infinitely many solutions to this equation. (Two more are $x = 5$, $y = 0$, and $x = 3$, $y = \sqrt{2}$.) We can represent each solution by a single point in the coordinate plane. For instance, we can represent the solution $x = 1$, $y = -2$ by the point $(1, -2)$. With most equations, if we string together all the points representing solutions, we get an elegant curve, called the *graph* of the equation.

GRAPH OF AN EQUATION

The **graph** of an equation in the two variables x and y consists of all points (x, y) in the plane whose coordinates are solutions of the equation.

▼   **EXAMPLE 3**

Sketch the graph of the equation $x + y = 4$.

SOLUTION   The graph consists of all points representing solutions of the equation, so we must first find the solutions. Since a solution of $x + y = 4$

consists of values for x and y that add up to 4, possible solutions are $(1, 3)$, $(0.5, 3.5)$, and $(100, -96)$.

To be more systematic about this, we proceed as follows. Notice that no matter what value we choose for x, we can always find a corresponding value for y so that the two numbers add up to 4. So we do the following.

1. Solve the equation $x + y = 4$ for y, getting $y = 4 - x$.

2. Choose values for x and substitute them in this formula to get the corresponding values for y.

This procedure gives a table of solutions.

x	-4	-3	-2	-1	0	1	2	3	4	5	...	411	...
$y = 4 - x$	8	7	6	5	4	3	2	1	0	-1	...	-407	...

The pairs (x, y) in this table, such as $(-4, 8)$, $(-3, 7)$, $(-2, 6)$, ... are solutions to the equation. Of course, there are infinitely many "in-between" solutions, such as $(-4.1, 8.1)$, $(-4.11, 8.11)$, and so on, that are not shown in this table. For now, you need only be aware that these other solutions exist. Reading the table from left to right gives the points

$$(-4, 8), (-3, 7), (-2, 6), (-1, 5), (0, 4), (1, 3), (2, 2),$$
$$(3, 1), (4, 0), (5, -1), \ldots, (411, -407), \ldots$$

on the graph of $x + y = 4$. Plotting several of them gives the picture on the left in Figure 6.

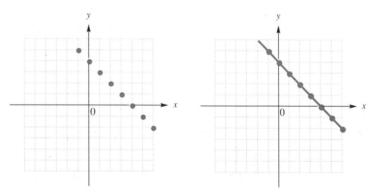

FIGURE 6

Now this figure is very suggestive of a straight line. If we decided to plot "in-between" points, such as $(0.5, 3.5)$, $(0.25, 3.75)$ and so on, we would find that they all lie precisely on the line shown on the right in Figure 6. This line is the graph of the equation $x + y = 4$.

Note that the line segment sketched gives only a *range* of solutions; the entire solution set would encompass an infinitely long line extending the segment shown in both directions. We can imagine what this may look like, but we can never actually see it.

Before we go on... Our first step, solving the equation for *y*, gives us *y* **as a function of** *x*. By this phrase we mean that the equation $y = 4 - x$ gives us a *rule* for calculating the value of *y* if we are given any value of *x*, as shown in the table. Through this rule, we think of *y* as **depending on** *x*. We shall discuss functions in the next section.

The graphing programs used by calculators and computers operate by simply plotting hundreds of solutions to a given equation and then joining them, usually with straight lines, creating the effect of a smooth curve. If you have a graphing calculator at hand, you can experiment by using it to plot the graph of $y = -x + 4$.* (On a TI-82, for example, you press [Y=] to get the display "$Y_1 =$" and enter

$$Y_1 = -X + 4$$

then press [GRAPH].) In order to view different portions of the graph, you must set the **viewing window coordinates** accordingly. These are usually denoted by *xMin*, *xMax*, *yMin*, and *yMax*, and they designate the portion of the *xy*-plane you will see. On the left in Figure 7 you see a rectangle drawn around the portion of the *xy*-plane defined by $xMin = -1$, $xMax = 5$, $yMin = -2$, and $yMax = 3$. On the right you see what the resulting graph would look like on a calculator.

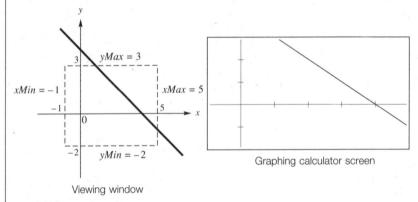

Viewing window

Graphing calculator screen

FIGURE 7

Once you have graphed the equation, you can use the trace feature to move around on the graph and find the coordinates of several points. Some of these points may correspond to the ones we calculated by hand in the table. (On the TI-82, you can obtain such a table automatically by pressing [2nd] [TABLE].)

Be warned, however, that the trace feature will not show all the points on the graph, only the finite number that are plotted on the screen. Also, the accuracy of the coordinates shown may be limited. (Check the

▼* See the first section of Appendix G for more details on using a calculator to graph.

device you are using: Do the x- and y-coordinates always add up to *exactly* 4?) You can use your calculator's zoom feature to see more of the "in-between" points on the graph.

Note finally that most graphing calculators and computer programs will only graph functions. Graphing a general equation (not solved for y) is difficult to do mechanically.

▼ **EXAMPLE 4**

Sketch the curve $xy = 1$.

SOLUTION   We start, as before, by solving for y to get $y = \frac{1}{x}$. We then make a table of values by choosing values for x and calculating the corresponding y-values from the equation. Notice that this means we cannot choose $x = 0$, since there is no such number as $\frac{1}{0}$. But that shouldn't stop us from getting as close to zero as we like. For example, we can choose $x = \pm\frac{1}{2}$, $x = \pm\frac{1}{100}$ or $x = \pm\frac{1}{100,000}$. Think of it as a dangerous exploration, very close to the "forbidden" zone. Here are some values:

x	-3	-2	-1	$-\frac{1}{2}$	$-\frac{1}{3}$	$-\frac{1}{100,000}$	0	$\frac{1}{100,000}$	$\frac{1}{3}$	$\frac{1}{2}$	1	2	3
$y = \frac{1}{x}$	$-\frac{1}{3}$	$-\frac{1}{2}$	-1	-2	-3	$-100,000$	✖	$100,000$	3	2	1	$\frac{1}{2}$	$\frac{1}{3}$

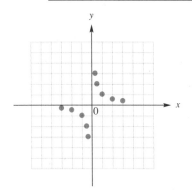

Plotting those points we can manage gives us Figure 8. Before joining up the points, note that the curve is not permitted to cross the y-axis.

Q Why not?

A Because if it did, then you would have a point on the curve with $x = 0$. But, as the "✖" reminds you, there is no possible y-value when $x = 0$. Thus, the y-axis should be regarded here as a wall that cannot be breached.

Q So how can the curve possibly get from the left-hand portion to the right-hand portion without crossing the y-axis?

A It can't, so there are two separate curves instead of one, as shown in Figure 9.

FIGURE 8

This curve is a **hyperbola.** Notice that it gets closer and closer to the axes as it extends outward in all directions, but never quite touches them.

As we've mentioned, graphing calculators and computer graphing programs draw graphs by plotting a large number of points (and possibly joining them up) to get an approximation to a curve. If you use a graphing calculator or computer to graph the equation $y = \frac{1}{x}$, then, depending on the type of graphing calculator or program you are using, you may not get the exact picture shown in Figure 9. Instead, you may see a near-vertical

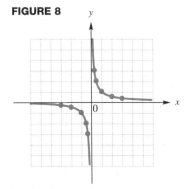

FIGURE 9

line down the *y*-axis joining the left-hand portion of the curve to the right-hand portion. The reason for this is that graphing software is not very good at spotting **singularities**—isolated points where the function of *x* is not defined. In the case we are looking at, there is a singularity at *x* = 0. When you instruct your graphing calculator to draw the graph, it may simply skip over the singularity and join the two portions of the curve with a near-vertical line.

INTERVALS ON AN AXIS

There are some sets of numbers that arise often enough to have their own notation. Suppose *a* and *b* are real numbers with *a* < *b*. Then by [*a*, *b*] we mean the set of all real numbers *x* with *a* ≤ *x* ≤ *b*. For instance, [1, 3] is the set of all real numbers between 1 and 3, inclusive. We can think of [1, 3] graphically as a segment of the real line, or an axis.

Here, the segment is shown as a heavy line. The solid dots at the ends indicate that we are including the two points 1 and 3. We call [*a*, *b*] a **closed interval.**

By (*a*, *b*) we mean the set of all real numbers *x* with *a* < *x* < *b*. Thus, for example, the set (1, 3) is the set of all real numbers *strictly between* 1 and 3. We represent (1, 3) graphically as follows.

Here, the hollow circles at 1 and 3 indicate that these two numbers are left out. We call (*a*, *b*) an **open interval.**

▶ CAUTION   The notation (*a*, *b*) is ambiguous. For instance, (1, 3) could mean the open interval as shown above, or it could refer to a point in the *xy*-plane with coordinates 1 and 3. Mathematical notation can sometimes be ambiguous. You are expected to infer its meaning from the context of the discussion. Here we are talking about intervals, not points in the plane.   ◀

It sometimes happens that we wish to consider an infinitely long interval, such as the set of all real numbers (the whole real line), or the set of positive real numbers or real numbers less than 6 (half-lines). For these, we use the notations $(-\infty, +\infty)$, $(0, +\infty)$, and $(-\infty, 6)$ respectively. This is consistent with our previous use of the notation.

Formally:

$(a, +\infty)$ is the set of all real numbers *x* with $a < x < +\infty$, or $x > a$.*

$(-\infty, b)$ is the set of all real numbers with $-\infty < x < b$, or $x < b$.

$(-\infty, +\infty)$ is just the whole real line, sometimes written as $\mathbb{R}$.

▼ * Saying $x < +\infty$ is redundant; *all* real numbers are less than infinity, so we could leave this out, and simply say $a < x$, or equivalently, $x > a$. There is a similar redundancy in the description of $(-\infty, b)$.

For example, $(-\frac{1}{4}, +\infty)$ is the set of all real numbers $x > -\frac{1}{4}$, as shown below.

Similarly, $(-\infty, \frac{1}{4})$ is the set of all real numbers $x < -\frac{1}{4}$.

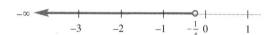

We also have four kinds of *half-open intervals.* Rather than defining each type formally, we'll simply give examples using diagrams.

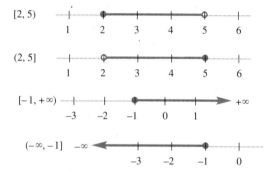

Just when you thought it was over, we mention one more thing: we'll also be considering *unions* of intervals of the various types just discussed. Again, we'll simply illustrate this with several examples.

This is the set of all points that are either in $(-\infty, 0]$ *or* in $(2, 3)$.

Here, the hollow dot at zero indicates that 0 is missing, as though we have "punctured" the line there. Had we wanted $(-3, 0) \cup [0, +\infty)$ instead, the diagram would have been exactly the same, except for the fact that 0 would not have been missing. This would, of course, be identical to the interval $(-3, +\infty)$.

▶ NOTE   When using a graphing calculator or computer program, we can use the interval notation

$$[xMin, xMax]$$

and

$$[yMin, yMax]$$

to refer to the viewing window coordinates. For instance, we can specify a viewing window by saying x is in $[-1, 5]$ and y is in $[2, 4]$. This is equivalent to saying that $-1 \le x \le 5$ and $2 \le y \le 4$, or that $xMin = -1$, $xMax = 5$, $yMin = 2$, and $yMax = 4$. ◄

▶ **1.1 EXERCISES**

1. Referring to the following figure, determine the coordinates of the indicated points as accurately as you can. (The grid lines are placed at intervals of one unit.)

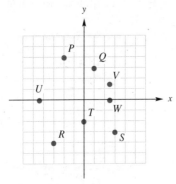

2. Referring to the following figure, determine the coordinates of the indicated points as accurately as you can. (The grid lines are placed at intervals of one unit.)

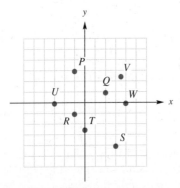

3. Graph the following points.

$P(4, 4)$, $Q(6, -5)$, $R(3, 0)$, $S(4, \frac{1}{4})$, $T(\frac{1}{2}, \frac{3}{4})$, $U(-2, 0)$, $V(-100, 0)$

4. Graph the following points.

$P(4, -2)$, $Q(2, -5)$, $R(1, -3)$, $S(-4, \frac{1}{4})$, $T(\frac{1}{4}, -1)$, $U(-\frac{1}{4}, 0)$, $V(0, 100)$

APPLICATIONS

Stock Market Index *Exercises 5 and 6 are based on the following table,
which shows the monthly highs and lows of the Dow Jones Industrial Average for
the year 1929.**

	Jan	Feb	Mar	Apr	May	Jun	Jul	Aug	Sep	Oct	Nov	Dec
High	318	322	321	319	327	334	348	380	381	353	258	264
Low	297	296	297	299	293	299	335	338	344	230	199	231

5. Graph the monthly highs. Choose the scales on your
axes carefully so that you can clearly see the upward
and downward moves of the Dow.

6. Repeat the previous exercise for the monthly lows.

7. *Deficits* The monthly U.S. trade balance for the
months from March 1992 to March 1993 is shown in
the following graph.[†] Use this graph to estimate the
approximate values (to the nearest $0.5 billion) of the
trade deficit for these months.

8. *Deficits* The U.S. federal deficit for the years 1982–
1993 is shown in the following graph.[‡] Use this graph
to estimate the values of the deficit (to the nearest $10
billion) for these years.

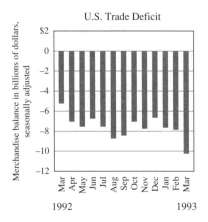

U.S. Trade Deficit

Source: U.S. Department of Commerce

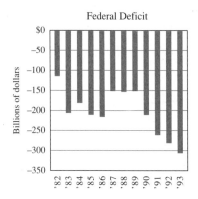

Federal Deficit

Source: Federal Reserve Board, Congressional
Budget Office

9. *Demand* Market research for Fullcourt Press
makes the following estimates of sales of a new book
at various possible prices:

Price ($)	20	30	40	50	60
Sales (1,000)	110	90	70	50	30

(a) Graph these data.
(b) What effect does a price increase of $10 per book
have on sales?

10. *Demand* For another book, Fullcourt Press's market
research projects the following sales at various possible
prices:

Price ($)	30	40	60	80	120
Sales (1,000)	150	110	75	55	37

(a) Graph these data.
(b) What is the effect on sales of doubling the price per
book?

▼ * Source: Standard & Poor's *Security Price Index Record,* Statistical Service, 1992 Edition.

† As printed in the *Chicago Tribune* business section, May 20, 1993.

‡ As printed in the *Chicago Tribune* business section, March 7, 1993.

11. *Cannons*   A cannon is fired horizontally from the top of a 500-ft cliff. The height of the cannonball after t seconds is shown in the following graph. Use this graph to estimate the cannonball's height at 0, 1, 2, 3, and 4 seconds.

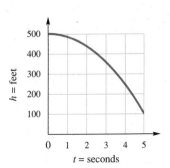

12. *Archery*   An arrow is shot straight up at a speed of 100 ft/s. Its height after t seconds is shown in the following graph. Use this graph to estimate the arrow's height at 0, 1, 2, 3, 4, 5, and 6 seconds.

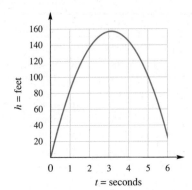

13. *Infant Height*   Alexander Costenoble's height during his first year of life was measured as follows:

Age (months)	0	2	3	4	5	6	9	12
Height (inches)	21	23	24	25	26	27	28	29

(a) Graph these data.
(b) Use the graph to estimate Alexander's approximate growth during his ninth month.

14. *Infant Weight*   Alexander Costenoble's weight during his first year of life was measured as follows:

Age (months)	0	2	3	4	5	6	9	12
Weight (pounds)	8	9	13	14	16	17	18	19

(a) Graph these data.
(b) Use the graph to estimate Alexander's approximate gain in weight during his tenth month.

Carefully sketch the graphs of each of the equations in Exercises 15–30.

15. $2x + y = 2$

16. $x - y = 4$

17. $2x - 3y = 4$

18. $2x + 3y = 4$

19. $y = -x^2 + 2$

20. $y = -2x^2 + 1$

21. $y = \dfrac{1}{x^2}$

22. $y = \dfrac{1}{x^3}$

23. $y = \dfrac{1}{x - 2}$

24. $y = \dfrac{1}{x + 1}$

25. $y = \dfrac{1}{(x + 1)^2}$

26. $y = \dfrac{1}{(x - 1)^2}$

27. $y(x^2 + 2x + 1) = x$

28. $y(x^2 + 3x + 2) = -x$

29. $y = \sqrt{x}(x - 1)$

30. $y = \sqrt{x}(x + 1)$

Exercises 31–36 require the use of a graphing calculator or computer software.

 31. (a) Graph the equation $y = x^3 - 2x - 5$ with $-5 \le x \le 5$ and $-10 \le y \le 10$.
(b) What happens if you change the range to $-1 \le x \le 1$ and $-1 \le y \le 1$? Explain the result.

 32. (a) Graph the equation $y = 1/(x^3 - 2x - 5)$ with $-5 \le x \le 5$ and $-10 \le y \le 10$.
(b) Find viewing window coordinates that show the interesting features of the graph more clearly.

Use a graphing calculator (or a graphing computer program) to display the graphs in Exercises 33–36. Answer the accompanying questions.

 33. (a) $y = x^{1/2}(x - 1)$ **(b)** $y - 1 = x^{1/2}(x - 1)$
(c) $y - 2 = x^{1/2}(x - 1)$ **(d)** $y + 1 = x^{1/2}(x - 1)$

What is the effect on the graph of an equation if you replace y with the quantity $(y - c)$?

 34. (a) $y = x^3 + x^2 + 4$ **(b)** $y - 1 = x^3 + x^2 + 4$
(c) $y - 2 = x^3 + x^2 + 4$ **(d)** $y + 1 = x^3 + x^2 + 4$

What is the effect on the graph of an equation if you replace y with the quantity $(y - c)$?

35. (a) $y = x^3 - 2x^2 + 4$ **(b)** $y = (x - 1)^3 - 2(x - 1)^2 + 4$
(c) $y = (x + 1)^3 - 2(x + 1)^2 + 4$ **(d)** $y = (x - 2)^3 - 2(x - 2)^2 + 4$

What is the effect on the graph of an equation if you replace x with the quantity $(x - c)$?

36. (a) $y = x^{1/2}(x - 1)$ **(b)** $y = (x - 1)^{1/2}(x - 2)$
(c) $y = (x + 1)^{1/2}(x)$ **(d)** $y = (x - 2)^{1/2}(x - 3)$

What is the effect on the graph of an equation if you replace x with the quantity $(x - c)$?

In Exercises 37–52, draw the given subsets of the real line.

37. $(-3, 4)$ **38.** $(-\infty, 4)$

39. $(-\infty, 0]$ **40.** $[4, 8]$

41. $[-1, +\infty)$ **42.** $(-\infty, +\infty)$

43. $(-4, 3) \cup (3, 4)$ **44.** $(-\infty, 1) \cup (1, +\infty)$

45. $(-4, 3) \cup [3, 4)$ **46.** $(-1, 0) \cup (1, 2) \cup (2, 3)$

47. $(-1, 0) \cup [1, 2] \cup (3, 4)$ **48.** $(-1, 1) \cup [1, 2] \cup (2, 4)$

49. $[2, 2]$ **50.** $[6.1, 6.1]$

51. $(2, 2)$ **52.** $(6, 6)$

In Exercises 53–58, represent the given line segments in interval notation.

53.

54.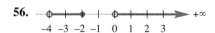

55.

56.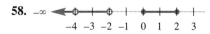

57.

58.

▶ ══════ **1.2** FUNCTIONS AND THEIR GRAPHS

FUNCTIONS

Simply speaking, a **function** is a *rule* or *set of instructions* for manufacturing a new object from an old one. The functions we deal with in this text manufacture new *real numbers* from old ones, and we call them *real-valued functions of a real variable*. The rule or set of instructions can be viewed in a number of ways, including *algebraically* when the rule is given by an algebraic formula, *numerically* using a table of values, and *geometrically* using a graph. We shall use all three perspectives to help us understand functions.

First, a formal definition. We shall explore the notation and terminology used in this definition in the examples below.

FUNCTION

A **real-valued function f of a real variable** is a rule that assigns to each real number x in a specified set of numbers a single real number $f(x)$. The set of numbers x for which $f(x)$ is defined is called the **domain** of f.

It is customary to give functions single-letter names, such as "f" or "g." In applications, we generally use a letter related to the quantities that we are studying.

▶ NOTE   Since a function f assigns a *single* number $f(x)$ to each x in the domain, it follows that $f(x)$ cannot have more than one value. For instance, we would not write $f(0) = \pm 1$, since $f(0)$ cannot equal both $+1$ and -1. ◀

A simple example of a function is the *squaring* function, by which we mean the rule that *squares* numbers. Thus, given an *input* number, say 4, this function produces the *output* 4^2, or 16. Similarly, if the input is -1, then the output is $(-1)^2 = 1$. If we call this function f, we would write

$$f(4) = 16,$$

meaning that "the function f, when applied to the input 4, yields the output 16," or, more concisely,

"f, when evaluated at 4, is 16,"

or even more concisely,

"f of 4 is 16."

Similarly, $f(-1) = 1, f(3) = 9, f(-11) = 121$, and so on. More succinctly, we can write the algebraic formula

$$f(x) = x^2.$$

This formula tells us that no matter what number x we use as the input, the output will be the square of that number, x^2. We can use this formula to evaluate f at any number we like. For example, if we wish to evaluate f at 6, we just take the formula and *replace x with the quantity 6 everywhere x occurs,* getting

$$f(6) = 6^2 = 36.$$

If we wish to evaluate f at -707, we replace x with the quantity -707 everywhere it occurs, getting

$$f(-707) = (-707)^2 = 499{,}849.$$

We say that we **substitute** the value -707 for x in the formula. Note that when we replace x with -707, we get $(-707)^2$ (which is positive), not -707^2 (which would be negative), on the right. Similarly, evaluating f at the unknown quantity a gives

$$f(a) = a^2.$$

This substitution rule also works for algebraic expressions. For instance, if we want to evaluate f at $x + h$ (which we shall want to when we start discussing calculus), then we substitute the quantity $x + h$ for x, getting

$$f(x + h) = (x + h)^2 = x^2 + 2xh + h^2.$$

In other words, we substitute the quantity $(x + h)$ for the quantity x. (We got the expression $x^2 + 2xh + h^2$ by expanding* the binomial $(x + h)^2$.)

There is a nice way of picturing the concept of a function: we can think of a function as a "black box," as shown in Figure 1. When the number x is fed into this particular function, out pops the square of that number, x^2. Thus, when the number -3 is fed in, we get the number 9 as the output.

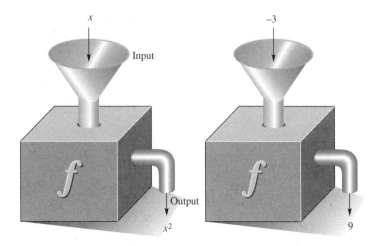

FIGURE 1

▼ *See the Algebra Review (Appendix A).

▼ | **EXAMPLE 1**

(a) Given that $f(x) = 3x^2 + 4x - 11$, find $f(0)$, $f(-1)$, $f\left(\frac{1}{2}\right)$, $f(\pi)$, and $f(x + h)$.

(b) With f as in part (a), calculate and simplify the quantity $\dfrac{f(x + h) - f(x)}{h}$.

SOLUTION

(a) Substituting,

$$f(0) = 3(0)^2 + 4(0) - 11 = -11.$$
$$f(-1) = 3(-1)^2 + 4(-1) - 11 = -12.$$
$$f\left(\frac{1}{2}\right) = 3\left(\frac{1}{2}\right)^2 + 4\left(\frac{1}{2}\right) - 11 = \frac{3}{4} + 2 - 11 = -\frac{33}{4}.$$
$$f(\pi) = 3\pi^2 + 4\pi - 11,$$

which, as good mathematicians, we'll simply leave like that. (Attempting decimal approximations will ruin the exactness of the answer!)
     Finally,

$$f(x + h) = 3(x + h)^2 + 4(x + h) - 11$$

We've replaced every occurrence of x with the quantity $x + h$.

$$= 3(x^2 + 2xh + h^2) + 4x + 4h - 11$$
$$= 3x^2 + 6xh + 3h^2 + 4x + 4h - 11.$$

(b) We have already calculated

$$f(x + h) = 3x^2 + 6xh + 3h^2 + 4x + 4h - 11.$$

Thus,

Note that we are subtracting the **whole quantity** $f(x)$ as required.

$$\frac{f(x + h) - f(x)}{h} = \frac{[3x^2 + 6xh + 3h^2 + 4x + 4h - 11] - [3x^2 + 4x - 11]}{h}$$

$$= \frac{3x^2 + 6xh + 3h^2 + 4x + 4h - 11 - 3x^2 - 4x + 11}{h}$$

$$= \frac{6xh + 3h^2 + 4h}{h} \qquad \text{All the other terms cancel.}$$

$$= \frac{h(6x + 3h + 4)}{h}$$

$$= 6x + 3h + 4.$$

You can enter a formula for a function into a graphing calculator (as well as certain other calculators and computer software) and have it evaluate the function at several numbers. For instance, you could have used a graphing calculator to find the first three answers above. You could also *approximate* the fourth answer (because a calculator or computer uses an approximation to π, not the exact value). On the other hand, it would require a more sophisticated calculator or computer software to evaluate $f(x + h)$ or the expression in part (b) since the calculation of $f(x + h)$ above is *symbolic,* and not *numerical.*

For details about using a graphing calculator to evaluate a function, see the corresponding section in Appendix B.

▼   **EXAMPLE 2**

If $f(x) = \frac{1}{x}$, find $f(-1), f(\frac{1}{2}), f(\pi), f(x + h)$, and $f(x + h) - f(x)$.

SOLUTION   Substituting, we get

$$f(-1) = \frac{1}{-1} = -1,$$

$$f\left(\frac{1}{2}\right) = \frac{1}{1/2} = 2,$$

$$f(\pi) = \frac{1}{\pi},$$

$$f(x + h) = \frac{1}{x + h}.$$

Finally,

$$f(x + h) - f(x) = \frac{1}{x + h} - \frac{1}{x}$$

$$= \frac{x - (x + h)}{x(x + h)} \qquad \text{by the rule for subtracting rational expressions*}$$

$$= \frac{x - x - h}{x(x + h)}$$

$$= \frac{-h}{x(x + h)}.$$

▼   * Here is the rule, as given in the Algebra Review (Appendix A):

$$\frac{A}{B} \pm \frac{C}{D} = \frac{AD \pm BC}{BD}.$$

▶ **NOTES**

1. When we write, say, $f(x) = \frac{1}{x}$, we call x the **independent variable,** since we think of it as a freely varying quantity. If we have an equation $y = f(x)$, we call y the **dependent variable,** and also say that y **depends on x** via the function f.

2. There is nothing magical about the letter x. We might just as well say $f(t) = \frac{1}{t}$ which means *exactly the same thing* as $f(x) = \frac{1}{x}$. For example, if we are told that $f(t) = \frac{1}{t}$, then, to get $f(3)$, we would substitute 3 for t, getting $f(3) = \frac{1}{3}$. This is exactly the same thing we would do, and exactly the same answer we would get, if we were told that $f(x) = \frac{1}{x}$. We replace x by t (or some other letter) if the independent variable stands for a particular quantity, such as time, and we want to use an appropriate letter for the quantity. ◀

▼ **EXAMPLE 3**   Income Tax

Each year millions of citizens of the United States are required to compute a function. The income tax they owe is a function of the taxable income (gross income less deductions) that they earned. If we write I for the taxable income, then we can write the income tax owed as $T(I)$.

It is interesting how the function T is defined. There is a formula for T, but at some point the government decided that people could more easily and accurately read a table than use the formula, so they publish *tax tables* in which you can look up the tax owed on a given taxable income. (Figure 2 shows a portion of the 1993 tax table.)

Curiously, if your income is high enough, the government trusts you to use the formulas, for the tax tables stop at a taxable income of $99,999. The government does publish the formula for T as well as the tax tables. For 1993, the formula appeared as in Figure 3.

1993 Tax Table

If your taxable income is		And you are
At least	But less than	Single
		Your tax is
23,000	23,050	3,574
23,050	23,100	3,588
23,100	23,150	3,602
23,150	23,200	3,616

FIGURE 2

Schedule X – Use if your filing status is **single**

If your taxable income is		Your tax is	of the amount over
Over	*But not over*		
$0	$22,100	15%	$0
22,100	53,500	$3,315.00 + 28%	22,100
53,500	115,000	12,107.00 + 31%	53,500
115,000	250,000	31,172.00 + 36%	115,000
250,000	....	79,772.0 + 39.6%	250,000

FIGURE 3

Before we go on... The tax table is an example of a **numerical definition** of a function. That is, a function can be defined by simply listing its values at all possible values of the independent variable, instead of giving a formula. On the other hand, even if we are given a formula for a function it may be useful to list some of its values (we shall see this when we discuss graphs below).

There are many functions, such as the value of the Dow Jones industrial average as a function of time, for which it is difficult or impossible to write down a formula, but for which at least some values may be easily listed. One of the goals of mathematical modeling is to find algebraic formulas for numerically specified functions, since a formula is easier to work with, and more amenable to the use of mathematical tools, than a table of values. We shall see examples of this in Section 4 and scattered throughout the book.

THE DOMAIN OF A FUNCTION

It sometimes happens that you can't evaluate a function at a specified real number. For example, the income tax $T(I)$ is not defined for a negative value of I. Similar difficulties may occur if a function is specified algebraically: if $f(x) = \sqrt{x}$, for instance, then what is the value of $f(-3)$? If you try to calculate the square root of -3 on a calculator, it will give you an error message (unless it is a sophisticated calculator that does complex number arithmetic). In the realm of real numbers, we just don't have square roots of negative numbers, so we must restrict the possible inputs to prevent them from being negative.* This is why we mentioned the *domain* of a function in the definition at the beginning of this section. The domain is the set of all numbers that we permit as inputs to the function.

For example, the square root function should be defined as

$$f(x) = \sqrt{x}, \text{ with domain } [0, +\infty).$$

Similarly, if $g(x) = \frac{1}{x}$, then we dare not attempt to evaluate $g(0)$. Thus, we must restrict the domain of g to consist of real numbers *other than* 0. This we can do by defining g as

$$g(x) = \frac{1}{x}, \text{ with domain } (-\infty, 0) \cup (0, +\infty).$$

If, for some reason, we did not want to consider negative numbers, we could restrict this function further, by defining

$$h(x) = \frac{1}{x}, \text{ with domain } (0, +\infty).$$

▼ * Some calculators will also give an error message if you try to take the *cube* root of a negative number. This is not because there is no such real number—cube roots of negative numbers always exist (for instance, the cube root of -8 is -2). This result is caused by the way the calculator has been programmed. An *ideal* calculator would always give an answer, but most calculators are not ideal.

Q What is the difference between *g* and *h*? After all, they are defined by the same formula.

A The two functions have different domains. For example, whereas $g(-11)$ is defined and is $-\frac{1}{11}$, $h(-11)$ is *not* defined. Also notice that *g* has the *largest possible domain* given its formula $\frac{1}{x}$, whereas *h* does not. No matter, there is no requirement that all functions must be given their largest possible domain! In fact, we shall see that it is often useful to consider functions with restricted domains. Remember that *the domain of a function is part of its definition*.

▼ **EXAMPLE 4**   Cost

Your electronics plant is equipped to manufacture up to 40 large-screen television sets per day. The cost in dollars for producing the sets is given by

$$C(x) = 400x + 10,000,$$

where *x* represents the number of large-screen sets manufactured in a day. How should you completely specify the function *C*?

SOLUTION   To completely specify a function, we need to know its domain as well as the rule that tells us how to evaluate it. But all we have been given is the rule, so we need to come up with a domain. Since the plant can make at most 40 large-screen television sets per day, we must restrict *x* to be less than or equal to 40 $(x \le 40)$. Furthermore, it is meaningless to talk about manufacturing a *negative* number of sets, so we must also restrict *x* to be greater than or equal to zero $(x \ge 0)$.

Thus, the domain is $[0, 40]$, and we can completely specify the function *C* by

$$C(x) = 400x + 10,000, \text{ with domain } [0, 40].$$

Before we go on... Actually, it does not make sense to evaluate *C* on any number other than an integer (unless you can make sense of what it means to manufacture, say, $\frac{1}{3}$ of a television in a day). Strictly speaking, then, the domain of *C* should be the set of integers $\{0, 1, 2, \ldots, 40\}$. There is, however, a great *mathematical* advantage to using a whole interval as the domain, and in particular the techniques of calculus require it. In practice, this will not be a problem.

You may have seen exercises such as this: "Find the domain of the function $g(x) = \sqrt{x + 3}$." This is not worded well. What is meant is that you should find the *largest possible domain* of *g*. Now, this is a convention we all agree to adopt: if a domain is not specified, we mean that it should be as large as possible. While we won't devote too much time to the calculation of largest possible domains, an example or two might be illuminating.

▼   **EXAMPLE 5**

Find the largest possible domain of the function $g(x) = \sqrt{x + 3}$.

SOLUTION   We are looking for all the values of x for which the given formula makes sense. For example, the formula clearly *doesn't* make sense when $x = -479$, for then we would be trying to evaluate the square root of -476. If we want the function to make sense, we had better make sure that *the quantity under the radical is at least* 0. Writing this phrase as a formula,

$$x + 3 \geq 0.$$

We solve this inequality by adding -3 to both sides.

$$x \geq -3$$

Thus, we must have $x \geq -3$ for the formula to make sense. In other words, x must be in the interval $[-3, +\infty)$. It follows that the largest possible domain for this function is $[-3, +\infty)$. Further, we can now specify the function precisely, as follows:

$$g(x) = \sqrt{x + 3}, \text{ with domain } [-3, +\infty).$$

▼   **EXAMPLE 6**

Find the largest possible domain of the function

$$H(t) = \frac{-3}{\sqrt{1 + t}}.$$

SOLUTION   We must find all values of t for which $H(t)$ makes sense. Since there is a denominator, a warning bell should be going off in our heads:

▮   *denominators can never be zero!*

Another restriction is imposed by the square root sign:

▮   *the quantity under the radical can never be negative.*

Writing these two restrictions mathematically,

$$1 + t \neq 0 \qquad \text{(or else its square root will be zero),}$$
$$1 + t \geq 0 \qquad \text{(or else we can't take its square root in the first place).}$$

These restrictions must hold simultaneously. Together, they say that $1 + t > 0$, or, adding -1 to both sides, that $t > -1$. Thus, t must be in the interval $(-1, +\infty)$, and this is the largest possible domain for the function H.

GRAPHS OF FUNCTIONS

It is very useful to visualize a function by means of its **graph.** Given a function f, we get an equation in x and y by setting $y = f(x)$. For instance, if the function f is specified by

$$f(x) = x^2 + 1,$$

then we get the corresponding equation by setting $y = f(x)$, getting

$$y = x^2 + 1.$$

The graph of f is then the graph of the equation $y = x^2 + 1$, as discussed in the previous section.

THE GRAPH OF A FUNCTION

The **graph of the function f** is the graph of the equation $y = f(x)$, where we restrict the values of x to lie in the domain of f. In other words, the graph of f is the set of all points $(x, f(x))$, where x lies in the domain of f.

▼   **EXAMPLE 7**

Let $f(x) = -x + 4$, with domain the set $\mathbb{R}$ of all real numbers. Then the graph of f is just the graph of the equation $y = -x + 4$, with no restriction on x since the domain is the set of all real numbers. As we know from Section 1, the graph of this equation is the straight line shown in Figure 4.

Before we go on... Recall how we obtained this graph in Section 1. We first constructed a table of values of the function:

x	-4	-3	-2	-1	0	1	2	3	4	5
$f(x) = 4 - x$	8	7	6	5	4	3	2	1	0	-1

We then plotted the points $(-4, 8)$, $(-3, 7)$, and so on. We then connected up these points in the most reasonable way possible, which looked like a straight line.

We now have three different ways of thinking about this function: algebraically, from the formula $f(x) = -x + 4$, numerically from the table above, and graphically from the graph above.

Q Of what use is the table of values in the last example?

A First, we needed it to draw the graph. But there are features of the function that show up most clearly in the table. For example, it is obvious from the table that the value of f decreases by 1 whenever x increases by 1.

Q Of what use is the graph?

A Again, there are features of the function that show up very clearly in the graph. For example, we can see once again the steady decrease by 1 in the value of f for every increase by 1 in x. Also, we can see clearly that $f(x)$ is positive precisely when $x < 4$ (this is where the graph lies above the x-axis) and that $f(x)$ is negative when $x > 4$ (this is where the graph lies below the x-axis).

Also, we can use the graph to find $f(x)$ for any value of x in the domain. For example, suppose we want to find $f(1)$, looking only at the graph and not at the formula. We can reason as follows: If $(1, y)$ is a point on the graph, then y is given by the equation $y = f(1)$. Therefore, we look for the y-value of the point on the graph above $x = 1$ on the x-axis. (See Figure 5.)

The point above $x = 1$ has y-coordinate 3, so $f(1) = 3$. In this way, the graph can be used as another way to *define* a function.

In general, the graph is a visual way of "taking in the whole function." A picture is worth a thousand words.

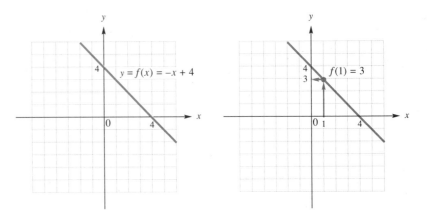

FIGURE 4   **FIGURE 5**

▼ **EXAMPLE 8**

Let $f(x) = x + \frac{1}{x}$, with domain $(0, +\infty)$. Its graph is the curve $y = x + \frac{1}{x}$, but with x restricted to be strictly positive. Figure 6 shows what the graph of f looks like.

Notice that as x increases, $\frac{1}{x}$ is small, so $x + \frac{1}{x}$ is very close to x. Thus, the graph of $y = x + \frac{1}{x}$ gets close to the line $y = x$. For values of x very close to 0, x is small, so the graph is close to the curve $y = \frac{1}{x}$. These dotted graphs $y = x$ and $y = \frac{1}{x}$ are called **asymptotes** of the function f. An *asymptote* is a line that the graph of a function approaches arbitrarily closely. Note that the function f also has the *vertical asymptote* $x = 0$ (why?).

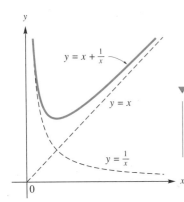

FIGURE 6

Graph of $f(x) = x + \dfrac{1}{x}$, with domain $(0, +\infty)$

Before we go on... It is obvious from the graph (but not from the formula) that this function has a minimum value at some smallish x (actually, at $x = 1$). Notice that the domain of f is not the largest possible domain. What would the graph look like if the domain were extended to $(-\infty, 0) \cup (0, +\infty)$?

 How did we get this curve? We could have plotted several points by hand by using a table of x- and y-values, or we could have instructed a graphing calculator to plot the curve $y = x + \frac{1}{x}$. (Actually, we used a computer graphing package something like the one available with this book.) If we use a graphing calculator, then we can "trace" the graph to obtain approximate coordinates of the lowest point of the curve.

Let us restrict the domain further.

▼   **EXAMPLE 9**

Let $g(x) = x + \frac{1}{x}$, with domain $(0, 2]$. Then its graph is the curve $y = x + \frac{1}{x}$, but this time x is restricted to be larger than 0 but less than or equal to 2. The graph is shown in Figure 7. Here, we get only a small segment of the graph described in Example 8.

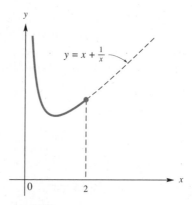

FIGURE 7

Graph of $g(x) = x + \dfrac{1}{x}$, with domain $(0, 2]$

▼   **EXAMPLE 10**

Let f be given by

$$f(x) = \begin{cases} -1 & \text{if } -4 \le x < -1 \\ x & \text{if } -1 \le x \le 1 \\ x^2 - 1 & \text{if } \ 1 < x \le 2 \end{cases}$$

with domain $[-4, 2]$. Sketch its graph, and use the graph to evaluate $f(1)$, $f(-1)$, and $f(-3)$.

SOLUTION This is an example of a **piecewise-defined function:** it is defined using different formulas for different intervals of its domain (another example is the income tax function in Example 3). To sketch its graph, we need to sketch the three graphs $y = -1$, $y = x$, and $y = x^2 - 1$, using the appropriate domain for each (Figure 8). From the graph, we see that $f(1) = 1$ (and not 0, since there is a hollow dot there), $f(-1) = -1$, and $f(-3) = -1$.

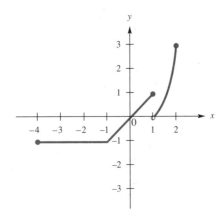

FIGURE 8

Before we go on... We didn't really need to specify that the domain of f is $[-4, 2]$, as that is implied by the description of $f(x)$ in the displayed formula. We could have simply let f be given by

$$f(x) = \begin{cases} -1 & \text{if } -4 \le x < -1 \\ x & \text{if } -1 \le x \le 1 \\ x^2 - 1 & \text{if } 1 < x \le 2. \end{cases}$$

The only values of x for which $f(x)$ is defined by this formula are those between -4 and 2, so the domain is $[-4, 2]$.

Some graphing calculators will not allow you to draw a piecewise-defined function very easily, since they apply the same x-range to all the graphs. Thus, you could plot the following three graphs with $-4 \le x \le 2$:

$$y_1 = -1,$$
$$y_2 = x,$$
$$y_3 = x^2 - 1.$$

You would then want to pay attention to the first graph only when $-4 \le x < -1$, the second one when $-1 \le x \le 1$, and the third when $1 < x \le 2$. (See Appendix B for other methods of graphing piecewise-defined functions.)

We have graphed functions by graphing corresponding equations. This raises the question, is the graph of every equation the graph of some function? To answer this, recall that if f is a function, there can be only a single value $f(x)$ assigned to each value of x. It follows that in the graph of a function, there should be only one y corresponding to any value of x, namely, $y = f(x)$. In other words, the graph of a function cannot contain two or more points with the same x-coordinate—that is, two or more points on the same vertical line. This gives us the following rule.

VERTICAL LINE TEST

For a graph to be the graph of a function, each vertical line must intersect the graph in *at most* one point.

▼ **EXAMPLE 11**

Which of the graphs shown in Figure 9 are the graphs of functions?

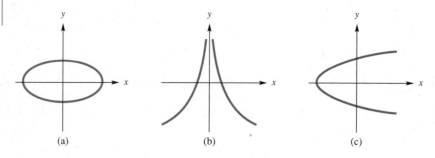

(a)   (b)   (c)

FIGURE 9

SOLUTION   As illustrated in Figure 10, only graph (b) passes the vertical line test and is the graph of a function.

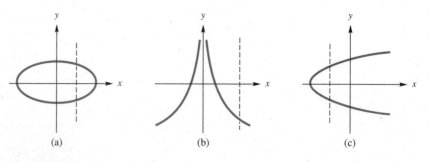

(a)   (b)   (c)

FIGURE 10

Following is a table listing some important functions with their graphs. Knowing these graphs will be *immensely* useful in the rest of this book.

Function	Equation for Graph	Comments	Graph
$f(x) = x$, with domain $\mathbb{R}$	$y = x$	The graph is a straight line passing through the origin and inclined at $45°$.	
$f(x) = \lvert x \rvert$, with domain $\mathbb{R}$	$y = \lvert x \rvert$	For x positive or zero, this agrees with $y = x$. For x negative or zero, it agrees with $y = -x$.	
$f(x) = x^2$, with domain $\mathbb{R}$	$y = x^2$	This is a **parabola** passing through the origin.	

Continued

Function	Equation for Graph	Comments	Graph
$f(x) = \sqrt{x}$, with domain $[0, +\infty)$	$y = \sqrt{x}$	The domain must be restricted to the nonnegative numbers, since the square root of a negative number is not real. Also note that the graph is the top half of a horizontally oriented parabola, since squaring both sides of the equation gives $x = y^2$.	
$f(x) = \frac{1}{x}$, with domain $(-\infty, 0) \cup (0, +\infty)$	$y = \frac{1}{x}$	This is a **hyperbola.** The domain excludes zero since there is no such thing as the reciprocal of 0.	

▶ __1.2 EXERCISES__

In Exercises 1–6, perform the given evaluations, simplifying the answer where appropriate.

1. Given $f(x) = x^2 + 2x + 3$, find
   (a) $f(0)$ (b) $f(1)$ (c) $f(-1)$ (d) $f(-3)$
   (e) $f(a)$,   (f) $f(x + h)$ (g) $\dfrac{f(x + h) - f(x)}{h}$ (if $h \neq 0$).

2. Given $g(x) = 2x^2 - x + 1$, find
   (a) $g(0)$ (b) $g(-1)$ (c) $g(r)$ (d) $g(x + h)$
   (e) $\dfrac{g(x + h) - g(x)}{h}$ (if $h \neq 0$).

3. Given $g(s) = s^2 + \frac{1}{s}$, find
   (a) $g(1)$ (b) $g(-1)$ (c) $g(4)$ (d) $g(x)$
   (e) $g(s + h)$ (f) $g(s + h) - g(s)$.

4. Given $h(r) = \dfrac{1}{r + 4}$, find
   (a) $h(0)$ (b) $h(-3)$ (c) $h(-5)$,   (d) $h(x^2)$
   (e) $h(x^2 + 1)$ (f) $h(x^2) + 1$.

5. Given $\phi(x) = \sqrt{x^2 + 3}$, find
   (a) $\phi(0)$ (b) $\phi(-2)$ (c) $\phi(x + h)$ (d) $\phi(x) + h$.

6. Given $\alpha(\gamma) = \gamma^{3/2} - \gamma$, find
   (a) $\alpha(9)$ (b) $\alpha(16)$ (c) $\alpha(\sigma + h)$ (d) $\alpha(\sigma) + h$.

In Exercises 7–14, calculate and simplify the quotient $\dfrac{f(x + h) - f(x)}{h}$ *$(h \neq 0)$.**

7. $f(x) = -x^2 - 2x - 1$ $\qquad\qquad$ **8.** $f(x) = 3x^2 - 2x - 1$

9. $f(x) = \dfrac{2}{x + 1}$ $\qquad\qquad$ **10.** $f(x) = \dfrac{1}{2 - x}$

11. $f(x) = x + \dfrac{1}{x}$ $\qquad\qquad$ **12.** $f(x) = x^2 - \dfrac{1}{x}$

13. $f(x) = \dfrac{1}{x^2}$ $\qquad\qquad$ **14.** $f(x) = \dfrac{1}{x^2 + 1}$

In Exercises 15–20, say whether $f(x)$ is defined for the given values of x. If it is defined, give its value.

15. $f(x) = x - \dfrac{1}{x^2}$, with domain $(0, +\infty)$
   (a) $x = 4$ (b) $x = 0$ (c) $x = -1$

16. $f(x) = \dfrac{2}{x} - x^2$, with domain $[2, +\infty)$
   (a) $x = 4$ (b) $x = 0$ (c) $x = 1$

17. $(x) = \sqrt{x + 10}$, with domain $[-10, 0)$
   (a) $x = 0$ (b) $x = 9$ (c) $x = -10$

18. $f(x) = \sqrt{9 - x^2}$, with domain $(-3, 3)$
   (a) $x = 0$ (b) $x = 3$ (c) $x = -3$

19. $f(x) = \sqrt{1 - x}$
   (a) $x = 0$ (b) $x = 2$ (c) $x = -3$

20. $f(x) = \dfrac{1}{\sqrt{x - 3}}$
   (a) $x = 0$ (b) $x = 3$ (c) $x = 4$

Calculate the largest possible domain of each of the functions in Exercises 21–32.

21. $f(x) = x^2 - 1$ $\qquad$ **22.** $f(x) = \sqrt{x}$ $\qquad$ **23.** $g(x) = \sqrt{3x}$

24. $g(x) = \sqrt{x^2}$ $\qquad$ **25.** $h(x) = \sqrt{x - 1}$ $\qquad$ **26.** $h(x) = \sqrt{x + 1}$

27. $f(x) = \dfrac{1}{x}$ $\qquad$ **28.** $f(x) = 4 - \dfrac{1}{x}$ $\qquad$ **29.** $g(x) = 4 - \dfrac{1}{x^2}$

30. $g(x) = \dfrac{1}{\sqrt{x}}$ $\qquad$ **31.** $h(x) = \dfrac{1}{x - 2}$ $\qquad$ **32.** $h(x) = \dfrac{3}{x - 1}$

▼ * This quotient, known as the *difference quotient,* measures how fast the value of the function changes as x changes. We shall study the difference quotient in detail when we study calculus.

In Exercises 33–40, use the graph of the function f to find approximations of the given values.

33.

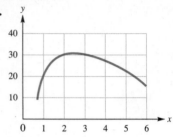

(a) $f(1)$ (b) $f(2)$
(c) $f(3)$ (d) $f(5)$
(e) $f(3) - f(2)$

34.

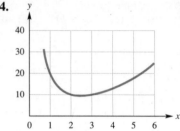

(a) $f(1)$ (b) $f(2)$
(c) $f(3)$ (d) $f(5)$
(e) $f(3) - f(2)$

35.

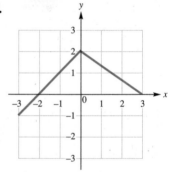

(a) $f(1)$ (b) $f(-2)$
(c) $f(0)$ (d) $f(3)$
(e) $f(3) - f(2)$

36.

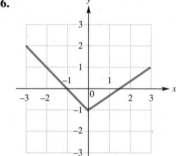

(a) $f(1)$ (b) $f(-2)$
(c) $f(0)$ (d) $f(3)$
(e) $f(3) - f(2)$

37.

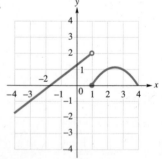

(a) $f(-3)$ (b) $f(0)$
(c) $f(1)$ (d) $f(2)$
(e) $\dfrac{f(3) - f(2)}{3 - 2}$

38.

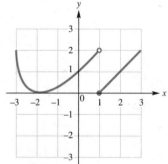

(a) $f(-2)$ (b) $f(0)$
(c) $f(1)$ (d) $f(3)$
(e) $\dfrac{f(3) - f(1)}{3 - 1}$

39.

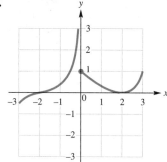

(a) $f(-3)$ (b) $f(-2)$
(c) $f(0)$ (d) $f(2)$
(e) $\dfrac{f(2) - f(0)}{2 - 0}$

40.

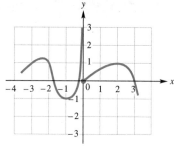

(a) $f(-2)$ (b) $f(0)$
(c) $f(1)$ (d) $f(3)$
(e) $\dfrac{f(2) - f(-1)}{2 - (-1)}$

In Exercises 41 and 42, match the functions to the graphs

41. (a) $f(x) = x$, with domain $[-1, 1]$ **(b)** $f(x) = -x$, with domain $[-1, 1]$
    (c) $f(x) = \sqrt{x}$, with domain $[0, 4]$ **(d)** $f(x) = x + \frac{1}{x} - 2$, with domain $(0, 4)$
    (e) $f(x) = |x - 1|$, with domain $(-2, 2)$ **(f)** $f(x) = x - 1$, with domain $(-1, 2)$

(I)

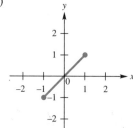

(II)

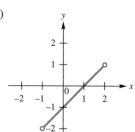

(III)

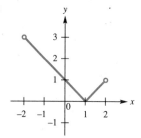

(IV)

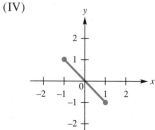

(V)

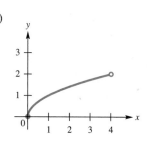

(VI)

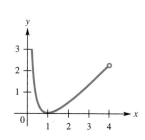

42. (a) $f(x) = -x + 1$, with domain $(-2, 2]$ **(b)** $f(x) = 2 - |x|$, with domain $[-2, 1)$
(c) $f(x) = \sqrt{x + 2}$, with domain $(-2, 2]$ **(d)** $f(x) = -x^2 + 3$, with domain $(-2, 2]$
(e) $f(x) = \frac{1}{x} - 1$, with domain $(0, 4)$ **(f)** $f(x) = x^2 - 1$, with domain $(-2, 2]$

(I)
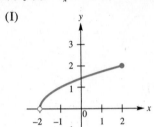

(II)
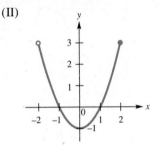

(III)
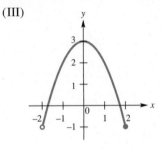

(IV)
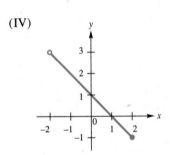

(V)
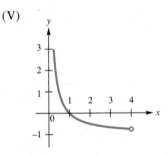

(VI)
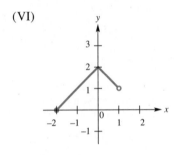

Graph the functions given in Exercises 43–48. We suggest that you become familiar with these graphs in addition to those in the chart at the end of the section.

43. $f(x) = x^3$, with domain $\mathbb{R}$

44. $f(x) = x^3$, with domain $[0, +\infty)$

45. $f(x) = x^4$, with domain $\mathbb{R}$

46. $f(x) = \sqrt[3]{x}$, with domain $\mathbb{R}$

47. $f(x) = \dfrac{1}{x^2}$, with domain $(-\infty, 0) \cup (0, +\infty)$

48. $f(x) = x + \dfrac{1}{x}$, with domain $(-\infty, 0) \cup (0, +\infty)$

In each of Exercises 49–54, sketch the graph of the given function. (The domain is implied by the formula. See the "Before we go on" discussion in Example 10.)

49. $f(x) = \begin{cases} x & \text{if} \quad -4 \le x < 0 \\ 2 & \text{if} \quad 0 \le x \le 4 \end{cases}$

50. $f(x) = \begin{cases} -1 & \text{if} \quad -4 \le x \le 0 \\ x & \text{if} \quad 0 < x \le 4 \end{cases}$

51. $f(x) = \begin{cases} x & \text{if} \quad -1 < x \le 0 \\ x + 1 & \text{if} \quad 0 < x \le 2 \\ x & \text{if} \quad 2 < x \le 4 \end{cases}$

52. $f(x) = \begin{cases} -x & \text{if} \quad -1 < x < 0 \\ x - 2 & \text{if} \quad 0 \le x \le 2 \\ -x & \text{if} \quad 2 < x \le 4 \end{cases}$

53. $f(x) = \begin{cases} x^2 & \text{if} \quad -2 < x \le 0 \\ \frac{1}{x} & \text{if} \quad 0 < x \le 4 \end{cases}$

54. $f(x) = \begin{cases} -x^2 & \text{if} \quad -2 < x \le 0 \\ \sqrt{x} & \text{if} \quad 0 < x < 4 \end{cases}$

55. The **greatest integer function** I is defined by taking $I(x)$ to be the greatest integer $\le x$. For example, $I(3.2) = 3$; $I(3.999) = 3$; $I(5) = 5$; $I(-4.1) = -5$ (since -5 is ≤ -4.1); $I(-4) = -4$.* Sketch the graph of this function. The domain of I is all of $\mathbb{R}$.

▼ * In other words, it drops the decimal part of positive numbers and rounds negative numbers down.

56. The **rounding** function R is defined by taking $R(x)$ to be the integer near-est to x, with the convention that a number midway between two integers is rounded up in magnitude. Thus, for example, $R(3.499) = 3$; $R(3.5) = 4$; $R(-2.4999) = -2$; $R(-2.5) = -3$. Sketch the graph of this function. The do-main of R is all of $\mathbb{R}$.

In Exercises 57–70, use either a graphing calculator or computer graphing soft-ware. Sketch the graphs of the indicated functions, and answer the additional questions.

57. $f(x) = (x - 1)(x - 2)(x - 3)(x - 4)$, with domain $\mathbb{R}$

58. $f(x) = (x - 1)(x - 2)(x - 3)(x - 4)(x - 5)$, with domain $\mathbb{R}$

59. $f(x) = \dfrac{1}{(x - 1)(x - 2)(x - 3)(x - 4)}$
    What is the largest possible domain of f?

60. $f(x) = \dfrac{1}{(x - 1)(x - 2)(x - 3)(x - 4)(x - 5)}$
    What is the largest possible domain of f?

61. $f(x) = \begin{cases} \dfrac{1}{x^2 + 1} & \text{if} \quad -2 \le x \le 0 \\[2mm] \dfrac{x}{x^2 + 1} & \text{if} \quad 0 < x \le 6 \end{cases}$

62. $f(x) = \begin{cases} \sqrt{x^2 + x} & \text{if} \quad -10 < x \le -1 \\[2mm] \sqrt{x^2 + 2x + 4} & \text{if} \quad -1 < x < 10 \end{cases}$

63. $f(x) = \dfrac{x^2}{(x - 1)(x - 2)(x - 3)}$, $\quad -5 \le x \le 5$
    Find the largest possible domain of f.

64. $f(x) = \dfrac{2x - 5}{(x^2 - 9)(x+1)}$, $\quad -5 \le x \le 5$
    Find the largest possible domain of f.

65. $f(x) = (x - 1)(x + 2)\sqrt{x} - \dfrac{1}{(x - 1)(x - 2)}$, $\quad 0 \le x \le 5$
    Find the largest possible domain of f.

66. $f(x) = \dfrac{(x - 1)(x + 2)}{x} - \dfrac{1}{(x - 1)(x - 2)}$, $\quad -2 \le x \le 5$
    Find the largest possible domain of f.

67. $f(x) = \sqrt{x}(x - 1)$, with domain $(0, 5]$. What is the lowest point on this graph?

68. $f(x) = \sqrt{x}(x + 1)$, with domain $(0, 5]$. Is this function increasing (getting larger) or decreasing (getting smaller)?

69. $f(x) = \dfrac{\sqrt{x}}{x - 1}$ Use your graphing calculator to find the largest possible domain. Is there any region in which this function is increasing?

70. $f(x) = \dfrac{x}{\sqrt{x - 1}}$ Use your graphing calculator to find the largest possible do-main. Is there any region in which this function is increasing?

APPLICATIONS

71. Demand   The demand for Sigma Mu Fraternity plastic brownie dishes is

$$q(p) = 361{,}201 - (p + 1)^2,$$

where q represents the number of brownie dishes Sigma Mu can sell per month at a price of $p¢$ each. Use this function to determine

(a) the number of brownie dishes Sigma Mu can sell per month if the price is set at 50¢,

(b) the number of brownie dishes they can unload per month if they give them away,

(c) the price at which Sigma Mu will be unable to sell any dishes.

72. Revenue   The total weekly revenue earned at Royal Ruby Retailers is given by

$$R(p) = -\frac{4}{3}p^2 + 80p,$$

where p was the price (in dollars) RRR charges per ruby. Use this function to determine

(a) the weekly revenue to the nearest dollar when the price is set at $20 per ruby,

(b) the weekly revenue to the nearest dollar when the price is set at $200 per ruby (interpret your result),

(c) the price RRR should charge in order to obtain a weekly revenue of $1200.

73. Investments in South Africa   The number of U.S. companies that invested in South Africa from 1986 through 1994 closely followed the function

$$n(t) = 5t^2 - 49t + 232.$$

Here, t is the number of years since 1986, and $n(t)$ is the number of U.S. companies that own at least 50% of their South African subsidiaries and employ 1,000 or more people.[*]

(a) Find the appropriate domain of n.

(b) Is $[0, +\infty)$ the appropriate domain? Give reasons for your answer.

74. Sony Net Income   The annual net income for *Sony Corp.* from 1989 through 1994 can be approximated by the function

$$I(t) = -77t^2 + 301t + 524.$$

Here, t is the number of years since 1989 and $I(t)$ is *Sony Corp.*'s net income in millions of dollars for the corresponding fiscal year.[†]

(a) Find the appropriate domain of t.

(b) Is $[0, +\infty)$ the appropriate domain? Give reasons for your answer.

75. Toxic Waste Treatment   The cost of treating waste by removing PCPs goes up rapidly as the quantity of PCPs removed goes up. Here is a possible model:

$$C(q) = 2{,}000 + 100q^2,$$

where q is the reduction in toxicity (in pounds of PCPs removed per day) and $C(q)$ is the daily cost (in dollars) of this reduction.

(a) Find the cost of removing 10 pounds of PCPs per day.

(b) Government subsidies for toxic waste cleanup amount to

$$S(q) = 500q,$$

where q is as above and $S(q)$ is the daily dollar subsidy. Calculate the net cost function (the cost after the subsidy is taken into account) $N(q)$, given the cost function and subsidy above, and find the net cost of removing 20 pounds of PCPs per day.

76. Dental Plans   A company pays for its employees' dental coverage at an annual cost C given by

$$C(q) = 1{,}000 + 100 \sqrt{q},$$

where q is the number of employees covered, and $C(q)$ is the annual cost in dollars.

(a) If the company has 100 employees, find its annual outlay for dental coverage.

(b) Assuming that the government subsidizes cover-

▼ * The model is the authors' (least squares quadratic regression with coefficients rounded to nearest integer). Source for raw data: Investor Responsibility Research Center Inc., Fleming Martin/New York Times, *The New York Times*, June 7, 1994, p. D1.

† The model is the authors' (least squares quadratic regression with coefficients rounded to nearest integer). Source for raw data: Sony Corporation/New York Times, *The New York Times*, May 20, 1994, p. D1.

age by an annual dollar amount of

$$S(q) = 200q,$$

calculate the net cost function, $N(q)$, to the company, and calculate the net cost of subsidizing its 100 employees. Comment on your answer.

77. *Cost* I want to fence in a 20-square-foot rectangular vegetable patch. For reasons too complicated to explain, the fencing for the east and west sides costs $4 per foot, while the fencing for the north and south sides costs only $2 per foot.

(a) Express the total cost of the project as a function of the length x of the east and west sides stretch of fencing.

(b) Graph this function, and, if you are using graphing technology, determine the approximate value of x that leads to the lowest total cost.

78. *Cost* Professor Gaunce Lewis is a keen gardener, specializing in the rarest exotic orchids. He is planning to start a new garden at the far end of the university's property, and he wishes to keep stray animals and students out. He decides to buy razor wire fencing at cost of $100 per yard and decides that he can afford to lay out enough money for 20 yards of fencing. His garden is planned as shown in the figure, although he can't quite decide on a value of x for its length.

Orchid garden

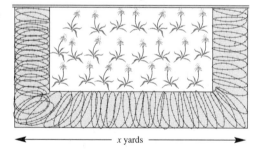

◄─────── *x* yards ───────►

(a) Express the area of the garden as a function A of x, being careful to state the domain.

(b) Graph this function, and, if you are using graphing technology, determine the approximate value of x that leads to the largest area.

79. *Volume* The volume of a cone is given by the formula $V = \frac{1}{3}\pi r^2 h$, where r is the radius of the base, and h is the height.

(a) If the ratio of h to r is given by $\frac{h}{r} = 3$, express V as a function of h only, and graph this function.

(b) Given the proportions of the cone in part (a), express h as a function of V, and graph this function.

80. *Surface Area* The surface area of a hollow cylinder of cross-sectional radius r and height h is given by $S = 2\pi rh$. Further, the surface area of a disc of radius r is given by $A = \pi r^2$.

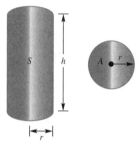

You are the design manager for Pebbles and Blips gourmet cat food, and you wish to design the shape of the cans that are to hold Pebbles and Blips Lobster Delight. Your expert consultant decides (after reading this book) that it would be most economical to design the cans with height equal to the diameter. Further, your cost analysis reveals that the gold/silver alloy you plan to use for the cans will cost the company $40 per square inch. Use this data to express the cost of a single can as a function of the height h. Graph this function.

81. *Biology—Reproduction* The Verhulst Model for population growth specifies the reproductive rate of an organism as a function of the total population according to the following formula:

$$R(p) = \frac{r}{1 + kp}.$$

Here, p is the total population, r and k are constants that depend on the particular circumstances and the organism being studied, and $R(p)$ is the reproduction

rate in new organisms per time period. Suppose that $r = 45$ and $k = 1/8000$ for a particular population.
(a) What is an appropriate domain for this function?
(b) Predict the reproduction rate when the population is 4,000.
(c) Graph R with the domain $0 < p \leq 10,000$.

82. Biology—Reproduction   Another model, the Predator Satiation Model for population growth, specifies the reproductive rate of an organism as a function of the total population, according to the following formula:

$$R(p) = \frac{rp}{1 + kp}.$$

Here, p is the total population, r and k are constants that depend on the particular circumstances and the organism being studied, and $R(p)$ is the reproduction rate in new organisms per time period.* Suppose $r = 1/125$ and $k = 1/5,000$ for a particular population.
(a) What is an appropriate domain for this function?
(b) Predict the reproduction rate when the population is 3,000.
(c) Graph R with the domain $0 \leq p \leq 50,000$.

83. Inflation Rates (GRE economics exam)[†]
Suppose the inflation rate in an economy is given by i (where, for example, a 20% inflation rate corresponds to $i = 0.20$). According to an economic model, inflation will cause the Gross National Product (GNP) Y to differ from the potential GNP Y_P according to the formula

$$\frac{Y - Y_P}{Y_P} = -9i.$$

Assuming a current inflation rate of 5%, use the formula to obtain a function G that specifies the GNP (Y) in terms of the potential GNP (Y_P), and use your function to calculate the GNP in the event that the potential GNP is $2 trillion. How does inflation affect GNP?

84. Inflation Rates (GRE economics exam)[†]
Suppose the inflation rate in an economy is given by i (where, for example, a 20% inflation rate corresponds to $i = 0.20$). According to an economic model, if an economy has a potential Gross National Product (GNP) of Y_P then the inflation rate j the following year is related to the present year's inflation rate i by the formula

$$Y_P(j - i) = 0.9(Y - Y_P),$$

where Y is this year's GNP. Assuming a current GNP of $2 trillion and a current inflation rate of 2.5%, specify j as a function of Y_P and use your function to estimate next year's inflation rate given a potential GNP of $2.2 trillion. Interpret your answer.

85. Acquisition of Language   The percentage $p(t)$ of children who are able to speak in at least single words by the age of t months can be approximated by the equation[‡]

$$p(t) = 100\left(1 - \frac{12,196}{t^{4.478}}\right) \qquad (t \geq 8.5).$$

(a) Graph p for $9 \leq t \leq 20$ and $0 \leq p \leq 100$.
(b) What percentage of children are able to speak in at least single words by the age of 12 months?
(c) By what age are 90% of children speaking in at least single words?

86. Acquisition of Language   The percentage $p(t)$ of children who are able to speak in sentences of five or more words by the age of t months can be approximated by the equation[‡]

$$p(t) = 100\left(1 - \frac{5.2665 \times 10^{17}}{t^{12}}\right) \qquad (t \geq 30).$$

(a) Graph p for $30 \leq t \leq 45$ and $0 \leq p \leq 100$.
(b) What percentage of children are able to speak in sentences of five or more words by the age of 36 months?
(c) By what age are 75% of children speaking in sentences of five or more words?

▼ * Source: *Mathematics in Medicine and the Life Sciences* by F. C. Hoppensteadt and C. S. Peskin (Springer-Verlag, New York, 1992) pp. 20–22.

[†] Based on a sample question in a GRE Economics exam. Source: *GRE Economics* by G. G. Gallagher, G. E. Pollock, W. J. Simeone, and G. Yohe, p. 48. (Research and Education Association, Piscataway, N.J., 1989).

[‡] The model is the authors' and is based on data presented in the article *The Emergence of Intelligence* by William H. Calvin, *Scientific American*, October, 1994, pp. 101–7.

87. *The Theory of Relativity*   In science fiction terminology, a speed of *warp 1* is the speed of light—about 3×10^8 meters per second. (Thus, for instance, a speed of warp 0.8 corresponds to 80% of the speed of light—about 2.4×10^8 meters per second.) According to Einstein's Special Theory of Relativity, a moving object appears to get shorter to a stationary observer as its speed approaches that of light. If a rocket ship whose length is l_0 meters at rest travels at a speed of warp p, its length in meters, as measured by a stationary observer, will be given by

$$l(p) = l_0\sqrt{1 - p^2}, \text{ with domain } [0, 1]$$

(a) Assuming that a rocket is 100 meters long at rest ($l_0 = 100$), estimate $l(0.95)$. What does this tell you?

(b) At what speed should a rocket ship be traveling in order that it appears to be one-half as long as when it is at rest?

88. *Newton's Law of Gravity*   The gravitational force exerted on a particle with mass m by another particle with mass M is given by the following function of distance:

$$F(r) = G\frac{Mm}{r^2}, \text{ with domain } (0, +\infty).$$

Here, r is the distance between the two particles in meters, the masses M and m are given in kilograms, $G \approx 0.0000000000667$, or 6.67×10^{-11}, and the resulting force is given in newtons.

(a) Given that M and m are both 1,000 kilograms, estimate $F(10)$. What does the answer tell you?

(b) How far apart should two 1,000-kilogram masses be in order that they experience an attractive force of 1 newton?

COMMUNICATION AND REASONING EXERCISES

89. Give a real-life scenario leading to a function with domain $[2, 100]$.

90. Give a real-life scenario leading to a function with domain $[-100, +\infty)$.

91. Why is the following assertion false?
"If $f(x) = x^2 - 1$, then $f(x + h) = x^2 + h - 1$."

92. Why is the following assertion false?
"If $f(x) = \sqrt{x}$, then $f(x + h) - f(x) = \sqrt{x + h} - \sqrt{x} = \sqrt{h}$."

93. How do the graphs of two functions differ if they are specified by the same formula but have different domains?

94. How do the graphs of two functions $f(x)$ and $g(x)$ differ if $g(x) = f(x) + 10$? (Try an example.)

95. How do the graphs of two functions $f(x)$ and $g(x)$ differ if $g(x) = f(x - 5)$? (Try an example.)

96. How do the graphs of two functions $f(x)$ and $g(x)$ differ if $g(x) = f(-x)$? (Try an example.)

▶ ══════ **1.3** LINEAR FUNCTIONS

The simplest interesting equations are the **linear equations.** The simplest interesting functions are the **linear functions.** Understanding linear functions and equations is crucial to understanding calculus, and we shall see some of the basic ideas of calculus in this and the next section.

LINEAR EQUATION

A **linear equation in the two unknowns** (or **variables**) x **and** y is an equation of the form

$$ax + by + c = 0$$

with a, b, and c being fixed numbers, and a and b not both zero.

▼   **EXAMPLE 1**

The following are linear equations.

(a) $2x + 3y + 4 = 0$ ($a = 2$, $b = 3$, $c = 4$)

(b) $x - y + 1 = 0$ ($a = 1$, $b = -1$, $c = 1$)

(c) $-2x + 1 = 0$ ($a = -2$, $b = 0$, $c = 1$)

(d) $\frac{3}{4}y = 0$ ($a = 0$, $b = \frac{3}{4}$, $c = 0$)

(e) $4 = 0$ is *not* a linear equation in two unknowns, since both a and b are zero.

(f) $x^2 + 2xy^2 + xy = 0$, while a perfectly respectable equation in x and y, is not a *linear* equation in x and y, since the definition does not allow expressions such as x^2, y^2, or xy in linear equations.

In the case of equations **(a)**, **(b)**, and **(d)** of the above example, we can solve for y as function of x and write

(a) $y = -\frac{2}{3}x - \frac{4}{3}$

(b) $y = x + 1$

(d) $y = 0$.

These are examples of **linear functions.**

LINEAR FUNCTION

A **linear function** is a function that can be written in the form

$$f(x) = mx + b$$

with m and b being fixed numbers (the names m and b are traditional).

Note that a linear function $f(x) = mx + b$ leads to the linear equation $y = mx + b$, so the graph of a linear function is the graph of a linear equation. Therefore, all that we say about the graph of linear equations below will also apply to linear functions.

In Section 1 we sketched the graph of the linear equation $x + y = 4$ and got a straight line. In fact, the graph of *any* linear equation is a straight line (hence the name *linear*). Before we discuss easy methods of drawing the graph of any linear equation, we need to introduce an extremely important concept, the **slope** of a straight line. The slope of a straight line is a way of measuring its steepness, similar to the "gradient" used by civil engineers and surveyors to measure the steepness of roads or railway tracks. Figure 1 shows a car proceeding from left to right up or down various inclines.

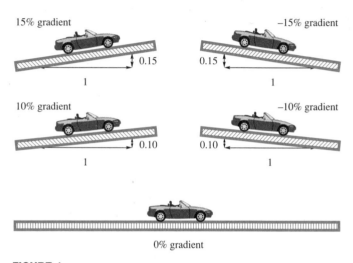

15% gradient   −15% gradient

0.15   0.15

1   1

10% gradient   −10% gradient

0.10   0.10

1   1

0% gradient

FIGURE 1

Look first at the 15% gradient at the top left. For every one unit across to the right, the road goes up 0.15 units, or 15% of the distance across. The −15% gradient at the top right goes *down* 0.15 units for every one unit across to the right. The ±10% gradients are similar: the road goes up or down 0.10 units for every unit across. The 0% gradient is horizontal: it goes up 0 units for every unit across.

We measure the steepness of a line in the same way, but instead of saying that the road at the top left of Figure 1 has a gradient of 15%, we shall say that, as a straight line, it has a **slope of 0.15.** Similarly, the line with a −10% gradient has a slope of −0.10. In other words, the **slope** of a line is the number of units it goes up for every unit across to the right. If it goes down instead, the line has negative slope. Consider the line of slope 2 shown in Figure 2. This line goes up two units for every unit across to the right. Thus, it goes up 6 units for every 3 units across, and it goes up 8 units for every 4 units across, and so on. We say that we have a **rise** of 2 for a **run** of 1, a rise of 6 for a run of 3, and so on.

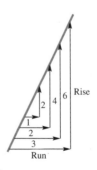

6 Rise

2   4

1

2

3

Run

FIGURE 2

Referring to Figure 2, we see that the ratios *rise/run* all reduce to the same answer, 2 (since the right triangles are all similar). Thus, another way of defining the slope is to say that the slope is the ratio *rise/run*.

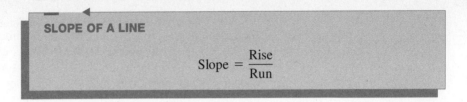

SLOPE OF A LINE

$$\text{Slope} = \frac{\text{Rise}}{\text{Run}}$$

Figure 3 gives several more examples. Notice that the larger the numerical value of the slope, the steeper the line.

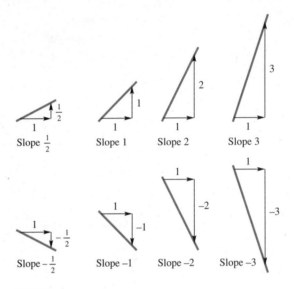

FIGURE 3

Q Suppose we have a straight line in the xy-plane, passing through two given points (x_1, y_1) and $(x_2\, y_2)$. Can we compute its slope in terms of these coordinates?

A According to Figure 4, the rise is $y_2 - y_1$, while the run is $x_2 - x_1$.

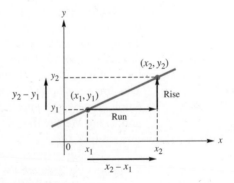

FIGURE 4

Thus, the slope, which is traditionally called* m, is given by the following formula.

SLOPE m OF THE LINE THROUGH THE POINTS (x_1, y_1) AND (x_2, y_2)

$$m = \frac{y_2 - y_1}{x_2 - x_1}$$

▶ NOTE   Here is a good way of thinking about the above formula. The quantity $y_2 - y_1$ in the numerator is the change in the y-coordinate as we go from the first point (x_1, y_1) to the second point (x_2, y_2). (Refer to Figure 4.) Similarly, the quantity $x_2 - x_1$ in the denominator is the change in the x-coordinate as we go from the first point to the second. Thus, we can write

$$m = \frac{y_2 - y_1}{x_2 - x_1}$$
$$= \frac{\text{Change in } y}{\text{Change in } x} = \frac{\Delta y}{\Delta x},$$

where we use the shorthand Δy and Δx (Δ is the Greek letter *delta*) for "change in y" and "change in x." ◀

So to get the slope of a line in the xy-plane, all we need to know are the coordinates of any two distinct points on that line.

▼ **EXAMPLE 2**

Find the slope of the line through $(1, 3)$ and $(5, 11)$.

SOLUTION   Take $(x_1, y_1) = (1, 3)$ and $(x_2, y_2) = (5, 11)$ and use the formula to get

$$m = \frac{\Delta y}{\Delta x} = \frac{y_2 - y_1}{x_2 - x_1} = \frac{11 - 3}{5 - 1} = \frac{8}{4} = 2.$$

Before we go on... What if we had chosen to list the two points in reverse order? That is, suppose we had taken $(x_1, y_1) = (5, 11)$ and $(x_2, y_2) = (1, 3)$. Then we would have obtained

$$m = \frac{\Delta y}{\Delta x} = \frac{y_2 - y_1}{x_2 - x_1} = \frac{3 - 11}{1 - 5} = \frac{-8}{-4} = 2,$$

the same answer. The order in which we take the points is not important, *as long as we use the same order on the top and the bottom.*

▼ * This is done for reasons that no one knows. There has actually been some research into this question lately, but still no one has found why the letter m is used.

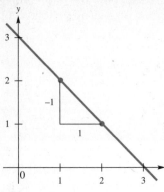

FIGURE 5

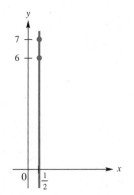

FIGURE 6

▼ EXAMPLE 3

Find the slope of the line through (1, 2) and (2, 1).

SOLUTION

$$m = \frac{\Delta y}{\Delta x} = \frac{y_2 - y_1}{x_2 - x_1} = \frac{1 - 2}{2 - 1} = \frac{-1}{1} = -1$$

Before we go on... Recall that lines with negative slope go downhill as you move from left to right. The line in this example is shown in Figure 5.

▼ EXAMPLE 4

Find the slope of the line through the points (2, 3) and (−5, 3).

SOLUTION

$$m = \frac{\Delta y}{\Delta x} = \frac{3 - 3}{-5 - 2} = \frac{0}{-7} = 0$$

Before we go on... A line of slope 0 has a 0 rise, so it is a *horizontal* line.

▼ EXAMPLE 5

Finally, find the slope of the line through ($\frac{1}{2}$, 6) and ($\frac{1}{2}$, 7).

SOLUTION

$$m = \frac{\Delta y}{\Delta x} = \frac{7 - 6}{\frac{1}{2} - \frac{1}{2}} = \frac{1}{0} \quad ✖$$

We have used the symbol ✖ to remind you that there is no such number as $\frac{1}{0}$. This line has an **undefined** or **infinite** slope. (Although there is no such number as $\frac{1}{0}$, we sometimes think of it as being infinite.) If we plot the two points in question, we see that the line passing through them is *vertical* (Figure 6).

Learn to recognize the approximate slope of a line by looking at it. For example, recognize a line of slope 0 as being horizontal, a line of slope 1 as going up at an angle of 45°, and a line of slope −1 as going downward at the same angle. A line rising at an angle steeper than 45° would have a positive slope larger than 1, a line falling at an angle shallower than 45° would have a slope between −1 and 0, and so on.

PARALLEL AND PERPENDICULAR LINES

From Figure 7, you can conclude the following.

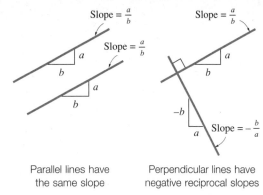

Parallel lines have   Perpendicular lines have
the same slope   negative reciprocal slopes

FIGURE 7

1.   Parallel lines have the same slope. For example, if a line has slope -3, then any line parallel to it also will have slope -3.
2.   Perpendicular lines (i.e., lines at right angles to each other) have negative reciprocal slopes. For example, if a line has slope 3, then any line perpendicular to it will have slope $-\frac{1}{3}$.

THE EQUATION OF THE STRAIGHT LINE WITH A GIVEN SLOPE THROUGH A GIVEN POINT

The slope by itself is not enough to completely specify a line. We need two pieces of information:

1.   the *slope m* (which specifies the direction of the line), and
2.   a *point* (x_0, y_0) on the line (which pins down its location in the plane).

Q What is the equation of the line through the point (x_0, y_0) with slope *m*?

A Before stating the answer in general, let us look at a specific example: suppose we want to find the equation of the line through $(1, 2)$ with slope 3. If (x, y) is a point on this line other than $(1, 2)$, then, since the line through the points $(1, 2)$ and (x, y) has slope 3, the slope formula tells us that

$$m = 3 = \frac{y - 2}{x - 1}.$$

Multiplying both sides by the quantity $(x - 1)$ gives

$$y - 2 = 3(x - 1),$$

which is an equation of this line, since it must be satisfied by any point (x, y) on the line. Thus, an equation of the line through $(1, 2)$ with slope 3 is $y - 2 = 3(x - 1)$.

We can easily generalize this derivation to get the following formula.

> **THE POINT-SLOPE FORMULA**
>
> An equation of the line through the point (x_0, y_0) with slope m is given by
>
> $$y - y_0 = m(x - x_0).$$

▶ CAUTION   The point (x_0, y_0) represents a given point, so the subscripted variables x_0 and y_0 will always be replaced by actual numbers, just as in the formula for the slope of a line passing through two given points: $m = (y_2 - y_1)/(x_2 - x_1)$. The terms x and y, on the other hand, remain as x and y, since they are the variables in the equation of the line.   ◀

▼ **EXAMPLE 6**

Find an equation of the line through $(1, 3)$ with slope 2.

SOLUTION   In order to apply the point-slope formula, we need:

- a point—given here by $(x_0, y_0) = (1, 3)$;
- the slope—given here by $m = 2$.

An equation of the line is therefore

$$y - y_0 = m(x - x_0).$$

In other words,

$$y - 3 = 2(x - 1)$$

and

$$y - 3 = 2x - 2, \quad \text{or} \quad y = 2x + 1.$$

Before we go on... It is a good idea to check our answers whenever possible. Here, the answer is $y = 2x + 1$. We can check that it is correct by making sure that it does pass through $(1, 3)$ and has slope 2. To check that it passes through $(1, 3)$, we substitute $x = 1$ and $y = 3$ into our equation, getting

$$3 = 2(1) + 1. ✔$$

To make certain that the line has slope 2, notice first that the formula $y - y_0 = m(x - x_0)$ tells us that the coefficient of x in the equation should be the slope m (since m is the only term multiplied by x in the point-slope formula). Thus, to check that the line has slope 2, we need only check that the coefficient of x in our answer is 2.

Another way you could check the answer is by using a graphing calculator or computer. Use it to plot the graph of $y = 2x + 1$, and then use the trace feature to check that it does in fact pass through the point $(1, 3)$ and that it rises 2 units for every unit to the right.

▶ CAUTION   If the axes are scaled differently on your graphing calculator, then lines won't appear to have the correct slope. Figure 8 shows the graph of $y = 2x + 1$ twice: on the xy-plane, with axes scaled the same, and on a graphing calculator, with axes scaled differently.

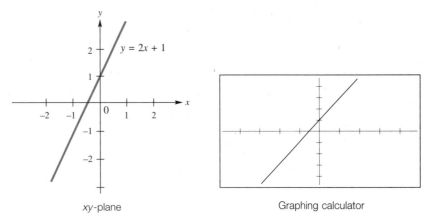

xy-plane   Graphing calculator

FIGURE 8

The graphing calculator line appears to have a slope slightly less than 2. This distortion is caused by the fact that (as with most calculators) the screen is not square. If, as is done above, you draw the graph for $-5 \le x \le 5$ and $-5 \le y \le 5$, the units on the x-axis will be different from the units on the y-axis, and lines will not appear to have their correct slopes.   ◀

▼   **EXAMPLE 7**

Find an equation of the line through the points $(1, 2)$ and $(3, -1)$.

SOLUTION   We need the following:

■   a point—we have two to choose from, so we take the first, $(x_0, y_0) = (1, 2)$;
■   the slope—not given *directly*, but we do have enough information to calculate it. Since we are given two points on the line, we can use the slope formula:

$$m = \frac{y_2 - y_1}{x_2 - x_1} = \frac{-1 - 2}{3 - 1} = \frac{-3}{2}.$$

An equation of the line is therefore

$$y - y_0 = m(x - x_0).$$

Thus, $$y - 2 = -\frac{3}{2}(x - 1)$$

and $$y - 2 = -\frac{3}{2}x + \frac{3}{2}, \quad \text{or} \quad y = -\frac{3}{2}x + \frac{7}{2}.$$

Before we go on... Once again we check the answer. All we need to verify is that it passes through the given points $(1, 2)$ and $(3, -1)$. Substituting each of these into the equation gives

$$2 = -\frac{3}{2}(1) + \frac{7}{2} \quad ✔$$

$$-1 = -\frac{3}{2}(3) + \frac{7}{2}. \quad ✔$$

As in the preceding example, you could check the answer by using a graphing calculator or computer to plot $y = -\frac{3}{2}x + \frac{7}{2}$, and then use the trace feature to check that it passes through the two points $(1, 2)$ and $(3, -1)$. Always bear in mind that the trace feature has limited accuracy (even if you use it in conjunction with the zoom feature).

▼ **EXAMPLE 8**

Find the points where the line in Example 7 crosses the x- and y-axes.

SOLUTION   The line is shown in Figure 9. The point where it crosses the x-axis has y-coordinate 0, so we substitute $y = 0$ in the equation and solve for x.

$$0 = -\frac{3}{2}x + \frac{7}{2}$$

$$\frac{3}{2}x = \frac{7}{2}$$

$$x = \frac{7}{3}$$

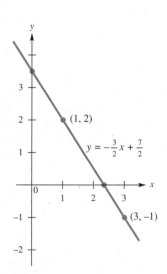

FIGURE 9

Therefore, the line crosses the x-axis at $(\frac{7}{3}, 0)$. The point where it crosses the y-axis has x-coordinate 0, so we substitute $x = 0$ in the equation.

$$y = -\frac{3}{2} \cdot 0 + \frac{7}{2} = \frac{7}{2}$$

Therefore, the line crosses the y-axis at $(0, \frac{7}{2})$.

Before we go on... These numbers $\frac{7}{3}$ and $\frac{7}{2}$ are called the x-intercept and the y-intercept of the line. In general, the x-coordinate of the point where a line crosses the x-axis is called the **x-intercept** of the line. Similarly, the y-coordinate of the point where it crosses the y-axis is the **y-intercept** of the line. Given an equation of a line, one easy way of sketching the line is to find its intercepts and draw the line connecting the corresponding points. This technique does not always work: try it with $2x - y = 0$. This equation has both x- and y-intercepts zero and thus gives only a single point—the origin—to plot. To draw the graph of such an equation, you could either find the coordinates of an additional point on the line or use the technique we are about to discuss.

▼ **EXAMPLE 9**

Find the equation of the line with slope m and y-intercept b.

SOLUTION   Although we have probably seen this answer somewhere before, let's pretend this is entirely new to us. We need two things:

- ■ a point—not given directly, but the point b on the y-axis has coordinates $(0, b)$, so take $(x_0, y_0) = (0, b)$;
- ■ the slope—given here simply as m.

The equation of the line is therefore

$$y - y_0 = m(x - x_0).$$

Thus,

$$y - b = m(x - 0)$$

and

$$y - b = mx, \quad \text{or} \quad y = mx + b.$$

Before we go on... To check that the graph of $y = mx + b$ does in fact pass through $(0, b)$, substitute this into the equation to get

$$b = m(0) + b. \quad ✔$$

◀

THE SLOPE-INTERCEPT FORMULA

An equation of the line with slope m and y-intercept b is

$$y = mx + b.$$

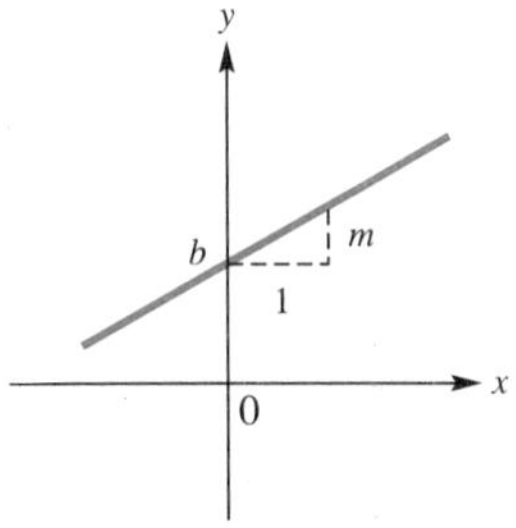

Figure 10 illustrates the graph of the line $y = mx + b$. Notice that the line crosses the y-axis at $y = b$, and that, since the slope is $m = \frac{m}{1}$, it goes up m units for every 1 unit in the x-direction.

Q What use is this formula if we already have the point-slope formula to give us an equation for any line?

A Suppose you are given the equation of a straight line in this form: for example, $y = \frac{3}{4}x - 1$. Then you can tell at a glance that the slope is $\frac{3}{4}$ and the y-intercept is -1.

The line $y = mx + b$

FIGURE 10

FINDING THE SLOPE AND y-INTERCEPT FROM THE EQUATION OF A LINE

In the equation $y = mx + b$, the coefficient m of x is the slope, and the constant term b is the y-intercept of the line.

◀

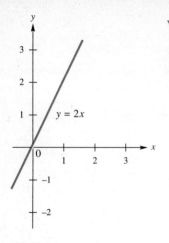

FIGURE 11

▼ **EXAMPLE 10**

Sketch the graph of $2x - y = 0$.

SOLUTION   Rewriting this equation in the form $y = mx + b$, we get

$$y = 2x.$$

We can now see that the slope of the line is 2 and the y-intercept is 0. Figure 11 shows the line.

Although you could use a graphing calculator or computer to sketch this line, it would really be overkill to do so. It may take you longer to use the device to draw the plot than it would take you to draw it yourself.

Note that everything we say about the graph of the equation $y = mx + b$ applies equally well to the graph of the general linear *function* $f(x) = mx + b$.

▼ **EXAMPLE 11**

Find the linear function whose graph is the straight line through $(2, -2)$ that is parallel to the line $3x + 4y = 5$.

SOLUTION   We begin by finding an equation of this line. As usual, we need

■ a point—given here as $(2, -2)$;
■ the slope. Since the required line is parallel to $3x + 4y = 5$, it must have the same slope. We can find the slope of this line by solving for y and then looking at the coefficient of x. Solving for y gives

$$y = -\frac{3}{4}x + \frac{5}{4},$$

so the slope is $-\frac{3}{4}$, since this is the coefficient of x.

Thus, an equation for the line is

$$y - y_0 = m(x - x_0)$$

$$y + 2 = -\frac{3}{4}(x - 2)$$

or $\qquad y + 2 = -\frac{3}{4}x + \frac{3}{2}, \quad$ or $\quad y = -\frac{3}{4}x - \frac{1}{2}.$

Notice that we have put the answer in $y = mx + b$ form by solving for y. The corresponding function is therefore

$$f(x) = -\frac{3}{4}x - \frac{1}{2}.$$

Before we go on... To check the solution, $f(x) = -\frac{3}{4}x - \frac{1}{2}$, we first test that the graph of this function passes through the point $(2, -2)$.

$$f(2) = -\frac{3}{4}(2) - \frac{1}{2} = -2 \quad \checkmark$$

To test that it is indeed parallel to the given line, all we need to do is make sure that the slope is $-\frac{3}{4}$. Since this is the coefficient of x, all is well.

With a graphing calculator or computer you can plot both $y = -\frac{3}{4}x + \frac{5}{4}$ and $y = -\frac{3}{4}x - \frac{1}{2}$ and look at the graphs to check that they do in fact appear to be parallel. As usual, you can use the trace feature to check (to within the accuracy of the display) that the second line does pass through the point $(2, -2)$.

▼ **EXAMPLE 12**

Find the linear function whose graph is the straight line through $(2, -2)$ that is perpendicular to the line $3x + 4y = 5$.

SOLUTION   Again, we start by finding an equation of the line. As usual, we need

- a point—given here as $(2, -2)$;
- the slope. Since the required line is perpendicular to $3x + 4y = 5$, its slope is the negative reciprocal of the slope of $3x + 4y = 5$. As in Example 11, we solve for y to get $y = -\frac{3}{4}x + \frac{5}{4}$. Thus, the slope we want is the negative reciprocal of $-\frac{3}{4}$, namely, $\frac{4}{3}$.

So the desired equation is

$$y - y_0 = m(x - x_0)$$
$$y + 2 = \frac{4}{3}(x - 2)$$

or $\qquad y + 2 = \frac{4}{3}x - \frac{8}{3}, \quad$ or $\quad y = \frac{4}{3}x - \frac{14}{3}.$

The corresponding function is

$$f(x) = \frac{4}{3}x - \frac{14}{3}.$$

Before we go on... First, we test that the graph passes through $(2, -2)$.

$$f(2) = \frac{4}{3}(2) - \frac{14}{3} = -2 \quad \checkmark$$

Finally, the coefficient of x tells us that the slope is $\frac{4}{3}$, which is what we wanted.

With a graphing calculator or computer, you can plot both $y = -\frac{3}{4}x + \frac{5}{4}$ and $y = \frac{4}{3}x - \frac{14}{3}$. Do they appear to cross at right angles? If not, why not? (See the caution after Example 6.)

▼ **EXAMPLE 13**

Sketch the lines found in Examples 11 and 12 above.

SOLUTION   The line in Example 11 has equation $y = -\frac{3}{4}x - \frac{1}{2}$. Thus, its slope is $-\frac{3}{4}$ and its y-intercept is $-\frac{1}{2}$. Since the slope is $-\frac{3}{4}$, it goes down 3 units for every 4 units to the right, so we can sketch it as in Figure 12. For the line in Example 12, the equation is $y = \frac{4}{3}x - \frac{14}{3}$, so we can sketch it as in Figure 13.

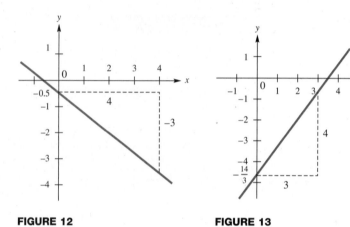

FIGURE 12   **FIGURE 13**

▼ **EXAMPLE 14**

Find the equation of **(a)** the horizontal line passing through $(-9, 5)$, and **(b)** the vertical line passing through the same point.

SOLUTION

(a)   Here the point is $(-9, 5)$ and the slope is 0, yielding $y = 5$. (Check this for yourself using the point-slope formula.)

(b)   Here, the point remains $(-9, 5)$, but the slope is undefined! So we can't use the point-slope formula, since there isn't a well-defined slope. (That formula only makes sense for a slope m that is a real value.) What can we do? Well, here are some points on the required line:

$$(-9, 1), (-9, 2), (-9, 3), \ldots,$$

so $x = -9$, and $y = anything$. If we simply say that $x = -9$, then these points are all solutions, so the equation is $x = -9$.

Before we go on... The horizontal line is the graph of the function $f(x) = 5$. However, the vertical line is not the graph of a function (why not?).

SUMMARY: SURE-FIRE METHOD OF FINDING AN EQUATION OF A
STRAIGHT LINE

1. All we need are two pieces of information: **(a)** a *point* on the line and **(b)** the *slope* of the line. We can then use the point-slope formula $y - y_0 = m(x - x_0)$.
2. Sometimes the problems are not posed so simply, but there must *always* be enough information to get these two quantities. Here are some instances that can occur:

    (a) We are given two points (x_1, y_1) and (x_2, y_2): use $m = (y_2 - y_1)/(x_2 - x_1)$ to get the slope. If the slope is undefined, see (d).
    (b) We are given a point and the equation of a parallel line: obtain the slope of the other line by solving its equation for y. This value is the slope to use. However, if the parallel line is vertical, and hence has no slope, see (d).
    (c) We are given a point and the equation of a perpendicular line: obtain the slope of the other line by solving its equation for y, and then take its negative reciprocal. This value is the slope to use. However, if the slope of the given line is zero, then the required line, being perpendicular to a horizontal line, is vertical, so see (d).
    (d) An equation of the vertical line passing through (x_0, y_0) is $x = x_0$.

3. The equations we will be using are these, and we strongly suggest you commit them to memory.

    (a) Point-Slope Formula An equation of the line through (x_0, y_0) with slope m is
    $$y - y_0 = m(x - x_0).$$

    (b) Slope-Intercept Formula An equation of the line with y-intercept b and slope m is
    $$y = mx + b.$$

    (c) Horizontal and Vertical Lines An equation of the horizontal line through (x_0, y_0) is
    $$y = y_0.$$
    An equation of the vertical line through (x_0, y_0) is
    $$x = x_0.$$

▶ 1.3 EXERCISES

In each of Exercises 1–18, find the slope of the straight line through the given pair of points. Try to do as many as you can without writing anything down.

1. $(0, 0)$ and $(1, 2)$

2. $(0, 0)$ and $(-1, 2)$

3. $(-1, -2)$ and $(0, 0)$

4. $(2, 1)$ and $(0, 0)$

5. $(4, 3)$ and $(5, 1)$

6. $(4, 3)$ and $(-1, -5)$

7. $(1, -1)$ and $(2, -2)$

8. $(-2, 2)$ and $(-1, -1)$

9. $(0, 1)$ and $(-\frac{1}{2}, \frac{3}{4})$

10. $(\frac{1}{2}, 1)$ and $(-\frac{1}{2}, \frac{3}{4})$

11. $(4, \sqrt{2})$ and $(5, \sqrt{2})$

12. $(1, 1)$ and $(\sqrt{2}, \sqrt{2})$

13. $(4, \sqrt{2})$ and $(5, 2\sqrt{2})$

14. $(4\sqrt{2}, 2\sqrt{2})$ and $(0, -2\sqrt{2})$

15. (a, a) and $(a, 3a)$ $(a \neq 0)$

16. $(a, 1)$ and $(a, 2)$

17. (a, b) and (c, d) $(a \neq c)$

18. (a, b) and (a, d) $(b \neq d)$

Referring to Exercises 1–18, say whether each of the following pairs of lines are parallel, perpendicular, or neither.

19. 1 and 3   **20.** 2 and 4   **21.** 5 and 7   **22.** 6 and 8

23. 11 and 15   **24.** 12 and 14   **25.** 5 and 9   **26.** 16 and 18

27. Referring to the following figure, match the lines with their slopes.
(I) $-\frac{1}{3}$; (II) 3; (III) $-\frac{1}{4}$; (IV) 1; (V) -1; (VI) undefined;
(VII) $\frac{1}{2}$; (VIII) -2; (IX) 0.

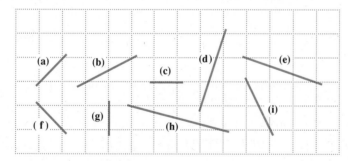

28. Referring to the following figure, match the lines with their slopes.
(I) $-\frac{1}{3}$; (II) 3; (III) 2; (IV) 1; (V) -1; (VI) undefined;
(VII) $-\frac{1}{2}$; (VIII) -2; (IX) 0.

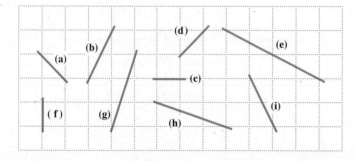

Find the linear functions whose graphs are the straight lines in Exercises 29–44.

29. Through $(1, 3)$ with slope 3

30. Through $(2, 1)$ with slope 2

31. Through $(1, -\frac{3}{4})$ with slope $\frac{1}{4}$

32. Through $(0, -\frac{1}{3})$ with slope $\frac{1}{3}$

33. Through $(2, -4)$ and $(1, 1)$

34. Through $(1, -4)$ and $(-1, -1)$

35. Through $(1, -\frac{3}{4})$ and $(\frac{1}{2}, \frac{3}{4})$

36. Through $(\frac{1}{2}, -\frac{3}{4})$ and $(\frac{1}{4}, \frac{3}{4})$

37. Through $(6, 6)$ and parallel to the line $x + y = 4$

38. Through $(\frac{1}{3}, -1)$ and parallel to the line $3x - 4y = 8$

39. Through $(\frac{1}{2}, 5)$ and parallel to the line $4x - 2y = 11$

40. Through $(\frac{1}{3}, 0)$ and parallel to the line $6x - 2y = 11$

41. Through $(0, 2)$ and perpendicular to the line $x + y = 4$

42. Through $(0, 4)$ and perpendicular to the line $2x + 4y = 4$

43. Through $(3, -2)$ and perpendicular to the line $y = -2$

44. Through $(3, -2)$ and perpendicular to the line $x = 3$

Sketch the straight lines with equations given in Exercises 45–56.

45. $2x + 3y = 6$

46. $x + 2y = 6$

47. $y + \frac{1}{4}x = -4$

48. $y - \frac{1}{4}x = -4$

49. $7x - 3y = 5$

50. $2x - 3y = 1$

51. $3x = 8$

52. $2x = -7$

53. $6y = 9$

54. $3y = 4$

55. $2x = 3y$

56. $3x = -2y$

Sketch each of the lines in Exercises 57–66 on a graphing calculator. In each case, use your graph to find the approximate value of the x-intercept (if any) of the given line. Check by finding the exact value of the intercept.

57. $y = 4.1x - 5.4$

58. $y = 2.3x + 5.5$

59. $y = -10.4x + 10$

60. $y = 20.3x - 31.2$

61. $y = 10,050x + 4,323$

62. $y = -5,300x - 2,000$

63. $13x - 15y = 23$

64. $7x + 19y = 43$

65. $100x + 50y = -1,020$

66. $1,000x - 3y = 66$

COMMUNICATION AND REASONING EXERCISES

67. To what linear function of x does the linear equation $ax + by = c$ $(b \neq 0)$ correspond?

68. Why did we specify $b \neq 0$ in Exercise 67?

69. Which linear equations $ax + by = c$ have graphs that are *not* the graphs of linear functions?

70. In terms of a, b, and c, what are the slope, x-intercept, and y-intercept of the line with equation $ax + by = c$ $(a \neq 0, b \neq 0)$?

71. Complete the following. The slope of the line with equation $y = mx + b$ is the number of units that _____ increases per unit increase in _____ .

72. Complete the following. If, in a straight line, y is increasing three times as fast as x, then its _____ is _____ .

73. A friend tells you that the line through the points $(3, 5)$ and $(4, 5)$ has no slope. Is she correct? (Explain your answer.)

74. Another friend tells you that the line through the points $(3, 5)$ and $(3, 6)$ has no slope. Is he correct? (Explain your answer.)

▶ ══════ **1.4** LINEAR MODELS

Linear functions can be used to represent a wide variety of situations in everyday life. Using linear functions to represent situations in real life is called **linear modeling,** and we illustrate linear modeling with several examples.

▼ **EXAMPLE 1**   Cost Function

The Yellow Cab Company charges $1 on entering the cab, plus an additional $2 per mile.*

(a) Find the cost C of an x-mile trip.

(b) Use your answer to calculate the cost of a 40-mile trip.

(c) What is the cost of the second mile?

(d) What is the cost of the tenth mile?

(e) Graph C as a function of x.

SOLUTION

(a) We are being asked to find how the cost C depends on the length x of the trip, or to find C as a function of x. Look at the cost in a few instances.

$$\text{When } x = 1, \text{ the cost is } C = 1 + 2(1) = 3.$$
$$\text{When } x = 2, \text{ the cost is } C = 1 + 2(2) = 5.$$
$$\text{When } x = 3, \text{ the cost is } C = 1 + 2(3) = 7.$$
$$\cdots$$
$$\text{When } x = x, \text{ the cost is } C = 1 + 2x.$$

Thus, the cost of an x-mile trip is given by the linear function

$$C(x) = 2x + 1.$$

▼ * This is equivalent to charging $3 for the first mile and $2 for each subsequent mile. In 1993, cab fare in Chicago cost $2.40 for the first mile and $1.40 for each subsequent mile. (We are allowing for inflation.)

Notice that the slope 2 is equal to the incremental cost per mile, which we call **marginal cost,** while the varying quantity $2x$ we call the **variable cost.** The y-intercept is equal to the basic fee, which we call the **fixed cost.** In general, a linear cost function has the following form.

$$\overbrace{}^{\text{Variable cost}}$$

$$C(x) = \underset{\underset{\text{Marginal cost}}{\uparrow}}{mx} + \underset{\underset{\text{Fixed cost}}{\uparrow}}{b}$$

(b) We can now use the cost function to calculate the cost of a 40-mile trip as

$$C(40) = 2(40) + 1 = \$81.$$

(c) To calculate the cost of the second mile, we *could* proceed as follows.

1. Find the cost of a one-mile trip: $C(1) = 2(1) + 1 = \$3$.
2. Find the cost of a two-mile trip: $C(2) = 2(2) + 1 = \$5$.
3. Therefore, the cost of the second mile is $\$5 - \$3 = \$2$.

But notice that this is just the marginal cost. In fact, the marginal cost is the cost of each additional mile, so we could have done this more simply as follows.

$$\text{Cost of second mile} = \text{Marginal cost} = \$2.$$

(d) Since the marginal cost is the cost of each additional mile, the answer once again is $2.

(e) Figure 1 shows the graph of the cost function, which we can interpret as a *cost vs. miles* graph. The fixed cost is the starting height on the left, while the marginal cost is the slope of the line.

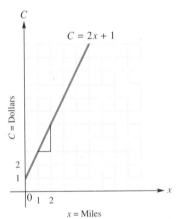

FIGURE 1

LINEAR COST FUNCTION

A **cost function** specifies the cost C as a function of the number of items x. Thus, $C(x)$ is the cost of x items. A cost function of the form

$$C(x) = mx + b$$

is called a **linear cost function.** The quantity mx is called the **variable cost,** and the intercept b is called the **fixed cost.** The slope m, which represents the **marginal cost,** measures the incremental cost per item.

▼ **EXAMPLE 2**   Cost Function

The manager of the FrozenAir Refrigerator factory notices that on Monday it cost them a total of $25,000 to build 30 refrigerators, and on Tuesday it cost them $30,000 to build 40 refrigerators. Assuming a linear cost function, what is their daily fixed cost, and what is the marginal cost?

SOLUTION   The secret in this kind of problem is to realize that we have been given two points on a line whose equation we want to find. The assumption is that the cost C will depend on x, the number of refrigerators, by an equation of the form

$$C = mx + b,$$

where m is the marginal cost and b is the fixed cost. We are told that $C = 25{,}000$ when $x = 30$, and this amounts to being told that $(30, 25{,}000)$ is a point on the line we get by graphing $C = mx+b$. Similarly, $(40, 30{,}000)$ is another point on the line. The graph of C vs. x is shown in Figure 2. To find the equation of this line we recall that you need two items of information: a point on the line, and the slope.

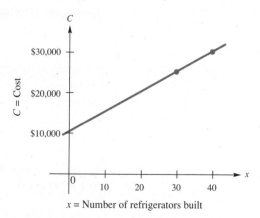

FIGURE 2

Point: We are given two of them, so you can choose the first: $(30, 25{,}000)$.
Slope: We use the slope equation (with C playing the role of y).

$$m = \frac{C_2 - C_1}{x_2 - x_1}$$

$$= \frac{30{,}000 - 25{,}000}{40 - 30}$$

$$= \frac{5{,}000}{10} = 500$$

In other words, the marginal cost is $500 per refrigerator. Put another way, each refrigerator adds $500 to the total cost. To complete the problem, we use the point-slope formula:

$$C - C_0 = m(x - x_0)$$
$$C - 25,000 = 500(x - 30) = 500x - 15,000$$
$$C = 500x + 10,000$$

or

$$C(x) = 500x + 10,000.$$

Since $b = 10,000$, the factory's fixed cost is $10,000 each day.

Before we go on... We can check our equation by making sure that it gives the known costs of making 30 and 40 refrigerators:

$$C(30) = 500 \cdot 30 + 10,000 = 15,000 + 10,000$$
$$= 25,000 \ ✔$$
$$C(40) = 500 \cdot 40 + 10,000 = 20,000 + 10,000$$
$$= 30,000. \ ✔$$

We should also think about the domain of C. Certainly it makes no sense to speak of a negative number of refrigerators here, so $x \geq 0$. On the other hand, there likely is a largest number of refrigerators that this factory can churn out. If, for example, they could make no more than 70 refrigerators in one day, we would say that the domain of C is $[0, 70]$.

We also wish to consider revenue and profit.

REVENUE AND PROFIT

The **revenue** resulting from one or more business transactions is the total payment received, sometimes called the gross proceeds. The **profit,** on the other hand, is the *net* proceeds, or what remains of the revenue when costs are subtracted. Profit, revenue, and cost are related by the following formulas.

$$\text{Profit} = \text{Revenue} - \text{Cost}$$
$$P = R - C$$

If the profit is negative, say $-$500, we refer to a **loss** (of $500 in this case).

▼ **EXAMPLE 3**   Revenue, Profit, and Break-Even Analysis

Lumber futures prices at the Chicago Mercantile Exchange reached a record high of $363.50 per lot of 1,000 board feet in February 1993.*

▼ *Source: *The Chicago Tribune*, February 10, 1993, Business Section, p. 1.

(a) Find the revenue that a seller of lumber at the Merc would earn if she sold x lots of lumber at the above price.

(b) Assuming that she has fixed costs of $1,000 per day and can buy lumber at $200 per lot, express her profit as a linear function of the number of lots she trades (buys and sells) per day.

(c) How many lots should she trade per day to break even?

SOLUTION

(a) Here, the revenue the seller obtains from the sale of a single lot is $363.50. To obtain the revenue from the sale of x lots, let us do a little accounting.

$$x \text{ lots @ } \$363.50 \text{ per lot gives } x \cdot 363.50 = 363.5x$$

Thus, the revenue function is

$$R(x) = 363.5x.$$

(b) We are now asked for the *profit*. In order to calculate profit, we use the formula

$$\text{Profit} = \text{Revenue} - \text{Cost}.$$

We already have the revenue function, so we must now find the cost function. Since the fixed daily cost is $1,000 per day and the marginal cost is $200 per lot, we can simply write down the total cost function.

$$C(x) = 200x + 1,000$$

We can now write the profit function.

$$\begin{aligned} P(x) &= R(x) - C(x) \\ &= 363.5x - (200x + 1,000) \\ &= 163.5x - 1,000 \end{aligned}$$

Here, $P(x)$ is the daily profit the seller can make by trading x lots per day (assuming prices do not vary).

(c) To "break even" means to make zero profit. Thus, the question can be rephrased as

$$\text{Find } x \text{ so that } P(x) = 0.$$

So, all we have to do is set $P(x) = 0$ and solve for x.

$$163.5x - 1,000 = 0$$
$$163.5x = 1,000$$
$$x = \frac{1,000}{163.5} \approx 6.1$$

Thus, in order to break even, the seller should trade approximately 6.1 lots per day.

Q How do we interpret this answer? In other words, how is it possible to trade 6.1 lots per day?

A We can interpret it as follows. To break even, she should trade an *average* of 6.1 lots per day. For instance, if she trades 61 lots in 10 days, she will have traded an average of $\frac{61}{10} = 6.1$ lots per day, and is thus breaking even.

Before we go on... As we saw, break-even occurs when the profit $P(x) = 0$. We can also look at it as the point where Revenue = Cost: $R(x) = C(x)$. We can interpret this by graphing the revenue and cost functions on the same axes (Figure 3).

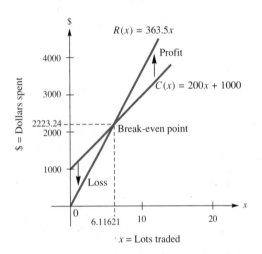

Break-even occurs at the point of intersection of the graphs of revenue and cost.

FIGURE 3

There are several things we can see from the figure.

1. Break-even occurs when the revenue and cost graphs intersect. The x-coordinate of the break-even point gives the number of lots that must be traded to break even: 6.1 per day. The y-coordinate gives the amount of money traded: the daily costs and revenue will each amount to $2,223 at break-even. (This figure was obtained from a calculator by taking $x = 1,000/163.5$ rather than the approximation $x \approx 6.1$.)

2. If she trades fewer than 6.1 lots per day, then she is to the left of 6.1 on the x-axis, where the graph of the cost function is above the graph of the revenue function, so she is making a loss amounting to the vertical distance between the graphs. Similarly, if she trades more than 6.1 lots per day, the revenue graph is above the cost graph, telling us that she is making a profit.

With a graphing calculator or computer, you can graph $y = 363.5x$ and $y = 200x + 100$ on the same axes. You can then use the trace and zoom features to locate the coordinates of the break-even point. Alternatively, you could graph the profit $P = 163.5x - 1000$ and locate the point where the profit changes from negative (a loss) to positive.

We summarize what we have gleaned from this example.

PROFIT FUNCTION AND BREAK-EVEN

The **profit function** P is given by

$$P(x) = R(x) - C(x)$$

(Profit = Revenue − Cost).

Break-even occurs when *the revenue equals the cost*, or

$$R(x) = C(x).$$

Equivalently, it occurs when *the profit equals zero*, or

$$P(x) = 0.$$

The **break-even quantity** is the number of items at which break-even occurs.

OBTAINING THE BREAK-EVEN QUANTITY

We can obtain the break-even quantity *graphically* by finding the coordinates of the point of intersection of the graphs of revenue and cost. We can calculate the break-even quantity *algebraically* by solving one of the equations $R(x) = C(x)$ or $P(x) = 0$ for x. The revenue (and cost) corresponding to the break-even quantity is then the value of either $R(x)$ or $C(x)$ for this value of x.

▼ **EXAMPLE 4**   Velocity

You are driving down the Ohio Turnpike watching the mileage markers to stay awake. Measuring time in hours after you see the 20-mile marker, you see the following markers each half hour:

Time (hrs)	0	0.5	1	1.5	2
Marker (mi)	20	47	74	101	128

Find your location s as a function of t, the number of hours you have been driving. (The number s is called your **displacement,** or **position**).

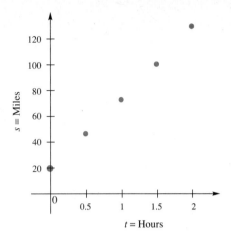

FIGURE 4

SOLUTION   If we plot the location s versus the time t, the 5 markers listed give us the graph in Figure 4.

These points appear to lie along a straight line. We can verify this by calculating how far you traveled in each half-hour. In the first half-hour you traveled $47 - 20 = 27$ miles. In the second half-hour you traveled $74 - 47 = 27$ miles also. In fact, you traveled exactly 27 miles each half-hour. The points we plotted lie on a straight line that rises 27 for every $\frac{1}{2}$ unit we go to the right, for a slope of $27 \div \left(\frac{1}{2}\right) = 54$.

To get the equation of that line, we notice that we have the s-intercept, which is the starting marker of 20. From the slope-intercept form (using s in place of y and t in place of x) we get

$$s(t) = 54t + 20.$$

Before we go on... Notice the significance of the slope: for every hour you travel, you drive a distance of 54 miles. In other words, you are traveling at a constant speed of 54 mph. We have uncovered one of the most important principles in applied mathematics:

In the graph of displacement vs. time, velocity is given by the slope.

In Example 1, the slope was the marginal cost (in dollars per mile). Here, the slope is velocity (in miles per hour). In both instances, the slope can be interpreted as *the speed or rate at which a quantity y (or s) changes per unit of x (or t)*. Be sure to spend some time thinking about this idea. It is one of the central concepts underlying calculus.

 A computer (particularly a spreadsheet program) is very useful in analyzing tables of data like this. By taking the differences in successive s values, we were able to notice a regularity: you traveled the same distance in each half-hour. This regularity indicated a linear relationship between s and t. You can analyze more complicated relations this way, too, and we shall occasionally return to this.

VELOCITY AND POSITION

If $s(t)$ is the position (or displacement) of a moving object at time t, and s is given by

$$s(t) = mt + b,$$

then the slope m is the **velocity** of the object, and b is its **starting position** or **initial position.**

▼ **EXAMPLE 5**   Fish Tanks

My 36-gallon tropical fish tank is leaking at a rate of 6 gallons per day. Assuming that it starts out full, how long will it take to empty?

SOLUTION   Before attempting to answer the question, let us first get an equation for the amount of water y left in the tank after t days, assuming that it is full when $t = 0$.

When $t = 1$, the amount of water left is $36 - 6(1)$.

When $t = 2$, the amount of water left is $36 - 6(2)$.

. . .

When $t = t$, the amount of water left is $36 - 6t$.

Thus,

$$y = -6t + 36.$$

Notice that the slope is negative. The reason: since the quantity y of water is *decreasing* at 6 gal/day, the rate at which y is changing is -6 gal/day. Thus, the slope is again measuring the rate of change of y, as it did in the previous examples. As usual, 36 is the initial value of y.

Now we can answer the question, "How long will it take to empty?" Mathematically, the question reads:

"Find that t for which y will equal 0."

We take the equation $y = -6t + 36$, put $y = 0$, and solve for t.

$$0 = -6t + 36$$
$$6t = 36, \text{ so } t = 6$$

Thus, my tropical fish tank will empty in exactly 6 days.

Before we go on... You might argue that you could have answered this by using "common sense" (36 gal ÷ 6 gal/day = 6 days). Well, yes, but this is because the numbers were simple to work with. What is important here is the interpretation of slope as the rate of change.

It is commonplace to observe that demand for a commodity goes down as its price goes up. It is traditional to use the letter q for the (quantity of) demand. Consider the following example.

▼ **EXAMPLE 6**   Setting Up a Demand Equation

You are the owner of the upscale Workout Fever Health Club and have been charging an annual membership fee of $600.* You are disappointed with the response: the club has been averaging only 10 new members per month. In order to remedy this, you decide to lower the fee to $500, and you notice that this boosts new membership to an average of 16 per month. Taking the demand q to be the average number of new members per month, express q as a linear function of the annual membership fee p.

SOLUTION   A **demand equation** or **demand function** expresses demand q (in this case, the number of new memberships sold per month) as a function of the unit price p (in this case, membership fees). Since we are asking for a *linear* demand function, we are looking for an equation of the form

$$q = mp + b,$$

where m and b are constants to be determined. (Thus, we take p to play the role of x and q to play the role of y.) In order to find these constants, notice that, just as in the case of Example 2, we are given two points on the graph of q versus p: (600, 10) and (500, 16) (see Figure 5).

We can therefore use this information to obtain the equation of the line.

Point: (600, 10) (As usual, we can choose either of the points we are given.)

Slope:

$$m = \frac{q_2 - q_1}{p_2 - p_1}$$

$$= \frac{16 - 10}{500 - 600}$$

$$= -\frac{6}{100} = -\frac{3}{50}$$

Thus, the equation we want is

$$q - q_0 = m(p - p_0)$$

$$q - 10 = -\frac{3}{50}(p - 600)$$

or

$$q = -\frac{3}{50}p + 46.$$

Before we go on... We should check that this equation gives us the correct membership figures for the prices of $600 and $500, but we leave this to you as an exercise. What we shall do instead is use the equation to make some predictions about membership.

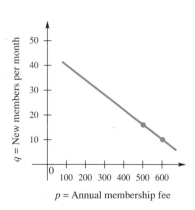

FIGURE 5

The graph axes: q = New members per month (vertical, marked 10, 20, 30, 40, 50), p = Annual membership fee (horizontal, marked 100 200 300 400 500 600).

▼ *This is $1 more than what the East Bank Club in Chicago was charging for individual memberships in 1993, according to the *Chicago Tribune*. ("What the Clubs Cost," Section 6, May 10, 1993, p. 15)

Q Suppose you decide to lower the annual fee to $100. How will this affect new membership?

A Here is where the demand equation is useful. Simply substitute $p = 100$ in the equation to obtain

$$q(100) = -\frac{3}{50}(100) + 46$$

$$= 40,$$

so you expect your new membership to increase to 40 new members per month.

Q Suppose, in a fit of greed, you decided to *increase* the price to $2,000. How would this affect membership?

A Our linear demand equation tells us that

$$q(2,000) = -\frac{3}{50}(2,000) + 46$$

$$= -74.$$

Our linear model seems to be telling us that -74 new members will join each month! In other words, the model has failed to give a meaningful answer.

This answer suggests the following further question.

Q Just how reliable is the linear model?

A Look at it this way: the *actual* demand graph could, in principle, be obtained by tabulating new membership figures for a large number of different annual fees. If the resulting points were plotted on the pq plane, they would probably suggest a curve and not a straight line. However, if you looked at a small enough portion of this curve, you could closely *approximate* it by a straight line. In other words, *over a small range of values of p, the linear model is accurate.* Linear models of real-life situations are generally reliable only for small ranges of the variables. (This point will come up again in some of the exercises.)

DEMAND EQUATION

A **demand equation** or **demand function** expresses demand q (the number of items demanded) as a function of the unit price p (the price per item). A **linear demand equation** has the form

$$q = mp + b.$$

It is usually the case that demand decreases as the unit price increases, so m is usually negative.

> ### INTERPRETATION OF m
> The (usually negative) slope m measures the change in demand per unit change in price. Thus for instance, if $m = -400$, p is measured in dollars and q in monthly sales, then each \$1 increase in the price per item will result in a drop in sales of 400 items per month.

> ### INTERPRETATION OF b
> The y-intercept b gives the demand if the items were given away.*

▼ **EXAMPLE 7**   Demand Equation

The demand for rubies at Royal Ruby Retailers (RRR) is given by the equation

$$q = -\frac{4}{3}p + 80,$$

where p is the price RRR charges (in dollars) and q is the number of rubies sold per week. If we plot demand q vs. price p, we get a graph with q-intercept 80 and slope $-\frac{4}{3}$. (See Figure 6.)

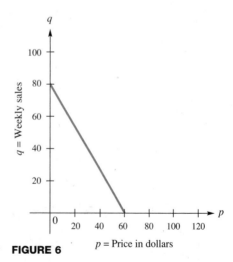

FIGURE 6

How should we interpret this? Since the slope is $-\frac{4}{3}$, we know that q, the quantity of rubies sold weekly, is decreasing by $\frac{4}{3}$ of a ruby per \$1 increase in price. In other words, a rise in price of \$3 results in a drop in sales of 4 rubies

---

▼ * This demand is not unlimited. For instance, campus newspapers are sometimes given away, and yet piles of them are often left unclaimed.

per week. Since the q-intercept is 80, RRR would hand out only 80 rubies per week if it were to give them away! (We see again what can go wrong with a linear model if we push it to extremes.)

Q At what price would sales stop completely?

A Translating into mathematics, "Find p when $q = 0$." We put $q = 0$ and solve for p, getting

$$0 = -\frac{4}{3}p + 80,$$

or

$$\frac{4}{3}p = 80,$$

and

$$p = \frac{3}{4}(80) = 60.$$

Thus, if RRR were to raise the price to $60, it would be stuck with them! Since $60 is ridiculously cheap for a good ruby, these stones must be very poor in quality!

▼ **EXAMPLE 8**   Using a Demand Function to Calculate Revenue

Referring to the demand equation $q = -\frac{4}{3}p + 80$ for rubies in Example 7, calculate the weekly revenue as a function of price.

SOLUTION   The weekly revenue is the total amount of money Royal Ruby Retailers will make from their sales of rubies. To calculate this as a function of p, we assume that they are selling them at $\$p$, and then do a little accounting.

q rubies @ $\$p$ per ruby gives a total revenue of $q \cdot p = pq$

In other words,

$$\text{Revenue} = \text{Price} \cdot \text{Quantity}$$
$$R = pq.$$

This formula gives R, but not as a function of p, since there is still the term q on the right-hand side. To get R in terms of p alone, we substitute the formula for q into this equation.

$$R = pq$$
$$= p(-\frac{4}{3}p + 80)$$

Thus,

$$R(p) = -\frac{4}{3}p^2 + 80p$$

is the function we are after.

Before we go on... Notice that R is not a *linear* function of p, it is a *quadratic* function of p because of the p^2 term. (We shall be studying quadratic expressions in the next section.)

Q If RRR gives rubies away, it makes no money. We saw in the last example that, if it sells them at $60, it still makes no money, because it doesn't sell any. Therefore, there should be a price somewhere *in between* $0 and $60 at which RRR makes the *largest* revenue. What is that price, and how much money does it bring in?

A We just saw that the weekly revenue is given by

$$R(p) = -\frac{4}{3}p^2 + 80p.$$

We now need to know: *What value of p gives the largest value of R(p)?* We'll see how to find the answer quickly in the next section. Calculus also provides an easy way to find it. In the meantime, we can make a pretty good guess at the answer by drawing the graph of R vs. p and then locating the highest point. We start by making a table.

p	0	10	20	30	40	50	60
$R(p) = -\frac{4}{3}p^2 + 80p$	0	667	1067	1200	1067	667	0

We suspect from this table that the best price is $30, which will bring RRR a total revenue of $1,200 per week. To support this claim, take a look at the graph in Figure 7.

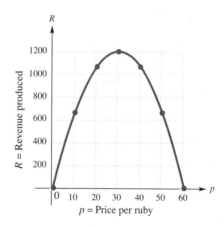

FIGURE 7

By the way, this curve is a parabola. (The next section is devoted to the study of parabolas.)

With a calculator or computer you can easily generate a table of revenues corresponding to lots of different prices, to get more points on this graph. Or, you can just use the calculator or computer to graph the equation $R = -\frac{4}{3}p^2 + 80p$ directly (whereupon it will generate its own very large table and plot large numbers of points to draw the graph). You will need to enter the equation as

$$Y_1 = -(4/3) \text{ X}^{\wedge}2 + 80\text{X}$$

Notice that graphing calculators usually expect y to be given as a function of x, rather than p as a function of q. (Also, why have we inserted the parentheses, and what would happen if we omitted them?) You can use the ranges $0 \le x \le 60$ and $0 \le y \le 1300$. (You know that y will be between 0 and 1200, since we have already computed the values in the above table. In general, these ranges will not be given to you, and you will need to experiment with different ranges for x and y to obtain a suitable graph.)

Once you have plotted the curve, you can then find the price that gives the largest revenue by using the trace feature to move to the highest point of the graph—zooming in if you want more accuracy—and reading the x-coordinate from the display. (The y-coordinate gives the corresponding largest revenue.)

All the preceding examples illustrate the following principle.

GENERAL LINEAR MODELS

If $y = mx + b$ is a linear model of changing quantities x and y, then the slope m is the rate at which y is increasing per unit increase in x, while the y-intercept b is the value of y that corresponds to $x = 0$.

▶ **1.4 EXERCISES**

APPLICATIONS

1. *Cost*   A piano manufacturer has a daily fixed cost of $1,200 and a marginal cost of $1,500 per piano. Find the daily cost $C(x)$ of manufacturing x pianos per day.

2. *Cost*   A soft-drink manufacturer has a weekly fixed cost of $1,000 and a marginal cost of $5 per case of soda. Find the weekly cost $C(x)$ if the company produces x cases of soda per week.

3. *Demand*   Sales figures show that your company sold 1,960 pen sets per week for $1 per pen set, and 1,800 pen sets per week for $5 per pen set. What is the linear demand function for your pen sets?

4. *Demand*   A large department store is prepared to buy 3,850 of your neon-colored shower curtains per month for $5 apiece, but only 3,700 shower curtains per month for $10 apiece. What is the linear demand function for your neon-colored shower curtains?

5. *Revenue* The annual revenue of *United Airlines* increased from $8.50 billion in 1988 by approximately $0.95 billion per year for several years.*

(a) Use these data to express *United's* annual revenue *R* (in billions of dollars) as a linear function of the number of years *t* since 1988.

(b) Use your model to predict *United's* revenue in the year 2000.

(c) Comment on the limitations of this model (if any).

6. *Profit* The annual profit of *United Airlines* decreased from $1,124 million in 1988 by approximately $520 million per year for several years.*

(a) Use these data to express *United's* annual profit *P* (in millions of dollars) as a linear function of the number of years *t* since 1988.

(b) Use your model to predict *United's* profit (or loss) in the year 2000.

(c) Comment on the limitations of this model (if any).

7. *Fahrenheit and Celsius* In the Fahrenheit temperature scale, water freezes at 32°F and boils at 212°F. In the Celsius (or Centigrade) scale, water freezes at 0°C and boils at 100°C. Assuming that the Fahrenheit temperature *f* and the Celsius temperature *c* are related by a linear equation, find *f* in terms of *c*. Use your equation to find the Fahrenheit temperatures corresponding to 30°C, 22°C, −10°C, and −14°C.

8. *Fahrenheit and Celsius* Use the relationship between degrees Celsius and Fahrenheit you obtained in the previous exercise to obtain a linear equation for *c* in terms of *f*, and use your equation to find the Celsius temperatures corresponding to 100°F, 70°F, 10°F, and −40°F.

9. *Cost* The cost of renting tuxes for the Choral Society's concert is $20 down, plus $88 per tux. Express the cost *C* as a function of *x*, the number of tuxes rented, and graph *C(x)*. Use your function to answer the following questions.

(a) What is the cost of renting 2 tuxes?

(b) What is the cost of the second tux?

(c) What is the cost of the 4,098th tux?

(d) What is the marginal cost per tux?

10. *Income* The Enormous State University (ESU) Information Office pays student aides $3.50 per hour in addition to a daily meal allowance of $5.00 (even though the cafeteria charges $6.50 for a rather tasteless hamburger). Express a student aide's daily earnings *q* in terms of the number of hours *h* of work per day, and graph it. Use your equation to answer the following questions:

(a) How much does a student aide earn in an eight-hour work day?

(b) How much does a student aide earn in the eighth hour of an eight-hour work day?

(c) What is the marginal earning rate for a student aide?

(d) How long (to the nearest minute) must a student aide work in order to afford two cafeteria hamburgers?

11. *Income* The well-known romance novelist, Celestine A. Lafleur (a.k.a. Bertha Snodgrass), has decided to sell the screen rights to her latest book, *Henrietta's Heaving Heart*, to Boxoffice Success Productions, Inc. for $50,000. In addition, the contract assures Ms. Lafleur royalties of 5% of the net profits.[†] Express her income *I* as a function of the net profit *N*, and determine the net profit necessary to bring her an income of $100,000. What is her marginal income (share of each dollar of net profit)?

12. *Income* Due to the enormous success of the movie *Henrietta's Heaving Heart* based on a novel by Celestine A. Lafleur (see the preceding exercise), Boxoffice Success Productions Inc. decides to film the sequel, "Henrietta, Oh Henrietta." At this point, Bertha Snodgrass (whose novels now top the bestseller lists) feels she is in a position to demand $100,000 for the screen rights, and royalties of 8% of

---

▼ * Based on data published in *The New York Times*, December 24, 1993, p. D1. Source: Company Reports.

† Percentages of net profit are commonly called "monkey points." Few movies ever make a net profit on paper, and anyone with any clout in the business gets a share of the *gross*, not the net.

the net profits. Express her income I as a function of the net profit N, and determine the net profit necessary to bring her an income of $1,000,000. What is her marginal income (share of each dollar of net profit)?

13. *Cost*   The RideEm Bicycles company can produce 100 bicycles in a day at a total cost of $10,500, and it can produce 120 bicycles in a day at a total cost of $11,000. What are the company's daily fixed costs, and what is the marginal cost to build a bicycle?

14. *Biology*   The Snowtree cricket behaves in a rather interesting way: the rate at which it chirps depends linearly on the temperature. One summer evening you hear a cricket chirping at a rate of 140 chirps per minute, and you notice that the temperature is 80°F. Later in the evening the cricket has slowed down to 120 chirps per minute, and you notice that the temperature has dropped to 75°F. Express the temperature, T, as a function of the cricket's rate of chirping, r. What is the temperature if the cricket is chirping at a rate of 100 chirps per minute?

15. *Tax Depreciation*   In calculating the value of its assets for tax purposes, a large law firm calculates that its office equipment, which cost them $60,000 new, is decreasing in value at a rate of 5% of the original price per year.
  (a) Express the total value v of the equipment as a function of n, the numbers of years after purchase.
  (b) How long will it take for the law firm's office equipment to depreciate to $1,000?
  (c) How long will it take for the law firm's office equipment to be worthless?
  (d) When does your linear model cease to be meaningful?

16. *Cost*   Joe Silly's overdue library books (which he can never remember to return) are costing him a small fortune. For each overdue book, the Enormous State University (ESU) Library charges a $5 late fee, plus an additional $10 for each week a book remains overdue. Joe had 47 overdue books and his parents had $50,000 in the bank when the fines started mounting.
  (a) Express the total fine f for a *single* book as a function of the number w of overdue weeks.

  (b) Now express the total fine F on Joe Silly's 47 books as a function of w.
  (c) How long does Joe have before his overdue library fees wipe out his family's entire life savings?

17. *Muscle Recovery Time*   Most workout enthusiasts will tell you that muscle recovery time is about 48 hours. But it is not quite as simple as that; the recovery time ought to depend on the number of sets you do involving the muscle group in question. For example, if you do no sets of biceps exercises, then the recovery time for your biceps is (of course) zero. To take a compromise position, let us assume that, if you do three sets of exercises on a muscle group, then its recovery time is 48 hours. Use this data to write a linear function that gives the recovery time (in hours) in terms of the number of sets affecting a particular muscle. Use this model to calculate how long it would take your biceps to recover if you did 15 sets of curls. Comment on your answer with reference to the usefulness of a linear model.

18. *Oil Reserves*   The total estimated crude oil reserves (i.e., oil in the ground) in Saudi Arabia on January 1, 1989, were 255 billion barrels.* Further, Saudi Arabia produces on the order of 10 million ($=$ .01 billion) barrels of crude oil per day. Use these data to give a function that estimates the crude oil reserves n years after the January 1, 1989 date, and use your equation to predict the Saudi Arabia crude oil reserves at the start of the year 2000. When (to the nearest year) will Saudi Arabia run out of crude?

19. *Profit Analysis—Aviation*   The operating cost of a *Boeing* 747-100, which seats up to 405 passengers, is estimated to be $5,132 per hour.† If an airline charges each passenger a fare of $100 per hour of flight, find the hourly profit P it earns operating a 747-100 as a function of the number of passengers x (be sure to specify the domain). What is the least number of passengers it must carry in order to make a profit?

20. *Profit Analysis—Aviation*   The operating cost of a *McDonnell Douglas DC* 10-10, which seats up to 295 passengers, is estimated to be $3,885 per hour.† If an airline charges each passenger a fare of $100 per

▼ *Source: *Oil and Gas Journal* (December 26, 1989).
  † In 1992. Source: Air Transportation Association of America

hour of flight, find the hourly profit P it earns operating a DC 10-10 as a function of the number of passengers x (be sure to specify the domain). What is the least number of passengers it must carry in order to make a profit?

21. *Insurance Losses* *Allstate* Insurance Company charged an average of $360 per year for household insurance in Florida in 1992 (the year Hurricane Andrew struck the East Coast), and it reported that it had covered 1.1 million homes during that year.* Assume that *Allstate* paid an average of $50,000 per damaged home.

   (a) Find the profit $P(x)$ giving the total profit if x homes were damaged.

   (b) Given that its total losses were $2 billion that year, how many of the homes it covered were damaged?[†] What percentage is this of the total number of homes it insured?

22. *Insurance Losses* After Hurricane Andrew, *Allstate* applied for a Florida state increase of 30% for household insurance premiums and was seeking to reduce the total number of Florida homes covered to 750,000.*Repeat the preceding exercise to see what would have happened if these changes had gone into effect before 1992, but *Allstate* still lost $2 billion.

23. *Break-Even Analysis (based on a question from a CPA exam)* The Oliver Company plans to market a new product. Based on its market studies, Oliver estimates that it can sell up to 5,500 units in 1992. The selling price will be $2 per unit. Variable costs are estimated to be 40% of total revenue. Fixed costs are estimated to be $6,000 for 1992. How many units should the company sell to break even?

24. *Break-Even Analysis (based on a question from a CPA exam)* The Metropolitan Company sells its latest product at a unit price of $5. Variable costs are estimated to be 30% of the total revenue, while fixed costs amount to $7,000 per month. How many units should the company sell per month in order to break even, assuming that it can sell at up to 5,000 units per month at the planned price?

25. *Break-Even Analysis (from a CPA exam)* Given the following notations, write a formula for the break-even sales level.

$$SP = \text{Selling price per unit}$$
$$FC = \text{Total fixed cost}$$
$$VC = \text{Variable cost per unit}$$

26. *Break-Even Analysis (based on a question from a CPA exam)* Given the following notation, write a formula for the total fixed cost.

$$SP = \text{Selling price per unit}$$
$$VC = \text{Variable cost per unit}$$
$$BE = \text{Break-even sales level in units}$$

27. *Profit Analysis—Wireless Communications* You are the CEO of a new communications company that is interested in boosting subscriptions. Your monthly operating costs are calculated as $20 per customer and $10,000 overheads (fixed monthly costs). Your monthly income averages $50 per customer each month.

   (a) Find the profit function $P(x)$, where x is the total number of subscribers.

   (b) The U.S. Government hopes to auction a large amount of the nation's airwaves to cellular telephone companies for about $10 billion.[‡] This translates into about $40 per potential subscriber—or "pop" in the industry jargon. Assuming that your company has 1,000 subscribers at present, how many new "pops" could you afford to buy from the government at the start of next month at the estimated price of $40 in order to break even by the end of the month? (Assume that due to your large waiting list you will be able to begin immediate service to up to 4,000 new customers.)

28. *Profit Analysis—Automobile Manufacturing* You are the CEO of an automobile company that has just spent $6 billion developing a new "world car" planned for worldwide sales at a wholesale price of $14,000 per car.[§] Manufacturing costs amount to $12,000 per car, plus monthly overheads of $200,000.

▼ * Source: *The Chicago Tribune*, May 28, 1993, Section 3, p. 3.

[†] Allstate actually reported a $2.7 billion loss resulting from Hurricane Andrew.

[‡] Such a plan was announced by the Federal Communications Commission in September, 1993. (Source: *The New York Times*, September 27, 1993, p. D1.)

[§] This was the actual manufacturing cost of the Ford "world car" known variously as the Ford Mondeo, the Ford Contour and the Mercury Mistique. The planned sticker price was $16,000–$17,000 in the U.S. (Source: *The New York Times*, September 27, 1993, p. D1.)

(a) Ignoring development costs, find the profit function $P(x)$, where x is the total number of automobiles you manufacture each month (assuming that you can sell every car your company manufactures).

(b) How many "world cars" should your company manufacture in the coming year in order to cover the development costs by the end of the year?

29. *Population Growth*   According to the U.S. Bureau of the Census, population growth in the U.S. has followed the declining pattern of most of the developed world, dropping from 55.2 live births per thousand in 1820 to 15.7 in 1987. Use these data to give a linear equation showing the annual number of live births b per thousand in terms of the number of years n since 1820, and use your model to predict when the rate will drop to zero (assuming the trend continues).

30. *Motion*   On the day I got my learner's driving permit, I wanted to impress my friends with the speed at which I could maneuver a car in reverse along an 800-ft country road (at the end of which was a large mud pool). Assuming that at time $t = 0$ seconds I was at the 10-ft mark and that I subsequently maintained a speed of 30 mph ($= 44$ ft/sec), express my position s as a function of the time t in seconds. At what point in time did I wind up in the mud pool?

31. *Pollution*   Radioactive pollutants are seeping into a lake at a rate of 3,000 gallons per year. At the start of 1992, it was estimated that the lake contained 2,500 gallons of pollutant. Find a function that gives the number of gallons p of pollutant in the lake n years since the start of 1992, and use your function to estimate the amount of pollutant at the start of 2001.

32. *Recycling*   In Sweden, the problem of abandoned cars was addressed by an innovative policy introduced in 1976: Swedish car owners got a government bonus of $60 for selling their jalopies to a registered scrap dealer. The cost of this bonus was subsidized by a $60 surcharge on new cars. Since not every car ultimately wound up at a registered scrap yard, the government stood to make a profit from this arrangement. Here is a test scenario: let n be the number of new cars sold in Sweden per year, and let us assume that half that number of cars were scrapped per year. Express the annual surplus s accruing to the government as a function of n, and estimate the annual surplus if 50,000 new cars were sold per year.

33. *Ecology—Logging*   In 1990, the U.S. Forest Service unveiled a plan calling for a reduction of logging in national forests from the then-current annual cut of 12.2 billion board feet to 10.8 billion board feet by the end of 1995.* Assuming that logging decreased linearly over this period, specify a function L, where $L(n)$ is the total cut (in billions of board feet) n years after 1990. Graph this function, and give the largest possible domain for which the model makes sense.

34. *Energy—Oil Reserves*   The U.S. Strategic Oil Reserve in 1990 amounted to 590 million barrels of oil.* Assuming that it was tapped at a rate of 25 million barrels per year, specify a function R, where $R(n)$ is the total oil reserve (in millions of barrels) n years after 1990. Graph this function, and give the largest possible domain for which the model makes sense.

35. *Break-even Analysis—Organized Crime*   The organized crime boss and perfume king Butch (Stinky) Rose has daily overheads (bribes to corrupt officials, motel photographers, wages for hitmen, explosives, etc.) amounting to $20,000 per day. On the other hand, he has a substantial income from his counterfeit perfume racket: he buys imitation French perfume (Chanel No. $22\frac{1}{2}$) at $20 per gram, pays an additional $30 per 100 grams for transportation, and sells it via his street thugs for $600 per gram. Specify Stinky's profit function, $P(x)$, where x is the quantity (in grams) of perfume he buys and sells, and use your answer to calculate how much perfume should pass through his hands per day in order that he break even.

36. *Break-even   Analysis—Disorganized   Crime*   Butch (Stinky) Rose's counterfeit Chanel No. $22\frac{1}{2}$ racket has run into difficulties; it seems that the *authentic* Chanel No. $22\frac{1}{2}$ perfume is selling at only $500 per gram, whereas his street thugs have been charging $600 per gram, and his costs amount to $400 per gram plus $3,000 per 100 grams transportation costs. (The perfume's smell is easily detected by specially trained Chanel Hounds, and this necessitates elaborate packaging measures.) He therefore decides to price it at $420 per gram in order to undercut the competition. Specify Stinky's profit function, $P(x)$, where x is the quantity (in grams) of perfume he buys and sells, and use your answer to calculate how much perfume should pass through his hands per day in order that he break even. Interpret your answer.

▼ *Source: Feb.-Mar. 1991 issue of *National Wildlife Magazine*.

37. *Demand and Revenue—Poultry*   A linear model of the demand for chicken in the U.S. predicts that if the average per capita disposable income is $30,000, then

$$q = 65.4 - 0.45p + 0.12b,$$

where q is the per capita demand for chicken in pounds per year, p is the wholesale price of chicken in cents per pound, and b is the wholesale price of beef in cents per pound.*

(a) If the wholesale price of beef is fixed at 45¢ per pound, recast the above formula as a demand function for chicken in terms of the wholesale price per pound.

(b) Use your demand function to predict the demand for chicken (to the nearest pound) if the wholesale price is set at 20¢ per pound.

(c) At what price (to the nearest cent) does the model predict that the demand will drop to zero? Use this result to specify a domain for the demand function so that the demand can never be negative.

(d) Use your demand function from parts (a) and (c) to calculate the revenue function R, which specifies annual per capita revenue in terms of the wholesale price of chicken per pound.

38. *Demand and Revenue—Poultry*   Repeat the preceding exercise, assuming that the wholesale price of beef is fixed at 60¢ per pound.

39. *Epidemics*   The following chart shows the number of new cases of tuberculosis per 100,000 people in New York City since the 1920s.[†]

(a) Use the data from 1920 and 1980 to model the incidence of tuberculosis per 100,000 New York City residents as a linear function of time since 1920.

(b) What year gives the greatest discrepancy between your model and the actual data?

(c) Comment on the reliability of the linear model.

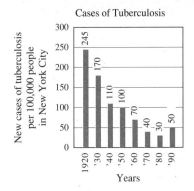

Cases of Tuberculosis

40. *Incidence of Melanoma*   The following chart shows the number of new cases of melanoma per 100,000 people in the U.S. since 1973.[‡]

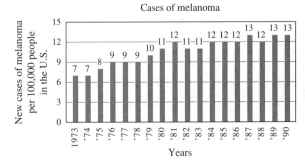

Cases of melanoma

(a) Use the data from 1975 and 1990 to model the incidence of melanoma per 100,000 U.S. residents as a linear function of time since 1975.

(b) What year gives the greatest discrepancy between your model and the actual data?

(c) Comment on the reliability of the linear model.

41. *Demand and Revenue*   You have been hired as a marketing consultant to Johannesburg Burger Supply, Inc., and you wish to come up with a demand equation for its hamburgers. In order to make life as simple as possible, you assume that the demand equation for Johannesburg hamburgers has the linear form

---

▼   * This equation is based on actual data from poultry sales in the period 1950–1984. In the "You're the Expert" section at the end of this chapter, we shall discuss this equation further. (Source: A. H. Studenmund, *Using Econometrics*, Second Edition (HarperCollins, 1992), pp. 180–181.)

† Sources: State Health Department, New York City Department of Health, Centers for Disease Control (New York Times, January 24, 1994, p. B1.)

‡ Source: American Cancer Society/New York Times (*The New York Times*, January 25, 1994, p. C3.)

$q = mp + b$, where p is the price per hamburger, q is the demand in weekly sales, and m and b are certain constants you'll have to figure out.

(a) Your market studies reveal the following sales figures: when the price is set at $2.00 per hamburger, the sales amount to 3,000 per week, but when the price is set at $4.00 per hamburger, the sales drop to zero. Use these data to calculate m and b and, hence, the actual demand equation.

(b) Use the demand equation obtained in (a) to estimate the number of hamburgers Johannesburg Burgers can unload per week if it gives them away.

(c) Use the equation to estimate Johannesburg's weekly revenue R if it were to sell the hamburgers at p dollars apiece.

(d) By tabulating values or otherwise, determine the price at which you would advise Johannesburg to sell its their hamburgers in order to maximize weekly revenue. What would this maximum revenue be?

42. *Demand and Revenue*   Johannesburg Burger Supply, Inc. decides to market a new product: the "Fatfree Slim Thin" pork lard patty. Once again, you assume that the demand equation for Johannesburg patties should have the linear form $q = mp + b$, where p is the price per meat patty, q is the demand in weekly sales, and m and b are unknown constants.

(a) Your market studies reveal the following sales figures: when the price is set at $2.00 per pork lard patty, Johannesburg sells 6,000 per week, but when the price is set at $10.00 per pork lard patty, the sales plummet to zero. Use these data to calculate m and b and, hence, the actual demand equation.

(b) Use the demand equation obtained in (a) to estimate the number of pork lard patties Johannesburg can unload per week if it gives them away.

(c) Use the equation to estimate Johannesburg Burger's weekly revenue R if it were to sell the pork lard patties at p dollars apiece.

(d) By tabulating values or otherwise, determine the price at which you would advise Johannesburg to sell its patties in order to maximize weekly revenue. What would this maximum revenue be?

*Exercises 43–46 use the idea of the **average cost**. If $C(x)$ is the total cost for x items, then the average cost per item for x items is*

$$\overline{C}(x) = \frac{C(x)}{x}.$$

43. *Cost and Average Cost*   A firm has monthly fixed costs of $30,000, variable (marginal) costs of $400 per item, and a manufacturing capacity of 1,000 items per month.

(a) Write the cost function $C(x)$, where x is the number of items produced per month.

(b) Write the average cost function (be sure to specify the domain).

(c) Find the production level (number of items produced per month) that gives an average cost of $450 per item.

44. *Cost and Average Cost*   The RideEm Bicycles company can produce 100 bicycles in a day at a total cost of $10,500, and it can produce 120 bicycles in a day at a total cost of $11,000. Its maximum production capacity is 200 bicycles per day.

(a) Construct a linear cost function for RideEm Bicycles (be sure to specify the domain).

(b) Write the average cost function.

(c) How many bicycles should RideEm manufacture per day in order to meet an average cost goal of $75 per bicycle?

45. *Break-Even Analysis (based on a GRE economics exam question*)* A firm's average cost function is presented as

$$\text{Average cost} = 350 + \frac{9000}{Q}.$$

The revenue from one unit of output is $500. How many units of output must this firm sell to break even?

46. *Break-Even Analysis (based on a GRE economics exam question*)*   A firm's average cost function is presented as

$$\text{Average cost} = 600 + \frac{900}{Q}.$$

The revenue from one unit of output is $700. How many units of output must this firm sell to make a profit of $100?

▼ *Source: *GRE Economics* by G. G. Gallagher, G. E. Pollock, W. J. Simeone and G. Yohe (Research and Education Association, Piscataway, N.J., 1989), p. 254.

COMMUNICATION AND REASONING EXERCISES

47. Describe some of the limitations of a linear demand model $q = mp + b$.

48. Describe some of the limitations of a linear cost function $C(x) = mx + b$.

49. If y and x are related by the linear expression $y = mx + b$, when should the quantity m be positive, when should it be negative, and when should it be zero?

50. If cost and revenue are expressed as linear expressions of the number of items x, what does the break-even quantity signify?

51. Suppose the cost function is $C(x) = mx + b$ (with m and b positive), the revenue function is $R(x) = kx$ ($k > m$) and the number of items is increased from the break-even quantity. Does this result in a loss, a profit, or is it impossible to say? Explain your answer.

52. You have been constructing a demand equation, and you obtained a (correct) expression of the form $p = mq + b$, whereas you would have preferred one of the form $q = mp + b$. Should you simply switch p and q in the answer, should you start again from scratch, using p in the role of x and q in the role of y, or should you solve your demand equation for q? Give reasons for your decision.

53. Come up with an interesting application leading to the linear model $s = -100k + 200$.

54. Come up with an interesting application leading to the linear model $s = 100k - 200$.

55. Explain intuitively why there is always a unit price that results in the maximum revenue, given a linear demand equation with negative slope.

56. Explain why there are always at least two unit prices that result in zero revenue, given a linear demand equation with negative slope.

▶ ═══ **1.5** QUADRATIC FUNCTIONS AND MODELS

We saw at the end of the preceding section that a linear demand function gives rise to a *quadratic* revenue function. Quadratic functions are the next most complicated functions after the linear ones, and they will be very useful as examples as we study calculus.

> **QUADRATIC FUNCTION**
>
> A **quadratic function** is a function of the form
>
> $$f(x) = ax^2 + bx + c \quad \text{(where } a \neq 0\text{)}.$$

▼ **EXAMPLE 1**

Sketch the graph of $y = x^2$.

SOLUTION We first construct a table with sufficiently many points to give us an idea of what the graph will look like.

x	-3	-2	-1	0	1	2	3
$y = x^2$	9	4	1	0	1	4	9

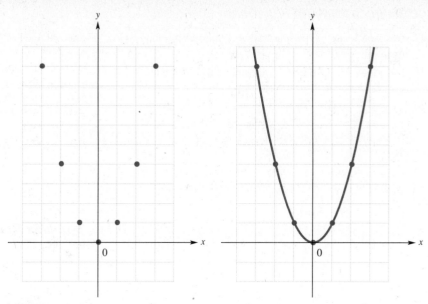

FIGURE 1

Plotting these points gives the picture on the left in Figure 1, suggesting the curve on the right.

This curve is called a **parabola,** and its lowest point, at the origin, is called its **vertex.**

 With a calculator or computer, you can generate this graph using the format

$$Y_1 = X{\char94}2.$$

(The caret symbol "^" is the standard graphing calculator and computer symbol for raising to a power.)

Any quadratic function $f(x) = ax^2 + bx + c \ (a \neq 0)$ has a parabola as its graph (possibly upside-down, and with its vertex not necessarily at the origin). These curves have the general shape shown in Figure 2.

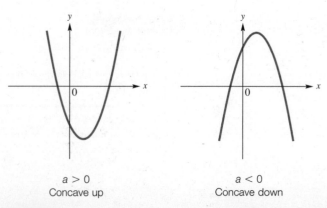

FIGURE 2

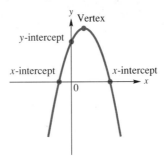

FIGURE 3

In this section, we present an easy method of sketching by hand the graph of any quadratic function, based on the features shown in Figure 3. If you are using a graphing calculator or computer program, think about how the following summary explains the picture the device gives you.

DETERMINING THE FEATURES OF A PARABOLA

The graph of $f(x) = ax^2 + bx + c$ is a parabola with the following features.

Vertex   The x-coordinate of the vertex is $-b/(2a)$. The y-coordinate is $f(-b/(2a))$.

x-Intercepts (if any)   These occur when $f(x) = 0$, or when

$$0 = ax^2 + bx + c.$$

We can solve this equation for x using the quadratic formula. Thus, the x-intercepts are

$$x = \frac{-b \pm \sqrt{b^2 - 4ac}}{2a}.$$

If the discriminant $\Delta = b^2 - 4ac$ is positive, there are two x-intercepts. If it is zero, there is a single x-intercept (at the vertex). If it is negative, there are no x-intercepts (so the parabola doesn't touch the x-axis at all).

y-Intercept   This occurs when $x = 0$. So

$$y = a \cdot 0^2 + b \cdot 0 + c = c.$$

Symmetry   The parabola is symmetric with respect to the vertical line through the vertex, which is the line $x = -b/(2a)$.

We shall not fully justify the formula for the vertex (and the axis of symmetry), but notice the following. If there are two x-intercepts, they are

$$x = -\frac{b}{2a} - \frac{\sqrt{b^2 - 4ac}}{2a}$$

and

$$x = -\frac{b}{2a} + \frac{\sqrt{b^2 - 4ac}}{2a}.$$

These should be symmetric around the axis of symmetry, but the formulas make it clear that they are symmetric around the line $x = -b/(2a)$. So this must be the axis of symmetry, and the vertex is located on this line as well. You can see the symmetry clearly in the next example. (We shall give a better justification after we have studied some calculus.)

▼   **EXAMPLE 2**

Sketch the parabola $f(x) = x^2 + 2x - 8$.

SOLUTION   Here, $a = 1, b = 2$, and $c = -8$. Since $a > 0$, the parabola is concave up (Figure 4).

Vertex   The x-coordinate of the vertex is $x = -b/(2a) = -2/2 = -1$. To get its y-coordinate, we substitute this value back into $f(x)$ to get $y = f(-1) = (-1)^2 + 2(-1) - 8 = 1 - 2 - 8 = -9$. Thus, the coordinates of the vertex are $(-1, -9)$.

x-Intercepts   To calculate the x-intercepts (if any), we solve the equation $x^2 + 2x - 8 = 0$. Luckily, this factors as $(x + 4)(x - 2) = 0$. Thus, the solutions are $x = -4$ and $x = 2$, giving these values as the x-intercepts.

y-Intercept   The y-intercept is given by $c = -8$.

Symmetry   The graph is symmetric around the vertical line $x = -1$.

Now we can sketch the curve as in Figure 5.

FIGURE 4

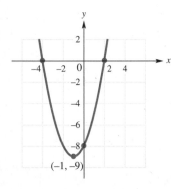

FIGURE 5

▼   **EXAMPLE 3**

Sketch the parabola $f(x) = -x^2 - 2x + 1$.

SOLUTION   Here, $a = -1, b = -2$, and $c = 1$. Since $a < 0$, the parabola is concave down (Figure 6).

Vertex   The x-coordinate of the vertex is $x = -b/(2a) = 2/-2 = -1$, and substitution gives its y-coordinate as $y = f(-1) = -(-1)^2 - 2(-1) + 1 = 2$. Thus, the vertex is the point $(-1, 2)$.

x-Intercepts   The x-intercepts are the solutions to $-x^2 - 2x + 1 = 0$. This quadratic does not factor nicely, so we use the quadratic formula to obtain

$$x = -\frac{1}{2}(2 \pm \sqrt{8}) = -\frac{1}{2}(2 \pm 2\sqrt{2}) = -1 \pm \sqrt{2}.$$

Thus, the x-intercepts are $-1 + \sqrt{2}$ and $-1 - \sqrt{2}$. In order to plot them, we approximate $\sqrt{2}$ by 1.4. (We don't need much accuracy when we sketch

FIGURE 6

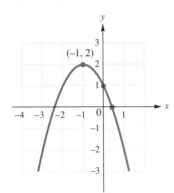

(-1, 2)

FIGURE 7

this curve, as we only want its general features.) Thus, the *x*-intercepts are approximately 0.4 and −2.4.

y-Intercept   The *y*-intercept is $c = 1$.

Symmetry   The graph is symmetric around the vertical line $x = -1$.

The graph is shown in Figure 7.

▼ **EXAMPLE 4**

Sketch the graph of $f(x) = 4x^2 - 12x + 9$.

SOLUTION   We have $a = 4$, $b = -12$, and $c = 9$. Since $a > 0$, this parabola is concave up.

Vertex   The *x*-coordinate of the vertex is given by $x = -b/(2a) = \frac{12}{8} = \frac{3}{2}$. Substituting to get the *y*-coordinate gives $y = f\left(\frac{3}{2}\right) = 4\left(\frac{3}{2}\right)^2 - 12\left(\frac{3}{2}\right) + 9 = 0$. Thus, the vertex is at the point $\left(\frac{3}{2}, 0\right)$.

x-Intercepts   To get the *x*-intercepts, we must solve $4x^2 - 12x + 9 = 0$. This quadratic factors as $(2x - 3)^2 = 0$, so the only solution is $2x - 3 = 0$, or $x = \frac{3}{2}$. Note that this coincides with the vertex, which also lies on the *x*-axis.

y-Intercept   The *y*-intercept is $c = 9$.

Symmetry   The graph is symmetric around the vertical line $x = \frac{3}{2}$.
The graph is the very narrow parabola shown in Figure 8.

Why is the parabola so narrow? One way you can find out experimentally is by varying the coefficients of the equation $y = 4x^2 - 12x + 9$ and plotting them on the same set of axes. For instance, you could vary the coefficient of x^2 by plotting

$$Y_1 = 4X^2 - 12X + 9$$
$$Y_2 = 3X^2 - 12X + 9$$
$$Y_3 = 2X^2 - 12X + 9$$

to examine the effect of this coefficient on the shape of the graph. You could then vary the coefficient of *x*, and finally, the constant.

$\left(\frac{3}{2}, 0\right)$

FIGURE 8

▼ **EXAMPLE 5**

Sketch the graph of $f(x) = -\frac{1}{2}x^2 + 4x - 12$.

SOLUTION   Here, $a = -\frac{1}{2}$, $b = 4$, and $c = -12$. Since $a < 0$, the parabola is concave down.

Vertex   The vertex has *x*-coordinate $-b/(2a) = -4/-1 = 4$, with corresponding *y*-coordinate $f(4) = -\frac{1}{2}(4)^2 + 4(4) - 12 = -4$. Thus, the vertex is at $(4, -4)$.

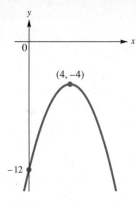

FIGURE 9

x-Intercepts   For the *x*-intercepts, we must solve $-\frac{1}{2}x^2 + 4x - 12 = 0$. But in doing so, we discover that the discriminant Δ is $b^2 - 4ac = 16 - 24 = -8$. Since this is negative, there are no solutions of the equation, so there are no *x*-intercepts.

y-Intercept   The *y*-intercept is given by $c = -12$.

Symmetry   The graph is symmetric around the vertical line $x = 4$.

Since there are no *x*-intercepts, the graph lies entirely below the *x*-axis, as shown in Figure 9.

We can now put our study of parabolas to use.

▼   **EXAMPLE 6**   Demand and Revenue

A publishing company predicts that the demand equation for the sale of its latest romance novel is

$$q = -2{,}000p + 150{,}000,$$

where q is the number of books it can sell per year at a price of $\$p$ per book. What price should it charge in order to obtain the maximum annual revenue?

SOLUTION   The total revenue depends on the price, as follows.

$$
\begin{aligned}
R &= pq \\
&= p(-2{,}000p + 150{,}000) \\
&= -2{,}000p^2 + 150{,}000p
\end{aligned}
$$

We are after the price p that gives the largest possible revenue. Notice that what we have is a quadratic function of the form $R(p) = ap^2 + bp + c$, where $a = -2{,}000$, $b = 150{,}000$, and $c = 0$. Since a is negative, the graph of the function is a parabola, concave down, and so its vertex is its highest point. The p-coordinate of the vertex is

$$p = -\frac{b}{2a} = -\frac{150{,}000}{-4{,}000} = 37.5.$$

This value of p gives the highest point on the graph, and thus gives the largest value of $R(p)$. We may conclude that the company should charge $\$37.50$ per book to maximize its annual revenue.

Before we go on...   You might ask what the maximum annual revenue is. The answer is supplied for us by the revenue function $R(p) = -2{,}000p^2 + 150{,}000p$. Since we have $p = 37.5$, we can substitute this value into the equation and obtain $R(37.5) = -2{,}000(37.5)^2 + 150{,}000(37.5) = 2{,}812{,}500$. In other words, the company will earn total annual revenues from this book amounting to $\$2{,}812{,}500$. Not bad!

▼ **EXAMPLE 7**   Break-Even Analysis

As the operator of Workout Fever Health Club (see Example 6 in the preceding section) you calculate your demand equation to be

$$q = -\frac{3}{50}p + 46,$$

where q is the number of new members who join the club per month, and p is the annual membership fee you charge.

(a)   Since you are running a shoestring operation, your annual operating costs amount to only $5,000 per year. At what price should you set annual memberships in order to break even?

(b)   How would the situation change if your operating costs went up to $10,000 per year?

SOLUTION

(a)   First recall from the preceding section that break-even occurs when total revenue equals total cost. The annual revenue is given by

$$R = pq$$

$$= p\left(-\frac{3}{50}p + 46\right)$$

$$= -\frac{3}{50}p^2 + 46p$$

while the annual cost C is fixed at $5,000. Thus, for break-even,

$$R = C,$$

or $$-\frac{3}{50}p^2 + 46p = 5,000$$

or $$-\frac{3}{50}p^2 + 46p - 5,000 = 0.$$

Note that the second equation above has the form

$$R - C = 0.$$

In other words,

$$\text{Profit} = 0.$$

We now have a quadratic equation in p, and we must solve it for the break-even price p. To simplify the equation, we can first multiply both sides by the bothersome 50 in the denominator, getting

$$-3p^2 + 2{,}300p - 250{,}000 = 0.$$

Now solve, using the quadratic formula to obtain

$$p = 635.55 \quad \text{or} \quad 131.12.$$

Although you may be tempted to choose the larger figure on the grounds that you will be getting more in membership fees, remember that both these options will result in your operation breaking even—there is no advantage to one or the other. Thus, to break even, you should charge either $635.55 or $131.12 per year.

(b)   If we repeat the above analysis with the $5,000 fixed cost replaced by $10,000, we wind up with the quadratic equation

$$-3p^2 + 2,300p - 500,000 = 0.$$

This equation has no real solutions since the discriminant, $b^2 - 4ac$, is negative. You simply cannot break even with these costs, so you might as well close down the operation.

You can answer both parts of the question by using a graphing calculator or computer program to sketch the revenue and cost functions. We had $R = -\frac{3}{50}p^2 + 46p$, and $C = 5,000$ for part (a) and $C = 10,000$ for part (b). If you plot all three functions on the same coordinate system, you get the picture shown in Figure 10.

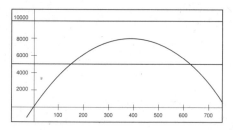

FIGURE 10

The horizontal lines represent the two fixed annual costs, while the parabola is the graph of the revenue function. The two intersection points are the break-even points we calculated in part (a). You can zoom and trace to obtain the coordinates of the break-even point and confirm the calculations we did. The higher cost line fails to touch the revenue curve, confirming our conclusion in part (b) that the health club cannot break even with annual costs of $10,000.

The graph also shows the price you should charge to obtain the largest profit. Trace to the highest point of the parabola (which has a p-coordinate midway between the break-even points) and you will find that in order to obtain the largest revenue, and hence profit, you should charge an annual fee of $383.33. You can then find the annual revenue predicted by the demand equation as the y-coordinate.

▶ **1.5 EXERCISES**

Sketch the graphs of the quadratic functions given in Exercises 1–12, indicating the coordinates of the vertex, the y-intercept and the x-intercepts (if any).

1. $f(x) = x^2 + 3x + 2$

2. $f(x) = -x^2 - x - 12$

3. $f(x) = x^2 + x - 1$

4. $f(x) = x^2 + \sqrt{2}x + 1$

5. $f(x) = \frac{1}{4}x^2 + \sqrt{2}x - 1$

6. $f(x) = -\frac{1}{3}x^2 + 3x - 1$

7. $f(x) = x^2 + 2x + 1$

8. $f(x) = -x(3x + 2)$

9. $f(x) = x^2$

10. $f(x) = -x^2$

11. $f(x) = x^2 + 1$

12. $f(x) = -x^2 + 5$

For each of the demand equations in Exercises 13–16, express the total revenue R as a function of the price p per item, sketch the graph of the resulting function, and determine the price p that maximizes total revenue in each case.

13. $q = -4p + 100$

14. $q = -3p + 300$

15. $q = -2p + 400$

16. $q = -5p + 1200$

APPLICATIONS

17. *Revenue* The market research department of the Better Baby Buggy Co. notices the following. When its buggies are priced at $80 each, it can sell 100 each month. However, when the price is raised to $100, it can only sell 90 each month. Assuming that the demand is linear, at what price should it sell the buggies to get the largest revenue? What is the largest monthly revenue?

18. *Revenue* The Better Baby Buggy Co. has just come out with a new model, the Turbo. The market research department now estimates that the company can sell 200 Turbos per month at $60, but only 120 per month at $100. Assuming that the demand is linear, at what price should it sell its buggies to get the largest revenue? What is the largest monthly revenue?

19. *Revenue* Pack-Em-In Real Estate is building a new housing development. The more houses it builds, the lower the price people will be willing to pay, due to the crowding and smaller lot sizes. In fact, if the company builds 40 houses in this particular development, it can sell them for $200,000 each, but if it builds 60 houses, it will only be able to get $160,000 each. Assuming that the demand is linear, how many houses should the company build in order to get the largest revenue? What is the largest possible revenue?

20. *Revenue* Pack-Em-In has another development in the works. If it builds 50 houses in this development, it will be able to sell them at $190,000 each, but if it builds 70 houses, it will only get $170,000 each. Assuming that the demand is linear, how many houses should it build in order to get the largest revenue? What is the largest possible revenue?

21. *Revenue (computer or graphing calculator recommended)* The wholesale price for chicken in the U.S. fell from 25¢ per pound in 1951 to 14¢ per pound in 1958. At the same time, per capita chicken consumption rose from 21.7 pounds per year to 28.1 pounds per year.*
  (a) Use these data to set up a linear demand equation for poultry. (Round all decimals to four places.)
  (b) Use your demand equation to express the wholesale revenue per capita as a function of the price p of poultry per pound.
  (c) Calculate, to the nearest cent, the price per pound that should result in the largest per capita wholesale revenue.

▼ *Source: U.S. Department of Agriculture, *Agricultural Statistics*. Also see the next footnote.

22. **Revenue** *(computer or graphing calculator recommended)* Repeat Exercise 21, given that the wholesale price for chicken in the U.S. fell from 10¢ per pound in 1962 to 9¢ per pound in 1968. At the same time, per capita chicken consumption rose from 30.8 pounds per year to 41.8 pounds per year.*

23. **Fuel Efficiency**   The fuel consumption of an automobile engine increases with speed. If the latest model Guzzler is driven at 30 mph, it burns a gallon of gas in $\frac{1}{2}$ hour. If it is driven at 90 mph it burns a gallon in $\frac{1}{6}$ hour. Assuming that the time it takes to burn a gallon of gas depends linearly on the speed, find the speed at which the fuel efficiency (in miles/gallon) is highest. What is the highest fuel efficiency that the Guzzler can reach?

24. **Fuel Efficiency**   If the latest model Sipper is driven at 25 mph, it takes $1\frac{1}{2}$ hours to burn a gallon of gas. If it is driven at 50 mph, it takes 1 hour to burn a gallon. Assuming that the time it takes to burn a gallon of gas depends linearly on the speed, find the speed at which the fuel efficiency (in miles/gallon) is highest. What is the highest fuel efficiency that the Sipper can reach?

25. **Break-Even Analysis**   You are the only supplier of beef in a town where the demand equation for beef is given by

$$q = 500 - 40p.$$

Here, q is the number of pounds of beef consumed per month in your town, and p is the price of beef in dollars per pound.[†] Assume that you have stockpiled a large amount of frozen beef and wish to set the price in order to break even. Your fixed costs for storage and refrigeration amount to $1,000 per month. What is the most you could charge per pound?

26. **Break-Even Analysis**   Repeat Exercise 25 using the demand equation $q = 400 - 30p$.

27. **Motion Under Gravity**   If a ball is tossed straight up from ground level ($h = 0$) at a velocity of v_0 feet per second, its height h (in feet) after t seconds will be

$$h = v_0 t - 16t^2.$$

(This formula neglects the effect of air resistance.)
(a) With $v_0 = 64$, sketch the graph of h as a function of t, and use your graph to determine how long it will take the ball to reach the ground.
(b) True or false: If v_0 is doubled, then the time the ball is airborne is also doubled. Justify your answer.

28. **Motion Under Gravity**   If a stone is thrown down a shaft from ground level ($d = 0$) at a velocity of v_0 feet per second, its depth d (in feet) after t seconds will be

$$d = v_0 t + 16t^2.$$

(This formula neglects the effect of air resistance.)
(a) With $v_0 = 2$, sketch the graph of h as a function of t, and use your graph to determine how long it will take the ball to reach a depth of 100 feet.
(b) True or false: If v_0 is doubled, then the time it takes the ball to reach a depth of 100 feet is halved. Justify your answer.

COMMUNICATION AND REASONING EXERCISES

29. Suppose the graph of revenue as a function of unit price is a parabola that is concave down. What are the significance of the coordinates of the vertex, the (possible) x-intercepts, and the y-intercept?

30. Suppose the height of a stone thrown vertically upward is given by a quadratic function of the time. What are the significance of the coordinates of the vertex, the (possible) x-intercepts, and the y-intercept?

▼ * Source: U.S. Department of Agriculture, *Agricultural Statistics*. If you happen to do both Exercises 21 and 22, you will notice that entirely different demand equations result. This is a further illustration of the limitations of applying linear modeling to every pair of data points in sight. It turns out that the case study from which these figures are quoted had to take into account not only the effect of the price of poultry, but also the effect of the price of *beef* on the demand for poultry. It also had to take into account the U.S. per capita disposable income! See "You're the Expert" at the end of this chapter.

[†] Source: *Using Econometrics: A Practical Guide,* by A. H. Studenmund (New York: HarperCollins 1992). The equation was obtained from actual statistics compiled by the U.S. Department of Agriculture. We have adapted the figures to match the scenario of the exercise.

31. Explain why, if demand is a linear function of unit price p (with negative slope) then there must be a *single value of p* that results in a maximum revenue.

32. Explain why, if the average cost of a commodity is given by $y = 0.1x^2 - 4x - 2$, where x is the number of units sold, there is a single choice of x that results in the lowest possible average cost.

33. If the revenue function for a particular commodity is $R(p) = -50p^2 + 60p$, what is the (linear) demand function? Give a reason for your answer.

34. If the revenue function for a particular commodity is $R(p) = -50p^2 + 60p + 50$, can the demand function be linear? What is the associated demand function?

▶ ═══════ **1.6** SOLVING EQUATIONS USING GRAPHING CALCULATORS OR COMPUTERS

In this section we'll have a short discussion on the use of graphing calculators to solve equations in a single unknown (such as $x^2 - 4\sqrt{x} = 0$, or $x^3 - 4x + 1 = 0$). In the language of functions, this amounts to **finding zeros of functions:** finding values of x for which $f(x) = 0$. (For instance, solving $x^3 - 4x + 1 = 0$ is the same as finding the zeros of $f(x) = x^3 - 4x + 1$.) If you will not be using a graphing calculator or similar computer software, you can safely skip this section.

Before we start, we should point out that there are two methods of solving an equation: analytical and numerical. To solve an equation **analytically** means to obtain exact solutions using algebraic techniques. (The Algebra Review in Appendix A has several sections dealing with the analytic solution of equations.) To solve an equation **numerically** means to use a graphing calculator or computer program to obtain *approximate* solutions. Although numerical solutions are only approximations of true solutions, we can calculate them as accurately as we want. Further, some equations can be solved analytically only with great difficulty, and some cannot be solved analytically at all. Often, numerical solution is the best we can do.

Most standard graphing calculators come equipped with "trace" and "zoom" features. The trace feature allows you to move a cursor along the displayed graph and gives you the coordinates of the points as you go. The zoom feature lets you magnify a portion of the curve. Now, in any process of approximation it is important to have some idea of how accurate your answer is. In this regard the trace feature is misleading—it can fool you into thinking that your answer is more accurate than it is. We shall rely more on the zoom feature to keep track of the accuracy of our answers.

▼ **EXAMPLE 1**

Use a graphing calculator to solve the equation $3x^3 - x + 1 = 0$. The solution(s) should be accurate to within ± 0.05 (that is, accurate to one decimal place).

SOLUTION   Begin by using your calculator to graph the equation $y = 3x^3 - x + 1$. If you need to specify x- and y-ranges, start by specifying x

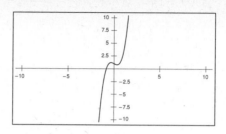

FIGURE 1

between -10 and 10, and do the same for y. Figure 1 shows the kind of picture you should get on your calculator display.

You are looking for a solution of $3x^3 - x + 1 = 0$. Since you have graphed $y = 3x^3 - x + 1$, you are looking for a point on the graph where $y = 0$. That is, you are looking for a point where the graph crosses the x-axis. Looking at the graph, notice that it crosses the x-axis at exactly one point, somewhere between -2 and 0. This observation tells you two things. First, there is only one real solution to the equation $3x^3 - x + 1 = 0$. (If there were another solution, you would see the curve crossing the x-axis again.) Second, this solution is somewhere between -2 and 0.

As your first estimate of the solution, take the point midway in this range $[-2, 0]$, that is, $x = -1$. Since you are not sure exactly where in this range the solution x lies, all you can say with certainty is that $x = -1$ with a possible error of 1 unit in either direction, that is, $x = -1 \pm 1$. (See Figure 2.)

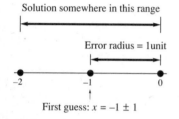

FIGURE 2

To get a more accurate estimate of where this solution occurs, "zoom in" by using your calculator's "zoom" feature or by redrawing the graph specifying the x-range as $-2 \le x \le 0$, since this is where you know the solution lies. As for the y-range, you can use any range that includes zero, say $-1 \le y \le 1$. Figure 3 shows what the output should look like.

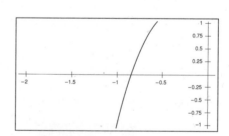

FIGURE 3

Now you see that the solution lies somewhere between -1 and -0.5, so as your next estimate, take the midpoint, -0.75, of this range. Since the width of the interval $[-1, -0.5]$ is 0.5 units, your estimate is accurate to within plus or minus half that, or 0.25. Thus, your second estimate is

$$x = -0.75 \pm 0.25.$$

Q Wait! My graphing calculator does not put numbers on the axes as shown in the diagram. How can I tell that the curve crosses the *x*-axis between −1 and −0.5?

A You can use the scale feature to place tick marks on the axes as shown in the figures. Alternatively, you can use the trace feature as follows.

1. Place the cursor on any point of the curve to the left of the point of intersection with the *x*-axis, and read the value of *x* from the display. Call this value *a*. If—as often occurs when you use trace—the point *a* is a messy decimal such as $a = -0.9874512$, you can choose any convenient *smaller* number as *a*, for instance, $a = -1$. (Why smaller?)

2. Now trace to any point of the curve to the *right* of the point of intersection with the *x*-axis, and again read the value of *x*. Call this value *b*. If the point *b* is a messy decimal such as $b = -0.0821348$, you can choose any convenient *larger* number as *b*, for instance, $b = -0.08$. (Why larger?)

   You can now see that the solution lies somewhere in the interval [*a*, *b*]. In fact, this technique can get you to the desired accuracy rather quickly.

The error is still too big, so you can't stop here. (Remember that you are looking for an accuracy of ± 0.05 or less.) So "zoom in" once again, using the *x*-range $-1 \le x \le -0.5$ and a smaller *y*-range if you like, say $-0.5 \le y \le 0.5$. (Figure 4)

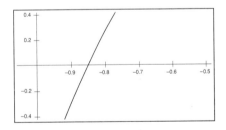

FIGURE 4

Now we are getting somewhere. According to the graph, the solution is somewhere between −0.9 and −0.8. Thus, as your next estimate, choose the midpoint −0.85, with a possible error of ± 0.05, since this is half the length of the interval [−0.9, −0.8]. This is the accuracy you needed, so you are done: the solution of $3x^3 - x + 1 = 0$ is $x = -0.85$, to within ± 0.05.

▼ **EXAMPLE 2**

Using a graphing calculator, find all zeros of the function $f(x) = x^5 - 2x^2 + 1$, to within ± 0.02.

SOLUTION   Recall that to find the zeros of $f(x)$ means to solve the equation $f(x) = 0$. Begin by having the calculator draw the graph of $y = x^5 - 2x^2 + 1$ in the range $-10 \le x \le 10$, and do the same for y. The output is shown in Figure 5.

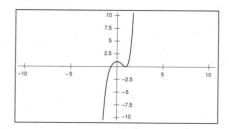

FIGURE 5

This time, it looks as though there are several solutions, one between -2 and 0, and one or two between 0 and 2. To get a better view, zoom in to $-2 \le x \le 2$, and do the same for y. (Figure 6)

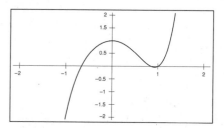

FIGURE 6

You can see now that there are three solutions. You will have to zoom in on each of the zeros separately to get the desired accuracy. We shall find the middle one, and leave the others to you! Looking at the last figure, notice that the solution second from the right is somewhere between 0.75 and 1. So zoom in there, getting Figure 7.

This figure shows the solution as slightly to the left of 0.85, and definitely in the range $0.8 \le x \le 0.9$. Thus, you get the estimate $x = 0.85 \pm 0.05$. Although you suspect it to be accurate to within ± 0.02 (why?) zoom in once more to confirm this (Figure 8).

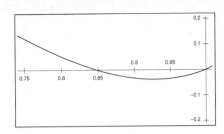

FIGURE 7

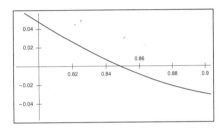

FIGURE 8

Now you can be absolutely certain that the solution is 0.85 ± 0.01, which is even more accurate than what was required!

$\boldsymbol{Q}$ We know how to locate the zeros of a function f—values of x for which $f(x) = 0$. What if we need to locate values of x for which $f(x) = g(x)$, where g is some other function?

$\boldsymbol{A}$ There are two ways to do this, and we illustrate with the example $f(x) = x^5 - x^2 + 2$, and $g(x) = \sqrt{4 - x^2}$.
    1. Notice that solving

$$x^5 - x^2 + 2 = \sqrt{4 - x^2}$$

is the same as solving

$$x^5 - x^2 + 2 - \sqrt{4 - x^2} = 0,$$

and we already know how to locate zeros of functions such as this.
    2. Alternatively, graph both $y = x^5 - x^2 + 2$ and $y = \sqrt{4 - x^2}$ on the same set of axes, and find values of x where the graphs cross. (This is what we did in our discussion of break-even analysis.)

▶ ___ **1.6 EXERCISES**

Use a graphing calculator to solve the equations in Exercises 1–8 to the specified accuracy.

1. $x^2 + 2x - 5 = 0$, to within ± 0.05

2. $-x^2 + 5x + 12 = 0$, to within ± 0.05

3. $-x^3 - 2x^2 + x - 1 = 0$, to within ± 0.01

4. $x^3 - 2x^2 + x - 1 = 0$, to within ± 0.01

5. $x^5 - 10x + 5 = 0$, to within ± 0.001

6. $x^5 - 16x - 1 = 0$, to within ± 0.001

7. $x^7 - x^5 + x - 2 = 0$, to within ± 0.02

8. $3x^7 - 2x^3 + x = 0$, to within ± 0.02

Use a graphing calculator to locate all zeros of each of the functions in Exercises 9–12 to the specified accuracy.

9. $f(x) = x^2 + \frac{1}{x} - 4x$, to within ± 0.001

10. $f(x) = x^2 - \frac{1}{x} - 5$, to within ± 0.001

11. $f(x) = x^5 - x - 3$, to within ± 0.05

12. $f(x) = x^4 - 3x^2 - x$, to within ± 0.05

Locate all values of x that satisfy the equations in Exercises 13–16.

13. $2^x = x$, to within ± 0.05

14. $3^x = 4x^2$, to within ± 0.05

15. $\dfrac{x^2 + 1}{x^2 - 1} = 2x - \sqrt{x}$, to within ± 0.05

16. $(x - 1)^{2/3} = 2^x$, to within ± 0.05

APPLICATIONS

17. *Dental Plans*   A company pays for its employees' dental coverage at an annual cost C given by

$$C(q) = 1{,}000 + 100\sqrt{q},$$

where q is the number of employees covered, and $C(q)$ is the annual cost in dollars. If the government subsidizes coverage by an annual dollar amount of

$$S(q) = 50q,$$

at what number of employees does the company actually start making a profit on dental coverage?

18. *Surface Area*   The surface area of a hollow cylinder of cross-sectional radius r and height h is given by $S = 2\pi rh$. Further, the surface area of a disc of radius r is given by $A = \pi r^2$. (See the diagram on page 39 for this exercise.

You are the design manager for Pebbles and Blips gourmet cat food, and you wish to design the shape of the cans that are to hold Pebbles and Blips Lobster Delight. Your can must hold 20 cubic inches (the volume of a can is given by $V = \pi r^2 h$). Further, the alloy you plan to use to manufacture the cans will cost 4¢ per square inch. Express the cost of a single can as a function of the radius r, and determine at what radius the can will use $2.50 worth of metal to build. (Give your answer to the nearest 0.01 in.)

19. *Investments*   If you invest $2,000 at interest rate r compounded monthly, the amount of money you will have after 10 years is given by the function

$$A(r) = 2{,}000\left(1 + \frac{r}{12}\right)^{120}.$$

What must the interest rate be in order for you to double your money in 10 years? (Give your answer to the nearest 0.01%.)

20. *Radioactive Decay*   Carbon 14 is radioactive and decays over time to nitrogen. If you start with 10 g of Carbon 14, after t years you will still have

$$A(t) = 10\left(\frac{1}{2}\right)^{t/5{,}730}$$

grams. How long will it take your original 10 g to decay to 1 g? (Give your answer to the nearest year.)

▶ You're the Expert

MODELING THE DEMAND FOR POULTRY

A government agency plans to regulate the price of poultry with the goal of increasing revenue to poultry producers. Economists developed the demand equation

$$q = 56.9 - 0.45p + 0.12b$$

in which q is the annual per capita demand for poultry in pounds per year, p is the wholesale price of poultry in cents per pound, and b is the wholesale price of beef in cents per pound.*

The agency has hired you to develop a pricing policy and to analyze the effect that fluctuations in the beef price would have on revenue to poultry producers under your proposed policy.

You observe immediately that the demand for poultry depends on the price of beef as well as the price of poultry. You decide that the pricing policy should "tie" the poultry price to the beef price in some way. You must do the following things.

1. Find the function f such that if b is the price of beef, then $p = f(b)$ is the price of poultry that will maximize revenue to poultry producers. (The price of beef is expected to be between $0.50 and $1.00 per pound.)
2. Estimate the revenue to poultry producers for the range of beef prices given.
3. Determine the increase in per capita annual revenue for poultry production for each one-cent increase in the price of beef, and find the largest value of this quantity over the range of beef prices given.

You then get to work. The first part of the project reminds you of the kind of calculation done in Sections 3 and 4. First, you recall that

$$\text{Revenue} = \text{Price per pound} \times \text{Number of pounds},$$

so you obtain

$$R(p) = p(56.9 - 0.45p + 0.12b).$$

You express this as a quadratic function in p, getting

$$R(p) = -0.45p^2 + (56.9 + 0.12b)p$$

▼ *Source: A. H. Studenmund, *Using Econometrics*, Second Edition (HarperCollins, 1992), pp. 180–81. This equation was calculated using the "least squares" method, from data collected from 1950 through 1984 (see the chapter on functions of several variables). The equation originally involved a parameter for per capita disposable income, which we have assumed to be $15,000.

for the annual per capita revenue (in cents) to the poultry industry. You now calculate the value of p that gives the largest revenue.

$$f(b) = \frac{56.9 + 0.12b}{0.90} = 63.2 + 0.13b$$

Since you are interested in values of b between 50¢ and 100¢, you take the domain to be $[50, 100]$, and you discover that you have completed the requirements for the first part of the assignment (and you take the rest of the day off).

Upon returning to your desk, you turn to the calculation of the total revenue. You notice that you already have the formula

$$R(p) = -0.45p^2 + (56.9 + 0.12b)p$$

and since $p = f(b)$, you calculate R as a function of b by substituting.

$$R(f(b)) = -0.45(63.2 + 0.13b)^2$$
$$+ (56.9 + 0.12b)(63.2 + 0.13b)$$

You give this new function of b a name, S, and simplify by expanding and combining terms.

$$S(b) = -0.45(3994.24 + 16.432b + 0.0169b^2)$$
$$+ (3596.08 + 14.981b + 0.0156b^2)$$
$$\approx 1800 + 7.6b + 0.008b^2$$

with domain $[50, 100]$ as before. This *almost* completes the second part of your assignment, but since the agency wants to *see* the revenue, you decide to draw the graph of S (Figure 1). (This is a tiny piece of a parabola, so small that you can hardly see the curvature.) You can now see at a glance, for example, that if the price of beef is set at 60¢ per pound, you can expect the average person to consume about $23 worth of poultry per year.

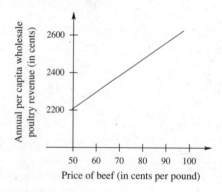

FIGURE 1

Q As the price of beef goes up, so does the price of poultry. Why?

A The higher the price of beef, the greater the demand for poultry. This allows the government to increase the price of poultry, and at the same time increase revenue from poultry.

Finally, the agency wants to know how much the revenue from poultry will go up for each 1¢ increase in the price of beef under your pricing policy. Now the revenue from poultry at a beef price of b is given by

$$S(b) = 1800 + 7.6b + 0.008b^2.$$

An increase of 1¢ in this price gives

$$S(b+1) = 1800 + 7.6(b + 1) + 0.008(b + 1)^2.$$

Thus, the increase in the revenue from poultry is

$$
\begin{aligned}
S(b + 1) - S(b) &= [1800 + 7.6(b + 1) + 0.008(b + 1)^2] \\
&\quad - [1800 + 7.6b + 0.008b^2] \\
&= [1800 + 7.6b + 7.6 + 0.008b^2 + 0.016b \\
&\quad + 0.008] - [1800 + 7.6b + 0.008b^2] \\
&\approx 7.6 + 0.016b.
\end{aligned}
$$

The change in S goes up as the beef price goes up. Thus, the largest value $S(b + 1) - S(b)$ can have occurs at a beef price of $b = 100$. So, if beef is priced at \$1.00 per pound, a 1¢ increase would result in an increase in poultry revenue of

$$7.6 + 0.016 \cdot 100 \approx 9.2¢ \text{ per capita per year.}$$

Exercises

1. If the wholesale price of beef is 82¢ per pound, what, according to your proposed policy, should the wholesale price of poultry be?

2. Under your proposal, by how much will the price of poultry rise under your proposal for each 1¢ increase in the price of beef?

3. Explain the significance of each of the constants in the demand equation given by the economists.

4. Explain in general terms why one expects the revenue for poultry producers to increase as the price of beef increases.

5. Repeat the analysis assuming a per capita disposable income of \$30,000, which makes the demand equation look like this:

$$q = 65.4 - 0.45p + 0.12b.$$

6. Just as you are about to submit your proposal to the agency, you get an urgent phone call: The term b in the demand equation should have been b^2. (The exponent got lost during faxing.) What adjustments will you need to make to your report?

7. Given the demand equation used in the discussion, assume that it costs poultry producers 35¢ to produce one pound of poultry. Repeat the first part of your analysis, but this time with a view to maximizing poultry producers' profit rather than revenue.

▶ ## Review Exercises

Find the equations of the lines described in Exercises 1–10, in the form
$ax + by + c = 0$.

1. Through the origin with slope 3

2. Through the origin with slope $-\frac{1}{2}$

3. Through the point $(1, -1)$ with slope 3

4. Through the point $(-1, -2)$ with slope -1

5. Through the points $(-3, -6)$ and $(1, -1)$

6. Through the points $(1, -2)$ and $(-1, 0)$

7. Through the point $(1, -2)$ and parallel to the line $x + 3y = 1$

8. Through the point $(1, -1)$ and parallel to the line $2x - 3y = 11$

9. Through the point $(1, -2)$ and perpendicular to the line $2x - y = 1$

10. Through the point $(0, -1)$ and perpendicular to the line $2x + 2y = 3$

Sketch the lines whose equations are given in Exercises 11–16.

11. $2x + y = 6$ **12.** $x - 2y = 8$ **13.** $2y = -3$

14. $3y = 0$ **15.** $2x + 1 = 0$ **16.** $4x - 3 = 0$

Sketch the parabolas whose equations are given in Exercises 17–22.

17. $y = x^2 - 3x + 2$ **18.** $y = -x^2 - 11x - 30$

19. $y = -5x^2 + x - 2$ **20.** $y = 6x^2 - 4x - 2$

21. $y = x^2 - x - 1$ **22.** $y = -x^2 + x - 1$

*In Exercises 23–28, sketch the graphs of the given functions without plotting
points, by referring to the graphs of the standard functions you have learned.
(Unless otherwise stated, the domain is the largest possible.)*

23. $f(x) = x^3$, with domain $[-\infty, 0)$ **24.** $f(x) = x^4$, with domain $[-1, 1]$

25. $f(x) = \sqrt{x}$, with domain $[0, 9]$ **26.** $f(x) = \sqrt[4]{x}$, with domain $[0, 16]$

27. $f(x) = 1/|x|$ **28.** $f(x) = |x| + 1/|x|$

Sketch the graph of each of the functions in Exercises 29–34.

29. $f(x) = x^2 + x + 2$ **30.** $f(x) = x^2 + \dfrac{1}{x}$

31. $f(x) = \dfrac{1}{(x + 1)^2}$ **32.** $f(x) = \sqrt{x} + \dfrac{1}{x^2}$

33. $f(x) = \sqrt{1 - x^2} - x$ **34.** $f(x) = \sqrt{1 - x^2 + 2x}$

*Use a graphing calculator to find at least one solution for each of the equations in
Exercises 35–40 to within ± 0.005.*

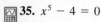 **35.** $x^5 - 4 = 0$ **36.** $x^5 - 2x = 0$ **37.** $x^2 - 5\sqrt{x} = 0$

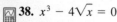 **38.** $x^3 - 4\sqrt{x} = 0$ **39.** $4 = \sqrt{x} + \dfrac{1}{x^3}$ **40.** $\dfrac{1}{(x + 1)^2} = 7x - 1$

41. Given $f(r) = \dfrac{1}{r + 2}$, find

    (a) $f(0)$ **(b)** $f(1)$ **(c)** $f(-1)$ **(d)** $f(x - 2)$ **(e)** $f(x^2 + x)$ **(f)** $f(x^2) + x$

42. Given $f(x) = x - \dfrac{1}{x}$, find

    (a) $f(1)$ **(b)** $f(-1)$ **(c)** $f(\pi)$ **(d)** $f(x + h) - f(x)$ **(e)** $f(x) + h$ **(f)** $f(x) + h - f(x)$

43. Given $g(x) = \sqrt{x^2 - 1}$, find

    (a) $g(1)$ **(b)** $g(-1)$ **(c)** $g(\sqrt{x + h})$ **(d)** $g(\sqrt{x}) + h.$

44. Given $g(y) = \sqrt[3]{y}$ find

    (a) $g(8)$ **(b)** $g(16)$ **(c)** $g((y + h)^3)$ **(d)** $g(y^3) + h.$

45. Given $f(x) = \dfrac{x^2 + 1}{x}$, with domain $(0, +\infty)$, find

    (a) $f(1)$ **(b)** $f(a^2)$ **(c)** $f(x + h) - h$ **(d)** $f(\sqrt{x}) + h \, (x > 0)$

46. Given $f(x) = \dfrac{6}{\sqrt{x}} - \sqrt{x}$, with domain $(0, +\infty)$, find

    (a) $f(x^2)$ **(b)** $(f(x))^2$ **(c)** $\sqrt{x}f(x)$ **(d)** $f(x\sqrt{x})$

In Exercises 47–50, calculate and simplify the quotient $\dfrac{f(x + h) - f(x)}{h}$ *$(h \neq 0)$.*

47. $f(x) = x^2 + x - 1$ **48.** $f(x) = -x^2 - 2x - 1$

49. $f(x) = \dfrac{2}{2x - 1}$ **50.** $f(x) = \dfrac{1}{3 - 2x}$

APPLICATIONS

51. *Cost*   Rock Solid Insurance, Inc.'s premium for a $100,000 life insurance policy is $96 for a 25-year-old nonsmoker, and $186 for a 45-year-old nonsmoker. If n represents the number of years since a policyholder was 25, and p the premium, express p as a linear function of n, and use your model to predict what the premium will be for a 90-year-old nonsmoker.

52. *Demand*   The market research department for Ultrafast Computers, Inc., finds that it can sell 10,000 computers at $5,000 apiece, but only 6,000 computers priced at $7,000. Express the demand q as a linear function of the price p.

53. *Motion*   As you start your cross-country drive, your odometer reads 45,000 miles. If you maintain a constant speed of 55 mph, find a linear function giving your odometer reading s after t hours.

54. *Motion*   Your brand new Corvette can accelerate from zero to 60 mph in 6 seconds. Write a linear function giving its speed v (mph) in terms of time t (seconds) What is your average acceleration over this period (that is, at what rate is your speed increasing)? Assuming you continue this rate of acceleration, how fast would your Corvette be going after 1 minute?

55. *Free Fall*   If you drop a cannonball from the top of the leaning tower of Pisa, it will start with a velocity of 0, but after 5 seconds will have a velocity of 160 ft/s. Assuming that the velocity increases linearly, find an equation for the velocity v in terms of the time t. What is the acceleration of the cannonball?

56. *Cost Equation*   Dirty Dudley's coin-operated clothes dryers require 50¢ to get started, and $1 for each fifteen minutes of drying time. Express the cost of drying a load of laundry as a function of the time t in hours. What is the marginal cost per hour? Because these dryers operate at close to room temperature, a typical load requires two hours of drying time. How much does this cost?

57. *Sales Commission (from the GMAT)*   An employee is paid a salary of $300 per month and earns a 6 percent commission on all her sales. What must her annual sales be in order for her to have a gross annual salary of exactly $21,600?

58. *Break Even Analysis (from the GMAT)*   Ken left a job paying $75,000 per year to accept a sales job paying $45,000 per year plus 15 percent commission. If each of his sales is for $750, what is the least number of sales he must make per year if he is not to lose money because of the change?

59. *Getting Ahead (from the GMAT)* In 1980 John's salary was $15,000 a year and Don's salary was $20,000 a year. If every year thereafter John receives a raise of $2,450 and Don receives a raise of $2,000, what is the first year in which John's salary will be more than Don's salary?

60. *Cost Equations (from the GMAT)* Marion rented a car for $18.00 plus $0.10 per mile driven. Craig rented a car for $25.00 plus $0.05 per mile driven. If each drove *d* miles and each was charged exactly the same amount for the rental, then what was *d*?

61. *Demand and Revenue* Your underground used book business is doing a booming trade. Your policy is to sell all used versions of *Calculus and You* at the same price (regardless of condition). When you set the price at $10, sales amounted to 120 volumes during the first week of classes. What was your total revenue that week? The following semester, you set the price at $30, and sold not a single book. Use these data to set up a demand equation, and use your equation to express the total revenue as a function of the price per book. What price gives you the maximum revenue, and what does that revenue amount to?

62. *Demand and Revenue* Banana Computers has just introduced a new model. The company estimates that it can sell 12,000 units at a price of $2,000, but only 11,000 units at a price of $2,250. Assuming a linear demand function, at what price will it get the largest revenue?

63. *Break-even Analysis (from the CPA exam)* At a break-even point of 400 units sold, the variable costs were $400 and the fixed costs were $200. What will the 401st unit contribute to profit before income taxes?

64. *Break-even Analysis (adapted from a CPA exam question)* At a break-even point of 200 units sold, the variable costs were $800 and the fixed costs were $100. What will the 401st unit contribute to profit before income taxes?

65. *Demand* The demand function for a commodity is
$$q(p) = 40 - 18p - p^2,$$
where q represents the number of items the manufacturer can sell per month at a price of p dollars each. Use this function to determine
(a) the number of items the manufacturer can sell per month if the price is set at $1;
(b) the price at which all sales will stop.

66. *Demand* The demand function for a commodity is
$$q(p) = 200 - 10p - p^2,$$
where q represents the number of items the manufacturer can sell per month at a price of p dollars each. Use this function to determine
(a) the number of items the manufacturer can sell per month if the price is set at $5;
(b) the price at which all sales will stop.

*Exercises 67 and 68 use the idea of the **average cost**. If $C(x)$ is the total cost for x items, then the average cost per item for x items is*
$$\overline{C}(x) = \frac{C(x)}{x}$$

67. *Cost and Average Cost* Lite Up My Life, Inc. can produce 1,000 light fixtures in a month at a total cost of $20,000, and it can produce 2,000 fixtures in a month at a total cost of $30,000.
(a) Construct a linear cost function for Lite Up My Life, Inc.
(b) Write the average cost function.
(c) How many light fixtures should Lite Up My Life, Inc. manufacture per month in order to meet an average cost goal of $12.50 per fixture?

68. *Cost and Average Cost* A firm has monthly fixed costs of $50,000 and variable (marginal) costs of $100 per item.
(a) Write the cost function $C(x)$, where x is the number of items produced per month.
(b) Write the average cost function.
(c) Find the production level (number of items produced per month) that gives an average cost of $200 per item.

69. *The Theory of Relativity* In science fiction terminology, a speed of *warp 1* is the speed of light—about 3×10^8 meters per second. (Thus, for instance, a speed of warp 0.8 corresponds to 80% of the speed of light—about 2.4×10^8 meters per second.) According to Einstein's Special Theory of Relativity, a moving object appears to get shorter to a stationary observer as its speed approaches that of light. If a rocket ship whose length is l_0 meters at rest travels at a speed of warp p, its length in meters, as measured by a stationary observer, will be given by
$$l(p) = l_0\sqrt{1 - p^2}, \text{ with domain } [0, 1].$$
(a) Assuming that a 100-meter rocket ship is traveling at warp 0.9, what will be its length as measured by a stationary observer?

(b) At what speed will it need to travel in order that it appears to be squashed to a length of 1 meter?

(c) What would happen at the speed of light (warp 1)?

70. *Newton's Law of Gravity*   The gravitational force exerted on a particle with mass m by another particle with mass M is given by the following function of distance:

$$F(r) = G\frac{Mm}{r^2}, \text{ with domain } (0, +\infty).$$

Here, r is the distance between the two particles in meters, the masses M and m are given in kilograms, $G \approx 0.0000000000667$, or 6.67×10^{-11}, and the resulting force is given in newtons.

(a) Given that M and m are both 1,000 kilograms, find $F(1)$ and $F(10)$.

(b) How much would a battleship have to weigh in order to attract a 1-kg mass situated 1,000 meters away with a force of 1 newton?

71. *Sales Commissions*   The Bigger the Better Publishing Company hires students to sell its 100-volume *Encyclopedia Galactica* (which also includes graphing software) for $3,025 per set. The sales staff are each paid a commission of 5% of the square root of total sales, plus a basic wage of $100 per month.

(a) Write a function that expresses the student's earnings on sales of x sets per month.

(b) Use your formula to calculate the earnings from the sale of 1 set per month and 100 sets per month.

(c) Approximately how many sets should a student sell in order to earn $200 per month? Comment on the company's policy.

72. *Sales Commissions*   The Smaller the Better Publishing Company hires students to sell its 100-page *Encyclopedia Miniscula* (which includes graphing hardware) for $25 per volume. The sales staff are each paid a commission of 5% of the square of total sales, plus a basic wage of $100 per month.

(a) Write down a function giving a student's earnings on sales of x volumes per month.

(b) Use your formula to calculate the earnings from 1 volume per month and 100 volumes per month.

(c) Approximately how many sets should a student sell in order to earn $2,000 per month? Comment on the company's policy.

73. *Salary Scales in Japan*   According to data in a *New York Times* article, the average annual salary for college graduates in Japan is approximately a linear function of the age, rising from an average of $30,000 per year for a 28-year-old to $65,000 per year for a 48-year-old worker.*

(a) Use this information to express the average salary S of a Japanese college graduate in thousands of dollars as a function of age x. (Also give the domain.)

(b) The same data also show the average salary increasing more slowly, at about $1,000 per year, from age 48 to 54, and then leveling off at $71,000 per year. Use these data to express annual salary as a piecewise-defined function of age x for $28 \le x \le 58$.

74. *Salary Scales in Japan*   According to data in the *New York Times* article mentioned in Exercise 73, the average annual salary for high school graduates in Japan is approximately a linear function of the age, and rises from an average of $28,000 per year for a 28-year-old to $55,000 per year for a 48-year-old worker.*

(a) Use this information to express the average salary S of a Japanese high school graduate in thousands of dollars as a function of age x. (Also give the domain.)

(b) The same data also show the average salary increasing more slowly, at about $1,000 per year, from age 48 to 54, and then leveling off at $61,000 per year. Use these data to express annual salary as a piecewise-defined function of age x for $28 \le x \le 58$.

75. *Demand*   The demand equation for donuts is

$$p = \frac{500}{q + 100},$$

where p is the price per donut and q is the quantity of demand. Solve for q as a function of p. What does this function signify?

76. *Demand*   The demand equation for bagels is

$$p\sqrt{q} = 400,$$

where p is the price per bagel and q is the quantity of demand. Find q as a function of p, and also find p as a function of q. What do these functions signify?

▼ * Source: *The New York Times*, Oct. 2, 1993, p. 6, and the Japan Federation of Employee Associations.

Cerebrospinal Meningitis Epidemics

A debilitating and often deadly disease, meningitis remains common in many developing countries. New insights may soon enable us to predict and control outbreaks

By Patrick S. Moore and Claire V. Broome

By the middle of April 1988 the meningitis epidemic in N'Djamena, the capital of Chad, was in full swing. The outbreak had begun with a few isolated cases in mid-February; within four weeks nearly 150 patients were being admitted to the city's Central Hospital every day. As the facility ran out of bed space, people were treated in huge army tents scattered throughout the inner courtyards. Despite the best efforts of the Ministry of Health and foreign volunteer agencies, the epidemic spread. A shortage of medicines burdened health workers straining under the fatigue of seemingly endless days. Although a massive vaccination campaign was being implemented that would eventually stem the epidemic, each day threatened to paralyze further the country's fragile health care system.

By the time the scourge ended, 4,500 people had acquired meningitis, according to official statistics. Hundreds, or even thousands more, however, were uncounted. In Chad, as in many African countries, medical care is generally not available to people outside major cities. Meningitis sufferers who lived more than a day's walk from the nearest health station generally did not receive antibiotic treatment. Many died or were left with permanent brain damage.

The chaos and misery caused by the epidemic in Chad characterize most outbreaks of meningococcal meningitis, commonly known as spinal meningitis. The hallmark of the disease is its extremely rapid onset, which has earned this illness an uncommon degree of respect among medical experts. A healthy person first develops fever and malaise similar to that associated with influenza. Within hours these symptoms evolve into severe headache, neck rigidity and aversion to bright lights. If untreated, the patient can lapse into coma and eventually a fatal form of shock. Although it is now rare in the U.S., intense epidemics still affect most of the developing world; within weeks an entire country can be stricken.

Why do such epidemics occur? What causes a disease like meningitis to simmer within a population for years and then suddenly erupt? While many mysteries continue to surround the potentially lethal illness, its peculiar epidemiology offers some clues about the causes of meningitis epidemics and how to prevent them. The disease includes cycles of incidence that may correspond to environmental changes, to unusual patterns of immunity as well as to an association with still other infectious diseases. Medical detective work and the application of new biological techniques have begun to unveil some of these deadly secrets.

Source: From Patrick S. Moore and Claire V. Broome, "Cerebrospinal Meningitis Epidemics," *Scientific American*, November 1994, p. 38.

Exponential and Logarithmic Functions

APPLICATION ▶ A mysterious epidemic is spreading through the population of the United States. An estimated 150,000,000 people are susceptible to this particular disease. There are 10,000 people already infected, and the number is doubling every two months. As advisor to the Surgeon General, it is your job to predict the course of the epidemic. In particular, the Surgeon General needs to know when the disease will have infected 1,000,000 people, when it will reach 10,000,000, and when it will affect 100,000,000 people. How do you answer these questions?

| INTRODUCTION | ▸Exponentials and logarithms are indispensable for an understanding of many processes in economics and nature. Examples include interest, inflation, population growth, spread of epidemics, and radioactive decay. All of these processes are modeled by exponential functions, and logarithms are necessary to answer many of the questions that naturally arise.

You are strongly urged to read through the review sections on the algebra of exponentials and radicals in Appendix A before you begin this chapter, because much of the material in this chapter assumes a familiarity with manipulating exponents.

In this chapter, we will discuss in some detail the use of graphing calculators in the examples and exercises. Even if you do not have access to graphing technology, however, you will benefit greatly by carefully reading *all* of the examples. If you *are* using a graphing calculator, refer to Appendix B and to the first section in Appendix A.

▸ 2.1 EXPONENTIAL FUNCTIONS AND APPLICATIONS

We have already seen examples of functions involving exponents, such as $f(x) = x^2$ or $g(x) = (x + 1)^{1/2}$. In each of these examples, the exponent is constant and the base is variable. We are now going to turn the tables and consider functions such as $f(x) = 2^x$, where the base is constant and the exponent is variable. This function is an example of an *exponential function*.

EXPONENTIAL FUNCTION

An **exponential function** is a function of the form

$$f(x) = Ca^x,$$

where C and a are constants and $a > 0$. (We call a the **base** of the exponential function.)

Examples of exponential functions are

$$f(x) = 2^x \qquad C = 1, a = 2$$

and
$$g(x) = 3 \cdot 2^{-4x}$$
$$= 3(2^{-4})^x$$
$$= 3\left(\frac{1}{16}\right)^x. \qquad C = 3, a = \frac{1}{16}$$

For reference, we repeat below the list of laws of exponents from the algebra review in the appendix. Which of the laws did we use in reformulating $g(x)$ above?

THE LAWS OF EXPONENTS

If a and b are positive and x and y are any real numbers, then the following laws hold.

Law	Example
1. $a^x a^y = a^{x+y}$	$2^3 2^2 = 2^5 = 32$
2. $\dfrac{a^x}{a^y} = a^{x-y}$	$\dfrac{4^3}{4^2} = 4^{3-2} = 4^1 = 4$
3. $a^{-x} = \dfrac{1}{a^x}$	$9^{-0.5} = \dfrac{1}{9^{0.5}} = \dfrac{1}{3}$
4. $a^0 = 1$	$(3.3)^0 = 1$
5. $(a^x)^y = a^{xy}$	$(3^2)^2 = 3^4 = 81$
6. $(ab)^x = a^x b^x$	$(4 \cdot 2)^2 = 4^2 2^2 = 64$
7. $\left(\dfrac{a}{b}\right)^x = \dfrac{a^x}{b^x}$	$\left(\dfrac{4}{3}\right)^2 = \dfrac{4^2}{3^2} = \dfrac{16}{9}$

▼ **EXAMPLE 1**

Let $f(x) = 2^x$, with domain the set $\mathbb{R}$ of real numbers. Then

$$f(3) = 2^3 = 8$$

$$f(-3) = 2^{-3} = \frac{1}{8}$$

$$f(0) = 2^0 = 1$$

$$f(x + h) = 2^{x+h}$$

$$f(20) = 2^{20} = 1,048,576$$

$$f(-20) = 2^{-20} = \frac{1}{1,048,567} \approx 0.000000953.$$

Before we go on... We didn't calculate all of these values by hand. You will find that the use of a scientific calculator is indispensable for most of what follows. A graphing calculator would be even better.

Notice something interesting about the values of $f(x)$ we obtained: *they are all positive.* In general,

$$a^x > 0 \text{ for every real number } x.$$

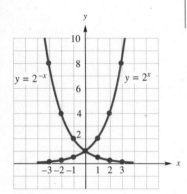

$y = 2^{-x}$ $y = 2^x$

FIGURE 1

▼ **EXAMPLE 2**

On the same set of axes, graph the functions $f(x) = 2^x$ and $g(x) = 2^{-x}$.

SOLUTION   Although we can graph these easily on a graphing calculator, we can also graph them easily by hand. Here is a table of values to start with.

x	-3	-2	-1	0	1	2	3
$f(x) = 2^x$	$\frac{1}{8}$	$\frac{1}{4}$	$\frac{1}{2}$	1	2	4	8
$g(x) = 2^{-x}$	8	-4	-2	1	$\frac{1}{2}$	$\frac{1}{4}$	$\frac{1}{8}$

Before we graph the curves, notice the symmetry in the table: values of $f(-x)$ correspond to values of $g(x)$. This is a consequence of the fact that

$$f(-x) = 2^{-x} = g(x)$$

and is reflected by a symmetry between their graphs (see Figure 1).

Notice also how the curve $y = 2^x$ goes "shooting up" for larger values of x. In fact, the y-coordinate doubles for each increase of 1 in x. (For example, if we'd plotted an extra point for $x = 4$, the coordinates would be (4,16), which is much higher than we've allowed for in the picture.) On the other hand, notice how the curve "levels off" for larger and larger negative values of x, since the sequence $2^{-1}, 2^{-2}, 2^{-3}, \ldots$ gets closer and closer to zero. The curve $y = 2^{-x}$ is the mirror image of $y = 2^x$. The curves meet on the y-axis at $y = 1$, since $2^0 = 2^{-0} = 1$.

 If you wish to reproduce these graphs on your graphing calculator, be sure to use the "range" menu to specify $-3 \le x \le 3$ and $0 \le y \le 8$, so that the graphs will use the entire screen. Also, the correct format for entering the equations corresponding to these functions on most graphing calculators is

$$Y_1 = 2^{\wedge}X$$
$$Y_2 = 2^{\wedge}(-X)$$

where the second formula specifies that y_2 is 2 raised to the *quantity* $(-x)$.

▼ **EXAMPLE 3**

On the same set of axes, sketch the functions $f_1(x) = (\frac{1}{2})^x$, $f_2(x) = 1^x$, $f_3(x) = 2^x$, $f_4(x) = 3^x$, and $f_5(x) = 4^x$.

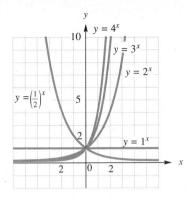

FIGURE 2

SOLUTION   We can save ourselves some work by noting the following things about these functions.

$y = (\frac{1}{2})^x$ is the same as $y = 2^{-x}$, by the laws of exponents, and we have already drawn that curve.

$y = 1^x$ is the same as $y = 1$, a horizontal line with y-intercept 1.

$y = 2^x$ we have already drawn.

$y = 3^x$ will look like $y = 2^x$, except that the y-values will triple, rather than double, for each increase of 1 in x.

$y = 4^x$ will behave like $y = 2^x$ and $y = 3^x$, except that the y-values will increase by a factor of 4 for each increase of 1 in x.

The graphs are shown in Figure 2. Notice that all the graphs pass through the point (0, 1) because of the identity $a^0 = 1$. If we were to sketch the curve $y = (\frac{3}{2})^x$, it would have the same general shape as $y = 2^x$ but lie between the line $y = 1^x$ and the curve $y = 2^x$, because $\frac{3}{2}$ lies between 1 and 2. Notice also that the graphs *cross* at (0, 1). For example, the curve $y = 3^x$ lies above $y = 2^x$ on the right of the y-axis, but lies below it to the left of the y-axis. (Why?)

▼   **EXAMPLE 4**

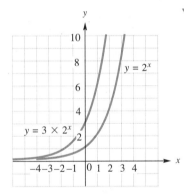

FIGURE 3

Sketch the graphs of $f(x) = 2^x$ and $g(x) = 3 \cdot 2^x$ on the same axes. How are the graphs related?

SOLUTION   The graph of $g(x) = 3 \cdot 2^x$ will look like that of $f(x) = 2^x$, except that each point on $y = 3 \cdot 2^x$ will be 3 times as high as the corresponding point on $y = 2^x$. The graphs are shown in Figure 3.

When entering the equation $y = 3 \cdot 2^x$ on your graphing calculator or computer graphing software, you can use the format

$$Y_1 = 3 \times 2\text{^}X$$

rather than the equivalent

$$Y_1 = 3 \times (2\text{^}X)$$

and not bother with parentheses, since calculators are programmed to respect the usual order of operations: exponents first, and then products. Notice that this is *not* the same as

$$Y_1 = (3 \times 2)\text{^}X$$

which is $y = (3 \cdot 2)^x = 6^x$.

EXAMPLE 5

▼

Examples of the values of two functions, f and g, are given in the following table.

x	-2	-1	0	1	2
$f(x)$	-7	-3	1	5	9
$g(x)$	2/9	2/3	2	6	18

One of these functions is linear and the other is exponential. Which is which?

SOLUTION   There are several ways of telling when a function is linear or exponential. We can start with a numerical approach. Remember that a linear function changes by the same amount every time x increases by 1. The values of f behave this way. Every time x increases by 1, the value of $f(x)$ increases by 4. Therefore, f could be a linear function with a *slope* of 4. Since $f(0) = 1$, we see that

$$f(x) = 4x + 1$$

is a linear formula fitting the data.

On the other hand, an exponential function $y = Ca^x$ is *multiplied* by the same amount every time x increases by 1:

$$Ca^{x+1} = Ca^x \cdot a^1 = (Ca^x) \cdot a.$$

The values of g behave this way. Every time x increases by 1, the value of $g(x)$ is multiplied by 3:

$$\frac{2/3}{2/9} = \frac{2}{2/3} = \frac{6}{2} = \frac{18}{6} = 3.$$

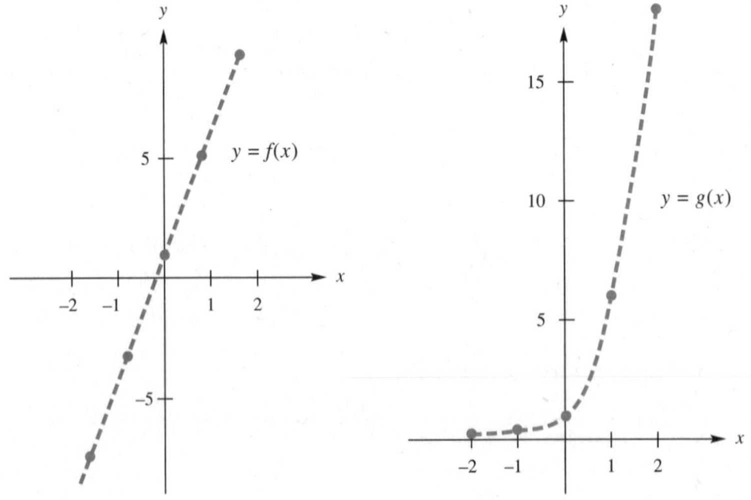

FIGURE 4

Since $g(0) = 2$, we can see that

$$g(x) = 2 \cdot 3^x$$

is an exponential formula fitting the data.

We could also examine the data graphically. Figure 4 shows the graphs of the given points $(x, f(x))$ and $(x, g(x)$. The given values of $f(x)$ clearly lie along a straight line, whereas the values of $g(x)$ lie along a curve. Therefore, f is linear and g is not. To see that g is actually exponential, we need to use the numerical approach above.

Before we go on... The way in which a linear function changes, *adding* the same amount each time x increases by 1, is called **arithmetic growth.** The way in which an exponential function changes, *multiplying* by the same amount each time x increases by 1, is called **geometric growth** or **exponential growth.**

APPLICATIONS

Exponential functions arise in finance and economics mainly through the idea of **compound interest.** We start with a typical scenario: you deposit a **principal** $P = \$10,000$ at Solid Savings & Loan in an account that pays 6% interest **compounded annually.** This means that 6% interest is paid at the end of each year, and *added back into the account*. This money then earns interest the next year along with the original principal.

After one year, you will have the original amount P plus 6% interest:

$$\$10,000 + 10,000(0.06) = \$10,000(1 + 0.06)$$
$$= \$10,600.$$

In other words, if we write $r = 0.06$ for the interest rate, the balance after one year is obtained by multiplying the principal by the quantity $1 + r = 1.06$. To obtain the balance after two years, we would multiply the $10,600 that is in the account at the beginning of the year by $1 + r$ again, since the whole amount earns interest in the second year. In other words, the amount of money in the account grows exponentially, being multiplied by $1 + r$ each year. After t years you would have a total of

$$A = P(1 + r)^t$$

in the account. Note that we can think of A as a function of t, and write

$$A(t) = P(1 + r)^t.$$

Then, for instance, $A(10)$ is the accumulated amount after 10 years. How should we interpret, say, $A(-10)$?

▼ | **EXAMPLE 6** | Mutual Funds

Fidelity Investments advertised that their Fidelity Equity Income II Fund yielded an average annual return of 23.35%.* Assuming that this rate of return will continue indefinitely, express the value of the investment as a function of the number of years t, and use your function to find the value of the investment after 4 years.

SOLUTION   Since the income is reinvested at the end of each year, this amounts to compounding the interest each year, so we use the formula

$$A(t) = P(1 + r)^t.$$

We now substitute the given information: $P = 10,000$ (initial investment) and $r = 0.2335$ (rate of return). This gives

$$A(t) = 10,000(1 + 0.2335)^t,$$

or

$$A(t) = 10,000(1.2335)^t,$$

specifying A as an exponential function of t.
    After 4 years, the value is

$$A(4) = 10,000(1.2335)^4 = \$23,150.30.$$

Note that the order of operations in calculating the value of $10,000(1.2335)^4$ is the standard one: first exponentiate, and then multiply. On a standard calculator, we use the following sequence.

| 1.2335 | x^y | 4 | = | × | 10000 | = |

On a graphing calculator or function-based calculator, we enter

$$10000 \times (1.2335^\wedge 4)$$

The parentheses are optional, as calculators are programmed to use the usual order of operations.
    We can use the graph of the function A to answer interesting questions. First, to graph $A(t) = 10,000(1.2335)^t$, enter the corresponding equation on your graphing calculator using one of the formats shown below.

$$Y_1 = 10000 \times 1.2335^\wedge X$$

or

$$Y_1 = 10000(1.2335^\wedge X)$$

▼ *Rate was current as of 6/30/93, as quoted in an advertisement in *The New York Times* on September 26, 1993.

You also need to set the viewing window ranges. To graph the value of the investment for a 10-year period, use $0 \leq x \leq 10$. Since the investment starts at $10,000 and goes up, start with the y-scale $10,000 \leq y \leq 100,000$. The graph is shown in Figure 5.

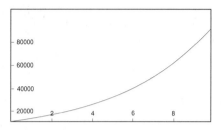

FIGURE 5

Q How long will it take before the investment is worth $75,000?

A The question is asking for the value of t (x on the graph) such that $A(t) = 75,000$. Since $A(t)$ is represented by y, we are asking for the x-coordinate of the point on the graph where $y = 75,000$.

To answer this graphically, graph the line

$$Y_2 = 75000$$

along with A, and use the trace feature to approximate the coordinates of the point of intersection (Figure 6).

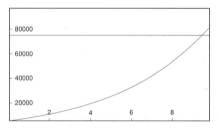

FIGURE 6

You will find that $t \approx 9.6$ years (to the nearest 0.1 year). Thus, it takes a little more than $9\frac{1}{2}$ years to accumulate $75,000.

In Section 4 we shall see how to answer this question algebraically, using logarithms.

Q Suppose that, instead of compounding the interest once a year, your bank compounds the interest four times a year (once per quarter). What formula should we use to calculate the accumulated amount after t years?

A Look once again at the formula $A = P(1 + r)^t$. We can interpret the exponent as the number of times the interest is added to the account, and r as the interest earned each time. This allows us to answer the question as follows: If the interest is added four times a year for t years, it will be added a total of $4 \times t = 4t$ times. Thus, the exponent for our formula should be $4t$ instead of t. Further, the bank is not going to give you a full year's interest every quarter, but only one-fourth of that. Thus, we should replace r with $r/4$. In other words, the formula we want is

$$A = P\left(1 + \frac{r}{4}\right)^{4t}.$$

This example leads us to the following general formula.

COMPOUND INTEREST

If an amount P (the **present value**) earns interest at an annual interest rate r, compounded m times per year, then the accumulated amount (or **future value**) after t years is

$$A = P\left(1 + \frac{r}{m}\right)^{mt}.$$

▶ **NOTES**

1. This formula does generalize the previous formula, $A = P(1 + r)^t$, which is the special case $m = 1$.
2. Once again, we can express A as a function of t by writing

$$A(t) = P\left(1 + \frac{r}{m}\right)^{mt}.$$

   If we write $C = P$ and $a = (1 + r/m)^m$, then we see that $A(t) = Ca^t$ is an *exponential* function of t.   ◀

▼ **EXAMPLE 7**   Interest Compounded Monthly

You invest $10,000 in CDs at the Park Avenue Bank, which pays 6% interest compounded monthly. Express the value of your investment as an exponential function of the number of years t that your capital remains invested. How much money will you have after 5 years?

SOLUTION   Here, the number of times per year that the interest is compounded is 12. In the compound interest formula

$$A = P\left(1 + \frac{r}{m}\right)^{mt}$$

we substitute $P = 10,000$, $r = 0.06$, and $m = 12$. This gives

$$A(t) = P\left(1 + \frac{0.06}{12}\right)^{12t}$$
$$= 10,000(1 + 0.005)^{12t}$$
$$= 10,000(1.005)^{12t}.$$

After five years, you will have accumulated

$$A(5) = 10,000(1.005)^{12(5)}$$
$$= \$13,488.50.$$

▼ **EXAMPLE 8**   Population Growth

The population of a town was 30,000 in 1990 and was increasing by one-third every year. Express the town's population as a function of the number of years since 1990.

SOLUTION   Although it is not money that is growing, the concept is similar. To say that the population is increasing by one-third every year is to say that each year the population is multiplied by $1 + \frac{1}{3}$. Therefore, the population is growing exponentially, as if it were accumulating "interest" at an annual rate of $33\frac{1}{3}\%$ per year. The population is given by the function

$$A(t) = 30,000\left(1 + \frac{1}{3}\right)^{t}$$
$$= 30,000\left(\frac{4}{3}\right)^{t}.$$

Before we go on... Why is the formula *not* $A(t) = 30,000(\frac{1}{3})^{t}$?

▼ **EXAMPLE 9**   Suntanning

When DeltaBlock Sunscreen is first applied, it has an SPF rating of 25. According to experimental data, its SPF rating decreases by 20% per hour during exposure to sunlight.

(a)   Find and graph DeltaBlock's SPF rating S as a function of x, the number of hours after exposure to the sun.

(b) Your dermatologist advises you not to stay in the sun with anything less than SPF-15 protection. If you use DeltaBlock Sunscreen and wish to follow your dermatologist's advice, how often should you reapply the cream?

SOLUTION

(a) Think of the SPF as a bank balance that is depreciating by 20% every hour. Since r stands for the rate of growth, we must use a negative value in this case: $r = -0.20$. In other words, the SPF rating is being multiplied by $1 - 0.20 = 0.80$ every hour. So

$$S(x) = 25(1 - 0.20)^x$$
$$= 25(0.80)^x.$$

Figure 7 shows a graphing calculator plot of the function S.

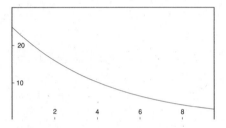

FIGURE 7

(b) We can use the graph to estimate the length of time it takes for the SPF-rating to drop to 15 by graphing the horizontal line $y = 15$ and locating the point of intersection with the curve (using the trace feature if you are using a graphing calculator). As shown in Figure 8, the SPF-rating drops to 15 after approximately 2.29 hours. Thus, you should reapply DeltaBlock after 2.29 hours in order to maintain an SPF of 15 or better.

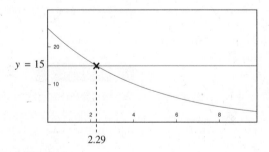

FIGURE 8

▶ **2.1 EXERCISES**

Sketch the graph of each of the functions given in Exercises 1–12.

1. $f(x) = 2^x$.

2. $f(x) = 3^x$

3. $f(x) = 3^{-x}$

4. $f(x) = 4^{-x}$

5. $g(x) = 2(2^x)$

6. $g(x) = 2(3^x)$

7. $h(x) = -3(2^{-x})$

8. $h(x) = -2(3^{-x})$

9. $r(x) = 1 - 2^x$

10. $r(x) = 2 + 2^{-x}$

11. $s(x) = 2^{x-1}$

12. $s(x) = 2^{1-x}$

13. How are the graphs in Exercises 1 and 11 related?

14. How are the graphs in Exercises 6 and 8 related?

15. How are the graphs in Exercises 1 and 9 related?

16. How are the graphs in Exercises 10 and 12 related?

In Exercises 17–22, sketch the graphs of each of the following pairs of functions on the same set of axes for $-3 \le x \le 3$. (Use a graphing calculator or computer program.)

 17. $f_1(x) = 1.6^x$, $f_2(x) = 1.8^x$

18. $f_1(x) = 2.2^x$, $f_2(x) = 2.5^x$

19. $f_1(x) = 300(1.1^x)$, $f_2(x) = 300(1.1^{2x})$

20. $f_1(x) = 100(1.01^{2x})$, $f_2(x) = 100(1.01^{3x})$

21. $f_1(x) = 1,000(1.045^{-3x})$, $f_2(x) = 1,000(1.045^{3x})$

22. $f_1(x) = 1,202(1.034^{-3x})$, $f_2(x) = 1,202(1.034^{3x})$

In Exercises 23–30:

(a) *Express the future value A of a $10,000 investment earning the given interest as a function of time t in years.*

(b) *Use your function to calculate, to the nearest cent, the future value of the investment after the stated time.*

23. 3% per year compounded annually, after 10 years

24. 4% per year compounded annually, after 8 years

25. 2.5% per year compounded quarterly (4 times per year), after 5 years

26. 1.5% per year compounded weekly, after 5 years

27. 6.5% per year compounded daily (365 times per year), after 10 years

28. 11.2% per year compounded monthly, after 12 years

29. 0.2% per month, after 10 years

30. 0.45% per month, after 20 years

APPLICATIONS

31. *Investments*   The balance in Susan's bank account is given by $A(t) = 1,000(1.056)^t$ after t years. Find the balance in her account after

(a) 11 years;

(b) 21 years.

32. *Investments*   The balance in Mike's bank account is given by $A(t) = 1,000(1.0225)^{2t}$ after t years. Find the balance in his account after

(a) 11 years;

(b) 21 years.

33. *Appreciation*   The value of my '68 Classic Pontiac in t years' time will be given by

$$V(t) = 6,000(1.1)^{2t}.$$

Use a graphing calculator to estimate when the car will be worth $15,000. [*Hint:* Graph the two equations $y_1 = 6,000(1.1)^{2x}$ and $y_2 = 15,000$ on the same set of axes.]

34. *Depreciation*   The value of my '88 Pontiac in t years' time will be given by

$$V(t) = 8,000(0.75)^{1.5t}.$$

Use a graphing calculator to estimate when the car will be worth \$4,000. [*Hint:* Graph the two equations $y_1 = 8.000(0.75)^{1.5x}$ and $y = 4,000$ on the same set of axes.]

35. *Investments*   You invest \$100 in the Lifelong Trust Savings and Loan Company, which pays 6% interest compounded quarterly. By how much will your investment have grown after four years?

36. *Investments*   You invest \$1,000 in Rapid Growth Funds, which appreciate by 2% per year, with yields reinvested quarterly. By how much will your investment have grown after 5 years?

37. *Depreciating Investments*   During the first nine months of 1993, mutual funds in health-related industries depreciated at a rate of 6.8% per year.* Assuming this trend were to continue, how much will a \$3,000 investment in this category of funds be worth in 5 years?

38. *Depreciating Stocks*   During the first nine months of 1993, the value of *IBM* stocks depreciated at a rate of 16.28%.† Assuming that this trend were to continue, what will a \$10,000 investment in *IBM* stocks be worth in 10 years?

39. *Depreciation*   During a prolonged recession, property values in Long Island depreciated by 5% every six months. If a new house cost \$200,000, express its future value as a function of its age in years. How much was the house worth 10 years after it was built?

40. *Depreciation*   Stocks in the health industry depreciated by 5.1% in the first nine months of 1993.* Assuming this trend were to continue, express the future value of a \$40,000 investment as a function of its age in years. How much will it be worth in nine years?

Use a graphing calculator or computer software for Exercises 41 through 48.

Compound Interest   Graph the accumulated amount A as a function of the number of years t in Exercises 41–44.

41. \$550 invested at 1.5% per year, compounded annually; $0 \le t \le 10$

42. \$600 invested at 2.2% per year, compounded annually; $0 \le t \le 10$

43. \$550 invested at 1.5% per year, compounded daily; $0 \le t \le 10$

44. \$600 invested at 2.2% per year, compounded daily; $0 \le t \le 10$

45. *Competing Investments*   Bob Carlton just purchased \$5,000 worth of municipal funds that are expected to yield 5.4% per year, compounded every six months. Susan Hessney just purchased \$6,000 worth of CDs that are expected to earn 4.8% per year, compounded every six months. Determine when, to the nearest year, the value of Bob's investment will be the same as Susan's, and what this value will be.

46. *Investments*   Determine when, to the nearest year, \$3,000 invested at 5% per year, compounded daily, will be worth \$10,000.

47. *Epidemics*   At the start of 1985, the incidence of AIDS was doubling every six months, and 40,000 cases had been reported in the U.S. Assuming this trend had continued, determine when, to the nearest 0.1 year, the number of cases would have reached 1 million.

48. *Depreciation*   Lee Anne Fisher's investment in Genetic Splicing Inc. is now worth \$4,354 and is depreciating by 5% every six months. For some reason, she is reluctant to sell the stocks and swallow her losses. Determine when, to the nearest year, her investment will drop below \$50.

▼ *Source: The New York Times*, October 9, 1993, p. 37
  †*Source: The New York Times*, October 9, 1993, p. 40.

49. *Inflation* If the percentage rate of inflation is r per year, then interpret the formula

$$A(r) = 1,000(1 + r)^{10}$$

by completing the following sentence: "$A(r)$ is the amount that an item costing \$ _____ now will cost in _____ years, given an annual rate of inflation _____." Evaluate $A(0.1)$ and $A(1.15)$ and interpret your answers.

50. *Depreciation* If the value of a car depreciates every six months at an annual percentage rate of r, interpret the formula

$$A(r) = 10,200\left(1 - \frac{r}{2}\right)^{40}$$

by completing the following sentence: "If the car is now worth \$ _____ and is depreciating every six months at an annual rate of _____, then $A(r)$ is the amount that it will be worth in _____ years." Evaluate $A(0.1)$ and $A(0.35)$ and interpret your answers.

COMMUNICATION AND REASONING EXERCISES

51. Model the following data using an exponential function $f(x) = Ca^x$.

x	0	1	2
$f(x)$	500	225	101.25

52. Model the following data using an exponential function $f(x) = Ca^x$.

x	-1	0	1
$f(x)$	-97.08737	-100	-103

53. Which of the following three functions will be largest for large values of x?
(a) $f(x) = x^2$
(b) $r(x) = 2^x$
(c) $h(x) = x^{10}$

54. Which of the following three functions will be smallest for large values of x?
(a) $f(x) = x^{-2}$
(b) $r(x) = 2^{-x}$
(c) $h(x) = x^{-10}$

55. What limitations are there to using an exponential function to model growth in real-life situations? Illustrate your answer with an example.

56. Describe two real-life situations in which a linear model would be more appropriate than an exponential one, and two in which an exponential model would be more appropriate than a linear one.

▶ ══════ **2.2** CONTINUOUS GROWTH AND DECAY AND THE NUMBER *e*

In the examples in the last section, capital grew (or depreciated) in discrete steps. For instance, in Example 7 interest was added at the end of each month.

In nature, we find examples of growth that occurs *continuously,* as though "interest" is being added more often than every second or fraction of a second. To model this, we need to see what happens to our compound interest formula as we let m (the number of times interest is added per year) become extremely large. Something very interesting happens: instead of getting a bulky formula with very large numbers, we instead get a compact and elegant formula. To begin to see why, let's look at a very simple situation.

Suppose we invest \$1 in the bank for 1 year at 100% interest, compounded m times per year. Then the accumulated capital is

$$A = 1\left(1 + \frac{1}{m}\right)^m = \left(1 + \frac{1}{m}\right)^m.$$

Now, we are interested in what this becomes for large values of m. So let's make a chart that shows how this quantity behaves as m increases.

m	$\left(1 + \dfrac{1}{m}\right)^m$
1	2
10	2.59374246
100	2.70481383
1000	2.71692393
10^4	2.71814593
10^5	2.71826824
10^6	2.71828047
10^7	2.71828169
10^8	2.71828182

Something interesting *does* seem to be happening! The numbers appear to be getting closer and closer to a specific value. In mathematical terminology, we say that the numbers *converge* to a fixed number, 2.7182818 . . . . This number is one of the most important in mathematics, and is referred to as e. The number e is irrational, just as the more familiar π is, so we cannot write down its exact numerical value. To 20 decimal places, $e = 2.71828182845904523536$ . . . .

▶   NOTE   **Evaluating Powers of e on a Calculator**

1. To obtain an approximation to the number e on your calculator, evaluate the quantity e^1. On traditional calculators, you can do this by first entering 1 and then pressing the "e^x" button. This button is sometimes labeled as "inv ln," where "ln" stands for the natural logarithm (which we shall discuss in the next section). On some of the newer calculators, you must enter expressions in formula form: first press the "e^x" button, and then enter 1, followed by "=". In most computer programs, you write exp(x) for e^x, so exp(1) would represent e.

2. To obtain a power of e, such as $e^{-4.5}$ on a traditional calculator, enter the following sequence:

    $$\boxed{4.5}\ \boxed{+/-}\ \boxed{e^x}$$

    On a formula-based calculator, enter

    $$\boxed{e^x}\ \boxed{(}\ \boxed{(-)}\ \boxed{4.5}\ \boxed{)}\ \boxed{=}$$

    remembering to enclose the -4.5 in parentheses. In a computer program, enter exp(-4.5)   ◀

We now say that if \$1 is invested for 1 year at 100% interest **compounded continuously,** the accumulated capital at the end of that year will amount to \$$e$ = \$2.72 (to the nearest cent). But what about the following more general question?

Q Suppose we invest an amount P for t years at an interest rate of r, compounded continuously. What will the accumulated capital A be at the end of that period?

A In the special case just discussed, we took the compound interest formula and let m get larger and larger. We do the same again, combined with a little of the algebra of exponentials.

$$A = P\left(1 + \frac{r}{m}\right)^{tm}$$

$$= P\left(1 + \frac{1}{(m/r)}\right)^{tm}$$

$$= P\left(1 + \frac{1}{(m/r)}\right)^{(m/r)rt}$$

$$= P\left[\left(1 + \frac{1}{(m/r)}\right)^{(m/r)}\right]^{rt}$$

For continuous compounding of interest, we let m, and hence m/r, get very large. This only affects the term in brackets, which converges to e, and we get the following formula.

COMPOUND INTEREST FORMULA—CONTINUOUS COMPOUNDING

If $\$P$ is invested at an annual interest rate r compounded continuously, then the accumulated amount after t years is

$$A = Pe^{rt}.$$

▶ NOTE   As in the previous section, we can interpret A as a function of t by writing $A(t) = Pe^{rt}$.

If we write $A(t) = P(e^r)^t$, we see that $A(t)$ is an exponential function of t, where the base is $a = e^r$.   ◀

▼ **EXAMPLE 1**   Continuous Compounding

Suppose we deposit $\$1$ in an account yielding 100% interest per year, compounded continuously for x years. Then we have $P = \$1, r = 1.0$, and $t = x$, so

$$A(x) = Pe^{rx}$$
$$= 1 \cdot e^{1.0x}$$
$$= e^x.$$

Thus, the function $A(x) = e^x$ represents something real: the value after x years of a \$1 investment that is growing continuously at 100% per year. We can graph this function by using the following table of values.

x	-3	-2	-1	0	1	2	3
$A(x) = e^x$	0.050	0.135	0.368	1	2.718	7.389	20.08

The graph is shown in Figure 1.

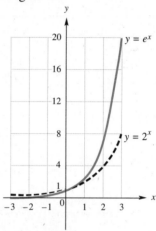

FIGURE 1

Notice the similarity of this graph to the graph of $f(x) = 2^x$ (shown dashed in the figure). Notice also that the investment is more than doubling each year; in fact, it grows by a factor of $e \approx 2.72$ every year.

▼ **EXAMPLE 2**   Continuous Appreciation

You invest \$10,000 at Fastrack Savings & Loan, which pays 6% compounded continuously. Express the balance in your account as a function of the number of years t, and calculate the amount of money you will have after 5 years.

SOLUTION   For the first part, we use the continuous growth formula with $P = 10,000$, $r = 0.06$, and t variable, getting

$$A(t) = Pe^{rt}$$
$$= 10,000e^{0.06t}.$$

To answer the second part of the question, we calculate

$$A(5) = 10,000e^{0.06(5)}$$
$$= 10,000e^{0.3}$$
$$\approx \$13,498.59.$$

Before we go on... Compare this answer to Example 7 from the last section. Continuous compounding earns more interest than monthly compounding, but not a lot more.

To graph A as a function of t, enter the function on your graphing calculator using the following format.

$$Y_1 = 10000e^{\wedge}(0.06X)$$

Most graphing calculators display "e^" when the [e^x] key is pressed.

▼ **EXAMPLE 3** Continuous Depreciation

A $100 investment in Constant Growth Funds is continuously declining at a rate of 4% per year. Specify the value of the investment as a function of the number of years, t, and predict the value of the investment in 10 years.

SOLUTION We use the continuous growth formula with $P = 100$, $r = -0.04$, and t variable, getting

$$A(t) = Pe^{rt}$$
$$= 100e^{-0.04t}.$$

After 10 years, the investment will be worth

$$A(10) = 100e^{-0.04(10)}$$
$$= 100e^{-0.4}$$
$$\approx \$67.03.$$

Thus, the investment with Constant Growth Funds will have declined to $67.03 in ten years.

▼ **EXAMPLE 4** Investments

(a) You have $100 invested in Quarterly Savings and Loan Company, which pays 5% interest compounded quarterly. By how much will your investment have grown after four years?

(b) Continuity Continental Corp. advertises that they can improve on Quarterly's offer by giving the same interest rate, but compounded continuously. How much more would you have earned had you invested with them?

SOLUTION Part (a) is an ordinary compound interest calculation. We use the compound interest formula $A = P(1 + \frac{r}{m})^{tm}$ with $P = 100$, $r = 0.05$, $m = 4$ and $t = 4$. This gives

$$A = 100(1 + \tfrac{0.05}{4})^{16}$$
$$= 100(1.0125)^{16}$$
$$\approx 100(1.2199) = 121.99.$$

Thus, you will earn $21.99 in interest on your $100 investment in four years.

For part (b), we use the continuous compounding formula $A = Pe^{rt}$ with P, t, and r as in part (a), giving

$$A = 100e^{0.20}$$
$$\approx 100(1.2214)$$
$$= 122.14.$$

This shows that, with Continuity Corp., the interest on your investment will amount to $22.14, which is 15¢ better than Quarterly S&L can do. A difference of 15¢ after four years is nothing to get excited about, so you may as well stick with Quarterly.

Before we go on... Would you make the same decision if you had $100,000 to invest?

The **effective annual yield** or **effective interest rate** of an investment is the actual percent by which an investment rises after one year. Another way of thinking of it is as the interest rate that would give the same yield if compounded only once per year. We can calculate the effective yield by calculating the interest earned on an investment of $1 for one year. This gives the following formulas.

EFFECTIVE YIELD FROM COMPOUNDING m TIMES PER YEAR

An investment at an annual interest rate of r, compounded m times per year, has an effective annual yield of

$$r_e = \left(1 + \frac{r}{m}\right)^m - 1.$$

EFFECTIVE YIELD FROM CONTINUOUS COMPOUNDING

An investment at an annual interest rate of r, compounded continuously, has an effective annual yield of

$$r_e = e^r - 1.$$

We can compare investments compounded with different frequencies by comparing their effective yields.

▼   **EXAMPLE 5**   Effective Yield

Which yields more, an investment paying 5% compounded continuously, or an investment paying 5.1% compounded quarterly?

SOLUTION   We compare the effective yields of the two investments. The investment paying 5% compounded continuously yields

$$r_e = e^{0.05} - 1 \approx 0.0513.$$

That is, the effective yield is 5.13%. Said another way, after one year the investment will have appreciated by 5.13%.

The investment paying 5.1% compounded quarterly yields

$$r_e = \left(1 + \frac{0.051}{12}\right)^{12} - 1 \approx 0.0522.$$

Thus the effective yield is 5.22%. This investment is slightly better than the first.

Few banks actually compound interest continuously (although some do!). However, there are other situations where we see exponential growth in which the continuous compounding formula is useful (see also Example 8 in the previous section). In applications beyond finance, it is customary to use the letter *k* rather than *r*, and to call *k* the **fractional rate of growth.** The formula is shown below.

> **EXPONENTIAL GROWTH AT FRACTIONAL GROWTH RATE *k***
> If a quantity *P* at time *t* = 0 grows continuously with a fractional rate of growth *k* per unit time, then, after a time *t*, the quantity will measure
>
> $$A = Pe^{kt}.$$

▼ **EXAMPLE 6**   Continuous Population Growth

The population of fleas on my cat Fluffy is increasing continuously at a growth rate of 15% per day, despite all my efforts at eradication! Today, I estimated that Fluffy's fur is host to about 100 fleas, and a week from today I'll be entertaining the Dean of the Faculties (who is fond of cats but is allergic to flea bites). Estimate the flea population on my cat when the Dean comes to visit.

SOLUTION   This is an example of exponential growth at a growth rate of $k = 0.15$. We also have $P = 100$ and $t = 7$ (note that time is given in *days*, with the rate of increase as a percentage *per day*). We have

$$A = Pe^{kt}$$
$$= 100e^{1.05}$$
$$= 100(2.8577)$$
$$= 285.77 \approx 286.*$$

Thus, one can expect the Dean of the Faculties to be unpleasantly surprised when she begins to cuddle Fluffy next week!

▼ * To the nearest flea.

The continuous compounding formula can also be used to describe quantities that are *decreasing*: for example, continuously depreciating assets. We say that such a quantity is showing **exponential decay.**

EXPONENTIAL DECAY AT FRACTIONAL DECAY RATE k
If a quantity P at time $t = 0$ decays continuously with a **fractional rate of decay** k per unit time, then, after a time t, the amount left is given by

$$A = Pe^{-kt}.$$

▼ **EXAMPLE 7**   Radioactive Decay

Carbon-14, an unstable isotope of carbon, decays continuously to nitrogen at a rate of about 0.0121% per year. If a sample originally contains 50 grams of Carbon-14, how much will be left after 20,000 years?

SOLUTION   We use the exponential decay formula with $P = 50$, $k = 0.000121$, and $t = 20,000$.

$$A = Pe^{-kt}$$
$$= 50e^{-0.000121(20,000)}$$
$$= 50e^{-2.42}$$
$$\approx 4.45$$

Thus, after 20,000 years, there will still be approximately 4.45 grams of Carbon-14 left in the sample.

Before we go on...   Because Carbon-14 decays so slowly, paleontologists use measurements of C-14 to date fossils, in a procedure known as "carbon dating." Carbon dating will be discussed in more detail in the next section.

The next example also involves continuous decay, and it requires either a graphing calculator or graphing software.

▼ **EXAMPLE 8**   Exponential Decay

After getting fired from the university two days after an unfortunate experience involving the Dean of the Faculties and my cat Fluffy, I decided to purchase a "Red Flag 1-in-6 Collar" from the pet store. The manufacturer of the collar claims that it works by continuously lowering the flea population at a rate of 1 in every 6 (or 16.67%) per day. By the time I managed to get close enough to Fluffy to put the thing on, his flea population had grown to about 500,000. Graph Fluffy's flea population as a function of time. When will the population be back down to 100 fleas?

SOLUTION   We again use the continuous decay formula, with $P = 500{,}000$ and $k = 0.1667$.

$$A = Pe^{-kt}$$
$$= 500{,}000e^{-0.1667t}$$

If you are using a graphing calculator, enter this in the usual format.

$$Y_1 = 500000e^{\wedge}(-0.1667X)$$

The graph of A vs. t, is shown in Figure 2, together with a "zoomed-in" portion of the curve (which you can obtain by using the "zoom" feature on a graphing calculator, or by respecifying the x- and y-ranges for the graph). From the graph (using "trace"), we see that the flea population will be back down to 100 in about 51 days. Maybe I'd better have the cat wear two Red Flag 1-in-6 Collars!

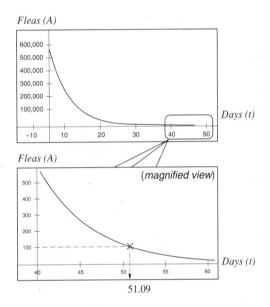

FIGURE 2

Before we go on... We can check our estimate of 51 days by substituting into the decay equation to see what the flea population should be then.

After 51 days: $A = 500{,}000e^{-0.1667 \cdot 51} = 102$ (to the nearest flea).
After 52 days: $A = 500{,}000e^{-0.1667 \cdot 52} = 86$ (to the nearest flea).

We therefore estimate that the flea population will be down to 100 some time on the 52nd day.

We'll be able to answer the question: "Exactly when will the flea population be down to 100 fleas?" without the need for a graphing calculator after we study logarithms in the next section.

▶ ## 2.2 EXERCISES

Compute each of the numbers in Exercises 1–10, rounded to four decimal places.

1. e^3

2. e^4

3. e^{-1}

4. e^{-2}

5. $e^{30 \times 0.001}$

6. $e^{0.014 \times 20}$

7. $100e^{0.0125}$

8. $200e^{-1.124}$

9. $10,200e^{-0.025 \times 20}$

10. $20,000e^{0.004 \times 40}$

Sketch the graphs of the functions in Exercises 11–16.

11. $f(x) = e^x$

12. $f(x) = e^{-x}$

13. $g(x) = e^{-2x}$

14. $g(x) = e^{2x}$

15. $h(x) = 2e^x$

16. $h(x) = 2e^{-x}$

17. How are the graphs in Exercises 11 and 15 related?

18. How are the graphs in Exercises 12 and 16 related?

19. How are the graphs in Exercises 11 and 13 related?

20. How are the graphs in Exercises 12 and 14 related?

In Exercises 21–26, calculate the value, to the nearest dollar, of an investment of $10,000 *after the given time earning the given interest rate compounded continuously.*

21. 5%, 10 years

22. 6%, 10 years

23. 2.5%, 50 years

24. 3.5%, 20 years

25. 2.5%, 11.5 years

26. 3.5%, 21.5 years

APPLICATIONS

27. *Investments* The Second Bank of Chicago offers 2.35%, compounded continuously, on its savings accounts. Express the value of a $1,000 deposit as a function of the number of years *t*, and use your function to find its value after 5 years.

28. *Investments* The Third Bank of Vancouver offers 3.25%, compounded continuously, on its savings accounts. Express the value of a $1,000 deposit as a function of the number of years *t*, and use your function to find its value after 6 years.

29. *Investments* Rock Solid Bank & Trust is offering a CD that pays 4% compounded continuously. How much interest would a $1,000 deposit earn over 10 years?

30. *Savings* SemiSolid Savings & Loan is offering a savings account that pays $3\frac{1}{2}$% interest compounded continuously. How much interest would a deposit of $2,000 earn over 10 years?

31. *Investments* Rock Solid Bank & Trust is offering a CD that pays 4% compounded continuously. What is the effective yield?

32. *Savings* SemiSolid Savings & Loan is offering a savings account that pays $3\frac{1}{2}$% interest compounded continuously. What is the effective yield?

33. *Loans* Your local loan shark makes an offer you can't refuse: a loan at 20% interest, compounded continuously. What is the effective interest rate?

34. *Credit Cards* Your local credit card company is charging 17% interest compounded continuously. What is the effective interest rate?

35. *Loans* Fifth Federal Bank is offering a loan at 9% compounded monthly, and Ninth National is offering a loan at 8.9% compounded continuously. Which is the better deal?

36. *Credit Cards* PassPort Credit Corp. offers a credit card charging 20% interest compounded monthly, while the competing Eureka card charges 19.9% compounded continuously. Which is the better deal?

37. *Investments* My investment portfolio is as follows:
(i) $5,000 in Shady Professorial Deals Inc., earning 5.5% per year, compounded continuously;

(ii) $1,000 in Op Art Treasures Funds, depreciating continuously at 30% per year;

(iii) $10,000 in Holiday Magic Cosmetics, Inc., earning a steady 2% per annum, compounded quarterly.

What will the total value of my portfolio be in ten years' time?

38. *Gold Stocks*   When I recently inherited my aunt's fortune in gold stocks, worth $300,000, I was told that the stocks had been continuously appreciating at 5.5% per year. My late aunt had bought them 25 years ago. What did she originally pay for the stocks?

39. *Carbon Dating*   Carbon-14 decays into nitrogen continuously at a rate of .0121% per year. Express the amount of Carbon-14 in a fossil originally containing 100 grams of Carbon-14 as a function of its age in years. Use your function to estimate the amount left in a 10,000-year-old fossil.

40. *Plutonium Dating*   Plutonium-239 decays continuously at a rate of 0.00284% per year. Express the amount of Plutonium-239 left if a sample originally containing 10 grams of the substance is stored for t years. Use your function to estimate the amount that will be left undecayed after 20,000 years.

41. *Bacteria Growth*   A dangerous strain of bacteria (Bug XXX) is reproducing in a petrie dish at a rate of 2 new bugs for every bug present per hour. Express this as a percentage growth, and estimate the size of the bacteria culture after 12 hours, assuming there were 1,000 organisms originally present. [This and the next exercise are *not* continuous compounding problems.]

42. *Population Growth*   The population of the U.S. in 1990 was about 250,000,000. Assume that for every 1000 citizens, 16 infants are born and 8 citizens die each year. Estimate the population in 2010 to 6 significant figures. [See comment in Exercise 41.]

43. *Varying Inflation*   At the start of 1988, prices in Argentina were increasing continuously at a rate of 158% per year. By the start of 1993, the rate was down to 17%.*

(a) Assuming that the inflation rate declined linearly, express the rate of inflation r as a linear function of t, the number of years since 1988.

(b) The average rate $\bar{r}$ of inflation over t years since 1988 is given by the formula

$$\bar{r} = \frac{r(0) + r(t)}{2},$$

where $r(0)$ is the inflation rate in 1988 and $r(t)$ is the inflation rate t years later. Find a formula for $\bar{r}$ in terms of t.

(c) Use your answer to part (b) to find the price $A(t)$ you would expect to pay after t years for an item that cost P at the start of 1988.

(d) If an item cost $100 on Jan. 1, 1988, what would you expect it to have cost on Jan. 1, 1992? (Use the function you obtained in part (c), and round your answer to the nearest dollar.)

44. *Varying Inflation*   Repeat Exercise 43 using the following figures for Mexico: 42% at the start of 1988, 12% at the start of 1992.*

45. *Inflation*   Inflation is running at the rate of 3% per year compounded continuously when you deposit $1,000 in an account earning 5% compounded continuously. Determine the purchasing power *in today's dollars* of the amount of money you will have in the account 4 years from now. (In other words, taking into account inflation, what amount of money today would buy the same amount of goods as the money you will have in your account 4 years from now?) Explain why this is the same amount you would get by depositing your money in an account paying 2% interest if there were no inflation. This is called the **real interest rate.**

46. *Inflation*   Inflation is running at the rate of 8% per year compounded continuously when you deposit $10,000 in an account earning 5% compounded continuously. Determine the purchasing power *in today's dollars* of the amount of money you will have in the account 4 years from now.

▼

*These figures are based on the actual rates of inflation for the years 1988 and 1992 and have been rounded to the nearest percentage point. Source: International Monetary Fund, U.N. Economic Commission for Latin America and the Caribbean (as cited in *The Chicago Tribune*, June 20, 1993).

In exercises 47–56 use a graphing calculator or computer software.
Plot the graphs of the functions given in Exercises 47–50.

47. $f(x) = 100e^{0.3x}$; $-10 \le x \le 10$

48. $f(x) = 900e^{0.125x}$; $-10 \le x \le 10$

49. $f(x) = 1,000e^{-0.005x}$; $-50 \le x \le 50$

50. $f(x) = 9,000e^{-0.252x}$; $-20 \le x \le 20$

51. *Carbon Dating*   Carbon-14 decays into nitrogen continuously at a rate of 0.0121% per year. If a fossil that originally contained 100 grams of Carbon-14 now contains only 1 gram of Carbon-14, how old (to the nearest year) is the fossil?

52. *Depreciation*   Determine when, to the nearest year, a $3,000 investment that is continuously depreciating at a rate of 5% per year will be worth $1,000.

53. *Competing Investments*   Mark Badgett just purchased $10,000 worth of municipal funds that are appreciating continuously at a rate of 3% per year. Kathi Callahan just purchased $12,000 worth of CDs that are appreciating continuously at a rate of 2% per year. Determine when, to the nearest year, the value of Mark's investment will be the same as Kathi's, and what this value will be.

54. *Competing Investments*   Rich just purchased $100,000 worth of gold coins whose value is appreciating continuously at a rate of 5% per year. Prudence has just purchased $200,000 worth of antiques that are appreciating continuously at a rate of 4.5% per year. Determine when, to the nearest year, the value of Rich's gold coins will pass the value of Prudence's antiques and what this value will be.

55. *Depreciation*   Sandra Groff's investment in Leveraged Buyout, Inc. is now worth $7,000 and is depreciating continuously at a rate of 5.5% per year. For some reason, she is reluctant to sell the stocks and swallow her losses. Determine when, to the nearest year, her investment will drop below $1,000.

56. *Bacteria*   A strain of bacteria is reproducing continuously at a rate of 0.35% per minute. Determine when, to the nearest minute, a culture containing 2,000 organisms will double in size.

57. *Present Value*   Determine the amount of money, to the nearest dollar, you must invest at 6% per year, compounded continuously, so that you will be a millionaire in 10 years' time.

58. *Present Value*   Determine the amount of money, to the nearest dollar, you must invest at 7% per year, compounded continuously, so that you will be a millionaire in 10 years' time.

Employment   Exercises 59 and 60 are based on the following chart, which shows the number of people employed by Northrop Grumman *in Long Island, New York for the period from 1987 to 1995.**

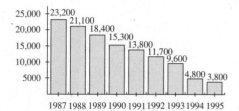

The method of least squares exponential regression[†] *gives the following model.*

$$P = 28,131.7e^{-0.220800t},$$

where t is the number of years since 1987.

59. According to the least squares model, when will employment by *Northrop Grumman* Long Island drop to 1,000 employees? (Give the answer to the nearest year.)

60. According to the least squares model, when will employment by *Northrop Grumman* Long Island drop to 500 employees? (Give the answer to the nearest year.)

▼ * The 1987–1994 figures show employment at Grumman Corp. before it was acquired by Northrop Corp. in 1994. The 1994 and 1995 figures are company projections. Source: Grumman Corp., Northrop Grumman Corp./Long Island, *Newsday*, September 23, 1994, p. A5.

[†] This method is described in the chapter on functions of several variables.

COMMUNICATION AND REASONING EXERCISES

61. Explain in words why 5% per year compounded continuously yields more interest than 5% per year compounded monthly.

62. Explain in words why the continuous growth model is appropriate for a population of algae.

63. Your local banker tells you that the reason his bank doesn't compound interest continuously is that it would be too demanding of computer resources, since the computer would need to spend a great deal of time keeping all accounts updated. Comment on his reasoning.

64. Your other local banker tells you that the reason *her* bank doesn't offer continuously compounded interest is that it is equivalent to offering a fractionally higher interest rate compounded daily. Comment on her reasoning.

▶ ═══ **2.3** LOGARITHMIC FUNCTIONS

LOGARITHMS

Logarithms were invented by John Napier (1550–1617) in the late sixteenth century as a means of aiding calculation. His invention made possible the prodigious hand calculations of astronomer Johannes Kepler (1571–1630), who was the first to describe accurately the orbits and the motions of the planets. Today electronic calculators have done away with that use of logarithms, but many other uses remain.

Consider the equation

$$2^3 = 8.$$

This tells us that the power to which you need to raise 2 in order to get 8 is 3. We shall abbreviate the phrase "*the power to which you need to raise 2 in order to get 8*" as "$\log_2 8$." Thus, another way of writing the equation $2^3 = 8$ is

$$\log_2 8 = 3.$$

This is read: "The logarithm of 8 with base 2 is 3."

Now here is the general definition. Let x be any positive number, and let a be a positive number other than 1.

LOGARITHM WITH BASE a

The **logarithm of x with base a, $\log_a x$,** is the power to which you need to raise a in order to get x. In other words, it is *the exponent that turns a into x.* Symbolically,

$$\log_a x = y$$

means

$$a^y = x.$$

▶ **NOTE**   If we raise a nonzero number a to any power, the result is always positive. (Recall that the graph of $f(x) = a^x$ lies entirely above the x-axis.) Thus, we cannot speak of the logarithm of a negative number. In other words,

the largest possible domain of $f(x) = \log_a x$ is $(0, +\infty)$.   ◀

▼ **EXAMPLE 1**

When we say "$\log_3 9 = 2$," we mean: "the power you raise 3 to in order to get 9 is 2." In other words, $3^2 = 9$. The statement $\log_3 9 = 2$ is called the **logarithmic form** of the statement $3^2 = 9$, while the statement $3^2 = 9$ is called the **exponential form** of the statement $\log_3 9 = 2$.

▼ **EXAMPLE 2**

Find $\log_{10} 1,000$.

SOLUTION   We are asking the question, "What power of 10 gives 1,000?" A moment's thought reveals that $10^3 = 1,000$, so $\log_{10} 1,000 = 3$. If you prefer a more mechanical way of doing this, proceed as follows.

1.   Let $\log_{10} 1,000 = x$.
2.   In exponent form, this is $10^x = 1,000$.
3.   But $1,000 = 10^3$.
4.   Thus, $x = 3$.

▼ **EXAMPLE 3**

Find $\log_4 64$.

SOLUTION   Let $\log_4 64 = x$.
Then, in exponent form, $4^x = 64$.
But $64 = 4^3$.
Thus, $x = 3$, so $\log_4 64 = 3$.

Before we go on... Again, you can also ask yourself: What power of 4 gives 64? The answer: $4^3 = 64$, so $\log_4 64 = 3$.

▼ **EXAMPLE 4**

Find $\log_4\left(\dfrac{1}{64}\right)$.

SOLUTION   Let $x = \log_4\left(\dfrac{1}{64}\right)$.

Then $4^x = \dfrac{1}{64}$.

But $\dfrac{1}{64} = \dfrac{1}{4^3} = 4^{-3}$

and so $x = -3$.

▼ **EXAMPLE 5**

Find $\log_{1/5} 25$.

SOLUTION   Let $\log_{1/5} 25 = x$.
In exponent form, this is $\left(\frac{1}{5}\right)^x = 25$.
Using the rules for exponents, this may be written as

$$\frac{1}{5^x} = 25,$$

or $\qquad\qquad\qquad\qquad 5^{-x} = 25.$

But $\qquad\qquad\qquad\qquad 25 = 5^2.$

Thus, $\qquad\qquad\qquad\qquad -x = 2.$

Hence, $\qquad x = -2, \quad \text{or} \quad \log_{1/5} 25 = -2.$

▼ **EXAMPLE 6**

Find $\log_5 1$.

SOLUTION   Let $x = \log_5 1$.
Thus, $\qquad\qquad 5^x = 1.$
But $\qquad\qquad 1 = 5^0$, whence $x = 0$
that is, $\qquad \log_5 1 = 0.$

Before we go on... We could repeat this calculation with any base to get $\log_a 1 = 0$ for any $a \neq 1$.

▼ **EXAMPLE 7**

Find $\log_2 \sqrt{2}$.

SOLUTION   Putting $\log_2 \sqrt{2} = x$, we get $2^x = \sqrt{2} = 2^{1/2}$,
and so $x = \frac{1}{2}$ (that is, $\log_2 \sqrt{2} = \frac{1}{2}$).

The following are standard abbreviations.

$$\log_{10} x = \log x$$
$$\log_e x = \ln x$$

We call $\log x = \log_{10} x$ the **common logarithm** of x, and $\ln x = \log_e x$ the **natural logarithm** of x.

To obtain $\log_{10} 5$ on a traditional scientific calculator, press $\boxed{5}$ followed by $\boxed{\text{log}}$. (You should get 0.6989 . . .) On a graphing calculator or other function-style calculator, press $\boxed{\text{log}}$ followed by $\boxed{5}$ and then $\boxed{=}$.*

To obtain $\log_e 5$ on a traditional scientific calculator, press $\boxed{5}$ followed by $\boxed{\text{ln}}$. (You should get 1.6094 . . .) On a graphing calculator or other function-style calculator, press $\boxed{\text{ln}}$ followed by $\boxed{5}$ and then $\boxed{=}$.

Q Suppose we want to calculate logs to bases other than 10 and e?

A We use the change-of-base formula.

CHANGE-OF-BASE-FORMULA

$$\log_a b = \frac{\log b}{\log a} = \frac{\ln b}{\ln a}$$

(Take our word for it now. We'll derive this formula later in the section.) We can use this formula to get the logarithm of any number to any base using a calculator. For example, to find $\log_{3.45} 2.261$, we divide log (2.261) by log(3.45), getting 0.6588 (to four significant digits). We get the same answer by dividing ln(2.261) by ln(3.45) (try it).

The following identities for logarithms are as important as the ones for exponents listed in Section 1. We shall discuss why they are true after using them in several examples.

LOGARITHM IDENTITIES

The following identities hold for any positive base $a \neq 1$ and any positive numbers x and y.

Identity	**Example**
(a) $\log_a (xy) = \log_a x + \log_a y$	$\log_2 16 = \log_2 8 + \log_2 2$
(b) $\log_a \left(\frac{x}{y}\right) = \log_a x - \log_a y$	$\log_2 \left(\frac{5}{3}\right) = \log_2 5 - \log_2 3$
(c) $\log_a (x^r) = r \log_a x$	$\log_2 (6^5) = 5 \log_2 6$
(d) $\log_a a = 1; \log_a 1 = 0$	$\log_2 2 = 1; \log_{11} 1 = 0$
(e) $\log_a \left(\frac{1}{x}\right) = -\log_a x$	$\log_2 \left(\frac{1}{3}\right) = -\log_2 3$
(f) $\log_a x = \dfrac{\log_b x}{\log_b a}$	$\log_2 5 = \dfrac{\log_{10} 5}{\log_{10} 2}$

▼ * We do not guarantee that this will work on *your* calculator—for instance, you might have to use an "Execute" button rather than an "=" button, depending on the brand of calculator—so you should check your instruction manual.

▶ NOTE   Since these rules hold for any base, they hold in particular for the bases 10 and e, so we can replace "$\log_a$" with either "log" or "ln", as long as we stick to the same base throughout an identity.

Some people like to remember these identities in words. For example, the first identity says that "multiplication on the *inside* corresponds to addition on the *outside.*" Another way of remembering them is to notice that "log" converts more difficult operations into simpler operations.   ◀

▶ CAUTION

1. In all of these identities except (f), the bases of the logarithms must match. For example, rule (a) gives $\log_3 11 + \log_3 4 = \log_3 44$, but does *not* apply to the expression $\log_3 11 + \log_5 4$.

2. People sometimes invent their own identities. Here is one of the most popular ones:

$$\log_a (x + y) = (\log_a x)(\log_a y). \qquad \text{WRONG!}$$

For instance, $\log(99 + 1) \neq \log(99)\log(1)$, since the left-hand side is $\log(100) = 2$, while the right-hand side is $\log(99) \times 0 = 0$.

The following formula is also wrong (we suggest that you give an example to show why).

$$\log_a (x + y) = \log_a x + \log_a y \qquad \text{WRONG!}$$

These just don't work! Just because they're popular doesn't mean they're right.*   ◀

Let us now use these identities to help solve some equations having unknowns in the exponent.

▼   **EXAMPLE 8**

Solve $3^{2x} = 4$ for x.

SOLUTION   The unknown is in a most inconvenient place up there in the exponent. To bring it down, we shall take the logarithm of both sides and then use rule (c). We could use the logarithm with any base, but we choose to use the natural log.

▼   * Mathematics is not a democracy. Mathematicians try to uncover the truth, and not create it. What is true is true, and what is false is false, and there's absolutely nothing we can do about it! There was an embarrassing incident a few decades ago when a certain state in the Union (that shall remain unnamed) seriously attempted to introduce a law declaring π to be *exactly* 3!

Actually, there is a philosophical controversy here. Platonists believe that mathematics already exists and the mathematician's job is one of exploration and discovery. Another camp believes that mathematicians *invent* mathematics. Both sides agree, however, that once the ground rules have been set, there is an objective criterion of mathematical truth. In other words, what is true is true and what is false is false.

Taking logs,

$$\ln(3^{2x}) = \ln 4.$$

By identity (c),

$$2x \cdot \ln 3 = \ln 4.$$

Dividing,

$$x = \frac{\ln 4}{2 \ln 3} \approx \frac{1.386294}{2.197225} = 0.63093.$$

It is a good idea to leave the actual evaluation of logarithms until the very end, as we did here. This eliminates the need to record long decimal numbers, and it tends to give more accurate answers.

Before we go on... Here is the sequence of calculator operations that will give the answer without having to store intermediate answers.

Traditional calculator: [4] [ln] [÷] [3] [ln] [=] [÷] [2] [=]

Function-style calculator:

[ln] [4] [÷] [(] [2] [×] [ln] [3] [)] [=]

As usual, we should check the anwer by substituting it back into the original equation.

$$3^{2(0.63093)} \approx 4 \text{ (to six significant digits)} \quad \checkmark$$

▼   **EXAMPLE 9**

Solve $3^{3t} = 2^{t+1}$ for t.

SOLUTION   We shall again take the natural log of both sides, as it worked so nicely last time.

$$\ln(3^{3t}) = \ln(2^{t+1})$$

By identity (c),

$$3t \ln 3 = (t + 1) \ln 2.$$

To solve for t, we must get all the terms with t in them together.

$$3t \ln 3 = t \ln 2 + \ln 2$$

Subtracting $t \ln 2$ from both sides,

$$3t \ln 3 - t \ln 2 = \ln 2.$$

Hence, $$t(3 \ln 3 - \ln 2) = \ln 2.$$

Thus, $$t = \frac{\ln 2}{3 \ln 3 - \ln 2} \approx 0.26632.$$

Before we go on... Checking the answer,

$$3^{3(0.26632)} \approx 2.40547$$
$$2^{0.26632+1} \approx 2.40547. \quad ✔$$

Q I did the above calculation on my graphing calculator and obtained a different answer: $t = -0.482837$. What did I do wrong?

A This wrong answer, $t = -0.482837$, will result from entering the following information.

$$\ln 2/3 \ln 3 - \ln 2 \qquad \text{WRONG!}$$

What is wrong is our representation of the fraction bar: a fraction bar indicates that one *quantity* is divided by another *quantity*. In other words,

$$t = \frac{\ln 2}{3 \ln 3 - \ln 2} = \frac{(\ln 2)}{(3 \ln 3 - \ln 2)}.$$

Thus, the correct calculator format is

$$(\ln 2)/(3 \ln 3 - \ln 2)$$

▼ **EXAMPLE 10**

Solve $49(1.01)^t = 59$ for t.

SOLUTION   Although you are again tempted to take the log of both sides, the problem is the 49 in front of the $(1.01)^t$. If you decide to go ahead and take the logs anyway, you must be careful not to make a common mistake. See if you can determine what is wrong with the following.

$$(1) \ \ln 49(1.01)^t = \ln 59;$$

therefore,

$$(2) \ t \ln(49(1.01)) = \ln 59.$$

Thus,   WRONG!

$$(3) \ t \ln(49(1.01)) = \ln 59.$$

Hence,

$$(4) \ t = \ln 59/\ln(49(1.01)) = 1.045.$$

Q How are we sure this is the wrong answer?

A Check it:

$$49(1.01)^{1.045} \approx 49.5 \ne 59.$$

$\boldsymbol{Q}$ What went wrong here?

$\boldsymbol{A}$ The error occurred in Step 2: $\ln 49(1.01)^t$ is *not* equal to $t \cdot \ln(49(1.01))$. It is true that $t \cdot \ln(49(1.01))$ is $\ln(49(1.01))^t$, but this is not what we're evaluating. The point is that $49(1.01)^t$ is *not* the same as $(49(1.01))^t$. The first expression is the *product* of 49 and $(1.01)^t$, while the second is the tth power of the product $49(1.01)$. What we should say is

$$\ln 49(1.01)^t = \ln 49 + \ln(1.01)^t \qquad \text{By identity (a)}$$
$$= \ln 49 + t \ln(1.01). \qquad \text{By identity (c)}$$

Going back to the beginning, instead of solving for t immediately by taking the log of both sides, we'll first divide by the annoying factor 49.

$$(1.01)^t = \frac{59}{49}$$

Now take the logs.

$$t \cdot \ln 1.01 = \ln\left(\frac{59}{49}\right)$$

Thus,

$$t = \frac{\ln\left(\dfrac{59}{49}\right)}{\ln 1.01} \approx 18.6644.$$

Before we go on... Check the answer.

$$49(1.01)^{18.6644} \approx 59 \quad \text{(to 5 decimal places)} \quad \checkmark$$

For practice, solve again for t by first taking the natural log of both sides.

▼   **EXAMPLE 11**

Solve $\dfrac{1}{2^x} = 2^{x^2}$ for x.

SOLUTION   Since 2 is the base of both expressions, let us take logs with base 2 for a change. (This will be more convenient, but you might try solving by taking natural logs of both sides, for practice.)
   We have

$$\log_2\left(\frac{1}{2^x}\right) = \log_2\left(2^{x^2}\right).$$

By identity (f),

$$-\log_2\left(2^x\right) = \log_2\left(2^{x^2}\right).$$

Applying identity (c),

$$-x \log_2 2 = x^2 \log_2 2.$$

But identity (d) says that $\log_2 2 = 1$. Thus, we get

$$-x = x^2.$$

This is a quadratic. To solve, we bring all the terms to the same side and factor:

$$x^2 + x = 0,$$

or

$$x(x + 1) = 0.$$

This gives

$$x = 0 \quad \text{or} \quad x = -1.$$

Done!

ALTERNATE SOLUTION   (without logarithms)
Rewrite the equation as

$$2^{-x} = 2^{x^2}.$$

Then we must have

$$-x = x^2,$$

giving

$$x^2 + x = 0.$$

Before we go on... Check the answers $x = 0$ and $x = -1$.

$$\frac{1}{2^0} = \frac{1}{1} = 1 \quad \text{and} \quad 2^{0^2} = 2^0 = 1 \quad ✔$$

$$\frac{1}{2^{-1}} = 2^1 = 2 \quad \text{and} \quad 2^{(-1)^2} = 2^1 = 2 \quad ✔$$

DERIVATION OF THE LOGARITHM IDENTITIES

We now pause to see where (at least some of) the rules for logarithms come from, as we promised. Roughly speaking, they are restatements in logarithmic form of the laws of exponents.

Q Why is $\log_a xy = \log_a x + \log_a y$?

A Let $s = \log_a x$ and $t = \log_a y$. In exponential form, these equations say that

$$a^s = x \quad \text{and} \quad a^t = y.$$

Multiplying these two equations together gives

$$a^s a^t = xy;$$

that is,

$$a^{s+t} = xy.$$

Rewriting this in logarithmic form gives

$$\log_a (xy) = s + t = \log_a x + \log_a y,$$

as claimed.

Here is an intuitive way of thinking about it: Since logs are exponents, this identity expresses the familiar law that the exponent of a product is the sum of the exponents.

Identity (b) is shown in almost the identical way, and we leave it for you for practice.

Q Why is $\log_a (x^r) = r \log_a x$?

A Let $t = \log_a x$. Writing this in exponential form gives

$$a^t = x.$$

Raising this equation to the rth power gives

$$a^{rt} = x^r.$$

Rewriting in logarithmic form gives

$$\log_a (x^r) = rt = r \log_a x,$$

as claimed.

Identity (d) we will leave for you to do as practice.

Q Why is $\log_a \left(\frac{1}{x}\right) = -\log_a x$?

A This follows from identities (b) and (d) (think about it).

Q Why is $\log_a x = \dfrac{\log_b x}{\log_b a}$?

A Let $s = \log_a x$. In exponential form, this says that

$$a^s = x.$$

Take the logarithm with base b of both sides, getting

$$\log_b a^s = \log_b x,$$

then use identity (c):

$$s \log_b a = \log_b x,$$

so
$$s = \frac{\log_b x}{\log_b a}.$$

GRAPHS OF LOGARITHMIC FUNCTIONS

It is useful to have a sense of what the graph of $f(x) = \log_a x$ looks like for various values of a. Recall that the largest possible domain of the function $f(x) = \log_a x$ is $(0, +\infty)$.

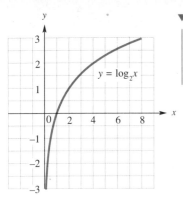

FIGURE 1

▼ **EXAMPLE 12**

Sketch the graph of $f(x) = \log_2 x$.

SOLUTION   To graph this curve, we could use a graphing calculator (as described below) or make a table, as follows. The values of x that we use must be positive, and we have chosen the values shown for convenience. Since $\log_2 x$ is not defined when $x = 0$, we chose several values of x close to zero.

x	$\frac{1}{8}$	$\frac{1}{4}$	$\frac{1}{2}$	1	2	4	8
$f(x) = \log_2 x$	-3	-2	-1	0	1	2	3

Graphing these points gives us Figure 1.

To enter this function in your graphing calculator, use the change-of-base formula

$$\log_2 x = \frac{\log x}{\log 2}.$$

Thus, you will need to use the format

$$Y_1 = (\log X)/(\log 2)$$

(Although the parentheses are not strictly necessary, it is good practice to use them to distinguish the quantities you are dividing, since the fraction bar indicates that one *quantity* is divided by another.) As for the ranges, you could use $0.001 \leq x \leq 8$ and $-5 \leq y \leq 5$ to get a facsimile of Figure 1.

▼ **EXAMPLE 13**

On the same axes, sketch the graphs of

$$f(x) = \log_2 x \quad \text{and} \quad g(x) = 2^x.$$

SOLUTION   We have already sketched both curves—the first curve in the last example, and the second curve back in the first section. If we superimpose them on the same set of axes, we notice something interesting (Figure 2).

There is a symmetry between the two curves: one is the mirror image of the other across the diagonal line $y = x$. It is as though the graph of $f(x) = \log_2 x$ was obtained from the graph $g(x) = 2^x$ by switching the x- and y-coordinates. For instance, the point $(2, 1)$ is on the graph of $f(x) = \log_2 x$, while the point $(1, 2)$ is on the graph of $g(x) = 2^x$. It is this property that makes the functions $f(x) = \log_2 x$ and $g(x) = 2^x$ **inverse functions**.

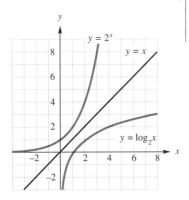

FIGURE 2

Q What does it mean for two functions to be inverse functions?

A Two functions f and g are **inverse functions** if

$$g(f(x)) = x$$

for every x in the domain of f, and

$$f(g(x)) = x$$

for every x in the domain of g. For example, $f(x) = x^2$ with domain $[0, +\infty)$ and $g(x) = \sqrt{x}$ are inverse functions.

Q Are the functions $f(x) = \log_2 x$ and $g(x) = 2^x$ inverse functions?

A Let us check the requirement directly.

$$g(f(x)) = g(\log_2 x)$$
$$= 2^{\log_2 x}$$

Now $\log_2 x$ is the power to which you raise 2 in order to get x. In other words, 2 raised to that power must be x! Thus,

$$g(f(x)) = x.$$

Next,

$$f(g(x)) = f(2^x)$$
$$= \log_2 (2^x)$$
$$= x \log_2 2$$
$$= x,$$

because $\log_2 2 = 1$. Thus, we have shown that the two functions are indeed inverse functions.

There is nothing special about the base 2 in the above discussion. So we can draw the following conclusion.

RELATIONSHIP OF THE FUNCTIONS $f(x) = \log_a x$ **AND** $g(x) = a^x$

If a is any positive number other than 1, then the functions $f(x) = \log_a x$ and $g(x) = a^x$ are inverse functions. This means that

$$a^{\log_a x} = x$$

for all positive x and

$$\log_a (a^x) = x$$

for all real x.

▼ **EXAMPLE 14**

Using a graphing calculator or graphing software, graph the functions

$$f(x) = \log_a x$$

with $a = \frac{1}{4}, \frac{1}{2}, 2$, and 4.

SOLUTION   Referring to the graphing calculator discussion at the end of Example 12, we can do this using the formula

$$y = \frac{\log x}{\log a},$$

where $a = \frac{1}{4}, \frac{1}{2}, 2$, and 4. This gives us Figure 3.

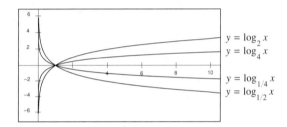

FIGURE 3

Notice that the graphs all pass through the point $(1, 0)$. (Why?) Notice further that the graphs of the logarithmic functions with bases smaller than 1 are upside-down versions of the others.

▶ **2.3 EXERCISES**

Rewrite the equations in Exercises 1–12 in logarithmic form.

1. $2^5 = 32$

2. $3^2 = 9$

3. $3^{-2} = \dfrac{1}{9}$

4. $2^{-4} = \dfrac{1}{16}$

5. $10^3 = 1,000$

6. $10^6 = 1,000,000$

7. $y = e^x$

8. $y = e^{-x}$

9. $y = x^{-3}$ $(x \neq 1)$

10. $y = x^4$ $(x \neq 1)$

11. $1 = e^0$

12. $1 = 10^0$

Rewrite the equations in Exercises 13–24 in exponential form.

13. $\log_6 36 = 2$

14. $\log_5 125 = 5$

15. $\log_2 \left(\dfrac{1}{4} \right) = -2$

16. $\log_3 \left(\dfrac{1}{27} \right) = -3$

17. $\log 100,000,000 = 8$

18. $\log\left(\dfrac{1}{10,000}\right) = -4$

19. $\ln\left(\dfrac{1}{e}\right) = -1$

20. $\ln(e^{-3}) = -3$

21. $\log_x y = 3$

22. $\log_3 x = y$

23. $y = \ln(-x) \ (x < 0)$

24. $y = \ln(x^{-1}) \ (x > 0)$

Let $a = \ln 2$, $b = \ln 3$, and $c = \ln 5$. Use the identities for logarithms to express the quantities in Exercises 25–36 in terms of a, b, and c.

25. $\ln 6$

26. $\ln 10$

27. $\ln\left(\dfrac{2}{3}\right)$

28. $\ln\left(\dfrac{3}{5}\right)$

29. $\ln 256$

30. $\ln 81$

31. $\ln\left(\dfrac{3}{10}\right)$

32. $\ln\left(\dfrac{25}{3}\right)$

33. $\ln 0.02$

34. $\ln 0.05$

35. $\ln\left(\dfrac{9}{e}\right)$

36. $\ln 25e$

Complete the equations in Exercises 37–44 by filling in the missing quantity.

37. $\log_a 3 + \log_a 4 = \log_a(\quad)$

38. $\log_a 3 - \log_a 4 = \log_a(\quad)$

39. $2\log_a x + \log_a y = \log_a(\quad)$

40. $2\ln x - 4\ln y = \ln(\quad)$

41. $2\ln x + 4\ln y - 6\ln z = \ln(\quad)$

42. $-(2\ln x + \ln y + 6\ln 1) = \ln(\quad)$

43. $\log_2(x) + 2^x\log_2 y = \log_2(\quad)$

44. $x^2\log 2 - \log(2x^2) = \log(\quad)$

Solve the equations in Exercises 45–56 for the indicated variable.

45. $4 = 2^x$; solve for x.

46. $9 = 3^{-x}$; solve for x.

47. $2^t = 4^{2t}$; solve for t.

48. $9^t = 3^{3t}$; solve for t.

49. $2^t = 2^{-t^2}$; solve for t.

50. $2^{-m} = 2^{m^2} \cdot 4^m$; solve for m.

51. $100 = 50e^{3t}$; solve for t.

52. $20 = 10e^{5t}$; solve for t.

53. $10 = 1{,}000e^{-2x}$; solve for x

54. $5 = 5{,}000e^{-2.2x}$; solve for x.

55. $200 = 5(2^y)$; solve for y.

56. $1{,}000 = 6(3^{2x})$; solve for x.

Sketch the graphs of the functions in Exercises 57–62.

57. $f(x) = \log_3 x;\ 0 < x \le 27$

58. $f(x) = \log_4 x;\ 0 < x \le 64$

59. $g(x) = x - \log_2 x;\ 0 < x \le 8$

60. $g(x) = \log_3 x - x;\ 0 < x \le 27$

61. $h(x) = (\log_2 x)^2;\ 0 < x \le 8$

62. $h(x) = (\log_3 x)^3;\ 0 < x \le 27$

Use a graphing calculator or graphing software to graph the functions in Exercises 63–66.

63. $f(x) = \log_2(x^2 + 1);\ -2 \le x \le 2$

64. $f(x) = \log_2((x - 1)^2 + 1);\ -10 \le x \le 10$

65. $f(x) = \ln|x|;\ -10 \le x \le 10 \quad (x \ne 0)$

66. $f(x) = \ln|x - 3|;\ -10 \le x \le 10 \quad (x \ne 3)$

APPLICATIONS

67. *Richter Scale* The **Richter scale** is used to measure the intensity of earthquakes. The Richter scale rating of an earthquake is given by the formula

$$R = \frac{2}{3}(\log E - 4.4),$$

where E is the energy released by the earthquake (measured in joules*).
(a) The Great San Francisco Earthquake of 1906 registered $R = 8.2$ on the Richter scale. How many joules of energy were released?
(b) In 1989, another San Francisco earthquake registered 7.1 on the Richter scale. What percentage of the energy released in the 1906 earthquake was released in the smaller 1989 earthquake?
(c) Show that if two earthquakes registering R_1 and R_2 on the Richter scale release E_1 and E_2 joules of energy, respectively, then

$$\frac{E_2}{E_1} = 10^{1.5(R_2 - R_1)}.$$

(d) Fill in the missing quantity: If one earthquake registers one point more on the Richter scale than another, then it releases _____ times the amount of energy.

68. *Sound Intensity* The loudness of a sound is measured in **decibels**. The decibel level of a sound is given by the formula

$$D = 10 \log \frac{I}{I_0},$$

where D is the decibel level (dB), I is its intensity in watts per square meter (W/m^2) and $I_0 = 10^{-12}$ W/m^2 is the intensity of a barely audible "threshold" sound. A sound intensity of 90dB or greater causes damage to the average human ear.
(a) Find the decibel levels of each of the following, rounding to the nearest dB:
   Whisper: 115×10^{-12} W/m^2
   T.V. (average volume from 10 feet):
   320×10^{-7} W/m^2
   Loud music: 900×10^{-3} W/m^2
   Jet aircraft (from 500 ft.): 100 W/m^2

(b) Which of the above sounds would cause damage to the average human ear?
(c) Show that if two sounds of intensity I_1 and I_2 register decibel levels of D_1 and D_2 respectively, then

$$\frac{I_2}{I_1} = 10^{0.1(D_2 - D_1)}.$$

(d) Fill in the missing quantity: If one sound registers one decibel more than another, then it is _____ times as intense.

69. *Sound Intensity* The decibel level of a T.V. set decreases with the distance from the set according to the formula

$$D = 10 \log\left(\frac{320 \times 10^7}{r^2}\right),$$

where D is the decibel level and r is the distance from the T.V. set in feet.
(a) Find the decibel level (to the nearest decibel) at distances of 10, 20, and 50 feet.
(b) Express D in the form $D = A + B \log r$ for suitable constants A and B.
(c) How far must a listener be from a T.V. so that the decibel level drops to 0?

70. *Acidity* The acidity of a solution is measured by its pH. The pH of a solution is given by the formula

$$pH = -\log(H^+),$$

where H^+ measures the concentration of hydrogen ions in moles per liter.[†] The pH of pure water is 7. A solution is referred to as *acidic* if its pH is below 7 and as *basic* if its pH is above 7.
(a) Calculate the pH of each of the following substances.
   Blood: 3.9×10^{-8} moles/liter
   Milk: 4.0×10^{-7} moles/liter
   Soap solution: 1.0×10^{-11} moles/liter
   Black coffee: 1.2×10^{-7} moles/liter
(b) How many moles of hydrogen ions are contained in a liter of acid rain with a pH of 5.0?
(c) Complete the following sentence: If the pH of a solution increases by 1.0, then the concentration of hydrogen ions _____ .

▼ *A joule is a unit of energy. 100 joules of energy would light up a 100-watt light bulb for a second.
[†] A mole corresponds to about 6.0×10^{23} hydrogen ions. (This number is also known as Avogadro's Number.)

71. *Modeling Demand*   You are the owner of a new fried chicken franchise, and you would like to construct a demand equation of the form $q = f(p)$, where q is the number of quarter-chicken servings you sell per hour, and p is the price you charge per serving. You have tried selling the servings at three different prices, and the resulting sales are shown in the following table:

Price per serving (p)	$3.00	$4.00	$5.00
Average hourly sales (q)	4.826	3.889	3.290

(a) Use the sales figures for the prices $3 and $5 to construct a linear demand function of the form $f(p) = mp + b$, where m and b are constants you will need to determine. (Round m and b to three significant digits.)

(b) Use the same sales figures to construct a demand function of the form $f(p) = kp^r$, where k and r are constants you will need to determine. (Round k and r to three significant digits.)

(c) Which of the two demand functions predicts the demand at $4 per serving most accurately?

(d) Use the better of the two demand functions to predict the demand when the price is set at $3.50 per portion.

72. *Modeling Supply*   You are the purchasing manager for a large drugstore and have been trying to purchase vitamin supplements at a bulk discount from Back to Nature Supplements, Inc. You have noticed that the number of cases of vitamin C supplement the manufacturer is willing to supply depends on the price you offer, as shown by the following table:

Price per case (p)	$30	$40	$50
Average number of cases supplied (q)	23.40	25.15	26.59

(a) Use the supply figures for the prices $30 and $50 to construct a linear supply function of the form $q = f(p) = mp + b$, where m and b are constants you will need to determine. (Round m and b to three significant digits.)

(b) Use the same sales figures to construct a supply function of the form $f(p) = kp^r$, where k and r are constants you will need to determine. (Round k and r to three significant digits.)

(c) Which of the two supply functions predicts the supply figure for $40 per case most accurately?

(d) Use the better of the two supply functions to predict the number of cases you can order at $60 per case.

COMMUNICATION AND REASONING EXERCISES

73. Explain in words why the logarithm of a product of two numbers is the sum of the logarithms of the individual numbers.

74. Explain in words why the logarithm of a number's reciprocal is the negative of the logarithm of that number.

75. Why is the logarithm of a negative number not defined?

76. Of what use are logarithms, now that they are no longer needed to perform complex calculations?

77. Complete the following: if $y = 4^x$, then $x =$ _____.

78. Complete the following: if $y = \log_6 x$, then $x =$ _____.

79. Complete the following sentence: If y is a linear function of $\log x$, then x is a _____ function of y.

80. If y is an exponential function of x, how are x and $\log y$ related?

▶ ▬▬▬ **2.4** APPLICATIONS OF LOGARITHMS

You may have noticed in the first section that when we had a compound interest problem in which the number of years was the unknown, we had to solve the problem graphically, not analytically. The reason was that the number of years t appears in the exponent in the expression for compound interest:

$$A = P\left(1 + \frac{r}{m}\right)^{mt},$$

and we knew no way of solving for it. As we saw in the previous section, the logarithm function makes solving such equations straightforward. We can now tackle a greater variety of questions about compound interest and also about exponential growth and decay.

▼ **EXAMPLE 1**   Investments

Tax-exempt bonds are yielding an average of 5.2% per year.* Assuming this rate continues, how long will it take a $1,000-dollar investment to be worth $1,500 if the interest is compounded monthly?

SOLUTION   Substituting $A = 1,500$, $P = 1,000$, $r = 0.052$, and $m = 12$ in the compound interest equation gives

$$1,500 = 1,000\left(1 + \frac{0.052}{12}\right)^{12t}$$

or

$$1,500 = 1,000(1.004333)^{12t},$$

and we must solve for t. We saw how to solve this kind of equation in the previous section. We first divide both sides by 1,000, getting

$$1.5 = 1.004333^{12t},$$

and then take the natural log of both sides.

$$\ln(1.5) = \ln(1.004333^{12t}) = 12t \ln(1.004333)$$

We can now solve for t.

$$t = \frac{\ln(1.5)}{12 \ln(1.004333)} \approx 7.8 \text{ years}$$

Thus, it will take approximately 7.8 years for a $1,000-dollar investment to be worth $1,500.

▼ **EXAMPLE 2**   Doubling Time

How long does it take to double P dollars invested at 5% interest, compounded continuously?

SOLUTION   We want to know when $\$P$ becomes $\$2P$. We substitute $A = 2P$, $P = P$, and $r = 0.05$ into the continuous compounding formula $A = Pe^{rt}$ to get

$$2P = Pe^{0.05t}.$$

▼ * In October 1993, according to an index of yields for long-term A-rated general obligation bonds compiled weekly by The Bond Buyer. (Source: *The New York Times*, October 18, 1993).

We need to solve for t. As a first step, we can divide both sides by P, yielding

$$2 = e^{0.05t}.$$

Taking the natural logarithm of both sides,

$$\ln 2 = \ln e^{0.05t},$$

or

$$\ln 2 = 0.05\, t \ln e.$$

But $\ln e = 1$, by identity (d). Thus, we have

$$\ln 2 = 0.05t.$$

Hence,

$$t = \frac{\ln 2}{0.05} = 13.863 \text{ years.}$$

Before we go on ... What we have just found is known as the **doubling time** t_D. Let's check our answer by substituting it back into the compound interest formula.

$$Pe^{(0.05)(13.863)} = 2P \quad ✔$$

In the above example, we obtained the doubling time t_D by dividing $\ln 2$ by the interest rate. We can carry through this calculation in general to get the following (the calculation for exponential growth is identical because the basic formula is identical).

DOUBLING TIME FOR CONTINUOUS COMPOUNDING

If $\$P$ is invested at an annual interest rate r compounded continuously, the doubling time t_D is given by

$$t_D = \frac{\ln 2}{r}.$$

DOUBLING TIME FOR EXPONENTIAL GROWTH

For a quantity growing exponentially with fractional growth rate k, the doubling time t_D is given by

$$t_D = \frac{\ln 2}{k}.$$

Another way of writing the doubling time formulas is

$$t_D \cdot r = \ln 2$$
$$t_D \cdot k = \ln 2.$$

This tells us that once we know one of the two quantities r and t_D (or k and t_D), we can solve for the other.

▼ EXAMPLE 3   Investments

I would like to double my money in 10 years. At what interest rate (compounded continuously) can this be accomplished?

SOLUTION   We have the formula

$$t_D \cdot r = \ln 2,$$

where

$$t_D = 10.$$

Thus,

$$r = \frac{\ln 2}{t_D} = \frac{\ln 2}{10} \approx 0.069315.$$

The required rate of interest is therefore 6.9315%.

Before we go on... Check:

$$e^{0.069315 \times 10} = 2. \quad ✔$$

▼ EXAMPLE 4   Epidemics

In the early stages of the AIDS epidemic, the number of people infected was doubling every six months. Assuming an exponential growth model, and given that there were an estimated 1 million persons infected at a certain time, estimate the number of infected people 2.5 years later.

SOLUTION   Since the doubling time is $t_D = 0.5$ years, this gives us

$$k = \frac{\ln 2}{t_D} = \frac{\ln 2}{0.5} = 2 \ln 2 \approx 1.3863.$$

Using the exponential growth model,

$$A = Pe^{tk},$$

we have

$$P = 1{,}000{,}000, \, t = 2.5, \, k = 2 \ln 2.$$

This gives

$$A = 1{,}000{,}000 e^{2.5}(2 \ln 2) = 32{,}000{,}000.$$

Thus, the model predicts 32 million infected people 2.5 years later.

Before we go on... Is the exponential growth model reliable here? There are really two questions: (1) Does the model produce the numbers we expect? (2) Are these reasonable numbers?

1. In order to test whether the exponential growth model gives the expected numbers, note that we were told that the number of infected people was doubling every 6 months. Thus, after 6 months, it will have doubled once, giving 2 million cases; after 1 year, it will have doubled twice, giving 4 million cases ($= 2^2$), and so on. After 2.5 years, it will have doubled 5 times, giving $2^5 = 32$ million cases, agreeing with the prediction of the model. Thus, the exponential model produces the expected numbers.

2. On the other hand, the doubling every six months couldn't continue for very long, and this is borne out by observations. If doubling every six months did continue, then in 20 years the number of infected people would be

$$2^{40} \text{ million} \approx 1,099,511,628,000,000,000,$$

a number that is considerably larger than the population of the earth! Thus the exponential model is unreliable for predicting long-term trends.

Epidemiologists use more sophisticated models to measure the spread of epidemics, and these models predict a "leveling-off" phenomenon as the number of cases becomes a significant part of the total population (see the "You're the Expert" section at the end of this chapter). However, the exponential growth model *is* fairly reliable in the early stages of an epidemic.

▼ **EXAMPLE 5**   Half-Life

Let's go back to the story of Fluffy's "Red Flag 1-in-6 Flea Collar." (Examples 6 and 8 in Section 2). We saw there that the flea population was decreasing at a fractional rate of $\frac{1}{6}$ per day. Thus, using the exponential decay formula,

$$A = Pe^{(-1/6)t}.$$

How long will it take for the flea population to be cut in half? This time is called the **half-life** t_H of the flea population.

SOLUTION   We want to know when the population P becomes $\frac{P}{2}$. Substituting $A = \frac{P}{2}$ gives

$$\frac{P}{2} = Pe^{(-1/6)t} = Pe^{-t/6}.$$

Dividing by P,

$$\frac{1}{2} = e^{-t/6}.$$

Taking the natural log of both sides gives

$$\ln \frac{1}{2} = \ln(e^{-t/6})$$

$$= -\frac{t}{6} \ln e$$

$$= -\frac{t}{6}.$$

In other words, $-\ln 2 = -\dfrac{t}{6}$

or $\qquad \ln 2 = \dfrac{t}{6},$

so $\qquad t = 6 \ln 2 \approx 4.1589$ (days).

Thus, the half-life of the flea population is $t_H \approx 4.2$ days.

Before we go on... Let's check our answer.

$$Pe^{-4.1589/6} \stackrel{?}{=} 0.5P \quad \checkmark$$

Before we leave this example, let's go back to the step

$$\ln 2 = \frac{1}{6}t.$$

Since $k = \frac{1}{6}$ is the fractional rate of decay, we can see from this equation that

$$\ln 2 = \text{Fractional rate of decay} \times \text{Half-life}$$

or

$$\ln 2 = k \cdot t_H.$$

HALF-LIFE FOR EXPONENTIAL DECAY

For a quantity decaying exponentially with fractional rate of decay k, the half-life t_H is given by

$$t_H = \frac{\ln 2}{k}.$$

Another way of writing this formula is

$$t_H \cdot k = \ln 2.$$

An important form of exponential decay is **radioactive decay.** Some forms of chemical elements, such as uranium, plutonium, cobalt, and carbon, are unstable and decay into more stable elements. The decay takes place on an essentially continuous basis and is modeled with great precision by the exponential decay equation $A = Pe^{-kt}$, where P is the original number of atoms (or the original weight) and A is the number of atoms (or weight) undecayed after time t. Different unstable elements can have vastly different values of k, and hence vastly different half-lives, ranging from millionths of a second to thousands of years.

Carbon dating, one of the methods used by archaeologists to determine the age of a fossil, is based on the following principle. There are two isotopes of carbon: Carbon-12, the stable form, and Carbon-14, which is unstable and gradually decays into nitrogen. Carbon-14 originates in the upper atmosphere when nitrogen is exposed to cosmic radiation and is absorbed in small amounts by all living organisms along with Carbon-12. When an organism dies, absorption of carbon stops, and the Carbon-14 in the organism slowly decays, with a half-life of 5730 years. When a fossil is analyzed, the ratio of Carbon-14 to Carbon-12 in the fossil is measured and compared with the original ratio.* In this way, it is possible to estimate the amount of Carbon-14 that has decayed since the organism was alive, and hence determine the age of the fossil.

▼   **EXAMPLE 6**   Carbon Dating

If tests on a fossilized skull reveal that 99.95% of the Carbon-14 has decayed, how old is the skull?

SOLUTION   We set this up as a standard decay problem, with $A = Pe^{-kt}$. We know that

A = amount of undecayed C-14 left = 0.05% of P, or 0.0005P.

$P = P$ (unspecified),

$t = ?$ (the unknown),

$k = \ldots$ well, we are told that the half-life is 5730 years. That is, $t_H = 5730$. To get k, we use the formula $t_H \cdot k = \ln 2$.

So

$$5730k = \ln 2,$$

and

$$k = \frac{\ln 2}{5730} \approx 0.00012097.$$

Now that we have all the constants we need, we substitute them in the decay formula to get

$$0.0005P = Pe^{-0.00012097t}.$$

▼   * The ratio of Carbon-14 to Carbon-12 in any living organism is the same as the ratio in the environment, and this ratio has stayed fairly constant over the millennia, according to geophysical studies of the earth.

As usual, the Ps cancel, giving

$$0.0005 = e^{-0.00012097t}.$$

Taking the natural log,

$$\ln(0.0005) = \ln(e^{-0.00012097t}) = -0.00012097t.$$

Thus,

$$t = -\frac{\ln(0.0005)}{0.00012097} \approx 62,833.$$

We conclude that the skull is approximately 63,000 years old.

Before we go on... Let us check our answer.

$$Pe^{-62,833 \cdot 0.00012097} = 0.0005P \quad \checkmark$$

▼ **EXAMPLE 7**   Continuous Population Decline

Given the situation in Example 5, how long will it take for the flea population of 500,000 to get down to 100? (Recall that this was the question we answered graphically in Section 2.)

SOLUTION   Because this is not about half-life, we go back to the original decay equation

$$A = Pe^{-t/6}.$$

Substituting $A = 100$ and $P = 500,000$ gives

$$100 = 500,000e^{-t/6}.$$

Divide by 500,000.

$$0.0002 = e^{-t/6}$$

Take natural logs.

$$\ln 0.0002 = \ln(e^{-t/6})$$
$$= -\frac{t}{6}$$

Thus,

$$t = -6 \ln 0.0002 \approx 51.103 \text{ days}.$$

Before we go on... This confirms the approximate result we got graphically in the last section. We can check that this is an accurate solution.

$$500,000e^{-51.103/6} \approx 100 \quad \checkmark$$

▶ ## 2.4 EXERCISES

APPLICATIONS

Government Bonds   *Exercises 1–10 are based on the following table, which lists annual percentage yields on government bonds in several countries.**

Country	U.S	Japan	Germany	Britain	Canada	Mexico
Yield (%)	5.16	3.86	5.97	6.81	6.58	13.7

1. Assuming that you invest $10,000 in U.S. government bonds, how long (to the nearest year) must you wait before your investment is worth $15,000 if the interest is compounded annually?

2. Assuming that you invest $10,000 in Japanese government bonds, how long (to the nearest year) must you wait before your investment is worth $15,000 if the interest is compounded annually?

3. If you invest $10,400 in German government bonds and the interest is compounded monthly, how many months will it take for your investment to grow to $20,000?

4. If you invest $10,400 in British government bonds and the interest is compounded monthly, how many months will it take for your investment to grow to $20,000?

5. How long, to the nearest year, will it take an investment in Mexico to double its value if the interest is compounded every 6 months?

6. How long, to the nearest year, will it take an investment in Canada to double its value if the interest is compounded every 6 months?

7. How long will it take a $1,000 investment in Canadian government bonds to be worth the same as a $800 investment in Mexican government bonds? (Assume all interest is compounded annually, and give the answer to the nearest year.)

8. How long will it take a $1,000 investment in Japanese government bonds to be worth the same as an $800

investment in U.S. government bonds? (Assume all interest is compounded annually, and give the answer to the nearest year.)

9. If the interest on U.S. government bonds is compounded continuously, how long will it take the value of an investment to double? (Give an answer correct to two decimal places.)

10. If the interest on Canadian government bonds is compounded continuously, how long will it take the value of an investment to double? (Give an answer correct to two decimal places)

11. *Investments*   *Berger Funds* advertised that its Berger 100 Mutual Fund yielded an average interest rate of 15.9% from 1974 to 1993[†]. Assuming that this rate of return continues, how long, to the nearest tenth of a year, will it take a $5,000 investment in the fund to triple in value?

12. *Investments*   *Berger Funds* also advertised that its Berger 101 Mutual Fund yielded an average interest rate of 14.1% from 1974 to 1993.[†] Assuming that this rate of return continues, how long, to the nearest tenth of a year, will it take a $5,000 investment in the fund to double in value?

13. *Influenza Epidemics*   If each infected student at Enormous State University infects one healthy student with flu every day, how long (to the nearest day) after the first student is infected will it take for the epidemic to spread to 500 students?

▼ * Information was current as of October 18, 1993. (Source: Salamon Brothers, Mexican Government, S.G. Wartburg & Company, J.P. Morgan Global Research, as quoted in *The New York Times,* October 18, 1993.)

[†] According to an ad placed in *The New York Times* on September 26, 1993.

14. *Cold Epidemics* If each infected student at Enormous State University infects one healthy student with a cold every three days, how long (to the nearest day) after the first student is infected will it take for the epidemic to spread to 500 students?

15. *Investments* How long will it take a $500 investment to be worth $700 if it is continuously compounded at 10% per year? (Give an answer to two decimal places.)

16. *Investments* How long will it take a $500 investment to be worth $700 if it is continuously compounded at 15% per year? (Give an answer to two decimal places.)

17. *Investments* How long, to the nearest year, will it take an investment to triple if it is continuously compounded at 10% per year?

18. *Investments* How long, to the nearest year, will it take me to become a millionaire if I invest $1,000 at 10% interest compounded continuously?

19. *Stocks* Professor Stefan Schwartzenegger invests $1,000 in Tarnished Trade (TT) stocks, which are depreciating continuously at a rate of 20% per year. Luckily for Professor Schwartzenegger, TT declares bankruptcy at the instant his investment has declined to $666. How long after the initial investment did this occur?

20. *Investments*

(a) The Hofstra choral society's investment of $100,000 in Tarnished Teak Enterprises is depreciating continuously at 16% per year. At this rate, how long, to the nearest year, will it take for the investment to be worth $1?

(b) After six agonizing months watching its savings dwindle, the choral society pulls out what is left of its investment and buys Sammy Solid Trust Bonds, which earn a steady 6% interest, compounded semi-annually. How long will it take the choral society to recover its losses?

21. *Carbon Dating* The half-life of Carbon-14 is 5730 years. It is found that in a fossilized math professor's skull, 99.875% of the Carbon-14 has decayed since the time it was part of a living math professor (if you call that living). How old is the skull?

22. *Carbon Dating* A fossilized math book has just been dug up, and archeologists find that 99.5% of the Carbon-14 in the paper has decayed. When were the trees cut down to make the paper for the book?

23. *Automobiles* The rate of auto thefts is tripling every 6 months. Find the doubling time.

24. *Televisions* The rate of television thefts is doubling every 4 months. Find the tripling time.

25. *Bacteria* A bacteria culture starts with 1,000 bacteria and doubles in size every 3 hours. Approximately how many bacteria will there be after 2 days?

26. *Bacteria* A bacteria culture starts with 1,000 bacteria. Two hours later there are 1,500 bacteria. Assuming exponential growth, approximately how many critters will there be after two days?

27. *Frogs* Frogs in Nassau County have been breeding like flies! Two years ago, the pledge class of Epsilon Delta was instructed by the brothers to tag all the frogs residing on the ESU campus (Nassau County Branch) as an educational exercise. After several agonizing days, they managed to tag all 50,000 of them (with little Epsilon Delta Fraternity tags). This year's pledge class discovered that the tags had all fallen off, and they wound up tagging a total of 75,000 frogs. Assuming exponential population growth, how many tags should Epsilon Delta order for next year's pledge class?

28. *Flies* Flies in Suffolk County have been breeding like frogs! Three years ago the Health Commission caught 4,000 flies in one hour in a trap. This year it caught 7,000 flies in one hour. Assuming exponential population growth, how many flies should it expect to catch next year?

29. *U.S. Population* The U.S. population was 180,000,000 in 1960 and 250,000,000 in 1990. Assuming exponential population growth, what will be the population in the year 2010?

30. *World Population* World population was estimated at 1.6 billion people in 1900 and 5.3 billion people in 1990. Assuming exponential growth, when were there only two people in the world? Comment on your answer.

31. *Membership* Membership in the Enormous State University Choral Society has been increasing exponentially, doubling every semester. Assuming that there were a total of 20 Choral Society members at the start of this semester, how long do you estimate it would take for its membership to exceed the total U.S. population of 250,000,000?

32. *Membership* Having reached a membership of 10,000,000, the ESU Choral Society began charging its members a $100 annual membership fee. From that point on, its membership started to decline exponentially, going down by half every year. How long did it take for the membership to reach its original total of 20?

33. *Radioactive Decay*   Uranium-235 is used as fuel for some nuclear reactors. It has a half-life of 710 million years. How long would it take 10 grams of Uranium-235 to decay down to 1 gram?

34. *Radioactive Decay*   Plutonium-239 is used as fuel for some nuclear reactors, and also as the fissionable material in atomic bombs. It has a half-life of 24,400 years. How long would it take 10 grams of Plutonium-239 to decay to 1 gram?

35. *Radioactive Decay*   You are trying to determine the half-life of a new radioactive element you have isolated. You start with 1 gram, and 2 days later determine that it has decayed down to 0.7 gram. What is its half-life?

36. *Radioactive Decay*   You have just isolated a new radioactive element. If you can only determine its half-life, you will win the Nobel prize in physics. You purify a sample of 2 grams. One of your colleagues steals half of it, but three days later you find that 0.1 gram of radioactive material is still left. What is the half-life?

Employment   *Exercises 37–42 are based on the following chart, which shows the number of people employed by* Northrop Grumman *in Long Island, New York for the period from 1987 to 1995.* *

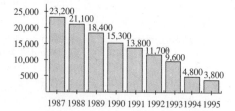

37. Use the 1987 and 1995 figures to construct an exponential decay model for employment at *Northrop Grumman*/Long Island, and use your model to predict

the number of people employed there in 1997. Your model should have the form

$$P = P_0 e^{-kt},$$

where t is the number of years since 1987. (Round your answer to the nearest 100 people.)

38. Repeat Exercise 37 using the 1987 and 1994 figures.

39. The method of least squares exponential regression[†] gives the model

$$P = 28,131.7 e^{-0.220800t},$$

where t is the number of years since 1987. Referring to the model you obtained in Exercise 37, which model predicts a higher employment figure for 1997?

40. Repeat Exercise 39 using your model from Exercise 38.

41. According to the least squares model in Exercise 39, when will employment by *Northrop Grumman*/Long Island drop to 1,000 employees? (Give the answer to the nearest year.)

42. According to the least squares model in Exercise 39, when will employment by *Northrop Grumman*/Long Island drop to 500 employees? (Give the answer to the nearest year.)

43. *Temperature*   You just bought a six-pack of non-alcoholic brew, but it is at room temperature (70°F). Since you desire a cold drink, you stick it in your refrigerator, which maintains 35°F. Now the brew will cool down in such a way that the difference between its temperature and the refrigerator's temperature will halve every hour. How long will it take to get down to a drinkable 40°F? [*Hint*: Use the difference D between the temperature of the brew and the refrigerator's temperature as the decaying quantity.]

44. *Temperature*   You are heating up grög for your winter party. The drink starts out at 50°F, and you are heating the pot to 400°F. If the difference between the temperature of the grög and the (fixed) temperature of the pot halves every 30 minutes, how long will it take for the grög to reach its ideal temperature of 200°F? [*Hint:* Use the difference D between the temperature of the grög and the temperature of the pot as the decaying quantity.]

▼ * The 1987–1994 figures show employment at Grumman Corp. before it was acquired by Northrop Corp. in 1994. The 1994 and 1995 figures are company projections. Source: Grumman Corp., Northrop Grumman Corp. *Newsday*, September 23, 1994, p. A5.

† This method is described in the chapter on functions of several variables.

COMMUNICATION AND REASONING EXERCISES

45. You have decided to use an exponential growth model to describe the student enrollment at your college. What information would you need, and how would you go about constructing the model?

46. Your friend has begun a project to use an exponential growth model to describe the student enrollment at his college. You come to the conclusion that an exponential model is inappropriate for modeling student enrollment. What arguments might you use to support this view?

47. You are midway through your project to describe student enrollment at your college, when a colleague argues that an exponential model is inappropriate. Are there any circumstances that might give you reasons to support the use of such a model?

48. Comment on the following statement: Ultimately, all exponential growth models fail.

▶ You're the Expert

EPIDEMICS

A mysterious epidemic is spreading through the population of the United States. An estimated 150,000,000 people are susceptible to this particular disease. There are 10,000 people already infected, and the number is doubling every two months. As advisor to the Surgeon General, it is your job to predict the course of the epidemic. In particular, the Surgeon General needs to know when the disease will have infected 1,000,000 people, when it will have reached 10,000,000, and when it will have affected 100,000,000 people.

Although the initial spread of an epidemic appears to be exponential, it cannot continue to be so, since the susceptible population is limited. A commonly used model for epidemics is the **logistic curve,** given by the function

$$A(t) = \frac{NP_0}{P_0 + (N - P_0)e^{-kt}},$$

where $A(t) = $ the infected population at time t, $P_0 = $ the population initially infected, and $N = $ the total susceptible population. The number k is a constant that governs the rate of spread of the epidemic. (Its exact meaning will be made clear in a moment.) The graph of this function is shown in Figure 1.

The initial part of this graph shows roughly exponential growth. To see why, multiply the top and bottom of the formula above by e^{kt}, which gives

$$A(t) = \frac{NP_0 e^{kt}}{P_0 e^{kt} + (N - P_0)}.$$

Now if P_0 is small compared to N, then for small t the denominator of this expression is approximately N. This gives

$$A(t) \approx P_0 e^{kt}$$

for small t; that is, $A(t)$ grows approximately exponentially in the early part of the epidemic. This also tells us that the number k is the fractional rate of growth that governs the early stages of the epidemic, when $A(t)$ is approximately exponential.

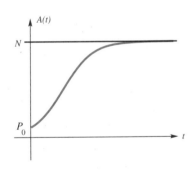

FIGURE 1

On the other hand, as t gets large, the term e^{-kt} in the original formula gets very small, and $A(t)$ gets close to the number N. You can see this in the rightmost part of the graph as $A(t)$ levels off under the line at height N.* Thus, the line $A = N$ is a horizontal asymptote.

The problems you now face are (1) to determine the constants to use in the logistic curve, and (2) to use the formula to predict the course of the epidemic. From the data you have, you know that $P_0 = 10,000$ and $N = 150,000,000$. The main problem is to find k. But remember that the initial spread of the epidemic is given by $A \approx P_0 e^{kt}$, so k is the fractional rate of growth for this exponential growth formula. You know that the infected population is doubling every two months, so the doubling time is $t_D = 2$ (months). Thus,

$$k = \frac{\ln 2}{t_D} = \frac{\ln 2}{2} = 0.3466.$$

Now we can write the logistic curve governing this particular epidemic:

$$A(t) = \frac{1,500,000,000,000}{10,000 + 149,990,000e^{-0.3466t}}$$

$$= \frac{150,000,000}{1 + 14,999^{-0.3466t}}.$$

The graph of this function is shown in Figure 2.

To check your calculation, calculate $A(2) \approx 20,000$, showing the expected doubling after 2 months.

Now you tackle the question of prediction: When will the disease infect 1,000,000 people? This is asking: When is $A(t) = 1,000,000$? Setting $A(t)$ equal to 1,000,000, you get the equation

$$1,000,000 = \frac{150,000,000}{1 + 14,999e^{-0.3466t}}.$$

Now you solve for t.

$$1,000,000(1+14,999e^{-0.3466t}) = 150,000,000$$

$$1 + 14,999e^{-0.3466t} = 150$$

$$14,999e^{-0.3466t} = 149$$

$$e^{-0.3466t} = \frac{149}{14,999} = 0.009934$$

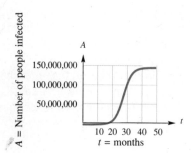

A

A = Number of people infected

150,000,000

100,000,000

50,000,000

10 20 30 40 50
t = months

t

FIGURE 2

▼ *If you look at $N - A(t) = \dfrac{N(N - P_0)e^{-kt}}{P_0 + (N - P_0)e^{-kt}} \approx \dfrac{(N - P_0)}{P_0}e^{-kt}$ for large t, you can see that the difference between A and N decays exponentially for large t.

To solve this equation you take the natural log of both sides.

$$-0.3466t = \ln 0.009934 = -4.612$$

$$t = \frac{-4.612}{-0.3466} = 13.3 \text{ months}$$

So, in just over 13 months 1,000,000 people are expected to be infected.
   When will 10,000,000 people be infected? Set $A(t) = 10,000,000$, and solve for t as before.

$$10,000,000 = \frac{150,000,000}{1 + 14,999e^{-0.3466t}}$$

$$1 + 14,999e^{-0.3466t} = 15$$

$$e^{-0.3466t} = \frac{14}{14,999} = 0.0009334$$

$$-0.3466t = \ln 0.0009334 = -6.977$$

$$t = \frac{-6.977}{-0.3466} = 20.1 \text{ months}$$

The epidemic will reach 10,000,000 people in just over 20 months. Finally, to determine when 100,000,000 will be infected, you solve

$$100,000,000 = \frac{150,000,000}{1 + 14,999e^{-0.3466t}}$$

and get $t = 29.7$ months. Notice that it took under 7 months to go from 1 million to 10 million infected people, but it takes over 9 months to from 10 million to 100 million. This is the slowing down of the epidemic from the exponential growth in its early stages.

Exercises

1. Track the later stages of the epidemic, by determining when 110 million people will be infected, when 120 million will be, when 130 million will be, and when 140 million will be infected.

2. Referring to Figure 2, when—to the nearest five months—would you estimate the epidemic to be spreading the fastest?

3. Give an estimate of the number of people you expect will be infected during the first year of the epidemic.

4. Explain the role of each of the constants in the logistic function

$$A(t) = \frac{NP_0}{P_0 + (N - P_0)e^{-kt}}.$$

5. Give a logistic model for the following: You have sold 100 "I ❤ Calculus" T-shirts, and sales are going up continuously at a rate of 30% per day. You estimate the total market for "I ❤ Calculus" T-shirts to be 3,000. Now use your model to predict when you will have sold 2,000 T-shirts.

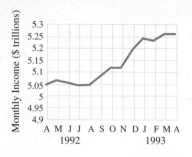

Monthly Income (\$ trillions)

A M J J A S O N D J F M A
1992   1993

6. In Russia, the average consumer drank two servings of *Coca-Cola*® in 1993. This amount appeared to be increasing exponentially with a doubling time of two years.* Given a long-range market saturation estimate of 100 servings per year, find a logistic model for the consumption of *Coca-Cola* in Russia, and use your model to predict when the average consumption will be 50 servings per year.

7. The graph to the left shows the monthly total U.S. personal income from April 1, 1992 to April 1, 1993, in trillions of dollars.† Obtain a rough logistic model for the income by estimating P_0 and N from the graph, and experimenting with several values of k.

▶ ▦ Review Exercises

Rewrite the equations in Exercises 1–6 in logarithmic form.

1. $2^{10} = 1024$

2. $5^{-2} = \frac{1}{25}$

3. $10^{-4} = 0.0001$

4. $y = e^{-2x}$

5. $y = x^e (x \neq 1)$

6. $y = 2x^{1/2}$

Rewrite the equations in Exercises 7–12 in exponential form.

7. $\log_3 81 = 4$

8. $\log_{1/3} 81 = -4$

9. $\log 1{,}000 = 3$

10. $\ln e^3 = 3$

11. $\log_x y = -1$

12. $y = -\ln x$

Solve the equations in Exercises 13–18 for x.

13. $4 = 3^x$

14. $10^x = 5^{2x}$

15. $9^x = 3^{-x^2}$

16. $10 = 2^{x^2}$

17. $1000 = 5e^{10x}$

18. $5 = 1000e^{-10x}$

Find the future value of the investments in Exercises 19–24.

19. \$2,000 invested at 4% compounded monthly for 4 years

20. \$3,000 invested at 5% compounded monthly for 3 years

21. \$2,000 invested at 6.75% compounded daily (365 times per year) for 4 years

22. \$3,000 invested at 8.25% compounded daily for 5 years

23. \$2,000 invested at 3.75% compounded continuously for 4 years

24. \$3,000 invested at 4.75% compounded continuously for 3 years

In each of Exercises 25–30, find the time required for the investment to reach the desired goal.

25. \$2,000 invested at 4% compounded monthly, goal = \$3,000

26. \$3,000 invested at 5% compounded monthly, goal = \$5,000

27. \$2,000 invested at 6.75% compounded daily, goal = \$3,000

28. \$3,000 invested at 8.25% compounded daily, goal = \$5,000

29. \$2,000 invested at 3.75% compounded continuously, goal = \$3,000

30. \$3,000 invested at 4.75% compounded continuously, goal = \$5,000

▼ * The doubling time is based on retail sales of Coca-Cola products in Russia. Sales in 1993 were double those in 1991, and were expected to double again by 1995. Source: *The New York Times*, September 26, 1994, p. D2.

† Source: National Association of Purchasing Management, Department of Commerce (*The New York Times*, June 2, 1993, Section 3, p. 1.) Our graph is a facsimile of the published graph (though accuracy may have suffered slightly).

APPLICATIONS

31. *Investments*   Diane Blake invested her life savings, $10,000, in five-year CDs that paid 6% per year, compounded monthly. After five years she took her proceeds and put it all in the stock market, which proceeded to lose 1% each month. When she took out her money one year later, how much did she have left?

32. *Investments*   Frank Capek invested his life savings, $7,500, in 10-year CDs that paid 7.25% per year, compounded monthly. After 10 years he took his proceeds and put it all in the bond market, which proceeded to lose 1.5% each month. When he took out his money one year later, how much did he have left?

33. *Inflation*   Inflation in East Camelot is running at 4% per month. If a car costs $10,000 now, how much will an equivalent car cost in two years?

34. *Inflation*   Inflation in West Camelot is running at 3% per month. How long will it take prices to double?

35. *Rabbits*   The rabbits on my farm are reproducing like . . . well, rabbits. I started out with 10, but 4 months later had 30. Assuming exponential growth, when will I be completely overrun with 1,000 bunnies?

36. *Investments*   The $1,000 you invested in mutual funds has grown to $1,200 in half a year. Assuming that this rate of return continues, how large will your investment be after 5 years total?

37. *Radioactive Decay*   Costenobelium is a highly unstable element. 10μg will decay down to 1μg in one second. What is its half-life?

38. *Radioactive Decay*   Wanerium is a highly stable element. 10μg will decay down to 9μg after 10,000 years. What is its half-life?

39. *Investments*   Utopia Investments offers an account paying 25% compounded annually, while Erewhon Investments offers an account paying 24% compounded continuously. Which is the better investment?

40. *Investments*   Pomegranate Computer's stock is depreciating at a rate of 2% per month, while Mega-Soft's stock dropped 12.5% in the last half year. Which is the worse investment?

41. *Radioactive Decay*   Potassium-40 has a half-life of 1.28×10^9 years. If 95% of the Potassium-40 present in a rock when it was formed has decayed, how old is the rock?

42. *Radioactive Decay*   A sample of 1 milligram of Einsteinium-246 will decay to 0.9 milligram in 1.11 minutes. How long will it take to decay to 0.1 milligram?

43. *Investments* (*from the GMAT*)   Each month for 6 months the amount of money in a benefit fund is doubled. At the end of the 6 months there is a total of $640 in the fund. How much money was in the fund at the end of 3 months?

44. *Savings* (*from the GMAT*)   On July 1, 1982, Ms. Fox deposited $10,000 in a new account at the annual interest rate of 12 percent compounded monthly. If no additional deposits or withdrawals were made, and if interest was credited on the last day of each month, what was the amount of money in the account on September 1, 1982?

45. *Bacteria* (*from the GMAT*)   The population of a bacteria culture doubles every 2 minutes. Approximately how many minutes will it take for the population to grow from 1,000 to 500,000 bacteria?

46. *Population* (*based on a question from the GMAT*)   The population of a small town doubles every 10 years. Approximately how many years will it take for the population to grow from 10,000 to 25,000 people?

47. *Interest* (*from the GRE economics exam*)   If the interest rate is 10 percent, what is the present discounted value of a dollar due two years from now?

48. *Exponential Decay* (*from the GRE economics exam*)   To estimate the rate at which new instruments will have to be retired, a telephone company uses the "survivor curve":

$$L_x = L_0 e^{-x/t},$$

where

L_x = number of survivors at age x,
L_0 = number of initial installations,
t = average life in years.

All of the following are implied by the curve *except*:
(a) Some of the equipment is retired during the first year of service.
(b) Some equipment survives three average lives.
(c) More than half the equipment survives the average life.
(d) Increasing the average life of equipment by using more durable materials would increase the number surviving at every age.
(e) The number of survivors never reaches zero.

Hard Times Dilute Enthusiasm for Clean-Air Laws

By ROBERT REINHOLD

LOS ANGELES, Nov. 25—One of the few things about life in Southern California that has improved in recent years is the air. The air is undisputably cleaner and clearer and the region's notorious lung-searing smog alerts are all but history.

But the prospects for further progress are clouded. Just when other states are emulating California's strict controls on auto emissions and other pollutants, the state may have begun to retreat on air quality.

The worst economic slump since the Depression has created an audience for the argument that pollution restrictions are luxuries that Southern California cannot afford and have begun to undermine their original opposite premise: that cleaned-up air is essential to the region's future economic health.

The local air-quality district here has scaled back enforcement because of severe budget cuts, and the chairwoman of the state's Air Resources Board has resigned under fire from industry. The complex politics of smog are shifting.

"California had been on the leading edge of air quality," said Mary D. Nichols, an Angeleno who is the new assistant administrator for air and radiation at the Environmental Protection Agency in Washington. "Now, it is leading the backlash."

The backlash comes as Southern Californians this year breathed the cleanest air in a generation. Peak levels of ozone, the most dangerous pollutant, have dropped in Los Angeles to a quarter of what they were in 1955, despite huge increases in population and traffic. Since 1985, there have been only two stage-two health alerts, when all people are advised to stay indoors.

Partly, the improvements have to do with an unusually cool summer and reduced traffic because of the recession. . . .

Source: From Robert Reinhold, "Hard Times Dilute Enthusiasm for Clean-Air Laws," *The New York Times,* November 26, 1993, pp. 1 and A30.

Introduction to the Derivative

APPLICATION ▶ The Environmental Protection Agency (EPA) wants to formulate a policy that will encourage utilities to reduce sulfur emissions. Its goal is to reduce annual emissions of sulfur dioxide by a total of 10 million tons from the current level of 25 million tons by imposing a fixed charge for every ton of sulfur released into the environment per year. The EPA has data showing the marginal cost to utilities of reducing sulfur emissions by 8, 10, and 12 million tons per year. As a consultant to the EPA, you must determine the amount to be charged per ton of sulfur emissions in light of these data.

| INTRODUCTION | ▶ With this chapter we begin to study calculus—one of the most important, most useful, most *used* parts of mathematics. In the world around us, everything is changing, and calculus, at its heart, is the study of *how* things change; how fast and in what direction. Is the Dow Jones average going up, and if so, how fast? If I raise my prices, how many customers will I lose? If I launch this missile here, how high will it go, and where will it come down?

Calculus is concerned first with the *rate of change* of a function. We have already discussed this for linear functions (straight lines), where the *slope* measures the rate of change. But this works only because a straight line maintains the same rate of change along its whole length. Other functions rise faster here than there—or rise in one place and fall in another—so the rate of change varies along the graph. The first and greatest achievement of calculus is that it provides a systematic and straightforward way of *calculating* (hence the name) these rates of change. To describe a changing world, we need a language of change, and that is what calculus gives us.

The history of calculus is an interesting story of personalities, intellectual movements, and controversy. Credit for its invention is given to two mathematicians: Isaac Newton (1642–1727) and Gottfried Leibniz (1646–1716). Newton, an English mathematician and scientist, was the first to invent calculus, probably in the 1660s. We say "probably" because, for various reasons, he did not publish his ideas until much later. This allowed Leibniz to publish his own version of calculus first, in 1684. Fifteen years later, stirred up by nationalist fervor in England and on the continent, controversy erupted over who should get the credit for its invention. The debate got so heated that the Royal Society (of which Newton and Leibniz were both members) set up a commission to investigate the question. The commission decided in favor of Newton, who just happened to be president of the Society at the time. The consensus today is that both mathematicians deserve credit, since they both came to the same conclusions working independently. This is not really surprising: both built on well-known work of other people, and it was almost inevitable that someone would put it all together at about that time.

The controversy did have one unfortunate consequence for English mathematics. Newton's and Leibniz's versions of calculus were not identical. In particular, Leibniz had by far the better notation. This was mainly a consequence of Leibniz's work in philosophy: he believed that all knowledge could be expressed and manipulated in symbolic form and that a good notation would make this easier. He was therefore very interested in finding good notations for his mathematics and worked particularly hard on a notation for calculus. In fact, we still use his notation and no longer use Newton's. The problem for English mathematicians was that they felt compelled to use Newton's version of calculus. Their nationalist fervor also caused them to cut themselves off from the explosion of work on calculus that took place on the continent in the early 1700s, setting back English mathematics for decades.

Another controversy associated with calculus was much more important in the development of mathematics and still influences the way we teach calculus today. Calculus is, first of all, a formal way of calculating the rate of change of a function if the formula for that function is known. It works, and you can use it without knowing why it works, just as you can do long division or punch the buttons of a calculator without worrying about why you get the right answer. But mathematicians and philosophers of the eighteenth century were not satisfied with this—they wanted to know *why* calculus works. (This was part of the intellectual climate of the time. Much of this controversy erupted around the time of the Enlightenment, beginning in the 1730s. We have inherited much of the rationalism of that age.) Nobody could deny that calculus worked. Newton had used it to explain the motions of the planets, and successful applications soon came fast and furiously. But unfortunately, neither Newton nor Leibniz had provided satisfactory justification for their work, and the annoying fact remained that mathematicians could not adequately explain *why* the techniques of calculus worked. It took another hundred years for this to be resolved. Not until the 1820s did Augustin Louis Cauchy (1789–1857) publish papers showing how calculus could be justified by the use of *limits*. He also gave the careful definition of limits that we use to this day. His work began the great nineteenth-century project of introducing logical rigor into mathematics, establishing a precedent that mathematicians follow to this day.

▶ 3.1 RATE OF CHANGE AND THE DERIVATIVE

There are many situations in which we would like to know how fast a quantity is changing. We saw examples of this when we talked about straight lines and linear functions. In that case, the rate of change is measured by the *slope* of the line. Our first goal is to see what we can use in place of the slope when we have a function that is not linear.

▼ **EXAMPLE 1**   Rate of Change

Suppose we monitor the stock market very closely during one rather active week, and we find that during that week the value of the Dow Jones Average could be accurately described by the function $A(t) = 2{,}500 + 500t - 100t^2$, where t is time in days. The graph of this function is shown in Figure 1 (this is, of course, a parabola).

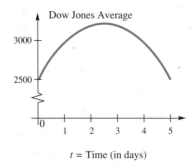

FIGURE 1

Looking at the graph, we can see that the Dow Jones Average ("Dow") rose rather rapidly at the beginning of the week, but by the middle of the week the rise had slowed, until the market faltered and the Dow began to fall more and more rapidly towards the end of the week. Can we calculate exactly how fast the Dow was rising at the beginning of the second day ($t = 1$), and how fast it was falling at the beginning of the fourth day ($t = 3$)?

SOLUTION   One way to answer this question is as follows. Let us first see how fast the Dow rose during the second day.

Start of day 2 ($t = 1$): $A(1) = 2{,}900$
End of day 2 ($t = 2$): $A(2) = 3{,}100$

Change during day 2: $A(2) - A(1) = 3{,}100 - 2{,}900 = 200$

Thus, the Dow increased by 200 points in one day, for a rate of change of 200 points per day. This was its **average rate of change** during the second day.

Now, if we look closely at the graph, we can see that it was actually rising faster at the beginning of the second day than at the end, so its rate of change must have been greater at the beginning of the day than we just calculated. To get a better idea of that rate, let us look at only the first half of the day, from $t = 1$ to $t = 1.5$.

Start of day 2 ($t = 1$):	$A(1) = 2,900$
Middle of day 2 ($t = 1.5$):	$A(1.5) = 3,025$

Change during first half of day 2: $A(1.5) - A(1) = 3,025 - 2,900 = 125$

Thus, the Dow rose 125 points in *half* a day. To translate this into points per day, we divide by the number of days—in this case, 0.5:

$$\text{Average rate of change of } A(t) = \frac{\text{Change in } A(t)}{\text{Time}}$$

$$= \frac{125}{0.5} = 250 \text{ points per day.}$$

As expected, this was greater than the average rate for the whole day.

Still, the market was rising faster at the beginning of the day than at noon, so the rate at which it was rising *right at the beginning of the day* must be even greater than the 250 we just calculated. To get closer to the rate we seek, we must do the same calculation over a smaller part of the day.

To organize our work a little, notice that we are looking at the time interval from time 1 to time $1 + h$, where h is the fraction of the day we are using. We have already calculated the rates for $h = 1$ and $h = 0.5$. The calculation we have been doing is as follows:

$$\text{Average rate of change of } A(t) \text{ from time 1}$$
$$\text{to time } 1 + h = \frac{A(1 + h) - A(1)}{h}.$$

Here is a table of the average rates of change we get when we choose smaller and smaller values for h:

h	1	0.5	0.1	0.01	0.001	0.0001
$A(1 + h) - A(1)$	200	125	29	2.99	0.2999	0.029999
Ave. rate of change $= \dfrac{A(1 + h) - A(1)}{h}$	200	250	290	299	299.9	299.99

As h decreases, the average rate of change seems to be getting closer and closer to 300 points per day. So it seems reasonable to say that the **instantaneous rate of change** at time $t = 1$ was 300 points per day. This is how fast the Dow was rising at the *instant* day 2 began ($t = 1$).

To see how fast the market was falling at the beginning of the fourth day ($t = 3$), we can do the same calculations with the time intervals from $t = 3$ to $t = 3 + h$, as follows.

$$\text{Average rate of change from time 3}$$
$$\text{to time } 3 + h = \frac{A(3 + h) - A(3)}{h}.$$

Here is a table of the average rates of change we get for various values of h.

h	1	0.5	0.1	0.01	0.001	0.0001
$A(3 + h) - A(3)$	−200	−75	−11	−1.01	−0.1001	−0.010001
Ave. rate of change $= \dfrac{A(3 + h) - A(3)}{h}$	−200	−150	−110	−101	−100.1	−100.01

$$\underset{1}{\uparrow\ A(4) - A(3)} \quad \underset{0.5}{\uparrow\ A(3.5) - A(3)} \quad \underset{0.1}{\uparrow\ A(3.1) - A(3)} \quad \underset{0.01}{\uparrow\ A(3.01) - A(3)} \quad \underset{0.001}{\uparrow\ A(3.001) - A(3)} \quad \underset{0.0001}{\uparrow\ A(3.0001) - A(3)}$$

Again, as h decreases, the average rate of change seems to be settling down to a "limiting value," here −100 points per day. Notice that this rate is negative, because the market is *falling* at the beginning of the third day.

Before we go on...

Q Why do we "sneak up" on the value $h = 0$, rather than just setting h equal to 0?

A Look at the formula $(A(1 + h) - A(1))/h$. We cannot simply substitute $h = 0$ because h appears in the denominator and we cannot divide by 0. More fundamentally, in order to see how the Dow is changing, we need to look at it at two different times, so that we can see some change. The single value $A(1) = 2,900$ tells us the Dow at the beginning of the second day, but it tells us absolutely nothing about how fast it is changing.

Notice also that we have measured the rates of change in points per day, that is, *units of A per unit of t*.

With a graphing calculator, you can automate these calculations. A computer spreadsheet is particularly useful for generating tables such as those above. Consult the Appendix for details and examples of the use of a graphing calculator or a spreadsheet program to calculate average rates of change.

We shall see computations similar to the previous example in many different situations. Let us summarize and introduce some terminology.

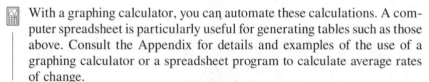

AVERAGE RATE OF CHANGE; DIFFERENCE QUOTIENT

The **average rate of change** of the function f over the interval $[x, x + h]$ from x to $x + h$ is

$$\text{Average rate of change} = \frac{f(x + h) - f(x)}{h}.$$

We also call this average rate of change the **difference quotient** of f over the interval $[x, x + h]$. Its units of measurement are units of f per unit of x.

The process of letting h get smaller and smaller is called taking the **limit** as h approaches 0. We write $h \to 0$ as shorthand for "h approaches 0." Taking the limit of the average rates of change gives us the instantaneous rate of change. Here is our notation for this limit.

INSTANTANEOUS RATE OF CHANGE; DERIVATIVE

The **instantaneous rate of change** of f at x is given by taking the limit of the average rates of change (given by the difference quotient) as h approaches 0. We write:

$$\text{Instantaneous rate of change} = \lim_{h \to 0} \frac{f(x + h) - f(x)}{h}.$$

We also call this instantaneous rate of change the **derivative** of f at x and write it as $f'(x)$ (read "f prime of x"). Thus,

$$f'(x) = \lim_{h \to 0} \frac{f(x + h) - f(x)}{h}.$$

The units of $f'(x)$ are units of f per unit of x.

▶ NOTES
1. For now, we shall trust our intuition when it comes to limits. We shall discuss limits in detail in the third section of this chapter.
2. $f'(x)$ is a number we can calculate, or at least approximate, for various values of x, as we have done in the example above. This means that it depends on x, or is a function of x. In other words, f' is another function of x. An old name for this is "the function *derived from f* '", which has been shortened to the *derivative* of f.
3. Finding the derivative of f is called **differentiating** f. ◀

▼ **EXAMPLE 2**   Demand

The economist Henry Schultz calculated the following demand function for corn:

$$q(p) = \frac{176{,}000}{p^{0.77}}.$$

Here, p is the price (in dollars) per bushel of corn and q is the number of bushels of corn that could be sold at the price p in one year.* How sensitive is demand to price when $p = \$2$ per bushel, and how sensitive is it when $p = \$8$ per bushel?

▼ * This demand function is based on data for the period 1915–1929. Source: Henry Schultz: *The Theory and Measurement of Demand* [as cited in *Introduction to Mathematical Economics* by A. L. Ostrosky, Jr., and J.V. Koch (Waveland Press, Prospect Heights, Illinois, 1979.)]

SOLUTION   When we ask how *sensitive* demand is to price, we would like to know how much the demand will change if the price changes. More precisely, we want to know the *rate* at which the demand will change as the price changes.

From the formula above, we have

$$\text{Rate of change of } q = \text{Derivative of } q, \text{ evaluated at } p = 2$$

$$= q'(2) = \lim_{h \to 0} \frac{q(2 + h) - q(2)}{h}.$$

The table below shows the average rate of change,

$$\frac{q(2 + h) - q(2)}{h} = \frac{\left[\dfrac{176{,}000}{(2 + h)^{0.77}} - \dfrac{176{,}000}{2^{0.77}} \right]}{h},$$

for various values of h. (On a graphing calculator, enter this function of h as follows, using x instead of h.

$$Y_1 = (176000/(2 + X)^{\wedge}0.77 - 176000/2^{\wedge}0.77)/X$$

and obtain the averages by having the calculator evaluate Y_1 at X = 1, 0.1, 0.01, 0.001, and 0.0001.)

h	1	0.1	0.01	0.001	0.0001
Ave. rate of change	−27,678	−38,055	−39,561	−39,718	−39,734

How do we interpret these figures? First, recall that we measure the rate of change in units of q per unit of p; that is, in (demand for) bushels per \$1 increase in price. Thus, the figure for h = 1 tells us that when the price of corn is raised from \$2 per bushel to 2 + h = \$3, demand will drop by an average of 27,678 bushels per \$1 increase in price. Similarly, the figure for h = 0.1 tells us that when the price of corn is raised from \$2 per bushel to 2 + h = \$2.10, demand will drop by an average of 38,055 bushels per \$1 increase in price.

To calculate the instantaneous rate of change q'(2), we need to know what number the average rate approaches as h approaches 0. Although this is not obvious from the table, we can say with some certainty that to three significant digits, $q'(2) \approx -39{,}700$ bushels per dollar. Thus, at a price of \$2 per bushel, the demand for corn will fall at a rate of approximately 39,700 bushels for each dollar increase in price.

We should also be interested in what happens if the price *decreases*. This means that we should also look at negative h:

h	−1	−0.1	−0.01	−0.001	−0.0001
Ave. rate of change	−72,791	−41,579	−39,912	−39,753	−39,737

This agrees with what we saw when we looked at positive h: as h approaches 0, the average rate of change approaches about $-39,700$. This means that, if the price were to *drop*, the demand would *rise* at a rate of 39,700 bushels per dollar drop in price (draw a graph and think about this for a while).

We can repeat all of these computations for a price of $8 per bushel. Notice, though, that we are really interested in the limit of the difference quotient, as we were above. We can *approximate* the value of the limit by substituting a small value of h. This is essentially what we did above: What we really paid attention to was the value of the difference quotient for $h = 0.0001$. So, let us compute

$$\frac{q(8 + 0.0001) - q(8)}{0.0001} \approx -3,416$$

Therefore, at a price of $8 per bushel, the demand for corn will fall by approximately 3,416 bushels per $1 increase in price.

Before we go on... When doing these calculations, it is important to record the calculations to enough significant digits to see the change. For example, in the last calculation the values of $q(8.0001)$ and $q(8)$ differ only in the fifth significant digit. If we write down only three or four significant digits, we would not notice any change at all. It is best to let your calculator or computer use as many digits as it calculates until the very end, when you may round the final answer to as many digits as are reasonable (usually three or four, depending on the accuracy of the data with which you are working).

Q What is the difference in meaning between $q(2)$ and $q'(2)$?

A Briefly, $q(2)$ is the *value of q* when $p = 2$, while $q'(2)$ is the *rate at which q is changing* when $p = 2$. In the above example, $q(2) = 176,000/2^{0.77} \approx 103,209$ bushels. This means that at a price of $2 per bushel, the demand for corn (measured in annual sales) is 103,209 bushels. On the other hand, $q'(2) \approx -39,700$ bushels per dollar. This means that at a price of $2 per bushel, the demand *is dropping by* 39,700 *bushels per* $1 *increase in price.*

Q Do we always need to make a table of difference quotients with decreasing values of h in order to calculate an approximate value for the derivative, or is there a quicker way?

A As we saw in the above example, we can *approximate* the value of the derivative by using a single, small value of h such as $h = 0.0001$. Graphing calculators do it this way (although they use a far smaller value). The only problem with using $h = 0.0001$ is that (1) we do not get an exact answer, and (2) it is not clear just how accurate our answer is.

Thus we have the following.

CALCULATING AN APPROXIMATE VALUE FOR THE DERIVATIVE

We can calculate an approximate value of $f'(x)$ by using the formula

$$f'(x) \approx \frac{f(x + 0.0001) - f(x)}{0.0001}.$$

▼ **EXAMPLE 3**

Use a graphing calculator to approximate $f'(5)$ if $f(x) = x^2 - x^{-0.4}$.

SOLUTION   The approximation given above is

$$f'(x) \approx \frac{f(x + 0.0001) - f(x)}{0.0001}$$

$$= \frac{(x + 0.0001)^2 - (x + 0.0001)^{-0.4} - (x^2 - x^{-0.4})}{0.0001}$$

$$= \frac{(x + 0.0001)^2 - (x + 0.0001)^{-0.4} - x^2 + x^{-0.4}}{0.0001}$$

We enter this on the graphing calculator as

$$Y_1 = ((X+0.0001)^2 - (X+0.0001)^{\wedge}(-0.4) - X^{\wedge}2 + X^{\wedge}(-0.4))/0.0001$$

Then set $x = 5$ (on the home screen) and evaluate Y_1 by entering*

$$5 \rightarrow X$$
$$Y_1$$

to obtain $f'(5) \approx 10.04212$. This answer is accurate to three decimal places; in fact, $f'(5) = 10.04202. \ldots$

DELTA NOTATION

We have used the notation $f'(x)$ for the derivative of f at x, but there is another interesting notation. We can write $\Delta f = f(x + h) - f(x)$ for the *change in* f. The Greek letter delta (Δ) is often used to stand for the phrase "the change in." Likewise, we can write $\Delta x = (x + h) - x = h$ for the change in x. Then the difference quotient is

$$\frac{\Delta f}{\Delta x} = \frac{f(x + h) - f(x)}{h},$$

▼ * This applies to TI models of graphing calculators.

which leads to the notation

$$\frac{df}{dx} = \lim_{\Delta x \to 0} \frac{\Delta f}{\Delta x}$$

for the derivative. That is, df/dx is just another notation for $f'(x)$. You should not really think of this as a quotient, but simply as a notation. We read df/dx as "the derivative of f with respect to x." It is necessary to include the phrase "with respect to x" when there are other variables around, and in any case it reminds us which variable is being used as the independent variable. As a last piece of notation, the phrase "the derivative with respect to x" is often abbreviated as d/dx, so that we could write

$$f'(x) = \frac{d}{dx}(f(x)) = \frac{df}{dx}.$$

This notation is most useful if we have a formula for a function that is still unnamed. For example, to refer to the derivative of the function with the formula x^3, we could write

$$\frac{d}{dx}(x^3).$$

We shall say more about this notation in Section 4.

Let us finish this section with another important application.

▼ **EXAMPLE 4** Velocity

If I throw a ball upward at a speed of 100 ft/s, our physicist friends tell us that its height t seconds later will be $s = 100t - 16t^2$. How fast will the ball be rising exactly 2 seconds after I throw it?

SOLUTION The graph of the ball's height as a function of time is shown in Figure 2. Asking "How fast will it be rising?" is really asking for the rate of change of height with respect to time. Why? Think about average velocity for a moment. If we wanted to compute the average velocity of the ball from time 2 to time 3, say, we would first compute the distance the ball rose in that time, or the change in height:

$$\Delta s = s(3) - s(2) = 156 - 136 = 20 \text{ ft.}$$

Since it rose 20 feet in $\Delta t = 1$ second, we use the formula $Speed = Distance/Time$ to get an average velocity of

$$\text{Average velocity} = \frac{\Delta s}{\Delta t} = \frac{20}{1} = 20 \text{ ft/s}$$

from time $t = 2$ to $t = 3$. Note that this is just the difference quotient. Now, to get the **instantaneous velocity** at $t = 2$, we take the limit. In other words, we need to calculate the derivative ds/dt at $t = 2$.

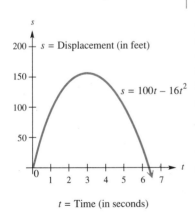

$s = $ Displacement (in feet)

$s = 100t - 16t^2$

$t = $ Time (in seconds)

FIGURE 2

Using the quick approximation described above, we obtain

$$\frac{ds}{dt} \approx \frac{s(2 + 0.0001) - s(2)}{0.0001}$$

$$= \frac{s(2.0001) - s(2)}{0.0001}$$

$$= \frac{100(2.0001) - 16(2.0001)^2 - (100(2) - 16(2)^2)}{0.0001}$$

$$= 35.9984.$$

In fact, the instantaneous velocity at $t = 2$ is exactly 36 ft/s. (Try an even smaller value of h to convince yourself!)

Before we go on... If you repeat the calculation at time $t = 5$, you will obtain

$$\frac{ds}{dt} = \lim_{h \to 0} \frac{s(5 + h) - s(5)}{h} = -60 \text{ ft/s}.$$

The negative sign tells us that the ball was *falling* at a rate of 60 ft/s at time $t = 5$. (How does the fact that it is falling at $t = 5$ show up on the graph?)

The last example gives another interpretation of the derivative:

VELOCITY

For an object moving in a straight line with position $s(t)$ at time t, the **average velocity** from time t to time $t + h$ is given by

$$v_{\text{average}} = \frac{s(t + h) - s(t)}{h} = \frac{\Delta s}{\Delta t}.$$

The **instantaneous velocity** at time t is given by

$$v = \lim_{h \to 0} \frac{s(t + h) - s(t)}{h} = \frac{ds}{dt}.$$

Velocity is really the rate of change of position with respect to time.

Here is one last bit of notation. In the last example, we could have written the velocity either as s' or as ds/dt, as we chose to do. In order to write the answer to the question, that the velocity at 2 seconds was 36 ft/s, we can write either

$$s'(2) = 36$$

or

$$\left. \frac{ds}{dt} \right|_{t=2} = 36.$$

The notation "$|_{t=2}$" is read "evaluated at $t = 2$." This notation is obviously more cumbersome than the functional notation $s'(2)$, but the notation ds/dt has other advantages, as we shall see. You should feel free to use whichever notation is most convenient at any given time. In fact, you should practice using both notations.

▶ **3.1 EXERCISES**

In Exercises 1–8, calculate the difference quotient for the given function f at the given value of x. Use the resulting formula to find the average rates of change corresponding to $h = \pm1, \pm0.1, \pm0.01, \pm0.001,$ and ±0.0001. (It will be easier to do this if you first simplify the difference quotient as much as possible.)

1. $f(x) = 2x^2;\quad x = 0$

2. $f(x) = \dfrac{x^2}{2};\quad x = 1$

3. $f(x) = \dfrac{1}{x};\quad x = 2$

4. $f(x) = \dfrac{2}{x};\quad x = 1$

5. $f(x) = x^3;\quad x = 1$

6. $f(x) = \dfrac{x^3}{3};\quad x = -1$

7. $f(x) = x^2 + 2x;\quad x = 3$

8. $f(x) = 3x^2 - 2x;\quad x = 0$

In Exercises 9–20, estimate the derivative of the given function at the indicated point.

9. $f(x) = 1 - 2x;\quad x = 2$

10. $f(x) = \dfrac{x}{3} - 1;\quad x = -3$

11. $f(x) = x^3;\quad x = -1$

12. $g(x) = x^4;\quad x = -2$

13. $h(x) = x^{1/4};\quad x = 16$

14. $s(x) = x^{2/3};\quad x = 8$

15. $f(x) = x^{-1/2};\quad x = 2$

16. $f(x) = x^{-1/3};\quad x = 1$

17. $g(t) = \dfrac{1}{t^5};\quad t = 1$

18. $s(t) = \dfrac{1}{t^3};\quad t = -2$

19. $f(x) = \dfrac{x^2}{4} - \dfrac{x^3}{3};\quad x = -1$

20. $f(x) = \dfrac{x^2}{2} + \dfrac{x}{4};\quad x = 2$

Consider the functions in Exercises 21–28 as representing the position (in miles) of a car t hours after starting to drive down a road. Find the average velocities of the car over the time intervals $[t, t + h]$, where t is as indicated and $h = 1$, 0.1, and 0.01 hours. Also, estimate the instantaneous velocity at time t (that is, the reading on the speedometer at that instant).

21. $s(t) = 50t - t^2;\quad t = 5$

22. $s(t) = 60t - 2t^2;\quad t = 3$

23. $s(t) = 100 + 20t^3;\quad t = 1$

24. $s(t) = 1{,}000 + 50t - t^3;\quad t = 2$

25. $s(t) = 60\left(t + \dfrac{1}{t}\right);\quad t = 10$

26. $s(t) = 50\left(t - \dfrac{1}{t^2}\right);\quad t = 5$

27. $s(t) = 40\sqrt{t};\quad t = 4$

28. $s(t) = 50(t - \sqrt{t});\quad t = 16$

Consider each of the functions in Exercises 29–34 as representing the cost to manufacture x items. Find the average costs of manufacturing h more items (i.e., the average rate of change of the total cost) at a production level of x, where x is as indicated and h = 50, 10, and 1. Also, estimate the rate of change of the total cost at the given production level x.

29. $C(x) = 10,000 + 5x - \dfrac{x^2}{10,000}$; $x = 1,000$

30. $C(x) = 20,000 + 7x - \dfrac{x^2}{20,000}$; $x = 10,000$

31. $C(x) = 1,000 + 10x - \sqrt{x}$; $x = 100$

32. $C(x) = 10,000 + 20x - \sqrt{x}$; $x = 10$

33. $C(x) = 15,000 + 100x + \dfrac{1,000}{x}$; $x = 100$

34. $C(x) = 20,000 + 50x + \dfrac{10,000}{x}$; $x = 100$

APPLICATIONS

35. *Demand*   Suppose the demand for a new brand of sneakers is given by
$$q = \frac{5,000,000}{p},$$
where p is the price per pair of sneakers, in dollars. Find $q(100)$ and estimate $q'(100)$, and interpret your answers.

36. *Demand*   Suppose the demand for an old brand of TV is given by
$$q = \frac{100,000}{p + 10},$$
where p is the price per TV set, in dollars. Find $q(190)$ and estimate $q'(190)$, and interpret your answers.

37. *Profit*   Your monthly profit (in dollars) from selling magazines is given by
$$P = 5n + \sqrt{n},$$
where n is the number of magazines you sell in a month. If you are currently selling $n = 50$ magazines per month, find P and estimate dP/dn. Interpret your answers.

38. *Profit*   Your monthly profit (in dollars) from your newspaper route is given by
$$P = 2n - \sqrt{n},$$
where n is the number of subscribers on your route. If you currently have 100 subscribers, find P and dP/dn. Interpret your answers.

39. *Popularity*   A study of enrollment at a certain university shows that there is a relationship between enrollment in each professor's classes and the average grade that the professor awards. Measuring grades g on a scale of 0 to 4, the relationship can be expressed by the equation
$$E = 20\sqrt[4]{g},$$

where E is the professor's average enrollment per class. For a professor whose average grade awarded is $g = 2.5$, find E and dE/dg, and interpret your answers.

40. *Popularity*   A study of enrollment at another university shows that, again, there is a relationship between enrollment in each professor's classes and the average grade that the professor awards. Measuring grades g on a scale of 0 to 4, the relationship this time can be expressed by the equation
$$E = 15\sqrt[3]{g^2},$$

where E is the professor's average enrollment per class. For a professor whose average grade awarded is $g = 2.5$, find E and dE/dg, and interpret your answers.

41. *Market Average*   Joe Downs runs a small investment company from his basement. Every week he

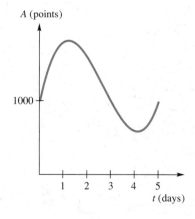

publishes a report on the success of his investments, including the progress of the infamous Joe Downs Average. At the end of one particularly memorable week he reported that the Average for that week had the value $A(t) = 1000 + 1500t - 800t^2 + 100t^3$ points, where t represents the number of days into the week; t ranges from 0 at the beginning of the week to 5 at end of the week. The graph of A is shown above. During the upswing at the beginning of the week, say halfway through the first day, approximately how fast was the value of the Average increasing?

42. *Market Average*   Referring to the Joe Downs Average in the previous exercise: During the downswing, say, at the end of the third day, how fast was the value of the Average decreasing?

43. *Learning to Speak*   Let $p(t)$ represent the percentage of children who are able to speak at the age of t months.
    (a) It is found that $p(10) = 60$ and $p'(10) = 18.2$.*
        What does this mean?
    (b) As t increases, what happens to $p(t)$ and $p'(t)$?

44. *Learning to Read*   Let $p(t)$ represent the number of children in your class who learned to read at the age of t years.
    (a) Assuming that everyone in your class could read by the age of 7, what does this tell you about $p(7)$ and $p'(7)$?
    (b) Assuming that 25% of the people in your class could read by the age of 5, and that 25.3% of them could read by the age of 5 years and one month, estimate $p'(5)$, remembering to give its units.

45. *Biology—Reproduction*   The Verhulst model for population growth specifies the reproductive rate of an organism as a function of the total population according to the following formula:

$$R(p) = \frac{r}{1 + kp}.$$

Here, p is the total population in thousands of organisms, r and k are constants that depend on the particular circumstances and the organism being studied, and $R(p)$ is the reproduction rate in thousands of organisms per hour.[†] Assume that $k = 0.125$ and $r = 45$ for a particular population, and estimate $R'(4)$. Interpret the result.

46. *Biology—Reproduction*   Another model, the predator satiation model for population growth, specifies that the reproductive rate of an organism as a function of the total population varies according to the following formula:

$$R(p) = \frac{rp}{1 + kp}.$$

Here, p is the total population in thousands of organisms, r and k are constants that depend on the particular circumstances and the organism being studied, and $R(p)$ is the reproduction rate in new organisms per hour.[†] Assume that $k = 0.2$ and $r = 0.08$ for a particular population, and estimate $R'(2)$. Interpret the result.

47. *The Theory of Relativity*   In science fiction terminology, a speed of *warp 1* is the speed of light—about 3×10^8 meters per second. (Thus, for instance, a speed of warp 0.8 corresponds to 80% of the speed of light—about 2.4×10^8 meters per second.) According to Einstein's Special Theory of Relativity, a moving object appears to get shorter to a stationary observer as its speed approaches that of light. If a rocket ship whose length is l_0 meters at rest travels at a speed of warp p, its length in meters, as measured by a stationary observer, will be given by

$$l(p) = l_0\sqrt{1 - p^2}, \text{ with domain } [0, 1].$$

Assuming that your rocket is 100 meters long ($l_0 = 100$), estimate $l(0.95)$ and $l'(0.95)$. What do these figures tell you?

48. *Newton's Law of Gravity*   The gravitational force exerted on a particle with mass m by another particle with mass M is given by the following function of distance:

$$F(r) = G\frac{Mm}{r^2}, \text{ with domain } (0, +\infty).$$

Here, r is the distance between the two particles in meters, the masses M and m are given in kilograms, $G \approx 0.0000000000667 = 6.67 \times 10^{-11}$, and the resulting force is given in newtons. Given that M and m are both 1,000 kilograms, estimate $F(10)$ and $F'(10)$. What do these figures tell you?

▼ * Based on data presented in the article *The Emergence of Intelligence* by William H. Calvin, *Scientific American*, October, 1994, pp. 101–7.

† Source: *Mathematics in Medicine and the Life Sciences* by F.C. Hoppensteadt and C.S. Peskin (Springer-Verlag, New York, 1992) pp 20-22.

Exercises 49–56 are based on logarithmic and exponential function.

In the following exercises, estimate the derivative of the given function at the indicated point.

E/L 49. $f(x) = e^x$; $x = 0$ **E/L 50.** $f(x) = 2e^x$; $x = 1$

E/L 51. $f(x) = \ln x$; $x = 1$ **E/L 52.** $f(x) = \ln x$; $x = 2$

E/L 53. Sales   Weekly sales of a new brand of sneakers are given by

$$S(t) = 200 - 150e^{-t/10},$$

pairs sold per week, where t is the number of weeks since the introduction of the brand. Estimate $S(5)$ and estimate $S'(5)$, and interpret your answers.

E/L 54. Sales   Weekly sales of an old brand of TV are given by

$$S(t) = 100e^{-t/5},$$

sets per week, where t is the number of weeks after the introduction of a competing brand. Estimate $S(5)$, and $S'(5)$ and interpret your answers.

E/L

55. Logistic Growth in Demand   The demand for a new product can be modeled by a **logistic** curve of the form

$$q(t) = \frac{N}{1 + ke^{-rt}},$$

where $q(t)$ is the total number of units sold t months after the introduction of the new product, and N, k, and r are constants that depend on the product and the market. Assume that the demand for video game units is determined by the above formula, with $N = 10,000$, $k = 0.5$, and $r = 0.4$. Estimate $q(2)$ and $q'(2)$, and interpret the results.

E/L 56. Information Highway   The amount of information transmitted each month on the National Science Foundation's Internet network for the years 1988–1994 can be modeled by the equation

$$q(t) = \frac{2e^{0.69t}}{3 + 1.5e^{-0.4t}},$$

where q is the amount of information transmitted each month in billions of data packets, and t is the number of years since the start of 1988.* Estimate the number of data packets transmitted during the first month of 1990, and also the rate at which this number was increasing.

COMMUNICATION AND REASONING EXERCISES

57. Give an algebraic explanation of the fact that if f is a linear function, then the average rate of change over any interval equals the instantaneous rate of change at any point.

58. Give a geometric explanation of the fact that, if f is a linear function, then the average rate of change over any interval equals the instantaneous rate of change at any point.

59. Explain why we cannot put $h = 0$ in the formula

$$f'(x) = \lim_{h \to 0} \frac{f(x + h) - f(x)}{h}$$

for the derivative of f.

60. A manufacturer has manufactured 10,000 surfboards, but the manufacturing level is decreasing at a rate of 4,000 per year. What does this tell you about the number $N(t)$ of surfboards produced as a function of time t (years)?

▶ ═══════ **3.2** GEOMETRIC INTERPRETATION OF THE DERIVATIVE

As we mentioned at the beginning of the previous section, the rate of change of a linear function is the slope of the corresponding line. For a function whose graph is not a straight line, there is no notion of *the* slope of the graph. Look at the line drawn in Figure 1.

This line is just as steep at the point P as it is at the point Q. Put another way, the line is rising just as fast at P as it is at Q. Now look at the curve shown in Figure 2.

▼ *This is the authors' model, based on figures published in *The New York Times*, Nov. 3, 1993.

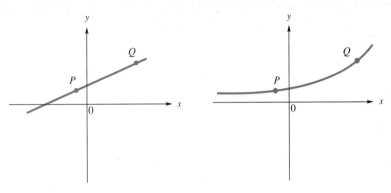

FIGURE 1   **FIGURE 2**

This graph is decidedly steeper at Q than it is at P. This suggests that the "slope" of the curve increases from left to right, and so we cannot assign a single number to measure the steepness of the whole graph. Instead, we shall have to assign a number *to each point of the graph* to measure its steepness *at that point*.

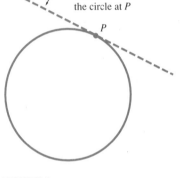

FIGURE 3

Q Just what do we *mean* by the steepness of a graph at a specified point? In other words, how can we measure it?

A Here is a way to measure steepness. Since we all know how to measure the steepness of a *line* (by its slope), and since we would like to build on what we already know, we will say that the steepness of a graph at a specified point is the slope of the *tangent line to the graph at that point*.

Remember that a tangent line to a circle is a line that touches the circle in just one point. A tangent line gives the circle "a glancing blow," as shown in Figure 3.

A tangent line to an arbitrary curve is similar. Figure 4 shows the lines tangent to the parabola $y = x^2$ at the five points $(-2, 4)$, $(-1, 1)$, $(0, 0)$, $(1, 1)$, and $(2, 4)$.

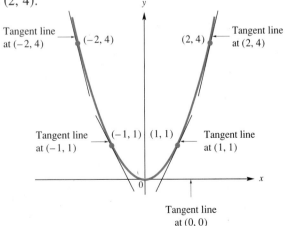

FIGURE 4

The steepness of the parabola at $(-2, 4)$ is the slope of the tangent line we have drawn there. Notice that it has a negative slope of large magnitude. Similarly, the steepness of the parabola at $(-1, 1)$ is the slope of the tangent line at $(-1, 1)$. This line has a negative slope of smaller magnitude. The steepness of the graph at $(0, 0)$ is 0, since the tangent line there is horizontal (it happens to be the x-axis). The steepness of the graph at $(1, 1)$ is positive, since the tangent line there has a positive slope, while the steepness of the graph at $(2, 4)$ is also positive, but larger.

In summary:

STEEPNESS OF A GRAPH

The steepness of a graph at a point P is measured by the slope of the tangent to the curve at P.

We shall often use the abbreviation m_{tan} to denote the slope of the tangent to a graph at a specified point.

The problem we are now faced with is *how to calculate the slope of a tangent line*. The first part of that problem is that we do not know exactly what we *mean* by the tangent line. For the moment we shall not attempt to define it, but continue to use our intuition. We shall return to the *definition* of the tangent line shortly. Now, assuming that we know what we mean by the tangent line, we return to the question of finding its slope. We *could* draw the graph very carefully, draw the tangent line very carefully, and then compute the slope using two points on the line. But can we really draw the graph carefully enough to trust our answers? There is a better way to approach this.

Figure 5 shows what we have in general: the graph of some function, a point P on the graph, and the line tangent to the graph at P. Our task is to find a way of *calculating* the slope of the tangent line.

If we're not told the slope of a line, the only way we can calculate it is by using two points on the line. But, as you can see from the figure, we start out knowing only *one* point—the point on the graph through which the tangent line passes.

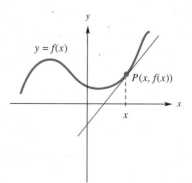

$y = f(x)$

$P(x, f(x))$

FIGURE 5

Q So how do we go about finding the slope of the tangent line knowing only one point?

A We first find the slope, not of the tangent line, but of an *approximate* tangent line, by selecting a second point Q on the curve close to P. Take a look at Figure 6.

Let us explain what is going on here: The line passing through the points P and Q is an *approximate* tangent line. We get the point Q as follows. Since the original point P has x-coordinate x, we choose a nearby value by adding a small quantity h, getting $x + h$. The point Q is then taken to be the point on the graph with x-coordinate $x + h$ ($h > 0$ in Figure 6, but h may also be negative, which would put Q to the left of P). We obtain the y-coordinate of Q by evaluating the function f there, getting $f(x + h)$. Thus, Q is the point $(x + h, f(x + h))$.

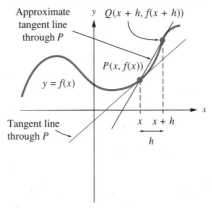

Approximate tangent line through P

y $Q(x + h, f(x + h))$

$P(x, f(x))$

$y = f(x)$

Tangent line through P

x $x + h$

h

FIGURE 6

The line through P and Q is called a **secant line** of the graph. Its slope can be calculated using the usual slope formula:

$$m_{\text{sec}} = \frac{y_2 - y_1}{x_2 - x_1} = \frac{f(x + h) - f(x)}{(x + h) - x} = \frac{f(x + h) - f(x)}{h}.$$

Do you recognize this? It is the difference quotient we saw in the first section.

Q So we know a formula for the slope of the *secant* line, which is an *approximate* tangent line. How do we get the slope of the *exact* tangent line?

A Here is the key idea: The closer the point Q gets to the point P, the more closely the secant line will approximate the tangent line. In Figure 7 is shown a graph with a tangent line drawn at P and three secant lines drawn through P and points Q_1, Q_2, and Q_3. As Q gets closer to P in the sequence Q_1, Q_2, Q_3, the secant line gets closer to the tangent line, so the slope of the secant line approaches the slope of the tangent line.

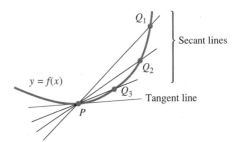

Q_1

Secant lines

Q_2

$y = f(x)$

Q_3

Tangent line

P

FIGURE 7

This tells us that if we want to find the slope of the tangent line, we need to take the limit of the slopes of the secant lines as h approaches 0. In other words, *the slope of the tangent line at a point is given by the derivative at that point.*

Thus we have the following important formulas:

SECANT AND TANGENT LINES

The slope of the secant line through $(x, f(x))$ and $(x + h, f(x + h))$ is given by the difference quotient:

$$m_{\text{sec}} = \frac{f(x + h) - f(x)}{h}$$

($=$ Average rate of change of f over the interval $[x, x + h]$).

The slope of the line tangent to the graph of f at $(x, f(x))$ is given by the derivative:

$$m_{\text{tan}} = f'(x) = \lim_{h \to 0} \frac{f(x + h) - f(x)}{h}$$

($=$ Instantaneous rate of change of f at x).

Q So how do we *define* the tangent line?

A Since we started the discussion above by assuming that there was a tangent line, what we have really shown is that *if* there is a tangent line at a point, its slope ought to be given by the derivative at that point. We turn this around and *define* the tangent line to f through the point $P = (x, f(x))$ to be the line through P with slope $f'(x)$.

It's about time for some examples.

▼   **EXAMPLE 1**

Let $f(x) = x^2$.

(a) Obtain a formula for the slope of the secant line through the points $(x, f(x))$ and $(x + h, f(x + h))$.

(b) Use the answer in part (a) to find the slope of the secant line through the points $(1, 1)$ and $(1 + h, (1 + h)^2)$.

SOLUTION

(a) The slope of any secant line is given by the difference quotient.

$$\begin{aligned}
m_{\text{sec}} &= \frac{f(x + h) - f(x)}{h} \\[2mm]
&= \frac{(x + h)^2 - x^2}{h} \\[2mm]
&= \frac{x^2 + 2xh + h^2 - x^2}{h} \\[2mm]
&= \frac{2xh + h^2}{h} \\[2mm]
&= \frac{h(2x + h)}{h} = 2x + h.
\end{aligned}$$

This is the slope of the secant line through (x, x^2) and $(x + h, (x + h)^2)$.

(b) For the secant line through $(1, 1)$ and $(1 + h, (1 + h)^2)$, we have $x = 1$, so

$$m_{\text{sec}} = 2x + h = 2 + h.$$

Before we go on... This single formula enables us to find the slope of many secant lines. For instance, choosing $h = 1$, the secant line through $(1, 1)$ and $(2, 4)$ has slope $2 + 1 = 3$. Choosing $h = 2$, the secant line through $(1, 1)$ and $(3, 9)$ has slope $2 + 2 = 4$. The secant line through $(1, 1)$ and $(0, 0)$ has slope $2 - 1 = 1$ (since $h = 0 - 1 = -1$). Figure 8 shows the graph of f, together with these three secant lines.

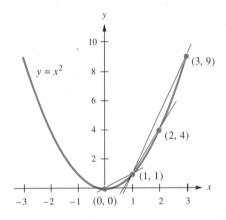

FIGURE 8

In the last instance, we used a negative value for h. There is nothing in our analysis that says h can't be negative. Of course, if h *is* negative, then $x + h$ is to the *left* of x.

▼ **EXAMPLE 2**

This continues Example 1, with $f(x) = x^2$.

(a) Find a formula for the slope of the tangent line at the point $(x, f(x))$.

(b) Use the answer in part (a) to find the slope of the tangent line at the point $(1, 1)$.

(c) Find an equation of the tangent line at the point $(1, 1)$.

SOLUTION

(a) In Example 1, we calculated the slope of the secant line through $(x, f(x))$ and $(x + h, f(x + h))$ to be

$$m_{\text{sec}} = \frac{f(x + h) - f(x)}{h} = 2x + h.$$

To obtain the slope of the *tangent* line, we let h approach 0:

$$m_{\text{tan}} = \lim_{h \to 0} \frac{f(x + h) - f(x)}{h}$$

$$= \lim_{h \to 0} 2x + h.$$

As h gets closer to 0, the sum $2x + h$ gets closer and closer to $2x + 0 = 2x$. Thus,

$$m_{\text{tan}} = \lim_{h \to 0} 2x + h = 2x$$

is the slope of the tangent line at $(x, f(x)) = (x, x^2)$.

(b)   For the tangent line at $(1, f(1)) = (1, 1)$, we have $x = 1$, so

$$m_{\text{tan}} = 2x = 2.$$

(c)   To find an equation of the tangent line, we may use the point-slope formula, because we know that the tangent line goes through the point $(1, 1)$ with slope 2:

$$y - 1 = 2(x - 1)$$

or

$$y = 2x - 1.$$

This tangent line is shown in Figure 9.

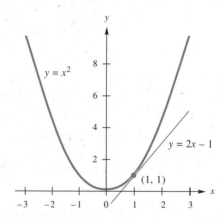

FIGURE 9

Before we go on... We did the following calculation in part (a): If $f(x) = x^2$, then $f'(x) = 2x$. This is our first complete calculation of a derivative function. After we talk about limits in the next section we shall do many more such calculations. In part (b), we calculated $f'(1) = 2$.

If you look back at Figure 4, we can now quantify what we saw there. The tangent line at the point where $x = -2$ has slope $f'(-2) = 2(-2) = -4$.

This is a steeply falling line, as appears in the figure. The tangent line at the point where $x = -1$ has slope $f'(-1) = 2(-1) = -2$. This is also a falling line, but it is less steep than the first. Similarly, the tangent lines at the points where $x = 0$, $x = 1$, and $x = 2$ have slopes $f'(0) = 0$, $f'(1) = 2$, and $f'(2) = 4$, respectively.

ESTIMATING DERIVATIVES GRAPHICALLY USING GRAPHING CALCULATORS OR GRAPHING SOFTWARE

We mentioned in the first section of this chapter that graphing calculators and computers were helpful in approximating derivatives numerically. We can also use graphing calculators or graphing software to approximate derivatives graphically.

▼ **EXAMPLE 3**

If $f(x) = x^{1.2} - \sqrt{x^{2.2} + x}$, calculate an approximate value for $f'(2)$ graphically.

SOLUTION   We shall exploit this idea: If we magnify a small portion of the graph of f near the point of interest, the result is almost indistinguishable from a straight line. In fact, it is almost indistinguishable from the *tangent* line. Therefore, if we calculate the slope of this "line," we will have a close approximation to the slope of the tangent line. Here is what we do.

First, we enter the function in the correct format:

$$Y_1 = X^{\wedge}1.2 - (X^{\wedge}2.2 + X)^{\wedge}0.5$$

Next, we set the viewing window by taking $0 \le x \le 5$ (to make sure that the point with $x = 2$ is on the graph) and $-0.6 \le y \le 0.6$. (We obtained the y-ranges by experimenting.) Figure 10 shows the graphing calculator output.

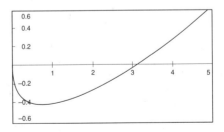

FIGURE 10

Since we are interested in the slope of the tangent line at the point where $x = 2$, we shall zoom into a small portion of the curve centered *exactly* at the point on the graph where x is 2. (This will improve our accuracy.) To do this, we choose a small value for h, say $h = 0.001$, and use

$Xmin = 2 - h = 1.999$, and $Xmax = 2 + h = 2.001$. To specify the y-coordinates of the window, we can use "trace" to give rough values for $f(1.999)$ and $f(2.001)$, and use these values as $Ymin$ and $Ymax$. Thus, let us take $Ymin = -0.271$ and $Ymax = -0.270$.* Figure 11 shows the zoomed-in view.

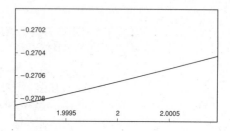

FIGURE 11

As we said, this small portion of the graph is approximately the tangent line. Now we need to measure its slope. To do this, recall that we need the coordinates of two points. *We shall use the two furthest points on the zoomed-in graph:* the points where $x = 1.999$ and $x = 2.001$. Using "trace" to find the y-coordinates, we obtain the points $(1.999, -0.2708359)$ and $(2.001, -0.2704366)$ (alternatively, we could calculate $f(1.999)$ and $f(2.001)$ directly using the calculator). This gives the slope as

$$m = \frac{y_2 - y_1}{x_2 - x_1}$$

$$= \frac{-0.2704366 - (-0.2708359)}{2.001 - 1.999} = 0.19965$$

Since the slope of the tangent is approximately 0.19965, we conclude that $f'(2) \approx 0.19965$.

Before we go on... Why did we use so many decimal places for the y-coordinates? How can we improve the accuracy of our calculation of the slope?

Q Just how accurate is the answer?

A The exact value of $f'(2)$ to five decimal places turns out to be 0.19966. Thus, the answer we obtained is *extremely* accurate—within 0.00001 of the exact answer. As to how we obtained the exact answer, you will find out how to take the derivative of functions as complicated as $f(x)$ in the next chapter.

▼ *Alternatively, we can run the "windows" program in Section 1 Appendix C. This program automatically sets $Ymin$ and $Ymax$ for us.

▶   **3.2 EXERCISES**

In each of the graphs in Exercises 1–8, say at which labeled point the slope of the tangent is **(a)** *greatest and* **(b)** *least (in the sense that −7 is less than 1).*

1.

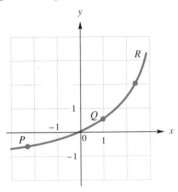

2.

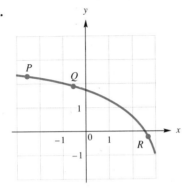

3.

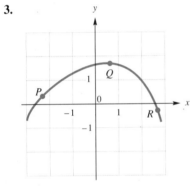

4.

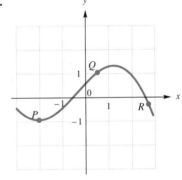

5.

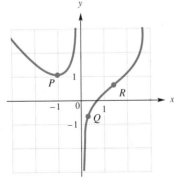

6.

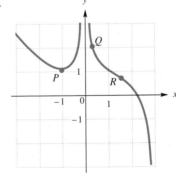

7.

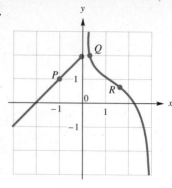

8.

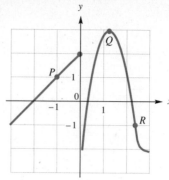

In each of Exercises 9–14, three slopes are given. For each slope, determine at which of the labeled points on the graph the tangent line has that slope.

9. (a) 0   **(b)** 4   **(c)** −1   **10. (a)** 0   **(b)** 1   **(c)** −1

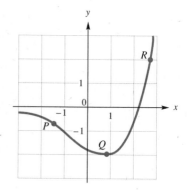

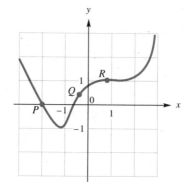

11. (a) 0   **(b)** 3   **(c)** −3   **12. (a)** 0   **(b)** 3   **(c)** 1

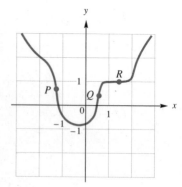

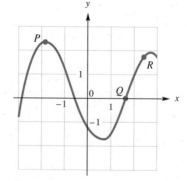

13. (a) 3   **(b)** −2.5   **(c)** −0.5   **14. (a)** 1.5   **(b)** −1.5   **(c)** −2.5

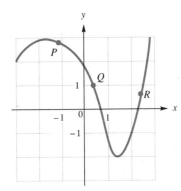
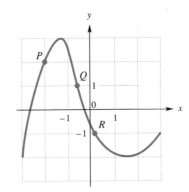

In each of Exercises 15–18, find the approximate coordinates of all points (if any) where the slope of the tangent is **(a)** 0,   **(b)** 1,   **(c)** −1.

15.

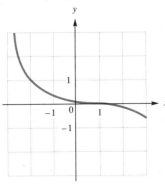

16.

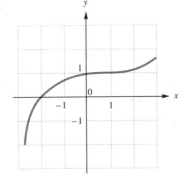

17.

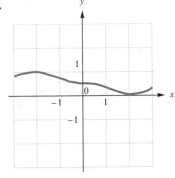

18.

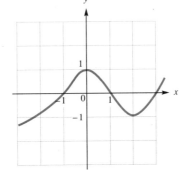

19. Complete the following sentence: The tangent to the graph of the function f at the general point where $x = a$ is the line passing through _____ with slope _____ .

20. Complete the following sentence: The difference quotient for f at the point where $x = a$ gives the slope of the _____ passing through the points _____ .

21. Let $f(x)$ have derivative $f'(x)$. Find a formula for the equation of the tangent to the graph of f through the point where $x = a$.

22. Find a formula for the equation of the secant line through the points on the graph of the function f corresponding to $x = a$ and $x = a + h$.

In each of Exercises 23–34:
(a) *find a formula for the slope of the secant line through $(x, f(x))$ and $(x + h, f(x + h))$;*
(b) *find a formula for the slope of the tangent to the graph of f at the point $(x, f(x))$;*
(c) *find the slope of the tangent to the graph of f at the indicated point.*

23. $f(x) = x^2 + 1$; $(2, 5)$

24. $f(x) = x^2 - 3$; $(1, -2)$

25. $f(x) = 2 - x^2$; $(-1, 1)$

26. $f(x) = -1 - x^2$; $(0, -1)$

27. $f(x) = 3x$; $(2, 6)$

28. $f(x) = x - 1$; $(2, 1)$

29. $f(x) = 1 - 2x$; $(2, -3)$

30. $f(x) = \dfrac{x}{3} - 1$; $(-3, -2)$

31. $f(x) = 3x^2 + 1$; $(-1, 4)$

32. $f(x) = 2x^2$; $(-2, 8)$

33. $f(x) = x - x^2$; $(2, -2)$

34. $f(x) = x^2 + x$; $(3, 12)$

In each of Exercises 35–40:
(a) *use any method to find the slope of the tangent to the graph of the given function at the point with the indicated x-coordinate;*
(b) *find an equation of the tangent line in part (a). In each case, sketch the curve together with the appropriate tangent line.*

35. $f(x) = x^3$; $x = -1$

36. $f(x) = x^2$; $x = 0$

37. $f(x) = x + \dfrac{1}{x}$; $x = 2$

38. $f(x) = \dfrac{1}{x^2}$; $x = 1$

39. $f(x) = \sqrt{x}$; $x = 4$

40. $f(x) = 2x + 4$; $x = -1$

Match each of the functions graphed in Exercises 41–46 to the graph of its derivative (shown below Exercise 46).

41.

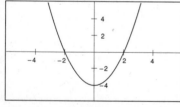

42.

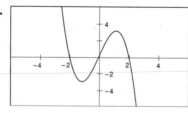

43.

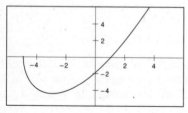

44.

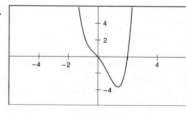

45.
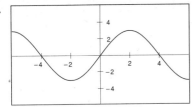

46.
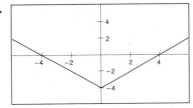

Graphs of derivatives:

(a)
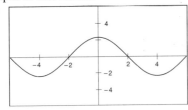

(b)
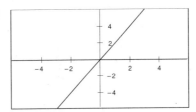

(c)
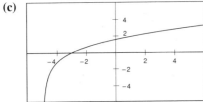

(d)
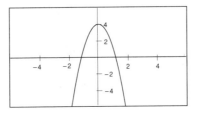

(e)
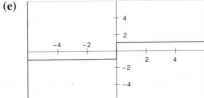

(f)
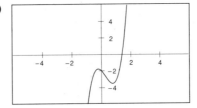

In Exercises 47–50, use a graphing calculator or graphing software to estimate the derivative of

$$f(x) = \frac{1}{x^{1.1} - 4}$$

graphically at the point with the given x-coordinate.

47. $x = 1$ **48.** $x = 2$ **49.** $x = 1.2$ **50.** $x = 4.3$

In Exercises 51–54, use a graphing calculator or graphing software to estimate the derivative of

$$f(x) = \sqrt{1 - x^2}$$

graphically at the point with the given x-coordinate.

51. $x = 0.5$ **52.** $x = 0.75$ **53.** $x = -0.25$ **54.** $x = -0.5$

In Exercises 55–58, use a graphing calculator to graph the given function $f(x)$ over the given range of values of x, and determine the approximate values of x in that range for which $f'(x) = 0$ (if any).

55. $f(x) = x^{3.4} - x^{1.2}$; $0 \le x \le 5$

56. $f(x) = 2x^{4.1} - x^{1.3}$; $0 \le x \le 1$

57. $f(x) = (x - 1)(x - 2)(x - 3)$; $-1 \le x \le 5$

58. $f(x) = (x - 1)^2(x - 2)^2(x - 3)$; $-1 \le x \le 5$

APPLICATIONS

Exercises 59–74 were given in the last section, but this time we ask you to use either a graphing calculator or graphing software to solve them graphically.

59. *Demand*   Suppose that the demand for a new brand of sneakers is given by

$$q = \frac{5,000,000}{p},$$

where p is the price per pair of sneakers in dollars. Graph q for $1 \le p \le 200$, use your graph to find $q(100)$ and estimate $q'(100)$, and interpret your answers.

60. *Demand*   Suppose that the demand for an old brand of TV is given by

$$q = \frac{100,000}{p + 10},$$

where p is the price per TV set, in dollars. Graph q for $0 \le p \le 400$, use your graph to find $q(190)$ and estimate $q'(190)$, and interpret your answers.

61. *Profit*   Your monthly profit (in dollars) from selling magazines is given by

$$P = 5n + \sqrt{n},$$

where n is the number of magazines you sell in a month. Graph P for $0 \le n \le 100$, and use your graph to answer this question: If you are currently selling $n = 50$ magazines per month, find P and estimate dP/dn. Interpret your answers.

62. *Profit*   Your monthly profit (in dollars) from your newspaper route is given by

$$P = 2n - \sqrt{n},$$

where n is the number of subscribers on your route. Graph P for $0 \le n \le 200$, and use your graph to answer this question: If you currently have 100 subscribers, find P and dP/dn. Interpret your answers.

63. *Popularity*   A study of enrollment at a certain university shows that there is a relationship between enrollment in each professor's classes and the average grade that the professor awards. Measuring grades g on a scale of 0 to 4, the relationship can be expressed by the equation

$$E = 20\sqrt[4]{g},$$

where E is the professor's average enrollment per class. Graph E, and use your graph to answer the following question: For a professor whose average grade awarded is $g = 2.5$, find E and dE/dg, and interpret your answers.

64. *Popularity*   A study of enrollment at another university shows that, again, there is a relationship between enrollment in each professor's classes and the average grade that the professor awards. Measuring grades g on a scale of 0 to 4, the relationship this time can be expressed by the equation

$$E = 15\sqrt[3]{g^2},$$

where E is the professor's average enrollment per class. Graph E, and use your graph to answer the following question: For a professor whose average grade awarded is $g = 2.5$, find E and dE/dg, and interpret your answers.

65. *Market Average*   Joe Downs runs a small investment company from his basement. Every week he publishes a report on the success of his investments, including the progress of the infamous Joe Downs Average. At the end of one particularly memorable week he reported that the Average for that week had the value $A(t) = 1000 + 1500t - 800t^2 + 100t^3$ points, where t represents the number of days into the week; t ranges from 0 at the beginning of the week to 5 at end of the week. Graph A, and use your graph to estimate how fast the Average was growing halfway through the first day of the week.

66. *Market Average*   Referring to the Joe Downs Average in the previous exercise: Graph A, and use the graph to estimate how fast the Average was increasing at the end of the third day.

67. *Biology—Reproduction*   The Verhulst model for population growth specifies the reproductive rate of an organism as a function of the total population according to the following formula:

$$R(p) = \frac{r}{1 + kp}.$$

Here, p is the total population in thousands of organisms, r and k are constants that depend on the particular circumstances and the organism being studied, and $R(p)$ is the reproduction rate in thousands of organisms per hour.* Graph this function for $0 \le p \le 10$, given that $k = 0.125$ and $r = 45$, and use your graph to find an approximate value of $R'(4)$. Interpret the result.

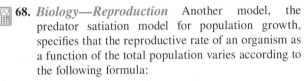 **68. Biology—Reproduction** Another model, the predator satiation model for population growth, specifies that the reproductive rate of an organism as a function of the total population varies according to the following formula:

$$R(p) = \frac{rp}{1 + kp}.$$

Here, p is the total population in thousands of organisms, r and k are constants that depend on the particular circumstances and the organism being studied, and $R(p)$ is the reproduction rate in new organisms per hour.* Graph this function for $0 \le p \le 10$, given that $k = 0.2$ and $r = 0.08$, and use your graph to find an approximate value of $R'(2)$. Interpret the result.

69. The Theory of Relativity In science fiction terminology, a speed of *warp 1* is the speed of light—about 3×10^8 meters per second. (Thus, for instance, a speed of warp 0.8 corresponds to 80% of the speed of light—about 2.4×10^8 meters per second.) According to Einstein's Special Theory of Relativity, a moving object appears to get shorter to a stationary observer as its speed approaches that of light. If a rocket ship whose length is l_0 meters at rest travels at a speed of warp p, its length in meters, as measured by a stationary observer, will be given by

$$l(p) = l_0\sqrt{1 - p^2}, \text{ with domain } [0, 1].$$

Graph l as a function of p for a 100-meter rocket ship ($l_0 = 100$), and use your graph to estimate $l(0.95)$ and $l'(0.95)$. What do these figures tell you?

70. Newton's Law of Gravity The gravitational force exerted on a particle with mass m by another particle with mass M is given by the following function of distance:

$$F(r) = G\frac{Mm}{r^2}, \text{ with domain } (0, +\infty).$$

Here, r is the distance between the two particles in meters, the masses M and m are given in kilograms, $G \approx 0.0000000000667$, or 6.67×10^{-11}, and the resulting force is given in newtons. Graph F as a function of r, given that M and m are both 1,000 kilograms, and use your graph to estimate $F(10)$ and $F'(10)$. What do these figures tell you?

Exercises 71–74 are based on logarithmic and exponential functions and also require the use of a graphing calculator or graphing computer software.

71. Sales Weekly sales of a new brand of sneakers are given by

$$S(t) = 200 - 150e^{-t/10},$$

pairs sold per week, *where t is the number of weeks* since the introduction of the brand. Graph S as a function of t, and use your graph to estimate $S(5)$ and estimate $S'(5)$. Interpret your answers.

72. Sales Weekly sales of an old brand of TV are given by

$$S(t) = 100e^{-t/5},$$

sets per week where t is the number of weeks after the introduction of a competing brand. Graph S as a function of t, and use your graph to estimate $S(5)$ and $S'(5)$. Interpret your answers.

73. Logistic Growth in Demand The demand for a new product can be modeled by a **logistic** curve of the form

$$q(t) = \frac{N}{1 + ke^{-rt}},$$

where $q(t)$ is the total number of units sold t months after the introduction of the new product, and N, k, and r are constants that depend on the product and the market. Assume that the demand for video game units

---

▼ *Source: *Mathematics in Medicine and the Life Sciences* by F. C. Hoppensteadt and C. S. Peskin (New York: Springer-Verlag, 1992), pp. 20–22.

is determined by the above formula, with $N = 10{,}000$, $k = 0.5$, and $r = 0.4$. Graph the demand curve using $0 \leq t \leq 10$ and $5{,}000 \leq q \leq 10{,}000$, use your graph to estimate $q(2)$ and $q'(2)$, and interpret the results.

 74. *Information Highway*   The amount of information transmitted each month on the National Science Foundation's Internet network for the years 1988–1994 can be modeled by the equation

$$q(t) = \frac{2e^{0.69t}}{3 + 1.5e^{-0.4t}},$$

where q is the amount of information transmitted each month in billions of data packets and t is the number of years since the start of 1988.* Graph q as a function of t with $0 \leq t \leq 4$ and $0 \leq q \leq 40$, and use your graph to estimate the number of data packets transmitted during the first month of 1990 and also the rate at which this number was increasing.

COMMUNICATION AND REASONING EXERCISES

75. If the derivative of f is zero at a point, what do you know about the graph of f near that point?

76. Sketch the graph of a function whose derivative never exceeds 1.

77. Sketch the graph of a function whose derivative exceeds 1 at every point.

78. Sketch the graph of a function whose derivative is exactly 1 at every point.

79. If the derivative of f is always positive, what do you know about the graph of f?

80. If the derivative of f is increasing, what do you know about the graph of f?

▶ ═══ **3.3** LIMITS AND CONTINUITY

The derivative is defined using a limit, and it is now time to say what that means. It is possible to speak of limits by themselves, rather than in the context of the derivative. The story of limits is a long one that we will try to make as concise as possible.

EVALUATING LIMITS NUMERICALLY

Start with a simple example: Look at the function $f(x) = 2 + x$ and ask yourself: "What happens to $f(x)$ as x approaches 3?" The following table shows the value of $f(x)$ for values of x close to, and on either side of 3:

x approaching 3 from the left $\rightarrow$ $\leftarrow$ x approaching 3 from the right

x	2.9	2.99	2.999	2.9999	3	3.0001	3.001	3.01	3.1
$f(x) = 2 + x$	4.9	4.99	4.999	4.9999		5.0001	5.001	5.01	5.1

We have left the entry under 3 blank to emphasize that when we are calculating the limit of $f(x)$ as x *approaches* 3, *we are not interested in its value when x equals* 3.

▼ * This is the authors' model, based on figures published in *The New York Times*, Nov. 3, 1993.

Notice from the table that the closer x gets to 3 from either side, the closer $f(x)$ gets to 5. We write this as

$$\lim_{x \to 3} f(x) = 5,$$

meaning

the limit of $f(x)$, as x approaches 3, equals 5.

Q Why all the fuss? Can't we simply put $x = 3$ and avoid having to use a table?

A This happens to work for *some* functions, but not for *all* functions. The following example illustrates this point.

▼   **EXAMPLE 1**

Use a table to evaluate $\lim_{x \to 2} \dfrac{x^3 - 8}{x - 2}$.

SOLUTION   Notice that we cannot simply substitute $x = 2$, because the function $f(x) = \dfrac{x^3 - 8}{x - 2}$ is not defined at $x = 2$. As above, we can use a table of values, with x approaching 2 from either side.

x approaching 2 from the left $\to$ $\leftarrow$ x approaching 2 from the right

x	1.9	1.99	1.999	1.9999	2	2.0001	2.001	2.01	2.1
$f(x) = \dfrac{x^3 - 8}{x - 2}$	11.41	11.9401	11.9940	11.9994		12.0006	12.0060	12.0601	12.61

We notice that as x approaches 2 from either side, $f(x)$ approaches 12. This suggests that the limit is 12, and we write

$$\lim_{x \to 2} \frac{x^3 - 8}{x - 2} = 12.$$

Before we go on... Although the table *suggests* that the limit is 12, it by no means establishes that fact conclusively. It is *conceivable* (though not in fact the case here) that putting $x = 1.99999987$ will result in $f(x) = 426$. Using a table can only suggest a value for the limit. We shall talk soon about algebraic techniques for finding limits exactly.

   See "Evaluating a Function Using a Table" in the Evaluating Functions section of Appendix B for a quick and easy method of obtaining tables such as the one above on your graphing calculator.

Before we continue, let us make a more formal definition.

DEFINITION OF A LIMIT

If $f(x)$ approaches the number L as x approaches (but is not equal to) a from both sides, then we say that the **limit** of $f(x)$ as $x \to a$ ("x approaches a") is L. We write

$$\lim_{x \to a} f(x) = L$$

or

$$f(x) \to L \quad \text{as} \quad x \to a.$$

If $f(x)$ *fails* to approach *a single fixed number* as x approaches a from both sides, then we say that $f(x)$ **has no limit** as $x \to a$, or

$$\lim_{x \to a} f(x) \quad \text{does not exist.}$$

▶ NOTES

1. It is important that $f(x)$ approach the same number as x approaches a from either side. For instance, if $f(x)$ approaches 5 for $x = 1.9$, 1.99, 1.999, . . ., but approaches 4 for $x = 2.1$, 2.01, 2.001, . . ., then the limit as $x \to 2$ does not exist.

2. It may happen that $f(x)$ does not approach any fixed number at all as $x \to a$ from either side. In this case, we also say that the limit does not exist.

3. We are deliberately suppressing the exact definition of "approaches," and instead shall trust to your intuition. The following phrasing of the definition of the limit is closer to the one used by mathematicians: *we can make $f(x)$ be as close to L as we like by making x be sufficiently close to a.* ◀

EVALUATING LIMITS GRAPHICALLY

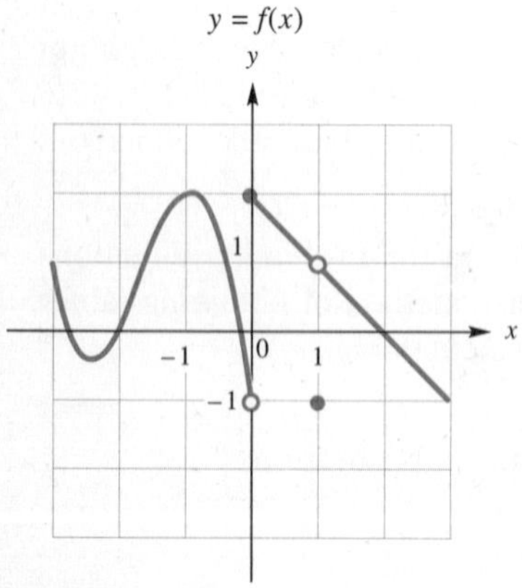

$y = f(x)$

FIGURE 1

▼ **EXAMPLE 2**

The graph of the function f is shown in Figure 1.
From the graph, analyze the following limits.

(a) $\lim_{x \to -2} f(x)$ **(b)** $\lim_{x \to 0} f(x)$ **(c)** $\lim_{x \to 1} f(x)$

SOLUTION   Since we are given only a graph of f, we must analyze these limits graphically.

(a)   Suppose that we had Figure 1 drawn on a graphing calculator.
   Graphing calculators are usually equipped with a "trace" feature that allows us to move a cursor along the graph and read the coordinates of points as we go. If we started at a point on the graph to the left of $x = -2$ and moved the cursor along the graph to the right, we could see numerically what the limit of $f(x)$ is. If we just have the graph drawn on paper, we can place a pencil

point on the graph to the left of $x = -2$ and move it along the curve to the right so that the x-coordinate approaches -2. (See Figure 2.)

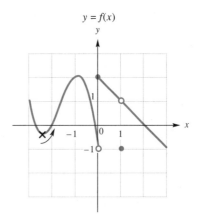

FIGURE 2

As the x-coordinate of the point approaches -2, we can see from the graph that the y-coordinate will approach 0. Similarly, if we place our pencil point to the right of $x = -2$ and move it along the graph to the left, the y-coordinate will again approach 0 (Figure 3).

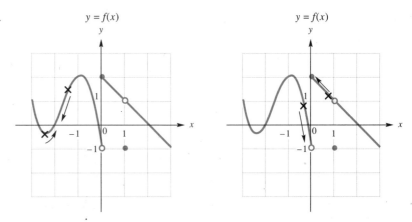

FIGURE 3   **FIGURE 4**

Therefore, as x approaches -2 from either side, $f(x)$ approaches 0, so

$$\lim_{x \to -2} f(x) = 0.$$

(b)   Here we move our pencil point toward $x = 0$. If we start from the left of $x = 0$ and approach 0 by moving right, the y-coordinate approaches -1 (Figure 4).

However, if we start from the right of $x = 0$ and approach 0 by moving left, the y-coordinate approaches 2 (see Figure 4 again). So there appear to be two *different* limits: the limit as we approach 0 from the left and the limit as we approach 0 from the right. We write

$$\lim_{x \to 0^-} f(x) = -1,$$

read "the limit as x approaches 0 from the left is -1," and

$$\lim_{x \to 0^+} f(x) = 2,$$

read "the limit as x approaches 0 from the right is 2." These are called the **one-sided limits** of $f(x)$. In order for the **two-sided limit** to exist (the one we are asked to compute), the two one-sided limits must be equal. Since they are not, we conclude that

$$\lim_{x \to 0} f(x) \text{ does not exist.}$$

Limits may or may not exist. In this case there is a "break" in the graph, and we say that the function is **discontinuous** there. We shall return to this concept in more detail later.

(c) Once more, we think about a pencil point (or cursor on a graphing calculator) moving along the graph with x-coordinate approaching $x = 1$ from the left and from the right (Figure 5).

  As the x-coordinate of the point approaches 1 from either side, the y-coordinate approaches 1 also. Therefore,

$$\lim_{x \to 1} f(x) = 1.$$

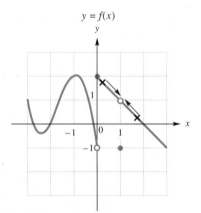

FIGURE 5

Before we go on... Notice that $f(1) = -1$. (Why?) Thus, $\lim\limits_{x \to 1} f(x) \neq f(1)$. In other words, the limit of $f(x)$ as x *approaches* 1 is not the same as the value of f at $x = 1$. Always bear in mind that when we evaluate a limit as $x \to a$ *we do not care about the value of the function at $x = a$. We only care about the value of $f(x)$ as x approaches a.* In other words, $f(a)$ may or may not equal $\lim\limits_{x \to a} f(x)$. Part (a) gives us an example where the limit and the value of the function are equal:

$$\lim_{x \to -2} f(x) = 0 = f(-2).$$

We can summarize the graphical method we used in this example as follows.

EVALUATING LIMITS GRAPHICALLY

To decide whether $\lim_{x \to a} f(x)$ exists, and to find its value if it does:

1. Draw the graph of $f(x)$ either by hand or using a graphing calculator.
2. Position your pencil point (or the graphing calculator "trace" cursor) on a point of the graph to the right of $x = a$.
3. Move the point *along the graph* toward $x = a$ from the right, and read the y-coordinate as you go. The value the y-coordinate approaches (if any) is then the limit $\lim_{x \to a^+} f(x)$.
4. Repeat Steps 2 and 3, but this time starting from a point on the graph to the left of $x = a$, and approach $x = a$ along the graph from the left. The value the y-coordinate approaches (if any) is then $\lim_{x \to a^-} f(x)$.
5. If the left and right limits both exist and have the same value L, then

$$\lim_{x \to a} f(x) \text{ exists and equals } L.$$

EVALUATING LIMITS ALGEBRAICALLY

In parts (a) and (c) of Example 2, we saw an example where $\lim_{x \to a} f(x) = f(a)$, and another where $\lim_{x \to a} f(x) \ne f(a)$. In the first case, the graph had no break; in the second, it did have a break. This motivates the following definition.

CONTINUOUS FUNCTION

A function f is **continuous at a** if $\lim_{x \to a} f(x) = f(a)$. The function f is said to be **continuous on its domain** if it is continuous at each point in its domain. If f is not continuous at a particular a, we say that f is **discontinuous** at a or that f has a **discontinuity** at a.

Besides describing a particularly nice kind of function, continuity is useful for evaluating many limits. If we know that a function is continuous at a point a, then we can compute $\lim_{x \to a} f(x)$ by simply substituting $x = a$ into $f(x)$. In order to use this, we need to know some continuous functions. Luckily, there is a large class of functions that are known to be continuous on their domains: roughly speaking, those that are specified by a single formula.

We can be more precise. A **closed-form function** is any function that can be obtained by combining constants, powers of x, exponential functions, radicals, logarithms (and some other functions we shall not encounter in this text) into a *single* mathematical expression by means of the usual arithmetic operations and composition of functions. Examples of closed-form functions are

$$3x^2 - x + 1, \quad \frac{\sqrt{x^2 - 1}}{6x - 1}, \quad e^{-\frac{4x^2 - 1}{x}}, \quad \sqrt{\log_3(x^2 - 1)}.$$

They can be as complicated as you like. The following is *not* a closed-form function.

$$f(x) = \begin{cases} -1 & \text{if } x \leq -1 \\ x^2 + x & \text{if } -1 \leq x \leq 1 \\ 2 - x & \text{if } 1 < x \leq 2 \end{cases}$$

The reason for this is that $f(x)$ is not specified by a *single* mathematical expression. What is nice about closed-form functions is the following.

CONTINUITY OF CLOSED-FORM FUNCTIONS

Every closed-form function is continuous on its domain. Thus, *the limit of a closed-form function at a point on its domain can be obtained by substitution.*

For example, if $f(h) = \sqrt{h^2 + 2h + 2}$, then f is continuous on its domain, and so

$$\lim_{h \to 0} f(h) = f(0) = \sqrt{2},$$

since 0 is in the domain of f.

▶ NOTES

1. The reason we refer to such functions as continuous on their domain is that their graphs do not break anywhere on their domain (although the graph of a continuous function may break if its domain is broken; see Example 3 below).
2. Mathematics majors spend a great deal of time proving results such as this. We shall ask you to accept it without proof. ◀

In the definition of the derivative, we must take the limit of the difference quotient $(f(x + h) - f(x))/h$ as $h \to 0$. Although the difference quotients we encounter are usually closed-form functions, we cannot evaluate them by substitution because of the following catch: Since h appears in the denominator, $h = 0$ is not in the domain of the difference quotient, and so we cannot evaluate the limit by substitution. However—and this is the the key to finding such limits—some preliminary algebraic simplification may allow us to obtain a closed-form function with $h = 0$ in its domain. We can find the limit by substituting $h = 0$ in the new function. Many limits can be computed by this technique.

▼ **EXAMPLE 3**

Evaluate $\lim\limits_{x \to 2} \dfrac{x^3 - 8}{x - 2}$.

SOLUTION We found this limit in Example 1 using a numerical approach, but we shall now find it algebraically. Notice that $x = 2$ is not in the domain of $f(x) = (x^3 - 8)/(x - 2)$, so we cannot obtain the limit by substitution. We first simplify $f(x)$ to obtain a new function with $x = 2$ in its domain. To do this, notice first that the numerator can be factored as

$$x^3 - 8 = (x - 2)(x^2 + 2x + 4).$$

Thus,

$$\frac{x^3 - 8}{x - 2} = \frac{(x - 2)(x^2 + 2x + 4)}{x - 2}$$
$$= x^2 + 2x + 4.$$

Once we have canceled the offending $(x - 2)$ in the denominator, we are left with a closed-form function *with 2 in its domain.* Thus,

$$\lim\limits_{x \to 2} \frac{x^3 - 8}{x - 2} = \lim\limits_{x \to 2} (x^2 + 2x + 4)$$
$$= 2^2 + 2(2) + 4 = 12.$$

This confirms the answer we found in Example 1.

Before we go on... If the given function fails to simplify, you can always evaluate the limit numerically. It may very well be that the limit does not exist in that case.

▼ **EXAMPLE 4**

Calculate

$$\lim\limits_{h \to 0} \frac{(1 + h)^2 - 1}{h}.$$

SOLUTION The given function is not defined when $h = 0$. So, we try to simplify the expression.

$$\lim\limits_{h \to 0} \frac{(1 + h)^2 - 1}{h} = \lim\limits_{h \to 0} \frac{1 + 2h + h^2 - 1}{h}$$
$$= \lim\limits_{h \to 0} \frac{h(2 + h)}{h}$$
$$= \lim\limits_{h \to 0} (2 + h) = 2$$

Before we go on... There is something suspicious about this example and the last one. If 0 was not in the domain before simplifying, but was in the domain after simplifying, we must have changed the function. In fact, when we say that

$$\frac{(1 + h)^2 - 1}{h} = 2 + h,$$

we are lying a little bit. What we really mean is that these two expressions are equal *where both are defined*. The functions $f(h) = ((1 + h)^2 - 1)/h$ and $g(h) = 2 + h$ are different functions. However, the only difference is that $h = 0$ is in the domain of g and is not in the domain of f. Since $\lim_{h \to 0} f(h)$ explicitly *ignores* any value that f may have at 0, this does not matter. Formally, we can say the following.

EQUALITY OF LIMITS

If $f(x) = g(x)$ as long as $x \neq a$, then

$$\lim_{x \to a} f(x) = \lim_{x \to a} g(x).$$

▼ **EXAMPLE 5**

If $f(x) = x^3$, find $f'(x)$.

SOLUTION

$$
\begin{aligned}
f'(x) &= \lim_{h \to 0} \frac{f(x + h) - f(x)}{h} \\
&= \lim_{h \to 0} \frac{(x + h)^3 - x^3}{h} \\
&= \lim_{h \to 0} \frac{x^3 + 3x^2 h + 3xh^2 + h^3 - x^3}{h} \\
&= \lim_{h \to 0} \frac{3x^2 h + 3xh^2 + h^3}{h} \\
&= \lim_{h \to 0} \frac{h(3x^2 + 3xh + h^2)}{h} \\
&= \lim_{h \to 0} (3x^2 + 3xh + h^2) \\
&= 3x^2 + 3x(0) + 0^2 \\
&= 3x^2
\end{aligned}
$$

Before we go on... When computing derivatives from the definition, as we just did, we always start by writing the definition. This helps get us started in the right direction.

▼ **EXAMPLE 6**

If $f(x) = \dfrac{1}{x}$, find $f'(x)$.

SOLUTION

$$f'(x) = \lim_{h \to 0} \frac{f(x + h) - f(x)}{h}$$

$$= \lim_{h \to 0} \frac{\left[\dfrac{1}{x + h} - \dfrac{1}{x}\right]}{h}$$

$$= \lim_{h \to 0} \frac{\left[\dfrac{x - (x + h)}{(x + h)x}\right]}{h}$$

$$= \lim_{h \to 0} \frac{x - x - h}{h(x + h)x}$$

$$= \lim_{h \to 0} \frac{-h}{h(x + h)x}$$

$$= \lim_{h \to 0} \frac{-1}{(x + h)x}$$

$$= -\frac{1}{x^2}$$

▼ **EXAMPLE 7**

If $f(x) = \sqrt{x}$, find $f'(x)$.

SOLUTION

$$f'(x) = \lim_{h \to 0} \frac{f(x + h) - f(x)}{h}$$

$$= \lim_{h \to 0} \frac{\sqrt{x + h} - \sqrt{x}}{h}$$

Now here we encounter a slight problem, since there is no way to "expand" the numerator as we did in Example 5, or combine terms as we did in Example 6. However, there is a technique that you may have seen before called "rationalizing the numerator": multiply top and bottom by the "conjugate," $\sqrt{x + h} + \sqrt{x}$, of the numerator.

$$f'(x) = \lim_{h \to 0} \frac{(\sqrt{x + h} - \sqrt{x})(\sqrt{x + h} + \sqrt{x})}{h(\sqrt{x + h} + \sqrt{x})}$$

The numerator is of the form $(a - b)(a + b) = a^2 - b^2$, so we get

$$f'(x) = \lim_{h \to 0} \frac{(\sqrt{x + h})^2 - (\sqrt{x})^2}{h(\sqrt{x + h} + \sqrt{x})}$$

$$= \lim_{h \to 0} \frac{(x + h) - x}{h(\sqrt{x + h} + \sqrt{x})}$$

$$= \lim_{h \to 0} \frac{h}{h(\sqrt{x + h} + \sqrt{x})}$$

$$= \lim_{h \to 0} \frac{1}{\sqrt{x + h} + \sqrt{x}}$$

$$= \frac{1}{\sqrt{x} + \sqrt{x}}$$

$$= \frac{1}{2\sqrt{x}}.$$

Obviously, you don't want to do too many of these by hand. In the next section we shall start to talk about some shortcuts for finding derivatives, and we will continue this discussion in the next chapter.

▶ ___ **3.3 EXERCISES**

Compute the limits in Exercises 1–4 numerically.

1. $\lim\limits_{x \to 0} \dfrac{x^2}{x + 1}$

2. $\lim\limits_{x \to 0} \dfrac{x - 3}{x - 1}$

3. $\lim\limits_{x \to 2} \dfrac{x^2 - 4}{x - 2}$

4. $\lim\limits_{x \to -1} \dfrac{x^2 + 2x + 1}{x + 1}$

In each of Exercises 5–20, the graph of f is given. Use the graph to compute the quantities asked for. If a particular quantity fails to exist, say why.

5. (a) $\lim\limits_{x \to 1} f(x)$ **(b)** $\lim\limits_{x \to -1} f(x)$

6. (a) $\lim\limits_{x \to -1} f(x)$ **(b)** $\lim\limits_{x \to 1} f(x)$

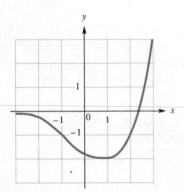

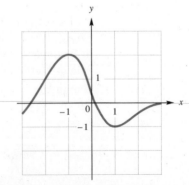

7. (a) $\lim\limits_{x\to 0} f(x)$ **(b)** $\lim\limits_{x\to 1} f(x)$

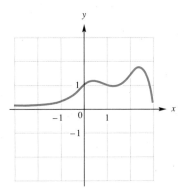

8. (a) $\lim\limits_{x\to -1} f(x)$ **(b)** $\lim\limits_{x\to 1} f(x)$

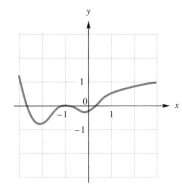

9. (a) $\lim\limits_{x\to 2} f(x)$ **(b)** $\lim\limits_{x\to 0^+} f(x)$
  (c) $\lim\limits_{x\to 0^-} f(x)$ **(d)** $\lim\limits_{x\to 0} f(x)$ **(e)** $f(0)$

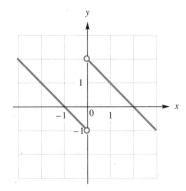

10. (a) $\lim\limits_{x\to 3} f(x)$ **(b)** $\lim\limits_{x\to 1^+} f(x)$
  (c) $\lim\limits_{x\to 1^-} f(x)$ **(d)** $\lim\limits_{x\to 1} f(x)$ **(e)** $f(1)$

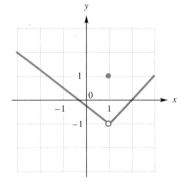

11. (a) $\lim\limits_{x\to -2} f(x)$ **(b)** $\lim\limits_{x\to -1^+} f(x)$
  (c) $\lim\limits_{x\to -1^-} f(x)$ **(d)** $\lim\limits_{x\to -1} f(x)$ **(e)** $f(-1)$

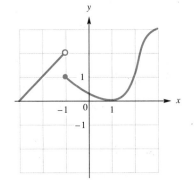

12. (a) $\lim\limits_{x\to -1} f(x)$ **(b)** $\lim\limits_{x\to 0^+} f(x)$
  (c) $\lim\limits_{x\to 0^-} f(x)$ **(d)** $\lim\limits_{x\to 0} f(x)$ **(e)** $f(0)$

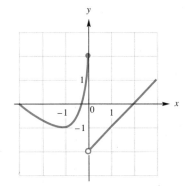

13. (a) $\lim\limits_{x \to -1} f(x)$ **(b)** $\lim\limits_{x \to 0^+} f(x)$
(c) $\lim\limits_{x \to 0^-} f(x)$ **(d)** $\lim\limits_{x \to 0} f(x)$ **(e)** $f(0)$

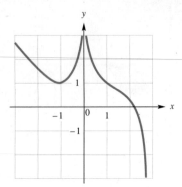

14. (a) $\lim\limits_{x \to 1} f(x)$ **(b)** $\lim\limits_{x \to 0^+} f(x)$
(c) $\lim\limits_{x \to 0^-} f(x)$ **(d)** $\lim\limits_{x \to 0} f(x)$ **(e)** $f(0)$

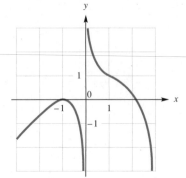

15. (a) $\lim\limits_{x \to -2} f(x)$ **(b)** $\lim\limits_{x \to 0^+} f(x)$
(c) $\lim\limits_{x \to 0^-} f(x)$ **(d)** $\lim\limits_{x \to 0} f(x)$
(e) $f(0)$ **(f)** $f(-2)$

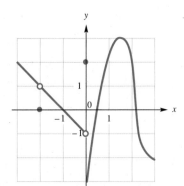

16. (a) $\lim\limits_{x \to 0^-} f(x)$ **(b)** $\lim\limits_{x \to 2^+} f(x)$
(c) $\lim\limits_{x \to 0} f(x)$ **(d)** $\lim\limits_{x \to 2} f(x)$
(e) $f(0)$ **(f)** $f(2)$

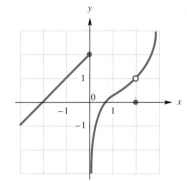

17. (a) $\lim\limits_{x \to 0} f(x)$ **(b)** $\lim\limits_{x \to 1} f(x)$ **(c)** $f(0)$

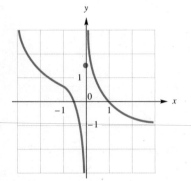

18. (a) $\lim\limits_{x \to 0} f(x)$ **(b)** $\lim\limits_{x \to 1} f(x)$ **(c)** $f(0)$

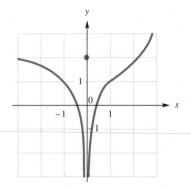

19. (a) $\lim\limits_{x \to 1} f(x)$ **(b)** $\lim\limits_{x \to 2} f(x)$

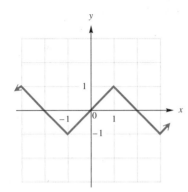

20. (a) $\lim\limits_{x \to 1} f(x)$ **(b)** $\lim\limits_{x \to 2} f(x)$

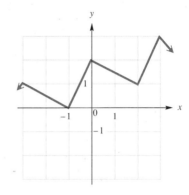

In Exercises 21–36, determine whether each of the functions given in Exercises 5–20 is continuous on its domain. If a particular function is not continuous on its domain, say why.

21. Graph from Exercise 5
22. Graph from Exercise 6
23. Graph from Exercise 7
24. Graph from Exercise 8
25. Graph from Exercise 9
26. Graph from Exercise 10
27. Graph from Exercise 11
28. Graph from Exercise 12
29. Graph from Exercise 13
30. Graph from Exercise 14
31. Graph from Exercise 15
32. Graph from Exercise 16
33. Graph from Exercise 17
34. Graph from Exercise 18
35. Graph from Exercise 19
36. Graph from Exercise 20

Calculate the limits in Exercises 37–44 mentally.

37. $\lim\limits_{x \to 0} (x + 1)$

38. $\lim\limits_{x \to 0} (2x - 4)$

39. $\lim\limits_{x \to 2} \dfrac{2 + x}{x}$

40. $\lim\limits_{x \to -1} \dfrac{4x^2 + 1}{x}$

41. $\lim\limits_{x \to -1} \dfrac{x + 1}{x}$

42. $\lim\limits_{x \to 4} (x + \sqrt{x})$

43. $\lim\limits_{x \to 8} (x - \sqrt[3]{x})$

44. $\lim\limits_{x \to 1} \dfrac{x - 2}{x + 1}$

Calculate each of the limits in Exercises 45–54.

45. $\lim\limits_{h \to 1} (h^2 + 2h + 1)$

46. $\lim\limits_{h \to 0} (h^3 - 4)$

47. $\lim\limits_{h \to 3} 2$

48. $\lim\limits_{h \to 0} -5$

49. $\lim\limits_{h \to 0} \dfrac{h^2}{h + h^2}$

50. $\lim\limits_{h \to 0} \dfrac{h^2 + h}{h^2 + 2h}$

51. $\lim\limits_{x \to 1} \dfrac{x^2 - 2x + 1}{x^2 - x}$

52. $\lim\limits_{x \to -1} \dfrac{x^2 + 3x + 2}{x^2 + x}$

53. $\lim\limits_{x \to 2} \dfrac{x^3 - 8}{x - 2}$

54. $\lim\limits_{x \to -2} \dfrac{x^3 + 8}{x^2 + 3x + 2}$

In each of Exercises 55–78, use the definition to calculate the derivative of the given function.

55. $f(x) = -14$
56. $f(x) = 5$
57. $f(x) = 2x - 3$
58. $f(x) = -3x + 5$
59. $g(x) = -4x - 1$
60. $g(x) = 10x - 100$
61. $g(x) = x^2 - 2x$
62. $g(x) = 3x^2 + 1$
63. $h(x) = -5x^2 + 2x - 1$

64. $h(x) = -3x^2 - x + 5$

65. $f(t) = t^3 + t$

66. $f(t) = 2t^3 - t^2$

67. $g(t) = t^4 - t$

68. $g(t) = 3t^4 + 2t^2$

69. $h(t) = \frac{6}{t}$

70. $h(t) = -\frac{1}{t}$

71. $f(x) = x + \frac{1}{x}$

72. $f(x) = 6 - \frac{6}{x}$

73. $f(x) = \dfrac{1}{x - 2}$

74. $f(x) = \dfrac{1}{2x + 1}$

75. $f(x) = \sqrt{x + 1}$

76. $f(x) = \sqrt{x - 2}$

77. $g(t) = \dfrac{1}{\sqrt{t}}$

78. $g(t) = t + \dfrac{1}{\sqrt{t}}$

COMMUNICATION AND REASONING EXERCISES

79. Describe the three methods of evaluating limits discussed in this section. Give at least one disadvantage of each.

80. Choose one of the limits in Exercises 45–54 and calculate it in three different ways.

81. What is wrong with the following statement? "If $f(a)$ is defined, then $\lim_{x \to a} f(x)$ exists and equals $f(a)$."

82. What is wrong with the following statement?

"$\lim_{x \to 4} \dfrac{\sqrt{x} - 2}{x - 4}$ does not exist, since substituting $x = 4$ yields $0/0$, which is undefined."

83. Give an example of a function f specified by means of algebraic formulas so that f is not continuous at $x = 2$.

84. Give an example of a function f with $\lim_{x \to 1} f(x) = f(2)$.

▶ ═══ **3.4** DERIVATIVES OF POWERS AND POLYNOMIALS

So far, we have calculated the derivatives of functions using the definition of the derivative as a limit. These calculations are tedious, so it would be nice to have a quicker way of doing them. In this section, we shall see how to quickly calculate the derivatives of many functions. By the end of the next chapter, we shall be able to find the derivative of almost any function we can write.

First, we review notation we mentioned in Section 1.

Differential Notation is based on an abbreviation for the phrase "the derivative with respect to x." For example, we learned in the preceding section (Example 5) that if $f(x) = x^3$, then $f'(x) = 3x^2$. When we say "$f'(x) = 3x^2$," we mean the following:

"The derivative with respect to x of x^3 equals $3x^2$."

You may wonder why we sneaked in the words "with respect to x." All this means is that the variable of the function is x, and nothing else.* Since

---

▼ * This may seem odd in the case of $f(x) = x^3$, since there are no other variables to worry about. But later, we shall see more complicated expressions than x^3 which involve variables other than x, and it will become necessary to specitfy just what the variable of the function is. This is the same reason that we write "$f(x) = x^3$" rather than just "$f = x^3$."

we shall be using the phrase "the derivative with respect to x" often, we use the following abbreviation:

DERIVATIVE WITH RESPECT TO x

$\dfrac{d}{dx}$ means "the derivative with respect to x."

Thus, for example, the statement

"the derivative with respect to x of x^3 is $3x^2$"

can now be written more compactly as:

$$\frac{d}{dx}(x^3) = 3x^2.$$

Similarly, by Example 7 of the preceding section,

the derivative with respect to x of $\sqrt{x}$ is $\dfrac{1}{2\sqrt{x}}$,

or

$$\frac{d}{dx}\left(\sqrt{x}\right) = \frac{1}{2\sqrt{x}}.$$

Sometimes, we don't want to spell out the function each time we use the phrase "d/dx." For instance, we might have $y = x^{100}$, and wish to talk about "the derivative (with respect to x) of y" rather than "the derivative (with respect to x) of x^{100}." Thus we can write:

$$\frac{d}{dx}(y)$$

or, more compactly,

$$\frac{dy}{dx}.$$

▶ **NOTE**   This notation is a little misleading; we must not think of dy/dx as a ratio of two quantities dy and dx, even though it looks like a ratio. Instead, dy/dx is the *limit* of a ratio, as we saw in Section 1:

$$\frac{dy}{dx} = \lim_{\triangle x \to 0} \frac{\triangle y}{\triangle x}. \quad ◀$$

Now we can state the first derivative rule.

POWER RULE

If n is any constant, then the derivative of $f(x) = x^n$ is

$$f'(x) = nx^{n-1}.$$

In differential notation, we write

$$\frac{d}{dx}(x^n) = nx^{n-1}.$$

We shall give reasons for believing this in a moment, but first let us do several quick examples.

▼   **EXAMPLE 1**

(a) $\dfrac{d}{dx}(x^3) = 3x^2$

(b) $\dfrac{d}{dx}(x^2) = 2x^1 = 2x$

(c) $\dfrac{d}{dx}\left(\dfrac{1}{x}\right) = \dfrac{d}{dx}(x^{-1}) = -x^{-2} = -\dfrac{1}{x^2}$

(d) $\dfrac{d}{dx}\sqrt{x} = \dfrac{d}{dx}(x^{1/2}) = \dfrac{1}{2}x^{-1/2} = \dfrac{1}{2\sqrt{x}}$

(e)   If $f(x) = 1/x^2$, then $f'(x) = -2x^{-3} = -2/x^3$.

(f)   If $f(x) = x$, then $f'(x) = 1x^0 = 1$.

(g)   If $f(x) = 1$, then $f'(x) = 0$ (think of $1 = x^0$).

Before we go on... We did (a), (c) and (d) using the definition of the derivative in Examples 5, 6 and 7 of the previous section and we did (b) in Section 2. You should think about (f) and (g) graphically to see why these are obvious.

Some of the derivatives in the last example are very useful to remember, so we summarize them in a table.

TABLE OF DERIVATIVE FORMULAS

$f(x)$	x^n	x	1	$\dfrac{1}{x}$	$\dfrac{1}{x^2}$	$\sqrt{x}$
$f'(x)$	nx^{n-1}	1	0	$-\dfrac{1}{x^2}$	$-\dfrac{2}{x^3}$	$\dfrac{1}{2\sqrt{x}}$

We suggest that you add to this table as you pick up more information. It is *extremely* helpful to remember the derivatives of common functions such as $1/x$ and $\sqrt{x}$.

▼ **EXAMPLE 2**

If $f(x) = x^{100}$, find $f'(0), f'(-1)$, and $f'(a)$.

SOLUTION $f'(x) = 100x^{99}$ by the power rule, so we get the required derivatives by substituting:

$$f'(0) = 100 \cdot 0^{99} = 0;$$
$$f'(1) = 100 \cdot 1^{99} = 100;$$
$$f'(a) = 100a^{99}.$$

Now let us see why the power rule is true. We have already checked the rule for x^3 and x^2. What follows is a rather slick proof that works for *any* positive integer. Don't worry if you might not think of doing it this way— neither did we at first! But mathematicians love to find the easiest or cleverest way of doing something. To follow this proof, you must recognize a nice little algebraic fact. First look at these identities.

$$a^1 - b^1 = (a - b)$$
$$a^2 - b^2 = (a - b)(a + b)$$
$$a^3 - b^3 = (a - b)(a^2 + ab + b^2)$$
$$a^4 - b^4 = (a - b)(a^3 + a^2b + ab^2 + b^2)$$
$$\cdots$$

(Use the distributive law to expand the right-hand side in each case.)

These examples generalize to give us the following formula.

DIFFERENCE OF TWO nTH POWERS

If a and b are real numbers, and n is a positive integer, then

$$a^n - b^n = (a - b)(a^{n-1} + a^{n-2}b + a^{n-3}b^2 + \ldots + ab^{n-2} + b^{n-1}).$$

PROOF OF THE POWER RULE FOR POSITIVE INTEGER POWERS

Write $f(x) = x^n$. Then

$$f'(x) = \lim_{h \to 0} \frac{f(x + h) - f(x)}{h}$$

$$= \lim_{h \to 0} \frac{(x + h)^n - x^n}{h}.$$

We now rewrite the numerator using the identity above with a replaced by the quantity $(x + h)$ and b by x, getting

$$= \lim_{h \to 0} \frac{[(x + h) - x][(x + h)^{n-1} + (x + h)^{n-2}x + \ldots + x^{n-1}]}{h}$$

$$= \lim_{h \to 0} \frac{h[(x + h)^{n-1} + (x + h)^{n-2}x + \ldots + x^{n-1}]}{h}$$

$$= \lim_{h \to 0} [(x + h)^{n-1} + (x + h)^{n-2}x + \ldots + x^{n-1}].$$

Now that we have canceled the h from the denominator, we have a closed-form function defined at $h = 0$, so we can evaluate the limit by substituting $h = 0$. Each term in the sum becomes x^{n-1} if we do that, so we get

$$f'(x) = x^{n-1} + x^{n-1} + \ldots + x^{n-1}.$$

There are n terms, so $f'(x) = nx^{n-1}$, completing the proof.

▶ **NOTE**   We have now proven the power rule only for positive integer powers. We have shown that the power rule also works in a few other specific cases, such as x^{-1} and $x^{1/2}$. In the next chapter we shall see why the power rule works for negative n, rational n (that is, $n = p/q$ with p and q integers) and irrational n as well. ◀

Never again will we have to evaluate a limit to find the derivative of x^n if n is a positive integer. We can now find the derivatives of more complicated functions, such as polynomials, using the following rules.

DERIVATIVES OF SUMS, DIFFERENCES AND CONSTANT MULTIPLES

If $f'(x)$ and $g'(x)$ exist, and c is a constant, then

$$\textbf{(A)} \ [f(x) \pm g(x)]' = f'(x) \pm g'(x),$$

and

$$\textbf{(B)} \ [cf(x)]' = cf'(x).$$

In differential notation, these rules are

(A) $\dfrac{d}{dx}[f(x) \pm g(x)] = \dfrac{d}{dx}[f(x)] \pm \dfrac{d}{dx}[g(x)]$

(B) $\dfrac{d}{dx}[cf(x)] = c\dfrac{d}{dx}[f(x)]$

In words:
(A) *The derivative of a sum is the sum of the derivatives, and the derivative of a difference is the difference of the derivatives.*
(B) *The derivative of c times a function is c times the derivative of the function.*

Before discussing why these rules are true, we illustrate their usefulness with some examples.

▼ **EXAMPLE 3**

Use the power rule and the rules above to find the derivative of $f(x) = 5x^3$.

SOLUTION   We need $\dfrac{d}{dx}(5x^3)$. Now

$$\frac{d}{dx}(5x^3) = 5\frac{d}{dx}(x^3) \qquad \text{by the rule for constant multiples}$$

$$= 5(3x^2) \qquad \text{by the power rule}$$

$$= 15x^2.$$

▼ **EXAMPLE 4**

Use the rules to find the derivative of $f(x) = 3x^2 + 2x - 4$.

SOLUTION

$$\frac{d}{dx}(3x^2 + 2x - 4) = \frac{d}{dx}(3x^2) + \frac{d}{dx}(2x - 4) \qquad \text{by the rule for sums}$$

$$= \frac{d}{dx}(3x^2) + \frac{d}{dx}(2x) - \frac{d}{dx}(4) \qquad \text{by the rule for differences}$$

$$= 3\frac{d}{dx}(x^2) + 2\frac{d}{dx}(x) - 4\frac{d}{dx}(1)$$

$$= 3(2x) + 2(1) - 0$$

$$= 6x + 2$$

Before we go on... Notice that we had three terms in the expression for $f(x)$, not just two. By applying the rule for sums and differences twice, we saw that the derivative of a sum or difference of three terms is the sum or difference of the derivatives of the terms. This works for sums and differences of any number of terms. We also saw two very common special cases that you should remember:

DERIVATIVE OF A CONSTANT TIMES x; DERIVATIVE OF A CONSTANT

If c is any constant, then

$$\frac{d}{dx}(cx) = c$$

and

$$\frac{d}{dx}(c) = 0.$$

Think about these graphically to see why they must be true.

▼ **EXAMPLE 5**

Find the derivative of a general polynomial

$$f(x) = a_0 + a_1 x + a_2 x^2 + \ldots + a_n x^n.$$

$$(a_0, a_1, \ldots, a_n \text{ constants})$$

SOLUTION   We take the derivative of each term and then add, getting

$$f'(x) = a_1 + 2a_2 x + \ldots + na_n x^{n-1}.$$

▼ **EXAMPLE 6**

Assuming the power rule is valid for all real powers, compute

$$\frac{d}{dx}\left(2 - \frac{3}{\sqrt{x}}\right).$$

SOLUTION

$$\frac{d}{dx}\left(2 - \frac{3}{\sqrt{x}}\right) = \frac{d}{dx}(2 - 3x^{-1/2})$$

$$= \frac{d}{dx}(2) - \frac{d}{dx}(3x^{-1/2})$$

$$= 0 - 3\frac{d}{dx}(x^{-1/2})$$

$$= -3\left(-\frac{1}{2}x^{-3/2}\right)$$

$$= \frac{3}{2x^{3/2}} = \frac{3}{2x\sqrt{x}}$$

Before we go on... Try to justify each of the steps above.

With practice, it should become second nature to you to use these rules. You will remember that if you see a sum of terms, you must take the derivative of each term separately, and that if a constant multiplies a function, then that constant will multiply the derivative of that function.

Here is something else to think about: If $y = mx + b$, then $y' = m$. It is not accidental that the derivative gives us the slope of the line. (Why does it happen?)

Finally, let us explain why the rule about sums is true.

PROOF OF THE RULE FOR SUMS

$$\frac{d}{dx}[f(x) + g(x)] = \lim_{h \to 0} \frac{[f(x + h) + g(x + h)] - [f(x) + g(x)]}{h}$$

$$= \lim_{h \to 0} \frac{[f(x + h) - f(x)] + [g(x + h) - g(x)]}{h}$$

$$= \lim_{h \to 0} \left[\frac{f(x + h) - f(x)}{h} + \frac{g(x + h) - g(x)}{h} \right]$$

$$= \lim_{h \to 0} \frac{f(x + h) - f(x)}{h} + \lim_{h \to 0} \frac{g(x + h) - g(x)}{h}$$

$$= \frac{d}{dx}[f(x)] + \frac{d}{dx}[g(x)]$$

The next-to-last step uses a property of limits: if $\lim_{x \to a} F(x) = L$ and $\lim_{x \to a} G(x) = M$, then $\lim_{x \to a}(F(x) + G(x)) = L + M$ (why should this be true?). The last step uses the definition of the derivative again.

We'll leave for you the cases of subtraction (which is essentially the same) and multiplication by a constant (which is easier).

▶ **3.4 EXERCISES**

In Exercises 1–14, calculate the derivative of the given function mentally.

1. $f(x) = x^3$ **2.** $f(x) = x^4$ **3.** $f(x) = 2x^{-2}$

4. $f(x) = 3x^{-1}$ **5.** $f(x) = -x^{1/4}$ **6.** $f(x) = -x^{-1/2}$

7. $f(x) = 2x^4 + 3x^3 - 1$ **8.** $f(x) = -x^3 - 3x^2 - 1$ **9.** $f(x) = -x + \frac{1}{x} + 1$

10. $f(x) = \frac{1}{x} + \frac{1}{x^2}$ **11.** $f(x) = 2\sqrt{x}$ **12.** $f(x) = \frac{2}{x}$

13. $f(x) = 3\sqrt[3]{x}$ **14.** $f(x) = \frac{2}{x^2} + \frac{3}{x^3}$

State the rules you use to obtain the indicated derivative in each of Exercises 15–30. You may use the power rule for any real exponent.

15. $y = 10;\ \frac{dy}{dx}$ **16.** $y = x^3;\ \frac{dy}{dx}$

17. $y = x^2 + x;\ \frac{dy}{dx}$ **18.** $y = x - 5;\ \frac{dy}{dx}$

19. $y = 4x^3 + 2x - 1$; $\dfrac{dy}{dx}$

20. $y = 4x^{-1} - 2x - 10$; $\dfrac{dy}{dx}$

21. $y = x^{104} - 99x^2 + x$; $\dfrac{dy}{dx}$

22. $y = \sqrt{x}(x + x^2)$; $\dfrac{dy}{dx}$

23. $s = \sqrt{t}(t - t^3) + \dfrac{1}{t^3}$; $\dfrac{ds}{dt}$

24. $s = 6t + \dfrac{6}{t}$; $\dfrac{ds}{dt}$

25. $V = \dfrac{4\pi r^3}{3}$; $\dfrac{dV}{dr}$

26. $A = 4\pi r^2 + 2\pi rh$ (h constant); $\dfrac{dA}{dr}$

27. $\dfrac{d}{dt}[t^2 + 4at^5]$ (a constant)

28. $\dfrac{d}{dt}[at^2 + bt + c]$ (a, b, c constant)

29. $\dfrac{d}{dx}\left(\sqrt{x}(1 + x)\right)$

30. $\dfrac{d}{dx}\left(\dfrac{1 + x}{x}\right)$

In Exercises 31–42, find the slope of the tangent to the graph of the given function at the indicated point.

31. $f(x) = x^3$; $(-1, -1)$

32. $g(x) = x^4$; $(-2, 16)$

33. $f(x) = 1 - 2x$; $(2, -3)$

34. $f(x) = \dfrac{x}{3} - 1$; $(-3, -2)$

35. $h(x) = x^{1/4}$; $(16, 2)$

36. $s(x) = x^{2/3}$; $(8, 4)$

37. $f(x) = x^{-1/2}$; $(2, 2^{-1/2})$

38. $f(x) = x^{-1/3}$; $(1, 1)$

39. $g(t) = \dfrac{1}{t^5}$; $(1, 1)$

40. $s(t) = \dfrac{1}{t^3}$; $(-2, -\frac{1}{8})$

41. $f(x) = \dfrac{x^2}{4} - \dfrac{x^3}{3}$; $(-1, \frac{7}{12})$

42. $f(x) = \dfrac{x^2}{2} + \dfrac{x}{4}$; $(2, \frac{5}{2})$

In Exercises 43–48, find the equation of the tangent line to the graph of the given function at the point with the indicated x-coordinate. In each case, sketch the curve together with the appropriate tangent line.

43. $f(x) = x^3$; $x = -1$

44. $f(x) = x^2$; $x = 0$

45. $f(x) = x + \dfrac{1}{x}$; $x = 2$

46. $f(x) = \dfrac{1}{x^2}$; $x = 1$

47. $f(x) = \sqrt{x}$; $x = 4$

48. $f(x) = 2x + 4$; $x = -1$

In Exercises 49–62, find the derivative of the given function.

49. $f(x) = x^2 - 3x + 5$

50. $f(x) = 3x^3 - 2x^2 + x$

51. $f(x) = x + \sqrt{x}$

52. $f(x) = x^{1/2} + 2x^{-1/2}$

53. $g(x) = \dfrac{1}{x^2} + \dfrac{2}{x^3}$

54. $g(x) = \dfrac{2}{x} - \dfrac{2}{x^3} + \dfrac{1}{x^4}$

55. $h(x) = \dfrac{1}{x} + \dfrac{1}{x^2} + \dfrac{1}{x^3}$

56. $h(x) = \dfrac{1}{\sqrt{x}}$

57. $r(x) = \sqrt{x} + \dfrac{1}{\sqrt{x}}$

58. $r(x) = x + \dfrac{7}{\sqrt{x}}$

59. $f(x) = x(x^2 - 1/x)$

60. $f(x) = x^{-1}(x - 2/x)$

61. $g(x) = \dfrac{x^2 - 2x^3}{x}$

62. $f(x) = \dfrac{2x + x^2}{x}$

In Exercises 63–70, evaluate the given expression.

63. $\dfrac{d}{dx}\left(x + \dfrac{1}{x^2}\right)$

64. $\dfrac{d}{dx}\left(2x - \dfrac{1}{x}\right)$

65. $\dfrac{d}{dx}(2x^{1.3} - x^{-1.2})$

66. $\dfrac{d}{dx}(2x^{4.3} + x^{0.6})$

67. $\dfrac{d}{dt}(at^3 - 4at);$ *(a constant)*

68. $\dfrac{d}{dt}(at^2 + bt + c);$ *(a, b, c constant)*

69. $\dfrac{d}{dx}(\sqrt{x}(1 + x))$

70. $\dfrac{d}{dx}\left(\dfrac{1 + x}{x}\right)$

In Exercises 71–76, find the indicated derivative.

71. $y = \dfrac{x^{10.3}}{2} + 99x^{-1};$ find $\dfrac{dy}{dx}$

72. $y = \dfrac{x^{1.2}}{3} - \dfrac{x^{0.9}}{2};$ find $\dfrac{dy}{dx}$

73. $s = 2.35 + \dfrac{2.1}{t^{1.1}} - \dfrac{t^{0.6}}{2};$ find $\dfrac{ds}{dt}$

74. $s = \dfrac{2}{t^{1.1}} + t^{-1.2};$ find $\dfrac{ds}{dt}$

75. $V = \dfrac{4}{3}\pi r^3;$ find $\dfrac{dV}{dr}$

76. $A = 4\pi r^2;$ find $\dfrac{dA}{dr}$

In Exercises 77–84, find all values of x (if any) where the tangent line to the graph of the given equation is horizontal.

77. $y = 2x^2 + 3x - 1$

78. $y = -3x^2 - x$

79. $y = 2x + 8$

80. $y = -x + 1$

81. $y = x + \dfrac{1}{x}$

82. $y = x + \sqrt{x}$

83. $y = \sqrt{x} - x$

84. $y = \sqrt{x} + \dfrac{1}{x}$

85. Write out the proof that $\dfrac{d}{dx}[x^4] = 4x^3$.

86. Write out the proof that $\dfrac{d}{dx}[x^5] = 5x^4$.

87. Write out the proof that $\dfrac{d}{dx}[3x^2] = 3\dfrac{d}{dx}[x^2]$.

88. Write out the proof that $\dfrac{d}{dx}\left[\dfrac{1}{2}x^2\right] = \dfrac{1}{2}\dfrac{d}{dx}[x^2]$.

89. Write out the proof that
$$\dfrac{d}{dx}[x^2 + x^3] = \dfrac{d}{dx}[x^2] + \dfrac{d}{dx}[x^3].$$

90. Write out the proof that
$$\dfrac{d}{dx}[2x^2 - 3x^2] = \dfrac{d}{dx}[2x^2] - \dfrac{d}{dx}[3x^3].$$

APPLICATIONS

91. **Cost** Consider the two cost functions $C_1(x) = 10{,}000 + 5x - x^2/10$ and $C_2(x) = 20{,}000 + 10x - x^2/5$. How do rates of change of these cost functions at the same production levels compare?

92. **Cost** The cost of making x teddy bears at the Cuddly Companion Company used to be $C_1(x) = 100 + 40x - 0.001x^2$. Due to rising health insurance costs, it now is $C_2(x) = 1{,}000 + 40x - 0.001x^2$. How does the rate of change of cost at a production level of x teddy bears compare to what it used to be?

93. **Profit** The cost to manufacture x cases of beer per week is $C(x) = 10{,}000 + 30x - 0.01x^2$, while the revenue from selling x cases is $R(x) = 20x$. How must the rate of change of cost and the rate of change of revenue be related when the rate of change of profit is 0? What can you conclude about the cost and revenue when the profit is zero?

94. **Profit** The cost to manufacture x cases of beer per week is $C(x) = 10{,}000 + 30x - 0.01x^2$, while the revenue from selling x cases is $R(x) = 20x$. How must the rate of change of cost and the rate of change of revenue be related when the rate of change of profit is positive? What can you conclude about the cost and revenue when the profit is positive?

95. **Market Average** Joe Downs runs a small investment company from his basement. Every week he publishes a report on the success of his investments, including the progress of the infamous Joe Downs Average. At the end of one particularly memorable week he reported that the Average for that week had the value $A(t) = 1000 + 1500t - 800t^2 + 100t^3$ points, where t represents the number of days into the week; t ranges from 0 at the beginning of the week to 5 at end of the week. The graph of A is shown here. During the upswing at the beginning of the week, say halfway through the first day, how fast was the value of the Average increasing?

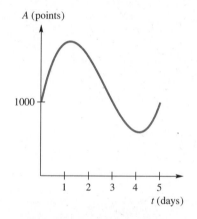

96. **Market Average** Referring to the Joe Downs Average in the previous exercise: During the downswing, say at the end of the third day, how fast was the value of the Average decreasing?

97. **Embryo Development** The oxygen consumption of a bird embryo increases from the time the egg is laid through the time the chick hatches. In the case of a typical galliform bird, the oxygen consumption (in milliliters per hour) can be approximated by

$$c(t) = -0.00271t^3 + 0.137t^2 - 0.892t + 0.149$$
$$(8 \le t \le 30),$$

where t is the time (in days) since the egg was laid.* (An egg will typically hatch at around $t = 28$.) Find $c'(15)$ and $c'(30)$. What do these results tell you about the embryo's oxygen consumption just prior to hatching?

▼ * The model approximates graphical data published in the article *The Brush Turkey* by Roger S. Seymour, *Scientific American*, December, 1991, pp. 108–14.

98. *Embryo Development*   The oxygen consumption of a turkey embryo increases from the time the egg is laid through the time the chick hatches. In the case of a brush turkey, the oxygen consumption (in milliliters per hour) can be approximated by

$$c(t) = -0.00118x^3 + 0.119x^2 - 1.83x + 3.972$$
$$(20 \le t \le 50),$$

where t is the time (in days) since the egg was laid.* (An egg will typically hatch at around $t = 50$.) Find $c'(30)$ and $c'(50)$. What do these results tell you about the embryo's oxygen consumption just prior to hatching?

99. *Velocity*   If a stone is dropped from a height of 100 feet, its height after t seconds is given by $s = 100 - 16t^2$.
   (a) Find its velocity at times $t = 0$, 1, 2, 3, and 4 seconds.
   (b) How long does it take to reach the ground, and how fast is it traveling when it hits the ground?

100. *Velocity*   If a stone is thrown down at 120 ft/s from a height of 1,000 feet, its height after t seconds is given by $s = 1,000 - 120t - 16t^2$.

   (a) Find its velocity at times $t = 0$, 1, 2, 3, and 4 seconds.
   (b) How long does it take to reach the ground, and how fast is it traveling when it hits the ground?

101. *Volume*   The volume, in cubic centimeters, of a spherical balloon is given by $V = \frac{4}{3}\pi r^3$, where r is its radius in centimeters, and $\pi \approx 3.141592$. Find $V'(10)$, and interpret the result.

102. *Volume*   The volume, in cubic centimeters, of an ellipsoid with a circular cross section of radius r centimeters. is given by $V = \frac{4}{3}\pi r^2 s$, where $\pi \approx 3.141592$, and s is as shown in the figure.

If s is fixed at 2 cm, find the rate of change of V with respect to r, and evaluate it at $r = 1$. Interpret the result.

COMMUNICATION AND REASONING EXERCISES

103. *Tangent Lines*   What instructions would you give to a fellow student who wanted to accurately graph the tangent line to the curve $y = 3x^2$ at the point $(-1, 3)$?

104. *Tangent Lines*   What instructions would you give to a fellow student who wanted to accurately graph a line at right angles to the curve $y = \frac{4}{x}$, at the point where $x = 0.5$?

105. *Tangent Lines*   Consider $f(x) = x^2$ and $g(x) = 2x^2$. How do the slopes of the tangent lines of f and g at the same x compare?

106. *Tangent Lines*   Consider $f(x) = x^3$ and $g(x) = x^3 + 3$. How do the slopes of the tangent lines of f and g compare?

107. *Tangent Lines*   Suppose that $f(x)$ and $g(x)$ are two functions, and $f(x) - g(x)$ has a horizontal tangent at $x = a$. How do the slopes of $f(x)$ and $g(x)$ at $x = a$ compare?

108. *Tangent Lines*   Suppose that $f(x)$ and $g(x)$ are two functions, and $f(x) + g(x)$ has a horizontal tangent at $x = a$. How do the slopes of $f(x)$ and $g(x)$ at $x = a$ compare?

109.   How would you respond to an acquaintance who says, "I finally understand what the derivative is: it is nx^{n-1}! Why weren't we taught that in the first place instead of the difficult way using limits?"

110.   Following is an excerpt from your friend's graded homework:

$$3x^4 + 11x^5 = 12x^3 + 55x^4 ✖ \mathcal{WRONG} -8$$

Why was it marked wrong?

▼ *The model approximates graphical data published in the article *The Brush Turkey,* by Roger S. Seymour, *Scientific American,* December 1991, pp. 108–14.

3.5 MARGINAL ANALYSIS

In Chapter 1 we considered linear *cost functions* of the form $C(x) = mx + b$, where C is the total cost, x is the number of items, and m and b are constants. The slope m was referred to as the *marginal cost*. It measured the *cost of producing one more item*. Notice that the derivative of $C(x) = mx + b$ is $C'(x) = m$. In other words, *the marginal cost is the derivative of the cost function*.

If $C(x)$ is any cost function—linear or not—then we shall define the **marginal cost function** to be the derivative $C'(x)$. Since the graph of C may be any curve, the marginal cost may have different values for different values of x. What $C'(x)$ measures is the steepness of the cost function curve at a specified production level x.

MARGINAL COST

A **cost function** specifies the total cost C as a function of the number of items x. Thus $C(x)$ is the total cost of x items. The **marginal cost function** is the derivative $C'(x)$ of the cost function $C(x)$. It measures the rate of change of cost with respect to x. The units of marginal cost are units of cost (dollars) per item.

▼ **EXAMPLE 1**   Marginal Cost

Assume that the cost to manufacture portable CD players is given by

$$C(x) = \$150{,}000 + 20x - \frac{x^2}{10{,}000},$$

where x is the number of CD players manufactured.* Find the marginal cost function $C'(x)$, and use it to estimate the cost of manufacturing the 50,001st CD player.

SOLUTION   Since

$$C(x) = \$150{,}000 + 20x - \frac{x^2}{10{,}000},$$

▼ * You might well ask where on earth this formula came from. There are two approaches to obtaining cost functions in real life: analytical and numerical. The analytical approach is to calculate the cost function from scratch. For example, in the above situation, we might have fixed costs of $150,000, plus a production cost of $20 per CD player. The term $x^2/10{,}000$ may reflect a cost saving for high levels of production, such as a bulk discount in the cost of electronic components. The numerical approach is as follows. First obtain the cost at several different production levels by direct observation. This gives several points on the (as yet unknown) cost versus production level graph. Then find the equation of the curve that best fits these points. This is called *curve-fitting* or *regression* and uses techniques of calculus. The resulting equation then gives the required cost function.

we have

$$C'(x) = 20 - \frac{x}{5,000}.$$

The units of $C'(x)$ are units of cost ($\$$) per unit of x (CD players). Thus, $C'(x)$ is measured in dollars per CD player.

The cost of the 50,001st CD player is the amount by which the total cost would rise if we increased production from 50,000 CD players to 50,001. Thus, we need to know the rate at which the total cost rises as we increase production. This rate of change is measured by the derivative, or marginal cost, which we just computed. At $x = 50,000$,

$$C'(50,000) = 20 - \frac{50,000}{5,000} = \$10 \text{ per CD player.}$$

In other words, we estimate that the 50,001st CD player will cost $\$10$.

Before we go on... The marginal cost is really only an approximation to the cost of the 50,001st CD player. The exact cost is

$$C(50,001) - C(50,000) = 150,000 + 20(50,001) - \frac{(50,001)^2}{10,000}$$

$$- \left(150,000 + 20(50,000) - \frac{(50,000)^2}{10,000} \right)$$

$$= \$9.9999.$$

So the marginal cost is a good approximation to the actual cost. One way of thinking about this calculation is that we are using the tangent line to approximate the cost function near a production level of 50,000. Figure 1 shows the graph of the cost function together with the tangent line at $x = 50,000$. Notice that the tangent line is essentially indistinguishable from the graph of the function for some distance on either side of 50,000.

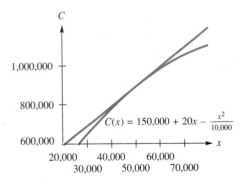

FIGURE 1

▶ NOTES

1. In general, the difference quotient $[C(x + h) - C(x)]/h$ gives the **average cost per item** to produce h more items at a current production level of x items.
2. Notice that $C'(x)$ is much easier to calculate than $[C(x + h) - C(x)]/h$ (try it). ◀

As the following examples show, the term "marginal" can apply to quantities other than cost.

▼ **EXAMPLE 2**   Marginal Revenue

Economist Henry Schultz calculated the following demand function for corn:

$$p = \frac{6{,}570{,}000}{q^{1.3}}.$$

Here, p is the price in dollars per bushel, and q is the number of bushels of corn that could be sold at the price p in one year.*

(a) Calculate the annual revenue as a function of the number of bushels q.
(b) Calculate $R(300{,}000)$, and interpret the result.
(c) Calculate the marginal revenue when $q = 300{,}000$ bushels, and interpret the result.

SOLUTION

(a) To calculate the annual revenue R, use $R = pq$ (revenue = price $\times$ quantity). Since we want revenue as a function of q only, we substitute for p using the demand equation.

$$R = pq$$
$$R(q) = \frac{6{,}570{,}000}{q^{1.3}}q$$
$$= \frac{6{,}570{,}000}{q^{0.3}}.$$

This gives the annual revenue as a function of the number of bushels of corn sold per year.

▼ * This demand function is based on data for the period 1915–1929. Notice that we have written p as a function of q, rather than the other way around, which would be more natural. Economists often specify demand functions this way, and we will find this convenient for the calculations. Source: Henry Schultz: *The Theory and Measurement of Demand* (as cited in *Introduction to Mathematical Economics* by A. L. Ostrosky, Jr., and J. V. Koch (Waveland Press, Prospect Heights, Illinois, 1979.)

(b) Since

$$R(q) = \frac{6{,}570{,}000}{q^{0.3}},$$

we have

$$R(300{,}000) = \frac{6{,}570{,}000}{(300{,}000)^{0.3}}$$
$$= 149{,}426.97.$$

Thus, if 300,000 bushels of corn were sold in one year, the total revenue for that year would be $149,426.97.

(c) To obtain the *marginal* revenue, we need to take the derivative of the revenue function. Since

$$R(q) = \frac{6{,}570{,}000}{q^{0.3}} = 6{,}570{,}000q^{-0.3},$$

we have

$$R'(q) = -(0.3)6{,}570{,}000q^{-1.3}$$
$$= -1{,}971{,}000q^{-1.3}.$$

This is the marginal revenue function. Its units are units of R per unit of q: that is, dollars per bushel. We are asked to find $R'(300{,}000)$.

$$R'(300{,}000) = -1{,}971{,}000(300{,}000)^{-1.3}$$
$$\approx -0.1494 \text{ dollars per bushel}$$
$$\approx -15¢ \text{ per bushel}$$

Since the derivative of a quantity measures its rate of change, we conclude that, at a sales level of 300,000 bushels per year, the annual revenue is dropping at a rate of 15¢ per additional bushel sold. In other words, each additional bushel a farmer sells will result in a decrease of 15¢ in annual revenue.

Before we go on ...

Q How is it possible for the revenue to decrease with increasing sales?

A In order to sell more corn, a farmer would have to lower his price. That lower price outweighs the increase in sales, and revenue goes down.

Q In order to *increase* revenue, what should a farmer do?

A The fact that the marginal revenue is negative implies that if q *decreases*, then R will *increase* by 15¢ per bushel. In other words, a farmer should raise the price to increase annual revenue, even though this will mean a decrease in the quantity sold.

▼ | **EXAMPLE 3** | Marginal Profit

In Chapter 1, we found that the demand equation for rubies at Royal Ruby Retailers (RRR) is given by

$$q = -\frac{4p}{3} + 80$$

where p is the retail price it charges per ruby and q is the number of rubies RRR can sell per week at $\$p$ per ruby. Assume that RRR pays $\$15$ per ruby. (These are rather cheap rubies.)

(a)   Calculate the weekly revenue and profit as functions of q.

(b)   Calculate the marginal profit function, $P'(q)$.

(c)   Calculate the profit and marginal profit for $q = 20$, $q = 30$, and $q = 40$, and interpret the results.

SOLUTION

(a)   The weekly revenue is given by $R = pq$. Since we need to express R as a function of q only, we must replace p in this equation by a function of q. Since the relationship between p and q is given in the demand equation, and we want p as a function of q, we need to first rewrite the demand equation by solving for p:

$$q = -\frac{4p}{3} + 80$$

gives

$$p = 60 - \frac{3q}{4}.$$

We now substitute this expression for p in $R = pq$ to obtain

$$R(q) = \left(60 - \frac{3q}{4}\right)q = 60q - \frac{3q^2}{4}.$$

This is the revenue as a function of q. For the profit function, recall that

$$P = R - C \text{ (Revenue } - \text{ Cost)}.$$

Since RRR pays $\$15$ per ruby, the cost of q rubies is $15q$. Thus,

$$P(q) = \left(60q - \frac{3q^2}{4}\right) - 15q$$

$$= 45q - \frac{3q^2}{4}.$$

(b)   The marginal profit function is the derivative,

$$P'(q) = 45 - \frac{3q}{2},$$

and its units are dollars per ruby. Thus, $P'(q)$ gives the rate of change of profit in dollars per ruby sold (per week).

(c)   We have

$$P'(20) = 45 - \frac{3(20)}{2} = \$15 \text{ per ruby.}$$

This means that at a demand level of 20 rubies per week, the profit is increasing by \$15 for every additional ruby RRR can sell. It would therefore pay RRR to sell more rubies, which it can do by lowering the price.

$$P'(30) = 45 - \frac{3(30)}{2} = \$0 \text{ per ruby}$$

This means that at a demand level of 30 rubies per week, the profit is neither increasing nor decreasing.

$$P'(40) = 45 - \frac{3(40)}{2} = -\$15 \text{ per ruby}$$

This means that at a demand level of 40 rubies per week, the profit is decreasing by \$15 for every additional ruby they sell. RRR should therefore sell fewer rubies, which they can do by increasing the price.

Before we go on... This analysis shows that RRR should sell more than 20 rubies per week, but less than 40 rubies per week, and the fact that $P'(30) = 0$ tells us that RRR should adjust the price to sell exactly 30 rubies per week. If it sells slightly fewer—say, 29 per week—then the marginal profit will be positive (as you can check by calculating $P'(29)$), indicating that it should sell more than 29. Similarly, $P'(31)$ is negative, indicating that it should sell fewer than 31. In general, *for maximum profit, $P'(q)$ must be* 0. Figure 2 shows the graph of $P(q)$ for this example, and you can see clearly in the graph that $P'(20) > 0$, $P'(40) < 0$, $P'(30) = 0$, and the largest value of $P(q)$ occurs at $q = 30$.

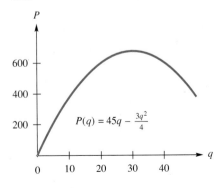

FIGURE 2

▼   **EXAMPLE 4**   Marginal Product

Precision Manufacturers, Inc. is informed by a consultant that its annual profit is given by

$$P = -200,000 + 4,000q - 0.5q^2,$$

where q is the number of surgical lasers it sells per year. The consultant also informs the company that the number of surgical lasers it can manufacture per year depends on the number n of assembly-line workers it employs according to the equation

$$q = 100n + 0.1n^2.$$

(a) Express P as a function of n, and find $P'(n)$. $P'(n)$ is called the **marginal product** at the employment level of n assembly line workers. What are its units?

(b) Calculate $P(10)$ and $P'(10)$, and interpret the results.

(c) Precision Manufacturers currently employs 10 assembly line workers and is considering laying off assembly line workers. What advice would you give the company's management?

SOLUTION

(a) Since P is given in terms of q, and q is given in terms of n, we can obtain P as a function of n by substituting the expression for q in the expression for P.

$$
\begin{aligned}
P &= -200,000 + 4,000q - 0.5q^2 \\
P(n) &= -200,000 + 4,000(100n + 0.1n^2) - 0.5(100n + 0.1n^2)^2 \\
&= -200,000 + 400,000n + 400n^2 - 0.5(10,000n^2 + 20n^3 + 0.01n^4) \\
&= -200,000 + 400,000n + 400n^2 - 5,000n^2 - 10n^3 - 0.005n^4 \\
&= -200,000 + 400,000n - 4,600n^2 - 10n^3 - 0.005n^4
\end{aligned}
$$

Then

$$P'(n) = 400,000 - 9,200n - 30n^2 - 0.02n^3.$$

The units of $P'(n)$ are profit (in dollars) per worker.

(b) The formula

$$P(n) = -200,000 + 400,000n - 4,600n^2 - 10n^3 - 0.005n^4$$

gives

$$
\begin{aligned}
P(10) &= -200,000 + 400,000(10) - 4,600(10)^2 - 10(10)^3 - 0.005(10)^4 \\
&= \$3,329,950.
\end{aligned}
$$

Thus, Precision Manufacturers will make an annual profit of $3,329,950 if it employs 10 assembly line workers. The formula

$$P'(n) = 400,000 - 9,200n - 30n^2 - 0.02n^3$$

gives

$$P'(10) = 400{,}000 - 9{,}200(10) - 30(10)^2 - 0.02(10)^3$$
$$= \$304{,}980 \text{ per worker.}$$

Thus, at an employment level of 10 assembly line workers, annual profit is increasing at a rate of $304,980 per additional worker. In other words, if the company were to employ one more assembly line worker, its annual profit would increase by approximately $304,980.

(c) Since the marginal product is positive, profits will increase if the company increases the number of workers and will decrease if it decreases the number of workers, so your advice would be to hire additional assembly line workers. Downsizing the assembly line work force would reduce the company's annual profits.

Before we go on... The algebra involved in finding P as a function of n was rather messy. In the next chapter we shall see a way to find $P'(n)$ that is not so algebraically intense.

 The following question might have occurred to you:

Q How many additional assembly line workers should the company hire to obtain the maximum annual profit?

A Since we have profit as a function of the number of workers n, we can graph the function $P(n)$. Figure 3 shows the graph of $P(n)$ for $0 \le n \le 80$, as drawn by a graphing calculator. (For the y-axis scale, we used $-2{,}000{,}000 \le P \le 8{,}000{,}000$.)

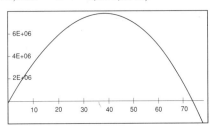

FIGURE 3

The horizontal axis represents the number of assembly line workers, while the vertical axis represents annual profit. At the point on the graph where $x = 10$, the slope is positive (as confirmed by our calculation of $P'(10)$). At approximately $n = 40$, the slope is zero (so the marginal product is zero) and the profit is largest. This tells us that the company should employ approximately 40 assembly line workers for a maximum profit. Thus, the company should hire approximately 30 additional assembly line workers.

Notice that Figure 3 also shows us that the company begins taking a loss at an employment level of about 75 assembly line workers.

▶ ___ **3.5 EXERCISES**

*Consider each of the functions in Exercises 1–6 as representing the cost to manu-
facture x items. Find the average costs of manufacturing h more items at a production level of x, where x is as indicated and h = 50, 10, and 1. Also, find the
marginal cost at the given production level x. (For the average costs, see the note
after Example 1.)*

1. $C(x) = 10,000 + 5x - \dfrac{x^2}{10,000}$; $x = 1,000$

2. $C(x) = 20,000 + 7x - \dfrac{x^2}{20,000}$; $x = 10,000$

3. $C(x) = 1,000 + 10x - \sqrt{x}$; $x = 100$

4. $C(x) = 10,000 + 20x - \sqrt{x}$; $x = 10$

5. $C(x) = 15,000 + 100x + \dfrac{1,000}{x}$; $x = 100$

6. $C(x) = 20,000 + 50x + \dfrac{10,000}{x}$; $x = 100$

*In Exercises 7 and 8, find the marginal cost, marginal revenue, and marginal
profit functions, and find all values of x for which the marginal profit is zero. Interpret your answer.*

7. $C(x) = 4x$; $R(x) = 8x - \dfrac{x^2}{1,000}$

8. $C(x) = 5x^2$; $R(x) = x^3 + 7x + 10$

APPLICATIONS

9. *Marginal Cost*   The cost of producing x teddy
bears per day at the Cuddly Companion Company is
calculated by their marketing staff to be given by the
formula

$$C(x) = 100 + 40x - .001x^2.$$

Find the marginal cost function, and use it to estimate
how fast the cost is going up at a production level of
100 Teddy bears. Compare this with the exact cost of
producing the 101st teddy bear.

10. *Marginal Cost*   Referring to the cost equation in
Exercise 9, find the production level for which the
marginal cost is zero.

11. *Marginal Net Profit*   Suppose that $P(x)$ represents
the net profit on the sale of x videocassettes. If
$P(1,000) = 3,000$, and $P'(1,000) = -3$, what does
this tell you?

12. *Marginal Loss*   An automobile retailer calculates
that its loss on the sale of Type M cars is given by
$L(50) = 5,000$ and $L'(50) = -200$, where $L(x)$ represents the loss on the sale of x Type M cars. What
does this tell you?

13. *Marginal Profit*   Your monthly profit (in dollars)
from selling magazines is given by

$$P = 5n + \sqrt{n},$$

where n is the number of magazines you sell in a

month. If you are currently selling $n = 50$ magazines
per month, find your profit and your marginal profit.
Interpret your answers.

14. *Marginal Profit*   Your monthly profit (in dollars)
from your newspaper route is given by

$$P = 2n - \sqrt{n},$$

where n is the number of subscribers on your route. If
you currently have 100 subscribers, find your profit
and your marginal profit. Interpret your answers.

15. *Marginal Product*   A car wash firm calculates that
its daily profit depends on the number n of workers it
employs according to the formula

$$P = 400n - 0.5n^2.$$

Calculate the marginal product at an employment
level of 50 workers, and interpret the result.

16. *Marginal Product*   Repeat Exercise 15 using the
formula

$$P = -100n + 25n^2 - 0.005n^4.$$

17. *Marginal Revenue*   Assume that the demand
function for tuna in a small coastal town is given by

$$p = \dfrac{50,000}{q^{1.5}},$$

where p is the price (in dollars) per pound of tuna, and q is the number of pounds of tuna that can be sold at the price p in one month.

(a) Calculate the price that the town's fishery should charge for tuna in order to produce a demand of 500 pounds of tuna per month.

(b) Calculate the annual revenue R as a function of the number of pounds of tuna q.

(c) Calculate $R(500)$, and interpret the result.

(d) Calculate the marginal revenue function and its value at $q = 500$ pounds, and interpret the result.

(e) If the town fishery's monthly tuna catch amounted to 600 pounds of tuna, and the price is at the level in part (a), would you recommend that the fishery raise or lower the price of tuna in order to increase its revenue?

18. Repeat Exercise 17 assuming a demand equation of

$$p = \frac{80{,}000}{q^{1.7}}.$$

19. *Marginal Revenue and Marginal Profit* The demand for poultry can be modeled as

$$q = 63.15 - 0.45p + 0.12b$$

where q is the per capita demand for poultry in pounds per year, p is the wholesale price of poultry in cents per pound, and b is the wholesale price of beef in cents per pound.* Assume that the wholesale price of beef is fixed at 45¢ per pound.

(a) Find the revenue as a function of q, and hence obtain the marginal revenue as a function of q. (Round constants to two decimal places.)

(b) Find the annual per capita revenue that will result at a demand level of 50 pounds of poultry per year, and estimate the change in revenue that will result if the price is raised to yield a demand level of 49 pounds of poultry per year.

(c) If a farmer breeds chickens at an average cost of 10¢ per pound, find the annual profit function P in terms of the per capita demand for poultry, evaluate $P'(50)$, and interpret the result.

20. *Demand for Poultry* Referring to the model for demand for poultry in Exercise 19, express the annual per capita revenue R as a function of the price of beef if the wholesale price of poultry is fixed at 40¢ per pound. Calculate $R(45)$ and $R'(45)$, and interpret the results.

21. *Revenue* In Chapter 1, we found that the demand equation for rubies at Royal Ruby Retailers (RRR) is given by

$$q = -\frac{4p}{3} + 80,$$

where p is the retail price RRR charges per ruby and q is the number of rubies RRR can sell per week at $\$p$ per ruby.

(a) Express the revenue as a function of p, and calculate the marginal revenue as a function of p.

(b) How fast is the revenue increasing as the price goes up at the following sales prices: (i) $20 per ruby; (ii) $30 per ruby; (iii) $40 per ruby?

(c) Interpret the results in part (b).

22. *Housing Costs* [†] The cost C of building a house is related to the number k of carpenters used and the number e of electricians used by the formula

$$C = 15{,}000 + 50k^2 + 60e^2.$$

(a) Assuming that 10 carpenters are currently being used, find the marginal cost as a function of e.

(b) If 10 carpenters and 10 electricians are currently being used, use your answer to part (a) to estimate the cost of hiring an additional electrician.

(c) If 10 carpenters and 10 electricians are currently being used, what is the cost of hiring an additional carpenter?

23. *Emission Control* The cost of controlling emissions at a firm goes up rapidly as the amount of emissions reduced goes up. Here is a possible model:

$$C(q) = 4{,}000 + 100q^2,$$

where q is the reduction in emissions (in pounds of pollutant per day) and C is the daily cost (in dollars) of this reduction.

(a) If a firm is currently reducing its emissions by 10 pounds each day, what is the marginal cost of reducing emissions further?

▼ * This equation is based on data from poultry sales in the period from 1950 to 1984. (Source: A.H. Studenmund, *Using Econometrics*, Second Ed. (New York: HarperCollins, 1992), pp. 180–81).

† Based on an exercise in *Introduction to Mathematical Economics* by A. L. Ostrosky, Jr., and J.V. Koch (Prospect Heights, IL: Waveland Press, 1979.)

(b) Government clean-air subsidies to the firm are based on the formula

$$S(q) = 500q,$$

where q is again the reduction in emissions in pounds and S is the subsidy. At what reduction level does the marginal cost surpass the marginal subsidy?

(c) Calculate the net cost function, $N(q) = C(q) - S(q)$, given the cost function and subsidy above, and find the value of q that gives the lowest net cost. What is this lowest net cost? Compare your answer to that for (b), and comment on what you find.

24. Taxation Schemes Here is a curious proposal for taxation rates based on income:

$$R(i) = \frac{\sqrt{i}}{1,000},$$

where i represents total annual income and $R(i)$ is the income tax rate as a percentage of total annual income. (Thus, for example, an income of $50,000 per year would be taxed at about 22%, while an income of double that amount would be taxed at about 32%.)*

(a) Calculate the after-tax (net) income $N(i)$ an individual can expect to earn as a function of income i.

(b) Calculate an individual's marginal after-tax income at income levels of $100,000 and $500,000.

(c) At what income does an individual's marginal after-tax income become negative? What is the after-tax income at that level, and what does this signify?

(d) What do you suspect is the most anyone can earn after taxes? (See the footnote.)

25. Fuel Economy Your Porsche's gas mileage (in miles per gallon) is given as a function $M(x)$ of speed x in mph. It is found that

$$M'(x) = \frac{3600x^{-2} - 1}{(3600x^{-1} + x)^2}.$$

Find $M'(10)$, $M'(60)$, and $M'(70)$. What do the answers tell you about your car?

26. Marginal Revenue The estimated marginal revenue for sales of ESU soccer team T-shirts is given by

$$R'(p) = \frac{(8 - 2p)e^{-p^2 + 8p}}{10,000,000}$$

where p is the price (in dollars) the soccer players charge for each shirt. Find $R'(3)$, $R'(4)$, and $R'(5)$. What do the answers tell you?

27. Transportation Costs Before the Alaskan pipeline was built, there was speculation as to whether it might be more economical to transport the oil by large tankers. The following cost equation was estimated by National Academy of Sciences:

$$C = 0.03 + \frac{10}{T} - \frac{200}{T^2},$$

where C is the cost in dollars of transporting one barrel of oil 1,000 nautical miles, and T is the size of an oil tanker in deadweight tons.[†]

(a) How much would it cost to transport 100 barrels of oil in a tanker weighing 1,000 tons?

(b) By how much is this cost increasing or decreasing as the weight of the tanker increases from 1,000 tons?

28. Transportation Costs Referring to the cost equation in Exercise 27, find the value of T so that $C'(T) = 0$. Interpret the result. By calculating values of $C(T)$ for T close to and on either side of this amount, what more can you say?

29. Marginal Cost *(from the GRE economics test)* In a multiple-plant firm in which the different plants have different and continuous cost schedules, if costs of production for a given output level are to be minimized, which of the following is essential?

(a) Marginal costs must equal marginal revenue.

(b) Average variable costs must be the same in all plants.

(c) Marginal costs must be the same in all plants.

(d) Total costs must be the same in all plants.

(e) Output per man-hour must be the same in all plants.

▼ *This model has the following interesting feature: an income of a million dollars per year would be taxed at 100%, leaving the individual penniless!

† Source: *Use of Satellite Data on the Alaskan Oil Marine Link,* Practical Applications of Space Systems: Cost and Benefits. (National Academy of Sciences, Washington, D.C., 1975, p. B–23)

30. *Study Time (from the GRE economics test)*   A student has a fixed number of hours to devote to study and is certain of the relationship between hours of study and the final grade for each course. Grades are given on a numerical scale (e.g., 0 to 100), and each course is counted equally in computing the grade average. In order to maximize his or her grade average, the student should allocate these hours to different courses so that

(a) the grade in each course is the same;

(b) the marginal product of an hour's study (in terms of final grade) in each course is zero;

(c) the marginal product of an hour's study (in terms of final grade) in each course is equal, although not necessarily equal to zero;

(d) the average product of an hour's study (in terms of final grade) in each course is equal;

(e) the number of hours spent in study for each course are equal.

31. *Marginal Product (from the GRE economics test)*   Assume that the marginal product of an additional senior professor is 50 percent higher than the marginal product of an additional junior professor and that junior professors are paid one-half the amount that senior professors receive. With a fixed overall budget, a university that wishes to maximize its quantity of output from professors should do which of the following?

(a) Hire equal numbers of senior professors and junior professors.

(b) Hire more senior professors and junior professors.

(c) Hire more senior professors and discharge junior professors.

(d) Discharge senior professors and hire more junior professors.

(e) Discharge all senior professors and half of the junior professors.

32. *Marginal Product (based on a question from the GRE economics test)*   Assume that the marginal product of an additional senior professor is twice the marginal product of an additional junior professor and that junior professors are paid two-thirds the amount that senior professors receive. With a fixed overall budget, a university that wishes to maximize its quantity of output from professors should do which of the following?

(a) Hire equal numbers of senior professors and junior professors.

(b) Hire more senior professors and junior professors.

(c) Hire more senior professors and discharge junior professors.

(d) Discharge senior professors and hire more junior professors.

(e) Discharge all senior professors and half of the junior professors.

COMMUNICATION AND REASONING EXERCISES

33. Carefully explain the difference between *cost* and *marginal cost* **(a)** in terms of their mathematical definition, **(b)** in terms of graphs, and **(c)** in terms of interpretation.

34. If your analysis of a manufacturing company yielded positive marginal profit but negative profit at the company's current production levels, what would you advise the company to do?

35. If a company's marginal average cost is zero at the current production level, positive for a slightly higher production level, and negative for a slightly lower production level, what should you advise the company to do?

36. The **acceleration** of cost is defined as the derivative of the marginal cost function: that is, the derivative of the derivative—or *second derivative*—of the cost function. What are the units of acceleration of cost, and how does one interpret this measure?

▶ ══════ **3.6** MORE ON LIMITS, CONTINUITY, AND DIFFERENTIABILITY

LIMITS

As we saw in Section 3, limits need not always exist. The next two examples show ways in which a limit can fail to exist.

▼ **EXAMPLE 1**

Let $f(x) = \dfrac{|x|}{x}$. Does $\lim\limits_{x \to 0} f(x)$ exist?

SOLUTION   We shall investigate this limit using both the numerical and the graphical approaches.

Numerical Approach   We construct a table of values, with x approaching 0 from both sides.

x approaches 0 from the left. $\rightarrow$ $\leftarrow$ x approaches 0 from the right.

x	−0.1	−0.01	−0.001	−0.0001	0	0.0001	0.001	0.01	0.1
$f(x) = \dfrac{|x|}{x}$	−1	−1	−1	−1		1	1	1	1

The table shows that $f(x)$ does not approach the same limit as x approaches 0 from both sides. In fact,

$$\lim_{x \to 0^-} f(x) = -1$$

and

$$\lim_{x \to 0^+} f(x) = 1.$$

Since the one-sided limits have different values, we conclude that $\lim\limits_{x \to 0} f(x)$ does not exist.

Graphical Approach   Consider the graph of $f(x)$. The domain of f consists of all real numbers except 0, and, as the table suggests,

$$f(x) = \begin{cases} 1 & \text{if } x > 0 \\ -1 & \text{if } x < 0. \end{cases}$$

(think about what $|x|$ means to see why). The graph of f is shown in Figure 1. As in Section 3, we imagine a pencil point or the "trace" cursor on a graphing calculator approaching $x = 0$ from either side (Figure 2).

Looking at the y-coordinates, we can see graphically that

$$\lim_{x \to 0^-} f(x) = -1$$

and

$$\lim_{x \to 0^+} f(x) = 1.$$

Since the one-sided limits have different values, we conclude that $\lim\limits_{x \to 0} f(x)$ does not exist.

If your graphing calculator does not have an absolute value function, you can still enter the function $f(x)$ by using the fact that

$$|x| = \sqrt{x^2},$$

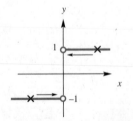

FIGURE 1

Graph of $f(x) = \dfrac{|x|}{x}$

FIGURE 2

so you can enter

$$Y_1 = ((X^2)^{\wedge}(0.5))/X$$

You can now use "trace" to move the cursor, reading the coordinates as you go.

▼ **EXAMPLE 2**

Does $\lim\limits_{x \to 0^+} \dfrac{1}{x}$ exist?

SOLUTION   We shall look at this limit using both the numerical and graphical approaches.

Numerical Approach   Since we are asked for only the right-hand limit, we need only list values of x approaching 0 from the right.

← x approaches 0 from the right.

x	0	0.0001	0.001	0.01	0.1
$f(x) = \dfrac{1}{x}$		10,000	1,000	100	10

What seems to be happening as x approaches 0 from the right is that $f(x)$ is increasing **without bound.** That is, if you name any number, no matter how large, $f(x)$ will be even larger if x is sufficiently close to zero. Since $f(x)$ is not approaching a specific real number, the limit does not exist. Since $f(x)$ is becoming arbitrarily large, we also say that the limit **diverges to $+\infty$,** and we write

$$\lim_{x \to 0^+} \frac{1}{x} = +\infty.$$

Graphical Approach   We again use the graphing calculator experiment. First, recall that the graph of $f(x) = \frac{1}{x}$ is the standard hyperbola shown in Figure 3.

The figure also shows the pencil point moving so that its x-coordinate approaches 0 from the right. Since the point moves along the graph, it is forced to go higher and higher. In other words, its y-coordinate becomes larger and larger, approaching $+\infty$. Thus, we conclude that

$$\lim_{x \to 0^+} \frac{1}{x} = +\infty.$$

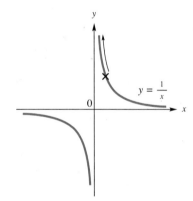

FIGURE 3

Before we go on... You should also check that

$$\lim_{x \to 0^-} \frac{1}{x} = -\infty.$$

We say that as x approaches 0 from the left, $\frac{1}{x}$ diverges to $-\infty$.

CONTINUITY

Recall that f is continuous at a if $\lim_{x \to a} f(x) = f(a)$. We saw in Section 3 that all closed-form functions are continuous on their domains. There are a number of ways in which a function can fail to be continuous at a point in its domain. The next example shows some of them.

▼ **EXAMPLE 3**

Let $f(x)$, $g(x)$, $h(x)$, and $k(x)$ be specified by the graphs in Figure 4. Determine which, if any, are continuous on their domains.

(a)

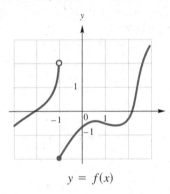

$y = f(x)$

(b)

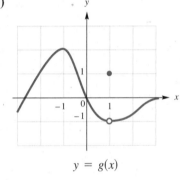

$y = g(x)$

(c)

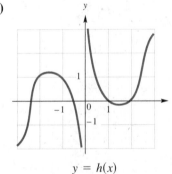

$y = h(x)$

(d)

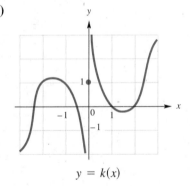

$y = k(x)$

FIGURE 4

SOLUTION   Looking at the graph of f, we notice that there is a break at the point where $x = -1$, so we suspect that it is not continuous there. If we use the graphical method to investigate the limit as $x \to -1$, we find that the limit does not exist. Thus, since -1 is a point in the domain of f ($f(-1) = -2$ by the graph) and the limit as $x \to -1$ fails to exist, f is not continuous at -1 and is not continuous on its domain. The particular kind of discontinuity we see here is called a **jump discontinuity.** Since the left and right limits are not equal, the value of f takes a sudden jump when x passes -1.

The graph of g has a "misplaced point" at $x = 1$. We can see graphically that

$$\lim_{x \to 1} g(x) = -1 \neq g(1) = 1.$$

Therefore, g is not continuous at 1, so it is not continuous on its domain. This type of discontinuity is called a **removable discontinuity,** since it could be removed by redefining g at one point: redefining $g(1) = -1$. If we made this change in g, the function would be continuous on its domain.

The graph of h has a break at $x = 0$, but notice that 0 is not in the domain of h: $h(0)$ is not defined. If you choose any a that *is* in its domain, h will be continuous at a, so we conclude that h is continuous on its domain. The reason that its graph is broken is that its domain is broken.

Finally, the function k is almost the same as the function h, except that it is defined at $x = 0$, since $k(0) = 1$. Since the limit of k as $x \to 0$ does not exist, we conclude that k is *not* continuous at 0 and thus not continuous on its domain.

DIFFERENTIABILITY

Since the derivative is defined using a limit, the derivative will not exist if the limit does not. If $f'(a)$ exists, we say that f is **differentiable at a.**

▼ EXAMPLE 4

If $f(x) = |x|$ find $f'(0)$.

SOLUTION   We compute

$$f'(x) = \lim_{h \to 0} \frac{f(x + h) - f(x)}{h}$$

$$f'(0) = \lim_{h \to 0} \frac{f(0 + h) - f(0)}{h}$$

$$= \lim_{h \to 0} \frac{|0 + h| - |0|}{h}$$

$$= \lim_{h \to 0} \frac{|h|}{h}.$$

In Example 1, we saw that this limit does not exist. We say that the derivative of f does not exist at 0, or that f is not differentiable at 0. You can see in the graph of f why this happens (Figure 5).

At $x = 0$ there is a sharp corner. What would be the tangent line there? There is no single reasonable answer. If you look at secant lines through $(0, 0)$, you see lines with slope -1 or $+1$, depending on whether you take the second point to the left or to the right of the origin (try it). We simply say that there is no tangent line, and thus no derivative, at 0.

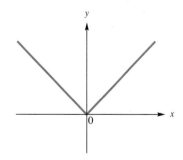

FIGURE 5

Graph of $f(x) = |x|$

Before we go on... Note that although the function $y = |x|$ is not differentiable at $x = 0$, it *is* continuous at every point on its domain. A function may be continuous at a point without being differentiable at that point. However, if a function is differentiable at a point, it can be shown that the function must be continuous there as well (think about what happens to the numerator in the difference quotient when you let $h \to 0$).

Recall that we were able to obtain an accurate estimate of the slope of the tangent at a point on a graph by "zooming in" to the graph in question until it appeared to be a straight line. If you use a graphing calculator to plot $f(x) = |x|$, you will notice that this approach fails at $(0, 0)$: no matter how many times you magnify the graph near $(0, 0)$, it will always appear just as sharp.*

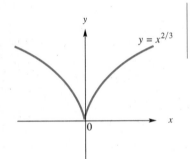

FIGURE 6

▼ **EXAMPLE 5**

The function $f(x) = x^{2/3}$ is continuous but not differentiable at 0.

The reason for this becomes clear if we find the derivative: $f'(x) = \frac{2}{3}x^{-1/3}$. This is defined except at $x = 0$, where it calls for dividing by 0. The derivatives near 0 are very large positive or negative numbers, approaching $+\infty$ as $x \to 0^+$ and $-\infty$ as $x \to 0^-$. This means that the tangent lines are approaching vertical as x gets close to 0. In other words, the graph comes to a very sharp point (called a **cusp**) at 0. The graph is shown in Figure 6.

If you try to graph the function $f(x) = x^{2/3}$ using the format

$$Y_1 = X^\wedge(2/3)$$

you will get only the right-hand portion of Figure 6, since calculators are (usually) not programmed to raise negative numbers to fractional exponents. To avoid this difficulty, you can take advantage of the identity

$$x^{2/3} = (x^2)^{1/3},$$

so that it is always a nonnegative number that is being raised to a fractional exponent. Thus, use the format

$$Y_1 = (X^\wedge 2)^\wedge(1/3)$$

to obtain both portions of the graph.

▼ *The claim is true provided you zoom in such a way that you preserve the width-to-height ratio of the window. For instance, you can start with the window $-1 \le x \le 1$, $-1 \le y \le 1$, and then $-0.1 \le x \le 0.1$, $-0.1 \le y \le 0.1$, and so on, keeping the width-to-height ratio 1:1.

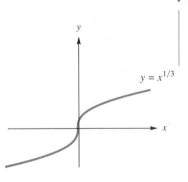

FIGURE 7

▼ **EXAMPLE 6**

The function $f(x) = x^{1/3}$ is continuous but not differentiable at 0.

Algebraically, this is similar to Example 5: $f'(x) = \frac{1}{3}x^{-2/3}$. Again, this is not defined at 0, and this time the derivative approaches $+\infty$ as $x \to 0$. If we look at Figure 7, we notice that instead of coming to a sharp point, the graph has a vertical tangent line. Since such a line has undefined slope, the derivative is undefined.

You can use the format

$$Y_1 = X^{(1/3)}$$

to draw this graph on most graphing calculators. These calculators recognize that it is legitimate to raise a negative number to a power $1/n$ if n is an odd integer. (Compare this comment to the comment in the previous example.)

Since a curve must not have a corner (a sudden change of direction) if it is to be differentiable, we sometimes say that a function is *smooth* rather than differentiable. Be warned, however, that this word is used by different people for different things, and it is often used for something even more stringent than being differentiable.

LIMITS AT INFINITY

In another useful kind of limit we let x approach either $+\infty$ or $-\infty$, by which we mean that we let x get arbitrarily large or let x become an arbitrarily large negative number. The next example illustrates this.

▼ **EXAMPLE 7**

Let $f(x) = \dfrac{2x^2 - 4x}{x^2 - 1}$, with domain all real numbers except 1 and -1. Find the following limits:

(a) $\lim\limits_{x \to +\infty} f(x)$ **(b)** $\lim\limits_{x \to -\infty} f(x)$.

SOLUTION

Numerical Approach   **(a)** By saying that x is "approaching $+\infty$," we mean that x is getting larger and larger without bound, so we make the following table.

x approaches $+\infty$. $\longrightarrow$

x	10	100	1,000	10,000	100,000
$f(x) = \dfrac{2x^2 - 4x}{x^2 - 1}$	1.6162	1.9602	1.9960	2.0000	2.0000

(Note that we are only approaching $+\infty$ from the left, as we can hardly approach it from the right!) What seems to be happening is that $f(x)$ is approaching 2. Thus, we write

$$\lim_{x \to +\infty} f(x) = 2.$$

(b)   Here, x is approaching $-\infty$, so we make a similar table, this time with x assuming negative values of greater and greater magnitude (read this table from right to left, as you would the graph).

$\longleftarrow$ x approaching $-\infty$.

x	$-100{,}000$	$-10{,}000$	$-1{,}000$	-100	-10
$f(x) = \dfrac{2x^2 - 4x}{x^2 - 1}$	2.0000	2.0004	2.0040	2.0402	2.4242

Once again, $f(x)$ is approaching 2. Thus,

$$\lim_{x \to -\infty} f(x) = 2$$

as well.

Graphical Approach   Figure 8 shows a graphing calculator plot of the function f.

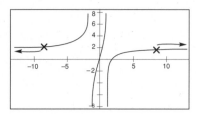

FIGURE 8

We have included the cursor marks for the graphing calculator experiment. To see what happens as $x \to +\infty$, start with your pencil point at the cursor mark on the right, and move to the right toward $+\infty$ while you read off the y-coordinate. The further you move to the right, the closer the y-coordinate gets to 2, showing once again that $f(x) \to 2$ as $x \to +\infty$. Similarly, if you start at the cursor mark on the left and move left toward $-\infty$, the y-coordinate also approaches 2. Thus, $f(x) \to 2$ as $x \to -\infty$.

Algebraic Approach   While calculating the values for the tables in the numerical approach, you might have noticed that the highest power of x in both the numerator and denominator dominated the calculations. For instance, when $x = 100{,}000$, the $2x^2$ in the numerator has the value of 20,000,000,000, whereas $4x$ has the comparatively insignificant value of 400,000. Similarly, the x^2 in the denominator overwhelms the -1. In other words, for large values of x (or negative values with large magnitude),

$$\frac{2x^2 - 4x}{x^2 - 1} \approx \frac{2x^2}{x^2} = 2.$$

Alternatively, do the following bit of algebra first.

$$\frac{2x^2 - 4x}{x^2 - 1} = \frac{(2x^2 - 4x)/x^2}{(x^2 - 1)/x^2} \qquad \text{Divide top and bottom by the highest power of } x.$$

$$= \frac{2 - \dfrac{4}{x}}{1 - \dfrac{1}{x^2}}.$$

Now, as x approaches $+\infty$ or $-\infty$, both $4/x$ and $1/x^2$ approach 0. Therefore,

$$\lim_{x \to \pm\infty} \frac{2x^2 - 4x}{x^2 - 1} = \frac{2}{1} = 2.$$

Before we go on... We say that the graph of f has a **horizontal asymptote** at $y = 2$ because of the limits we have just calculated. This means that the graph approaches the horizontal line $y = 2$ far to the right or left (in this case, both to the right and left). Figure 9 shows the graph of f together with the line $y = 2$.

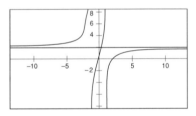

FIGURE 9

The graph also reveals additional interesting information: as $x \to 1^+$, $f(x) \to -\infty$, and as $x \to 1^-$, $f(x) \to +\infty$. Thus,

$$\lim_{x \to 1} f(x) \text{ does not exist.}$$

See if you can determine what is happening as $x \to -1$.

In the above example, $f(x)$ was a **rational function:** a quotient of polynomial functions. In the algebraic approach to that example, we calculated the limit of $f(x)$ at $\pm\infty$ by ignoring all powers of x in both the numerator and denominator except for the largest. It is possible to prove that this procedure is valid for any rational function (using the idea of dividing top and bottom by the highest power of x present).

EVALUATING THE LIMIT OF A RATIONAL FUNCTION AT $\pm\infty$

If $f(x)$ has the form

$$f(x) = \frac{c_n x^n + \ldots + c_2 x^2 + c_1 x + c_0}{d_m x^m + \ldots + d_2 x^2 + d_1 x + d_0}$$

with the c_i and d_i constants ($c_n \neq 0$ and $d_m \neq 0$), then we can calculate the limit of $f(x)$ as $x \to \pm\infty$ by ignoring all powers of x except the highest in both the numerator and denominator. Thus,

$$\lim_{x \to \pm\infty} f(x) = \lim_{x \to \pm\infty} \frac{c_n x^n}{d_m x^m}.$$

▼ **EXAMPLE 8**

Calculate

    (a) $\displaystyle\lim_{x \to +\infty} \frac{3x^4 - x^3 + 1}{x^3 + 40x^2}$ **(b)** $\displaystyle\lim_{x \to -\infty} \frac{x^3 + 40x^2}{10x^4}.$

SOLUTION

(a)   Ignoring all but the highest powers of x, we have

$$\lim_{x \to +\infty} \frac{3x^4 - x^3 + 1}{x^3 + 40x^2} = \lim_{x \to +\infty} \frac{3x^4}{x^3}$$

$$= \lim_{x \to +\infty} 3x.$$

Now we have a far simpler limit to evaluate—we can even say what the limit is without a table: as $x \to +\infty$, then $3x \to +\infty$ as well. Thus,

$$\lim_{x \to +\infty} \frac{3x^4 - x^3 + 1}{x^3 + 40x^2} = +\infty.$$

(b) $\displaystyle\lim_{x \to -\infty} \frac{x^3 + 40x^2}{10x^4} = \lim_{x \to -\infty} \frac{x^3}{10x^4}$

$$= \lim_{x \to -\infty} \frac{1}{10x}$$

At this stage, a table would be helpful, but once again we can manage without it. If x is, say, $-10,000$, then $1/(10x) = -1/100,000 = -0.00001$, extremely close to zero. In fact, the larger x gets in magnitude, the smaller $1/(10x)$ must get. Thus,

$$\lim_{x \to -\infty} \frac{x^3 + 40x^2}{10x^4} = 0.$$

▶ **3.6 EXERCISES**

In each of Exercises 1–12, the graph of f is given. Compute the asked-for limits. If a particular limit fails to exist, say why (for example, it might diverge to +∞).

1. (a) $\lim\limits_{x \to +\infty} f(x)$ **(b)** $\lim\limits_{x \to -\infty} f(x)$ **2. (a)** $\lim\limits_{x \to +\infty} f(x)$ **(b)** $\lim\limits_{x \to -\infty} f(x)$

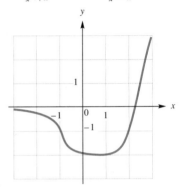
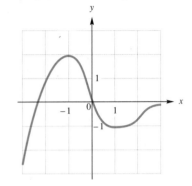

3. (a) $\lim\limits_{x \to +\infty} f(x)$ **(b)** $\lim\limits_{x \to -\infty} f(x)$ **4. (a)** $\lim\limits_{x \to \infty} f(x)$ **(b)** $\lim\limits_{x \to -\infty} f(x)$

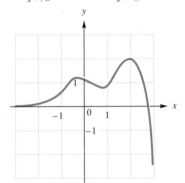
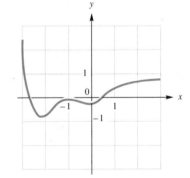

5. (a) $\lim\limits_{x \to 0^-} f(x)$ **(b)** $\lim\limits_{x \to 0^+} f(x)$ **(c)** $\lim\limits_{x \to 0} f(x)$ **6. (a)** $\lim\limits_{x \to 0^-} f(x)$ **(b)** $\lim\limits_{x \to 0^+} f(x)$ **(c)** $\lim\limits_{x \to 0} f(x)$

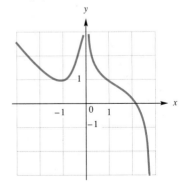
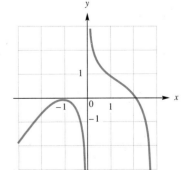

7. (a) $\lim\limits_{x\to 0^-} f(x)$ **(b)** $\lim\limits_{x\to 0^+} f(x)$ **(c)** $\lim\limits_{x\to 0} f(x)$ **8. (a)** $\lim\limits_{x\to 0^-} f(x)$ **(b)** $\lim\limits_{x\to 0^+} f(x)$ **(c)** $\lim\limits_{x\to 0} f(x)$

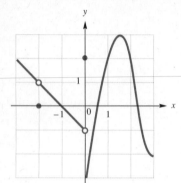

9. (a) $\lim\limits_{x\to 0} f(x)$ **(b)** $\lim\limits_{x\to -\infty} f(x)$ **(c)** $\lim\limits_{x\to +\infty} f(x)$ **10. (a)** $\lim\limits_{x\to 0} f(x)$ **(b)** $\lim\limits_{x\to -\infty} f(x)$ **(c)** $\lim\limits_{x\to +\infty} f(x)$

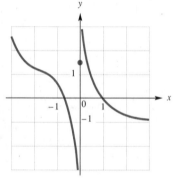

11. (a) $\lim\limits_{x\to -\infty} f(x)$ **(b)** $\lim\limits_{x\to +\infty} f(x)$ **12. (a)** $\lim\limits_{x\to -\infty} f(x)$ **(b)** $\lim\limits_{x\to +\infty} f(x)$

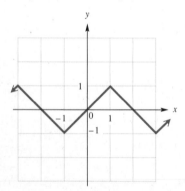
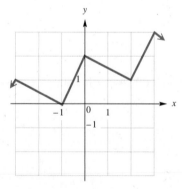

Calculate the limits in Exercises 13–38. If a limit fails to exist, say why. (Remember that you can use the numerical or graphical approaches if an algebraic approach will not work.)

13. $\lim\limits_{x\to 0^+} \dfrac{1}{x^2}$

14. $\lim\limits_{x\to 0^+} \dfrac{1}{x^2 - x}$

15. $\lim\limits_{x\to -1} \dfrac{x^2 + 1}{x + 1}$

16. $\lim\limits_{x\to -1^-} \dfrac{x^2 + 1}{x + 1}$

17. $\lim\limits_{x \to \infty} \dfrac{3x^2 + 10x - 1}{2x^2 - 5x}$

18. $\lim\limits_{x \to \infty} \dfrac{6x^2 + 5x + 100}{3x^2 - 9}$

19. $\lim\limits_{x \to \infty} \dfrac{x^5 - 1{,}000x^4}{2x^5 + 10{,}000}$

20. $\lim\limits_{x \to \infty} \dfrac{x^6 + 3{,}000x^3 + 1{,}000{,}000}{2x^6 + 1{,}000x^3}$

21. $\lim\limits_{x \to \infty} \dfrac{10x^2 + 300x + 1}{5x + 2}$

22. $\lim\limits_{x \to \infty} \dfrac{2x^4 + 20x^3}{1{,}000x^3 + 6}$

23. $\lim\limits_{x \to \infty} \dfrac{10x^2 + 300x + 1}{5x^3 + 2}$

24. $\lim\limits_{x \to \infty} \dfrac{2x^4 + 20x^3}{1{,}000x^6 + 6}$

25. $\lim\limits_{x \to -\infty} \dfrac{3x^2 + 10x - 1}{2x^2 - 5x}$

26. $\lim\limits_{x \to -\infty} \dfrac{6x^2 + 5x + 100}{3x^2 - 9}$

27. $\lim\limits_{x \to -\infty} \dfrac{x^5 - 1{,}000x^4}{2x^5 + 10{,}000}$

28. $\lim\limits_{x \to -\infty} \dfrac{x^6 + 3{,}000x^3 + 1{,}000{,}000}{2x^6 + 1{,}000x^3}$

29. $\lim\limits_{x \to -\infty} \dfrac{10x^2 + 300x + 1}{5x + 2}$

30. $\lim\limits_{x \to -\infty} \dfrac{2x^4 + 20x^3}{1{,}000x^3 + 6}$

31. $\lim\limits_{x \to -\infty} \dfrac{10x^2 + 300x + 1}{5x^3 + 2}$

32. $\lim\limits_{x \to -\infty} \dfrac{2x^4 + 20x^3}{1{,}000x^6 + 6}$

33. $\lim\limits_{x \to 0} |x|$

34. $\lim\limits_{x \to 1} |x - 1|$

35. $\lim\limits_{x \to 1} \dfrac{|x - 1|}{x - 1}$

36. $\lim\limits_{x \to -2} \dfrac{|x + 2|}{x + 2}$

E/L 37. $\lim\limits_{x \to 2} e^{x - 2}$

E/L 38. $\lim\limits_{x \to +\infty} e^{-x}$

E/L 39. $\lim\limits_{x \to +\infty} xe^{-x}$

E/L 40. $\lim\limits_{x \to -\infty} xe^{x}$

E/L 41. $\lim\limits_{x \to 2^+} \dfrac{e^{-1/(x-2)}}{x - 2}$

E/L 42. $\lim\limits_{x \to 2} \dfrac{e^{-1/(x-2)}}{x - 2}$

In each of Exercises 43–50, find all points of discontinuity of the given function. Classify discontinuities as removable, jump, or other.

43. $f(x) = \begin{cases} x + 2 & \text{if } x < 0 \\ 2x - 1 & \text{if } x \ge 0 \end{cases}$

44. $f(x) = \begin{cases} 1 - x & \text{if } x \le 1 \\ x + 2 & \text{if } x > 1 \end{cases}$

45. $g(x) = \begin{cases} x + 2 & \text{if } x < 0 \\ 2x + 2 & \text{if } x \ge 0 \end{cases}$

46. $g(x) = \begin{cases} 1 - x & \text{if } x \le 1 \\ x - 1 & \text{if } x > 1 \end{cases}$

47. $h(x) = \begin{cases} x + 2 & \text{if } x < 0 \\ 0 & \text{if } x = 0 \\ 2x + 2 & \text{if } x > 0 \end{cases}$

48. $h(x) = \begin{cases} 1 - x & \text{if } x < 1 \\ 1 & \text{if } x = 1 \\ x + 2 & \text{if } x > 1 \end{cases}$

49. $f(x) = \begin{cases} 1/x & \text{if } x < 0 \\ x & \text{if } x \ge 0 \end{cases}$

50. $f(x) = \begin{cases} x^2 & \text{if } x < 0 \\ x & \text{if } x \ge 0 \end{cases}$

In each of Exercises 51–56, find all points where the given function is not differentiable.

51. $f(x) = \begin{cases} x^2 & \text{if } x < 0 \\ x & \text{if } x \ge 0 \end{cases}$

52. $f(x) = \begin{cases} x^2 & \text{if } x < 0 \\ x^3 & \text{if } x \ge 0 \end{cases}$

53. $g(x) = x^{4/3}$

54. $g(x) = x^{4/9}$

55. $h(x) = |x - 1|$

56. $h(x) = |x + 2|$

APPLICATIONS

57. Social Ills   The number of DWI arrests in New Jersey during the period from 1990 to 1993 can be modeled by the equation*

$$n(t) = \frac{18,000}{(t + 1)^{0.4}}.$$

Here, $n(t)$ is the number of DWI arrests in year t, with $t = 0$ representing 1990. Calculate $\lim\limits_{t \to +\infty} n(t)$ and interpret your answer.

58. Social Ills   Repeat the previous exercise using the linear model

$$n(t) = -2.4t + 19.5.$$

59. Acquisition of Language   The percentage $p(t)$ of children who are able to speak in at least single words by the age of t months can be approximated by the equation[†]

$$p(t) = 100\left(1 - \frac{12,196}{t^{4.478}}\right) \quad (t \geq 8.5).$$

Calculate $\lim\limits_{t \to +\infty} p(t)$ and $\lim\limits_{t \to +\infty} p'(t)$ and interpret the results.

60. Acquisition of Language   The percentage $q(t)$ of children who are able to speak in sentences of five or more words by the age of t months can be approximated by the equation[†]

$$q(t) = 100\left(1 - \frac{5.2665 \times 10^{17}}{t^{12}}\right) \quad (t \geq 30).$$

If p is the function referred to in the previous exercise, calculate $\lim\limits_{t \to +\infty} [p(t) - q(t)]$ and interpret the result.

E/L 61. Logistic Growth in Demand   The demand for a new product can be modeled by a **logistic** curve of the form

$$q(t) = \frac{N}{1 + ke^{-rt}},$$

where $q(t)$ is the total number of units sold t months after the introduction of the new product, and N, k, and r are constants that depend on the product and the market. Assume that the demand for video game units is determined by the above formula, with $N = 10,000$, $k = 0.5$, and $r = 0.4$. Find $\lim\limits_{t \to +\infty} q(t)$ and interpret the result.

E/L 62. Information Highway   The amount of information transmitted each month on the National Science Foundation's Internet network for the years 1988–1994 can be modeled by the equation

$$q(t) = \frac{2e^{0.69t}}{3 + 1.5e^{-0.4t}},$$

where q is the amount of information transmitted each month in billions of data packets, and t is the number of years since the start of 1988.[‡] Find $\lim\limits_{t \to +\infty} q(t)$ and interpret the result.

COMMUNICATION AND REASONING EXERCISES

63. Find a function that is continuous everywhere except at two points.

64. Find a function that is continuous everywhere except at three points.

65. Find a function that is continuous everywhere but not differentiable at two points.

66. Find a function that is continuous everywhere except at one point and differentiable everywhere except at two points.

▼ * This is a regression model based on data gleaned from a graph. Source: New Jersey Administrative Office of the Courts/*The New York Times*, September 26, 1994, p. B1.

† The model is the authors' and is based on data presented in the article "The Emergence of Intelligence" by William H. Calvin, *Scientific American*, October, 1994, pp. 101–7.

‡ This is the authors' model, based on figures published in *The New York Times*, Nov. 3, 1993.

▶ You're the Expert

Reducing Sulfur Emissions

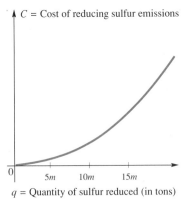

C = Cost of reducing sulfur emissions

q = Quantity of sulfur reduced (in tons)

FIGURE 1

Cost Curve before Emission Charge

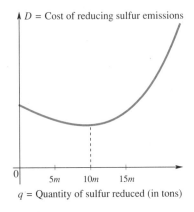

D = Cost of reducing sulfur emissions

q = Quantity of sulfur reduced (in tons)

FIGURE 2

Cost Curve with Emission Charge

The Environmental Protection Agency (EPA) wishes to formulate a policy that will encourage utilities to reduce sulfur emissions. Its goal is to reduce annual emissions of sulfur dioxide by a total of 10 million tons from the current level of 25 million tons by imposing a fixed charge for every ton of sulfur released into the environment per year. As a consultant to the EPA, you must determine the amount to be charged per ton of sulfur emissions.

You have the following data, which show the marginal cost to the utility industry of reducing sulfur emissions at several levels of reduction.*

Reduction (millions of tons)	8	10	12
Marginal Cost ($ per ton)	270	360	779

If $C(q)$ is the cost of removing q tons of sulfur dioxide, the table tells you that $C'(8,000,000) = \$270$ per ton, $C'(10,000,000) = \$360$ per ton, and $C'(12,000,000) = \$779$ per ton. Recalling that $C'(q)$ is the slope of the tangent to the graph of the cost function, you see from the table that this slope is positive and increasing as q increases, so the graph of this cost function has the general shape shown in Figure 1. (Notice that the slope is increasing as you move to the right.)

Thus the utility industry has no cost incentive to reduce emissions. What you would like to do—if the goal of reducing total emissions by 10 million tons is to be accomplished—is to alter this cost curve so that it has the general shape shown in Figure 2.

In this curve, the cost D to utilities is lowest at a reduction level of 10 million tons, so if the utilities act to minimize cost, they can be expected to reduce emissions by 10 million tons, which is the EPA goal. From the graph, you can see that $D'(10,000,000) = \$0$ per ton, while $D'(q)$ is negative for $q < 10,000,000$ and positive for $q > 10,000,000$.

At first, you are bothered by the fact that you were not given a cost function. Only the marginal costs were supplied, but you decide to work as best you can without knowing the original cost function $C(q)$.

You now assume that the EPA will impose an annual emission charge of $\$k$ per ton of sulfur released into the environment. It is your job to calculate k. Since you are working with q as the independent variable, you decide that it would be best to formulate the emission charge as a function of q, where q represents the amount by which sulfur emissions are *reduced*. The relationship between the annual sulfur emissions and the amount q by which emis-

▼ * These figures were produced in a computerized study of reducing sulfur emissions from the 1980 level by the given amounts. Source: Congress of the United States, Congressional Budget Office, *Curbing Acid Rain: Cost, Budget and Coal Market Effects* (Washington, DC: Government Printing Office, 1986), xx, xxii, pp. 23, 80.

sions are reduced from the original 25 million tons is given by

$$\text{Annual sulfur emissions} = \text{Original emissions}$$
$$- \text{Amount of reduction}$$
$$= 25,000,000 - q.$$

Thus, the total annual emission charge to the utilities is

$$k(25,000,000 - q) = 25,000,000k - kq.$$

This results in a total cost to the utilities of

$$\text{Total cost} = \text{Cost of reducing emissions} +$$
$$\text{Emission charge}$$
$$D(q) = C(q) + 25,000,000k - kq.$$

Even though you have no idea of the form of $C(q)$, you remember that the derivative of a sum is the sum of the derivatives, and so you differentiate both sides with respect to q and obtain

$$D'(q) = C'(q) + 0 - k$$
$$= C'(q) - k.$$

Remember that you want

$$D'(10,000,000) = 0.$$

Thus,

$$C'(10,000,000) - k = 0.$$

Referring to the table, this says that

$$360 - k = 0,$$

so

$$k = \$360 \text{ per ton.}$$

In other words, all you need to do is set the emission charge at $k = \$360$ per ton of sulfur emitted. Further, to ensure that the resulting curve will have the general shape shown in Figure 2, you would like to have $D'(q)$ negative for $q < 10,000,000$ and positive for $q > 10,000,000$. To check this, write

$$D'(q) = C'(q) - k$$
$$= C'(q) - 360$$

and refer to the table to obtain

$$D'(8,000,000) = 270 - 360 = -90 < 0. \quad ✔$$

and

$$D'(12,000,000) = 779 - 360 = 419 > 0. \quad ✔$$

Thus, based on the given data, the resulting curve will have the shape you require. You therefore inform the EPA that an annual emissions charge of $360 per ton of sulfur released into the environment will create the desired incentive: to reduce sulfur emissions by 10 million tons per year.

One week later, you are informed that this cost would be unrealistic, as the utilities cannot possibly afford such a cost, and are asked whether there is an alternative plan that accomplishes the 10-million-ton reduction goal, and yet is cheaper to the utilities by $5 billion dollars per year. You then look at your expression for the emission charge,

$$25,000,000k - kq,$$

and notice that, if you decrease this amount by $5 billion, the derivative will not change at all, since the derivative of a constant is zero. Thus, you propose the following revised formula for the emission charge:

$$25,000,000k - kq - 5,000,000,000$$
$$= 25,000,000(360) - 360q - 5,000,000,000$$
$$= 4,000,000,000 - 360q.$$

At the expected reduction level of 10 million tons, the total amount paid by the utilities will then be

$$4,000,000,000 - 360(10,000,000) = \$400,000,000.$$

Thus, your revised proposal is the following: Impose an annual emissions charge of $360 per ton of sulfur released into the environment, and hand back $5 billion in the form of subsidies. This policy will cause the utilities industry to reduce sulfur emissions by 10 million tons per year, and it will result in $400 million in annual revenues to the government.

Notice that this policy also provides incentive for the utilities to search for cheaper ways to reduce emissions. For instance, a reduction level of 12 million tons will result in an additional cost to the industry of

$$4,000,000,000 - 360(12,000,000) = -\$320,000,000.$$

The fact that this is negative means that the government would be paying the utilities $320 million in annual subsidies.

Exercises

1. Excluding subsidies, what should the annual emission charge be if the goal is to reduce sulfur emissions by 8 million tons?

2. Excluding subsidies, what should the annual emission charge be if the goal is to reduce sulfur emissions by 12 million tons?

3. What is the *marginal emission charge* in your revised proposal (as stated before the exercise set)? What is the relationship between the marginal cost of reducing sulfur emissions (before emissions charges are implemented) and the marginal emission charge, at the optimal reduction?

4. We said that the revised proposal provided an incentive for utilities to find cheaper ways to reduce emissions. How would $C(q)$ have to change to make 12 million tons the optimum reduction under your revised proposal?

5. What change in $C(q)$ would make 8 million tons the optimum reduction?

6. If the scenario in Exercise 5 took place, what would the EPA have to do in order to make 10 million tons the optimal reduction once again?

7. Due to intense lobbying by the utility industry, you are asked to revise the proposed policy so that the utility industry will pay no charge if sulfur emissions are reduced by the desired 10 million tons. How can you accomplish this?

8. Suppose that instead of imposing a fixed charge per ton of emission, you decide to use a sliding scale, so that the total charge to the industry for annual emissions of x tons will be $\$ kx^2$ for some k. What must k be to again make 10 million tons the optimum reduction?

9. Given the total charge for sulfur emissions used in Exercise 8, what would be the *marginal* charge to the industry at a reduction level of q tons? How does this marginal charge compare to the original proposal at a reduction level of 10 million tons? At 8 million tons? At 12 million tons?

▶ ## Review Exercises

In each of Exercises 1–4, several slopes are given. For each slope, determine at which of the labeled points on the graph the tangent line has that slope.

1. (a) 0   **(b)** 2   **(c)** -1

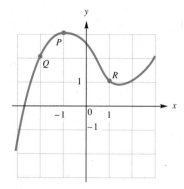

2. (a) 0   **(b)** 1

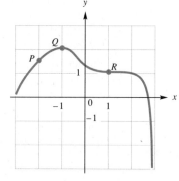

3. (a) 0   **(b)** -1

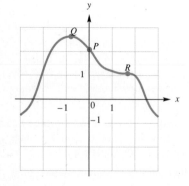

4. (a) 0   **(b)** 3   **(c)** 1

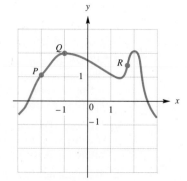

In each of Exercises 5–8, find the approximate coordinates of all points (if any) where the slope of the tangent is: (a) 0,   (b) 2,   (c) −2.

5.

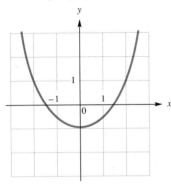

6.

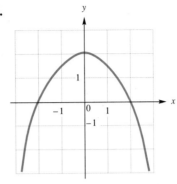

7.

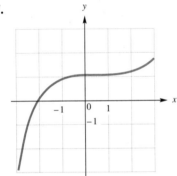

8.

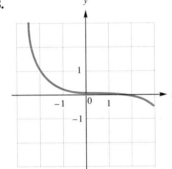

In Exercises 9–30, calculate the derivative of the given function mentally.

9. $f(x) = 3x^2 - 4x^{-1}$

10. $f(x) = 2x^4 + 3x^{-2}$

11. $f(x) = 1 + \dfrac{x^2}{2}$

12. $f(x) = x - \dfrac{x^3}{3}$

13. $g(x) = x^{1.2} + x^{2.3}$

14. $g(x) = 2x^{1.5} - x^{0.5}$

15. $g(s) = 3s^{0.2} - 44s$

16. $g(s) = 2s^{-0.1} - 5s$

17. $h(t) = at^3 - bt^2 + c$ (a, b, c constants)

18. $h(t) = \dfrac{t^2}{a} - \dfrac{t}{b}$ (a, b nonzero constants)

19. $r(x) = 2\sqrt{x} + \sqrt[3]{x}$

20. $r(x) = \dfrac{1}{x} - \dfrac{1}{x^2}$

21. $f(x) = 3x + \dfrac{1}{x^3}$

22. $f(x) = 2\sqrt{x} + x$

23. $h(x) = \dfrac{x}{5} - \dfrac{5}{x}$

24. $h(x) = \dfrac{x^2}{4} - \dfrac{4}{x^2}$

25. $r(x) = \sqrt{9x}$

26. $r(x) = 3(x - \sqrt{x})$

27. $g(x) = x^2(2x + 1)$

28. $g(x) = \dfrac{x^2 - 1}{x}$

29. $g(r) = \dfrac{1}{\sqrt{r}} + \dfrac{2}{r}$

30. $g(r) = \dfrac{2}{\sqrt{r}} - \dfrac{\sqrt{r}}{2}$

In each of Exercises 31–40, find the equation of the tangent line at the point on the graph of the given function with the indicated first coordinate.

31. $f(x) = x^2 + 2x - 1; \quad x = -2$

32. $g(x) = 3x^3 - x; \quad x = -2$

33. $g(t) = \dfrac{1}{5t^4}; \quad t = 1$

34. $s(t) = \dfrac{1}{3t^3}; \quad x = -2$

35. $h(s) = \dfrac{1}{s} + s; \quad s = 2$

36. $h(s) = \sqrt{s} - s; \quad s = 9$

37. $r(t) = \dfrac{t^2}{3} - \dfrac{2t^3}{6}; \quad t = -1$

38. $r(t) = \dfrac{2t^3}{9} - t; \quad t = -1$

39. $h(t) = \dfrac{t^2 - 1}{t}; \quad t = 2$

40. $h(t) = \sqrt{t} + 10t; \quad t = 4$

In each of Exercises 41–48, find all values of x (if any) where the tangent line to the graph of the given equation is horizontal.

41. $y = -x^2 - 3x - 1$

42. $y = x^2 - x + 4$

43. $y = x^2 + \dfrac{1}{x^2}$

44. $y = \sqrt{x}\,(x - 1)$

45. $y = \sqrt{x} - 1$

46. $y = \sqrt{x} + \dfrac{1}{x}$

47. $y = x - \dfrac{1}{x^2} + 4$

48. $y = 3x^2 - \dfrac{1}{x} + 3$

Calculate the limits in Exercises 49–56 mentally.

49. $\lim\limits_{x \to 0} 3x - 2$

50. $\lim\limits_{x \to 0} 2x + x^2$

51. $\lim\limits_{x \to -1} \dfrac{1 + x}{x}$

52. $\lim\limits_{x \to -1} \dfrac{x + 4}{x}$

53. $\lim\limits_{x \to 9} (x - \sqrt{x})$

54. $\lim\limits_{x \to 1} \dfrac{\sqrt{x} + x^2}{x}$

55. $\lim\limits_{x \to -\infty} (x^2 - 3x + 1)$

56. $\lim\limits_{x \to +\infty} \dfrac{x^2}{2x^2 + 1}$

Investigate each limit in Exercises 57–64 using the numerical approach. If the limit exists, state its value, and if it fails to exist, say why.

57. $\lim\limits_{x \to -1} \dfrac{x^3 + 1}{x + 1}$

58. $\lim\limits_{x \to -6} \dfrac{x^2 + 6x}{x + 6}$

59. $\lim\limits_{x \to 0^-} \dfrac{1}{x^2 - 2x}$

60. $\lim\limits_{x \to 0^+} \dfrac{x}{x^2 - x}$

61. $\lim\limits_{x \to -1^+} \dfrac{x^2 + 1}{x + 1}$

62. $\lim\limits_{x \to -1} \dfrac{x^2 + 1}{x + 1}$

63. $\lim\limits_{x \to +\infty} x^2 e^{-x}$

64. $\lim\limits_{x \to 0^+} \dfrac{x}{e^{-1/x}}$

In Exercises 65–72, the graph of f is given. Compute the indicated quantities graphically. If a particular limit exists, state its value, and if it fails to exist, say why.

65. (a) $\lim\limits_{x\to 1} f(x)$ **(b)** $\lim\limits_{x\to -1^+} f(x)$
(c) $\lim\limits_{x\to -1} f(x)$ **(d)** $\lim\limits_{x\to 3} f(x)$
(e) $f(1)$ **(f)** $f(-1)$

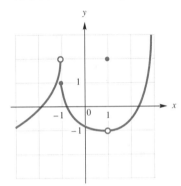

66. (a) $\lim\limits_{x\to -2} f(x)$ **(b)** $\lim\limits_{x\to 0^+} f(x)$
(c) $\lim\limits_{x\to 0^-} f(x)$ **(d)** $\lim\limits_{x\to 0} f(x)$
(e) $f(0)$ **(f)** $f(-2)$

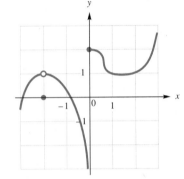

67. (a) $\lim\limits_{x\to 0} f(x)$ **(b)** $\lim\limits_{x\to +\infty} f(x)$
(c) $\lim\limits_{x\to -\infty} f(x)$

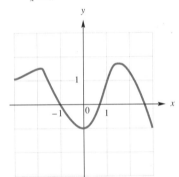

68. (a) $\lim\limits_{x\to 0} f(x)$ **(b)** $\lim\limits_{x\to +\infty} f(x)$
(c) $\lim\limits_{x\to -\infty} f(x)$

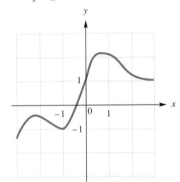

69. (a) $\lim\limits_{x\to -1} f(x)$ **(b)** $\lim\limits_{x\to -3^+} f(x)$
(c) $\lim\limits_{x\to +\infty} f(x)$ **(d)** $f(-1)$

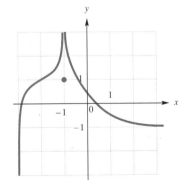

70. (a) $\lim\limits_{x\to 1} f(x)$ **(b)** $\lim\limits_{x\to -\infty} f(x)$
(c) $\lim\limits_{x\to +\infty} f(x)$ **(d)** $f(1)$

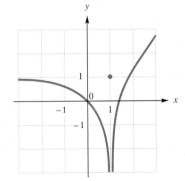

71. (a) $\lim\limits_{x \to 1^-} f(x)$ **(b)** $\lim\limits_{x \to -\infty} f(x)$

(c) $\lim\limits_{x \to +\infty} f(x)$

72. (a) $\lim\limits_{x \to 1^+} f(x)$ **(b)** $\lim\limits_{x \to -\infty} f(x)$

(c) $\lim\limits_{x \to +\infty} f(x)$

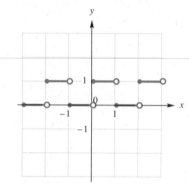

73–80. *Investigate the continuity of the functions whose graphs are given in Exercises 65–72. In each case, say whether f is continuous on its domain. If f fails to be continuous on its domain, say why.*

Calculate each limit in Exercises 81–90. If the limit fails to exist, say why.

81. $\lim\limits_{h \to -1^+} 4$

82. $\lim\limits_{h \to 0^+} -5$

83. $\lim\limits_{h \to 0} \dfrac{h^2 - 2h}{h + h^3}$

84. $\lim\limits_{h \to 0} \dfrac{h^2 + h - 1}{h^2 + h}$

85. $\lim\limits_{x \to -3} \dfrac{2x^2 + 5x - 3}{x + 3}$

86. $\lim\limits_{x \to -1} \dfrac{x^2 + x}{2x^2 + x - 1}$

87. $\lim\limits_{x \to 0} |-x|$

88. $\lim\limits_{x \to 0} \dfrac{1}{|x|}$

E/L 89. $\lim\limits_{x \to +\infty} \dfrac{1}{e^x - e^{-x}}$

E/L 90. $\lim\limits_{x \to +\infty} \dfrac{e^{-x}}{2}$

APPLICATION EXERCISES

91. *Velocity* If a stone is thrown upward at 100 ft/s, its height after t seconds is given by
$$s = 100t - 16t^2.$$
(a) Find its average velocity over the time intervals [1, 2], [1, 1.1], [1, 1.01], and [1, 1.001].
(b) Find its velocity at time $t = 1$ second.
(c) How long does it take to reach the ground, and how fast is it traveling when it hits the ground?

92. *Velocity* If a stone is thrown up at 120 ft/s from a height of 1,000 feet, its height after t seconds is given by
$$s = 1,000 + 120t - 16t^2.$$
(a) Find its average velocity over the time intervals [2, 3], [2, 2.1], [2, 2.01], and [2, 2.001].
(b) Find its velocity at time $t = 2$ seconds.

(c) How long does it take to reach the ground, and how fast is it traveling when it hits the ground?

93. *Marginal Cost* The cost of producing x soccer balls per day at the Taft Sports Company is calculated by its marketing staff to be given by the formula
$$C(x) = 100 + 60x - 0.001x^2.$$
Find the marginal cost function, and use it to estimate how fast the cost is going up at a production level of 50 soccer balls. Compare this with the exact cost of producing the 51st soccer ball.

94. *Marginal Cost* Referring to the cost equation in Exercise 93, find the production level for which the marginal cost is zero.

95. *Rates of Change* A cube is growing in such a way that each of its edges is growing at a rate of 1 centimeter per second. How fast is its volume growing at the instant when the cube has a volume of 1,000 cubic centimeters? [The volume of a cube with edge a is given by $V = a^3$.]

96. *Rates of Change*   The volume of a cone with a circular base of radius r and height h is given by $V = \frac{1}{3}\pi r^2 h$. (See the figure.) Find the rate of increase of V with respect to r (assuming that h is fixed). If the quantity r is growing at a uniform rate of 1 cm/s, how fast is the volume growing at the instant when $r = 1$, assuming that $h = 2$ does not change?

97. *Sales Incentives*   The Volume Sales Company pays its salespersons a weekly salary of $1,000 plus a sales commission based on the formula

$$C(s) = \frac{\sqrt{s}}{10},$$

where s represents a salesperson's total weekly sales (in dollars), and $C(s)$ is the percentage of total weekly sales paid to the salesperson as a commission.

(a) Calculate the net revenue $R(s)$ the company earns per week from a single salesperson, as a function of sales.

(b) Calculate the company's marginal revenue per salesperson for weekly sales of $100,000 and $500,000.

(c) For what weekly sales figure is the company's marginal net revenue positive? For what sales figures is it negative?

(d) What do you suspect is the most the company can earn in a week from a single salesperson?

98. *Medical Subsidies*

(a) The annual cost C of medical coverage is given by the formula

$$C(q) = 10,000 + 2q^2,$$

where q is the number of employees covered. Find the marginal cost of medical coverage for a 100-employee firm.

(b) The annual government subsidy to the firm is given by the formula

$$S(q) = 200q.$$

At what employee level does the marginal cost surpass the marginal subsidy?

(c) Calculate the net cost function, $N(q)$, given the cost function and subsidy above, graph it, and use your graph to estimate the value of q that gives the lowest net cost? What is this lowest net cost? Compare your answer to that of (b), and comment on what you find.

E/L 99. *Frogs*   (use of a graphing calculator or computer software required) The population P of frogs in Nassau County is given by the formula

$$P = 10,000e^{0.5t},$$

where t is the time in years since 1995.

(a) Roughly how fast will the frog population be growing in the year 2,000?

(b) In what year will the frog population be growing at a rate of one million per year?

(c) What is the significance of this for the frog leg industry?

E/L 100. *Epidemics*   (use of a graphing calculator or computer software required)   According to the logistic model discussed in the chapter on exponentials, the number of people P infected in an epidemic often follows a curve of the following type:

$$P = \frac{NP_0}{P_0 + (N - P_0)e^{-bt}},$$

where N is the total susceptible population, P_0 is the number of infected individuals at time $t = 0$, and b is a constant that governs the rate of spread. Taking t to be the time in years, and assuming that $b = 0.1$, $N = 1,000,000$, and $P_0 = 1,000$, answer the following questions.

(a) Roughly how fast is the epidemic spreading at the start ($t = 0$), after 10 years, and after 100 years?

(b) When does the spread of the disease slow to 1,000 new cases per year?

(c) What percentage of the population is infected by this time?

4 MORTGAGES TO SUIT YOUR BEST INTERESTS!

1 YEAR ADJUSTABLE

Your mortgage starts at an introductory below market rate. It is adjustable but it can't go up or down more than 2% every 12 months and it has a 12.95% lifetime cap. Convertible feature also available at a slightly higher interest rate. *LOANS UP TO $350,000 AVAILABLE.*

5.25% INTEREST RATE

6.48% ANNUAL PERCENTAGE RATE

FIXED RATE 15 YEAR BI-WEEKLY

Pays off in 13.04 years. 339 payments of $4.64 per thousand borrowed. FNMA-FHLMC limits apply.

7.50% INTEREST RATE

7.87% ANNUAL PERCENTAGE RATE

FIXED RATE 30 YEAR

360 payments of $7.34 per thousand borrowed. FNMA-FHLMC limits apply.

Conditional commitments on all mortgages in 72 hours. All Rates may be locked in at application. Rates, terms and conditions subject to change without notice

8.00% INTEREST RATE

8.21% ANNUAL PERCENTAGE RATE

SONYMA

*For 5 Years — then 8-1/8% $6.82 per $1,000 Borrowed

7.25% * INTEREST RATE

8.06% ANNUAL PERCENTAGE RATE

ALSO AVAILABLE:
■ Second Mortgages
■ Equity Credit Lines ■ Refinancing Available

CALL TOLL FREE 800-649-4005

Farmingdale:
516/249-0900

Riverhead:
516/727-2587

SUNRISE
FEDERAL SAVINGS BANK

Your Future Begins At Sunrise!

Source: Courtesy Sunrise Federal Savings Bank.

Techniques of Differentiation

APPLICATION ▶ You plan to go to the bank tomorrow to apply for a 30-year, $90,000 mortgage. The interest rate was advertised as 7.5%, but rates are volatile and you know that the rate you get could vary by as much as 0.5%. Your budget is tight, and your spouse is counting on you to make sure that you can afford the monthly payment. Also, you don't trust the loan officer to get the calculation right, so you would like to be able to calculate the monthly payment yourself once you know the actual rate.

What you would like is a simple linear formula that gives a good estimate of the monthly payment as a function of the interest rate for values close to 7.5%. How would you find such a formula?

| INTRODUCTION | ▶ In Chapter 3, we studied the concept of the derivative of a function, and we saw some of the things for which it is useful. However, the only functions we could differentiate easily were sums of terms of the form ax^n, where a and n are constants.

In this chapter, we develop techniques that will enable us to differentiate any closed-form function—that is, any function that can be specified by a formula involving powers, radicals, exponents, and logarithms. We also develop techniques for differentiating functions that are only specified *implicitly*—that is to say, functions for which we are not given an explicit formula, but only an equation relating x and y.

The chapter ends with an in-depth retrospective look at the concept of the derivative and approximation of a function by a linear function.

▶ ═══ **4.1** THE PRODUCT AND QUOTIENT RULES

We know how to find the derivatives of functions that are sums of powers, such as polynomials. However, that leaves many interesting functions whose derivatives we do not know how to find, such as $(x + 1)/(x - 1)$. We have discussed differentiation of sums and differences of functions. We now consider products and quotients.

▼ **EXAMPLE 1**

The following calculations are wrong.

$$\frac{d}{dx}\left(\frac{x^3}{x}\right) = \frac{3x^2}{1} = 3x^2 \qquad \text{WRONG}$$

$$\frac{d}{dx}(x^3 \cdot x) = 3x^2 \cdot 1 = 3x^2 \qquad \text{WRONG}$$

After all, $x^3/x = x^2$, and we know that $d/dx\,(x^2) = 2x$, not $3x^2$. Our error was the assumption that the derivative of a quotient is the quotient of the derivatives. Similarly, $x^3 \cdot x = x^4$, and its derivative is $4x^3$, not $3x^2$. In other words, the derivative of a product is *not* the product of the derivatives.

$\mathcal{Q}$ If this is not how we find the derivatives of products and quotients, how *do* we find them?

$\mathcal{A}$ We use the following rules to differentiate products and quotients.

PRODUCT RULE

$$\frac{d}{dx}[f(x)g(x)] = f'(x)g(x) + f(x)g'(x)$$

QUOTIENT RULE

$$\frac{d}{dx}\left(\frac{f(x)}{g(x)}\right) = \frac{f'(x)g(x) - f(x)g'(x)}{[g(x)]^2}$$

Don't try to remember these rules by the symbols we've used here, but remember them in words. The following slogans are easy to remember, even if the terminology is not precise.

PRODUCT AND QUOTIENT RULES IN WORDS

The derivative of a product is the derivative of the first times the second, plus the first times the derivative of the second.
    The derivative of a quotient is the derivative of the top times the bottom, minus the top times the derivative of the bottom, all over the bottom squared.

▶ CAUTION   One more time: *the derivative of a product is* not *the product of the derivatives, and the derivative of a quotient is* not *the quotient of the derivatives.* To find the derivative of a product, you must use the product rule, and to find the derivative of a quotient, you must use the quotient rule. Forgetting this is a mistake everyone makes from time to time.* ◀

    We shall see why the product rule works at the end of the section and why the quotient rule works in the next section. First, let us try some examples.

▼ *Leibniz made this mistake at first, too, so you are in good company.

▼ **EXAMPLE 2**

Calculate $\dfrac{d}{dx}(x^3 \cdot x^2)$ two different ways: by multiplying first, and by the product rule.

SOLUTION   If we multiply before taking the derivative, we get

$$\frac{d}{dx}(x^3 \cdot x^2) = \frac{d}{dx}(x^5)$$
$$= 5x^4.$$

If we use the product rule, we get

(Derivative of first) (Second) + (First) (Derivative of second)

$$\frac{d}{dx}(x^3 \cdot x^2) = 3x^2 \cdot x^2 + x^3 \cdot 2x$$

$$= 5x^4.$$

Before we go on... This example shows us how the product rule is consistent with the power rule—it leads to the same answer. The first of the two calculations above is obviously easier, so there is no necessity to use the product rule in this example. On the other hand, the product rule will prove indispensable when we study derivatives of more complicated functions, so we cannot do without it.

▼ **EXAMPLE 3**

Find $\dfrac{d}{dx}[(x^3 + 2x)(\sqrt{x} + 1)]$.

SOLUTION

$$\frac{d}{dx}[(x^3 + 2x)(\sqrt{x} + 1)]$$

$$= \text{(Derivative of first) (Second)} + \text{(First) (Derivative of second)}$$

$$= (3x^2 + 2)(\sqrt{x} + 1) + (x^3 + 2x)\left(\frac{1}{2\sqrt{x}}\right)$$

Before we go on... Notice that we could also do this example by expanding the whole expression first and *then* taking the derivative. This would enable us to avoid using the product rule. (You should do this for practice.)

▼ **EXAMPLE 4**

Find $\dfrac{d}{dx}[(x+1)(x^2+1)(x^3+1)]$.

SOLUTION   Here we have a product of three functions, not just two. We can find the derivative by using the product rule twice.

$$\frac{d}{dx}[(x+1)(x^2+1)(x^3+1)]$$

$$= \frac{d}{dx}(x+1) \cdot [(x^2+1)(x^3+1)] +$$

$$(x+1) \cdot \frac{d}{dx}[(x^2+1)(x^3+1)]$$

$$= (1)(x^2+1)(x^3+1) +$$

$$(x+1)[(2x)(x^3+1) + (x^2+1)(3x^2)]$$

$$= (1)(x^2+1)(x^3+1) + (x+1)(2x)(x^3+1) +$$

$$(x+1)(x^2+1)(3x^2)$$

We can see here a more general product rule: the derivative of a product of three functions is found by taking the derivatives of each function in turn and adding the results together. The general formula is

$$(fgh)' = f'gh + fg'h + fgh'.$$

There are similar formulas for products of more than three functions.

▼ **EXAMPLE 5**

Find $\dfrac{d}{dx}\left(\dfrac{x+1}{x-1}\right)$.

SOLUTION   We must use the quotient rule.

(Derivative of top)(Bottom) − (Top)(Derivative of bottom)

$$\frac{d}{dx}\left(\frac{x+1}{x-1}\right) = \frac{(1)(x-1) - (x+1)(1)}{(x-1)^2}$$

Bottom squared

$$= \frac{-2}{(x-1)^2}$$

▼ | EXAMPLE 6

Find $\dfrac{d}{dx}\left[\dfrac{(x+1)(x+2)}{x-1}\right]$.

SOLUTION   We seem to have both a product and a quotient here. Which rule do we use, the product or the quotient rule? Here is a way to decide.

Let us think about how we would calculate (by hand or with a scientific calculator) $(x+1)(x+2)/(x-1)$ for a specific value of x, say $x=11$. *What would be the last operation we would perform?* Here is how we would probably do the calculation:

1. Calculate $(x+1)(x+2) = (11+1)(11+2) = 156$.
2. Calculate $x-1 = 11-1 = 10$.
3. Divide 156 by 10 to get 15.6.

Thus, the last operation we would perform is division, and so we can regard the whole expression as a quotient—that is, as $(x+1)(x+2)$ *divided by* $x-1$. Therefore, we will use the quotient rule. The first thing the quotient rule tells us to do is take the derivative of the top. If we examine the top, we notice that it is a product, so we must use the product rule to take its derivative. Here is the calculation.

(Derivative of top) (Bottom) − (Top)(Derivative of bottom)

$$\frac{d}{dx}\frac{(x+1)(x+2)}{x-1} = \frac{[(1)(x+2)+(x+1)(1)](x-1)-[(x+1)(x+2)](1)}{(x-1)^2}$$

$$= \frac{(2x+3)(x-1)-(x+1)(x+2)}{(x-1)^2}$$

$$= \frac{x^2-2x-5}{(x-1)^2}$$

What was important here was to determine the *order of operations,* and in particular to determine the last operation to be performed. Pretending to do an actual calculation reminds us of the order of operations, and we shall call this technique the **calculation thought experiment.**

Before we go on... We had to use the product rule to calculate the derivative of the top because the top was $(x+1)(x+2)$, which is a product. Get used to this: differentiation rules often must be used in combination. Once you have determined one rule to use, do not assume that you can forget the others.

Now here is another way we could have done the problem: suppose that our calculation thought experiment took the following form.

1. Calculate $(x+1)/(x-1) = (11+1)/(11-1) = 1.2$.
2. Calculate $x+2 = 11+2 = 13$.
3. Multiply 1.2 by 13 to get 15.6.

Then we would have regarded the expression as a *product*—the product of $(x + 1)/(x - 2)$ and $(x + 2)$—so we could have used the product rule instead. We can't escape the quotient rule, however: we need to use it to take the derivative of the first factor, $(x + 1)/(x - 2)$. You should try this approach as an exercise, and check that you get the same answer.

In the last chapter we proved the power rule only for positive integers. We can now prove it for negative integers as well.

▼ **EXAMPLE 7**   Power Rule for Negative Integers

Show that if n is a positive integer, then

$$\frac{d}{dx}(x^{-n}) = -nx^{-n-1}.$$

SOLUTION   Since we are required here to justify the power rule for negative integers, we can't simply go ahead and *use* it! "Officially," all we can use are the power rule for *positive* integers, and the product and quotient rules (which are also unfinished business at the moment, but we'll attend to them shortly).

What we *can* do is this: write x^{-n} as $1/x^n$ and use the quotient rule. Applying the quotient rule to $1/x^n$ gives

$$\frac{d}{dx}\left(\frac{1}{x^n}\right) = \frac{(0)(x^n) - (1)(nx^{n-1})}{(x^n)^2}$$

$$= \frac{-nx^{n-1}}{x^{2n}}$$

$$= -nx^{n-1-2n} = -nx^{-n-1},$$

and we are done.

Before we go on... Notice that we did use the power rule, but only for a positive integer power, the case that we justified in the previous chapter.

▶ NOTE   In practice, we should not use the quotient rule for expressions like $3/x^2$. It is much simpler to first rewrite $3/x^2 = 3x^{-2}$, and then use the power rule. ◀

▼ **EXAMPLE 8**

Find $\frac{d}{dx}\left(6x^2 + 5\left(\frac{x}{x - 1}\right)\right)$.

SOLUTION   If this seems to be a little complicated at first sight, all we need to do is turn to the calculation thought experiment, which tells us that the expression we are asked to differentiate is a *sum*. Since the derivative of a sum is the sum of the derivatives, we get

$$\frac{d}{dx}\left(6x^2 + 5\left(\frac{x}{x-1}\right)\right) = \frac{d}{dx}(6x^2) + \frac{d}{dx}\left(5\left(\frac{x}{x-1}\right)\right).$$

In other words, we must take the derivatives of $6x^2$ and $5\left(\frac{x}{x-1}\right)$ separately, and then add the answers. The derivative of $6x^2$ is $12x$, and there are two ways of taking the derivative of $5\left(\frac{x}{x-1}\right)$. We could either first multiply the expression $\left(\frac{x}{x-1}\right)$ by 5 to get $\left(\frac{5x}{x-1}\right)$, and then take its derivative using the quotient rule, or we could proceed as follows.

$$\frac{d}{dx}\left(6x^2 + 5\left(\frac{x}{x-1}\right)\right) = \frac{d}{dx}(6x^2) + \frac{d}{dx}\left(5\left(\frac{x}{x-1}\right)\right)$$

$$= 12x + 5\frac{d}{dx}\left(\frac{x}{x-1}\right)$$

We use the quotient rule for the second expression, and we get

$$= 12x + 5\left(\frac{(1)(x-1) - (x)(1)}{(x-1)^2}\right)$$

$$= 12x + 5\left(\frac{-1}{(x-1)^2}\right)$$

$$= 12x - \frac{5}{(x-1)^2}.$$

Now let us see why the product rule works.

PROOF OF PRODUCT RULE

We wish to calculate the derivative of a function $f(x)g(x)$, so we proceed as usual, using the definition of the derivative:

$$\frac{d}{dx}[f(x)g(x)] = \lim_{h \to 0} \frac{f(x+h)g(x+h) - f(x)g(x)}{h}.$$

Q How can we rewrite the numerator so that we can evaluate the limit?

A There are several ways to do this. We choose to use one that is motivated by Figure 1.

The area of the whole rectangle represents $f(x+h)g(x+h)$, while the area of the rectangle labeled ① represents $f(x)g(x)$. The difference between these areas (which is the numerator we are trying to rewrite) is represented by the sum of the areas of the other three rectangles. Rectangle ② has a width of $f(x+h) - f(x)$ and a height of $g(x)$, so its area is $[f(x+h) - f(x)]g(x)$. Similarly, the area of rectangle ③ is

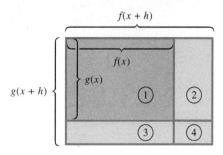

FIGURE 1

$f(x)[g(x + h) - g(x)]$, and the area of rectangle ④ is $[f(x + h) - f(x)]$ $[g(x + h) - g(x)]$. In other words,

$$f(x + h)g(x + h) - f(x)g(x)$$
$$= [f(x + h) - f(x)]g(x) + f(x)[g(x + h) - g(x)]$$
$$+ [f(x + h) - f(x)][g(x + h) - g(x)].$$

You should expand the right-hand side and simplify to convince yourself that this is really just an algebraic fact. Returning to the derivative, this gives

$$\frac{d}{dx}[f(x)g(x)]$$

$$= \lim_{h \to 0} \frac{[f(x + h) - f(x)]g(x) + f(x)[g(x + h) - g(x)] + [f(x + h) - f(x)][g(x + h) - g(x)]}{h}$$

$$= \lim_{h \to 0}\left(\frac{f(x + h) - f(x)}{h}\right)g(x) + \lim_{h \to 0} f(x)\left(\frac{g(x + h) - g(x)}{h}\right)$$

$$+ \lim_{h \to 0}\left(\frac{f(x + h) - f(x)}{h}\right)[g(x + h) - g(x)].$$

We already know the following:

$$\lim_{h \to 0} \frac{f(x + h) - f(x)}{h} = f'(x),$$

$$\lim_{h \to 0} \frac{g(x + h) - g(x)}{h} = g'(x),$$

and

$$\lim_{h \to 0} [g(x + h) - g(x)] = 0.$$

The last limit is 0 because if g is differentiable at x, it must be continuous there, so $\lim_{h \to 0} g(x + h) = g(x)$.

Putting this all together,

$$\frac{d}{dx}[f(x)g(x)]$$
$$= f'(x)g(x) + f(x)g'(x) + f'(x) \cdot 0$$
$$= f'(x)g(x) + f(x)g'(x),$$

as we claimed. ◀

It is possible to prove the quotient rule in a very similar way, but this proof would be completely unenlightening. Instead, we shall prove it in the next section using techniques we shall discuss there.

▶ **4.1 EXERCISES**

In each of Exercises 1 through 10,
(a) *calculate the derivative of the given function mentally without using either the product or quotient rule, and*
(b) *use the product or quotient rule to find the derivative, and check that you obtain the same answer.*

1. $f(x) = 3x$

2. $f(x) = 2x^2$

3. $g(x) = x \cdot x^2$

4. $g(x) = x \cdot x$

5. $h(x) = x(x + 3)$

6. $h(x) = x(1 + 2x)$

7. $r(x) = \dfrac{x^2}{3}$

8. $r(x) = \dfrac{x}{5}$

9. $s(x) = \dfrac{2}{x}$

10. $s(x) = \dfrac{3}{x^2}$

Evaluate $\dfrac{dy}{dx}$ *in each of Exercises 11–34.*

11. $y = (x + 1)(x^2 - 1)$

12. $y = (x^2 + x)(x - x^2)$

13. $y = (2x^{1/2} + 4x - 5)(x - x^{-1})$

14. $y = (x^{3/4} - 4x - 5)(x^{-1} + x^{-2})$

15. $y = (2x^2 - 4x + 1)^2$

16. $y = (2\sqrt{x} - x^2)^2$

17. $y = (x^2 - \sqrt{x})\left(\sqrt{x} + \dfrac{1}{\sqrt{x}}\right)$

18. $y = (4x^2 - \sqrt{x})\left(\sqrt{x} - \dfrac{2}{\sqrt{x}}\right)$

19. $y = (\sqrt{x} + 1)\left(\sqrt{x} + \dfrac{1}{x^2}\right)$

20. $y = (4x^2 - \sqrt{x})\left(\sqrt{x} - \dfrac{2}{x^2}\right)$

21. $y = \dfrac{2x + 4}{3x - 1}$

22. $y = \dfrac{3x - 9}{2x + 4}$

23. $y = \dfrac{2x^2 + 4x + 1}{3x - 1}$

24. $y = \dfrac{3x^2 - 9x + 11}{2x + 4}$

25. $y = \dfrac{x^2 - 4x + 1}{x^2 + x + 1}$

26. $y = \dfrac{x^2 + 9x - 1}{x^2 + 2x - 1}$

27. $y = \dfrac{\sqrt{x} + 1}{\sqrt{x} - 1}$

28. $y = \dfrac{\sqrt{x} - 1}{\sqrt{x} + 1}$

29. $y = \dfrac{\left(\dfrac{1}{x} + \dfrac{1}{x^2}\right)}{x + x^2}$

30. $y = \dfrac{\left(\dfrac{1}{x} - \dfrac{1}{x^2}\right)}{x^2 + 1}$

31. $y = \dfrac{(x + 3)(x + 1)}{3x - 1}$

32. $y = \dfrac{3x^2 - 9x + 11}{(x - 5)(x - 4)}$

33. $y = \dfrac{(x + 3)(x + 1)(x + 2)}{3x - 1}$

34. $y = \dfrac{3x^2 - 9x + 11}{(x - 5)(x - 4)(x - 1)}$

In Exercises 35 through 42, evaluate the derivatives.

35. $\dfrac{d}{dx}[(x^2 + x)(x^2 - x)]$

36. $\dfrac{d}{dx}[(x^2 + x^3)(x^3 - 2x + 1)]$

37. $\dfrac{d}{dx}[(x^3 + 2x)(x^2 - x)]\big|_{x=2}$

38. $\dfrac{d}{dx}[(x^2 + x)(x^2 - x)]\big|_{x=1}$

39. $\dfrac{d}{dt}\left((t^2 - \sqrt{t})\left(\sqrt{t} + \dfrac{1}{\sqrt{t}}\right)\right)\Big|_{t=1}$

40. $\dfrac{d}{dt}\left((t^2 + \sqrt{t})\left(\sqrt{t} - \dfrac{1}{\sqrt{t}}\right)\right)\Big|_{t=1}$

41. $\dfrac{d}{dt}\left(\dfrac{t^2 - \sqrt{t}}{\sqrt{t} + \dfrac{1}{\sqrt{t}}}\right)$

42. $\dfrac{d}{dt}\left(\dfrac{t^2 + \sqrt{t}}{\sqrt{t} - \dfrac{1}{\sqrt{t}}}\right)$

In each of Exercises 43 through 48, find the equation of the line tangent to the graph of the given function at the point with the indicated x-coordinate.

43. $f(x) = (x^2 + 1)(x^3 + x), \quad x = 1$

44. $f(x) = (\sqrt{x} + 1)(x^2 + x), \quad x = 1$

45. $f(x) = \dfrac{x + 1}{x + 2}, \quad x = 0$

46. $f(x) = \dfrac{\sqrt{x} + 1}{\sqrt{x} + 2}, \quad x = 4$

47. $f(x) = \dfrac{x^2 + 1}{x}, \quad x = -1$

48. $f(x) = \dfrac{x}{x^2 + 1}, \quad x = 1$

APPLICATIONS

49. *Revenue* The monthly sales of Sunny Electronics' new stereo system are given by $S(x) = 20x - x^2$ hundred units per month, x months after its introduction. The retail price Sunny charges is $p(x) = \$1{,}000 - x^2$, x months after introduction. The revenue Sunny earns must then be $R(x) = 100S(x)p(x)$. Find, five months after the introduction, the rate of change of monthly sales, the rate of change of the price, and the rate of change of revenue. Interpret your answers.

50. *Revenue* The monthly sales of Sunny Electronics' new portable tape player is given by $S(x) = 20x - x^2$ hundred units per month, x months after its introduction. The retail price Sunny charges is $p(x) = \$100 - x^2$, x months after introduction. The revenue

Sunny earns must then be $R(x) = 100S(x)p(x)$. Find, six months after the introduction, the rate of change of monthly sales, the rate of change of the price, and the rate of change of revenue. Interpret your answers.

51. *Revenue* Cyndi Keen is currently selling 20 "I ❤ Calculus" T-shirts per day, but sales are dropping at a rate of 3 per day. She is currently charging $7 per T-shirt, but to compensate for dwindling sales, she is increasing the unit price by $1 per day. How fast, and in what direction, is her daily revenue currently changing?

52. *Pricing Policy* Let us turn Exercise 51 around a little: Cyndi Keen is currently selling 20 "I ❤ Calculus" T-shirts per day, but sales are dropping at a rate of 3 per day. She is currently charging $7 per T-shirt, and she wishes to increase her daily revenue by $10 per day. At what rate should she increase the unit price to accomplish this (assuming the price increase does not affect sales)?

53. *Bus Travel* The Thoroughbred Bus Company finds that its monthly costs for one particular year were given by $C(t) = \$10,000 + t^2$ after t months. On the other hand, after t months, the company had $P(t) = 1,000 + t^2$ passengers per month. How fast is its cost per passenger changing after 6 months?

54. *Bus Travel* The Thoroughbred Bus Company finds that its monthly costs for one particular year were given by $C(t) = \$100 + t^2$ after t months. On the other hand, after t months, the company had $P(t) = 1,000 + t^2$ passengers per month. How fast is its cost per passenger changing after 6 months?

Some of the following exercises are variations of exercises you have already seen in Chapter 3.

55. *Fuel Economy* Your Porsche's gas mileage (in miles per gallon) is given as a function $M(x)$ of speed x in mph, where

$$M(x) = \frac{1}{\left(x + \dfrac{3,600}{x}\right)}.$$

Calculate $M'(x)$, and hence $M'(10)$, $M'(60)$, and $M'(70)$. What do the answers tell you about your car?

56. *Fuel Economy* Your used Chevy's gas mileage (in miles per gallon) is given as a function $M(x)$ of speed x in mph, where

$$M(x) = \frac{10}{\left(x + \dfrac{3,025}{x}\right)}.$$

Calculate $M'(x)$, and hence determine *the sign* of each of the following: $M'(40)$, $M'(55)$, $M'(60)$. Interpret your results.

57. *Expansion* The number of *Toys "R" Us*® stores (including *Kids "R" Us* stores) worldwide increased from 171 at the start of 1984 to 1,032 at the start of 1994.* If the annual revenue at each store was $600,000 in 1984 and was increasing by $50,000 per year, how fast would the company's worldwide revenue have been increasing at the start of 1990? (Use a linear model for the number of stores.)

58. *Investments* The price of *GTE*® stock rose from about $22 per share in January 1989 to $35 per share in January 1994.† If you had purchased 100 shares of *GTE* in January 1989 and steadily purchased additional shares at a rate of 10 shares per month, how fast would the value of your investment have been increasing in January 1994? (Use a linear model for the share price.)

59. *Biology—Reproduction* The Verhulst model for population growth specifies the reproductive rate of an organism as a function of the total population according to the following formula:

$$R(p) = \frac{r}{1 + kp}.$$

Here, p is the total population in thousands of organisms, r and k are constants that depend on the particular circumstances and the organism being studied, and $R(p)$ is the reproduction rate in thousands of organisms per hour.‡ If $k = 0.125$ and $r = 45$, find $R'(p)$ and hence $R'(4)$. Interpret the result.

60. *Biology—Reproduction* Another model, the predator satiation model for population growth, specifies that the reproductive rate of an organism as a function of the total population varies according to the following formula:

$$R(p) = \frac{rp}{1 + kp}.$$

Here, p is the total population in thousands of organisms, r and k are constants that depend on the particular circumstances and the organism being studied, and $R(p)$ is the reproduction rate in new organisms per hour.††† Given that $k = 0.2$ and $r = 0.08$, find $R'(p)$ and $R'(2)$. Interpret the result.

▼ *Source: Company Reports/Associated Press (*The New York Times*, Jan. 12, 1994, p. D4.)

†Source: Company Reports, Datastream (*The New York Times*, Jan. 14, 1994, p. D1.)

‡Source: *Mathematics in Medicine and the Life Sciences* by F. C. Hoppensteadt and C. S. Peskin (New York: Springer-Verlag, 1992) pp. 20–22.

61. *Embryo Development* Bird embryos consume oxygen from the time the egg is laid through the time the chick hatches. In the case of a typical galliform bird, the total oxygen consumption (in milliliters) t days after the egg was laid can be approximated by*

$$C(t) = -0.0163t^4 + 1.096t^3 - 10.704t^2$$
$$+ 3.576t \quad (t \le 30).$$

(An egg will usually hatch at around $t = 28$.) Suppose that at time $t = 0$ you have a collection of 30 newly hatched eggs and that the number of eggs is decreased linearly to zero at time $t = 30$ days. How fast is the total oxygen consumption of your collection of embryos changing after 25 days? (Answer to the nearest whole number.) Interpret the result.

62. *Embryo Development* Turkey embryos consume oxygen from the time the egg is laid through the time the chick hatches. In the case of a brush turkey, the total oxygen consumption (in milliliters) t days after the egg was laid can be approximated by*

$$C(t) = -0.00708t^4 + 0.952t^3$$
$$- 21.96t^2 + 95.328t \quad (t \le 50).$$

(An egg will typically hatch at around $t = 50$.) Suppose that at time $t = 0$ you have a collection of 100 newly hatched eggs and that the number of eggs is decreased linearly to zero at time $t = 50$ days. How fast is the total oxygen consumption of your collection of embryos changing after 40 days? (Answer to the nearest whole number.) Interpret the result.

COMMUNICATION AND REASONING EXERCISES

63. You have come across the following in a newspaper article: "Revenues of HAL Home Heating Oil Inc. are rising by $4.2 million per year. This is due to an annual increase of 70¢ per gallon in the price HAL charges for heating oil and an increase in sales of 6 million gallons of oil per year." Comment on this analysis.

64. Your friend says that since average cost is obtained by dividing the cost function by the number of units x, it follows that the derivative of average cost is the same as marginal cost, since the derivative of x is 1. Comment on this analysis.

65. Find a demand function $q(p)$ such that at a price per item of $p = \$100$, revenue will rise if the price per item is increased.

66. What must be true about a demand function $q(p)$ so that at a price per item of $p = \$100$, revenue will decrease if the price per item is increased?

67. *Marginal Product* (*from the GRE economics test*) Which of the following statements about average product and marginal product is correct?
(a) If average product is decreasing, marginal product must be less than average product.

(b) If average product is increasing, marginal product must be increasing.
(c) If marginal product is decreasing, average product must be less than marginal product.
(d) If marginal product is increasing, average product must be decreasing.
(e) If marginal product is constant over some range, average product must be constant over that range.

68. *Marginal Cost* (*based on a question from the GRE economics test*) Which of the following statements about average cost and marginal cost is correct?
(a) If average cost is increasing, marginal cost must be increasing.
(b) If average cost is increasing, marginal cost must be decreasing.
(c) If average cost is increasing, marginal cost must be more than average cost.
(d) If marginal cost is increasing, average cost must be increasing.
(e) If marginal cost is increasing, average cost must be larger than marginal cost.

▼ * The model is derived from graphical data published in the article "The Brush Turkey" by Roger S. Seymour, *Scientific American*, December, 1991, pp. 108–14.

▶ ═══════ **4.2** THE CHAIN RULE

We can now find the derivatives of sums, products and quotients of powers of x, but we still cannot take the derivative of an expression such as $\sqrt{3x + 1}$. For this we need one more rule, but we need to talk about an idea behind it first. Look at $h(x) = \sqrt{3x + 1}$. This function is not a sum, difference, product, or quotient. We can use the calculation thought experiment to find the last operation we would perform in calculating $h(x)$.

1.   Calculate $3x + 1$,
2.   Take the square root of the answer.

Thus, the last operation is the "square root." We do not yet have a rule for finding the derivative of the square root of a quantity other than x.

There is a way of building $h(x)$ out of two simpler functions: $f(x) = \sqrt{x}$ and $g(x) = 3x + 1$.

$$h(x) = \sqrt{3x + 1}$$
$$= f(3x + 1)$$
$$= f(g(x))$$

We say that h is the **composite** of f and g. We read $f(g(x))$ aloud as "f of g of x."

In order to compute $h(1)$, say, you would first compute $3 \cdot 1 + 1 = 4$ and then take the square root of 4, giving $h(1) = 2$. In order to compute $f(g(1))$, you would follow exactly the same steps: first compute $g(1) = 4$ then $f(g(1)) = f(4) = 2$. Remember that you must always compute $f(g(x))$ numerically from the inside out: given x, first compute $g(x)$ and then $f(g(x))$.

The reason for writing h as the composite of f and g is that f and g are functions *whose derivatives we know.* Formally, the derivative of the composite function $h(x) = f(g(x))$ is given by the following rule.

FORMAL STATEMENT OF THE CHAIN RULE

If f has derivative f' and g has derivative g', then

$$\frac{d}{dx}[f(g(x))] = f'(g(x)) \cdot g'(x).$$

A more convenient way of writing the chain rule is to introduce a new variable u and let $u = g(x)$. Then we can write $f(g(x)) = f(u)$, $f'(g(x)) = f'(u)$, and $g'(x) = du/dx$.

> **CHAIN RULE**
>
> If u is a function of x, then
>
> $$\frac{d}{dx}[f(u)] = f'(u)\frac{du}{dx},$$
>
> provided both derivatives on the right exist.
> In words: *The derivative of f(quantity) is the derivative of f, evaluated at that quantity, times the derivative of the quantity.*

For every function f whose derivative we know, we now get a "generalized" differentiation rule. For example, with $f(x) = x^3$ we get the following.

Original Rule	Generalized Rule
$\dfrac{d}{dx}x^3 = 3x^2$	$\dfrac{d}{dx}u^3 = 3u^2\dfrac{du}{dx}$

In words:

> *The derivative of a quantity cubed is 3 times that quantity squared times the derivative of the quantity.*

▼ **EXAMPLE 1**

Compute $\dfrac{d}{dx}(2x^2 + x)^3$.

SOLUTION Using the calculation thought experiment, we find that the last operation we would perform in calculating $(2x^2 + x)^3$ is that of cubing the quantity $(2x^2 + x)$. Thus, we think of $(2x^2 + x)^3$ as "a quantity cubed." We'll calculate its derivative in two ways: using the formula above, and using the verbal form.

Method 1: *Using the formula.* We think of $(2x^2 + x)^3$ as u^3, where $u = 2x^2 + x$. By the formula,

$$\frac{d}{dx}u^3 = 3u^2\frac{du}{dx}.$$

Now substitute for u.

$$\frac{d}{dx}(2x^2 + x)^3 = 3(2x^2 + x)^2\frac{d}{dx}(2x^2 + x)$$
$$= 3(2x^2 + x)^2(4x + 1)$$

Method 2: *Using the verbal form.*
   If we prefer to use the verbal form, we get

> *The derivative of $(2x^2 + x)$ cubed is three times $(2x^2 + x)$ squared, times the derivative of $(2x^2 + x)$.*

In other words,

$$\frac{d}{dx}(2x^2 + x)^3 = 3(2x^2 + x)^2(4x + 1),$$

as we obtained before.

For the next example, recall that the derivative with respect to x of the function $f(x) = \sqrt{x}$ is $1/(2\sqrt{x})$. Now suppose that instead of x under the square root sign, we had some other quantity. The chain rule would then take the following form.

The derivative of the square root of a quantity is one over twice the square root of the quantity, times the derivative of the quantity.

In terms of a formula, if we again let u stand for "the quantity," we can write the following.

Original Rule	Generalized Rule
$\dfrac{d}{dx}\sqrt{x} = \dfrac{1}{2\sqrt{x}}$	$\dfrac{d}{dx}\sqrt{u} = \dfrac{1}{2\sqrt{u}}\dfrac{du}{dx}$

▼ **EXAMPLE 2**

Compute $\dfrac{d}{dx}\sqrt{3x + 1}$.

SOLUTION   The steps in the calculation thought experiment for $\sqrt{3x + 1}$ are

1. Calculate $3x + 1$
2. Take the square root of the answer.

Thus, we are dealing here with *the square root of a quantity.*

Method 1: *Using the formula.*   Think of $\sqrt{3x + 1}$ as $\sqrt{u}$, where $u = 3x + 1$. The formula is

$$\frac{d}{dx}\sqrt{u} = \frac{1}{2\sqrt{u}}\frac{du}{dx}.$$

If we substitute for u, we get

$$\frac{d}{dx}\sqrt{3x + 1} = \frac{1}{2\sqrt{3x + 1}}\cdot\frac{d}{dx}(3x + 1)$$

$$= \frac{1}{2\sqrt{3x + 1}}(3)$$

$$= \frac{3}{2\sqrt{3x + 1}}.$$

Method 2: *Using the verbal form.* The verbal form is

> *The derivative of the square root of a quantity is one over twice the square root of that quantity, times the derivative of the quantity.*

In symbols, this is

$$\frac{d}{dx} \sqrt{3x + 1} = \frac{1}{2\sqrt{3x + 1}} \; (3),$$

giving the answer in one step!

Before we go on... The formula for $\frac{d}{dx} \sqrt{x}$ is part of the power rule:

$$\frac{d}{dx} \sqrt{x} = \frac{d}{dx} (x^{1/2}) = \frac{1}{2} x^{-1/2} = \frac{1}{2\sqrt{x}}.$$

We can rewrite the calculations we did above as

$$\frac{d}{dx} \sqrt{3x + 1} = \frac{d}{dx} [(3x + 1)^{1/2}]$$

$$= \frac{1}{2} (3x + 1)^{-1/2} \; (3)$$

$$= \frac{3}{2\sqrt{3x + 1}}.$$

These examples show one of the most common uses of the chain rule: to take the derivative of an expression raised to a power. It is worth writing this "generalized power rule" as another rule, recognizing that it is a special case of the chain rule.

GENERALIZED POWER RULE

Power Rule	Generalized Power Rule
$\dfrac{d}{dx} x^n = nx^{n-1}$	$\dfrac{d}{dx} u^n = nu^{n-1} \dfrac{du}{dx}$

In words:
The derivative of a quantity raised to the power n is n times that quantity raised to the power (n − 1), times the derivative of that quantity.

▼ **EXAMPLE 3**

Find $\dfrac{d}{dx}(x^3 + x)^{100}$.

SOLUTION   First, the calculation thought experiment: If we were computing $(x^3 + x)^{100}$, the last operation we would perform is *raising a quantity to the power* 100. Thus we are dealing with *a quantity raised to the power* 100, and so we must use the generalized power rule.

According to the generalized power rule, the derivative of a quantity raised to the power 100 is 100 times that quantity to the power 99, times the derivative of that quantity. In other words,

$$\frac{d}{dx}(x^3 + x)^{100} = 100(x^3 + x)^{99}(3x^2 + 1).$$

▶ **CAUTION**   The following are examples of common errors.

$$\frac{d}{dx}(x^3 + x)^{100} = 100(3x^2 + 1)^{99} \qquad \text{WRONG}$$

$$\frac{d}{dx}(x^3 + x)^{100} = 100(x^3 + x)^{99} \qquad \text{WRONG}$$

Remember that the generalized power rule says that the derivative of a quantity to the power 100 is 100 times *that same quantity* raised to the power 99, *times the derivative of that quantity*.   ◀

Q It seems that there are now two formulas for the derivative of an nth power:

$$(1) \quad \frac{d}{dx}x^n = nx^{n-1}$$

and

$$(2) \quad \frac{d}{dx}u^n = nu^{n-1}\frac{du}{dx}.$$

Which one do I use?

A Formula (1) is the original power rule, and it only applies to a power of x. Thus, for instance, it does not apply to $(2x + 1)^{10}$, since the quantity that is being raised to a power is not x. Formula (2) applies to a power of any *function of* x, such as $(2x + 1)^{10}$. It can even be used in place of the original power rule. For example, if we take $u = x$ in Formula (2), we obtain

$$\frac{d}{dx}x^n = nx^{n-1}\frac{dx}{dx}$$

$$= nx^{n-1},$$

since the derivative of x with respect to x is 1. Thus, the generalized power rule is really a generalization of the original power rule, as its name suggests.

▼ **EXAMPLE 4**

Find **(a)** $\dfrac{d}{dx} (2x^5 + x^2 - 20)^{-2/3}$, **(b)** $\dfrac{d}{dx}\left(\dfrac{1}{\sqrt{x + 2}}\right)$, and **(c)** $\dfrac{d}{dx}\left(\dfrac{1}{x^2 + x}\right)$.

SOLUTION

(a) $\dfrac{d}{dx}(2x^5 + x^2 - 20)^{-2/3} = -\dfrac{2}{3}(2x^5 + x^2 - 20)^{-5/3}(10x^4 + 2x)$

(b) $\dfrac{d}{dx}\left(\dfrac{1}{\sqrt{x + 2}}\right) = \dfrac{d}{dx}(x + 2)^{-1/2} = -\dfrac{1}{2}(x + 2)^{-3/2}(1) = -\dfrac{1}{2(x + 2)^{3/2}}$

(c) $\dfrac{d}{dx}\left(\dfrac{1}{x^2 + x}\right) = \dfrac{d}{dx}(x^2 + x)^{-1} = -(x^2 + x)^{-2}(2x + 1) = -\dfrac{2x + 1}{(x^2 + x)^2}$

Before we go on... In the last instance, we could have used the quotient rule instead of the generalized power rule. The reason for this is that we could think of the quantity $1/(x^2 + x)$ in two ways by using the calculation thought experiment:

(1) as 1 divided by something—in other words, as a quotient;
(2) as something raised to the -1 power.

There are two morals to be had from the last example. One is that there are usually several ways to find the right answer. More subtle, though, is that if there are several ways to the answer, those ways may be related. In fact that is the case here, and we can finally give you a reason for believing the quotient rule, if you will believe the chain and product rules.

PROOF OF THE QUOTIENT RULE

By the chain rule, with $u = g(x)$,

$$\frac{d}{dx}\left(\frac{1}{g(x)}\right) = \frac{d}{dx}[g(x)^{-1}]$$

$$= -[g(x)^{-2}]g'(x) = -\frac{1}{g(x)^2}g'(x) = -\frac{g'(x)}{g(x)^2}.$$

We use this to get the derivative of any quotient.

We have

$$\frac{d}{dx}\left(\frac{f(x)}{g(x)}\right) = \frac{d}{dx}\left(f(x)\,\frac{1}{g(x)}\right) \qquad \text{Derivative of a product}$$

$$= f'(x)\frac{1}{g(x)} + f(x)\left(-\frac{g'(x)}{g(x)^2}\right) \qquad \text{Product rule}$$

$$= \frac{f'(x)}{g(x)} - \frac{f(x)g'(x)}{g(x)^2}$$

$$= \frac{f'(x)g(x) - f(x)g'(x)}{g(x)^2}$$

which is, of course, the quotient rule. ◄

Thus, if you believe the chain rule, you *must* believe the quotient rule. Of course, we still owe you some reason for believing the chain rule in the first place. Truly rigorous proofs of the chain rule require very careful analysis of limits. Here is a rough outline of a real proof.

IDEA OF PROOF OF THE CHAIN RULE

From the definition of the derivative,

$$\frac{g(x + h) - g(x)}{h} \approx g'(x) \quad \text{for small } h.$$

(Remember that the difference quotient is an approximation to the derivative.) If we multiply both sides by h and add $g(x)$ to both sides, we find

$$g(x + h) \approx g(x) + g'(x)h \quad \text{for small } h.$$

Now the same is true for f.

$$f(y + k) \approx f(y) + f'(y)k \quad \text{for small } k.$$

What we are after is the derivative of $f(g(x))$. Thus, we want to calculate the limit of

$$\frac{f(g(x + h)) - f(g(x))}{h}.$$

If we approximate $g(x + h)$ by $g(x) + g'(x)h$, we get

$$\frac{f(g(x + h)) - f(g(x))}{h} \approx \frac{f(g(x) + g'(x)h) - f(g(x))}{h}.$$

Now we approximate $f(g(x) + g'(x)h)$.

$$f(g(x) + g'(x)h) \approx f(g(x)) + f'(g(x))g'(x)h$$

(Take $f(y + k) \approx f(y) + kf'(y)$ and replace y with $g(x)$ and k with $g'(x)h$.)

Substituting for $f(g(x) + g'(x)h)$, we now get

$$\frac{f(g(x+h)) - f(g(x))}{h} \approx \frac{f(g(x) + g'(x)h) - f(g(x))}{h}$$

$$\approx \frac{f(g(x)) + f'(g(x))g'(x)h - f(g(x))}{h}.$$

The terms $f(g(x))$ cancel to give

$$\frac{f(g(x+h)) - f(g(x))}{h} \approx \frac{f'(g(x))g'(x)h}{h} = f'(g(x))g'(x),$$

which is the chain rule.*   ◄

We now look at several more complicated examples.

▼ **EXAMPLE 5**

Find $\dfrac{dy}{dx}$ if $y = (\sqrt{x+1} + 3x)^{-3}$.

SOLUTION   The calculation thought experiment tells us that the last operation we would perform in calculating y is raising the quantity $(\sqrt{x+1} + 3x)$ to the power -3. Thus, we use the generalized power rule.

$$\frac{dy}{dx} = -3(\sqrt{x+1} + 3x)^{-4} \frac{d}{dx}(\sqrt{x+1} + 3x)$$

Notice that we are not yet done, since this equation indicates that we must still find the derivative of $\sqrt{x+1} + 3x$. We need not do everything in one step. Finding the derivative of a complicated function in several steps helps to keep the problem manageable. Continuing,

$$\frac{dy}{dx} = -3(\sqrt{x+1} + 3x)^{-4} \frac{d}{dx}(\sqrt{x+1} + 3x)$$

$$= -3(\sqrt{x+1} + 3x)^{-4}\left(\frac{d}{dx}\sqrt{x+1} + \frac{d}{dx}(3x)\right).$$

Now we have two derivatives left to calculate. The second of these we know to be 3, and the first is the derivative of the square root of a quantity. Thus,

$$\frac{dy}{dx} = -3(\sqrt{x+1} + 3x)^{-4}\left(\frac{1}{2\sqrt{x+1}} + 3\right).$$

▼ *You might wonder why we never seemed to take the limit as $h \to 0$. In fact, we did this when we replaced quantities by their *approximations*. These approximations become equalities only in the limit as $h \to 0$, so, in effect, we *did* take the limit.

▼ **EXAMPLE 6**

Find $\dfrac{d}{dx}\left((x + 10)^3\sqrt{1 - x^2}\right)$.

SOLUTION   The expression $(x + 10)^3\sqrt{1 - x^2}$ is a product, so we use the product rule.

$$\frac{d}{dx}\left((x + 10)^3\sqrt{1 - x^2}\right) = \left(\frac{d}{dx}(x + 10)^3\right)\sqrt{1 - x^2}$$
$$+ (x + 10)^3\frac{d}{dx}\sqrt{1 - x^2}$$

Notice that we left the calculation of the derivatives of the factors for the next step.

$$= 3(x + 10)^2\sqrt{1 - x^2} + (x + 10)^3\frac{1}{2\sqrt{1 - x^2}}(-2x)$$
$$= 3(x + 10)^2\sqrt{1 - x^2} - \frac{x(x + 10)^3}{\sqrt{1 - x^2}}$$

The next example is a new treatment of an exercise from the last chapter.

▼ **EXAMPLE 7**   Marginal Product

Precision Manufacturers, Inc. is informed by a consultant that its annual profit is given by

$$P = -200{,}000 + 4{,}000q - 0.5q^2,$$

where q is the number of surgical lasers it sells per year. The consultant also informs Precision that the number of surgical lasers it can manufacture per year depends on the number n of assembly-line workers it employs, according to the equation

$$q = 100n + 0.1n^2.$$

Use the chain rule to find the *marginal product* $\dfrac{dP}{dn}$.

SOLUTION   Recall that we calculated the marginal product in Chapter 3 by substituting the expression for q in the expression for P to obtain P as a function of n and then finding dP/dn. Alternatively—and this will simplify the calculation—we can use the chain rule. To see how the chain rule applies, notice that P is a function of q, where q in turn is given as a function of n. Thus, by the chain rule,

$$\frac{dP}{dn} = P'(q)\frac{dq}{dn}$$

$$= \frac{dP}{dq} \cdot \frac{dq}{dn}.$$

Now we compute

$$\frac{dP}{dq} = 4{,}000 - q$$

and

$$\frac{dq}{dn} = 100 + 0.2n.$$

Substituting into the equation for $\dfrac{dP}{dn}$ gives

$$\frac{dP}{dn} = (4{,}000 - q)(100 + 0.2n).$$

Notice that the answer has both q and n as variables. We can express this as a function of n alone by substituting for q.

$$\frac{dP}{dn} = [4{,}000 - (100n + 0.1n^2)](100 + 0.2n)$$

$$= (4{,}000 - 100n - 0.1n^2)(100 + 0.2n)$$

The equation

$$\frac{dP}{dn} = \frac{dP}{dq} \cdot \frac{dq}{dn}$$

is an appealing way of writing the chain rule. In general, we can write the chain rule as follows.

CHAIN RULE IN DIFFERENTIAL NOTATION

If y is a function of u, and u is a function of x, then

$$\frac{dy}{dx} = \frac{dy}{du}\frac{du}{dx}.$$

If y is a function of u, and x is a function of u, then

$$\frac{dy}{dx} = \frac{dy}{du} \bigg/ \frac{dx}{du}.$$

This is one of the reasons we still use Leibniz's differential notation. In this notation the chain rule looks like a simple "cancellation" of du terms.

▼ ▐ **EXAMPLE 8** ▌   Marginal Revenue

Suppose that a company's weekly revenue R is given as a function of the unit price p, and that p in turn is given as a function of weekly sales q (by means of a demand equation). If

$$\frac{dR}{dp}\bigg|_{q=1000} = \$40 \text{ per } \$1 \text{ increase in price, and}$$

$$\frac{dp}{dq}\bigg|_{q=1000} = -\$20 \text{ per additional item sold per week,}$$

find the marginal revenue when sales are 1,000 items per week.

SOLUTION   The marginal revenue is $\dfrac{dR}{dq}$. By the chain rule,

$$\frac{dR}{dq} = \frac{dR}{dp}\frac{dp}{dq}.$$

Since we are interested in the marginal revenue at a demand level of 1,000 items per week, we have

$$\frac{dR}{dq}\bigg|_{q=1000} = (40)(-20) = -\$800 \text{ per additional item demanded.}$$

Thus, if the price is lowered to increase the demand from 1,000 to 1,001 items per week, the weekly revenue will drop by approximately \$800.

So far, we have *proved* the (original) power rule only for integer exponents. (Positive integer exponents were dealt with in the previous chapter and negative integer exponents were dealt with in the previous section.) We end this section by proving that the power rule also works for *rational* exponents, those of the form $\frac{p}{q}$, with p and q integers.

▼ ▐ **EXAMPLE 9** ▌   Power Rule for Rational Exponents

Show that $\dfrac{d}{dx}(x^{p/q}) = \dfrac{p}{q}x^{p/q-1}$.

SOLUTION   We first let $y = x^{p/q}$. Thus, the problem is to calculate dy/dx without assuming the power rule for anything but *integer* exponents. Before we do anything, we raise both sides to the power q in order to get integer exponents everywhere.

$$y^q = (x^{p/q})^q = x^p$$

Now we take the equation $y^q = x^p$ and differentiate both sides with respect to x. By the chain rule,

$$\frac{d}{dx}(y^q) = qy^{q-1}\frac{dy}{dx},$$

whereas

$$\frac{d}{dx}(x^p) = px^{p-1}.$$

Equating these derivatives gives

$$qy^{q-1}\frac{dy}{dx} = px^{p-1}.$$

Now remember that we want dy/dx by itself. We can solve for dy/dx by dividing both sides by the quantity qy^{q-1}:

$$\frac{dy}{dx} = \frac{px^{p-1}}{qy^{q-1}}.$$

But $y = x^{p/q}$, and so

$$\frac{d}{dx}(x^{p/q}) = \frac{px^{p-1}}{q(x^{p/q})^{q-1}}$$

$$= \frac{px^{p-1}}{qx^{p-p/q}}$$

$$= \frac{p}{q}x^{p-1-(p-p/q)} = \frac{p}{q}x^{p/q-1}.$$

Done!

Before we go on... This calculation of dy/dx is an example of a technique called **implicit differentiation.** We shall discuss this technique in more detail in Section 4.

With some effort we have now proven the power rule for rational powers, but this still leaves the irrational powers. At the end of the next section we shall see a completely different proof of the power rule that applies to *all* real powers.

▶ ___ **4.2 EXERCISES**

Mentally calculate the derivatives of the functions in Exercises 1 through 10.

1. $f(x) = (2x + 1)^2$ **2.** $f(x) = (3x - 1)^2$ **3.** $f(x) = (x - 1)^{-1}$

4. $f(x) = (2x - 1)^{-2}$ **5.** $f(x) = (2 - x)^{-2}$ **6.** $f(x) = (1 - x)^{-1}$

7. $f(x) = \sqrt{2x + 1}$ **8.** $f(x) = \sqrt{3x - 2}$ **9.** $f(x) = \dfrac{1}{3x - 1}$

10. $f(x) = \dfrac{1}{(x + 1)^2}$

Calculate the derivatives of the functions in Exercises 11 through 32.

11. $f(x) = (x^2 + 2x)^4$

12. $f(x) = (x^3 - x)^3$

13. $f(x) = (2x^2 - 2)^{-1}$

14. $f(x) = (2x^3 + x)^{-2}$

15. $g(x) = (x^2 - 3x - 1)^{-5}$

16. $g(x) = (2x^2 + x + 1)^{-3}$

17. $h(x) = \dfrac{1}{(x^2 + 1)^3}$

18. $h(x) = \dfrac{1}{(x^2 + x + 1)^2}$

19. $s(t) = (t^2 - \sqrt{t})^4$

20. $s(t) = (2t + \sqrt{t})^{-1}$

21. $f(x) = \sqrt{1 - x^2}$

22. $f(x) = \sqrt{x + x^2}$

23. $f(x) = \sqrt{3\sqrt{x} - \dfrac{1}{\sqrt{x}}}$

24. $f(x) = \sqrt{2\sqrt{x} + \dfrac{1}{2\sqrt{x}}}$

25. $r(x) = (\sqrt{2x + 1} - x^2)^{-1}$

26. $r(x) = (\sqrt{x + 1} + \sqrt{x})^3$

27. $s(x) = \left(\dfrac{2x + 4}{3x - 1}\right)^2$

28. $s(x) = \left(\dfrac{3x - 9}{2x + 4}\right)^3$

29. $h(r) = [(r + 1)(r^2 - 1)]^{-1/2}$

30. $h(r) = [(2r - 1)(r - 1)]^{-1/3}$

31. $f(x) = (x^2 - 3x)^{-2}\sqrt{1 - x^2}$

32. $f(x) = (3x^2 + x)\sqrt{1 - x^2}$

Find the indicated derivative in each of Examples 33 through 40. In each case, the independent variable is an (unspecified) function of t.

33. $y = x^{100} + 99x^{-1}$. Find $\dfrac{dy}{dt}$.

34. $y = \sqrt{x}(1 + x)$. Find $\dfrac{dy}{dt}$.

35. $s = \dfrac{1}{r^3} + \sqrt{r}$. Find $\dfrac{ds}{dt}$.

36. $s = r + r^{-1}$. Find $\dfrac{ds}{dt}$.

37. $V = \dfrac{4}{3}\pi r^3$. Find $\dfrac{dV}{dt}$.

38. $A = 4\pi r^2$. Find $\dfrac{dA}{dt}$.

39. $y = x^3 + \dfrac{1}{x}$, $x = 2$ when $t = 1$, $\left.\dfrac{dx}{dt}\right|_{t=1} = -1$. Find $\left.\dfrac{dy}{dt}\right|_{t=1}$.

40. $y = \sqrt{x} + \dfrac{1}{\sqrt{x}}$, $x = 9$ when $t = 1$, $\left.\dfrac{dx}{dt}\right|_{t=1} = -1$. Find $\left.\dfrac{dy}{dt}\right|_{t=1}$.

APPLICATIONS

41. *Marginal Revenue* We saw in previous chapters that the weekly sales of rubies by Royal Ruby Retailers (RRR) is given by

$$q = -\dfrac{4p}{3} + 80.$$

(a) Express RRR's weekly revenue as a function of p, and hence calculate $\left.\dfrac{dR}{dp}\right|_{q=60}$.

(b) Use the demand equation to calculate $\left.\dfrac{dp}{dq}\right|_{q=60}$.

(c) Use the answers to parts (a) and (b) to find the marginal revenue at a demand level of 60 rubies per week, and interpret the result.

42. *Marginal Revenue* Repeat Exercise 41 for a demand level of $q = 52$ rubies per week.

43. *Marginal Product* Paramount Electronics, Inc. has an annual profit given by

$$P = -100,000 + 5,000q - 0.25q^2,$$

where q is the number of laptop computers it sells per year. The number of laptop computers it can manufacture per year depends on the number n of electrical engineers Paramount employs, according to the equation

$$q = 30n + 0.01n^2.$$

Use the chain rule to find $\left.\dfrac{dP}{dn}\right|_{n=10}$, and interpret the result.

44. *Marginal Product*   Referring to Exercise 43, give a formula for the average profit per computer

$$\bar{P} = \frac{P}{q}$$

as a function of q, and hence determine the **marginal average product,** $\dfrac{d\bar{P}}{dn}$, at an employee level of 10 engineers. Interpret the result.

45. *Ecology*   Manatees are grazing sea mammals sometimes referred to as sea sirens. Increasing numbers of manatees have been killed by boats off the Florida coast. Since 1976, the number M of manatees killed by boats each year is roughly linear, with

$$M(t) \approx 2.27t + 11.5 \qquad (0 \le t \le 16).$$

(t is the number of years since 1976*.) Over the same period, the total number B of boats registered in Florida has also been increasing at a roughly linear rate, given by

$$B(t) \approx 19{,}500t + 436{,}000 \qquad (0 \le t \le 16).$$

Use the chain rule to give an estimate of $\dfrac{dM}{dB}$. What does the answer tell you about manatee deaths?

46. *Ecology*   The linear models used in the previous exercise are rough, as shown in the following graphs of the actual data.*

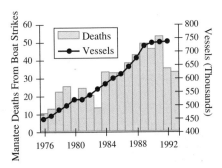

As you can see, the annual number of deaths was declining by 1992, despite an increase in the number of registered boats. This suggests that a linear model is not appropriate for the most recent period. If we consider only the data from 1986 onwards, use a quadratic model for manatee deaths, and use a linear model for the number of vessels, we obtain

$$M(t) \approx -1.35t^2 + 9.64t + 30.92$$

$$\text{and}\quad B(t) \approx 16{,}700t + 641{,}000,$$

where t is the number of years since 1986. Use the chain rule to obtain $\dfrac{dM}{dB}$. What does the answer tell you about manatee deaths? Give a possible explanation of this result.

47. *Pollution*   An offshore oil well is leaking oil and creating a circular oil slick. If the radius of the slick is growing at a rate of 2 mph, find the rate at which the area is increasing when the radius is 3 miles. (Remember that the area of a disc of radius r is $A = \pi r^2$.)

48. *Mold*   A mold culture in a dorm refrigerator is circular and growing in size. The radius is growing at a rate of 3 cm/day. How fast is the area growing when the culture is 4 centimeters in radius? (The area of a disc of radius r is $A = \pi r^2$.)

49. *Budget Overruns*   The Pentagon is planning to build a new satellite that will be spherical in shape. As is typical in these cases, the specifications keep changing, so that the size of the satellite keeps growing. In fact, the radius of the planned satellite is growing 0.5 feet per week. Now its cost will be \$1,000 per cubic foot. At the point when the plans call for a satellite 10 feet in radius, how fast is the cost growing? (Remember that the volume of a solid sphere of radius r is $V = \frac{4}{3}\pi r^3$.)

50. *Soap Bubbles*   The soap bubble I am blowing has a surface area that is growing at the rate of 4 cm²/sec. How fast is its radius growing when its surface area is 40 cm²? (The radius of a sphere of surface area S is $r = \sqrt{S/(4\pi)}$.)

51. *Revenue Growth (graphing calculator or computer software required)*   The demand for the Cyberpunk II arcade video game is modeled by the logistic curve

$$q(t) = \frac{10{,}000}{1 + 0.5e^{-0.4t}},$$

▼ * These models are best-fit linear functions based on graphical data in the article "Manatees," by Thomas J. O'Shea, *Scientific American*, July, 1994, pp. 66-72. Source: Florida Department of Environmental Protection.

where $q(t)$ is the total number of units sold t months after its introduction.

(a) Graph this function for $0 \le t \le 10$ using a range for q of $5{,}000 \le q' \le 10{,}000$, and estimate $q'(4)$.

(b) Assume that the manufacturers of Cyberpunk II sell each unit for $800. What is the company's marginal revenue?

(c) Use the chain rule to estimate the rate at which revenue is growing 4 months after the introduction of the video game.

52. *Information Highway* (*graphing calculator or computer software required*) The amount of information transmitted each month on the National Science Foundation's Internet network for the years from 1988 to 1994 can be modeled by the equation

$$q(t) = \frac{2e^{0.69t}}{3 + 1.5e^{-0.4t}},$$

where q is the amount of information transmitted each month in billions of data packets and t is the number of years since the start of 1988.*

(a) Graph this function with $0 \le t \le 5$ and $0 \le q \le 40$, and estimate $q'(2)$.

(b) Assume that it costs $5 to transmit a million packets of data. What is the marginal cost?

(c) How fast was the cost increasing at the start of 1990?

Money Stock Exercises 53–56 are based on the following demand function for money (taken from a question on the GRE economics test):

$$M_d = (2) \times (y)^{0.6} \times (r)^{-0.3} \times (p),$$

where

     M_d = *demand for nominal money balances;*

     y = *real income;*

     r = *an index of interest rates;*

     p = *an index of prices.*

These exercises also use the idea of **percentage rate of growth:** the percentage rate of growth of f at x is $f'(x)/f(x)$ (this is discussed again in Section 5).

53. (*from the GRE economics test*) If the interest rate and price level are to remain constant while real income grows at 5 percent per year, the money stock must grow at what percent per year?

54. (*from the GRE economics test*) If real income and the price level are to remain constant while the interest rate grows at 5 percent per year, the money stock must change by what percent per year?

55. (*from the GRE economics test*) If the interest rate is to remain constant while real income grows at 5 percent per year and the price level rises at 5 percent per year, the money stock must grow at what percent per year?

56. (*from the GRE economic test*) If real income grows by 5 percent per year, the interest rate grows by 2 percent per year, and the price level drops by 3 percent per year, the money stock must change by what percent per year?

COMMUNICATION AND REASONING EXERCISES

57. How is the graph of $y = f(x)$ related to the graph of $y = g(x) = f(x + 1)$? How is the slope of the graph of g at $x = a$ related to the slope of the graph of f at $x = a + 1$? Illustrate with $f(x) = x^2 + 2x$.

58. How is the graph of $y = f(x)$ related to the graph of $y = g(x) = f(2x)$? How is the slope of the graph of g at $x = a$ related to the slope of the graph of f at $x = 2a$? Illustrate with $f(x) = x^2 + 2x$.

59. Formulate a simple procedure for deciding whether to first apply the chain rule, the product rule, or the quotient rule when finding the derivative of a function.

60. Give an example of a function f with the property that calculating $f'(x)$ requires use of the following rules in the given order: (1) the chain rule, (2) the product rule, (3) the quotient rule, (4) the chain rule.

61. Give an example of a function f with the property that calculating $f'(x)$ requires use of the chain rule five times in succession.

62. What can you say about composites of linear functions?

▼ * This is the authors' model, based on figures published in *The New York Times*, Nov. 3, 1993.

▶ ═══ 4.3 DERIVATIVES OF LOGARITHMIC AND EXPONENTIAL FUNCTIONS

At this point, we know how to take the derivative of any algebraic expression in x (involving powers, radicals, and so on). We now turn to the derivatives of logarithmic and exponential functions.

> **DERIVATIVE OF THE LOGARITHM FUNCTION**
>
> $$\frac{d}{dx} \log_b x = \frac{1}{x \ln b}$$
>
> An important special case is this:
>
> **DERIVATIVE OF THE NATURAL LOGARITHM**
>
> $$\frac{d}{dx} \ln x = \frac{1}{x}$$

It is because $\ln x = \log_e x$ has the simplest-looking derivative that it is called the *natural* logarithm.

Q Where do these formulas come from?

A We shall show you at the end of this section. For now, let us look at the graphs of $y = \ln x$ and $y = 1/x$ to see that it is reasonable that the derivative of $\ln x$ should be $1/x$ (Figure 1).

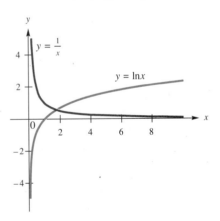

FIGURE 1

When x is close to 0, the graph of $\ln x$ is rising steeply, so the derivative should be a large positive number. And indeed, $1/x$ is positive and large. As x increases, the graph of $\ln x$ becomes less steep, although it continues to rise. Therefore, the derivative should become smaller but remain positive, and this is exactly what happens to $1/x$. So $1/x$ at least behaves the way the derivative of $\ln x$ should.

▼ **EXAMPLE 1**

Find $\dfrac{d}{dx}[x \ln x]$.

SOLUTION   By the calculation thought experiment, we see that $x \ln x$ is a product, so we need to use the product rule.

$$\frac{d}{dx}[x \ln x] = (1)\ln x + x\left(\frac{1}{x}\right) = \ln x + 1$$

(As we cautioned earlier, you must be prepared to use any of the rules in any example.)

If we were to take the derivative of the natural logarithm of a quantity, rather than just x, we would have to use the chain rule. In words, it would read like this:

> *The derivative of* ln*(quantity) is one over that quantity times the derivative of that quantity.*

We can also write this as a formula.

Original Rule	Generalized Rule
$\dfrac{d}{dx}\ln x = \dfrac{1}{x}$	$\dfrac{d}{dx}\ln u = \dfrac{1}{u}\dfrac{du}{dx}$

▼ **EXAMPLE 2**

Evaluate $\dfrac{d}{dx}\ln(x^2 + 1)$.

SOLUTION   If we were to evaluate $\ln(x^2 + 1)$, the last operation we would perform is taking the natural logarithm of something. Thus, the calculation thought experiment tells us that we are dealing with ln *of a quantity*, and so we need the generalized logarithm rule as stated above. Thus,

$$\frac{d}{dx}\ln (x^2 + 1) = \frac{1}{x^2 + 1}\frac{d}{dx}(x^2 + 1)$$

$$= \frac{1}{x^2 + 1}(2x) = \frac{2x}{x^2 + 1}.$$

▼ **EXAMPLE 3**

Find $\dfrac{d}{dx}\ln\sqrt{x + 1}$.

SOLUTION   The calculation thought experiment again tells us that we have the natural logarithm of something, so

$$\frac{d}{dx} \ln\sqrt{x + 1} = \frac{d}{dx} \ln[(x + 1)^{1/2}]$$

$$= \frac{1}{\sqrt{x + 1}}\left(\frac{1}{2}(x + 1)^{-1/2}\right)$$

$$= \frac{1}{\sqrt{x + 1}} \cdot \frac{1}{2\sqrt{x + 1}}$$

$$= \frac{1}{2(x + 1)}.$$

Before we go on... What happened to the square root? As with many problems involving logarithms, we could have done this one differently and with less bother had we simplified the expression $\ln\sqrt{x + 1}$ using the rules of logarithms *before* differentiating. Doing this, we get

$$\ln\sqrt{x + 1} = \ln(x + 1)^{1/2} = \frac{1}{2}\ln(x + 1).$$

Thus,

$$\frac{d}{dx} \ln\sqrt{x + 1} = \frac{d}{dx}\left(\frac{1}{2}\ln(x + 1)\right)$$

$$= \frac{1}{2}\left(\frac{1}{x + 1}\right) = \frac{1}{2(x + 1)},$$

the same answer as above.

▼ **EXAMPLE 4**

Evaluate $\dfrac{d}{dx} \ln[(x + 1)(x + 2)]$.

SOLUTION This time, we simplify the expression $\ln[(x + 1)(x + 2)]$ before taking the derivative.

$$\ln[(x + 1)(x + 2)] = \ln(x + 1) + \ln(x + 2)$$

Thus,

$$\frac{d}{dx} \ln[(x + 1)(x + 2)] = \frac{d}{dx} \ln(x + 1) + \frac{d}{dx} \ln(x + 2)$$

$$= \frac{1}{x + 1} + \frac{1}{x + 2}$$

$$= \frac{2x + 3}{(x + 1)(x + 2)}.$$

Before we go on... For practice, try doing this example without simplifying first. What other differentiation rule do you have to use?

▼ **EXAMPLE 5**

Find $\dfrac{d}{dx} \log_2(x^3 + x)$.

SOLUTION   Rather than work out a "very generalized logarithm rule" for logs with any base, it is easier to remember that

$$\log_b x = \frac{\ln x}{\ln b},$$

where $\ln b$ is a *constant*. This gives us

$$\frac{d}{dx} \log_2(x^3 + x) = \frac{d}{dx}\left(\frac{\ln(x^3 + x)}{\ln 2}\right)$$

$$= \frac{1}{\ln 2}\frac{d}{dx}\ln(x^3 + x) \qquad \text{because ln 2 is a constant}$$

$$= \frac{3x^2 + 1}{(x^3 + x)\ln 2}.$$

▼ **EXAMPLE 6**

Find $\dfrac{d}{dx}\ln|x|$.

SOLUTION   Before we start, you might ask why we are bothering with the natural log of the absolute value of x to begin with. The reason is this: $\ln x$ is defined only for positive values of x, so its domain is $(0, +\infty)$. The function $\ln|x|$, on the other hand, is defined for *all* values of x other than zero. For example, $\ln|-2| = \ln 2 \approx 0.6931$. Thus the domain of $\ln|x|$ is $(-\infty, 0) \cup (0, +\infty)$. For this reason, it often turns out to be more useful than the ordinary logarithm function.

Now we'll get to work. We might be tempted to use the chain rule here (with $u = |x|$), but there is a slight catch: the function $|x|$ is, as we saw in the last chapter, not differentiable at $x = 0$. We *could* work our way around this, since the logarithm is not defined there anyway, but we'll use the following approach instead.

First,

$$|x| = \begin{cases} x & \text{if } x \geq 0 \\ -x & \text{if } x < 0. \end{cases}$$

Thus,

$$\ln|x| = \begin{cases} \ln x & \text{if } x > 0 \\ \ln(-x) & \text{if } x < 0. \end{cases}$$

Hence,

$$\frac{d}{dx}\ln|x| = \begin{cases} \dfrac{d}{dx}\ln x = \dfrac{1}{x} & \text{if } x > 0 \\ \dfrac{d}{dx}\ln(-x) = \dfrac{1}{-x}(-1) = \dfrac{1}{x} & \text{if } x < 0. \end{cases}$$

In other words,

$$\frac{d}{dx}\ln|x| = \frac{1}{x}.$$

Before we go on... Figure 2 shows the graphs of $y = \ln|x|$ and $y = 1/x$. Figure 3 shows the graphs of $y = \ln|x|$ and $y = 1/|x|$. You should be able to see from these graphs why the derivative of $\ln|x|$ is $1/x$ and not $1/|x|$.

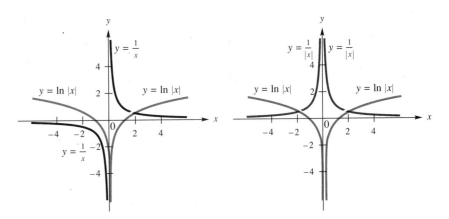

FIGURE 2 **FIGURE 3**

As we said, we could also find the derivative of $\ln|x|$ using the chain rule and the fact that

$$\frac{d}{dx}|x| = \begin{cases} 1 & \text{if } x > 0 \\ -1 & \text{if } x < 0. \end{cases}$$

Try this approach to see how well you understand the chain rule.

This example, in conjunction with the chain rule, gives us the following formulas.

DERIVATIVES OF LOGARITHMS OF ABSOLUTE VALUES

Original Rule	Generalized Rule
$\dfrac{d}{dx} \ln|x| = \dfrac{1}{x}$	$\dfrac{d}{dx} \ln|u| = \dfrac{1}{u}\dfrac{du}{dx}$
$\dfrac{d}{dx} \log_b|x| = \dfrac{1}{x \ln b}$	$\dfrac{d}{dx} \log_b|u| = \dfrac{1}{u \ln b}\dfrac{du}{dx}$

In other words, when taking the derivative of the logarithm of the absolute value of a quantity, we can simply ignore the absolute value sign!

▼ **EXAMPLE 7**

Find $\dfrac{d}{dx} \ln|x^2 - x + 1|$.

SOLUTION By the generalized logarithm rule,

$$\frac{d}{dx} \ln|x^2 - x + 1| = \frac{2x - 1}{x^2 - x + 1}.$$

We now turn to the derivatives of exponential functions—that is, functions of the form $f(x) = a^x$. We begin by showing how *not* to differentiate them.

▶ **CAUTION** The derivative of a^x is *not* xa^{x-1}. The power rule applies only to *constant* exponents. In this case, the exponent is decidedly *not* constant. Thus, the power rule does not apply, and we shall need a new rule. ◀

DERIVATIVE OF a^x

If a is any positive number, then

$$\frac{d}{dx} a^x = a^x \ln a.$$

In particular:

DERIVATIVE OF e^x

$$\frac{d}{dx} e^x = e^x$$

Thus, e^x has the amazing property that its derivative is itself! (There is another—very simple—function that is its own derivative. What is it?)

▼ **EXAMPLE 8**

Find $\dfrac{d}{dx}\left(\dfrac{e^x}{x}\right)$.

SOLUTION   Using the calculation thought experiment, we see that the expression e^x/x is a quotient, so we use the quotient rule.

$$\frac{d}{dx}\left(\frac{e^x}{x}\right) = \frac{e^x x - e^x(1)}{x^2} = \frac{e^x(x-1)}{x^2}$$

If we were to take the derivative of e raised to a quantity, not just x, we would have to use the chain rule. In words, it would read like this:

The derivative of e raised to a quantity is e raised to that quantity times the derivative of that quantity.

We can also write this as a formula.

DERIVATIVES OF EXPONENTIAL FUNCTIONS

Original Rule	Generalized Rule
$\dfrac{d}{dx}e^x = e^x$	$\dfrac{d}{dx}e^u = e^u\dfrac{du}{dx}$
$\dfrac{d}{dx}a^x = a^x \ln a$	$\dfrac{d}{dx}a^u = a^u \ln a\dfrac{du}{dx}$

▼ **EXAMPLE 9**

Find $\dfrac{d}{dx}e^{x^2+1}$.

SOLUTION   Here the calculation thought experiment says that we have e raised to a quantity. Thus, we will have to use the generalized exponential rule.

$$\frac{d}{dx}e^{x^2+1} = e^{x^2+1}(2x) = 2xe^{x^2+1}$$

▼ **EXAMPLE 10**

Find $\dfrac{d}{dx} 2^{3x}$.

SOLUTION Using the generalized exponential rule, we get

$$\frac{d}{dx} 2^{3x} = 2^{3x} \ln 2 \frac{d}{dx} (3x)$$
$$= 2^{3x} (\ln 2)(3) = (3 \ln 2)2^{3x}.$$

Before we prove the formulas for the derivatives of logarithmic and exponential functions, here are some applications.

▼ **EXAMPLE 11**   Epidemics

At the start of 1990, the number of U.S. residents infected with HIV was estimated to be 0.4 million. This number was growing exponentially, doubling every six months. Had this trend continued, how many new cases per month would have been occurring by the start of 1994?

SOLUTION   To find the answer, we model this exponential growth using the methods of Chapter 2: after n years, the number of cases is

$$A = 0.4e^{kn},$$

where k is given by

$$k = \frac{\ln 2}{doubling\ time} = \frac{\ln 2}{0.5} \approx 1.3863.$$

Thus,

$$A = 0.4e^{1.3863n}.$$

We are asking for the number of new cases per month. In other words, we want the rate of change, dA/dn.

$$\frac{dA}{dn} = 0.4(1.3863)e^{1.3863n}$$
$$= 0.55452e^{1.3863n}$$

At the start of 1994, $n = 4$, so the number of new cases per year is

$$\left.\frac{dA}{dn}\right|_{n=4} = 0.55452e^{1.3863(4)} \approx 141.96 \text{ million}$$

Thus, the number of new cases *per month* would be $141.96/12 = 11.83$ million!

Before we go on... The reason this figure is astronomically large is that we assumed that exponential growth—the doubling every six months—would continue. A more realistic model for the spread of a disease is the logistic model. (See the "You're the Expert" section in Chapter 2, as well as the next example.)

▼ **EXAMPLE 12** Sales Growth

The demand for the Cyberpunk II arcade video game is modeled by the logistic curve

$$q(t) = \frac{10,000}{1 + 0.5e^{-0.4t}},$$

where $q(t)$ is the total number of units sold t months after its introduction. How fast were the sales increasing two years after its introduction?

SOLUTION   We are asking for $q'(24)$. We can find the derivative of $q(t)$ using the quotient rule, or we can first write

$$q(t) = 10,000(1 + 0.5e^{-0.4t})^{-1},$$

and then use the generalized power rule.

$$q'(t) = -10,000(1 + 0.5e^{-0.4t})^{-2}(0.5e^{-0.4t})(-0.4)$$
$$= \frac{2,000e^{-0.4t}}{(1 + 0.5e^{-0.4t})^2}$$

Thus,

$$q'(24) = \frac{2,000e^{-0.4(24)}}{(1 + 0.5e^{-0.4(24)})^2} \approx 0.135 \text{ units per month.}$$

Thus, after two years, sales are increasing quite slowly.

You can check this answer graphically. If you plot the sales curve for $0 \le t \le 30$ and $6{,}000 \le q \le 10{,}000$, you will get the curve shown in Figure 4.

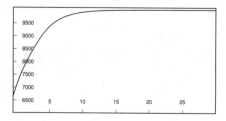

FIGURE 4

Notice that sales level off at about 10,000 units.* We computed $q'(24)$, which is the slope of the curve at the point with t-coordinate 24. If you zoom in to the portion of the curve near $t = 24$, you obtain the graph shown in Figure 5.

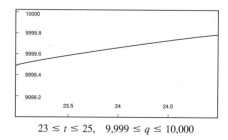

$23 \le t \le 25,\quad 9{,}999 \le q \le 10{,}000$

FIGURE 5

As you can see, the curve is almost linear. If we pick two points on this segment of the curve, say $(23, 9{,}999.4948)$ and $(25, 9{,}999.7730)$, we can approximate the derivative as

$$\frac{9{,}999.7730 - 9{,}999.4948}{25 - 23} = 0.1391,$$

which is accurate to within 0.004.

It is now time to explain where the formulas for the derivatives of $\ln x$ and e^x come from. We start with $\ln x$.

▼ * We can also say this using limits: $\lim\limits_{t \to +\infty} q(t) = 10{,}000$.

Q Why is $\dfrac{d}{dx}\ln x = \dfrac{1}{x}$?

A To compute $\dfrac{d}{dx}\ln x$, we need to use the definition of the derivative. We shall also use properties of the logarithm to help evaluate the limit.

$$\frac{d}{dx}\ln x = \lim_{h \to 0} \frac{\ln(x + h) - \ln x}{h}$$

$$= \lim_{h \to 0} \frac{1}{h}\left[\ln(x + h) - \ln x\right]$$

$$= \lim_{h \to 0} \frac{1}{h}\ln\left(\frac{x + h}{x}\right)$$

$$= \lim_{h \to 0} \frac{1}{h}\ln\left(1 + \frac{h}{x}\right)$$

$$= \lim_{h \to 0} \ln\left(1 + \frac{h}{x}\right)^{1/h}$$

$$= \ln\left(\lim_{h \to 0}\left(1 + \frac{h}{x}\right)^{1/h}\right)$$

(Switching ln and lim in the last step needs some justification that we'll glide over.*)

$$= \ln\left[\lim_{h \to 0}\left(1 + \frac{1}{(x/h)}\right)^{x/h}\right]^{1/x}$$

As $h \to 0$, the quantity x/h is getting large, and so the limit in brackets is approaching e. (Actually, we should say that $x/h \to +\infty$ as $h \to 0^+$, and $x/h \to -\infty$ as $h \to 0^-$ since x is positive for $\ln x$ to be defined. We should discuss both cases, but we shall leave the second to you.)

$$= \ln e^{1/x}$$

$$= \frac{1}{x}\ln e = \frac{1}{x}$$

The rule for the derivative of $\log_b x$ follows from the fact that $\log_b x = \ln x/\ln b$.

▼ * It is actually justified by the fact that the logarithm function is continuous.

Q Why is $\dfrac{d}{dx} e^x = e^x$?

A To find the derivative of e^x we use a shortcut*. Write $g(x) = e^x$. Then

$$\ln g(x) = x.$$

Take the derivative of both sides of this equation to get

$$\frac{g'(x)}{g(x)} = 1,$$

or

$$g'(x) = g(x) = e^x.$$

In other words, the exponential with base e is its own derivative. The rule for exponential functions with other bases follows from the equality $a^x = e^{x \ln a}$ (why?) and the chain rule (try it).

At the end of the last section, we promised to give a proof of the power rule,

$$\frac{d}{dx} x^n = nx^{n-1},$$

that works for all real n. This proof relies on the equality $x^n = e^{n \ln x}$. (You can check this by taking the natural logarithm of both sides.)

$$\frac{d}{dx} x^n = \frac{d}{dx} e^{n \ln x}$$

$$= e^{n \ln x} \frac{d}{dx} [n \ln x]$$

$$= e^{n \ln x} \left(\frac{n}{x} \right)$$

$$= x^n \left(\frac{n}{x} \right)$$

$$= nx^{n-1}$$

This proves the power rule for all real powers.

The power rule is a good example of something that often happens in mathematics. Something like the power rule might first be noticed in simple cases, like $(x^2)' = 2x$ and $(x^3)' = 3x^2$. In those simple cases, simple proofs may be found. Other cases can also be proved by simple means, but the straightforward approach starts to get pretty tough (try finding the derivative of $x^{1/4}$ straight from the definition). As mathematicians try to extend a result

▼ * This shortcut is an example of a technique called *logarithmic differentiation*, which is occasionally useful. We will see it again in the next section.

to cover more and more cases, first the proofs become tricky (for example, the proof of the power rule for rational powers). But a change of viewpoint often leads to greater generality and a simpler proof at the same time (as happened when we rewrote x^n as $e^{n \ln x}$). Often whole new areas of mathematics are opened up in the search for generalizations of the original simple cases. This is one of the ways that mathematics grows.

▶ **4.3 EXERCISES**

Find the derivatives of the functions in Exercises 1–14 mentally.

1. $f(x) = \ln(x - 1)$
2. $f(x) = \ln(x + 3)$
3. $f(x) = \log_2 x$

4. $f(x) = \log_3 x$
5. $g(x) = \ln|x^2 + 3|$
6. $g(x) = \ln|2x - 4|$

7. $h(x) = e^{x+3}$
8. $h(x) = e^{x^2}$
9. $f(x) = e^{-x}$

10. $f(x) = e^{1-x}$
11. $g(x) = 4^x$
12. $g(x) = 5^x$

13. $h(x) = 2^{x^2-1}$
14. $h(x) = 3^{x^2-x}$

Find the derivatives of the functions in Exercises 15–58.

15. $x \ln x$
16. $3 \ln x$

17. $f(x) = (x^2 + 1)\ln x$
18. $f(x) = (4x^2 - x)\ln x$

19. $f(x) = (x^2 + 1)^5 \ln x$
20. $f(x) = \sqrt{x + 1} \ln x$

21. $g(x) = \ln|3x - 1|$
22. $g(x) = \ln|5 - 9x|$

23. $g(x) = \ln|2x^2 + 1|$
24. $g(x) = \ln|x^2 - x|$

25. $g(x) = \ln(x^2 - \sqrt{x})$
26. $g(x) = \ln\left(x + \dfrac{1}{x}\right)$

27. $h(x) = \log_2(x+1)$
28. $h(x) = \log_3(x^2 + x)$

29. $r(t) = (t^2 + 1)\log_3(t + 1/t)$
30. $r(t) = (t^2 - t)\log_3(t + \sqrt{t})$

31. $f(x) = (\ln|x|)^2$
32. $f(x) = \dfrac{1}{\ln|x|}$

33. $r(x) = \ln(x^2) - (\ln(x - 1))^2$
34. $r(x) = (\ln(x^2))^2$

35. $f(x) = xe^x$
36. $f(x) = 2e^x - x^2e^x$

37. $r(x) = \ln(x + 1) + 3x^3e^x$
38. $r(x) = \ln|x + e^x|$

39. $f(x) = e^x \ln|x|$
40. $f(x) = e^x \log_2|x|$

41. $f(x) = e^{2x+1}$
42. $f(x) = e^{4x-5}$

43. $h(x) = e^{x^2-x+1}$
44. $h(x) = e^{2x^2-x+\sqrt{x}}$

45. $s(x) = x^2e^{2x-1}$
46. $s(x) = \dfrac{e^{4x-1}}{x^3 - 1}$

47. $r(x) = (e^{2x-1})^2$
48. $r(x) = (e^{2x^2})^3$

49. $g(x) = \dfrac{e^x + e^{-x}}{e^x - e^{-x}}$
50. $g(x) = \dfrac{1}{e^x + e^{-x}}$

51. $f(x) = \dfrac{1}{x(\ln x)^{1/2}}$

52. $f(x) = \dfrac{e^{-x}}{x(e^x)^{1/2}}$

53. $r(x) = \dfrac{\sqrt{\ln x}}{x}$

54. $r(x) = \dfrac{\sqrt{\ln x}}{x^2 - 1}$

55. $f(x) = \ln|\ln x|$

56. $f(x) = = \ln|\ln|\ln x||$

57. $s(x) = \ln \sqrt{\ln x}$

58. $s(x) = \sqrt{\ln(\ln x)}$

Find the equations of the straight lines described in Exercises 59–64.

59. Tangent to the curve $y = e^x \log_2 x$ at the point $(1, 0)$

60. Tangent to the curve $y = e^x + e^{-x}$ at the point $(0, 2)$

61. Tangent to the curve $y = \ln\sqrt{2x + 1}$ at the point on the curve where $x = 0$

62. Tangent to the curve $y = \ln\sqrt{2x^2 + 1}$ at the point where $x = 1$

63. At right angles to the curve $y = e^{x^2}$ at the point where $x = 1$

64. At right angles to the curve $y = \log_2(3x + 1)$ at the point where $x = 1$

APPLICATIONS

65. *Investments* If $10,000 is invested in a savings account yielding 4% per year, compounded continuously, how fast is the balance growing after 3 years?

66. *Investments* If $20,000 is invested in a savings account yielding 3.5% per year, compounded continuously, how fast is the balance growing after 3 years?

67. *Population Growth* The population of Lower Anchovia was 4,000,000 at the start of 1995 and doubling every 10 years. How fast was it growing per year at the start of 1995?

68. *Population Growth* The population of Upper Anchovia was 3,000,000 at the start of 1996 and doubling every 7 years. How fast was it growing per year at the start of 1996?

69. *Radioactive Decay* Plutonium-239 has a half-life of 24,400 years. How fast is a lump of 10 grams decaying after 100 years?

70. *Radioactive Decay* Carbon-14 has a half-life of 5,730 years. How fast is a lump of 20 grams decaying after 100 years?

71. *Investments* If $10,000 is invested in a savings account yielding 4% per year, compounded semiannually, how fast is the balance growing after 3 years?

72. *Investments* If $20,000 is invested in a savings account yielding 3.5% per year, compounded semiannually, how fast is the balance growing after 3 years?

73. *Life Span in Ancient Rome* The percentage $P(t)$ of people surviving to age t years in ancient Rome can be approximated by*

$$P(t) = 92e^{-0.0277t}.$$

Calculate $P'(22)$, and explain what the result indicates.

74. *Communication Among Bees* The audible signals honey bees use to communicate are in the frequency range 0–500 hertz and are generated by their wings. The speed with which bees' wings must move the air to generate these signals depends on the frequency of the signal, and this relationship can be approximated by[†]

$$V(f) = 95.6e^{0.0049f}.$$

▼ * Based on graphical data in Marvin Minsky's article "Will Robots Inherit the Earth?" *Scientific American*, October, 1994, pp. 109–13.

† Based on graphical data in the article "The Sensory Basis of the Honeybee's Dance Language," by Wolfgang H. Kirchner and William F. Towne, *Scientific American*, June 1994, pp. 74–80.

Here, $V(f)$ is the speed of the air near the bees' wings in millimeters per second, and f is the frequency of the communication signal in hertz. Calculate $V'(200)$, and explain what the result indicates.

75. *Epidemics* The epidemic described in the "You're the Expert" section of the chapter on exponentials followed the curve

$$A = \frac{150,000,000}{1 + 14,999e^{-0.3466t}},$$

where A is the number of people infected and t is the number of months after the start of the disease. How fast is the epidemic growing after 20 months? After 30 months? After 40 months?

76. *Epidemics* Another epidemic follows the curve

$$A = \frac{200,000,000}{1 + 20,000e^{-0.549t}},$$

where t is in years. How fast is the epidemic growing after 10 years? After 20 years? After 30 years?

77. *Information Highway* The amount of information transmitted each month on the National Science Foundation's Internet network for the years from 1988 to 1994 can be modeled by the equation

$$q(t) = \frac{2e^{0.69t}}{3 + 1.5e^{-0.4t}},$$

where q is the amount of information transmitted each month in billions of data packets, and t is the number of years since the start of 1988. How fast was this quantity growing at the start of 1994?

78. *Information Highway* Repeat Exercise 77 using the revised equation

$$q(t) = \frac{2e^{0.72t}}{3 + 1.4e^{-0.4t}}.$$

The following exercises require the use of a graphing calculator or graphing computer software.

79. *Diffusion of New Technology* Numeric control is a technology whereby the operation of machines is controlled by numerical instructions on disks, tapes,

or cards. In a study, W. Mansfield modeled the growth of this technology using the equation

$$p(t) = \frac{0.80}{1 + e^{4.46 - 0.477t}},$$

where $p(t)$ is the percentage of firms using numeric control in year t.*
  (a) Graph this function for $0 \le t \le 20$, and estimate $p'(10)$ graphically. Interpret the result.
  (b) Use your graph to estimate $\lim_{t \to +\infty} p(t)$, and interpret the result.
  (c) Compute $p'(t)$, graph it, and again find $p'(10)$.
  (d) Use your graph to estimate $\lim_{t \to +\infty} p'(t)$, and interpret the result.

80. *Diffusion of New Technology* Repeat Exercise 79 using the revised formula

$$p(t) = \frac{0.90e^{-0.1t}}{1 + e^{4.50 - 0.477t}},$$

which takes into account that in the long term, this new technology will eventually become outmoded and will be replaced by a newer technology.

81. *Growth of HMOs* The enrollment in health maintenance organizations (HMOs) in the years from 1975 to 1992 can be modeled by the equation

$$n(t) = 5 + \frac{40e^{0.002x}}{1 + 25e^{3 - 0.5x}},$$

where $n(t)$ represents the total number (in millions) of U.S. residents enrolled in HMOs t years after 1975.[†] Calculate $n'(t)$, graph it, and use your graph to determine the value of t ($0 \le t \le 20$) when $n'(t)$ was a maximum. Interpret your result.

82. *Demand for Poultry* The demand for poultry can be modeled as

$$q = -60.5 - 0.45p + 0.12b + 12.21 \ln(d),$$

where q is the per capita demand for chicken in pounds per year, p is the wholesale price of chicken in cents per pound, b is the wholesale price of beef in cents per pound, and d is the per capita annual dispos-

▼ * Source: "The Diffusion of a Major Manufacturing Innovation," in *Research and Innovation in the Modern Corporation* (W.W. Norton and Company, Inc., New York, 1971, pp. 186–205)

[†] The authors' model, based on data supplied by the Group Health Association of America (published in *The New York Times*, Oct. 18, 1993).

able income in dollars per year.* Assume that the wholesale prices of chicken and beef are fixed at 25¢ per pound and 50¢ per pound respectively and that the mean disposable income in t years' time will be

$25,000 + 1,000t$. Calculate and graph $q'(t)$ for $0 \le t \le 20$, and use your graph to estimate the value of t for which $q'(t) = 0.30$. Interpret the result.

COMMUNICATION AND REASONING EXERCISES

83. A quantity P is growing exponentially with time. Explain the difference between $P(10)$ and $P'(10)$. If P is measured in kilograms and t is measured in days, what are the units of $P'(10)$?

84. The number N of graphing calculators sold on campus is increasing by 3,000 per year at the present time $(t = 0)$. Does this mean that $N(0) = 3,000$? Explain your answer.

85. Make the correct selections: If $Q = 100e^{-0.3t}$, then Q is [**(a)** increasing, **(b)** decreasing] with increasing t, and Q' is [**(a)** increasing, **(b)** decreasing] with increasing t.

86. If $Q = 2,000 - e^{0.3t}$, then
   (a) both Q and Q' are increasing with increasing t;
   (b) both Q and Q' are decreasing with increasing t;
   (c) Q is increasing and Q' is decreasing with increasing t;
   (d) Q is decreasing and Q' is increasing with increasing t.

*The **percentage, or fractional, rate of change** of a function is defined to be the ratio $f'(x)/f(x)$. (It is customary to express this as a percentage when speaking about percentage rate of change.)*

87. Show that the fractional rate of change of the exponential function e^{kx} is its fractional rate of growth k.

88. Show that the fractional rate of change of $f(x)$ is the rate of change of $\ln(f(x))$.

89. Let $A(t)$ represent a quantity growing exponentially. Show that the percentage rate of growth, $A'(t)/A(t)$, is constant.

90. Let $A(t)$ be the amount of money in an account paying interest compounded some number of times per year. Show that the percentage rate of growth, $A'(t)/A(t)$, is constant. What might this constant represent?

▶ ═══════ **4.4** IMPLICIT DIFFERENTIATION

An equation in two variables x and y may or may not determine y as a function of x. For instance, the equation

$$2x^2 + y = 2$$

determines y as a function of x (solve for y to obtain $y = 2 - 2x^2$, a function of x). On the other hand, the equation

$$2x^2 + y^2 = 2$$

does not determine y as a function of x: solving for y yields $y = \pm\sqrt{2 - 2x^2}$. The "$\pm$" sign reminds us that for some values of x there are two corresponding values for y. In other words, y cannot be specified as a single

---

▼ * This equation is based on actual data from poultry sales in the period 1950–1984. Source: A. H. Studenmund, *Using Econometrics*, Second Edition, (New York: HarperCollins, 1992) pp. 180–81.

function of x. Even though y is not a function of x, we refer to y as an **implicit function** of x.

The best way to justify the term "implicit function" is to look at the graph of the equation $2x^2 + y^2 = 2$. We can graph it in two ways: by constructing a table along the lines described in Chapter 1, or by using a graphing calculator to superimpose the graphs of

$$y = \sqrt{2 - 2x^2} \quad \text{and} \quad y = -\sqrt{2 - 2x^2}.$$

The graph, an *ellipse,* is shown in Figure 1.

The curve $y = \sqrt{2 - 2x^2}$ constitutes the top half of the ellipse, and $y = -\sqrt{2 - 2x^2}$ constitutes the bottom half. Notice that the graph fails the "vertical line test": vertical lines between $x = -1$ and $x = 1$ pass through two points on the graph, and so the graph of $2x^2 + y^2 = 2$ is not the graph of a function. On the other hand, the graph is made up of the graphs of two very respectable functions, $f(x) = \sqrt{2 - 2x^2}$ and $g(x) = -\sqrt{2 - 2x^2}$. If we choose a point on the top half of the ellipse, then the part of the curve near that point is the graph of the function f. If we choose a point on the lower half, then the part of the curve near that point is the graph of the function g.

Figure 2 shows two selected points on the graph—P on the top half, and Q on the bottom half—together with segments of the graph through those points. Each of these segments is part of the graph of a function, as shown. Thus, given the equation $2x^2 + y^2 = 2$ and a point on its graph, there is a function whose graph is part of the graph of the equation near the point. (There are actually two points on the above graph for which this does not work. Can you see them?) The reason we use the word *implicit* is that we need not—and for many equations cannot—find an explicit formula for a function by solving for y.

We now find the slopes of the tangent lines to the ellipse at P and Q. We could do this by taking the derivative of f or g as appropriate. But there is a far simpler way of finding the derivative of an implicit function of x *without*

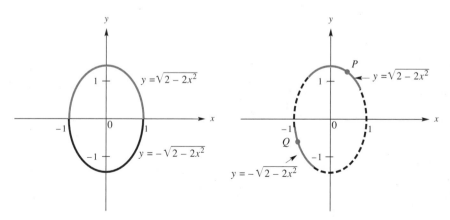

FIGURE 1   **FIGURE 2**

having to solve the equation for y. (We have already seen an example of this technique when we proved the power rule for rational exponents.)

Q How do we find $\dfrac{dy}{dx}$ if $2x^2 + y^2 = 2$ without first solving for y?

A We can do this by using the chain rule and a little cleverness, as follows. We think of y as a function of x and take the derivative with respect to x of both sides of the equation:

$$\frac{d}{dx}\left(2x^2 + y^2\right) = \frac{d}{dx}(2).$$

Recalling the rules for derivatives, we get

$$2\frac{d}{dx}(x^2) + \frac{d}{dx}(y^2) = 0.$$

Now we must be careful. The derivative with respect to x of x^2 is $2x$, but the derivative *with respect to x* of y^2 is *not* $2y$. Rather, since y is a function of x, we must use the chain rule, which tells us that $d/dx\,(y^2) = 2y\,dy/dx$. Thus, we obtain

$$2(2x) + 2y\frac{dy}{dx} = 0.$$

We want to find dy/dx, so we treat it as an unknown *and solve for it:* first take the expression not involving dy/dx over to the right-hand side.

$$2y\frac{dy}{dx} = -4x$$

Next, divide both sides by $2y$ to obtain

$$\frac{dy}{dx} = -\frac{4x}{2y} = -\frac{2x}{y},$$

and we have found $\dfrac{dy}{dx}$.

Q We calculated the derivative of the implicit function of x to be $-2x/y$. But this is not a function of x, as there is a "y" in the formula. Is this of any use to us?

A First, notice that we can hardly expect to obtain a function of x if y was not a function of x to begin with. The result is still useful, since we can evaluate the derivative at any point on the graph. For instance, if we choose P to be the point $(\frac{1}{\sqrt{2}}, 1)$ (which you can check is a point on the graph, since it satisfies $2x^2 + y^2 = 2$), we obtain

$$\frac{dy}{dx} = -\frac{2x}{y} = -\frac{2}{\sqrt{2}} = -\sqrt{2}\ .$$

Thus, the slope of the tangent to the curve $2x^2 + y^2 = 2$ at the point $P = (\frac{1}{\sqrt{2}}, 1)$ is $-\sqrt{2} \approx -1.414$.

This procedure—finding $\dfrac{dy}{dx}$ without first solving an equation for y—is called **implicit differentiation.**

▼ **EXAMPLE 1**

Use implicit differentiation to find dy/dx if $x^2 + 2xy = y$.

SOLUTION Taking the derivative with respect to x of both sides gives

$$\frac{d}{dx}(x^2) + \frac{d}{dx}(2xy) = \frac{d}{dx}(y).$$

Looking at the right-hand side, notice that $\dfrac{d}{dx}(y) = \dfrac{dy}{dx}$. Also notice that $2xy$ is the product of $2x$ and y, and so we must apply the product rule to find $\dfrac{d}{dx}(2xy)$.

$$2x + (2)(y) + (2x)\frac{dy}{dx} = \frac{dy}{dx}$$

That is,

$$2x + 2y + 2x\frac{dy}{dx} = \frac{dy}{dx}.$$

We have marked the term dy/dx in color to remind us that it is the unknown. To solve for it, we bring all the terms containing dy/dx to the left-hand side and all terms not containing it to the right-hand side.

$$2x\frac{dy}{dx} - \frac{dy}{dx} = -2x - 2y$$

Now factor out the common term dy/dx, obtaining

$$\frac{dy}{dx}(2x - 1) = -2x - 2y,$$

and finally, divide by $(2x - 1)$ to obtain

$$\frac{dy}{dx} = \frac{-2x - 2y}{2x - 1} = -\frac{2(x + y)}{2x - 1}.$$

Before we go on... Once again, the derivative is not an explicit function of x. Notice that we could have solved the original equation for y as a function of x and then obtained its derivative. You should do this as an exercise and check that you obtain the same derivative.

▼ **EXAMPLE 2**

Given that $\ln y = xy$, find the equation of the tangent line to the graph at the point where $y = 1$.

SOLUTION   Once again, we differentiate both sides of the equation with respect to x:

$$\frac{d}{dx}(\ln y) = \frac{d}{dx}(xy).$$

Using the chain rule on the left and the product rule on the right, we obtain

$$\frac{1}{y}\frac{dy}{dx} = (1)y + x\frac{dy}{dx}.$$

Bringing the terms with $\dfrac{dy}{dx}$ to the left gives

$$\frac{1}{y}\frac{dy}{dx} - x\frac{dy}{dx} = y,$$

and so

$$\frac{dy}{dx}\left(\frac{1}{y} - x\right) = y,$$

or

$$\frac{dy}{dx}\left(\frac{1 - xy}{y}\right) = y,$$

so that

$$\frac{dy}{dx} = y\left(\frac{y}{1 - xy}\right) = \frac{y^2}{1 - xy}.$$

The derivative gives us the slope of the tangent line, so we must next evaluate the derivative at the point where $y = 1$. Note that the formula for dy/dx requires values for both the x- and y-coordinates. We get the x-coordinate by substituting $y = 1$ in the original equation.

$$\ln y = xy$$
$$\ln(1) = x \cdot 1.$$

But $\ln(1) = 0$, and so $x = 0$ for this point. Thus,

$$\frac{dy}{dx}\bigg|_{y=1} = \frac{1^2}{1 - (0)(1)} = 1.$$

Therefore, the tangent line is the line through $(x, y) = (0, 1)$ with slope 1, which is

$$y = x + 1.$$

Before we go on... This is an example where it is simply not possible to solve for *y*. Try it.

▼ **EXAMPLE 3**

Find $\dfrac{d}{dx}\left[\dfrac{(x+1)^{10}(x^2+1)^{11}}{(x^3+1)^{12}}\right]$ without using the product or quotient rules.

SOLUTION   We can do this using a technique called **logarithmic differentiation.** We write

$$y = \frac{(x+1)^{10}(x^2+1)^{11}}{(x^3+1)^{12}}$$

and then take the natural logarithm of both sides.

$$\ln y = \ln\left[\frac{(x+1)^{10}(x^2+1)^{11}}{(x^3+1)^{12}}\right]$$

We can use properties of the logarithm to simplify the right-hand side.

$$\ln y = \ln(x+1)^{10} + \ln(x^2+1)^{11} - \ln(x^3+1)^{12}$$
$$= 10\ln(x+1) + 11\ln(x^2+1) - 12\ln(x^3+1)$$

Now we can find *dy/dx* using implicit differentiation.

$$\frac{1}{y}\frac{dy}{dx} = \frac{10}{x+1} + \frac{22x}{x^2+1} - \frac{36x^2}{x^3+1}$$

$$\frac{dy}{dx} = y\left(\frac{10}{x+1} + \frac{22x}{x^2+1} - \frac{36x^2}{x^3+1}\right)$$

$$= \frac{(x+1)^{10}(x^2+1)^{11}}{(x^3+1)^{12}}\left(\frac{10}{x+1} + \frac{22x}{x^2+1} - \frac{36x^2}{x^3+1}\right)$$

Before we go on... You should redo this example using the product and quotient rules instead of logarithmic differentiation, and compare the answers. Compare also the amount of work involved in both methods.

Now for an application. Productivity usually depends on both labor and capital. Suppose, for example, you are managing an automobile assembly plant. You can measure its productivity by counting the number of automobiles the plant produces each year. As a measure of labor, you can use the number of employees, and as a measure of capital you can use its operating budget. The **Cobb-Douglas production function** then has the form

$$P = Kx^a y^{1-a},$$

where *P* stands for the number of automobiles produced per year, *x* is the number of employees, and *y* is the operating budget. The numbers *K* and *a* are constants that depend on the particular factory studied, with a between 0 and 1.*

▼ **EXAMPLE 4** Cobb-Douglas Production Function

The automobile assembly plant you manage has the Cobb-Douglas production function

$$P = x^{0.3}y^{0.7},$$

where P is the number of automobiles it produces per year, x is the number of employees, and y is the daily operating budget (in dollars). Assume a production level of 1,000 automobiles per year.

(a) Find $\dfrac{dy}{dx}$.

(b) Evaluate this derivative when $x = 80$.

SOLUTION Since the production level is 1,000 automobiles per year, we have $P = 1,000$, and so the equation becomes

$$1,000 = x^{0.3}y^{0.7}.$$

(a) We find $\dfrac{dy}{dx}$ by implicit differentiation:

$$0 = \frac{d}{dx}(x^{0.3}y^{0.7})$$

$$= (0.3)x^{-0.7}y^{0.7} + x^{0.3}(0.7)y^{-0.3}\frac{dy}{dx}.$$

Thus,

$$0.7x^{0.3}y^{-0.3}\frac{dy}{dx} = -0.3x^{-0.7}y^{0.7},$$

giving

$$\frac{dy}{dx} = -\frac{0.3x^{-0.7}y^{0.7}}{0.7x^{0.3}y^{-0.3}}$$

$$= -\frac{3y}{7x}.$$

▼ * We shall be studying the Cobb-Douglas production function in detail in the chapter on functions of several variables. In particular, we shall see how to construct a production function to model a real-life situation.

(b) To evaluate this derivative at $x = 80$, we must first find the corresponding value of y. To obtain y, we substitute $x = 80$ in the original equation $1{,}000 = x^{0.3}y^{0.7}$ and solve for y.

$$1{,}000 = (80)^{0.3}y^{0.7} \approx 3.72329y^{0.7}$$

$$y^{0.7} \approx \frac{1{,}000}{3.72329} \approx 268.580.$$

To obtain y on its own, raise both sides to the power $1/0.7$ to obtain

$$y = (y^{0.7})^{1/0.7} = (268.580)^{1/0.7} \approx 2{,}951.92.$$

Now that we have the corresponding value for y, we evaluate the derivative at $x = 80$ and $y = 2{,}951.92$.

$$\left.\frac{dy}{dx}\right|_{x=80} = -\frac{3y}{7x} \approx -\frac{3(2{,}951.92)}{7(80)} \approx -15.81$$

Before we go on... How do we interpret this result? The first clue is to look at the units of the derivative: recall that the units of dy/dx are units of y per unit of x. Since y is the daily budget, its units are dollars, and since x is the number of employees, its units are employees. Thus,

$$\left.\frac{dy}{dx}\right|_{x=80} = -\$15.81 \text{ per employee.}$$

Next, recall that dy/dx measures the rate of change of y as x changes. Thus (since the answer is negative), the daily budget to maintain production of 1,000 automobiles is decreasing by approximately \$15.81 per additional employee at an employment level of 80 employees. Thus, increasing the work force by one worker will result in a saving of about \$15.81 per day. Roughly speaking, *a new employee is worth \$15.81 per day* at the current levels of employment and production.

▶ **4.4 EXERCISES**

In Exercises 1–10, find dy/dx using implicit differentiation. In each case, compare your answer with the result obtained by first solving for y as a function of x and then taking the derivative.

1. $2x + 3y = 7$ **2.** $4x - 5y = 9$ **3.** $x^2 - 2y = 6$ **4.** $3y + x^2 = 5$ **5.** $2x + 3y = xy$

6. $x - y = xy$ **7.** $e^x y = 1$ **8.** $e^x y - y = 2$ **9.** $y \ln x + y = 2$ **10.** $\dfrac{\ln x}{y} = 2 - x$

In Exercises 11–30 Find the indicated derivative using implicit differentiation.

11. $x^2 + y^2 = 5$. Find $\dfrac{dy}{dx}$.   **12.** $2x^2 - y^2 = 4$. Find $\dfrac{dy}{dx}$.

13. $x^2y - y^2 = 4$. Find $\dfrac{dy}{dx}$.

14. $xy^2 - y = x$. Find $\dfrac{dy}{dx}$.

15. $3xy - \dfrac{y}{3} = \dfrac{2}{x}$. Find $\dfrac{dy}{dx}$.

16. $\dfrac{xy}{2} - y^2 = 3$. Find $\dfrac{dy}{dx}$.

17. $x^2 - 3y^2 = 8$. Find $\dfrac{dx}{dy}$.

18. $(xy)^2 + y^2 = 8$. Find $\dfrac{dx}{dy}$.

19. $p^2 - pq = 5p^2q^2$. Find $\dfrac{dp}{dq}$.

20. $q^2 - pq = 5p^2q^2$. Find $\dfrac{dp}{dq}$.

21. $xe^y - ye^x = 1$. Find $\dfrac{dy}{dx}$.

22. $x^2e^y - y^2 = e^x$. Find $\dfrac{dy}{dx}$.

23. $e^{st} = s^2$. Find $\dfrac{ds}{dt}$.

24. $e^{s^2t} - st = 1$. Find $\dfrac{ds}{dt}$.

25. $\dfrac{e^x}{y^2} = 1 + e^y$. Find $\dfrac{dy}{dx}$.

26. $\dfrac{x}{e^y} + xy = 9y$. Find $\dfrac{dy}{dx}$.

27. $\ln(y^2 - y) + x = y$. Find $\dfrac{dy}{dx}$.

28. $\ln(xy) - x \ln y = y$. Find $\dfrac{dy}{dx}$.

29. $\ln(xy + y^2) = e^y$. Find $\dfrac{dy}{dx}$.

30. $\ln(1 + e^{xy}) = y$. Find $\dfrac{dy}{dx}$.

In Exercises 31–34, use logarithmic differentiation to find dy/dx.

31. $y = (x^3 + x)\sqrt{x^3 + 2}$

32. $y = \sqrt{\dfrac{x - 1}{x^2 + 2}}$

33. $y = x^x$

34. $y = x^{-x}$

In Exercises 35–46, use implicit differentiation to evaluate dy/dx at the indicated point of the graph (if only the x-coordinate is given, you must also find the y-coordinate).

35. $4x^2 + 2y^2 = 12$, $(1, -2)$

36. $3x^2 - y^2 = 11$, $(-2, 1)$

37. $2x^2 - y^2 = xy$, $(-1, 2)$

38. $2x^2 + xy = 3y^2$, $(-1, -1)$

39. $3x^{0.3}y^{0.7} = 10$, $x = 20$

40. $2x^{0.4}y^{0.6} = 10$, $x = 50$

41. $x^{0.4}y^{0.6} - 0.2x^2 = 100$, $x = 20$

42. $x^{0.4}y^{0.6} - 0.3x^2 = 10$, $x = 10$

43. $e^{xy} - x = 4x$, $x = 3$

44. $e^{-xy} + 2x = 1$, $x = -1$

45. $\ln(x + y) - x = 3x^2$, $x = 0$

46. $\ln(x - y) + 1 = 3x^2$, $x = 0$

APPLICATIONS

47. *Demand* The demand equation for soccer tournament T-shirts is

$$pq - 2{,}000 = q,$$

where q is the number of T-shirts the Enormous State University soccer team can sell for $\$p$ apiece.

(a) How many T-shirts can the team sell at $5 apiece?

(b) Find $\dfrac{dq}{dp}\bigg|_{p=5}$, and interpret the result.

48. *Cost Equations* The cost c (in cents) of producing x gallons of Ectoplasm hair gel is given by the cost equation

$$c^2 - 10cx = 200.$$

(a) Find the cost of producing 1 gallon and 3.5 gallons.

(b) Evaluate dc/dx at $x = 1$ and $x = 3.5$, and interpret the results.

49. *Housing Costs** The cost C of building a house is related to the number k of carpenters used and the number e of electricians used by the formula

$$C = 15{,}000 + 50k^2 + 60e^2.$$

If the cost of the house comes to $200,000, find $\dfrac{dk}{de}\bigg|_{e=15}$, and interpret your result.

50. *Employment* An employment research company estimates that the value of a recent MBA graduate to an accounting company is

$$V = 3e^2 + 5g^3,$$

where V is the value of the graduate, e is the number of years of prior business experience, and g is the graduate school grade point average. If $V = 200$, find de/dg when $g = 3.0$, and interpret the result.

51. *Grades** A production formula for a student's performance on a difficult English examination is

$$g = 4tx - 0.2t^2 - 10x^2 \quad (x < 30),$$

where g is the grade the student can expect to obtain, t is the number of hours of study for the examination, and x is the student's grade point average.
(a) For how long should a student with a 3.0 grade point average study in order to score 80 on the examination?

COMMUNICATION AND REASONING EXERCISES

55. Use logarithmic differentiation to give another proof of the product rule.

56. Use logarithmic differentiation to give another proof of the quotient rule.

57. If y is a specified function of x, then is finding dy/dx directly (that is, explicitly) the same as finding it by implicit differentiation? Explain.

58. Explain why one should not expect dy/dx to be a function of x if y is not a function of x.

(b) Find dt/dx for a student who earns a score of 80, evaluate it when $x = 3.0$, and interpret the result.

52. *Grades* Repeat Exercise 51 using the following production formula for a basket-weaving examination:

$$g = 10tx - 0.2t^2 - 10x^2 \quad (x < 10)$$

Comment on the result.

Exercises 53 and 54 are based on the following demand function for money (taken from a question on the GRE economics test):

$$M_d = (2) \times (y)^{0.6} \times (r)^{-0.3} \times (p),$$

where
  M_d = *demand for nominal money balances;*
  y = *real income;*
  r = *an index of interest rates;*
  p = *an index of prices.*

53. *Money Stock* If real income grows while the money stock and the price level remain constant, the interest rate must change at what rate? (First find dr/dy, then dr/dt; your answers will be expressed in terms of r and y.)

54. *Money Stock* If real income grows while the money stock and the interest rate remain constant, the price level must change at what rate?

59. True or false? If y is a function of x and $dy/dx \neq 0$ at some point, then, regarding x as an implicit function of y,

$$\frac{dx}{dy} = \frac{1}{dy/dx}.$$

Explain your answer.

60. If you are given an equation in x and y such that dy/dx is a function of x only, what can you say about the graph of the equation?

▼ * Based on an exercise in *Introduction to Mathematical Economics* by A. L. Ostrosky, Jr., and J. V. Koch (Prospect Heights, IL: Waveland Press, 1979).

▶ ━━━━ **4.5** LINEAR APPROXIMATION AND ERROR ESTIMATION

One of the central themes in this and the previous chapter has been the concept of the derivative as the slope of the line tangent to the graph of a function at a point. When we described how to use a graphing calculator to estimate derivatives, we pointed out that if you zoom in to a portion of a smooth curve near a specified point, it becomes indistinguishable from the tangent line at that point. In other words, the values of the function are close to the values of the linear function whose graph is the tangent line. In this section, we take a careful look at this idea, called the linear approximation of a function.

Let us start with a point $(a, f(a))$ on the graph of a function f. If the curve is smooth at that point—that is, if $f'(a)$ exists—then we have

$$f'(a) = \lim_{h \to 0} \frac{f(a + h) - f(a)}{h}.$$

This means that the smaller h becomes, the closer $(f(a + h) - f(a))/h$ approximates $f'(a)$. Thus, for small values of h (close to zero),

$$f'(a) \approx \frac{f(a + h) - f(a)}{h}.$$

Multiplying both sides by h and solving for $f(a + h)$ gives

$$hf'(a) \approx f(a + h) - f(a),$$

so

$$f(a + h) \approx f(a) + hf'(a).$$

Now, since h is small, $a + h$ is a number close to a. For our purposes it will be more useful to call this number x, so $x = a + h$. This also gives us $h = x - a$. Substituting gives us the approximation

$$f(x) \approx f(a) + (x - a)f'(a).$$

This formula for approximating $f(x)$ is referred to as the **linear approximation of $f(x)$ near $x = a$.** (We saw this formula once before: in our proof of the chain rule.)

━━ ◀

LINEAR APPROXIMATION OF $f(x)$ NEAR $x = a$

If x is close to a, then

$$f(x) \approx f(a) + (x - a)f'(a).$$

The right-hand side,

$$L(x) = f(a) + (x - a)f'(a),$$

which is a linear function of x, is called the **linear approximation of $f(x)$ near $x = a$.**

Q What is this function $L(x)$?

A Its graph is the line tangent to the graph of f at the point $(a, f(a))$. Indeed, the tangent line through this point has slope $f'(a)$, so the point-slope form of its equation is

$$y - f(a) = f'(a)(x - a)$$

or

$$y = f(a) + (x - a)f'(a).$$

Figure 1 shows the graphs of f and L.

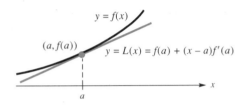

FIGURE 1

▼ **EXAMPLE 1**

Find the linear approximation of $f(x) = \sqrt{x}$ near $x = 4$.

SOLUTION   Since we are interested in $f(x)$ near $x = 4$, we take a to be 4.

$$L(x) = f(a) + (x - a)f'(a)$$

$$= \sqrt{a} + (x - a)\frac{1}{2\sqrt{a}}$$

$$= \sqrt{4} + \frac{x - 4}{2\sqrt{4}}$$

$$= 2 + \frac{x - 4}{4} = 2 + \frac{x}{4} - \frac{4}{4} = \frac{x}{4} + 1$$

Thus, the linear approximation to $\sqrt{x}$ near $x = 4$ is $L(x) = \frac{x}{4} + 1$.

Before we go on... The graphs of $y = \sqrt{x}$ and $y = \frac{x}{4} + 1$ are shown in Figure 2.

We can use $L(x)$ to approximate the square root of any number close to 4 very easily without using a calculator. For example,

$$\sqrt{4.1} \approx L(4.1) = \frac{4.1}{4} + 1 = 2.025$$

(the actual value of $\sqrt{4.1}$ is 2.02485. . . .), and

$$\sqrt{3.9} \approx L(3.9) = \frac{3.9}{4} + 1 = 1.975$$

(the actual value of $\sqrt{3.9}$ is 1.9748. . . .).

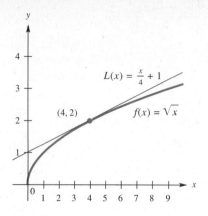

FIGURE 2

▶ **NOTE**   Don't forget that in the formula $L(x) = f(a) + (x - a) f'(a)$, $L(x)$ is a function of x, and a is a constant.   ◀

▼   **EXAMPLE 2**

Use linear approximation to approximate $\ln(1.134)$.

SOLUTION   Here, we are not given a value for a. The key is to use a value close to 1.134 whose natural logarithm we know. Since we know that $\ln(1) = 0$, we take a to be 1.

$$L(x) = f(a) + (x - a) f'(a)$$

$$= \ln a + (x - a)\frac{1}{a}$$

$$= \ln(1) + x - 1 = x - 1$$

Thus, the linear approximation of $\ln x$ near $x = 1$ is $L(x) = x - 1$. In particular,

$$\ln(1.134) \approx L(1.134) = 1.134 - 1 = 0.134.$$

The actual value is $\ln(1.134) = 0.12575 \ldots$ Thus, our approximation is off by $0.134 - 0.12575 \ldots = 0.00825 \ldots$, which is not bad for a quick estimation.

Before we go on... You can use $L(x) = x - 1$ to find approximations to the natural logarithm of *any* number close to 1: for instance,

$$\ln(0.843) \approx 0.843 - 1 = -0.157,$$
$$\ln(0.999) \approx 0.999 - 1 = -0.001.$$

Q $L(x) = f(a) + (x - a) f'(a)$ is an approximation of $f(x)$ near $x = a$. How accurate is this approximation?

A The accuracy depends on how close x is to a. The closer x is to a, the better the approximation. In fact, it can be shown that the error is of the order $(x - a)^2$, meaning that it is no more than a constant times $(x - a)^2$ (the constant depends on the function).

APPROXIMATING RELATIVE CHANGE

Suppose you are a screen printer, and the weekly sales of your T-shirts is given by a demand equation of the form $q = f(p)$. If you are currently charging $\$a$ per shirt, then you sell $f(a)$ T-shirts per week. If you change the price to $\$b$ per shirt, then your weekly sales will change to $f(b)$. Thus, if you change the price from a to b, the change in weekly sales is given by the difference, $f(b) - f(a)$.

Now if the new price b is close to the old price a, our linear approximation formula tells us that

$$f(b) \approx f(a) + (b - a) f'(a).$$

Thus, if you change the price from a to b, the change in weekly sales is

$$f(b) - f(a) \approx (b - a) f'(a),$$

or

$$\text{Change in sales} \approx \text{Change in price times } f'(a).$$

Now we would like to calculate the *percentage* change in f. We do so by dividing by the current sales level $f(a)$.

$$\text{Percentage change in sales}$$
$$= \frac{f(b) - f(a)}{f(a)} \approx (b - a) \frac{f'(a)}{f(a)}$$
$$= \text{Change in price} \times \frac{f'(a)}{f(a)}$$

ESTIMATING THE PERCENTAGE CHANGE IN A FUNCTION *f*

If x changes from a to b, then the **percentage** or **relative** change in f is approximated by

$$\text{Relative change in } f \approx (b - a) \frac{f'(a)}{f(a)}.$$

The ratio $f'(a)/f(a)$ is called the **percentage rate of change** or **relative rate of change.**

▼ EXAMPLE 3   Sales

The demand equation for your new fraternity T-shirts is given by

$$q = \frac{2{,}000}{p},$$

where q represents the weekly sales of T-shirts at a price of p. You are currently charging $5 per T-shirt. If you raise the price to $5.50, by what percentage will your sales drop?

SOLUTION   Notice first that q is given as a function of p.

$$q = f(p) = \frac{2{,}000}{p}$$

Let us take $a = \$5$ (the old price) and $b = \$5.50$ (the new price). Then $f(a) = f(5) = 400$ shirts per week. The derivative of f is

$$f'(p) = -\frac{2{,}000}{p^2},$$

so $f'(a) = f'(5) = -2{,}000/25 = -80$ shirts per $1 increase in price. The percentage rate of change is

$$\frac{f'(a)}{f(a)} = \frac{-80}{400} = -0.2,$$

or -20% per $1 increase in price. The relative change in f is approximated by

$$(b - a)\frac{f'(a)}{f(a)} = -(5.5 - 5)\frac{80}{400} = -0.1$$

In other words, you can expect your sales to drop by approximately 10% if you raise the price 50¢.

▼ EXAMPLE 4   Measurement Error

Precision Corp. manufactures ball bearings with a radius of 1 millimeter, varying by ±0.01 millimeters. By what percentage can the volume of its ball bearings vary?

SOLUTION   We can rephrase the question as follows:

(1)   If the radius of a ball bearing changes from 1 millimeter to 1.01 millimeters, by what percentage does the volume change?

(2)   If the radius of a ball bearing changes from 1 millimeter to 0.99 millimeter, by what percentage does the volume change?

We can answer both questions using our formula for relative change. Since we are asking for the relative change in volume, we use the formula for the volume of a sphere,

$$V(r) = \frac{4}{3}\pi r^3.$$

Note that this is a function of the radius r. To answer the first question, let us take $a = 1$ millimeter and $b = 1.01$ millimeters. Then $V(1) = 4\pi/3 \approx 4.1888$. To find $V'(a)$, we first take the derivative of V,

$$V'(r) = 4\pi r^2,$$

so $V'(1) = 4\pi \approx 12.5664$. Thus, the percentage rate of change is

$$\frac{V'(1)}{V(1)} = \frac{12.5664}{4.1888} = 3,$$

or 300% per millimeter when the radius is 1 millimeter. The percentage change in V is

$$(b - a)\frac{V'(a)}{V(a)} = (1.01 - 1)\frac{12.5664}{4.1888} = 0.03,$$

or 3%. Similarly, if we *decrease* the radius from 1 millimeter to 0.99 millimeter, the only change in the calculation is that $(b - a)$ becomes $(0.99 - 1) = -0.01$, so that the percentage change in V is -3%. Thus, the volume of the ball bearings will vary by approximately $\pm 3\%$.

We now state some of our results in delta notation. If x is close to a, then we saw that

$$f(x) - f(a) \approx (x - a)f'(a).$$

The quantity $f(x) - f(a)$ represents the change in f corresponding to a change in the independent variable from a to x. In other words,

$$\text{Change in } f \approx \text{Change in } x \times f'(a).$$

Using the delta notation, this becomes

$$\Delta f \approx \Delta x f'(x),$$

which is just another way of saying that $f'(x)$ is approximately $\Delta f/\Delta x$. This formula is sometimes written as

$$df = f'(x)\, dx,$$

where df and dx are thought of as very small changes in f and x, respectively, and are referred to as **differentials**. (Actually, $dx = \Delta x$, but df is the linear approximation to Δf. Thus, we can write $\Delta f \approx df = f'(x)\, dx$.)

You can think of df as the change in the linear approximation we called $L(x)$ earlier, while Δf is the actual change in f. Figure 3 shows these quantities on a graph.

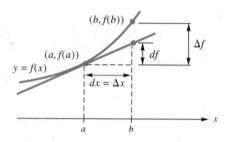

FIGURE 3

▼ **EXAMPLE 5**

Let $f(x) = \sqrt[3]{x}$. Using the differential notation, approximate $\sqrt[3]{990}$ and $\sqrt[3]{1,010}$.

SOLUTION   As in Example 1, we find the linear approximation to f near a point a where $f(a)$ is easy to compute. The obvious choice is $a = 1,000$, since $f(1,000) = 10$. Now the linear approximation, in differential form, is

$$df = \frac{1}{3}x^{-2/3}\,dx = \frac{1}{3\sqrt[3]{x^2}}\,dx.$$

When $x = 1,000$, we get

$$df = \frac{1}{300}\,dx.$$

Now, to approximate $\sqrt[3]{990}$, we take $dx = 990 - 1,000 = -10$, which gives

$$df = \frac{1}{300}(-10) = -0.0333.\ldots$$

This is an approximation to Δf, so

$$\sqrt[3]{990} = f(1,000) + \Delta f \approx f(1,000) + df$$
$$\approx 10 - 0.03333 = 9.96667.$$

Similarly, to approximate $\sqrt[3]{1,010}$ we take $dx = 1,010 - 1,000 = 10$, which gives

$$df = \frac{1}{300}(10) = 0.0333\ldots$$

and

$$\sqrt[3]{1,010} = f(1,000) + \Delta f \approx f(1,000) + df$$
$$\approx 10 + 0.03333 = 10.03333.$$

Before we go on... The actual cube roots are

$$\sqrt[3]{990} = 9.96665$$

and

$$\sqrt[3]{1,010} = 10.03322$$

to 5 decimal places (according to a calculator, which is using its own method of approximation . . .).

▶ ## 4.5 EXERCISES

Find the linear approximation near the indicated value for each of the functions in Exercises 1–20.

1. $f(x) = 3x + 5, \quad x = 2$

2. $f(x) = -x - 3, \quad x = 5$

3. $f(x) = 3x^2 - 4x + 5, \quad x = -1$

4. $f(x) = -x^2 + x - 1, \quad x = 0$

5. $f(x) = \dfrac{x}{2x + 1}, \quad x = 0$

6. $f(x) = \dfrac{2x - 1}{x + 1}, \quad x = -2$

7. $f(x) = e^x, \quad x = 0$

8. $f(x) = e^{-x}, \quad x = 0$

9. $f(x) = \ln(1 + x), \quad x = 0$

10. $f(x) = \ln(1 - x), \quad x = 0$

11. $f(x) = \sqrt{1 + x}, \quad x = 0$

12. $f(x) = \sqrt{1 - x}, \quad x = 0$

13. $f(x) = x^{1.3}, \quad x = 1$

14. $f(x) = x^{2.7}, \quad x = 1$

15. $f(x) = \dfrac{e^x + e^{-x}}{2}, \quad x = 2$

16. $f(x) = \dfrac{e^x - e^{-x}}{2}, \quad x = 2$

17. $f(x) = \dfrac{1}{1 + e^{0.2x}}, \quad x = 0$

18. $f(x) = \dfrac{e^{0.5x}}{x}, \quad x = 1$

19. $S(r) = 4\pi r^2, \quad r = 10$

20. $l(p) = \sqrt{1 - p^2}, \quad p = \frac{1}{2}$

Use linear approximation to estimate the numbers in Exercises 21–30.

21. $\sqrt{16.3}$

22. $\sqrt{36.1}$

23. $\sqrt{48.82}$

24. $\sqrt{24.73}$

25. $(8.1)^{2/3}$

26. $(3.9)^{3/2}$

27. $e^{0.3}$

28. $e^{-0.2}$

29. $\ln(0.95)$

30. $\ln(1.05)$

In each of Exercises 31–36, use linear approximation to estimate the percentage change in f(x) that results from a change from x = a to x = b.

31. $f(x) = 2x^2 - x; \quad a = 3, b = 3.5$

32. $f(x) = 5x^2 + x; \quad a = 1, b = 1.2$

33. $f(x) = e^x; \quad a = 0, b = -0.3$

34. $f(x) = e^{-x}; \quad a = 0, b = 0.3$

35. $f(x) = \dfrac{1}{x}; \quad a = 5, b = 6$

36. $f(x) = \dfrac{1}{x^2}; \quad a = 5, b = 7$

APPLICATIONS

Some of the themes of the following application exercises may already be familiar to you.

37. *Cost Analysis*   The daily cost of manufacturing x camcorders at Consumer Electronics, Inc. is calculated to be

$$C(x) = 1,000 + 150x - 0.01x^2.$$

(a) Find the linear approximation to $C(x)$ near $x = 100$, and use it to estimate $C(105)$.

(b) Estimate the percentage increase in daily costs if production is increased from 100 to 105 camcorders per day.

38. *Cost Analysis*   The daily cost of manufacturing x compact discs at the Techno Plus Recording Studio is given by

$$C(x) = 400 + 7x - 0.0001x^3.$$

(a) Find the linear approximation to $C(x)$ near $x = 200$, and use it to estimate $C(210)$.

(b) Estimate the percentage increase in daily costs if production is increased from 200 to 210 CD's per day.

39. *Cost Analysis (based on a GRE economics exam question*)*   A firm's average cost function is presented as

$$\text{Average cost} = 350 + \frac{9000}{Q}.$$

Estimate the percentage change in average cost if production increases from 1,000 units to 1,050 units.

40. *Cost Analysis (based on a GRE economics exam question)*   A firm's average cost function is presented as

$$\text{Average cost} = 600 + \frac{900}{Q}.$$

Estimate the percentage change in average cost if production increases from 5,000 units to 5,100 units.

41. *Toxic Waste Treatment*   The cost of treating waste by removing PCPs is given by

$$C(q) = 5,000 + 120q^2,$$

where q is the reduction in toxicity (in pounds of PCPs removed per day) and $C(q)$ is the daily cost (in dollars) of this reduction. Government subsidies for toxic waste cleanup amount to

$$S(q) = 600q,$$

where q is as above and $S(q)$ is the dollar subsidy.

(a) Calculate the net cost function, $N(q)$, given the cost function and subsidy above.

(b) If a company is currently removing 15 pounds of PCPs per day and decides to increase this amount by 10%, what will be the effect on the daily net cost?

42. *Transportation Costs*   Before the Alaskan pipeline was built, there was speculation as to whether it might be more economical to transport the oil by large tankers. The following cost equation was estimated by the National Academy of Sciences:

$$C = 0.03 + \frac{10}{T} - \frac{200}{T^2},$$

where C is the cost in dollars of transporting one barrel of oil 1,000 nautical miles and T is the size of an oil tanker in deadweight tons.[†]

(a) How much would it cost to transport 100 barrels of oil in a tanker weighing 1,000 tons?

(b) If the weight of the tanker is increased by 5%, how will this affect the cost?

43. *Sales Analysis*   British statistician Richard Stone published a demand equation for beer in Great Britain that had the form

$$q = Kp^{-1.040},$$

where q is the amount of beer demanded and p is the price of beer. K is a quantity depending on the average consumer's income and the price of other commodi-

▼ * Source: *GRE Economics* by G. G. Gallagher, G. E. Pollock, W. J. Simeone, and G. Yohe, (Piscataway, N.J.: Research and Education Association, 1989), p. 254.

† Source: "Use of Satellite Data on the Alaskan Oil Marine Link," *Practical Applications of Space Systems: Cost and Benefits* (Washington, D.C.: National Academy of Sciences, 1975), p. B–23.

ties.* You are the sales manager of a small brewery and would like to use Stone's equation to set an appropriate pricing policy.

(a) Your company has been charging $10 per case of beer and has orders for 2,000 cases per day. Use this information to determine a value for K, and hence a demand equation for your company.

(b) Use the demand equation in part (a) and linear approximation to estimate the percentage drop in sales if you increase the price to $12 per case.

(c) It costs your company $7.50 to manufacture and ship one case of beer. What effect will the price increase have on your profits?

44. *Sales Analysis*   Economist Henry Schultz calculated the demand function for corn to be

$$p = \frac{6,570,000}{q^{1.3}},$$

where p is the price of corn (in dollars per bushel) and q is the number of bushels of corn that could be sold at the price p in one year.[†]

(a) Rewrite the given demand equation to obtain q as a function of p.

(b) Use the equation obtained in part (a) and linear approximation to estimate the percentage drop in demand if the price of corn is raised from $1 per bushel to $1.10 per bushel.

(c) If a bushel of corn costs farmers 75¢ to produce,

what is the effect of this price increase on the annual profits?

45. *Quality Control*   Silicon Valley, Inc. manufactures blank compact discs for sale to recording studios. Its CDs have a radius of 5 centimeters and a thickness of 0.1 centimeter. A disc whose radius is off by more than 1% is automatically rejected. By what percentage can the volume of the discs that pass inspection vary? [Assume that all its CDs are exactly 0.1 centimeter thick. The volume of a CD (neglecting the hole) is $\pi r^2 t$, where r is its radius and t is its thickness.]

46. *Quality Control*   Precision Drills, Inc. manufactures diamond-tipped drills that use industrial diamonds with a radius of 0.5 mm. It will accept any diamond whose radius is within $\pm 10\%$ of 0.5 mm. By what percentage can the volume of the diamonds vary?

COMMUNICATION AND REASONING EXERCISES

47. For what functions is linear approximation exact?

48. Sketch a graph of a function for which a linear approximation will always be an overestimate. Sketch a graph of another function for which a linear approximation will always be an underestimate. What feature of the graph is important here?

▸ **You're the Expert**

ESTIMATING
MORTGAGE RATES

You plan to go to the bank tomorrow to apply for a 30 year, $90,000 mortgage. The interest rate was advertised as 7.5%, but rates are volatile and you know that the rate you get could vary by as much as 0.5%. Your budget is tight and your spouse is counting on you to make sure that you can afford the monthly payment. Also, you don't trust the loan officer to get the calculation right, so you would like to be able to calculate the monthly payment yourself once you know the actual rate. Now, you could bring a calculator and do the calculation on the spot, but that would be laborious and possibly insulting to

▼   * Source: Richard Stone, "The Analysis of Market Demand," *Journal of the Royal Statistical Society* 108 (1945): 286–382.

† Based on data for the period 1915–1929. Source: Henry Schultz: *The Theory and Measurement of Demand* (as cited in *Introduction to Mathematical Economics* by A. L. Ostrosky, Jr., and J. V. Koch (Prospect Heights, IL: Waveland Press, 1979).

the loan officer. What you need is a simple formula that will allow you to estimate the payment.

Your monthly payment M depends on the interest rate r according to the formula*

$$M(r) = 90{,}000\left(\frac{\dfrac{r}{12}}{1 - \left(1 + \dfrac{r}{12}\right)^{-360}} \right).$$

This is not a calculation you want to try to do in your head. However, you remember something from calculus: the tangent line to a curve lies close to that curve near the point of contact. (See Figure 1, which shows a plot of $M(r)$ and the tangent line at $r = 0.075$.)

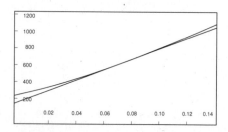

FIGURE 1

The equation of the tangent line is linear, and that would be easy to calculate in your head. The night before your appointment at the bank, you recall that the equation of the tangent line is given by the linear approximation to the curve at $r = 0.075$, with formula

$$L(r) = M(0.075) + (r - 0.075)M'(0.075).$$

First, you calculate

$$M(0.075) = \$629.29,$$

which will be your monthly payment if the interest rate is 7.5%. This gives you the point of contact of the tangent line with the curve: $(0.075, 629.29)$. Now you calculate the derivative, which will give you the slope of the line.

$$M(r) = 90{,}000\,\frac{\dfrac{r}{12}}{1 - \left(1 + \dfrac{r}{12}\right)^{-360}} = \frac{7{,}500r}{1 - \left(1 + \dfrac{r}{12}\right)^{-360}}$$

▼ *See a text that discusses the mathematics of finance, as most finite mathematics books do.

This gives, by the quotient rule,

$$M'(r) = \frac{7{,}500\left(1 - \left(1 + \dfrac{r}{12}\right)^{-360}\right) - 225{,}000r\left(1 + \dfrac{r}{12}\right)^{-361}}{\left(1 - \left(1 + \dfrac{r}{12}\right)^{-360}\right)^2}.$$

So

$$M'(0.075) = 6{,}162.77.$$

Thus, the linear approximation is

$$L(r) = M(0.075) + (r - 0.075)M'(0.075)$$
$$= 629.29 + (r - 0.075)6{,}162.77.$$

This is a fairly easy equation to use. For example, if the interest rate were 7.6%, this equation would estimate a monthly payment of

$$M \approx 629.29 + (0.076 - 0.075)6{,}162.77$$
$$= 629.29 + 6{,}162.77(0.001) = \$635.45.$$

Put another way,

$$\Delta M \approx 6{,}162.77\,\Delta r,$$

where $\Delta r = r - 0.075$ and $\Delta M = M - 629.29$. Thus, each 0.1% increase in the interest rate results in a $6{,}162.77(0.001) = \$6.16$ increase in the monthly payments. Similarly, each 0.1% decrease results in a \$6.16 decrease in the monthly payments. This is something you can remember and use tomorrow at the bank.

Remember, though, that this is only an approximation. To check your work and to see how good this approximation is, you look at the extremes of the possible interest rates: 7.0% and 8.0%. Your rough estimates, using the \$6.16 change per 0.1% change in the interest rate, are

$$M \approx 629.29 - 6.16(5) = \$598.49 \text{ at } 7.0\%$$

and

$$M \approx 629.29 + 6.16(5) = \$660.09 \text{ at } 8.0\%.$$

The actual monthly payments are given by the original function:

$$M(7.0) = \$598.77$$

and

$$M(8.0) = \$660.39.$$

So your estimates are quite reasonable, though not exact, as expected. You can go to the bank with some confidence that you know what you're doing.

Exercises

1. Suppose that the interest rate might be anything between 5% and 10%. How good is the linear approximation at the extremes now?

2. Suppose that instead of a $90,000 loan, you are going to take out a $180,000 loan. How will this change the linear approximation?

3. Suppose that the bank is willing to adjust the interest rate to produce whatever monthly payment you wish (within reason). If you want a monthly payment near $629.29, use implicit differentiation to find a linear approximation for Δr in terms of ΔM that you can use to approximate the interest rate that will produce a given monthly payment.

4. Compare the equation you found in Exercise 3 to the equation found in the discussion. How could you have answered Exercise 3 without using implicit differentiation?

The following two exercises use the general formula

$$M = P\frac{r/m}{1 - (1 + r/m)^{-mn}},$$

where M is the periodic payment on a loan of $\$P$ at the yearly interest rate r for n years, with payments made m times per year.

5. Suppose that in the scenario of the discussion, you know that the interest rate will be exactly 7.5%, but the bank is willing to be flexible in the length of the loan (i.e., n in the formula above). If the length of the loan will be something near 30 years, roughly how will the monthly payments change with n?

6. Suppose that in the scenario of the discussion, you know that the interest rate will be exactly 7.5%, but the bank is willing to be flexible in the number of payments made per year (i.e., m in the formula above). If the number of payments will be something near 12 per year, roughly how will the monthly payments change with m?

▶ ## Review Exercises

In each of Exercises 1–16, calculate the derivative of the given function mentally.

1. $f(x) = (x - 1)^3$

2. $f(x) = (x + 1)^{-1}$

3. $f(x) = (2x + 4)^{-1}$

4. $f(x) = (1 - 3x)^4$

5. $f(x) = \sqrt{x + 1}$

6. $f(x) = \sqrt{1 - x}$

7. $f(x) = \dfrac{1}{2x + 1}$

8. $f(x) = \dfrac{3}{x + 1}$

9. $f(x) = \ln|x^2 + 1|$

10. $f(x) = \ln|x^2 + x|$

11. $f(x) = \log_2|x^2 + x|$

12. $f(x) = \log_3|x^2 + 7|$

13. $f(x) = e^{-x}$

14. $f(x) = e^{2x+1}$

15. $f(x) = xe^x$

16. $f(x) = xe^{-x}$

In each of Exercises 17–38, find the derivative of the given function.

17. $f(x) = x^2(3x - 1)$

18. $f(x) = x^{-1}(2x + 1)$

19. $h(x) = (2x^2 - 1)\sqrt{x}$

20. $s(x) = x(\sqrt{x} + 1)$

21. $g(x) = \dfrac{2x + 1}{2x - 1}$

22. $g(x) = \dfrac{3x + 2}{x^2 - 1}$

23. $r(x) = (x^2 - 2)e^x$

24. $r(x) = x^2\ln x$

25. $f(x) = (2x^2 - 1)\ln x$

26. $f(x) = (x^2 - x)e^x$

27. $f(x) = \dfrac{x^2 - 2x^{-3} + 1}{x + 1}$

28. $f(x) = \dfrac{2x + x^{-2}}{x - 1}$

29. $r(x) = \dfrac{e^x - e^{-x}}{e^x}$

30. $s(x) = \dfrac{e^{2x} - e^{-2x}}{e^{2x}}$

31. $f(x) = (2x^2 - 2x + 1)^{-4}$

32. $f(x) = (2x^3 + x + 1)^{-3}$

33. $s(t) = (t^2 + e^{3t} + 2\sqrt{t})^4$

34. $s(t) = (2t^2 + e^{-t} + \sqrt{t})^{-1}$

35. $f(x) = \sqrt{e^x - x^2}$

36. $f(x) = \sqrt{e^x + x^2}$

37. $h(x) = \sqrt{1 - \ln x}$

38. $h(x) = \sqrt{e^x + \ln x}$

In each of Exercises 39–50, evaluate the given expression.

39. $\dfrac{d}{dx}\left(\dfrac{x^{1.3}}{1 + x}\right)$

40. $\dfrac{d}{dx}\left(\dfrac{x}{x^{0.1} + 1}\right)$

41. $\dfrac{d}{dx}(e^{0.1x})$

42. $\dfrac{d}{dx}\left(1 + \dfrac{1}{e^{2.1x}}\right)$

43. $\dfrac{d}{dx}\left(\dfrac{1}{1 + 2e^x}\right)$

44. $\dfrac{d}{dx}\left(\dfrac{e^{-x}}{1 + x}\right)$

45. $\dfrac{d}{dt}(b\ln(at^2))$, with a and b constant

46. $\dfrac{d}{dt}(ae^{bt})$, with a and b constant

47. $\dfrac{d}{dx}((\ln x)^2)$

48. $\dfrac{d}{dx}((e^x)^2)$

49. $\dfrac{d}{dx}(e^{x^2 - 3x + 1})$

50. $\dfrac{d}{dx}\left(\dfrac{2e^{x^2+1}}{x}\right)$

In each of Exercises 51–60, find the equation of the tangent line to the graph of the given function at the indicated point.

51. $r(s) = (2s - s^3)(s^2 + 1)$, at $(1, 2)$

52. $r(x) = (x^3 + x)(x - x^2)$, at $(1, 0)$

53. $h(x) = (x^2 - x + 1)e^{-x}$, at $(0, 1)$

54. $h(x) = (2\sqrt{x} - x^2)e^{2x}$, at $(1, e^2)$

55. $s(t) = \dfrac{2t + 5}{t - 1}$, at $(2, 9)$

56. $r(t) = \dfrac{5t - 2}{2t + 4}$, at $(1, \frac{1}{2})$

57. $s(t) = \dfrac{e^t \ln|t|}{t^2}$, at $(1, 0)$

58. $r(t) = \dfrac{e^{-t} \ln|t|}{t}$, at $(1, 0)$

59. $f(x) = (x^2 - 2)^{-2}$, at $(2, \frac{1}{4})$

60. $f(x) = (x^3 + x)^{-1}$, at $(2, \frac{1}{10})$

In each of Exercises 61–68, find $\dfrac{dy}{dx}$.

61. $x^2 - y^2 = x$

62. $2xy + y^2 = y$

63. $e^{xy} + xy = 1$

64. $xe^y - x^2y = 0$

65. $\dfrac{x}{x + y} - x = 1$

66. $\dfrac{x}{xy - 1} = y$

67. $y = x^{x+1}$

68. $y = (x + 1)^x$

In each of Exercises 69–74, find all values of x (if any) where the tangent line to the graph of the given equation is horizontal.

69. $y = x - e^{2x-1}$

70. $y = e^{x^2}$

71. $y = \dfrac{x}{x+1}$

72. $y = \sqrt{x}(x - 1)$

73. $f(x) = \sqrt{e^x - x}$

74. $f(x) = \sqrt{e^x + x}$

APPLICATIONS

75. *Revenue* You are currently able to sell 100 quarts of ice cream per day at $5 per quart. Increasing the price will cause demand to fall by 15 quarts per dollar increase in price. Should you raise your price?

76. *Revenue* You are currently able to sell 50 quarts of ice cream per day at $5 per quart. Increasing the price will cause demand to fall by 15 quarts per dollar increase in price. Should you raise your price?

77. *Fuel Economy* You are accelerating to enter a highway while, unbeknownst to you, your engine is starting to fail and your gas mileage is decreasing. If you are accelerating at a rate of 5 mph/s at the moment you are driving 40 mph, and your gas mileage is 20 mpg and decreasing at a rate of 2 mpg/s, how fast is your gas use (in gallons per hour) changing?

78. *Fuel Economy* You are decelerating as you leave a highway while, unbeknownst to you, your engine is starting to fail and your gas mileage is decreasing. If you are decelerating at a rate of 5 mph/s at the moment you are driving 40 mph, and your gas mileage is 20 mpg and decreasing at a rate of 2 mpg/s, how fast is your gas use (in gallons per hour) changing?

79. *Marginal Product* In your bicycle factory, your marginal profit at your current production level is $50 per bicycle. Moreover, your production would increase by 10 bicycles per employee you hire. At what rate would your profit change as you hire more employees?

80. *Marginal Product* In your bicycle factory, your marginal profit at your current production level is $60 per bicycle. Moreover, your production would fall by 5 bicycles per employee you hire (due to overcrowding). At what rate would your profit change as you hire more employees?

81. *Rates of Increase* A cube is growing in such a way that each of its edges is growing at a rate of 2 cm/s. How fast is its volume growing at the instant when the cube has a volume of 1,000 cubic centimeters? (The volume of a cube with edge a is given by $V = a^3$.)

82. *Rates of Increase* The volume of a cone with a circular base of radius r and height h is given by $V = \frac{1}{3}\pi r^2 h$. (See the figure.)

Find the rate of increase of V with respect to r. If the quantity r is growing at a uniform rate of 3 cm/s, how fast is the volume growing at the instant when $r = 1$, assuming that $h = 2$ does not change?

83. *Investments* If $5,000 is invested in a mutual fund whose value is growing at a rate of 5% per year (compounded once per year), how fast is the investment growing at the end of two years?

84. *Investment* If $5,000 is invested in a mutual fund whose value is declining at a rate of 5% per year (compounded once per year), how fast is the investment declining at the end of two years?

85. *Population Growth* The population P of frogs in Nassau County is given by the formula

$$P = 10,000e^{0.5t},$$

where t is the time in years since 1995.
(a) How fast will the frog population be growing in the year 2,000?
(b) In what year will the frog population be growing at a rate of one million per year?
(c) What is the significance of this for the frog leg industry?

86. *Business Growth* The accompanying graph shows the number of transactions handled by *Western*

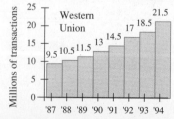

Union for the years 1987-1994.*
These data can be approximated by

$$Q(t) = 9.284e^{0.117t},$$

where t is time in years since 1987, and $Q(t)$ is the number of Western Union transactions (in millions) each year.
(a) According to the model, how fast was Western Union's business (as measured by annual transactions) growing in 1990?
(b) What is the first year that the actual increase (from that year to the next) was exceeded by the rate of increase that year as predicted by the model?

87. *Sulfur Emissions*   Worldwide industrial sulfur emissions since 1860 have followed the pattern shown in the accompanying graph.

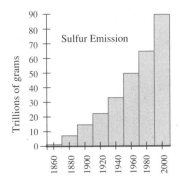

These data are approximately modeled by the function

$$Q(t) = 3.449e^{0.02576t},$$

where t is time in years since 1860, and $Q(t)$ is the amount of sulfur released into the atmosphere by industry each year, in trillions of grams.[†]
(a) According to the model, how fast was industrial sulfur emission growing in 1960?
(b) When (to the nearest year) did the level of industrial emission of sulfur surpass the earth's natural sulfur emissions level of 28 trillion grams?

88. *Epidemics*   According to the "logistic" model, the number of people P infected in an epidemic often follows a curve of the following type:

$$P = \frac{NP_0}{P_0 + (N - P_0)e^{-bt}},$$

where N is the total susceptible population, P_0 is the number of infected individuals at time $t = 0$, and b is a constant that governs the rate of spread. Taking t to be the time in years, and assuming that $b = 0.1$, $N = 1,000,000$, and $P_0 = 1,000$, answer the following questions.
(a) How fast is the epidemic spreading at the start ($t = 0$), after 10 years, after 100 years?
(b) When does the spread of the disease slow to 1,000 new cases per year?
(c) What percentage of the population is infected by this time?

89. *Investments*   You are considering depositing some money in an account earning 7% compounded continuously. You would like to end up with $20,000 in the account after 10 years.
(a) How much money would you have to invest?
(b) If you were able to increase your investment, how much time could you save in getting to $20,000? Express your answer as a rate, in years per dollar. (Suggestion: use implicit differentiation.)

90. *Investment*   You are considering depositing some money in an account earning 5% compounded continuously. You would like to end up with $30,000 in the account after 20 years.
(a) How much money would you have to invest?
(b) If you shorten the amount of time you want to wait, how much more money would you have to invest? Express your answer as a rate, in dollars per year. (Suggestion: use implicit differentiation.)

91. *Demand*   The price p of a new video game is related to the demand q by the equation

$$100pq + q^2 = 5,000,000.$$

Suppose that the price is set at $40, which will make the demand be 1,000 copies.
(a) Using implicit differentiation and linear approximation, estimate the demand if the price is raised to $42.
(b) Should the price be raised or lowered to increase revenue?

92. *Demand*   According to the demand equation in the previous exercise, if the price is set at $95, then the demand will be 500.
(a) Estimate the demand if the price is raised to $98.
(b) Should the price be raised or lowered to increase revenue?

▼ * Source: Company Reports/*The New York Times*, September 24, 1994.
† The exponential model is the authors'. The graphical data was obtained from "Sulfate Aerosol and Climatic Change," Robert J. Charlson and Tom M. L. Wigley, *Scientific American*, February 1994, pp. 48–57.

Source: Courtesy HarperCollins College Publishers.

Applications of the Derivative

APPLICATION ▶ Your book publishing company is planning the production of its latest best-seller, which is predicted to sell 100,000 copies per month over the coming year. The book will be printed in several batches of the same number evenly spaced throughout the year. Each printing run has a setup cost of $5,000, a single book costs $1 to produce, and monthly storage costs for books awaiting shipment average 1¢ per book. In order to minimize total cost to your company, how many printing runs should you plan in order to meet the anticipated demand?

INTRODUCTION ▶ In this chapter we begin to see the power of calculus as an optimization tool. For instance, if we are given the demand and cost functions for some item we are selling, we wish to price the item so as to get the largest possible profit. We have already seen how to do this in a restricted setting (linear demand and cost functions). However, not all the functions we encounter are linear. The true force of calculus comes into play in finding a maximum or minimum value of a *nonlinear* function.

We begin this chapter with this goal in mind: find the values of a variable that lead to a maximum (or minimum) value of a given function. Once we are familiar with the mechanics, we go on in the second section to apply these techniques to realistic situations.

There is a second, no less important, goal we address in this chapter: using calculus to assist you in understanding the graph of a function. By the time you have completed the material in the first section, you will be well on your way to being able to sketch the important features of a graph. For our section on curve-sketching, we have adopted a six-step "sketch-as-you-go" approach, the aim being to draw the graph of a function as quickly and efficiently as possible. If you use a graphing calculator or computer to draw graphs, the techniques we discuss are still necessary to locate and explain the important features of a graph.

We have also included sections on related rates and elasticity of demand. The first of these examines further the concept of the derivative as a rate of change, while the second discusses optimization of revenue based on the demand equation.

Throughout this chapter, you will notice our emphasis on a systematic approach, not only to the applications, but also to the computational and mechanical problems. This will help demystify a lot of the material and also sharpen your ability to extract the basic patterns underlying the examples.

▶ ══ **5.1** MAXIMA AND MINIMA

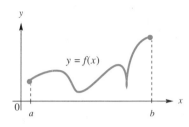

FIGURE 1

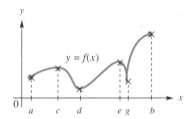

FIGURE 2

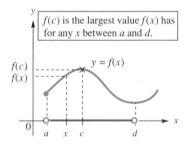

FIGURE 3

Take a look at the graph shown in Figure 1.

This is the graph of a function whose domain is the closed interval $[a, b]$. From the viewpoint of a mathematician, all kinds of exciting things are going on here. There are hills and valleys, and even a small chasm (called a "cusp") toward the right. For many purposes, the important features of this curve are the highs and lows. Suppose, for example, you knew beforehand that the stock price of some company would follow this graph during the course of a week. What would be your best strategy for buying and selling that stock? While you would certainly make a handsome profit if you bought at time a and sold at time b, you would do even better if you followed the old adage to "buy low and sell high," buying at all the lows and selling at all the highs.

Figure 2 shows the graph once again with these important points marked. The points marked on the graph of f give the lows and the highs, illustrating the best stock-trading strategy: buy at a, sell at c, buy at d, sell at e, buy at g, sell at b.

Mathematicians like to give these highs and lows Latin names: the highs (c, e, and b) are referred to as **local maxima**, and the lows (a, d, and g) are referred to as **local minima.** Collectively, these highs and lows are referred to as **local extrema.** (A point of language: the singular forms of minima, maxima, and extrema are minimum, maximum, and extremum.)

Why do we refer to these points as "local" extrema? Take a look at the point corresponding to $x = c$. Compared to nearby portions of the graph, it is the highest point of the graph *in the vicinity*. In other words, if you were an extremely near-sighted mountaineer and were positioned at this point, you would *think* that you were at the highest point of the graph, as you would be totally oblivious of the distant peaks at $x = e$ and $x = b$.

Let us translate this description into mathematical terms. We are talking about the heights of various points on the curve. Now the height of the curve at $x = c$ is measured by $f(c)$, so we are saying that $f(c)$ is larger than any neighboring $f(x)$. More specifically, $f(c)$ *is the largest value that $f(x)$ has for all choices of x between a and d.* (See Figure 3.)

Formally:

LOCAL EXTREMA

f has a **local** or **relative maximum at** c if there is some interval (r, s) (even a very small one) containing c for which $f(c) \geq f(x)$ for all choices of x between r and s for which $f(x)$ is defined.

f has a **local** or **relative minimum** at c if there is an interval (r, s) (even a very small one) containing c for which $f(c) \leq f(x)$ for all choices of x between r and s for which $f(x)$ is defined.

Figure 4 shows the location of all the local extrema on the graph of f.

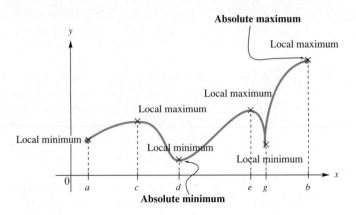

FIGURE 4

$f(a)$ is the smallest value $f(x)$ has for any x between p and c for which the function is defined.

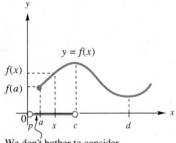

We don't bother to consider an x here, since $f(x)$ is not defined here.

FIGURE 5

You should try to find an interval containing the x-coordinate of each local extreme point as in the definition. For instance, let us show how our definition allows us to say that f has a local minimum at a. In Figure 5 is shown an interval (p, c) containing a. The only values of x in that interval for which $f(x)$ is defined are the numbers in $[a, c)$, and so these are the only ones we need to look at. From the picture, it is clear that $f(a)$ is the smallest value of f in this interval.

In Figure 4 we labeled one of the points an *absolute* minimum and one an absolute maximum. These simply correspond to the smallest and largest values of f overall. Formally:

> **ABSOLUTE EXTREMA**
>
> f has an **absolute maximum** at c if $f(c) \geq f(x)$ for every x in the domain of f.
>
> f has an **absolute minimum** at c if $f(c) \leq f(x)$ for every x in the domain of f.

Q What if $f(x)$ is constant, so the graph of f is a horizontal line?

A According to the definition of a local extremum, *every* point on the line would qualify as both a local minimum and a local maximum (because we use the inequalities $\leq$ and $\geq$ in the definition, rather than strict inequalities). Similarly, by the definition of an absolute extremum, every point would also be an absolute maximum and an absolute minimum.

We have already seen why local maxima and minima are interesting in the context of investment; your best strategy is to buy at the local minima and

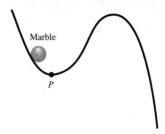

Marble

P

FIGURE 6

sell at the local maxima. (We suggest you spend some time thinking about why this is in fact the best strategy.) Here is another example that illustrates the importance of local extrema. Suppose you have a marble on an incline with the shape shown in Figure 6.

If you place that marble at the point P, it will remain there. Although there are lower points on the hill than P, there is a barrier between the marble and those lower points. The marble remains at the local minimum, which may not be an absolute minimum. Like our near-sighted mountaineer, the marble cannot "see" the lower points. This situation is an example of the scientific principle that systems tend toward the lowest possible energy state. We should be glad that systems can get stuck in local minima—otherwise, we would not stick to the surface of the Earth but all fall into the sun!

This principle also applies to some economic models. Economies are driven by consumers' desire to pay the lowest prices and producers' desire to get the largest profits. But it is possible for economies to fall into local extrema that are not globally best.

Q How do we find these local extreme points?

A We can find them using two approaches:

1. **graphically,** using a graphing calculator or computer graphing software to locate approximate numerical values for the local extrema, or
2. **analytically,** using calculus to obtain exact values for the local extrema. (Sketching the graph or using a graphing calculator is also very helpful in the analytical approach.)

At the end of this section, we shall also see how to combine these two approaches: we shall use analytic methods from calculus to help us improve the accuracy of the graphical method.

GRAPHICAL APPROACH (GRAPHING CALCULATOR OR COMPUTER)

To locate the local extrema graphically, you need some form of graphing technology.

▼ **EXAMPLE 1**

Use a graphing calculator or computer graphing software to locate the local extrema of

$$f(x) = 3x^5 - 25x^3 - 15x^2 + 60x.$$

SOLUTION   First, we must graph this function with a wide enough range of x-values to see all of the local extrema. If you experiment a little, you will find that the ranges

$$-5 \leq x \leq 5$$

and

$$-100 \leq y \leq 100$$

will show you the graph nicely. (See Figure 7.)

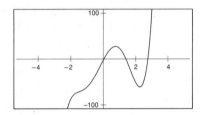

FIGURE 7

There appear to be two local extrema: a local maximum with x somewhere between 0 and 1, and a local minimum a little to the right of $x = 2$. To find the coordinates more precisely, we can use the trace or zoom features. For example, zooming in on the local maximum, we can obtain the graph shown in Figure 8, which uses an x-range of $0.7 \leq x \leq 0.8$ and a y-range of $26.6 \leq y \leq 26.8$.

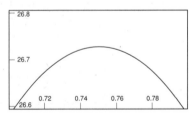

FIGURE 8

We can see from Figure 8 that the local maximum of $f(x)$ is at approximately $x = 0.75$, and $f(0.75) \approx 26.7$ (to three significant digits*). Similarly, zooming in on the local minimum shows that it is at approximately $x = 2.26$, and $f(2.26) \approx -52.7$.

Before we go on... If we want to locate the local extrema more accurately, we will need to zoom in closer. Repeated use of the zoom feature can cause difficulties, because the graph may appear flat near a local extremum, making its location difficult to find. How can we locate the extrema more accurately? Can we find the *exact* location of all local extrema? Using calculus, we shall see how to locate extrema accurately, and often exactly.

▼ *The number of significant digits we use is somewhat arbitrary. Three or four are sufficient for most purposes and are all that are justified in many cases when using real data.

▶ CAUTION   Many calculators are equipped with built-in procedures to locate absolute maxima and minima. (For instance, the TI-82 has one in the CALC menu and also in the MATH menu.) But these procedures sometimes behave unpredictably and may even miss absolute extrema completely!* In this text, we shall try not to rely on these features but rather on mathematical intuition and graphing technology. ◀

ANALYTICAL APPROACH

In Figure 9 we see once again the graph from Figure 1, but we have now classified the extrema into three types.

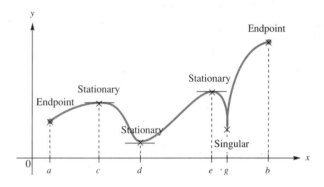

FIGURE 9

Look at the extrema we have labeled as "stationary." Notice that the tangent to the curve at each of these points is horizontal and thus has slope zero. Because the slope of the tangent at a specific value of x is equal to the derivative evaluated there, we conclude that the derivative of f is zero at each of these points. In other words,

$$f'(c) = 0, \quad f'(d) = 0, \quad \text{and} \quad f'(e) = 0.$$

We should be able to find the exact location of each of these extrema by solving the equation $f'(x) = 0$. We call points where $f'(x) = 0$ **stationary points** because the rate of change of f is zero there. We shall call an *extremum* that is a stationary point a **stationary extremum.**

▼   * Here is a glaring example: on a TI-82, the authors plotted the curve $y = x^3 - x^2 - 3x$ with $-5 \le x \le 2$ and $-135 \le y \le 20$. They then used the CALC menu to locate the minimum with a lower bound of -2.3 and an upper bound of 2, and the calculator located the *local* minimum at approximately 1.39 (whereas the absolute minimum in this range was at -2.3). Yet the same procedure gave the correct absolute minimum (of -5) if the lower bound of -5 was used instead.

To locate all stationary points (among which are all possible stationary extrema), we solve the equation $f'(x) = 0$ for x and we make sure that x is in the domain of f.

There is a local minimum at $x = g$, but something slightly different happens there: there is no horizontal tangent at that point. In fact, there is no tangent line at all, since the derivative is not defined at $x = g$. (Recall a similar situation with the graph of $f(x) = |x|$ at $x = 0$.) Thus, to locate g, we must look for the values of x (in the domain of f) for which $f'(x)$ does not exist. We call such points **singular points,** and we shall call an extremum that is a singular point a **singular extremum.**

LOCATING SINGULAR POINTS

To locate all singular points (among which are all possible singular extrema), we look for values of x such that $f'(x)$ does not exist, and we make sure that x is in the domain of f.

We call all points of the domain where either $f'(x) = 0$ or $f'(x)$ does not exist **critical points.** The critical points give us *all the possible candidates for stationary and singular local extrema.*

The remaining two extrema are at the endpoints of the domain.* As we see in the picture, they are (almost) always either local maxima or local minima.

LOCATING ENDPOINTS

If the domain of f has any endpoints, these are almost always local extrema.

▶ CAUTION   A critical point need not be an extreme point. For example, the graph shown in Figure 10 has two critical points, one of each type, but no local extrema at all!

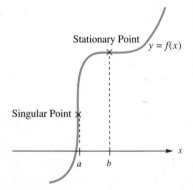

FIGURE 10

Stationary Point

$y = f(x)$

Singular Point

x

a b

▼ *Bear in mind that many calculus texts do not count endpoints of the domain as local extrema, although more advanced (analysis) texts do. In view of examples like our stock market investing strategy, we see no good reason not to count endpoints as extrema.

The point at $x = a$ in Figure 10 is a point in the domain at which the tangent is vertical, and so it is a point where the derivative $f'(x)$ is not defined, i.e., a singular point. But $x = a$ is neither a maximum nor a minimum: the graph is higher to the right and lower to the left. Similarly, the point at $x = b$ has a horizontal tangent and thus is a stationary point—yet it is neither a maximum nor a minimum. Thus, solving $f'(x) = 0$ for x and finding points in the domain where $f'(x)$ does not exist only gives us *candidates* for stationary extrema and singular extrema. These candidates need not be actual extrema. ◄

LOCATING CANDIDATES FOR LOCAL EXTREMA

To find all candidates for local extrema, we do the following.

Stationary Points: Solve the equation $f'(x) = 0$ for x, and make sure that x is in the domain of f.

Singular Points: Find all values of x such that $f'(x)$ does not exist and x is in the domain of f.

Endpoints: List all endpoints of the domain (if any).

Q Are there any other types of local extrema?

A The answer is a *qualified* "no." If the function we are looking at happens to be continuous on its domain and differentiable at every point except for a few isolated points, then these are the only kinds of local extrema we need consider. If the function has a discontinuity at some value of x, we need to look at the graph near this value to find out if there are any other local extrema. Because these cases rarely arise in practice, we can consider them on a case-by-case basis.

Q Now we know how to find the candidates for *local* extrema. What about the *absolute* extrema? (Recall that these are the highest and lowest points on the whole graph.)

A Finding these is a little more tricky. We shall see in some of the examples that there need not be any at all. There is, however, a useful theorem which tells us that if the function is continuous and the domain happens to be a closed interval (such as $[1, 5]$), then the function must have an absolute maximum and an absolute minimum. The absolute maximum is just the highest local maximum, and the absolute minimum is just the lowest local minimum. If the domain is not a closed interval, anything can happen. We see some of the possibilities in the following table (we shall see in the following examples exactly how we determine the extrema).

Function	Graph	Extrema
$f(x) = x^2$, with domain all real numbers		Absolute minimum at $x = 0$; no local or absolute maximum
$f(x) = \frac{1}{x}$, with domain $(0, +\infty)$		No extrema
$f(x) = x^3 - x^2 - 5x$, with domain $(-3, 4)$		A local minimum at $x = \frac{5}{3}$ and a local maximum at $x = -1$, but no absolute extrema

We shall see that no matter what the domain is, a rough sketch of the curve using the methods we shall discuss will often allow us to locate the absolute extrema, if any.

We now turn to several examples of finding maxima and minima analytically. In all of these examples we will follow this procedure: First we find the derivative, then we find the stationary points and singular points. Next, we make a table listing the critical points and endpoints, together with the value of the function at these points, and plot them. From this table and rough sketch we will usually have enough data to be able to say where the extreme points are.

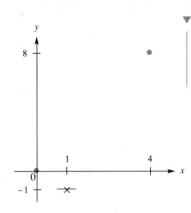

FIGURE 11

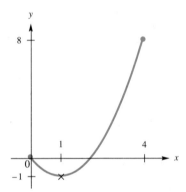

FIGURE 12

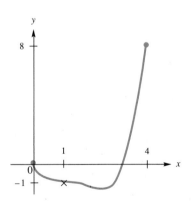

FIGURE 13

▼ | **EXAMPLE 2**

Find the relative and absolute maxima and minima of $f(x) = x^2 - 2x$ on the interval $[0, 4]$.

SOLUTION   We begin by locating the stationary points.

Stationary Points   To locate these points, we solve the equation $f'(x) = 0$. Because

$$f'(x) = 2x - 2,$$

we solve

$$2x - 2 = 0,$$

getting $x = 1$. The domain of the function is $[0, 4]$, so $x = 1$ is a point of the domain. Thus, $x = 1$ is the only candidate for a stationary local extremum.

Singular Points   We look for points where the derivative is not defined. Because the derivative is $2x - 2$, it is defined for every x. Thus, there are no singular points and hence no candidates for singular local extrema.

Endpoints   Because the domain is $[0, 4]$, the endpoints are $x = 0$ and $x = 4$.

We record these points in a table, together with the corresponding values of f.

x	0	1	4
$f(x)$	0	-1	8

Plotting these points gives us Figure 11.

The points at $x = 0$ and $x = 4$ are the endpoints, which we show as heavy dots, while the point at $x = 1$ is a stationary point, so that it has a horizontal tangent, and we remind ourselves of this by drawing a horizontal line segment through the point. Connecting these points gives us a graph that must look something like the curve shown in Figure 12.

Notice that the tangent is horizontal at the point $(1, -1)$ but not at the endpoints (because they are not stationary points).

Q All we have are three points! How do we know that the graph doesn't look something like the one in Figure 13, for instance?

A If it did, then there would be an extra stationary point at about $x = 3$. Because stationary points are found by solving $f'(x) = 0$, and we saw that there was only *one* solution, namely, $x = 1$, there can be no other stationary point. In other words, there can't be any maxima or minima not in our table!

Notice that from the correct graph (Figure 12), we see immediately that there is an absolute maximum of 8 at $x = 4$ and an absolute minimum of -1 at $x = 1$. Thus, the extrema are as follows:

local maximum at $(0, 0)$,

absolute minimum at $(1, -1)$,

absolute maximum at $(4, 8)$.

We can also state the result as follows: f has a local maximum of 0 at $x = 0$, an absolute minimum of -1 at $x = 1$, and an absolute maximum of 8 at $x = 4$.

Before we go on... Look once again at the derivative, $f'(x) = 2x - 2$. We know that $f'(1) = 0$ at the minimum. What about $f'(x)$ for values of x on either side of 1? We obtain the following table by choosing a value of x on either side of the stationary point at $x = 1$.

x	0	1	2
$f'(x) = 2x - 2$	-2 *(negative)*	0	2 *(positive)*
	↘		↗

Because $f'(0) = -2 < 0$, the graph has negative slope, or f is **decreasing,** for values of x to the left of 1. Because $f'(2) = 2 > 0$, the graph has positive slope, or f is **increasing,** for values of x to the right of 1. We have drawn arrows below the table to show where f is increasing and where it is decreasing, confirming the graph in Figure 12. This table also confirms the fact that there is a local minimum at $x = 1$, since it shows that f decreases approaching $x = 1$ from the left and then increases to the right. Formally, we say that f is decreasing on the interval $[0, 1]$ and increasing on the interval $[1, 4]$.

▼ **EXAMPLE 3**

Locate and classify the extrema of $g(t) = t^3$ on $[-2, 2]$.

SOLUTION   By "classifying" the extrema, we mean listing whether each extremum is a local or absolute maximum or minimum as we did in the last example. As before, we consider each type separately.

Stationary Points   Solve the equation $g'(t) = 0$. Here,

$$g'(t) = 3t^2,$$

so we solve the equation

$$3t^2 = 0,$$

whose only solution is $t = 0$, which is the only stationary point.

Singular Points   Because $g'(t) = 3t^2$ is defined for every t, there are no singular points.

Endpoints   Because the domain is given as $[-2, 2]$, the endpoints are $t = -2$ and $t = 2$.

We record these points in a table.

t	-2	0	2
$g(t)$	-8	0	8

Plotting these three points gives us Figure 14(a), suggesting the curve in Figure 14(b).

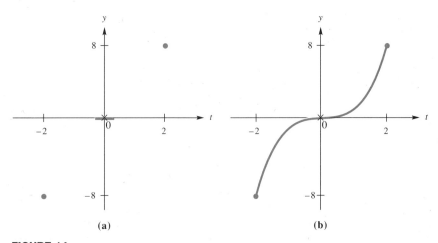

(a)　　　　　　　　　**(b)**

FIGURE 14

Notice something interesting at $t = 0$: we have a stationary point, so we know that the tangent line is horizontal. However, there is a lower point to the left and a higher point to the right, so that this critical point is neither a local maximum nor a local minimum.

Thus, we find the following extrema:

$$\text{absolute minimum of } -8 \text{ at } t = -2, \text{ and}$$
$$\text{absolute maximum of } 8 \text{ at } t = 2.$$

Before we go on... Notice that the shape of the curve is dictated by the requirement that the tangent be horizontal at $t = 0$. It would not be accurate to simply draw a straight line joining the three plotted points.

As in the last example, let us see where g is increasing and where it is decreasing.

t	-1	0	1
$g'(t) = 3t^2$	3 (*positive*)	0	3 (*positive*)
	↗		↗

We find that g is never decreasing, confirming that there can be no local extremum at $t = 0$. Once again, notice how the curve in Figure 14 follows the upward direction of the arrows on either side of the stationary point at $x = 0$.

▼ EXAMPLE 4

Locate and classify the maxima and minima of $f(t) = t^4 - 2t^2$, with domain $[0, +\infty)$.

SOLUTION

Stationary Points   We know $f'(t) = 4t^3 - 4t$, so we solve

$$4t^3 - 4t = 0,$$

giving

$$t^3 - t = 0,$$

or

$$t(t - 1)(t + 1) = 0.$$

Thus, $t = 0$, 1, or -1. We now have three stationary points. However, the point -1 is not in the domain $[0, \infty)$, so we discard it and keep only the two points 0 and 1.

Once again, there are no singular points.

Endpoints   The only endpoint is 0, which is also one of the stationary points.

Our table looks like this:

t	0	1
$f(t)$	0	-1

FIGURE 15

Plotting these points, we get Figure 15.

From the graph we can see that f decreases as t goes from 0 to 1. But what happens after that? We saw in the previous example that we cannot *assume* that it goes up again. Therefore, we try a "test point" farther to the right, at 2, say. We add this point to our table.

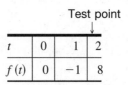

			Test point
t	0	1	2
$f(t)$	0	-1	8

After plotting these points in Figure 16(a), we can sketch the curve, as shown in Figure 16(b).

Now we can see that f does increase as t goes from 1 to ∞. Thus, f has a local maximum of 0 at 0 and an (absolute) minimum of -1 at 1. There is no absolute maximum, because f continues to increase forever to the right. Notice again that because $(0, 0)$ is a stationary point (in addition to being an endpoint), the tangent to the curve at the origin is horizontal, and so the curve departs from the origin with zero slope as it starts to dip.

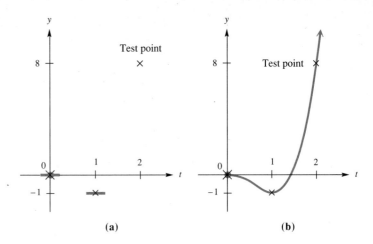

FIGURE 16

Before we go on... Notice two things about the graph in Figure 16.

1. The point (2, 8), is not an endpoint of the graph. The graph continues through that point as t goes to $+\infty$. The point (2, 8) was just a test point and is not a local extremum.
2. There is *no* absolute maximum. The graph climbs without bound as $t \to +\infty$. In mathematical terms, $\lim_{t \to +\infty} f(t) = +\infty$.

Let us once again examine where f is increasing and where it is decreasing, using the derivative.

t	0	$\frac{1}{2}$	1	2
$f'(t) = 4t^3 - 4t$	0	$-\frac{3}{2}$ (*negative*)	0	24 (*positive*)

The function is decreasing on $[0, 1]$ and increasing on $[1, +\infty)$.

Figure 17 shows graphing calculator plots of f and its derivative f' for $0 \le t \le 3$ and $-2 \le y \le 8$.

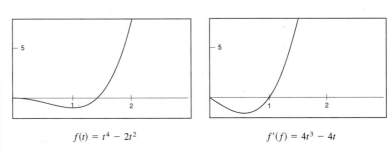

$f(t) = t^4 - 2t^2$ $f'(f) = 4t^3 - 4t$

FIGURE 17

Notice several things from these plots:

1. The stationary points at $t = 0$ and $t = 1$ in the graph of f correspond to the points where the graph of f' crosses the t-axis [since this is where $f'(t) = 0$].
2. From the graph of f', we see that $f'(t) \leq 0$ if $0 \leq t \leq 1$, which tells us that the original function f is decreasing for $0 \leq t \leq 1$.
3. Similarly, we see that $f'(t) \geq 0$ if $t \geq 1$, which tells us that the original function f is increasing for $t \geq 1$.

You can have your graphing calculator automatically calculate the derivative of any function numerically and plot it along with the original function. On a TI-82, enter the original function as Y_1 and its derivative as Y_2 using the following format.

$$Y_1 = X^4 - 2X^2$$
$$Y_2 = nDeriv\ (Y_1, X, X)$$

▼ **EXAMPLE 5**

Locate and classify the extrema of $f(x) = \sqrt[3]{x^2}$ on $[-1, 1]$.

SOLUTION
Stationary Points $f(x) = x^{2/3}$, so $f'(x) = \frac{2}{3}x^{-1/3}$. How do we solve the equation

$$\frac{2}{3}x^{-1/3} = 0?$$

First, *get rid of negative exponents*. Move the $x^{-1/3}$ to the denominator, where its exponent becomes positive, and the equation becomes

$$\frac{2}{3x^{1/3}} = 0.$$

Multiplying by $3x^{1/3}$ yields

$$2 = 0.*$$

But this is absurd! There is no solution to the equation, which means that there are no stationary points.

▼ * It is useful to remember the following rule:

If $a/b = 0$, then a must be zero.

(We see this by multiplying both sides by b.) The equation given above had $\frac{2}{3x^{1/3}} = 0$, which would imply that 2 would have to be zero, which it isn't. Thus, the equation represents a false statement. In other words, there is no x such that $\frac{2}{3\,x^{1/3}} = 0$.

Singular Points Look once again at the derivative, written in the form with no negative exponents.

$$f'(x) = \frac{2}{3x^{1/3}}$$

Because the derivative has an $x^{1/3}$ in the denominator, it is not defined when $x = 0$. Since $x = 0$ is in the domain of f, $x = 0$ is a singular point.

Endpoints The endpoints are 1 and -1.

Thus, our table looks like this:

x	-1	0	1
$f(x)$	1	0	1

As usual, we shall plot these points and sketch the curve as best we can. Notice the following, however. The point at $(0, 0)$, being a singular point, is a point at which the derivative is not defined. The derivative is not defined there because we have an x in the denominator of $f'(x)$. However, we can take the *limit* of $f'(x)$ as $x \to 0$ instead of trying to evaluate it at 0.

$$\lim_{x \to 0^+} f'(x) = \lim_{x \to 0^+} \frac{2}{3x^{1/3}} = +\infty$$

$$\lim_{x \to 0^-} f'(x) = \lim_{x \to 0^-} \frac{2}{3x^{1/3}} = -\infty$$

What these limits tell us is that close to $x = 0$ the curve is very steep (because the slope of the tangent is approaching $\pm\infty$). At that point, the tangent must be *vertical*. This will help us draw the curve in Figure 18(b). [Notice the vertical tangent at $(0, 0)$.]

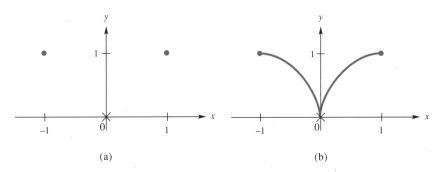

(a)　　　　　　　　　　　(b)

FIGURE 18

Summarizing, we have an

> absolute maximum of 1 at $x = -1$,
> absolute minimum of 0 at $x = 0$,
> absolute maximum of 1 at $x = 1$.

 Figure 19 shows graphing calculator plots of $f(x)$ and f' with $-1 \le x \le 1$ and $-2 \le y \le 2$.

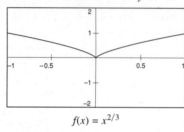

$f(x) = x^{2/3}$

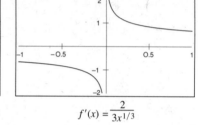

$f'(x) = \dfrac{2}{3x^{1/3}}$

FIGURE 19

Notice several things:

1. At the singular point $x = 0$, the derivative has a vertical asymptote, since it diverges to infinity.
2. For $x < 0$, the derivative is negative, so that f is decreasing, and for $x > 0$, the derivative is positive, so that f is increasing.

▼ **EXAMPLE 6**

Locate and classify the extrema of $g(x) = x + \dfrac{1}{x}$.

SOLUTION   Because no domain was specified, we take the domain to be as large as possible. Since the function cannot be defined when $x = 0$, the largest possible domain consists of all real numbers except 0. In other words, the domain is $(-\infty, 0) \cup (0, +\infty)$.

Stationary Points   We have $g'(x) = 1 - \dfrac{1}{x^2}$, so we solve

$$1 - \frac{1}{x^2} = 0.$$

Moving the $1/x^2$ over to the other side gives

$$1 = \frac{1}{x^2},$$

and multiplying by x^2 gives

$$x^2 = 1, \quad \text{so } x = \pm 1.$$

Singular Points $g'(x)$ is not defined when $x = 0$. However, $x = 0$ is not in the domain. Thus, $x = 0$ is disqualified as a singular point, so there are no singular points.

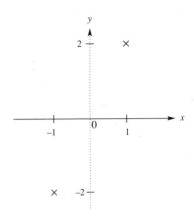

FIGURE 20

Endpoints   Since the domain is $(-\infty, 0) \cup (0, +\infty)$, there are no endpoints of the domain that also lie within the domain.

Because $g(0)$ is not defined, there will be a break in the graph where $x = 0$. We shall thus include $x = 0$ in our table as a reminder of this fact. Our table then looks like this:

x	-1	0	1
$g(x)$	-2	✖	2

These points are shown in Figure 20.

We have dotted the vertical line $x = 0$ to indicate where the graph breaks. In other words, the two points we plotted *cannot be joined*. This fact, together with the fact that we know nothing about what happens to the left of -1 and to the right of 1, means that we cannot yet tell whether these points are maxima, minima, or neither. To decide, we shall use test points. This time, we will take them on either side of *both* stationary points because we really have two separate curves. Thus we enlarge our table with some test points.

	test points				test points	
x	-2	-1	$-\frac{1}{2}$	0	$\frac{1}{2}$	1	2
$g(x)$	$-\frac{5}{2}$	-2	$-\frac{5}{2}$	✖	$\frac{5}{2}$	2	$\frac{5}{2}$

We plot these points in Figure 21(a), and connect them in the only way possible in Figure 21(b).

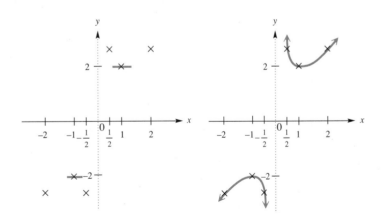

FIGURE 21

Thus we have the following classification:

local maximum of -2 at $x = -1$ and

local minimum of 2 at $x = 1$.

Figure 22 shows a graphing calculator plot of $f'(x) = 1 - \dfrac{1}{x^2}$.

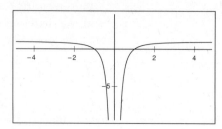

FIGURE 22

How much can you say about the graph of f by looking only at the graph of f'?

COMBINING THE GRAPHING CALCULATOR APPROACH WITH THE ANALYTICAL APPROACH

In the next example, we redo Example 1, this time using calculus together with a graphing calculator to find the local extrema.

▼ **EXAMPLE 7**

Use a graphing calculator or computer graphing software to locate and classify the local extrema of

$$f(x) = 3x^5 - 25x^3 - 15x^2 + 60x.$$

SOLUTION   In Example 1, we graphed this function (Figure 7) and saw by zooming in on the graph that there was one local maximum at approximately $x = 0.75$ and one local minimum at approximately $x = 2.26$. This time, we shall locate the extrema by using the fact that $f'(x) = 0$ at these points. First, we compute

$$f'(x) = 15x^4 - 75x^2 - 30x + 60.$$

Thus, the stationary points occur when

$$15x^4 - 75x^2 - 30x + 60 = 0.$$

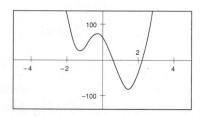

FIGURE 23

Instead of trying to solve this equation analytically, we solve it graphically. Figure 23 shows the graph of f' with the ranges $-5 \le x \le 5$, and $-125 \le y \le 125$.

Corresponding to the two local extrema we see two places where $f'(x) = 0$. It is somewhat easier to see places where a graph crosses the x-axis than it is to see local extrema. (See the "Before we go on" discussion below.) In this case, we can zoom in on, say, the first such point, getting a picture like Figure 24, with ranges $0.7504 \le x \le 0.7506$ and $-0.01 \le y \le 0.01$.

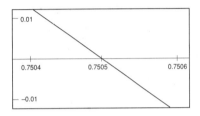

FIGURE 24

Figure 24 shows that the local maximum is at approximately $x = 0.7505$. We obtain the corresponding maximum value of f by computing $f(0.7505) = 26.7276$. Similarly, we can zoom in on the second crossing of the x-axis to find $x \approx 2.2586$. The corresponding minimum value of f is $f(2.2586) = -52.7199$.

Before we go on... If you try to avoid calculus completely by repeatedly zooming in on the local maximum for greater accuracy, you may find (as we did) that after three or four repetitions your graph resembles Figure 25.

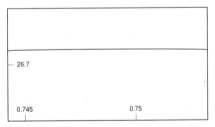

Graph of $f(x) = 3x^5 - 25x^3 - 15x^2 + 60x$ near $x = 75$.

FIGURE 25

As you see, it is difficult to pinpoint the location of the local maximum with the same accuracy we obtained using calculus.

▶ ## 5.1 EXERCISES

*Locate and classify all extrema in each of the graphs in Exercises 1–8 and indi-
cate the intervals on which the associated function is increasing or decreasing.
(By "classifying" the extrema, we mean listing whether each extremum is a local
or absolute maximum or minimum.) Also, locate any stationary points or singular
points that are not local extrema.*

1.

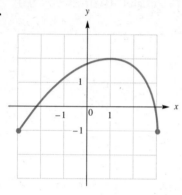

2.

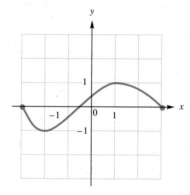

3.

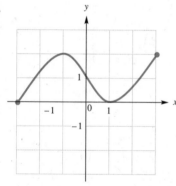

4.

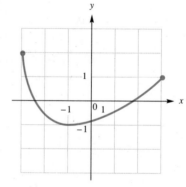

5.

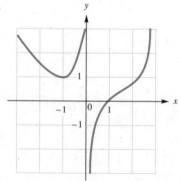

6.

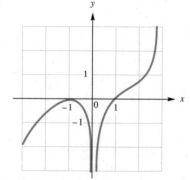

7.

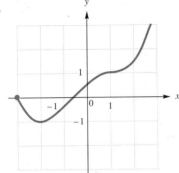

8.

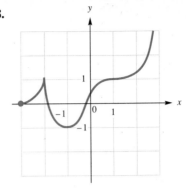

Use a graphing calculator or computer graphing software to find the approximate coordinates (correct to one decimal place) of all extrema for each of the functions in Exercises 9–16.

9. $f(x) = (x - 1)(x - 2)$ with domain $[0, +\infty)$

10. $f(x) = x(x + 3)$ with domain $[-4, 0]$

11. $f(x) = x^x$ with domain $[0.1, +\infty)$

12. $f(x) = \sqrt{x}\,\ln x$ with domain $[1.1, 5]$

13. $f(x) = x(x - 1)^{2/3}$ with domain all real numbers*

14. $f(x) = x + (x - 1)^{2/3}$ with domain all real numbers*

15. $f(x) = \dfrac{e^{-x}}{1 + e^{-x}}$ with domain all real numbers

16. $f(x) = \dfrac{\sqrt{x}}{1 + \sqrt{x}}$ with domain $[0, +\infty)$

Use calculus to find the exact location of all the local and absolute extrema of each of the functions in Exercises 17–40. In each case, give a rough sketch of the curve or use a graphing calculator to help you along.

17. $f(x) = x^2 - 4x + 1$ with domain $[0, 3]$

18. $f(x) = 2x^2 - 2x + 3$ with domain $[0, 3]$

19. $g(x) = x^3 - 12x$ with domain $[-4, 4]$

20. $g(x) = 2x^3 - 6x + 3$ with domain $[-2, 2]$

21. $f(t) = t^3 + t$ with domain $[-2, 2]$

22. $f(t) = -2t^3 - 3t$ with domain $[-1, 1]$

23. $h(t) = 2t^3 + 3t^2$ with domain $[-2, +\infty)$

24. $h(t) = t^3 - 3t^2$ with domain $[-1, +\infty)$

25. $f(x) = x^4 - 4x^3$ with domain $[-1, +\infty)$

26. $f(x) = 3x^4 - 2x^3$ with domain $[-1, +\infty)$

27. $g(t) = \frac{1}{4}t^4 - \frac{2}{3}t^3 + \frac{1}{2}t^2$ with domain $(-\infty, +\infty)$

28. $g(t) = 3t^4 - 16t^3 + 24t^2 + 1$ with domain $(-\infty, +\infty)$

29. $f(t) = (t^2 + 1)/(t^2 - 1)$; $-2 \le t \le 2, t \ne \pm 1$

30. $f(t) = (t^2 - 1)/(t^2 + 1)$ with domain $[-2, 2]$

31. $f(x) = \sqrt{x}(x - 1)$, $x \ge 0$

32. $f(x) = \sqrt{x}(x + 1)$, $x \ge 0$

33. $g(x) = x^2 - 4\sqrt{x}$

34. $g(x) = \dfrac{1}{x} - \dfrac{1}{x^2}$

▼ *Use the format $(X - 1)^{2\wedge}(1/3)$ for $x - 1^{2/3}$.

35. $g(x) = x^3/(x^2 + 3)$

36. $g(x) = x^3/(x^2 - 3)$

37. $f(x) = x - \ln x$ with domain $(0, +\infty)$

38. $f(x) = x - \ln(x^2)$ with domain $(0, +\infty)$

39. $g(t) = e^t - t$ with domain $[-1, 1]$

40. $g(t) = e^{-t^2}$ with domain $(-\infty, +\infty)$

In each of Exercises 41–44, a function is specified together with a computer-generated sketch of its graph. Use calculus to locate and classify all extrema.

41. $f(x) = \dfrac{2x^2 - 24}{x + 4}$

42. $f(x) = \dfrac{x - 4}{x^2 + 20}$

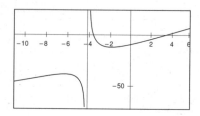

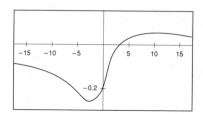

43. $f(x) = xe^{1 - x^2}$

44. $f(x) = x \ln x$ with domain $(0, +\infty)$

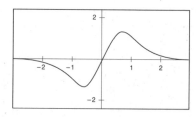

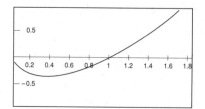

In each of Exercises 45–52, a graphing calculator plot of the derivative *of a function is shown. In each case, determine the x-coordinates of all stationary and singular points of the original function, and classify each one as a local maximum, minimum, or neither. (Assume that the function is defined for every x in the viewing window.)*

45.

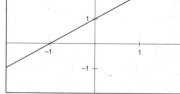

46.

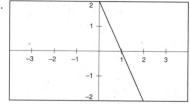

47.

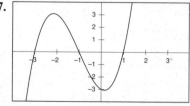

48.

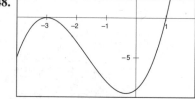

49.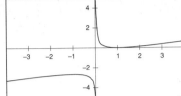

50.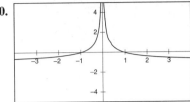

51.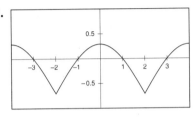

52.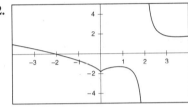

In each of Exercises 53–56, use a graphing calculator or computer to graph both the given function and its derivative, and hence locate all local and absolute extrema, with x-coordinates correct to two decimal places.

53. $y = x^2 + \dfrac{1}{x - 2}$ with domain $(-3, 2) \cup (2, 6)$

54. $y = x^2 - 10(x - 1)^{2/3}$ with domain $(-4, 4)$

55. $f(x) = (x - 5)^2(x + 4)(x - 2)$ with domain $[-5, 6]$

56. $f(x) = (x + 3)^2(x - 2)^2$ with domain $[-5, 5]$

COMMUNICATION AND REASONING EXERCISES

57. Draw the graph of a function f with domain all real numbers, such that f is not linear and has no local extrema.

58. Draw the graph of a function g with domain all real numbers, such that g has a local maximum and minimum but no absolute extrema.

59. Draw the graph of a function that has stationary and singular points but no local extrema.

60. Draw the graph of a function that has local, not absolute, maxima and minima, but has no stationary or singular points.

61. If a stationary point is not a local maximum, then must it be a local minimum? Explain your answer.

62. If one endpoint is a local maximum, must the other be a local minimum? Explain your answer.

▶ ═══════ **5.2** APPLICATIONS OF MAXIMA AND MINIMA

There are many times that we would like to find the largest or smallest possible value of some quantity—for instance, the largest possible profit or the lowest cost. We call this the *optimal* (best) value. We can often use calculus to find the optimal value.

In all applications, the first step is to translate a written description into a mathematical problem. The mathematical problem will have the following form. There will be some *unknowns* that we are asked to find, there will be an expression involving those unknowns that must be made as large or as small as possible—the **objective function**—and there may be **constraints**—equations or inequalities relating the variables.*

▼ **EXAMPLE 1**   Average Cost

Gymnast Clothing, Inc. manufactures expensive hockey jerseys for sale to college bookstores in runs of up to 500. Its cost function is

$$C(x) = 2000 + 10x + 0.2x^2,$$

where x is the number of hockey jerseys it manufactures. How many jerseys should Gymnast Clothing Inc. produce per run in order to minimize average cost?

SOLUTION   Why don't we seek to minimize total cost? The answer would be trivial: to minimize total cost, we would make *no* jerseys at all. Minimizing the average cost is a more practical objective. Here is the procedure we will follow to solve this problem.

1. *Identify the unknown(s) here.*   The only unknown is the number x of hockey jerseys Gymnast should manufacture (we know this because the question is "how many jerseys . . .").
2. *Identify the objective function.*   The objective function is the quantity that must be made as small (in this case) as possible. In this example, it is the average cost, given by

$$\overline{C}(x) = \frac{C(x)}{x} = \frac{2000 + 10x + 0.2x^2}{x}$$

$$= \frac{2000}{x} + 10 + 0.2x.$$

3. *Identify the constraints (if any).*   At most 500 jerseys can be manufactured in a run. Also, $\overline{C}(0)$ is not defined. Thus, our constraint is

$$0 < x \le 500.$$

   Another way of saying this is that the domain of the objective function $\overline{C}(x)$ is $(0, 500]$.
4. *State and solve the resulting optimization problem.*   Our optimization problem is to

$$minimize\ \overline{C}(x) = \frac{2000}{x} + 10 + 0.2x$$

$$subject\ to\ 0 < x \le\ 500.$$

We now proceed to solve this problem as in the previous section.

▼ *If you have studied linear programming, you will notice a similarity here, but unlike the situation in linear programming, neither the objective function nor the constraints need to be linear.

Stationary Points $\overline{C}'(x) = -\dfrac{2000}{x^2} + 0.2$. We set this to zero and solve for x:

$$-\frac{2000}{x^2} + 0.2 = 0,$$

giving

$$0.2 = \frac{2000}{x^2}.$$

Multiplying both sides by x^2,

$$0.2x^2 = 2000,$$

so

$$x^2 = \frac{2000}{0.2} = 10{,}000.$$

Thus, $x = \pm\sqrt{10{,}000} = \pm 100$.

We reject $x = -100$, since it is not in the domain, and so $x = 100$ is the only stationary point.

There are no singular points, since $\overline{C}'(x)$ is defined for all x in the domain $(0, 500]$. The only endpoint is $x = 500$.

Because there is no candidate for an extremum to the left of $x = 100$, we include the testpoint $x = 10$, and we obtain the following table:

x	10	100	500
$\overline{C}(x)$	212	50	114

We see from this table that $\overline{C}(x)$ has an absolute minimum at $x = 100$, and so Gymnast Clothing should manufacture 100 hockey shirts per run in order to minimize average cost. The average cost will be $\overline{C}(100) = \$50$ per jersey.

To obtain the solution graphically, plot the objective function,

$$\overline{C}(x) = \frac{2000}{x} + 10 + 0.2x,$$

with x-range $0.001 \le x \le 500$ (Figure 1).

We notice right away that there is an absolute minimum at about $x = 100$. You can check the accuracy of this answer by plotting $\overline{C}'(x)$ and determining where its graph crosses the x-axis. (See Example 7 of the previous section for a description of this method.)

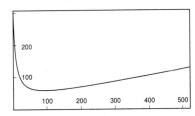

FIGURE 1

We now consider a constrained optimization problem in which the objective function is a function of two variables.

▼ **EXAMPLE 2**

Find x and y minimizing $f = x^2 + y^2$ and satisfying $x + y = 4$.

SOLUTION   Let us follow the procedure used in Example 1.

1.   We first identify the unknown(s). In this problem, we are asked to find x and y, so these are the unknowns, and they are already named for us.

2.   Next, we find the objective function. This is the quantity that we are required to minimize (or maximize). In this case, it is the function $f = x^2 + y^2$. Note that this is a function of *two* variables, so we can't simply go ahead and set the derivative equal to zero. (Would we take the derivative with respect to x or with respect to y?)

3.   Next, we locate the constraint(s). After all, if there were no restrictions on x and y, we could make $x^2 + y^2$ as small as possible by simply choosing x and y to be zero. Looking at the problem, we see the phrase "satisfying $x + y = 4$." This restriction gives us our only constraint: $x + y = 4$.

Our problem is now to

$$minimize\ f = x^2 + y^2\ subject\ to\ x + y = 4.$$

Now that we've restated the problem, we can solve it in two simple steps.
     First, solve the constraint equation for one of the variables (whichever is convenient), and substitute into the objective function. Doing so will eliminate one variable, giving us the objective as a function of a *single variable*.
     Here, our constraint is $x + y = 4$. We solve it for y, getting $y = 4 - x$. Substituting into the objective function gives

$$f = x^2 + (4 - x)^2,$$

a function of the single variable x.
     Second, locate the absolute maximum (or minimum) of the objective function as in the previous section.

Stationary Points $f'(x) = 2x + 2(4 - x)(-1) = 2x - 8 + 2x = 4x - 8$. Setting this equal to 0 gives

$$4x - 8 = 0,$$

so that

$$x = 2.$$

Thus, $x = 2$ is the only critical point, as there are no singular points. Also, there are no endpoints.
     Plotting this single point will tell us very little, so we choose test points on either side.

x	0	2	4
$f(x)$	16	8	16

Without even drawing the graph,* we can see that f has its minimum value of 8 when $x = 2$.

▶ CAUTION At this point in the problem, y has been eliminated, so we forget completely about y. We must be careful not to think of $f(x)$ as y as we usually do. The letter y is not playing its customary role here. Instead, y is one of the two unknowns. We are finding the minimum value of f, not y. When you draw the graph of the function f, you should label the vertical axis the f-axis, not the y-axis. ◀

To complete the problem, we must make sure that we answer the question, which was to find both x and y. To get y, we go back to the constraint $y = 4 - x$. Substituting $x = 2$ gives $y = 4 - 2 = 2$ also.

Thus, the minimum value of f is $2^2 + 2^2 = 8$ when x and y are both 2.

▼ **EXAMPLE 3**

Find x and y maximizing $A = xy$ and satisfying $y = 1 - x^2$ and $0 \leq x \leq 1$.

SOLUTION Here again, the unknowns are clearly stated, as is the objective function $A = xy$. There are two constraints: a constraint equation $y = 1 - x^2$ and an inequality $0 \leq x \leq 1$. Thus, our problem is to

$$\text{maximize } A = xy \text{ subject to } y = 1 - x^2 \text{ and } 0 \leq x \leq 1.$$

The constraint equation is already solved for y. We substitute this expression for y into the objective function, getting

$$A = x(1 - x^2) = x - x^3, \quad 0 \leq x \leq 1.$$

Notice that the second constraint does nothing more than specify that the domain of A is $[0, 1]$.

Now we locate the absolute maximum of A in the usual way. The stationary points are the solutions to $A'(x) = 0$, or

$$1 - 3x^2 = 0,$$

that is,

$$x^2 = \frac{1}{3}, \text{ so that } x = \pm\frac{1}{\sqrt{3}}.$$

We reject the negative solution because it is not in the domain of A, leaving us with a single stationary point at $x = 1/\sqrt{3}$. There are also the endpoints 0 and 1 of the domain.

▼ * You should still try to visualize the graph using the table as a reference—a good mental exercise.

Thus, we get the following table:

x	0	$\dfrac{1}{\sqrt{3}}$	1
$A(x)$	0	$\dfrac{2}{3\sqrt{3}}$	0

We can see from this table that we have an absolute maximum of $A = 2/(3\sqrt{3})$ at $x = 1/\sqrt{3}$. To answer the question, we also need y, which we get from the constraint equation: $y = 1 - x^2 = 1 - \frac{1}{3} = \frac{2}{3}$. Therefore, the maximum value of A is $2/(3\sqrt{3})$ and is achieved when $x = 1/\sqrt{3}$ and $y = \frac{2}{3}$.

Before turning to further applications, we summarize the steps we used in these examples.

PROCEDURE FOR SOLVING AN OPTIMIZATION PROBLEM

1. First, identify the unknown(s) (the quantities asked for in the problem).
2. Determine the objective function, the quantity that we are required to minimize (or maximize). The objective function may be a function of one, two, or more variables.
3. Determine the constraint(s). These can take the form of equations or inequalities.
4. Restate the problem mathematically in the form

   minimize [maximize] the objective function subject to the constraint(s)

5. If the objective function depends on several variables, rewrite it as a function of a single variable. This can be done by solving each constraint equation for one of the variables and substituting into the objective function.
6. Now you can locate the absolute maximum (or minimum) of the objective function as in the previous section. (Use the inequality constraints to specify the domain of the objective function.)

▼ **EXAMPLE 4**   Maximizing Area

Sam wants to build a rectangular enclosure as shown in Figure 2 for his pet rabbit, Killer, and he has bought 100 feet of fencing. What are the dimensions of the largest area that he can enclose?

SOLUTION   We must first identify the unknown(s), and for this, we go to the question: "what are the *dimensions* of the largest area he can enclose?" Thus,

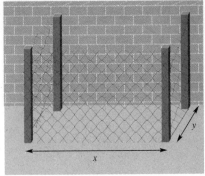

FIGURE 2   **FIGURE 3**

we are asked for the dimensions of the fence. We call these x and y, as shown in Figure 3.

To find the objective function, we look for what it is that we are trying to maximize (or minimize). The phrase "largest area" tells us that our object is to *maximize the area,* which is the product of length and width, so our objective function is

$$A = xy.$$

What about the constraints? If there were no constraints, Sam could simply make the area as large as he wanted by choosing x and y to be enormous. However, he has only 100 feet of electrified fencing to work with. This fact means that the sum of the lengths of the three sides must equal 100, or

$$x + 2y = 100.$$

One further point: because x and y represent lengths of sides of the enclosure, neither can be a negative number. Thus, we can rephrase our problem as follows.

Maximize $A = xy$ subject to $x + 2y = 100$,

$x \geq 0$, *and $y \geq 0$.*

Now we solve the constraint for one of the variables. We shall solve for x for a change.

$$x = 100 - 2y$$

Substituting this into the objective function gives

$$A = (100 - 2y)y = 100y - 2y^2,$$

and we have eliminated x. What about the inequalities? One says that $x \geq 0$, but we want to eliminate x from this as well. So again, we substitute for x, getting

$$100 - 2y \geq 0.$$

Solving this inequality for y gives $y \leq 50$. The second inequality says that $y \geq 0$. Thus, the constraints give us

$$A = 100y - 2y^2, \quad 0 \leq y \leq 50.$$

So we have A as a function of y this time, and the domain is $[0, 50]$. We now maximize in the usual way. We have

$$A'(y) = 100 - 4y,$$

so we set

$$100 - 4y = 0, \quad \text{giving } y = 25.$$

Adding in the endpoints, we get this table:

y	0	25	50
$A(y)$	0	1,250	0

Thus, the maximum area of 1,250 square feet occurs when $y = 25$ feet. The length of the other fence is obtained by substituting $y = 25$ into the constraint equation $x + 2y = 100$, giving $x = 50$ feet. Thus, the enclosure with the largest area is 50 feet across and 25 feet deep.

▼ **EXAMPLE 5**   Revenue

The Cuddly Carriage Co. builds baby strollers. Market Research estimates that if it sets the price at p dollars, then the company can sell $q = 300{,}000 - 10p^2$ strollers per year. What price will bring in the largest annual revenue?

SOLUTION   We first identify the unknowns by going to the question. We see that p is our main unknown, but there is also another variable we don't know: the demand q. Thus, we really have two unknowns, p and q. As for the objective function, we look for the quantity that we are trying to maximize (or minimize). We see in the last sentence that our objective is to maximize the annual *revenue,* which is the product of the price per stroller and the number of strollers sold per year. Thus, our objective function is

$$R = pq.$$

We are given the constraint in the form of a demand equation

$$q = 300{,}000 - 10p^2,$$

which is already solved for q. All we have to do, then, is substitute into the objective function.

$$R(p) = (300{,}000 - 10p^2)p = 300{,}000p - 10p^3$$

We should also think about the domain. Price cannot be negative (unless the company is going to *pay* people to take these strollers off its hands!) and neither can demand. Demand will become negative if p gets larger than the point where $q = 0$, that is, $300{,}000 - 10p^2 = 0$, which has a solution $p = 173$. So p must be between 0 and 173, and the domain of $R(p)$ is $[0, 173]$. Now for the calculus:

$$\frac{dR}{dp} = 300{,}000 - 30p^2,$$

so the critical points will be the solutions to $300{,}000 - 30p^2 = 0$. Solving, $30p^2 = 300{,}000$, $p^2 = 10{,}000$, $p = \pm 100$. Now, -100 is outside the domain, so 100 is the only critical point we can use. Together with the endpoints 0 and 173, this gives

p	0	100	173
R	0	20,000,000	170

So, the largest possible annual revenue is \$20,000,000, which is achieved by pricing the strollers at \$100 each.

The following problem is a classic one that shows the power of calculus.

▼ **EXAMPLE 6**   Maximizing Volume

The Cardboard Box Co. is going to make open-topped boxes out of squares of cardboard 30″ on a side by cutting squares out of the corners and folding up the sides. What is the largest volume box it can make this way?

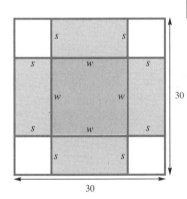

FIGURE 4

SOLUTION   Start by drawing a picture (Figure 4).
    We are asked to find the largest volume, so this will be our objective, but the real unknowns here are the dimensions of the squares cut out and the dimensions of the resulting box. In the picture we have labeled the sides of the squares cut out as s. This will be the height of the box once the sides are folded up. The bottom edges are labeled w in the picture. It is s and w that we need to find. Our objective function, the volume, is

$$V = s \cdot w \cdot w = sw^2.$$

Our constraint comes from the known dimensions of the original square of cardboard, 30″ on a side. Looking at the picture, you can see that

$$2s + w = 30.$$

Now we solve the constraint for one of the variables. It seems easiest to solve for w.

$$w = 30 - 2s$$

Substituting in the objective function,

$$V = s(30 - 2s)^2.$$

Finally, we think about the domain. Looking at the figure, we see that the smallest s can be is 0, and the largest it can be is 15 (half the width of the square). So the domain of $V(s)$ is $[0, 15]$. Using the product rule,

$$\frac{dV}{ds} = (30 - 2s)^2 + 2s(30 - 2s)(-2)$$
$$= (30 - 2s)(30 - 2s - 4s)$$
$$= (30 - 2s)(30 - 6s).$$

We must therefore solve $(30 - 2s)(30 - 6s) = 0$, which gives two critical points, $s = 5$ and $s = 15$. Taking these points together with the endpoints 0 and 15, we get the following table.

s	0	5	15
V	0	2,000	0

We see that the largest volume is 2,000 cubic inches and is achieved when 5-inch squares are cut out of each corner of the original piece of cardboard.

▼ **EXAMPLE 7**   Minimizing Resources

The Metal Can Co. has an order to make cans with a volume of 250 cubic centimeters. What should the dimensions of the cans be in order to use the least metal in their production?

SOLUTION   The unknown quantities are the dimensions of the cans. It is traditional to take as the dimensions of a cylinder the height h and the radius of the base r, so we take these as the unknowns. (See Figure 5.)

As for the objective function, we seek to minimize the total metal used in the can, which is the area of the surface of the cylinder. To calculate this, we imagine removing the circular top and bottom, then cutting vertically and flattening out the hollow cylinder to get a rectangle, as shown in Figure 6. The area of each disc is πr^2, while the area of the rectangular piece is $2\pi rh$. Thus, our objective function is given by the total surface area of a cylinder, which is

$$S = 2\pi r^2 + 2\pi rh.$$

As usual, there is a constraint: the volume must be exactly 250 cubic centimeters. The formula for the volume of a cylinder is $V = \pi r^2 h$, so we get the constraint

$$\pi r^2 h = 250.$$

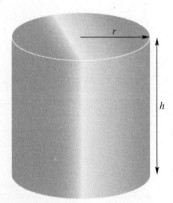

FIGURE 5

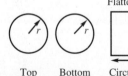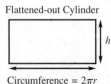

Flattened-out Cylinder

Top   Bottom   Circumference $= 2\pi r$

FIGURE 6

It is easiest to solve this constraint for h in terms of r:

$$h = \frac{250}{\pi r^2}.$$

Substituting in the objective function,

$$S = 2\pi r^2 + 2\pi r \frac{250}{\pi r^2} = 2\pi r^2 + \frac{500}{r}.$$

Now r cannot be negative or 0, but it can become very large (a very wide but very short can could have the right volume). We therefore take the domain of $S(r)$ to be $(0, +\infty)$. Now,

$$S'(r) = 4\pi r - \frac{500}{r^2}.$$

We set this equal to 0 and solve.

$$4\pi r - \frac{500}{r^2} = 0$$

$$4\pi r = \frac{500}{r^2}$$

$$4\pi r^3 = 500$$

$$r^3 = \frac{125}{\pi}$$

So

$$r = \sqrt[3]{\frac{125}{\pi}} = \frac{5}{\sqrt[3]{\pi}} \approx 3.41.$$

There are no other critical points, and there are no endpoints, so we need to choose test points on either side of this critical point.

r	1	3.41	5
S	506	220	257

From this table we see that the minimum amount of metal needed is 220 square centimeters, which occurs when the radius is 3.41 centimeters. We also want to know the height, and for this we go back to the constraint equation.

$$h = \frac{250}{\pi r^2} = \frac{250}{\pi \left(\dfrac{5}{\sqrt[3]{\pi}}\right)^2} = \frac{10}{\sqrt[3]{\pi}} \approx 6.82 \text{ cm}$$

Before we go on... More interesting than the actual numbers is the fact that the height must be exactly twice the radius. This means that the cans will look square when viewed from the side. Are most cans manufactured with those proportions, and if not, why not?

▼ **EXAMPLE 8**   Maximizing Volume

Picky Parcel Service is finicky about the size of the boxes it will accept: The perimeter of the base must be no more than 20 inches, while the perimeter of one side must be no more than 10 inches. What is the largest volume box it will accept?

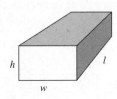

FIGURE 7

SOLUTION   We select the dimensions of the box as our unknowns, even though these are not mentioned explicitly. Thus, we have three unknowns in this problem: the three dimensions of the box l, w, and h as labeled in Figure 7.

Our objective is to maximize the volume

$$V = lwh.$$

Notice that we now need to eliminate *two* variables. To do this, we note that there are two constraints given, on the perimeters of the base and one side, which we will take to be the frontmost side in the picture. Because we want the volume to be as large as possible, we shall take these perimeters to be as large as possible. The stated constraints then translate into the equations

$$2l + 2w = 20$$
$$2h + 2w = 10.$$

The best thing to do now is to write two of the variables in terms of the third. Since w appears in both constraints, it will be easiest to write l and h in terms of w.

$$l = 10 - w$$
$$h = 5 - w$$

Substituting into the objective function, we get

$$V = (10 - w)w(5 - w) = 50w - 15w^2 + w^3.$$

Because none of the dimensions can be negative, the second constraint limits w to be no larger than 5, and so the domain of $V(w)$ is $[0, 5]$.

$$\frac{dV}{dw} = 50 - 30w + 3w^2$$

To solve $3w^2 - 30w + 50 = 0$, we use the quadratic formula, which gives

$$w = \frac{30 \pm \sqrt{900 - 600}}{6} = 5 \pm \frac{5\sqrt{3}}{3}.$$

Of these two solutions, only one is in the correct interval, $w = 5 - 5\sqrt{3}/3 \approx 2.11$.

Together with the endpoints, this gives

w	0	2.11	5
V	0	48.1	0

So the largest acceptable package has a volume of 48.1 cubic inches, with $w = 5 - 5\sqrt{3}/3 \approx 2.11$ inches. The other dimensions will then be $l = 10 - w = 5 + 5\sqrt{3}/3 \approx 7.88$ inches and $h = 5 - w = 5\sqrt{3}/3 \approx 2.88$ inches.

The next example is done with the aid of a graphing calculator.

▼ **EXAMPLE 9**   Labor Resource Allocation

The Gym Sock Company manufactures cotton athletic socks. Production is partially automated through the use of robots. Daily operating costs amount to $50 per laborer and $30 per robot. The number of pairs of socks q the company can manufacture in a day is given by a Cobb-Douglas* production formula

$$q = 50n^{0.6}r^{0.4},$$

where n is the number of laborers and r is the number of robots. Assuming that the company wishes to produce 1,000 pairs of socks per day at a minimum cost, how many laborers and how many robots should it use?

SOLUTION   The unknowns are n, the number of laborers, and r, the number of robots. The objective is to minimize the daily cost,

$$C = 50n + 30r.$$

The constraints are given by the daily quota,

$$1,000 = 50n^{0.6}r^{0.4},$$

and the fact that n and r are nonnegative.
  We solve the constraint equation for one of the variables—let us solve for n.

$$n^{0.6} = \frac{1,000}{50r^{0.4}} = \frac{20}{r^{0.4}}$$

Taking the $1/0.6$ power of both sides gives

$$n = \left(\frac{20}{r^{0.4}}\right)^{1/0.6} = \frac{20^{1/0.6}}{r^{0.4/0.6}} \approx \frac{147.36}{r^{2/3}}.$$

Substituting in the objective equation gives us the cost as a function of r.

$$C = 50\left(\frac{147.36}{r^{2/3}}\right) + 30r$$

$$= 7,368r^{-2/3} + 30r$$

▼ * Cobb-Douglas production formulas were discussed in the section on implicit differentiation in the preceding chapter.

Before minimizing, we graph this cost function on a graphing calculator, obtaining the graph shown in Figure 8. (The ranges of the coordinates are shown in the graph.)

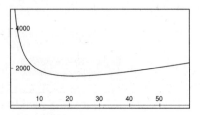

FIGURE 8

From the graph, we see that the cost is minimized when r is approximately 20, and that the minimum occurs at a stationary point. To obtain a more accurate answer, we could either zoom in or set the derivative equal to zero and solve for r. We choose the latter approach. Since

$$C = 7,368r^{-2/3} + 30r,$$

$$\frac{dC}{dr} = -4,912r^{-5/3} + 30.$$

So we must solve

$$-4,912r^{-5/3} + 30 = 0.$$

We can now either solve for r analytically or use our graphing calculator to solve it numerically. Graphing $-4,912r^{-5/3} + 30$ gives us Figure 9.

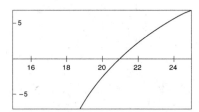

FIGURE 9

The value of r we desire is given by the point where this graph crosses the r-axis. Since r represents numbers of robots, and we are not interested in fractions of a robot, we need only estimate r to the nearest whole number, so we see that $r = 21$.

Now that we know r, we can obtain n by going back to the constraint equation.

$$1,000 = 50n^{0.6}r^{0.4} = 50n^{0.6}(21)^{0.4}.$$

Thus,

$$n^{0.6} = \frac{1,000}{50(21)^{0.4}} \approx 5.9176.$$

Taking reciprocal powers gives

$$n = 5.9176^{(1/0.6)} \approx 19.360.$$

Thus, to minimize daily operating costs, the company should use 19 laborers and 21 robots.

Before we go on... To find the resulting daily operating cost, we can either substitute the values $n = 19$ and $r = 21$ into the equation $C = 50n + 30r$ for cost, or find the cost from the graph in Figure 8. We leave this as an exercise for you.

▶ ___5.2 EXERCISES

Solve the optimization problems in Exercises 1–8.

1. Maximize $P = xy$
with $x + y = 10$.

2. Maximize $P = xy$
with $x + 2y = 40$.

3. Minimize $S = x + y$
with $xy = 9$ and both x and $y > 0$.

4. Minimize $S = x + 2y$
with $xy = 2$ and both x and $y > 0$.

5. Minimize $F = x^2 + y^2$
with $x + 2y = 10$.

6. Minimize $F = x^2 + y^2$
with $xy^2 = 16$.

7. Maximize $P = xyz$
with $x + y = 30$ and $y + z = 30$,
and x, y, and $z \geq 0$.

8. Maximize $P = xyz$
with $x + z = 12$ and $y + z = 12$
and x, y, and $z \geq 0$.

9. For a rectangle with perimeter 20 to have the largest area, what dimensions should it have?

10. For a rectangle with area 100 to have the smallest perimeter, what dimensions should it have?

APPLICATIONS

11. *Fences*   I want to fence in a rectangular vegetable patch. The fencing for the east and west sides costs $4 per foot, while the fencing for the north and south sides costs only $2 per foot. I want to spend $80 on the entire project. What is the largest area that I can enclose?

12. *Fences*   Actually, my vegetable garden abuts my house, so that the house itself forms the northern boundary. The fencing for the southern boundary costs $4 per foot, while the fencing for the east and west sides costs $2 per foot. If I want to spend $80 on the project, what is the largest area that I can enclose this time?

13. *Revenue*   Hercules Films is deciding on the price of the video release of their film "Son of Frankenstein." They estimate that at a price of p dollars, they can sell a total of $q = 200,000 - 10,000p$ copies. At what price will they bring in the largest revenue?

14. *Profit*   Hercules Films is also deciding on the price of the video release of their film "Bride of the Son of Frankenstein." Again, at a price of p dollars they can sell $q = 200,000 - 10,000p$ copies, but each copy costs them $4 to make. What should the price be to bring them the largest profit?

15. *Revenue* (Here we revisit Royal Ruby Retailers— see Chapter 1). The demand for rubies at RRR is given by the equation

$$q = -\frac{4}{3}p + 80,$$

where p is the price RRR charges (in dollars) and q is the number of rubies RRR sells per week. At what price should RRR sell its rubies in order to maximize its weekly revenue? (Try not to look at the answer we obtained in Chapter 1 until you have worked through this exercise.)

16. *Revenue* The consumer demand curve for tissues is given by

$$q = (100 - p)^2, \quad 0 \le p \le 100$$

where p is the price per case of tissues and q is the demand in weekly sales. At what price should tissues be sold in order to maximize the revenue?

17. *Revenue* Assume that the demand function for tuna in a small coastal town is given by

$$p = \frac{500,000}{q^{1.5}},$$

where p is the price (in dollars) per pound of tuna and q is the number of pounds of tuna that can be sold at the price p in one month. Assume that the town's fishery wishes to sell at least 5,000 pounds of tuna per month.

(a) How much should the town's fishery charge for tuna in order to maximize monthly revenue?
(b) How much tuna will it sell per month at that price?
(c) What will its resulting revenue be?

18. *Revenue* Economist Henry Schultz calculated the demand function for corn to be given by

$$p = \frac{6,570,000}{q^{1.3}},$$

where p is the price (in $) per bushel of corn, and q is the number of bushels of corn that could be sold at the price p in one year.[*] Assume that at least 10,000 bushels of corn per year must be sold.

(a) How much should farmers charge per bushel of corn in order to maximize annual revenue?
(b) How much corn can farmers sell per year at that price?
(c) What will the farmers' resulting revenue be?

19. *Revenue* The wholesale price for chicken in the U.S. fell from 25¢ per pound to 14¢ per pound, and at the same time, per capita chicken consumption rose from 22 pounds per year to 27.5 pounds per year.[†] Assuming that the demand for chicken depends linearly on the price, what wholesale price for chicken maximizes revenues for poultry farmers, and what does that revenue amount to?

20. *Revenue* Your underground used book business is doing a booming trade. Your policy is to sell all used versions of *Calculus and You* at the same price (regardless of condition). When you set the price at $10, sales amounted to 120 volumes during the first week of classes. The following semester, you set the price at $30 and sold not a single book. Assuming that the demand for books depends linearly on the price, what price gives you the maximum revenue, and what does that revenue amount to?

21. *Profit* As we have seen on several occasions, the demand for rubies at RRR is given by the equation

$$q = -\frac{4}{3}p + 80,$$

where p is the price RRR charges (in dollars) and q is the number of rubies RRR sells per week. Assuming that due to extraordinary market conditions RRR can obtain rubies for $25 each, how much should it charge per ruby to make the largest possible weekly profit, and what will that profit be?

22. *Profit* The consumer demand curve for tissues is given by

$$q = (100 - p)^2, \quad 0 \le p \le 100,$$

where p is the price per case of tissues and q is the demand in weekly sales. If tissues cost $30 per case, at what price should tissues be sold for the largest possible weekly profit, and what will that profit be?

▼ [*] Based on data for the period 1915–1929. Source: Henry Schultz, *The Theory and Measurement of Demand* (as cited in *Introduction to Mathematical Economics* by A. L. Ostrosky, Jr., and J.V. Koch (Prospect Heights, Ill.: Waveland Press, 1979.)

[†] Data are provided for the years 1951–1958. Source: U.S. Department of Agriculture, *Agricultural Statistics.*

23. *Profit* A demand equation for your company's virtual reality video headsets is given by

$$p = \frac{1,000}{q^{0.3}},$$

where q is the total number of headsets that your company can sell in a week at a price of p dollars. The total manufacturing and shipping cost amounts to $100 per headset.

(a) What is the largest profit your company can make in a week, and how many headsets will your company sell at this level of profit? (Give answers to the nearest whole number.)

(b) How much, to the nearest $1, should your company charge per headset for the maximum profit?

24. *Profit* Due to sales by a competing company, your company's sales of virtual reality video headsets have dropped, and your financial consultant revises the demand equation to

$$p = \frac{800}{q^{0.35}},$$

where q is the total number of headsets that your company can sell in a week at a price of p dollars. The total manufacturing and shipping cost still amounts to $100 per headset.

(a) What is the largest profit your company can make in a week, and how many headsets will your company sell at this level of profit? (Give answers to the nearest whole number.)

(b) How much, to the nearest $1, should your company charge per headset for the maximum profit?

25. *Box Design* The Chocolate Box Co. is going to make open-topped boxes out of $6'' \times 16''$ rectangles of cardboard by cutting squares out of the corners and folding up the sides. What is the largest-volume box it can make this way?

26. *Box Design* A packaging company is going to make open-topped boxes with square bases that hold 108 cubic centimeters. What are the dimensions of the box that can be built with the least material?

27. *Asset Appreciation* As the financial consultant to a classic auto dealership, you estimate that the total value of its collection of 1959 Chevrolets and Fords is given by the formula

$$v = 300,000 + 1,000t^2,$$

where t is the number of years from now. You anticipate an inflation rate running continuously at 5% per year, so that the discounted (present) value of an item that will be worth v in t years' time is given by

$$p = ve^{-0.05t}.$$

When would you advise the dealership to sell the vehicles in order to maximize their discounted value?

28. *Plantation Management* The value of a fir tree in your plantation increases with the age of the tree according to the formula

$$v = \frac{20t}{1 + 0.05t},$$

where t is the age of the tree in years. Given an inflation rate running continuously at 5% per year, the discounted (present) value of a newly planted seedling is given by

$$p = ve^{-0.05t}.$$

At what age (to the nearest year) should you harvest your trees in order to ensure the greatest possible discounted value?

29. *Marketing Strategy* The Feature Software Co. has a dilemma. Its new program, Doors 3.0, is almost ready to go on the market. However, the longer the company works on it, the better they can make it and the more they can charge for it. The company's marketing analysts estimate that if they delay t days they can set the price at $100 + 2t$. On the other hand, the longer they delay, the more market share they will lose to their main competitor (see the next exercise) so that if they delay t days they will be able to sell $400,000 - 2,500t$ copies of the program. How many days should it delay the release in order to get the largest revenue?

30. *Marketing Strategy* Feature Software's main competitor is Newton Software, and Newton is in a similar predicament. Its product, Walls 5.0, could be sold now for $200, but for each day they delay they could increase the price by $4. On the other hand, they could sell 300,000 copies now, but each day they wait will cut their sales by 1,500. How many days should it delay the release in order to get the largest revenue?

31. *Agriculture* The fruit yield per tree in an orchard containing 50 trees is 100 pounds per tree each year. Due to crowding, the yield decreases by 1 pound per season for every additional tree planted. How many additional trees should be planted for a maximum total annual yield?

32. *Agriculture*   Two years ago, your orange orchard contained 50 trees and yielded 75 bags of oranges. Last year, you sold ten of the trees and noticed that the total yield increased to 80 bags. Assuming that the yield of each tree depends linearly on the number of trees in the orchard, what should you do this year in order to maximize your yield?

33. *Average Cost*   A cost function for the manufacture of portable CD players is given by

$$C(x) = \$150,000 + 20x + \frac{x^2}{10,000},$$

where x is the number of CD players manufactured. Interpret each term in this formula. How many CD players should be manufactured in order to minimize average cost? What is the resulting average cost of a CD player? (Give your answer to the nearest dollar.)

34. *Average Cost*   Repeat the preceding exercise using the revised cost function

$$C(x) = \$150,000 + 20x + \frac{x^2}{100}.$$

35. *Pollution Control*   The cost of controlling emissions at a firm goes up rapidly as the amount of emissions reduced goes up. Here is a possible model:

$$C(q) = 4,000 + 100q^2,$$

where q is the reduction in emissions (in pounds of pollutant per day) and C is the daily cost (in dollars) of this reduction. Government clean-air subsidies amount to $500 per pound of pollutant removed. How many pounds of pollutant should the company remove each day in order to minimize *net* cost (cost minus subsidy)?

36. *Pollution Control*   Repeat the preceding exercise using the following data:

$$C(q) = 2,000 + 200q^2,$$

with government subsidies amounting to $100 per pound of pollutant removed per day.

37. *Luggage Dimensions*   Fly-by-Night Airlines has a peculiar rule about luggage: the length and width of a bag must add to 45 inches, while the width and height must also add to 45 inches. What are the dimensions of the bag with largest volume that it will accept?

38. *Luggage Dimensions*   Fair Weather Airlines has a similar rule. It will accept only bags for which the sum of the length and width is 36 inches, while the sum of length, height, and twice the width is 72 inches. What are the dimensions of the bag with largest volume that it will accept?

39. *Package Dimensions*   The U.S. Postal Service (USPS) will accept only packages with a length plus girth no more than 108 inches.* (See figure.)

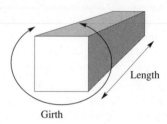

Girth

Length

Assuming that the front face of the package (as shown in the figure) is square, what is the largest-volume package that the USPS will accept?

40. *Package Dimensions*   The *United Parcel Service* (UPS) will only accept packages with a length no more than 108 inches and length plus girth no more than 130 inches*. (See figure above.) Assuming that the front face of the package (as shown in the figure) is square, what is the largest volume package that UPS will accept?

41. *Average Profit*   The Feature Software Company sells its graphing program, Dogwood, with a volume discount: if a customer buys x copies, then they pay[†] $\$500\sqrt{x}$. It cost the company $10,000 to develop the program and $2 to manufacture each copy. If just one customer buys all the copies of Dogwood, how many copies must the customer buy for Feature Software's average profit per copy to be maximized? How are average profit and marginal profit related at this number of copies?

42. *Average Profit*   Repeat the preceding exercise with the charge to the customer being $\$600\sqrt{x}$ and the cost to develop the program being $9,000.

▼   * The data were current at the time of this writing.
   [†] This is similar to the site license charge for the program Maple®.

43. *Prison Population* The prison population of the U.S. followed the curve

$$N(t) = 0.028234t^3 - 1.0922t^2 + 13.029t$$
$$+ 146.88 \quad (0 \le t \le 39)$$

in the years 1950–1989. Here t is the number of years since 1950 and N is the number of prisoners in thousands.* When, to the nearest year, was the prison population decreasing most rapidly, and when was it increasing most rapidly?

44. *Test Scores* Combined SAT scores in the United States can be approximated by

$$T(t) = -0.01085t^3 + 0.5804t^2 - 10.12t$$
$$+ 962.4 \quad (0 \le t \le 22)$$

in the years 1967–1991. Here t is the number of years since 1967 and T is the combined SAT score average for the U.S.[†] Based on this model, when (to the nearest year) was the average SAT score decreasing most rapidly? When was it increasing most rapidly?

45. *Embryo Development* The oxygen consumption of a bird embryo increases from the time the egg is laid through the time the chick hatches. In the case of a typical galliform bird, the oxygen consumption (in milliliters per hour) can be approximated by

$$c(t) = -0.00271t^3 + 0.137t^2 - 0.892t$$
$$+ 0.149 \quad (8 \le t \le 30),$$

where t is the time (in days) since the egg was laid.[‡] (An egg will typically hatch at around $t = 28$.) When, to the nearest day, is $c'(t)$ a maximum? What does the answer tell you?

46. *Embryo Development* The oxygen consumption of a turkey embryo increases from the time the egg is laid through the time the chick hatches. In the case of

a brush turkey, the oxygen consumption (in milliliters per hour) can be approximated by

$$c(t) = -0.00118t^3 + 0.119t^2 - 1.83t$$
$$+ 3.972 \quad (20 \le t \le 50)$$

where t is the time (in days) since the egg was laid.[‡] (An egg will typically hatch at around $t = 50$.) When, to the nearest day, is $c'(t)$ a maximum? What does the answer tell you?

47. *Minimizing Resources* Basic Buckets, Inc., has an order for plastic buckets holding 5,000 cubic centimeters. The buckets are open-topped cylinders, and the company wants to know what dimensions will use the least plastic per bucket. (The volume of an open-topped cylinder with height h and radius r is $\pi r^2 h$, while the surface area is $\pi r^2 + 2\pi rh$.)

48. *Optimizing Capacity* Basic Buckets would like to build a bucket with a surface area of 1,000 square centimeters. What is the volume of the largest bucket it can build? (See the previous exercise.)

The use of either a graphing calculator or graphing computer software is required for Exercises 49–54.

49. *Education* In 1991, the expected income of an individual depended on his or her educational level, according to the following formula.

$$I(n) = 2,928.8n^3 - 115,860n^2 + 1,532,900n$$
$$- 6,760,800 \quad (12 \le n \le 15)$$

Here, n is the number of school years completed, and $I(n)$ is the individual's expected income.[§] Using [12, 15] as the domain, use technology to locate and classify the absolute extrema of $I'(n)$. Interpret the results.

▼ * The model is the authors'. Source for data: *Sourcebook of Criminal Justice Statistics,* 1990, p. 604.

[†] The model is the authors'. Source for data: Educational Testing Service.

[‡] The model approximates graphical data published in the article "The Brush Turkey" by Roger S. Seymour, *Scientific American,* December, 1991, pp. 108–14.

[§] The model is a best-fit cubic based on Table 358, U.S. Department of Education, *Digest of Education Statistics, 1991* (Washington, DC: Government Printing Office, 1991).

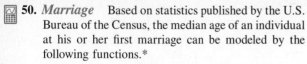

50. *Marriage* Based on statistics published by the U.S. Bureau of the Census, the median age of an individual at his or her first marriage can be modeled by the following functions.*

Females: $F(n) = 0.000023453n^3 - 0.0026363n^2 + 0.050582n + 21.766$

Males: $M(n) = 0.000023807n^3 - 0.0025184n^2 + 0.015754n + 0.015754n + 25.966$

n = number of years since 1890; $0 \le n \le 102$

(a) Using $[1, 102]$ as the domain, use technology to locate the local extrema of both functions. Round answers to the nearest integer, and interpret the results.

(b) What do the local extrema of $F'(n)$ and $M'(n)$ tell you?

(c) What do the local extrema of $M(n) - F(n)$ tell you?

51. *Asset Appreciation* You manage a small antique store that owns a collection of Louis XVI jewelry boxes. Their value v is increasing according to the formula

$$v = \frac{10000}{1 + 500e^{-0.5t}},$$

where t is the number of years from now. You anticipate an inflation rate of 5% per year, so that the present value of an item that will be worth $\$v$ in t years' time is given by

$$p = v(1.05)^{-t}.$$

When (to the nearest year) should you sell the jewelry boxes in order to maximize their present value? How much (to the nearest constant dollar) will they be worth at that time?

52. *Harvesting Forests* The following equation models the approximate volume in cubic feet of a typical Douglas fir tree of age t years:[†]

$$v = \frac{22514}{1 + 22514t^{-2.55}}.$$

The lumber will be sold at $\$10$ per cubic foot, and you

do not expect the price of lumber to appreciate in the foreseeable future. On the other hand, you anticipate a general inflation rate of 5% per year, so that the present value of an item that will be worth $\$v$ in t years' time is given by

$$p = v(1.05)^{-t}.$$

At what age (to the nearest year) should you harvest a Douglas fir tree in order to maximize its present value? How much (to the nearest constant dollar) will a Douglas fir tree be worth at that time?

53. *Resource Allocation* Your automobile assembly plant has a Cobb-Douglas production function given by

$$q = x^{0.4}y^{0.6},$$

where q is the number of automobiles it produces per year, x is the number of employees, and y is the daily operating budget (in dollars). Annual operating costs amount to an average of $\$20,000$ per employee plus the operating budget of $\$365y$. Assume that you wish to produce 1,000 automobiles per year at a minimum cost. How many employees should you hire?

54. *Resource Allocation* Repeat the preceding exercise using the production formula

$$q = x^{0.5}y^{0.5}.$$

55. *Revenue (based on a question in the GRE economics test[‡])* If total revenue (TR) is specified by $TR = a + bQ - cQ^2$, where Q is quantity of output and a, b, and c are positive parameters, then TR is maximized for this firm when it produces Q equal to:
(a) $b/2ac$ **(b)** $b/4c$ **(c)** $(a + b)/c$
(d) $b/2c$ **(e)** $c/2b$

56. *Revenue (based on a question in the GRE economics test)* If total demand (Q) is specified by $Q = -aP + b$, where P is unit price and a and b are positive parameters, then total revenue is maximized for this firm when it charges P equal to:
(a) $b/2a$ **(b)** $b/4a$ **(c)** a/b **(d)** $a/2b$
(e) $-b/2a$.

▼ *Source: U.S. Bureau of the Census, "Marital Status and Living Arrangements: March 1992," Current Population Reports, *Population Characteristics*, Series P-20, No. 468, March 1992, p. vii.

[†] The model is the authors' and is based on data in *Environmental and Natural Resource Economics*, Third Edition, by Tom Tietenberg (New York: HarperCollins, 1992), p. 282.

[‡] Source: GRE Economics Test, by G. Gallagher, G. E. Pollock, W. J. Simeone, G. Yohe (Piscataway, N.J.: Research and Education Association, 1989).

COMMUNICATION AND REASONING EXERCISES

57. Explain why the following problem is uninteresting: "A packaging company wishes to make cardboard boxes with open tops by cutting square pieces from the corners of a square sheet of cardboard and folding up the sides. What is the box with the least surface area they can make this way?"

58. Explain why finding the production level that minimizes a cost function is frequently uninteresting.

59. If demand q decreases as price p increases, what does the minimum value of dq/dp measure?

60. Explain why the following problem is uninteresting: "A cost function for the manufacture of portable CD players is given by

$$C(x) = \$150{,}000 + 20x + \frac{x^2}{10{,}000},$$

where x is the number of CD players manufactured. How many CD players should be manufactured in order to minimize total cost?"

61. Explain how you would solve an optimization problem of the following form. "Maximize $P = f(x, y, z)$ subject to $z = g(x, y)$ and $y = h(x)$."

62. Explain how you would solve an optimization problem of the following form. "Maximize $P = f(x, y, z)$ subject to $z = g(x, y)$, $x^2 + y^2 = 1$."

▶ ═══ **5.3** THE SECOND DERIVATIVE: ACCELERATION AND CONCAVITY

THE SECOND DERIVATIVE AS ACCELERATION

Now that we have seen some of the power of the derivative, we take a look at the derivative of the derivative. For example, if $f(x) = x^3 + 2x^2$, then its derivative is $f'(x) = 3x^2 + 4x$. Since this is also a function of x, we can take the derivative once again. When we do this, we have the derivative of the derivative, or the **second derivative,**

$$f''(x) = 6x + 4.$$

Q What does the second derivative mean?

A The answer to that question is the subject of this entire section.

First, let us go back to the interpretation of the derivative in terms of velocity. Suppose, for example, that the position (i.e., odometer reading) of a car is given by $s(t) = t^3 + 2t^2$ miles, where t is the time in hours. Then, as we saw, the derivative $s'(t)$ gives us the velocity—a measure of *how fast the odometer reading is increasing*—in miles per hour. Thus,

$$v(t) = s'(t) = 3t^2 + 4t.$$

Notice that v also shows up as the reading on the speedometer. Now how fast is *that* reading changing? To measure that, we must take the derivative of v: in other words, the second derivative of s:

$$a(t) = v'(t) = s''(t) = 6t + 4.$$

Here, $a(t)$ is the rate of change of the velocity $v(t)$, and is called the *acceleration*. Thus, just as the velocity measures how fast the odometer is going up (in miles per hour), the acceleration measures how fast the speedometer is changing (in miles per hour per hour).

> **ACCELERATION**
>
> The **acceleration** of a moving object is the derivative of its velocity, or the second derivative of its position.

Although cars don't usually come equipped with "accelerometers," we could easily set one up by hanging an enclosed pendulum somewhere in the car. As the car accelerates, the pendulum would move away from the vertical position. The greater the angle, the greater the acceleration.*

▶ NOTE   Since acceleration is the derivative of velocity, its units are given by units of velocity per unit of time: for example, (miles/hour)/hour, or miles/hour². ◀

▼ **EXAMPLE 1**   Acceleration Due to Gravity

According to the laws of physics, a particle falling in a vacuum from an initial rest position under the influence of gravity will travel a distance of

$$s = 16t^2 \text{ feet}$$

in t seconds. Find its acceleration.

SOLUTION   First, the velocity is given by

$$v(t) = s'(t) = 32t \text{ ft/s}.$$

Hence, the acceleration is given by

$$a(t) = s''(t) = 32 \text{ ft/s}^2.$$

This means that the velocity is increasing at a rate of 32 ft/s every second. We refer to 32 ft/s² as the **acceleration due to gravity**. If we ignore air resistance, then all falling bodies, no matter what their weight, will fall with this acceleration.[†]

Before we go on... This was one of Galileo's most important discoveries. In very careful experiments using balls rolling down inclined planes, he discovered first that acceleration due to gravity is constant, and second that it does not depend on the weight or composition of the object falling.[‡] A famous, though probably apocryphal, story about his experiments has him

---

▼ * Note that we also *feel* the effect of acceleration by being pushed back against the seat. We do not, however, feel the effect of velocity. Riding on a smooth road at 30 mph feels about the same as flying in a plane at 500 mph. This observation is one of the ideas that leads to the Theory of Relativity.

[†] There is nothing unique about the number 32. On other planets the number is different. For example, on Jupiter, the acceleration due to gravity is about three times as large.

[‡] An interesting aside: Galileo's experiments depended on getting extremely accurate timings. Since the timepieces of the day were very inaccurate, he used the most accurate time measurement he could: he sang and used the beat as his stopwatch.

dropping cannonballs of different weights off the leaning tower of Pisa to prove his point.*

▼ **EXAMPLE 2** Acceleration in Air

Since we do not live in a vacuum, we need a more realistic model of the acceleration of falling bodies. The distance traveled by a falling body in air might be given by a formula something like this:

$$s = 2(e^{-4t} + 4t - 1) \text{ feet.}$$

Find the acceleration as a function of time t.

SOLUTION   We have

$$v = s'(t) = 2(4 - 4e^{-4t}) = 8(1 - e^{-4t}) \text{ ft/s}$$

and so

$$a = 32e^{-4t} \text{ ft/s}^2.$$

Before we go on... Notice that the acceleration starts at 32 ft/sec², but then decreases exponentially to zero as time goes on. Also notice that as $t \to +\infty$, $v \to 8$ ft/s. This is called the **terminal** or **limiting velocity.** When a body falls in a medium such as air, its velocity increases more and more slowly until it approaches terminal velocity, which it never exceeds. The above functions do not apply to every object that happens to be falling in air. For example, the terminal velocity of a falling rock is far larger than the terminal velocity of a falling piece of paper. (If this weren't true, there would be no point to parachutes.)

 Figure 1 shows a graphing calculator plot of the velocity v as a function of time.

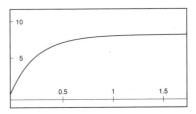

FIGURE 1

Notice that the object is close to its terminal velocity after about 1 second.

▼ *A true story: The point was made again during the Apollo 15 mission to the moon (July 1971) when astronaut David R. Scott dropped a feather and a hammer from the same height. The moon has no atmosphere, so the two hit the surface of the moon simultaneously.

▼ | **EXAMPLE 3** | Acceleration of Demand

For the first 15 months after its introduction, the total sales of a new video game grows exponentially and can be modeled by the curve

$$S(t) = 20e^{0.4t}.$$

Later, after about 25 months, the total sales follows more closely the curve

$$S(t) = 100,000 - 20e^{17-0.4t}.$$

How fast is the total sales accelerating after 10 months, and how fast is it accelerating after 30 months?

SOLUTION   To obtain the acceleration of a quantity, we must take its second derivative. During the first 15 months, the derivative of sales will be

$$S'(t) = 8e^{0.4t}$$

and so the second derivative will be

$$S''(t) = 3.2e^{0.4t}.$$

Thus, after 10 months the total sales will be

$$S(10) = 20e^4 \approx 1,092,$$

the rate of change of sales will be

$$S'(10) = 8e^4 \approx 437,$$

and the acceleration of sales will be

$$S''(10) = 3.2e^4 \approx 175.$$

What do these numbers mean? By the end of the 10th month, 1,092 video games will have been sold. The game will continue to sell at the rate of 437 games per month. This rate of sales is increasing by 175 sales per month per month.

To analyze the sales after 30 months, we must use the second formula for sales,

$$S(t) = 100,000 - 20e^{17-0.4t}.$$

The derivative is

$$S'(t) = 8e^{17-0.4t}$$

and the second derivative is

$$S''(t) = -3.2e^{17-0.4t}.$$

After 30 months,

$$S(30) = 100,000 - 20e^{17-12} \approx 97,032,$$
$$S'(30) = 8e^{17-12} \approx 1,187,$$

and

$$S''(30) = -3.2e^{17-12} \approx -475.$$

By the end of the 30th month, 97,032 video games will have been sold, the game will be selling at a rate of 1,187 games per month, and the rate of sales will be *decreasing* by 475 games per month per month.

GEOMETRIC INTERPRETATION OF THE SECOND DERIVATIVE

We now look at what the second derivative shows us about the graph of a function. We already know that the ordinary derivative, or **first derivative,** of a function measures the steepness of its graph as the slope of the tangent. The larger the absolute value of the first derivative, the steeper the curve. In order to interpret the second derivative, let us look at the two graphs shown in Figure 2.

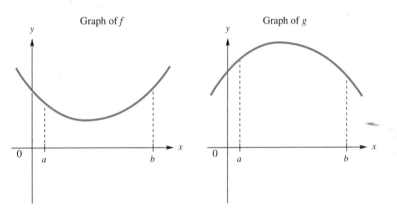

FIGURE 2

    The main difference between these two graphs is in the directions that they curve. We say that the graph of f is **concave up** and the graph of g is **concave down.**

Q How can we determine whether the graph of a function is concave up or concave down?

A If the *second derivative $f''(x)$* is *positive* for x between a and b, then the graph of f is concave up. If the second derivative is *negative* for x between a and b, then the graph of f is concave down.

    To see why, look at the graph of f. As x increases from a to b, the slope of the tangent to the graph *increases* from approximately -1 to approximately $+1$. On the other hand, the slope of the tangent to the graph of g *decreases* from $+1$ to -1. Thus, if we plot the slope as a function of x for each

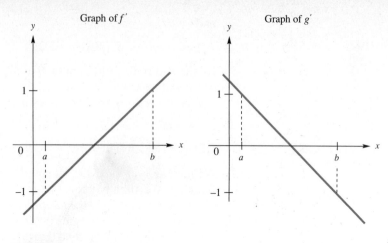

FIGURE 3

of these graphs (that is, the derivative as a function of x) we obtain graphs similar to those shown in Figure 3.

Now the second derivatives $f''(x)$ and $g''(x)$ measure the slopes of the graphs drawn in Figure 3. The graph of f' on the left has positive slope, so $f''(x)$ will be positive for every value of x between a and b. On the other hand, the graph of g' has negative slope, so $g''(x)$ will be negative for every x between a and b.

In short, we have the following:

Concave up; slope increasing

Concave down; slope decreasing

FIGURE 4

THE SECOND DERIVATIVE AND CONCAVITY

A curve is concave up if its slope is increasing, in which case the second derivative will be positive.

A curve is concave down if its slope is decreasing, in which case the second derivative will be negative.

◄

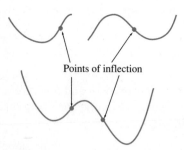

Points of inflection

FIGURE 5

Figure 4 shows a few more curves that are concave up and concave down.

Now we look at points where the curve switches from being concave up to concave down, or *vice versa*. (See Figure 5.) Such points are called **points of inflection.**

To locate these, we observe the following: when the graph switches from concave up to concave down, the second derivative must switch from being positive to being negative. This change can occur only if the second derivative is either zero or not defined at such a point. (The same argument applies to the case of switching from concave down to concave up.) Thus, we have the following.

> **LOCATING POINTS OF INFLECTION**
>
> To locate candidates for points of inflection,
>
> **(a)** set the second derivative $f''(x)$ equal to zero and solve for x (find the stationary points of the derivative f'), and
> **(b)** locate all points in the domain of f where f'' is not defined (find the singular points of the derivative f').

In Example 4 below we shall see how to determine whether a candidate really is a point of inflection.

Before getting to that example, we observe the following. Suppose that we have a stationary point at $x = a$ and do not want to look at other points to see whether it is a maximum or minimum (or neither). If the curve happens to be concave down in the vicinity of $x = a$, then there must be a local maximum at $x = a$ (draw yourself a picture). Thus, if the second derivative is negative when evaluated at $x = a$, we must have a minimum there. Similarly, if the second derivative is positive when evaluated at $x = a$, then the curve must be concave up at $x = a$, so we must have a local minimum there. This gives us another use for the second derivative:

> **SECOND-DERIVATIVE TEST FOR MAXIMA AND MINIMA**
>
> If $x = a$ is a stationary point, evaluate $f''(a)$. If the answer is positive, then there is a local minimum at $x = a$. If the answer is negative, then there is a local maximum at $x = a$. If the answer is zero, then the test is inconclusive, and you will need to look at test points or use a graphing calculator to obtain a conclusive result.

Let us put these discussions to use.

▼ **EXAMPLE 4**

Locate all the local extrema and points of inflection of the curve $y = 3x^4 - 16x^3 + 18x^2$.

SOLUTION   First, we locate the candidates for local extrema in the usual way:

$$f(x) = 3x^4 - 16x^3 + 18x^2,$$

so

$$f'(x) = 12x^3 - 48x^2 + 36x = 12x(x^2 - 4x + 3)$$
$$= 12x(x - 1)(x - 3).$$

Thus,

$$f'(x) = 0 \quad \text{when } x = 0, 1, \text{ or } 3,$$

giving us three stationary points. There are no singular points or endpoints. Let us use the second-derivative test to find out which are maxima and which of the stationary points are minima.

$$f''(x) = 36x^2 - 96x + 36 = 12(3x^2 - 8x + 3)$$

To find out whether a stationary point is a local maximum or minimum, we evaluate the second derivative at that point.

$x = 0$:
$f''(0) = 12(3(0)^2 - 8(0) + 3) = 36$, a positive number. Thus, we have a local minimum at $x = 0$.

$x = 1$:
$f''(1) = 12(3(1)^2 - 8(1) + 3) = -24$, a negative number. Thus, we have a local maximum at $x = 1$.

$x = 3$:
$f''(3) = 12(3(3)^2 - 8(3) + 3) = 72$, a positive number. Thus, we have another local minimum at $x = 3$.

Before looking for the points of inflection, we calculate the y-coordinates of the local extrema and sketch the curve to get a rough idea of what it looks like. (See Figure 6. We already know the minima and maxima, so we don't need any test points.)

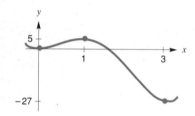

FIGURE 6

x	0	1	3
$f(x)$	0	5	-27

Looking at the curve, we see that there must be at least two points of inflection: one between $x = 0$ and $x = 1$ and another between $x = 1$ and $x = 3$. Alternatively, we know these points exist from the calculations $f''(0) > 0$, $f''(1) < 0$, and $f''(3) > 0$. We locate the points of inflection by setting the second derivative equal to zero and solving for x (there aren't any points where f'' is not defined).

$$f''(x) = 12(3x^2 - 8x + 3)$$

The second derivative is zero when

$$3x^2 - 8x + 3 = 0.$$

This quadratic does not factor with integer coefficients, so we need to use the quadratic formula. We get

$$x = \frac{8 \pm \sqrt{28}}{6} = \frac{4}{3} \pm \frac{\sqrt{7}}{3}.$$

Now we have two candidates for points of inflection: one at $x = 4/3 - \sqrt{7}/3 \approx 0.451$, and another at $x = 4/3 + \sqrt{7}/3 \approx 2.215$. These are the two points of inflection we expected to find.

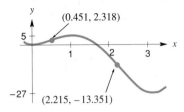

FIGURE 7

To find the y-coordinates of these points, we must, as usual, substitute into the original function $f(x)$: $f(0.451) \approx 2.318$ and $f(2.215) \approx -13.351$. These points are shown in Figure 7.

Before we go on... Recall that setting $f'(x) = 0$ and solving for x only gives us *candidates* for local extrema. Similarly, setting $f''(x) = 0$ and solving for x only gives us candidates for the points of inflection. In this example, we knew by looking at the graph that there had to be two points of inflection: one between 0 and 1 and another between 1 and 3. In general, we need to verify that a candidate for a point of inflection is indeed a point of inflection by checking that f'' changes sign at that point. Actually, testing for a point of inflection is very much like testing for extrema. There is a third-derivative test we could use, but it is unnecessary in most examples. Usually we have plenty of information about the graph by the time we begin testing for points of inflection, or we can obtain enough information with a graphing calculator.

Q Now that we have the second-derivative test for maxima and minima, we can forget about those other, more primitive, approaches (such as using test points). Right?

A Wrong. For example, try the second-derivative test on $f(x) = x^3$ and on $g(x) = x^4$. Both functions have stationary points at $x = 0$, but the second-derivative test tells us nothing about either function, since $f''(0) = 0$ and $g''(0) = 0$. In fact, g has an absolute minimum at $x = 0$, while f has neither. When the second derivative is 0, we need the other approaches.

The second derivative can be written in differential notation as

$$y'' = \frac{d^2 y}{dx^2} = \frac{d}{dx}\left(\frac{dy}{dx}\right) = \frac{d^2}{dx^2}(y).$$

We shall use this notation in the next example.

▼ **EXAMPLE 5**

Find all points of inflection of the curve $y = xe^{-x}$ for x in $[0, +\infty)$.

SOLUTION   Before taking derivatives, we can use a graphing calculator or a computer to get an idea of where any points of inflection might be. Figure 8 shows a graphing calculator plot of the curve for $0 \le x \le 5$.

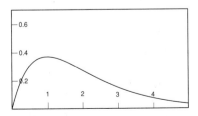

FIGURE 8

It looks as if there is one point of inflection, around $x = 2$. Now let us see what the derivative reveals.

$$\frac{dy}{dx} = e^{-x} - xe^{-x} \qquad \text{by the product rule}$$

$$= (1 - x)e^{-x}$$

(Using this derivative we can say that the maximum we can see in the graph is located at $x = 1$.)

$$\frac{d^2y}{dx^2} = -e^{-x} - (1 - x)e^{-x}$$

$$= (x - 2)e^{-x}$$

Setting the second derivative equal to 0 gives

$$(x - 2)e^{-x} = 0.$$

But, if a product of two quantities is zero, at least one of them must be zero. Since e^{-x} is never zero, it must be that the other quantity, $x - 2$, is zero. Thus,

$$x - 2 = 0,$$

so

$$x = 2.$$

This point is the only candidate for a point of inflection. But is this point really a point of inflection? Looking at Figure 8, we can see that the curve is concave down at the maximum at $x = 1$ and concave up at $x = 3$. (We could confirm this by evaluating the second derivative at these points, but let us trust our eyes and our calculator.) Therefore, there is a single point of inflection at $x = 2$.

Before we go on... Another way of telling if there is really a point of inflection is to see if $\dfrac{d^2y}{dx^2}$ changes sign at $x = 2$, which we can do easily by graphing this second derivative, as was done to get Figure 9.

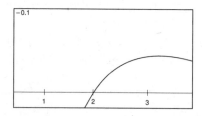

FIGURE 9

This graph (the graph of $(x - 2)e^{-x}$) clearly shows that $\dfrac{d^2y}{dx^2}$ changes sign at $x = 2$, verifying that there is a point of inflection at $x = 2$ in the graph of y.

▼ ■ **EXAMPLE 6**   Demand for a New Product

The demand for a new product over time often follows a logistic curve of the form

$$Q = \frac{a}{1 + e^{-k(t-c)}},$$

where Q is the total number of items sold up to time t (months, say), and a, c, and k are constants that depend on the particular market. For simplicity, let us look at the example with $a = c = k = 1$:

$$Q = \frac{1}{1 + e^{-(t-1)}}.$$

A graphing calculator plot of this demand equation is shown in Figure 10.

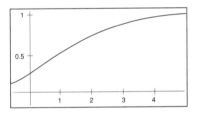

FIGURE 10

From the graph, we see that the sales rise fairly rapidly at first and then level off by about $t = 5$. Notice also that there is a point of inflection at about $t = 1$. Exactly where is this point of inflection, and what does it tell us about sales?

SOLUTION   We locate the point of inflection by setting $\dfrac{d^2Q}{dt^2} = 0$ and solving for t.

$$\frac{dQ}{dt} = \frac{e^{-(t-1)}}{(1 + e^{-(t-1)})^2} \qquad \text{by the quotient rule}$$

Taking the second derivative,

$$\frac{d^2Q}{dt^2} = \frac{(-e^{-(t-1)})(1 + e^{-(t-1)})^2 - e^{-(t-1)} \cdot 2(1 + e^{-(t-1)})(-e^{-(t-1)})}{(1 + e^{-(t-1)})^4}$$

$$= \frac{e^{-(t-1)}(1 + e^{-(t-1)})(-1 - e^{-(t-1)} + 2e^{-(t-1)})}{(1 + e^{-(t-1)})^4}$$

$$= \frac{e^{-(t-1)}(e^{-(t-1)} - 1)}{(1 + e^{-(t-1)})^3}.$$

Equating the second derivative to zero gives

$$\frac{e^{-(t-1)}(e^{-(t-1)} - 1)}{(1 + e^{-(t-1)})^3} = 0,$$

so that

$$e^{-(t-1)}(e^{-(t-1)} - 1) = 0.$$

Thus, either the first or second factor must be zero. Since the first factor can never be zero (why?) it follows that the second factor must be zero:

$$e^{-(t-1)} - 1 = 0,$$

or

$$e^{-(t-1)} = 1.$$

Because the unknown is in the exponent, we solve by taking the natural logarithm of both sides.

$$\ln(e^{-(t-1)}) = \ln(1)$$

That is,

$$-(t - 1)\ln(e) = 0,$$

or

$$-(t - 1) = 0.$$

Thus, $t = 1$ is the only solution, so the (only) point of inflection is the point on the curve where $t = 1$. By substituting into the equation for Q, we obtain the Q-coordinate.

$$Q = \frac{1}{1 + e^{-(1-1)}} = \frac{1}{1 + 1} = \frac{1}{2}.$$

Thus the point of inflection occurs at $(1, \frac{1}{2})$. What does this tell us about the sales? First, since Q represents total sales, the derivative of Q represents sales per month. Thus, dQ/dt measures *how fast the new product is selling*—a measure of the *demand* for the product. (We have often called this quantity q.)

If you look at the graph in Figure 10, you will notice that the derivative (slope) increases as t goes from 0 to 1 and then begins to decrease. In other words, the rate at which the product is selling increases from $t = 0$ to $t = 1$ and then starts to decrease. Thus, *the rate at which the product is selling reaches a maximum at $t = 1$*. After that, the rate of sales begins to decrease. (It is for this reason that we refer to $t = 1$ as *the point of diminishing returns*.) We can see directly how the rate dQ/dt reaches a maximum at $t = 1$ by plotting it (Figure 11).

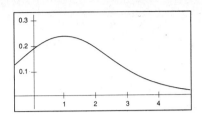

FIGURE 11

Before we go on... Notice something that came up in this discussion: *at a point of inflection, the derivative is either a local maximum or a local minimum.* In fact, if we were to look for extrema of f', we would do so by setting its derivative, f'', equal to zero—which is exactly what we do to find the points of inflection of f.

If we take the derivative of the second derivative, we get the *third derivative*, which can be written as

$$f'''(x), \quad f^{(3)}(x), \quad \text{or} \quad \frac{d^3y}{dx^3}.$$

Similarly, we can go on differentiating and obtain $f^{(n)}(x)$, or $\frac{d^ny}{dx^n}$, the *n*th **derivative** of f. These derivatives, $f''(x), f'''(x), \ldots, f^{(n)}(x), \ldots$ are referred to as **higher-order derivatives** of f. All of them convey subtle information about the graph of f. (We shall use them when we discuss numerical integration in the next chapter.)

▶ **5.3 EXERCISES**

Calculate $\dfrac{d^2y}{dx^2}$ *in each of Exercises 1–10.*

1. $y = 3x^2 - 6$ **2.** $y = -x^2 + x$ **3.** $y = \dfrac{2}{x}$ **4.** $y = -\dfrac{2}{x^2}$

5. $y = 4x^{0.4} - x$ **6.** $y = 0.2x^{-0.1}$ **7.** $y = e^{-(x-1)} - x$ **8.** $y = e^{-x} + e^x$

9. $y = \dfrac{1}{x} - \ln x$ **10.** $y = x^{-2} + \ln x$

*In Exercises 11–16, the position s of a point (in feet) is given as a function of time t (in seconds). Find **(a)** its acceleration as a function of t, and **(b)** its acceleration at the specified time.*

11. $s = 12 + 3t - 16t^2; \quad t = 2$ **12.** $s = -12 + t - 16t^2; \quad t = 2$ **13.** $s = \dfrac{1}{t} + \dfrac{1}{t^2}; \quad t = 1$

14. $s = \dfrac{1}{t} - \dfrac{1}{t^2}; \quad t = 2$ **15.** $s = \sqrt{t} + t^2; \quad t = 4$ **16.** $s = 2\sqrt{t} + t^3; \quad t = 1$

Find the approximate coordinates of all points of inflection (if any) in each of Exercises 17–24.

17.

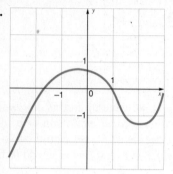

18.

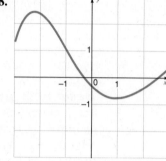

19.

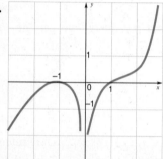

20.

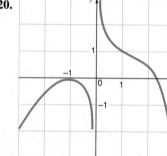

21.

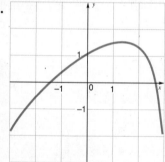

22.

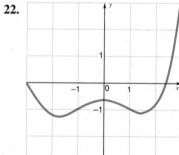

23.

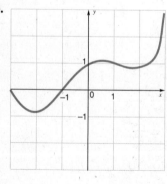

24.

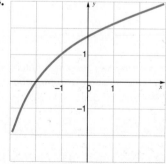

In Exercises 25–36, use the second derivative test to classify critical points where possible. Also, locate all points of inflection.

25. $f(x) = x^2 + 2x + 1$

26. $f(x) = -x^2 - 2x - 1$

27. $f(x) = 2x^3 + 3x^2 - 12x + 1$

28. $f(x) = 4x^3 + 3x^2 + 2$

29. $f(x) = -4x^3 - 3x^2 + 1$

30. $f(x) = -2x^3 - 3x^2 + 12x + 1$

31. $g(x) = (x - 3)\sqrt{x}$

32. $g(x) = (x + 3)\sqrt{x}$

33. $f(x) = x - \ln x$

34. $f(x) = x - \ln(x^2)$

35. $f(x) = x^2 + \ln x^2$

36. $f(x) = 2x^2 \ln x$

In Exercises 37–44, use a graphing calculator to plot the graph of f'' and hence find the approximate coordinates of all the points of inflection (if any) of the given function f. (All coordinates should be correct to two decimal places.)

 37. $f(x) = x^3 - 2.1x^2 + 4.3x$

38. $f(x) = 2x^3 + 4.2x^2 - 5.2x$

39. $f(x) = e^{-x^2}$

40. $f(x) = x + e^{-2x^2}$

41. $f(x) = x^4 - 2x^3 + x^2 - 2x + 1$

42. $f(x) = x^4 + x^3 + x^2 + x + 1$

43. $f(x) = x^2 - \ln x$

44. $f(x) = x^2 + \ln x$

APPLICATIONS

45. *Epidemics* The following graph shows the total number, n, of people (in millions) infected in an epidemic as a function of time t (in years).

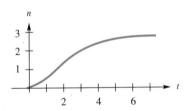

(a) When, to the nearest year, was the rate of infection largest?

(b) When could the Centers for Disease Control announce that the rate of infection was beginning to drop?

46. *Sales* The following graph shows the total number of Pomegranate II computers sold since their release.

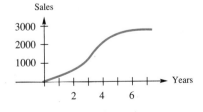

(a) When were the computers selling fastest?

(b) Explain why this graph might look as it does.

47. *Industrial Output* The following graph shows the yearly industrial output of a developing country (mesured in billions of dollars) over a seven-year period.

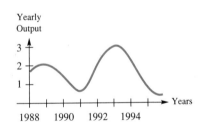

(a) When (to the nearest year) did the rate of change of yearly industrial output reach a maximum?

(b) When (to the nearest year) did the rate of change of yearly industrial output reach a minimum?

(c) When (to the nearest year) did the rate of change of yearly industrial output first start to increase?

48. Profits   The following graph shows the yearly profits of Gigantic Conglomerate, Inc. (GCI), from 1980 to 1995.

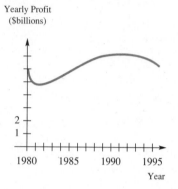

Yearly Profit
($billions)

1980   1985   1990   1995

Year

(a) When were the profits rising most rapidly?
(b) When were the profits falling most rapidly?
(c) When could GCI's board of directors legitimately tell stockholders that they had "turned the company around"?

49. Prison Population   The prison population of the United States followed the curve

$$N(t) = 0.028234t^3 - 1.0922t^2 + 13.029t$$
$$+ 146.88 \quad (0 \le t \le 39)$$

in the years 1950–1989. Here t is the number of years since 1950 and N is the number of prisoners in thousands.* Locate all points of inflection on the graph of N, and interpret the result.

50. Test Scores   Combined SAT scores in the U.S. can be approximated by

$$T(t) = -0.01085t^3 + 0.5804t^2 - 10.12t$$
$$+ 962.4 \quad (0 \le t \le 22)$$

in the years 1967–1991. Here t is the number of years since 1967 and T is the combined SAT score average for the U.S.† Locate all points of inflection on the graph of T, and interpret the result.

51. Education and Crime   The following graph shows a striking relationship between the total prison population and the average combined SAT score in the U.S.

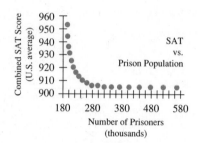

Combined SAT Score
(U.S. average)

960
950
940
930
920
910
900

SAT
vs.
Prison Population

180   280   380   480   580

Number of Prisoners
(thousands)

These data can be accurately modeled by

$$S(n) = 904 + \frac{1326}{(n - 180)^{1.325}} \quad (192 \le n \le 563).$$

Here, $S(n)$ is the combined U.S. average SAT score at a time when the total U.S. prison population is n thousand.‡

(a) Are there any points of inflection on the graph of S?
(b) What does the concavity of the graph of S tell you about prisons and SAT scores?

52. Education and Crime   Referring to the model in the previous exercise,

(a) Are there any points of inflection on the graph of S'?
(b) When is S'' a maximum? Interpret your answer in terms of prisoners and SAT scores.

53. Patents   In 1965, the economist F.M. Scherer modeled the number, n, of patents produced by a firm as a function of the size, s, of the firm (measured in annual sales in millions of dollars). He came up with the following equation based on a study of 448 large firms:§

$$n = -3.79 + 144.42s - 23.86s^2 + 1.457s^3.$$

▼ *The model is the authors'. Source for data: *Sourcebook of Criminal Justice Statistics,* 1990, p. 604.

† The model is the authors'. Source for data: Educational Testing Service.

‡ The model is the authors' based on data for the years 1967–1989. Sources: *Sourcebook of Criminal Justice Statistics,* 1990, p. 604; Educational Testing Service.

§ Source: F. M. Scherer, "Firm Size, Market Structure, Opportunity, and the Output of Patented Inventions," *American Economic Review* 55 (December 1965): 1097–1125.

(a) Find $\dfrac{d^2n}{ds^2}$ and evaluate it at $s = 3$. Is the rate at which patents are produced as the size of a firm goes up increasing or decreasing with size when $s = 3$? Comment on Scherer's words, " we find diminishing returns dominating."

(b) Find $\dfrac{d^2n}{ds^2}\bigg|_{s=7}$ and interpret the answer.

(c) Find the s-coordinate of any points of inflection and interpret the result.

54. *Returns on Investments* A company finds that the number of new products it develops per year depends on the size of its annual R&D budget, x (in thousands of dollars), according to the formula

$$n(x) = -1 + 8x + 2x^2 - 0.4x^3.$$

(a) Find $n''(1)$ and $n''(3)$, and interpret the results.

(b) Find the size of the budget that gives the largest rate of returns as measured in new products per dollar. (Again called the point of diminishing returns.)

55. *Modeling Demand* Your marketing group is launching a new 900 telephone service that supplies callers with the correct spelling of any word. You anticipate that the number of calls per day will initially be increasing at a rate of 10 new calls per day and that this rate will drop by 2 calls per day each day. Model this by an equation of the form

$$n = at + bt^2,$$

where n is the total number of phone calls you anticipate, t is time in days and a and b are constants you must determine. [*Hint:* The given information tells you something about $n'(0)$ and $n''(t)$.]

56. *Modeling Cost* You would like to construct a cost equation for your small tie-dye operation, so you decide on a general cubic equation of the form

$$C = a + bx + cx^2 + dx^3,$$

where C is the daily cost of producing x tie-dye T-shirts. Your daily overheads are $200. The marginal cost at a production level of zero T-shirts is $4.00 per shirt and is decreasing at a rate of 60¢ per T-shirt. The marginal cost reaches a minimum at a production level of 10 T-shirts. What is your cost equation?

57. *Modeling Revenue* As consultant to a medical research company, you would like to model the antici-

pated sales of its new antiviral drug Virastat using a logistic equation of the form

$$R = \frac{a}{1 + be^{-kt}},$$

where R is the total revenue from sales of Virastat in millions of dollars, t is time (in months), and a, b, and k are constants that you will need to determine. You have the following information to work with. First, the company estimates that it can sell a total of $10 million worth of the drug in the long term. Next, at the present time $(t = 0)$ the total sales revenue amounts to $0.5 million and is growing at a rate of $0.02375 million per month.

(a) Use the given information to find a, b, and k and hence the revenue as a function of time. [*Hint:* The given information tells you something about $\lim_{t \to +\infty} R(t)$, $R(0)$ and $R'(0)$.]

(b) Find $R''(0)$, and interpret the result.

Exercises 58 and 59 require the use of either a graphing calculator or graphing computer software.

58. *Asset Appreciation* You manage a small antique store that owns a collection of Louis XVI jewelry boxes. Their value v is increasing according to the formula

$$v = \frac{10000}{1 + 500e^{-0.5t}},$$

where t is the number of years from now. You anticipate an inflation rate of 5% per year, so that the present value of an item that will be worth v in t years' time is given by

$$p = v(1.05)^{-t}.$$

(a) Graph p as a function of t with $0 \le t \le 40$, $0 \le p \le 6{,}000$ and determine the approximate values of t for all points of inflection (to within ± 2 years).

(b) Now calculate dp/dt, graph it, and hence obtain more accurate estimates of the locations of the points of inflection (to the nearest year).

(c) What is the largest rate of increase of the value of your antiques, and when is this rate attained?

59. *Harvesting Forests*   You are considering harvesting a stand of Douglas fir trees. The following equation models the approximate volume in cubic feet of a typical Douglas fir tree of age t years.*

$$V = \frac{22{,}514}{1 + 22{,}514t^{-2.55}}$$

The lumber will be sold at \$10 per cubic foot, and you do not expect the price of lumber to appreciate in the foreseeable future. On the other hand, you anticipate a general inflation rate of 5% per year, so that the present value of an item that will be worth \$$v$ in t years' time is given by

$$p = v(1.05)^{-t}$$

(a) Graph p as a function of t using the following ranges: $0 \le t \le 80$, $0 \le p \le 1{,}500$ and determine the approximate values of t for all points of inflection (to within ± 5 years).

(b) Now calculate dp/dt, graph it, and hence obtain more accurate estimates of the locations of the points of inflection (to the nearest year).

(c) What is the largest rate of increase of the value of a fir tree, and when is this rate attained?

COMMUNICATION AND REASONING EXERCISES

60. Complete the following sentence. If the graph of a function is concave up on its entire domain, then its _____ derivative is never _____ .

61. Regarding position, s, as a function of time, t, what is the significance of the *third* derivative, $s'''(t)$? Describe an everyday scenario in which it arises.

62. Explain geometrically why the derivative of a function has a local extremum at a point of inflection. Which points of inflection give rise to local maxima in the derivative?

▶ ════ **5.4** CURVE SKETCHING

In Chapters 1 and 2 we sketched several curves by plotting points. Plotting points is precisely how most graph-generating software works: the computer or graphing calculator plots hundreds of points and connects them (usually with straight lines), creating the illusion of a smooth curve. There are, however, some drawbacks to this approach. First, it may be difficult to locate the local extrema simply by looking at a curve, but we know already how useful it is to know where they are located. Second, there are many other interesting features you might miss if you were plotting points by hand, such as points of inflection and behavior near points where the function is not defined.

In this section, we use our knowledge about derivatives and limits to analyze a variety of curves. First, we discuss a technique for sketching the graph of a function by hand. Later, we show how these techniques can be used to analyze a curve plotted on a graphing calculator or by computer graphing software. (You do not require graphing technology to benefit from this discussion.)

▼ *The model is the authors' and is based on data in *Environmental and Natural Resource Economics,* Third Edition, by Tom Tietenberg (New York: HarperCollins, 1992), p. 282.

GRAPHING FUNCTIONS BY HAND

Consider the curve $y = 2x - 1 + 2/(2x - 1)$. If we plot the points corresponding to the x-values $-3, -2, -1, 0, 1, 2, 3$, we get the points in the xy-plane shown in Figure 1.

This seems to suggest a curve like the one drawn in Figure 2. But this is not the right graph at all! The actual graph is shown in Figure 3. In other words, we completely missed two local extrema, as well as the fact that the curve breaks into two pieces.

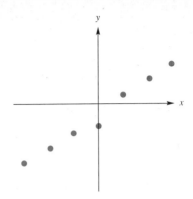

FIGURE 1

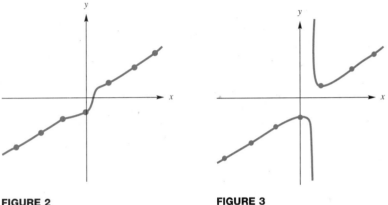

FIGURE 2   **FIGURE 3**

We now describe a simple six-step strategy that enables us to sketch virtually any curve that we encounter and to pinpoint all of its essential features. Most of these steps will not be new: we already know how to locate local extrema and points of inflection. Further, we already know from Chapter 1 what to do near points where the function is not defined. We will simply combine all these techniques and add a few more. Bear in mind the following guiding principle: the graph of a mathematical function is always elegant, so drawing a graph is an artistic exercise as well as a mathematical one.

▼ **EXAMPLE 1**

Sketch the graph of $f(x) = (x - 3)\sqrt{x}$.

SOLUTION   Here is our six-step approach:

Step 1: Find the domain of the function if it is not supplied.   Here, the domain of the function is not given, so we take its domain to be as large as possible. The domain will be all real numbers unless

(1)   there is an expression that can be zero in a denominator, or

(2)   there is a square root (or some other even root) of a quantity that can be negative.

Our function has a square root. We know that the function won't be defined where the quantity under the square root sign—namely x—is nega-

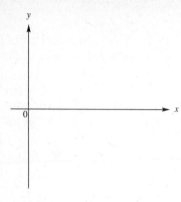

FIGURE 4

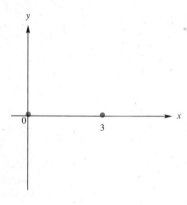

FIGURE 5

tive. Thus, x must be greater than or equal to zero for our function to make sense. So the domain is $[0, +\infty)$.

Because we are concerned only with values of $x \geq 0$, we can place the y-axis on the left of our drawing, as we shall draw nothing to the left of it (see Figure 4). (Let us not get impatient. We are quite aware that there is nothing on the graph yet!)

Step 2: Find the x- and y-intercepts.   The intercepts are the points where the curve touches or crosses the two axes. To find these points, we first replace $f(x)$ with y, getting

$$y = (x - 3)\sqrt{x}.$$

To find the y-intercept, we substitute $x = 0$ (points on the y-axis have $x = 0$), which gives us

$$y = (0 - 3)0 = 0.$$

Thus, the y-intercept is 0, so we can mark the point 0 on the y-axis. Similarly, we find the x-intercept by setting $y = 0$ and solving for x.

$$0 = (x - 3)\sqrt{x}$$

From this equation we see that either $x = 3$ or $x = 0$. Thus, there are two x-intercepts, 0 and 3. We plot these points.* We now have the beginnings of a graph (Figure 5).

Step 3: Locate and classify the local extrema, sketching them in.   We proceed as we have many times before.

Stationary Points   If we take the derivative of the function as it stands, we shall have to use the product rule. Instead, we can rewrite the radical in exponent form and multiply out.

$$y = (x - 3)\sqrt{x} = (x - 3)x^{1/2} = x^{3/2} - 3x^{1/2}$$

Thus,

$$\frac{dy}{dx} = \frac{3}{2}x^{1/2} - \frac{3}{2}x^{-1/2}.$$

We set

$$\frac{3}{2}x^{1/2} - \frac{3}{2}x^{-1/2} = 0.$$

To solve for x, we rewrite the equation with no negative exponents.

$$\frac{3x^{1/2}}{2} - \frac{3}{2x^{1/2}} = 0$$

▼ * In some examples it may be extremely difficult or even impossible to solve for the x-intercepts. On these occasions we either skip this step, or use graphing technology to help us.

We now clear the denominators by multiplying by $2x^{1/2}$, getting

$$3x - 3 = 0, \quad \text{so } x = 1.$$

Singular Points There is a singular point at $x = 0$, because the derivative is not defined there (why not?).

Endpoints $x = 0$ is an endpoint of the domain.

We set up our table, including the intercepts, which make convenient test points.

x	0	1	3
y	0	-2	0

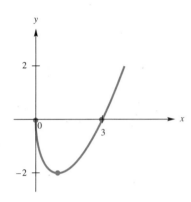

We can now sketch in the local extrema, getting Figure 6. (Remember to pay special attention to the singular point: the tangent there is vertical.)

FIGURE 6

 We seem to have the whole curve already after only three steps! Are we done?

A No. There may still be one or two hidden features, such as points of inflection. Also, we are not yet sure what happens towards the extreme right, as x goes off to infinity.

Step 4: Locate points of inflection, if any. We take $f''(x)$ and set it equal to zero.

$$\frac{dy}{dx} = \frac{3}{2}x^{1/2} - \frac{3}{2}x^{-1/2}$$

$$\frac{d^2y}{dx^2} = \frac{3}{4}x^{-1/2} + \frac{3}{4}x^{-3/2}$$

Setting the second derivative equal to zero gives

$$\frac{3}{4x^{1/2}} + \frac{3}{2x^{3/2}} = 0.$$

But the quantity on the left is positive, so it cannot be zero. Thus the equation has no solutions. The only other candidate for a point of inflection occurs at $x = 0$ where $f''(x)$ is not defined. But this cannot be a point of inflection. (Why?) We conclude that there are no points of inflection. (Our sketch so far suggests this.)

Step 5: Show behavior near points where the function is not defined. The points we have in mind are points where the function is not defined, but where the function is defined just to the left or right. There are no such points in this example, but there is one in the next example.

Step 6: Show behavior as $x \to +\infty$ and $x \to -\infty$. We should see what happens to the y-coordinate for large positive and negative values of x. In other words, *how high does the curve get as we move to the extreme right and left?* We answer this question by taking two limits,

$$\lim_{x \to -\infty} f(x) \quad \text{and} \quad \lim_{x \to +\infty} f(x).$$

In this example, we need only take the limit as $x \to +\infty$ because the domain does not include negative values of x. Now

$$\lim_{x \to +\infty} f(x) = \lim_{x \to +\infty} (x - 3)\sqrt{x}.$$

To calculate this limit, we do a quick mental version of the tabular approach: if x is a very large positive number, both $(x - 3)$ and $\sqrt{x}$ are large positive numbers, and hence, so is their product. Thus,

$$\lim_{x \to +\infty} \sqrt{x}(x - 3) = +\infty.$$

So, as we move farther and farther to the right, the curve gets higher and higher without bound.

Figure 7 shows a picture of an *incorrect* graph.

The graph in Figure 7 seems to be leveling off at about $y = 2$ for large values of x, which would contradict Step 6, which tells us that the graph's height increases without bound. The graph also contradicts Step 4, which says that there are no points of inflection. The curve in Figure 7 seems to have a point of inflection near $x = 4$.

The correct graph was shown in Figure 6.

 To check our work, Figure 8 shows a graphing calculator-generated plot of the curve $y = (x - 3)\sqrt{x}$.

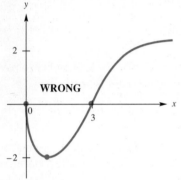

FIGURE 7

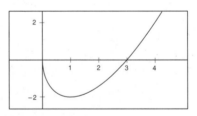

FIGURE 8

Before going on to the next example, we should say that there is nothing wrong with plotting a few additional points if they will help you visualize the curve or confirm your suspicions about its shape. As a rule, however, we'll always try to plot as few as possible.

▼ **EXAMPLE 2**

Sketch the graph of the function $f(x) = \dfrac{1}{x} - \dfrac{1}{x^2}$.

SOLUTION   We repeat the six-step process.

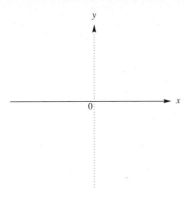

FIGURE 9

Step 1: Find the domain of the function if it is not supplied. Because the domain is not given, we assume the largest possible domain. There is an x in both denominators, so the function is not defined when $x = 0$. Thus, the domain consists of all real numbers except 0, or $(-\infty, 0) \cup (0, +\infty)$.

We can now begin to set up the sketch. Since x is not allowed to be zero, we mark the vertical line $x = 0$ (i.e., the y-axis) as a "forbidden zone"—the curve cannot touch this line. (See Figure 9. In Step 5 we shall see that $x = 0$ is a vertical asymptote.)

Step 2: Find the x- and y-intercepts. To calculate the intercepts, first replace $f(x)$ with y, getting

$$y = \frac{1}{x} - \frac{1}{x^2}.$$

To find the y-intercept, we would have to substitute $x = 0$. But $x = 0$ is not in the domain of f. Thus, there is no y-intercept.

We get the x-intercept(s) by setting $y = 0$ and solving for x.

$$0 = \frac{1}{x} - \frac{1}{x^2}$$

Multiplying by x^2 to clear denominators gives

$$0 = x - 1,$$

so

$$x = 1$$

is the only x-intercept. We mark this point on the x-axis.

Step 3: Locate and classify the local extrema, sketching them in. To find stationary points, we take the derivative and set it equal to zero:

$$f'(x) = -\frac{1}{x^2} + \frac{2}{x^3},$$

so we solve

$$-\frac{1}{x^2} + \frac{2}{x^3} = 0.$$

Multiplying by x^3 to clear denominators gives

$$-x + 2 = 0,$$

so

$$x = 2$$

is our only stationary point. There are no singular points or endpoints. Thus, we need a test point to the right of 2. (We already have one to the left: the x-intercept of 1.)

x	1	2	3
y	0	$\dfrac{1}{4}$	$\dfrac{2}{9}$

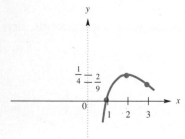

FIGURE 10

We can now sketch these points, showing the local maximum at $x = 2$ (Figure 10).

Step 4: Locate points of inflection, if any.   To locate the points of inflection, set the second derivative equal to zero and solve for x.

$$f''(x) = \frac{2}{x^3} - \frac{6}{x^4}$$

We thus solve

$$\frac{2}{x^3} - \frac{6}{x^4} = 0.$$

Multiplying by x^4,

$$2x - 6 = 0, \quad \text{so } x = 3.$$

So we have a *possible* point of inflection at $x = 3$ (which happened to be the test point we have already plotted). Now, there may or may not be a point of inflection at $x = 3$. To see that there really is a point of inflection there, we check the concavity to the right of $x = 3$, say, at $x = 4$. We find

$$f''(4) = \frac{2}{64} - \frac{6}{256} > 0,$$

so the curve is concave up to the right of $x = 3$. Because it is concave down to the left, there must be a point of inflection at $x = 3$.

Step 5: Show behavior near points where the function is not defined.   Recall that we are interested in points where f is not defined, but is defined just to the left or right. We have such a point: $x = 0$. Thus, we must look at points very close to $x = 0$ on *both* sides and determine what happens to the y-coordinates as x approaches 0. We do so by taking two limits,

$$\lim_{x \to 0^-} f(x) \quad \text{and} \quad \lim_{x \to 0^+} f(x).$$

We can evaluate these limits using the tabular approach. We calculate $f(x)$ for values of x close to 0 (and on either side).

x	$-\dfrac{1}{100}$	$\dfrac{1}{100}$
$f(x)$	$-10{,}100$	$-9{,}900$

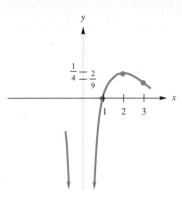

FIGURE 11

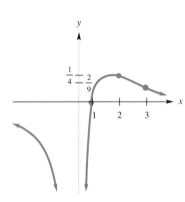

FIGURE 12

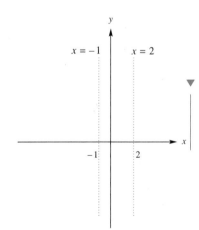

FIGURE 13

This table suggests that

$$\lim_{x \to 0^-} f(x) = -\infty, \text{ and } \lim_{x \to 0^+} f(x) = -\infty.$$

From these limits we see the following.

(1) Immediately to the *left* of $x = 0$, the graph plunges down toward $-\infty$.

(2) Immediately to the *right* of $x = 0$, the graph also plunges down toward $-\infty$.

Thus, the line $x = 0$ is a vertical asymptote. Figure 11 shows our sketch so far.

Step 6: Show behavior as $x \to +\infty$ and $x \to -\infty$. As in the last example, we calculate

$$\lim_{x \to +\infty} f(x) = \lim_{x \to +\infty} \left(\frac{1}{x} - \frac{1}{x^2} \right)$$
$$= 0 - 0 = 0,$$

and

$$\lim_{x \to -\infty} f(x) = \lim_{x \to -\infty} \left(\frac{1}{x} - \frac{1}{x^2} \right)$$
$$= 0 - 0 = 0.$$

Thus, on the extreme left and right of our picture the height of the curve levels off toward zero. Figure 12 shows the completed graph.

Before we go on... Notice that because the curve levels off as $x \to +\infty$, there must be a point of inflection *somewhere* to the right of $x = 2$. In Step 4 we saw that the only place there could be a point of inflection was at $x = 3$. Again, we have verified that there must be a point of inflection at $x = 3$.

Notice another thing: we haven't plotted a single point to the left of the y-axis, and yet we have a pretty good idea of what the curve looks like there!

▼ **EXAMPLE 3**

Sketch the graph of

$$f(x) = \frac{x^2}{(x + 1)(x - 2)}.$$

SOLUTION

Step 1: Find the domain of the function if it is not supplied. Again, the domain is not given to us. Because there is a denominator, we exclude those values of x that make the denominator equal to zero. That is, we exclude both $x = -1$ and $x = 2$. As usual, we begin by sketching in the "forbidden zones": the vertical lines $x = -1$ and $x = 2$ (Figure 13).

Step 2: Find the x- and y-intercepts.   We first replace $f(x)$ with y, getting

$$y = \frac{x^2}{(x + 1)(x - 2)}.$$

We then find the y-intercept by setting $x = 0$, getting

$$y = \frac{0}{(1)(-2)} = 0,$$

so the y-intercept is 0. For the x-intercept, we set $y = 0$ and solve for x.

$$0 = \frac{x^2}{(x + 1)(x - 2)}$$

Now recall that if a quotient is zero, the numerator must be zero, so that

$$x^2 = 0, \text{ whence } x = 0.$$

Thus, the only intercept we have is the origin $(0, 0)$. We mark it on the graph.

Step 3: Locate and classify the local extrema, sketching them in.   We calculate the derivative. To make this calculation easier, we first expand the denominator, getting

$$f(x) = \frac{x^2}{x^2 - x - 2}.$$

Thus,

$$f'(x) = \frac{2x(x^2 - x - 2) - x^2(2x - 1)}{(x^2 - x - 2)^2}$$

$$= \frac{-x^2 - 4x}{(x^2 - x - 2)^2}.$$

Setting $f'(x)$ equal to zero, it must be the numerator that is zero, so

$$-x - 4x = 0,$$

or

$$x(x + 4) = 0,$$

hence,

$$x = 0 \text{ or } x = -4.$$

We now have two stationary points. There are no singular points or endpoints. Notice that 0 and -4 are separated by the dividing line at $x = -1$, so they will be on different pieces of the graph. Therefore, we need to look at test points near 0 and -4 but not separated from them by the dividing lines.

x	-5	-4	-3	$-\dfrac{1}{2}$	0	1
y	$\dfrac{25}{28}$	$\dfrac{8}{9}$	$\dfrac{9}{10}$	$-\dfrac{1}{5}$	0	$-\dfrac{1}{2}$

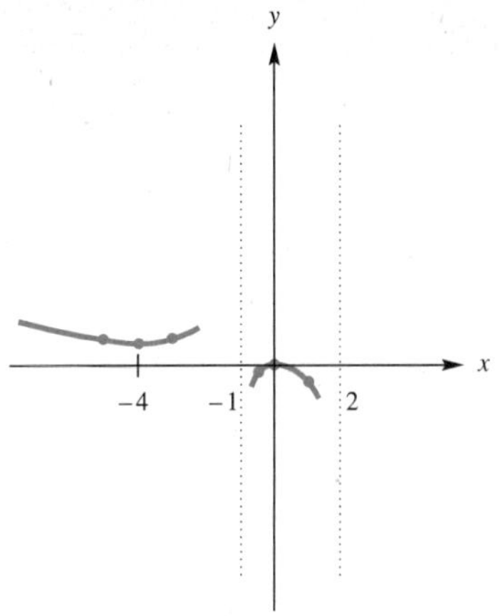

FIGURE 14

Plotting these points gives us Figure 14.

Step 4: Locate points of inflection, if any.   The second derivative is going to be very messy, and it will be difficult to solve for the points of inflection. We'll skip this step and hope that there are no points of inflection. If necessary, we can come back to this step later.

Step 5: Show behavior near points where the function is not defined.   We have two such points: $x = -1$ and $x = 2$. We must therefore calculate *four* limits,

$$\lim_{x \to -1^-} f(x), \qquad \lim_{x \to -1^+} f(x), \qquad \lim_{x \to 2^-} f(x), \quad \text{and} \quad \lim_{x \to 2^+} f(x).$$

We could estimate these limits numerically by choosing values of x on each side of (and close to) -1 and 2. Alternatively, if we remember that the reciprocal of a small number is a large number, we can argue as follows.

(1)   If we choose x very close to -1 and on its *left,* we get

$$y \approx \frac{1}{(small\ negative)(-3)}$$

$$= big\ positive, \text{ giving a limit of } +\infty.$$

(Notice that the denominator is positive, being the product of two negatives.) Thus, the graph shoots up to the left of -1.

(2)   If we choose x very close to -1 and on its *right,* we get

$$y \approx \frac{1}{(small\ positive)(-3)}$$

$$= big\ negative, \text{ giving a limit of } -\infty.$$

Thus, the graph plunges down to the right of -1.

(3)   If we choose x very close to 2 and on its *left,* we get

$$y \approx \frac{4}{(3)(small\ negative)}$$

$$= big\ negative, \text{ giving a limit of } -\infty.$$

Thus, the graph plunges down to the left of 2.

(4)   If we choose x very close to 2 and on its *right,* we get

$$y \approx \frac{4}{(3)(small\ positive)}$$

$$= big\ positive, \text{ giving a limit of } +\infty.$$

Thus, the graph shoots up to the right of 2.

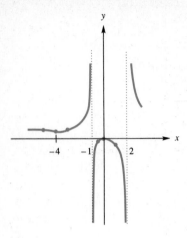

FIGURE 15

We can now sketch the partial graph shown in Figure 15. The lines $x = -1$ and $x = 2$ are vertical asymptotes.

Step 6: Show behavior as $x \to +\infty$ and $x \to -\infty$.
We find

$$\lim_{x \to -\infty} f(x), = \lim_{x \to -\infty} \frac{x^2}{x^2 - x - 2}$$

$$= \lim_{x \to -\infty} \frac{x^2}{x^2} = 1,$$

using the technique of ignoring all but the highest powers of x. Similarly,

$$\lim_{x \to +\infty} f(x) = 1.$$

Thus, on both the extreme left and right, the graph levels off at $x = 1$, so the line $y = 1$ is a horizontal asymptote. The completed graph is shown in Figure 16.

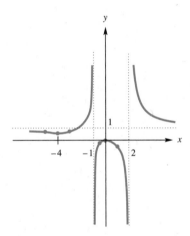

FIGURE 16

From the graph we now discover that there has to be a point of inflection somewhere to the left of the minimum at $x = -4$. Well, if we were forced to find out exactly where it is, we know what we would do. However, instead of trying to find the *exact* location of the point of inflection, let us use technology to locate it numerically.

 To locate the point of inflection, we need to solve $f''(x) = 0$. We can use technology to do this in at least two ways.

Method (a)   Graph $f''(x)$ near $x = -5$, and determine where the graph crosses the x-axis.

Method (b)   Have your calculator or computer solve $f''(x) = 0$ numerically.

Here is a version of (b) for the TI-82. First enter $f'(x)$ under Y_1 as

$$Y_1 = -(X^2+4X) / (X^2-X-2)^2.$$

Then let Y_2 be its derivative, $f''(x)$.

$$Y_2 = \text{nDeriv}(Y_1, X, X)$$

Now solve $f''(x) = 0$ using the "Solve" command on the Home screen.

$$\text{solve } (Y_2, X, -5) \boxed{\text{ENTER}}$$

This tells the calculator to solve the equation $Y_2 = 0$ numerically for x, searching for a solution near $x = -5$. A few seconds later, we find that $x \approx -6.1072$, which is the approximate x-coordinate of the point of inflection. How would you now find the y-coordinate?

SUMMARY: SKETCHING THE GRAPH OF A FUNCTION BY HAND

Step 1:   Find the domain of the function if it is not supplied.
If the function is not defined at one or more isolated points, then sketch in vertical dashed lines at those values. These lines break up the curve into disconnected pieces.

Step 2:   Find the x- and y-intercepts.
These are the points where the curve touches or crosses the two axes. To find these points, we first replace $f(x)$ with y, getting $y = $ function of x.

y-intercepts: Set $x = 0$ to obtain y.

x-intercepts: Set $y = 0$ and solve for x;

Step 3:   Locate and classify the local extrema, sketching them in.
To find the extrema we follow the methods of Section 1 in this chapter.

Step 4:   Locate points of inflection, if any.
We set $f''(x) = 0$ and solve for x.

Step 5:   Show behavior near points where the function is not defined.
If $x = a$ is such a point, we calculate

$$\lim_{x \to a^-} f(x) \text{ and } \lim_{x \to a^+} f(x).$$

(Remember that $1/small = $ big and $1/big = $ small.)

Step 6:   Show behavior as $x \to +\infty$ and $x \to -\infty$.
In other words: *How high does the curve get as we move to the extreme right and left?* To answer, we calculate the limits

$$\lim_{x \to -\infty} f(x) \quad \text{and} \quad \lim_{x \to +\infty} f(x).$$

ANALYZING THE GRAPH OF A FUNCTION PRODUCED BY A GRAPHING CALCULATOR

If we use a graphing calculator to plot a graph, the actual drawing is done for us, but we still need to analyze its features, such as the exact location of the x- and y-intercepts, the extrema, and points of inflection. We carry out this analysis in the same way as we did above.

▼ **EXAMPLE 4**

Figure 17 shows a graphing calculator plot of

$$y = 2(x - 4)^{2/3} + \frac{x}{2}.$$

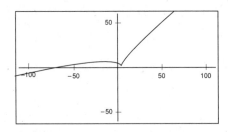

FIGURE 17

(We used the following format:

$$Y_1 = 2((X - 4)^{\wedge}2) ^{\wedge} (1/3) + X/2$$

with the x- and y-ranges as shown in the figure.)

Analyze this curve. (Round all values to one decimal place.)

SOLUTION First, notice that the domain consists of all real numbers, so we turn to the intercepts and see that there is only a single x-intercept. (We can check that there are no intercepts outside the range plotted by enlarging the x-range.)

y-intercept Set $x = 0$ in the equation of the curve, obtaining

$$y = 2(-4)^{2/3} + 0 = \sqrt[3]{16} \approx 2.5.$$

x-intercept Set $y = 0$ and solve for x.

$$0 = 2(x - 4)^{2/3} + \frac{x}{2}$$

If we take the term $x/2$ to the other side of the equation, multiply by 2, and cube both sides, we obtain the cubic equation

$$-x^3 = 64(x - 4)^2,$$

or

$$x^3 + 64(x - 4)^2 = 0.$$

Unfortunately, there is no easy analytical method for solving cubic equations: the easiest method of solving a general cubic equation is the graphical method—see Chapter 1. Thus, we decide that it is far simpler to find the x-intercept directly from the graph by zooming in for greater accuracy. Figure 18 shows a close-up of the curve.

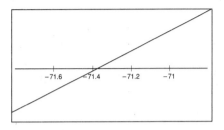

FIGURE 18

We see that the x-intercept occurs at approximately $x = -71.4$ (to the nearest one decimal place).

Extrema We notice in Figure 17 that there appears to be a stationary local maximum at around $x = -20$ and a singular local minimum at around $x = 5$. Analytically,

$$\frac{dy}{dx} = \frac{4}{3}(x - 4)^{-1/3} + \frac{1}{2}$$

$$= \frac{4}{3(x - 4)^{1/3}} + \frac{1}{2}.$$

Equating this to zero gives

$$\frac{4}{3(x - 4)^{1/3}} = -\frac{1}{2}.$$

Cross-multiplying (or clearing the denominators) gives

$$8 = -3(x - 4)^{1/3},$$

or

$$(x - 4)^{1/3} = -\frac{8}{3}.$$

Cubing both sides,

$$x - 4 = -\left(\frac{8}{3}\right)^3,$$

so

$$x = 4 - \left(\frac{8}{3}\right)^3 \approx -15.0 \text{ (to one decimal place)}.$$

This number is the x-coordinate of the single local maximum. To obtain its y-coordinate we substitute the value of x in the equation of the curve, getting $y \approx 6.7$. Thus, the stationary local maximum occurs at approximately $(-15.0, 6.7)$.

For the singular local minimum, we notice that the derivative is not defined when $x = 4$, so this is where the singular minimum must be. Its y-coordinate (obtained by substituting in the equation of the curve) is 2, so the coordinates of this point are $(4, 2)$.

Points of Inflection   Looking at the graph, we suspect that there are no points of inflection, and we can verify this by looking at the second derivative,

$$\frac{d^2y}{dx^2} = -\frac{4}{12}(x - 4)^{-4/3} = -\frac{4}{12(x - 4)^{4/3}}.$$

The second derivative is always negative (where it is defined), so the graph is always concave down, and there are no points of inflection.

Points Where the Function Is Not Defined   There are no such points, since the domain of the function f is the set of all real numbers.

Behavior as $x \to \pm\infty$ Because

$$f(x) = 2(x - 4)^{2/3} + \frac{x}{2},$$

a mental version of the tabular method will convince you that, as $x \to +\infty$, $f(x) \to +\infty$ as well (because we are adding larger and larger numbers). Evaluating the limit as $x \to -\infty$ is a little harder because if x is a large negative number, then $f(x)$ is the difference of two large numbers. However, because the graph is concave down everywhere, it must be the case that $\lim_{x \to -\infty} f(x) = -\infty$. (Think about it.)

▶ **5.4 EXERCISES**

Sketch the graphs of the functions in Exercises 1–26 by hand, analyzing all important features.

1. $f(x) = x^2 - 4x + 1$ with domain $[0, 3]$

2. $f(x) = 2x^2 - 2x + 3$ with domain $[0, 3]$

3. $g(x) = x^3 - 12x$ with domain $[-4, 4]$

4. $g(x) = 2x^3 - 6x$ with domain $[-4, 4]$

5. $f(t) = t^3 + t$ with domain $[-2, 2]$

6. $f(t) = -2t^3 - 3t$ with domain $[-1, 1]$

7. $h(t) = 2t^3 + 3t^2$ with domain $[-2, +\infty)$

8. $h(t) = t^3 - 3t^2$ with domain $[-1, +\infty)$

9. $f(x) = x^4 - 4x^3$ with domain $[-1, +\infty)$

10. $f(x) = 3x^4 - 2x^3$ with domain $[-1, +\infty)$

11. $g(t) = \frac{1}{4}t^4 - \frac{2}{3}t^2 + \frac{1}{2}t^2$ with domain $(-\infty, +\infty)$

12. $g(t) = 3t^4 - 16t^3 + 24t^2 + 1$ with domain $(-\infty, +\infty)$

13. $f(t) = \dfrac{t^2 + 1}{t^2 - 1}$ with $-2 \le t \le 2$

14. $f(t) = \dfrac{t^2 - 1}{t^2 + 1}$ with domain $[-2, 2]$

15. $f(x) = (x - 1)\sqrt{x}$

16. $f(x) = (x + 1)\sqrt{x}$

17. $g(x) = x^2 - 4\sqrt{x}$

18. $g(x) = 1/x + 1/x^2$

19. $g(x) = x^3/(x^2 + 3)$

20. $g(x) = x^3/(x^2 - 3)$

21. $f(x) = x - \ln x$ with domain $(0, +\infty)$

22. $f(x) = x - \ln x^2$ with domain $(0, +\infty)$

23. $g(t) = e^t - t$ with domain $[-1, 1]$

24. $g(t) = e^{-t^2}$ with domain $(-\infty, +\infty)$

25. $f(x) = x + \dfrac{1}{x}$

26. $f(x) = x^2 + \dfrac{1}{x^2}$

APPLICATIONS

27. *Asset Appreciation (graphing calculator or computer required)* You are the financial consultant to a classic auto dealership. You estimate that the total value of its collection of 1959 Chevrolets and Fords is given by the formula

$$v = 300,000 + 1,000t^2,$$

where t is the number of years from now. You anticipate an inflation rate running continuously at 5% per year, so that the discounted (present) value of an item that will be worth $\$v$ in t years' time is given by

$$p = ve^{-0.05t}.$$

Sketch the graph of the discounted value as a function of the time at which the vehicles are sold.

(a) Calculate the t-coordinates of the extrema analytically, and use a graphing calculator plot of the derivative of the discounted value function to locate the approximate location of points of inflection. What is an appropriate domain for this function?

(b) Use your graphs [or the information from part (a)] to determine when the value of the collection of classic cars is increasing most rapidly. When is it decreasing most rapidly?

(c) What is the significance of the asymptote? Is this reasonable?

28. *Plantation Management (graphing calculator or computer required)* The value of a fir tree in your plantation increases with the age of the tree according to the formula

$$v = \frac{20t}{1 + 0.05t},$$

where t is the age of the tree in years. Given an inflation rate running continuously at 5% per year, the

discounted (present) value of a newly planted seedling is given by

$$p = ve^{-0.05t}.$$

(a) Sketch the graph of the discounted value of a newly planted seedling as a function of the time at which the tree is harvested. Calculate the t-coordinates of the extrema analytically, and use a graphing calculator plot of the derivative of the discounted value function to locate the approximate location of points of inflection. What is an appropriate domain for this function?

(b) Use your graphs [or the information from part (a)] to determine when the value of a tree is increasing most rapidly. When is it decreasing most rapidly?

(c) What is the significance of the asymptote? Is this reasonable?

29. *Average Cost* A cost function for the manufacture of portable CD players is given by

$$C(x) = \$150,000 + 20x + \frac{x^2}{10,000},$$

where x is the number of CD players manufactured. Using an appropriate domain, sketch the graph of the average cost $\bar{C}$ to manufacture x CD players, indicating all absolute extrema and points of inflection (if there are any). What is the significance of the following features?

(a) absolute minimum

(b) $\lim\limits_{x \to +\infty} \bar{C}(x)$

(c) $\lim\limits_{x \to 0^+} \bar{C}$

30. *Average Cost* Repeat Exercise 29 using the revised cost function

$$C(x) = \$150,000 + 20x + \frac{x^2}{100}.$$

31. *Average Profit* The Feature Software Company sells its graphing program, Dogwood, with a volume discount. If a customer buys x copies, then the customer pays $500\sqrt{x}$. It cost the company $10,000 to develop the program, and it costs $2 to manufacture each copy. Assuming that just one customer buys all the copies of Dogwood, sketch the graph of the average profit obtained by selling x copies. On the same set of axes, sketch the graph of the marginal profit function. Locate the point where the two graphs cross each other, and explain its significance.

32. *Average Profit* Repeat Exercise 31 with the charge to the customer being $600\sqrt{x}$ and the cost to develop the program being $9,000.

33. *Minimizing Resources* Basic Buckets, Inc., has an order for plastic buckets holding 5,000 cubic centimeters. Their buckets are open-topped cylinders. Sketch the graph of the amount of plastic used as a function of the radius. Give a rationale for the behavior of the limits at 0 and infinity. (The volume of an open-topped cylinder with height h and radius r is $\pi r^2 h$, and the surface area is $\pi r^2 + 2\pi rh$.)

34. *Optimizing Capacity* Basic Buckets would like to build a bucket with a surface area of 1,000 square centimeters. Sketch the graph of the volume as a function of the radius. What is an appropriate domain for this function? Why? (See the previous exercise.)

Exercises 35–42 require the use of a graphing calculator or computer software.

35. *Resource Allocation* Your automobile assembly plant has a Cobb-Douglas production function given by

$$q = x^{0.4}y^{0.6},$$

where q is the number of automobiles it produces per year, x is the number of employees, and y is the daily operating budget (in dollars). Annual operating costs amount to an average of $20,000 per employee plus the operating budget of $365y$. Assume that you wish to produce 1,000 automobiles per year. Sketch the graph of the cost C as a function of the number of employees you hire.

(a) How would you use your graph to estimate the incremental cost per employee at an employment level of 50 employees?

(b) By looking at the graph of C, what can you say about $\lim_{x \to +\infty} C'(x)$? Interpret the answer.

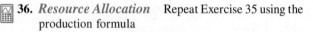 **36. *Resource Allocation*** Repeat Exercise 35 using the production formula

$$q = x^{0.5}y^{0.5}.$$

37. *Patents* In 1965, the economist F. M. Scherer modeled the number, n, of patents produced by a firm as a function of the size, s, of the firm (measured in annual sales of in millions of dollars). He came up with the following equation based on a study of 448 large firms.*

$$n = -3.79 + 144.42s - 23.86s^2 + 1.457s^3$$

Sketch the graph of n for firms of sizes up to $10 million in sales. Where on the graph is the smallest rate of increase in numbers of patents produced as a firm's size increases?

38. *Returns on Investments* A company finds that the number of new products it develops per year depends on the size of its annual R&D budget, x (in thousands of dollars), according to the formula

$$n(x) = -1 + 8x + 2x^2 - 0.4x^3.$$

Sketch the graph of n for R&D budgets up to $400,000. Where is the point of diminishing returns on the graph?

The Normal Curve *One of the most useful curves in statistics is the so-called normal distribution curve, which models the distributions of data in a wide range of applications. This curve is given by the function*

$$p(x) = \frac{1}{\sigma\sqrt{2\pi}} e^{-(x-\mu)^2/2\sigma^2},$$

where $\pi = 3.14159265 \ldots$ and μ and σ are constants called the mean and the standard deviation. Exercises 39 and 40, which require the use of a graphing calculator or computer graphing software, illustrate its use.

▼ *Source: F. M. Scherer, "Firm Size, Market Structure, Opportunity, and the Output of Patented Inventions," American Economic Review 55 (December 1965): 1097–1125.*

39. *Test Scores* Enormous State University's Calculus I test scores are modeled by the normal distribution with $\mu = 72.6$ and $\sigma = 5.2$. The quantity $p(x)$ represents an approximation of the percentage of students who obtained a score of between $x - 0.5$ and $x + 0.5$ on the test.

(a) Graph the function p on a graphing calculator or computer and determine all its features as outlined in the text. (Find the coordinates of points accurate to one decimal place.)

(b) What percentage of students scored between 89.5 and 90.5 on the test?

(c) What is the most likely test score?

(d) What interesting feature of the curve occurs at the values $x = \mu - \sigma$ and $x = \mu + \sigma$?

40. *Consumer Satisfaction* In a survey, consumers were asked to rate a new toothpaste on a scale of 1–10. The resulting data are modeled by a normal distribution with $\mu = 4.5$ and $\sigma = 1.0$. The quantity $p(x)$ represents an approximation of the percentage of consumers who rated the toothpaste with a score of between $x - 0.5$ and $x + 0.5$ on the test.

(a) Graph the function p on a graphing calculator or computer and determine all its features as outlined in the text. (Find the coordinates of points accurate to one decimal place.)

(b) What percentage of consumers rated the product with scores of between 0.5 and 1.5?

(c) What rating gives the maximum value of p?

(d) What interesting feature of the curve occurs at the values $x = \mu - \sigma$ and $x = \mu + \sigma$?

41. *The Sine Function* Check that your graphing calculator is in "radian mode" and then graph the function $y = \sin x$ for $-10 \le x \le 10$ and $-1.5 \le y \le 1.5$. (You can graph this using the "sin" button: $Y = \boxed{\sin} X$.) Determine the approximate coordinates of two maxima, two minima, and two points of inflection. What can you say about $\lim_{x \to \pm\infty} \sin x$? [Note on the sine function: if you rotate a wheel of radius 1 ft centered at the origin and keep track of the y-coordinate of a point on its rim, then the resulting graph of y versus time, t, will have this shape.]

42. *The Tan Function* Check that your graphing calculator is in "radian mode" and then graph the function $y = \tan x$ for $-10 \le x \le 10$ and $-10 \le y \le 10$. (You can graph this using the "tan" button: $Y = \boxed{\tan} X$.) Determine the approximate coordinates of two vertical asymptotes and two points of inflection. What can you say about $\lim_{x \to \pm\infty} \tan x$?

COMMUNICATION AND REASONING EXERCISES

43. How can you use the graph of $y = f'(x)$ to locate points of inflection on the graph of $y = f(x)$?

44. How can you use the graph of $y = f'(x)$ to locate local extrema on the graph of $y = f(x)$?

45. How can you use the graph of $y = f''(x)$ to calculate points of inflection on the graph of $y = f(x)$?

46. How can you use the graph of $y = f''(x)$ to calculate local extrema on the graph of $y = f'(x)$?

In each of Exercises 47 and 48, a graphing calculator plot of $y = f'(x)$ is shown. Use this graph to give a rough sketch of the graph of $y = f(x)$, assuming the graph passes through the origin. Your sketch should show as many of the features discussed in this section as possible.

47.

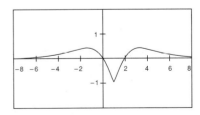

48.

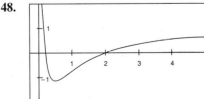

▶ ══════ **5.5** RELATED RATES

Suppose you are interested in a quantity, Q, that is varying with time. For example, you might be considering the mileage reading on your odometer, the volume of water in a tank, or the height of a space shuttle as it is taking off. Whatever Q stands for, we know that the derivative measures how fast Q is increasing or decreasing. Specifically, if we let t stand for time, then we know the following.

RATE OF CHANGE OF A QUANTITY

$\dfrac{dQ}{dt}$ is the rate of change of Q.

▼ **EXAMPLE 1**

If the height (in feet) of a space shuttle is given by the formula

$$h = 3t^2 + 2t,$$

where t is the time in seconds after blast-off, then the rate of ascent of the shuttle at time t seconds is given by

$$\frac{dh}{dt} = 6t + 2 \text{ ft/s.}$$

In particular, after 3 seconds it is ascending at a rate of $6(3) + 2 = 20$ ft/s.

▼ **EXAMPLE 2**

If the volume (in cm³) of gas in a canister is given by

$$V = e^{2t} + t^2,$$

where t is the time in minutes, then at time t minutes, the rate of change of the volume V is given by

$$\frac{dV}{dt} = 2e^{2t} + 2t \text{ cm}^3/\text{min.}$$

Thus, at the instant $t = 4$ minutes, the volume is increasing at a rate of $2e^8 + 8 \approx 5970$ cm³/min.

In this section, we will be concerned with problems called **related rates** problems. A typical related rates problem is the following.

The radius of a circle is increasing at a rate of 10 cm/sec. How fast is the area increasing at the instant when the radius has reached 5 cm?

If we look carefully at the problem, we see that there are *two* quantities that are varying with time: the radius, r, and the area, A. Moreover, these two quantities are *related* by the formula

$$A = \pi r^2.$$

The first sentence tells us that r is increasing at a certain rate. When we see a sentence like this referring to speed or rate of change, it is very helpful to rephrase the sentence using the phrase "the rate of change of."

REPHRASING A RELATED RATES PROBLEM

Rephrase all information regarding change using the words "The rate of change of . . . ."

◀ ―

Thus, the first sentence can be rephrased as follows:

The rate of change of r is equal to 10 cm/s.

The sentence can now easily be translated into the equation

$$\frac{dr}{dt} = 10 \text{ cm/s.}$$

The second sentence can be rephrased as follows:

Find the rate of change of A when $r = 5$.

Translating into mathematics, this becomes

Find $\dfrac{dA}{dt}$ when $r = 5$.

Thus, the problem can be read as follows.

$$\frac{dr}{dt} = 10. \text{ Find } \frac{dA}{dt} \text{ when } r = 5.$$

In other words, you are given one rate of change—namely, dr/dt—and are required to find the rate of change dA/dt of the related quantity A. Thus, we call this problem a "related rates" problem.

We shall now describe a systematic way of solving such problems, and we will illustrate our method by means of several examples. First, let's deal with the problem posed above.

▼ | **EXAMPLE 3**

The radius of a circle is increasing at a rate of 10 cm/s. How fast is the area increasing at the instant when the radius has reached 5 centimeters?

SOLUTION   Having done the initial analysis above, we now solve this problem using a *tableau* (that is, a fixed arrangement of all the data in the form of a kind of "spreadsheet"). We shall use three headings: The Related Quantities, The Problem, and Solution. We set up the tableau as follows

1. The Related Quantities
Under this heading, we do four things.

(1) State the quantities that are changing. In this example, they are

the radius r and the area A.

(2) Draw a diagram showing these quantities.

▶ CAUTION   Although the problem asks about $r = 5$, *do not put the 5 in the picture.* If a quantity is changing, it *must* be represented by a *letter* and not a number. Otherwise, you will be tempted to treat the quantity as constant and unchanging. On the other hand, if there was some *fixed* quantity, a quantity that is *not* changing (there is no such quantity here), then feel free to put the fixed number in the diagram.   ◀

(3) Give a formula relating these quantities.

$$A = \pi r^2$$

(4) Take the derivative *with respect to time, t,* of both sides. Before we do, notice that d/dt of the left-hand side is just dA/dt. On the right-hand side, we do not simply get $2\pi r$, as this would be the derivative *with respect to r*. Instead, we must use the chain rule.

The derivative of a quantity squared is twice that quantity times the derivative of that quantity.

Thus, the derivative of the right-hand side is $2\pi r \dfrac{dr}{dt}$.* We get

$$\frac{dA}{dt} = 2\pi r \frac{dr}{dt}.$$

▼ * Recall the generalized power rule: $\dfrac{d}{dx}[u^n] = nu^{n-1}\dfrac{du}{dx}$. Here, instead of x, the variable is t and the role of u is played by r.

Notice that whereas the formula $A = \pi r^2$ gives the relationship between the varying quantities, the derived formula gives the relationship between their *rates of change.*

2. The Problem

Under this heading, we go through the whole problem and restate it in terms of the quantities and their rates of change. Remember to rephrase all statements regarding changing quantities using the phrase "the rate of change of . . . ." We repeat the rephrasings we found in the discussion above.

The first sentence can be rephrased as follows:

The rate of increase of r is equal to 10 cm/s.

In mathematical terms,

$$\frac{dr}{dt} = 10.$$

The second sentence can be rephrased as follows:

Find the rate of increase of A when $r = 5$.

In mathematical terms,

$$\text{Find } \frac{dA}{dt} \text{ when } r = 5.$$

Thus, the whole problem can be stated as follows.

$$\frac{dr}{dt} = 10. \; \textit{Find } \frac{dA}{dt} \textit{ when } r = 5.$$

3. Solution

First copy the relationship between the rates of increase obtained under the first heading.

$$\frac{dA}{dt} = 2\pi r \frac{dr}{dt}$$

Now substitute the data given under heading 2, and solve for the unknown quantity $\dfrac{dA}{dt}$.

$$\frac{dA}{dt} = 2\pi(5)(10) = 100\pi \approx 314.15 \text{ cm}^2\text{/s}$$

(Note that the units, centimeters and seconds, are specified in the problem.)

Thus, the answer is that the area is increasing at a rate of approximately 314.15 cm²/s.

That's basically all there is to it! (There will sometimes be need for a little more work, as we'll see in the next few examples.)

This is what your tableau in Example 3 should look like when you are done:

FINISHED TABLEAU

1. The Related Quantities

    (1) Changing quantities:

        the radius r and the area A.

    (2) Diagram:

    (3) Formula relating the changing quantities: $A = \pi r^2$
    (4) Derived formula:

$$\frac{dA}{dt} = 2\pi r \frac{dr}{dt}$$

2. The Problem

$$\frac{dr}{dt} = 10. \text{ Find } \frac{dA}{dt} \text{ when } r = 5.$$

3. Solution

$$\frac{dA}{dt} = 2\pi r \frac{dr}{dt}$$

Substituting from heading 2,

$$\frac{dA}{dt} = 2\pi(5)(10) = 100\pi \approx 314.15 \text{ cm}^2/\text{s}.$$

Thus, the area is increasing at a rate of approximately 314.15 cm²/s.

▼ **EXAMPLE 4**   Average Cost

The cost to manufacture x portable pagers is

$$C(x) = \$10{,}000 + 3x + \frac{x^2}{10{,}000}.$$

The production level is currently $x = 5{,}000$ and is increasing by 100 units per day. How is the average cost changing?

SOLUTION

---

1. The Related Quantities

(1) The changing quantities are the production level, x, and the average cost, $\overline{C}$.

(2) In this example, the changing quantities cannot easily be depicted geometrically.

(3) The formula relating the changing quantities must feature both x and the average cost $\overline{C}$. We are given a formula for the *total* cost, so we can obtain one for the *average* cost by dividing by x.

$$\overline{C} = \frac{aC}{x}$$

So

$$\overline{C} = \frac{10{,}000}{x} + 3 + \frac{x}{10{,}000}.$$

(4) Taking derivatives with respect to t of both sides, we obtain the derived formula

$$\frac{d\overline{C}}{dt} = \left(-\frac{10{,}000}{x^2} + \frac{1}{10{,}000}\right)\frac{dx}{dt}.$$

2. The Problem

We see in the second sentence that $x = 5{,}000$ and is increasing by 100 units per day. We rephrase the second fact as follows:

The rate of change of x is 100 units/day.

In mathematical terms,

$$x = 5{,}000 \text{ and } \frac{dx}{dt} = 100.$$

The last sentence asks how the average cost is changing. In other words, we need to

find the rate of increase of $\overline{C}$.

In mathematical terms,

$$\text{Find } \frac{d\overline{C}}{dt}.$$

Thus, the problem reads

$$x = 5{,}000 \text{ and } \frac{dx}{dt} = 100. \text{ Find } \frac{d\overline{C}}{dt}.$$

3. Solution

We first copy the derived formula from heading 1:

$$\frac{d\overline{C}}{dt} = \left(-\frac{10,000}{x^2} + \frac{1}{10,000}\right)\frac{dx}{dt}.$$

Substituting from heading 2,

$$\frac{d\overline{C}}{dt} = \left(-\frac{10,000}{5,000^2} + \frac{1}{10,000}\right)100 = -0.03 \text{ dollars/day}$$

Thus, the average cost is decreasing by 3¢ per day.

FINISHED TABLEAU

1. The Related Quantities

   (1) The changing quantities are the production level, x, and the average cost, $\overline{C}$.
   (2) (No diagram)
   (3) Formula relating the varying quantities:

$$\overline{C} = \frac{10,000}{x} + 3 + \frac{x}{10,000}$$

   (4) Derived formula:

$$\frac{d\overline{C}}{dt} = \left(-\frac{10,000}{x^2} + \frac{1}{10,000}\right)\frac{dx}{dt}$$

2. The Problem

$$x = 5,000 \text{ and } \frac{dx}{dt} = 100. \text{ Find } \frac{d\overline{C}}{dt}.$$

3. Solution

Derived formula:

$$\frac{d\overline{C}}{dt} = \left(-\frac{10,000}{x^2} + \frac{1}{10,000}\right)\frac{dx}{dt}$$

Substituting from heading 2,

$$\frac{d\overline{C}}{dt} = \left(-\frac{10,000}{5,000^2} + \frac{1}{10,000}\right)100 = -0.03 \text{ dollars/day}$$

Thus, the average cost is decreasing by 3¢ per day.

The following is one of the "ladder problems" found in most traditional textbooks.

▼ EXAMPLE 5   Ladders

Jane is at the top of a 5-foot ladder when it starts to slide down the wall at a rate of 3 feet per minute. Jack is standing on the ground behind her. How fast is the base of the ladder moving when it hits him, if Jane is 4 feet from the ground at that instant?

SOLUTION   Instead of doing the tableau twice, we'll start with some discussion and move to the finished tableau right away.

Notice that the varying quantities are not clearly stated. If we force a rephrasing of the first sentence in our usual way, one of these quantities becomes clear.

> The rate of change of the height of (the top of) a 5 foot ladder is −3 ft/min.

Thus, one of the changing quantities is h, the height of the top of the ladder. Note the negative sign: the ladder is falling *down* at 3 ft/min.

The third sentence reveals the second changing quantity. Rephrased, it becomes:

> Find the rate of change of the distance of the base from the wall when $h = 4$ ft.

Thus, the second quantity that is changing is x, the distance of the base from the wall.

We can now move to the finished tableau.

---

1. The Related Quantities

    (1) The changing quantities are

        h, the height of the top of the ladder and

        x, the distance of the base from the wall.

    (2) Diagram:

(Note that the length of the ladder is *fixed* at 5, so we can include the 5 in the diagram.)

---

(3) Formula relating the changing quantities:

$h^2 + x^2 = 25$ (Pythagorean theorem for a right triangle)

(4) Derived formula:

$$2h\frac{dh}{dt} + 2x\frac{dx}{dt} = 0$$

2. The Problem

$$\frac{dh}{dt} = -3. \text{ Find } \frac{dx}{dt} \text{ when } h = 4.$$

3. Solution

Derived formula:

$$2h\frac{dh}{dt} + 2x\frac{dx}{dt} = 0$$

Substituting from heading 2,

$$2(4)(-3) + 2x\frac{dx}{dt} = 0.$$

(But wait! We cannot solve for the unknown dx/dt until we know the value of x. But we *do* know what h is, and moreover, h and x are related to each other by the original formula $h^2 + x^2 = 25$. Thus, we can get x by substituting the known value of h in this formula:

$$(4)^2 + x^2 = 25$$

which we solve to find $x = 3$.) Thus,

$$2(4)(-3) + 2(3)\frac{dx}{dt} = 0$$

$$-24 + 6\frac{dx}{dt} = 0,$$

$$\frac{dx}{dt} = \frac{24}{6} = 4 \text{ ft/min.}$$

Thus, the base of the ladder is moving away from the wall at a rate of 4 ft/min when it hits Jack.

The scenario in the next example comes from an example in the section on applications of maxima and minima.

▼ **EXAMPLE 6**   Automation

The Gym Sock Company manufactures cotton athletic socks. Production is partially automated through the use of robots. The number of pairs of socks q it can manufacture in a day is given by a Cobb-Douglas production formula

$$q = 50n^{0.6}r^{0.4},$$

where n is the number of laborers and r is the number of robots. The company currently produces 1,000 pairs of socks per day and employs 20 laborers. It is bringing one new robot on line every month. At what rate are laborers being laid off, assuming that daily productivity remains constant?

SOLUTION   Here is the tableau:

---

1. The Related Quantities

  (1) The changing quantities are

   the number of laborers, n,
   the number of robots, r.

  (2) (No diagram)
  (3) Formula relating the changing quantities:

$$1,000 = 50n^{0.6}r^{0.4}$$

   or

$$20 = n^{0.6}r^{0.4}.$$

   (Note that productivity is constant at 1,000 pairs of socks per day.)

  (4) Derived formula:

$$0 = 0.6n^{-0.4}\left(\frac{dn}{dt}\right)r^{0.4} + 0.4n^{0.6}r^{-0.6}\left(\frac{dr}{dt}\right)$$

$$= 0.6\left(\frac{r}{n}\right)^{0.4}\left(\frac{dn}{dt}\right) + 0.4\left(\frac{n}{r}\right)^{0.6}\left(\frac{dr}{dt}\right)$$

   (We solve this equation for dn/dt because we shall want to find dn/dt below, and because the equation becomes simpler when we do this.)

$$0.6\left(\frac{r}{n}\right)^{0.4}\left(\frac{dn}{dt}\right) = -0.4\left(\frac{n}{r}\right)^{0.6}\left(\frac{dr}{dt}\right)$$

$$\frac{dn}{dt} = -\frac{0.4}{0.6}\left(\frac{n}{r}\right)^{0.6}\left(\frac{n}{r}\right)^{0.4}\left(\frac{dr}{dt}\right)$$

$$= -\frac{2}{3}\left(\frac{n}{r}\right)\left(\frac{dr}{dt}\right)$$

2. The Problem

$$\frac{dr}{dt} = 1. \text{ Find } \frac{dn}{dt} \text{ when } r = 20.$$

3. Solution
Derived formula:

$$\frac{dn}{dt} = -\frac{2}{3}\left(\frac{n}{r}\right)\left(\frac{dr}{dt}\right)$$

Substituting from heading 2,

$$\frac{dn}{dt} = -\frac{2}{3}\left(\frac{20}{r}\right)(1).$$

(We can compute r by substituting the known value of n in the original formula:

$$20 = n^{0.6}r^{0.4}$$
$$20 = 20^{0.6}r^{0.4}$$
$$r^{0.4} = \frac{20}{20^{0.6}} = 20^{0.4},$$

giving $r = 20$.)
    Thus,

$$\frac{dn}{dt} = -\frac{2}{3}\left(\frac{20}{20}\right)(1) = -\frac{2}{3} \text{ laborer per month}$$

Thus, the company is laying off laborers at a rate of $\frac{2}{3}$ per month, or two every three months.

Before we go on... We can interpret this result as saying that at the current level of production and number of laborers, one robot is as productive as $\frac{2}{3}$ of a laborer, or that 3 robots are as productive as 2 laborers.

▶   **5.5 EXERCISES**

Translate the statements in Exercises 1–8 into mathematical terms.

1. The population P is currently 10,000 and growing at a rate of 1,000 per year.

2. There are presently 400 cases of Bangkok flu, and the number is growing by 30 new cases every month.

3. The annual revenue of your tie-dye T-shirt operation is currently $7,000 and growing by 10% each year. How fast are annual sales increasing?

4. A ladder is sliding down a wall so that the distance between the top of the ladder and the floor is decreasing at a rate of 3 ft/s. How fast is the base of the ladder receding from the wall?

5. The price of shoes is rising $5 per year. How fast is the demand changing?

6. Stock prices are rising $1,000 per year. How fast is the value of your portfolio increasing?

APPLICATIONS

9. *Doggie Puddles*   The area of a circular doggie pud-dle is growing at a rate of 12 cm²/s.
   (a) How fast is the radius growing at the instant when it equals 10 cm?
   (b) How fast is the radius growing at the instant when the puddle has an area of 49 cm²?

10. *Doggie Puddles*   The radius of a circular doggie puddle is growing at a rate of 5 cm/sec.
   (a) How fast is its area growing at the instant when the radius is 10 cm?
   (b) How fast is the area growing at the instant when it equals 36 cm²?

11. *Sliding Ladders*   The base of a 50-foot ladder is being pulled away from a wall at a rate of 10 feet per second. How fast is the top of the ladder sliding down the wall at the instant when the base of the ladder is 30 ft from the wall?

12. *Sliding Ladders*   The top of a 5-foot ladder is slid-ing down a wall at a rate of 10 feet per second. How fast is the base of the ladder sliding away from the wall at the instant when the top of the ladder is 3 feet from the ground?

13. *Demand*   Assume that the demand function for tuna in a small coastal town is given by

$$p = \frac{50,000}{q^{1.5}},$$

where p is the price (in $) per pound of tuna, and q is the number of pounds of tuna that can be sold at the price, p, in one month. The town's fishery finds that the demand for tuna is currently 900 pounds per month and is increasing at a rate of 100 pounds per month. How fast is the price changing?

14. *Demand*   In Chapter 1, we found that the demand equation for rubies at Royal Ruby Retailers (RRR) is

7. The average global temperature is 60°F and rising by 0.1°F per decade. How fast are annual sales of Bermuda shorts increasing?

8. The country's population is now 260,000,000 and is increasing by 1,000,000 people per year. How fast is the annual demand for diapers increasing?

given by

$$q = -\frac{4p}{3} + 80,$$

where p is the price RRR charges per ruby and q is the number of rubies it can sell per week at p dollars per ruby. RRR finds that the demand for its rubies is cur-rently 20 rubies per week and is dropping at a rate of one per week. How fast is the price changing?

15. *Demand*   Demand for your tie-dyed T-shirts is given by the formula

$$p = 5 + \frac{100}{\sqrt{q}},$$

where p is the price (in dollars) you can charge to sell q T-shirts per month. If you currently sell T-shirts for $15 each, and you raise your price by $2 per month, how fast will the demand drop?

16. *Supply*   The number of portable CD players you are prepared to supply to the local retail outlet every week is given by the formula

$$p = 0.1q^2 + 3q,$$

where p is the price it offers. The retail outlet is cur-rently offering you $40 per CD player. If the price it offers decreases at a rate of $10 per week, how will this affect the rate of supply?

17. *Revenue*   You can now sell 50 cups of lemonade per week at 30¢ per cup, but demand is dropping at a rate of 5 cups per week each week. Assuming that raising the price does not affect demand, how fast do you have to raise your price if you want to keep your weekly revenue constant?

18. *Revenue*   You can now sell 40 cars per month at $20,000 per car, and demand is increasing at a rate of 3 cars per month each month. What is the fastest you could drop your price before your monthly revenue starts to drop?

19. *Production* The automobile assembly plant you manage has a Cobb-Douglas production function given by

$$P = 10x^{0.3}y^{0.7},$$

where P is the number of automobiles it produces per year, x is the number of employees, and y is the daily operating budget (in dollars). You maintain a production level of 1,000 automobiles per year. If you currently employ 150 workers and are hiring new workers at a rate of 10 per year, how fast is your daily operating budget changing?

20. *Production* Referring to the Cobb-Douglas production formula in Exercise 19, assume that you maintain a work force of 200 workers and wish to increase production in order to meet a demand that is increasing by 100 automobiles per year. The current demand is 1,000 automobiles per year. How fast should your daily operating budget be increasing?

21. *Balloons* A spherical party balloon is being inflated by pumping in helium at a rate of 3 cubic feet per minute. How fast is the radius growing at the instant when the radius has reached 1 foot? (The volume of a sphere of radius r is $V = \frac{4}{3}\pi r^3$.)

22. *More Balloons* A rather flimsy spherical balloon is designed to pop at the instant its radius has reached 10 cm. Assuming the balloon is filled with helium at a rate of 10 cm³/s, calculate how fast the diameter is growing at the instant it pops. (The volume of a sphere of radius r is $V = \frac{4}{3}\pi r^3$.)

23. *Movement along a Graph* A point on the graph of $y = 1/x$ is moving along the curve in such a way that its x-coordinate is increasing at a rate of 4 units per second. What is happening to the y-coordinate at the instant the y-coordinate is equal to 2?

24. *Motion around a Circle* A point is moving along the circle $x^2 + (y - 1)^2 = 8$ in such a way that its x-coordinate is decreasing at a rate of 1 unit per second. What is happening to the y-coordinate at the instant when the point has reached $(-2, 3)$?

25. *Ships Sailing Apart* The H.M.S. Dreadnought is 40 miles north of Montauk and steaming due north at 20 mph, while the U.S.S. Mona Lisa is 50 miles east of Montauk and steaming due east at an even 30 mph. How fast is their distance apart increasing?

26. *Near Miss* My aunt and I were approaching the same intersection, she from the south and I from the west. She was traveling at a steady speed of 10 mph, while I was approaching the intersection at 60 mph. At a certain instant in time, I was one-tenth of a mile from the intersection, while she was one-twentieth of a mile from it. How fast were we approaching each other at that instant?

27. *Education* In 1991, the expected income of an individual depended on his or her educational level according to the following formula.

$$I(n) = 2,928.8n^3 - 115,860n^2 + 1,532,900n$$
$$-6,760,800 \quad (11.5 \le n \le 15.5)$$

Here, n is the number of school years completed, and $I(n)$ is the individual's expected income.* You have completed 13 years of school and are currently a part-time student. Your schedule is such that you will complete the equivalent of one year of college every three years. Assuming that your salary is linked to the above model, how fast is your income going up? (Round your answer to the nearest $1.)

28. *Education* Referring to the model in the previous exercise, assume that someone has completed 14 years of school and that her income is increasing by $10,000 per month. How much schooling per year is this rate of increase equivalent to?

29. *Employment* An employment research company estimates that the value of a recent MBA graduate to an accounting company is estimated as

$$V = 3e^2 + 5g^3,$$

where V is the value of the graduate, e is the number of years of prior business experience, and g is the graduate school grade point average. A company that currently employs graduates with a 3.0 average wishes to maintain a constant employee value of $V = 200$ but finds that the grade point average of its new employees is dropping at a rate of 0.2 per year. How fast must the experience of its new employees be growing in order to compensate for the decline in grade point average?

▼ * The model is a best-fit cubic based on Table 358, U.S. Department of Education, *Digest of Education Statistics, 1991,* Washington, D.C.: Government Printing Office, 1991.

30. *Grades** A production formula for a student's performance on a difficult English examination is given by

$$g = 4hx - 0.2h^2 - 10x^2,$$

where g is the grade the student can expect to obtain, h is the number of hours of study for the examination, and x is the student's grade point average. The instructor finds that students' grade point averages have remained constant at 3.0 over the years, and that students currently spend an average of 15 hours studying for the examination. However, scores on the examination are dropping at a rate of 10 points per year. At what rate is the average study time decreasing?

31. *Cones* A right circular conical vessel is being filled with green industrial waste at a rate of 100 cubic meters per second. How fast is the level rising after 200π cubic meters have been poured in? (The cone has height 50 m and radius of 30 m at its brim. The volume of a cone of height h and cross-sectional radius r at its brim is given by $V = \frac{1}{3}\pi r^2 h$.)

32. *More Cones* A circular conical vessel is being filled with ink at a rate of 10 cm³/s. How fast is the level rising after 20 cm³ have been poured in? (The cone has height 50 cm and radius of 20 cm at its brim. The volume of a cone of height h and cross-sectional radius r at its brim is given by $V = \frac{1}{3}\pi r^2 h$.)

33. *Cylinders* The volume of paint in a right cylindrical can is given by $V = 4t^2 - t$, where t is time in seconds and V is the volume in cm³. How fast is the level rising when the height is 2 cm? The can has a height of 4 cm and a radius of 2 cm. (*Hint:* To get h as a function of t, first solve the volume $V = \pi r^2 h$ for h.)

34. *Cylinders* A cylindrical bucket is being filled with

paint at a rate of 6 cm³ per minute. How fast is the level rising when the bucket starts to overflow? The bucket has radius 30 cm and height 60 cm.

Education and Crime *The following graph shows a striking relationship between the total prison population and the average combined SAT score in the United States.*

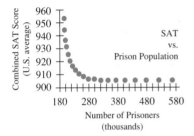

SAT
vs.
Prison Population

Exercises 35 and 36 are based on the following model for these data:

$$S(n) = 904 + \frac{1326}{(n - 180)^{1.325}} \quad (192 \le n \le 563).$$

Here, $S(n)$ is the combined average SAT score at a time when the total prison population is n thousand.[†]

35. In 1985, the U.S. prison population was 475,000 and increasing at a rate of 35,000 per year. What was the average SAT score, and how fast, and in what direction, was it changing? (Round your answers to two decimal places.)

36. In 1970, the U.S. combined SAT average was 940 and dropping by 10 points per year. What was the U.S. prison population, and how fast, and in what direction, was it changing? (Round your answers to the nearest 100.)

COMMUNICATION AND REASONING EXERCISES

37. If you know how fast one quantity is changing and need to compute how fast a second quantity is changing, what kind of information would you need to know?

38. If three quantities are related by a single equation, how would you go about computing how fast one of them is changing based on a knowledge of the other two?

▼ *Based on an exercise in *Introduction to Mathematical Economics* by A. L. Ostrosky, Jr., and J. V. Koch (Prospect Heights, Ill.: Waveland Press, 1979).

[†] The model is the authors', based on data for the years 1967–1989. Sources: *Sourcebook of Criminal Justice Statistics, 1990,* p. 604; Educational Testing Service.

39. Why is this section titled "related rates?"

40. In a recent exam, you were given a related rates problem based on an algebraic equation relating two variables x and y. Your friend told you that the correct relationship between dx/dt and dy/dt was given by

$$\left(\frac{dx}{dt}\right) = \left(\frac{dy}{dt}\right)^2.$$

Could he have been correct?

41. Transform the following into a mathematical statement about derivatives: "If my grades are improving at twice the speed of yours, then your grades are improving at half the speed of mine."

42. If two quantities x and y are related by a linear equation, how are their rates of change related?

▶ ══ 5.6 ELASTICITY OF DEMAND

Suppose you are manufacturing an extremely popular brand of sneakers, and you are trying to establish how the demand will be affected by an increase in price. Common sense tells you that the demand will decrease as you raise the price. However, it may be the case that the percentage drop in demand is smaller than the percentage increase in price. For example, if you raise the price by 1%, you might suffer only a 0.5% loss in sales. In this case, the loss in sales would be more than offset by the increase in price. Your overall revenue will go up, so a price increase would be in your best interests. In such a situation we say that the demand is **inelastic,** because it is not very sensitive to the increase in price. On the other hand, if your 1% price increase results in a 2% drop in demand, then raising the price will cause a drop in revenues. We would then say that the demand is **elastic** because it is sensitive to a price increase. The wise business executive would do well to *decrease* the price of sneakers in this case.

We can use calculus to predict the response of demand to price changes, provided we have a demand equation for the item we are selling.* To gauge the effect on revenue, we need to look at the percentage drop in demand corresponding to a 1% increase in price. In other words, we want the *percentage drop in demand per percentage increase in price*. This ratio is called the **elasticity of demand**, or **price elasticity of demand**, because it measures the degree to which demand responds to changes in price. The elasticity is usually denoted by the letter E. We will now show you how to get a formula for E starting from a demand equation.

So, let us assume that you are given a demand equation

$$q = f(p),$$

where q stands for the number of items you would sell (per week, per month or what have you) if you set the price per item at p. Now suppose you increase

▼ * Coming up with a good demand equation is not always easy. We saw in Chapter 1 that it is possible to find a linear demand equation if we know the sales figures at two different prices. However, such an equation is only a first approximation. To come up with a more accurate demand equation, we might need to gather data corresponding to sales at several different prices and use "curve-fitting" techniques. Another approach would be an analytic one, based on mathematical modeling techniques an economist might use.

the price per item p by a very small amount Δp. Then your percentage increase in price is $\Delta p/p \times 100$ (we multiply by 100 to get a percentage). Now this increase in p will presumably result in a decrease in demand q. Let us denote this corresponding decrease in q by $-\Delta q$ (we use the minus sign because, by convention, Δq stands for the *increase* in demand). Thus, the percentage decrease in demand is $-\Delta q/q \times 100$.

Now recall that E is the ratio

$$E = \frac{\text{Percentage decrease in demand}}{\text{Percentage increase in price}}.$$

Substituting gives

$$E = \frac{\dfrac{-\Delta q}{q} \times 100}{\dfrac{\Delta p}{p} \times 100}.$$

Canceling the 100s and reorganizing, we get

$$E = -\frac{\Delta q}{\Delta p} \cdot \frac{p}{q}.$$

Q This seems fine in theory, but what value should we use for the increase in price, Δp?

Here is an extreme example. Suppose that, delirious with greed, I increase the price of my latest-model sneakers by $1,000,000 per pair. Then I would be very lucky if I sold a single pair! In other words, I should expect the sales to drop to zero. Now this is hardly telling me how the market is going to respond to a modest price increase. In fact, it tells me nothing at all.

A In order to measure the effect most accurately, it seems best to choose the increase in p to be as *small* as possible. The smaller the value of Δp, the more accurately we can gauge the response to a price increase at the current pricing level. In other words, we are interested in the limit as Δp approaches 0.

Now the above equation for E includes the term $\Delta q/\Delta p$, which is precisely the difference quotient associated with the function q of p. Thus, as the value of Δp approaches 0, this quotient approaches the *derivative, dq/dp,* of q with respect to p. It follows that the most *useful* definition of E is the following.

ELASTICITY OF DEMAND

The **elasticity of demand, E,** is the percentage rate of decrease of demand per percentage increase in price. E is given by the formula

$$E = -\frac{dq}{dp} \cdot \frac{p}{q}.$$

▶ **NOTE**   We can also think of E another way: as *the percentage rate of increase of demand per percentage drop in price.* ◀

For our first example, we revisit Royal Ruby Retailers.

▼   **EXAMPLE 1**   Pricing Policy

The demand for rubies at RRR is given by the equation

$$q = -\frac{4p}{3} + 80,$$

where p is the price RRR charges (in dollars) and q is the number of rubies it sells per week. Find a formula for the elasticity of demand, E, evaluate it at the price level of \$40 per ruby, interpret the answer, and hence determine whether RRR should increase or decrease the price in order to increase revenue.

SOLUTION   The first part involves applying the formula

$$E = -\frac{dq}{dp} \cdot \frac{p}{q}.$$

Taking the derivative of q, we see that $\frac{dq}{dp} = -\frac{4}{3}$, so that

$$E = \frac{4p}{3q}.$$

We get E in terms of the price p alone by substituting for q.

$$E = \frac{4p}{3\left(-\dfrac{4p}{3} + 80\right)}$$

$$= \frac{4p}{240 - 4p} = \frac{p}{60 - p}$$

Now we have the formula for E that we want. When $p = \$40$ per ruby, we get

$$E = \frac{40}{60 - 40} = 2.$$

The percentage drop in demand is twice the percentage increase in price at the current pricing level, so RRR would do well not to increase the price. In fact it should *decrease* the price, because then the percentage *increase* in demand will be twice the percentage decrease in price.

As we said at the start of this section, the demand in the above example is elastic. More precisely:

ELASTIC AND INELASTIC DEMAND AND UNIT ELASTICITY

We say that the demand is
elastic if $E > 1$,
inelastic if $E < 1$, and
has **unit elasticity** if $E = 1$.

◄

Thus, if the demand is elastic, you can increase revenue by decreasing the price, and if the demand is inelastic you can increase revenue by increasing the price.

Q What about unit elasticity?

A Suppose that you have chosen the unit price exactly right so that the revenue is as large as possible. Then the demand cannot be elastic; otherwise, you would be able to increase the revenue by lowering the unit price. At the same time, the demand cannot be *inelastic*; otherwise, you would be able to increase the revenue by *raising* the unit price. Thus, it must be that the demand is neither elastic nor inelastic. That is, it must have unit elasticity, $E = 1$.

Before we make this more precise, let us apply this reasoning to the previous example.

▼ **EXAMPLE 2**   Revenue

Referring to Example 1, find the price RRR should charge in order to maximize revenue.

SOLUTION   We are seeking the price level that gives unit elasticity, $E = 1$. We already have an expression for E as a function of price p from the last example,

$$E = \frac{p}{60 - p}.$$

Because we want E to equal 1, we set this expression equal to 1 and obtain

$$1 = \frac{p}{60 - p}.$$

Multiplying by the denominator on the right gives

$$60 - p = p,$$

and so $2p = 60$, or $p = 30$.

Thus we conclude that RRR should sell rubies at \$30 apiece in order to maximize revenue. Notice that this is exactly the answer we got in Chapter 1 (when we didn't know any calculus!).

Before going on to the next example, let's look at why the revenue is maximized when $E = 1$. We'll start by treating this as we would any optimization problem. To do this, we must find an expression for the total revenue R in terms of p and then set its derivative with respect to p equal to 0. Recall that

Total revenue = Price per item × Total number of items sold

or $\qquad\qquad R = pq.$

We know that q is a function of p. Thus, R is a function of p although we don't have an explicit formula for this function. We proceed to take the derivative of R with respect to p and set it equal to 0. Because R is a product, we use the product rule.

$$\frac{dR}{dp} = \frac{dq}{dp} p + q \cdot 1 = \frac{dq}{dp} p + q$$

Now, in order to get a maximum revenue, dR/dp must be zero. Thus,

$$\frac{dq}{dp} p + q = 0.$$

Instead of solving for p, we fiddle with the above equation a little. First subtract q from both sides, getting

$$\frac{dq}{dp} p = -q$$

Now divide both sides by $-q$, to get

$$-\frac{dq}{dp} \cdot \frac{p}{q} = 1.$$

Recognize the expression on the left? That's E. Thus, in order to have maximum revenue, we must have $E = 1$, as we claimed.

DETERMINING THE PRICE THAT GIVES MAXIMUM REVENUE

The maximum revenue occurs when $E = 1$. Thus, to determine the price that gives maximum revenue, express elasticity of demand E as a function of p, equate it to 1, and solve for p. (If there is more than one solution for p, choose the solution yielding the largest revenue.)

Q Why does setting $E = 1$ *guarantee* finding the maximum revenue? This seems suspicious in view of the fact that many functions do not have absolute maxima at all.

A Well, we can argue as follows. Setting $p = 0$ always results in zero revenue. On the other hand, in all realistic situations, setting p too high will also result in zero revenue. Thus, there must be *some price p* in between such that the revenue is not zero. One of these intermediate prices will give the *maximum* revenue. Thus, there *is* a value of p that gives a maximum revenue, and setting the derivative of R equal to zero *must* yield that value!* As we saw above, this corresponds to setting $E = 1$.

▼ **EXAMPLE 3**   Revenue

Suppose that the demand equation for Bobby Dolls is given by $q = 216 - p^2$, where p is the price per doll in dollars and q represents weekly sales. Find the range of prices for which **(a)** the demand is elastic, **(b)** the demand is inelastic, and **(c)** the weekly revenue is maximized. Also, calculate the maximum weekly revenue.

SOLUTION   As in the previous examples, we must first calculate E in terms of p. We have

$$E = -\frac{dq}{dp} \cdot \frac{p}{q}$$

$$= 2p \cdot \frac{p}{216 - p^2}$$

$$= \frac{2p^2}{216 - p^2}.$$

We answer part (c) first. Setting $E = 1$, we get

$$\frac{2p^2}{216 - p^2} = 1,$$

or $2p^2 = 216 - p^2,$

so that $3p^2 = 216,$

or $p^2 = 72.$

Thus, we conclude that the maximum revenue occurs when $p = \sqrt{72} \approx \$8.49$. We can now answer parts (a) and (b) without further calculation: the demand is elastic when $p > \$8.49$ (the price is too high), and the demand is inelastic when $p < \$8.49$ (the price is too low).

Finally, we calculate the maximum weekly revenue, which equals the revenue at the price of $8.49.

$$R = qp = (216 - p^2)p = (216 - 72)\sqrt{72}$$
$$= 144\sqrt{72} \approx \$1,221.88$$

▼ * We are assuming that the demand equation is differentiable (smooth) so that we don't need to worry about singular points or other strange behavior.

▶ **5.6 EXERCISES**

APPLICATIONS

1. *Demand for Oranges* Given that the weekly sales of Honolulu Red Oranges is given by $q = 1,000 - 20p$, calculate the elasticity of demand for a price of $30 per orange. Interpret your answer and calculate the price that gives a maximum weekly revenue. Also find this maximum revenue.

2. *Demand for Oranges* Repeat Exercise 1 for weekly sales of $1,000 - 10p$.

3. *Tissues* The consumer demand curve for tissues is given by $q = (100 - p)^2$, where p is the price per case of tissues and q is the demand in weekly sales.
   (a) Determine the elasticity of demand E when the price is set at $30, and interpret your answer.
   (b) At what price should tissues be sold in order to maximize the revenue?
   (c) Approximately how many cases of tissues would be demanded at that price?

4. *Bodybuilding* The consumer demand curve for Professor Stefan Schwartzenegger dumbbells is given by $q = (100 - 2p)^2$, where p is the price per dumbbell, and q is the demand in weekly sales. Find the price Professor Schwartzenegger should charge for his dumbbells in order to maximize revenue.

5. *College Tuition* A study of about 1800 colleges and universities in the U.S. resulted in the demand equation $q = 9859.39 - 2.17p$, where q is the enrollment at a college or university, and p is the average annual tuition (plus fees) it charges.*
   (a) The study also found that the average tuition charged by universities and colleges was $2,867. What is the corresponding elasticity of demand? Interpret your answer.
   (b) Based on the study, what would you advise a college to charge its students in order to maximize total revenue, and what would the revenue be?

6. *Demand for Fried Chicken* A fried chicken franchise finds that the demand equation for its new roast chicken product, "Roasted Rooster," is given by

$$p = \frac{40}{q^{1.5}},$$

where p is the price (in dollars) per quarter-chicken serving, and q is the number of quarter-chicken servings that can be sold per hour at this price. Express q as a function of p and find the elasticity of demand when the price is set at $4 per serving. Interpret the result.

7. *Linear Demand Functions* A general linear demand function has the form $f(p) = mp + b$ (m, b constants, $m \neq 0$).
   (a) Obtain a formula for the elasticity of demand at a unit price of p.
   (b) Obtain a formula for the price that maximizes revenue.

8. *Exponential Demand Functions* A general exponential demand function has the form $f(p) = Ae^{-bp}$ (A, b nonzero constants).
   (a) Obtain a formula for the elasticity of demand at a unit price of p.
   (b) Obtain a formula for the price that maximizes revenue.

9. *Hyperbolic Demand Functions* A general hyperbolic demand function has the form $f(p) = \dfrac{k}{p^r}$ (r, k nonzero constants).
   (a) Obtain a formula for the elasticity of demand at unit price p.
   (b) How does E vary with p?
   (c) What does the answer to (b) say about the model?

10. *Quadratic Demand Functions* A general quadratic demand function has the form $f(p) = ap^2 + bp + c$ (a, b, c constants with $a \neq 0$).
    (a) Obtain a formula for the elasticity of demand at a unit price p.
    (b) Obtain a formula for the price or prices that could maximize revenue.

11. *Exponential Demand Functions* The estimated monthly sales of Mona Lisa paint-by-number sets is given by the formula $q = 100e^{-3p^2+p}$, where q is the demand in monthly sales and p is the retail price in ¥.
    (a) Determine the elasticity of demand, E, when the retail price is set at ¥3 and interpret your answer.
    (b) At what price will revenue be a maximum?
    (c) Approximately how many paint-by-number sets will be sold per week at that price?

▼ * Based on a study by A. L. Ostrosky, Jr., and J. V. Koch, as cited in their book, *Introduction to Mathematical Economics* (Prospect Heights, Ill.: Waveland Press, 1979), p. 133.

12. *Exponential Demand Functions* Repeat the previous exercise using the demand equation $q = 100e^{p-3p^2 12/p}$.

13. *Modeling Linear Demand (a new look at an old exercise from Chapter 1)*   You have been hired as a marketing consultant to Johannesburg Burger Supply, Inc., and you wish to come up with a unit price for its hamburgers in order to maximize the company's weekly revenue. In order to make life as simple as possible, you assume that the demand equation for Johannesburg hamburgers has the linear form $q = mp + b$, where p is the price per hamburger, q is the demand in weekly sales, and m and b are certain constants you'll have to figure out.
  (a) Your market studies reveal the following sales figures: when the price is set at $2.00 per hamburger, the sales amount to 3,000 per week, but when the price is set at $4.00 per hamburger, the sales drop to zero. Use this data to calculate the demand equation.
  (b) Now estimate the unit price in order to maximize weekly revenue and predict what the weekly revenue will be at that price.
  (Compare your answer with the corresponding exercise in Chapter 1, Section 4.)

14. *Modeling Linear Demand*   You have been hired as a marketing consultant to Big Book Publishing, Inc., and you have been approached to determine the best selling price for the hit calculus text by Whiner and Istanbul entitled *Fun With Derivatives*. You decide to make life easy and assume that the demand equation for *Fun With Derivatives* has the linear form $q = mp + b$, where p is the price per book, q is the demand in annual sales, and where m and b are certain constants you'll have to figure out.
  (a) Your market studies reveal the following sales figures: when the price is set at $50.00 per book, the sales amount to 10,000 per year; when the price is set at $80.00 per book, the sales drop to 1,000 per year. Use these data to calculate the demand equation.
  (b) Now estimate the unit price in order to maximize annual revenue and predict what Big Book Publishing, Inc.'s annual revenue will be at that price.

15. *Modeling Exponential Demand*   As the new owner of a supermarket, you have inherited a large inventory of unsold imported Limburger cheese and would like to set the price so that your revenue from selling it is as large as possible. Previous sales figures

of the cheese are shown in the following table:

Price per pound (p)	$3.00	$4.00	$5.00
Monthly sales in pounds (q)	407	287	223

  (a) Use the sales figures for the prices $3 and $5 per pound to construct a demand function of the form $f(p) = Ae^{-bp}$, where A and b are constants you must determine.
  (b) Use your demand function to find the elasticity of demand at each of the prices listed.
  (c) At what price should you sell the cheese in order to maximize monthly revenue?
  (d) If your total inventory of cheese amounts to only 200 pounds, and it will spoil one month from now, how should you price it in order to make the largest revenue? Is this the same answer you got in (c)? If not, give a brief explanation.

16. *Modeling Exponential Demand* Repeat Exercise 15, but this time use the sales figures for $4 and $5 per pound to construct the demand function.

17. *Income Elasticity of Demand (based on a question on the GRE economics test)*   If $Q = aP^\alpha Y^\beta$ is the individual's demand function for a commodity, where P is the price of the commodity, Y is the individual's income, and a, α, and β are parameters, explain why β can be interpreted as the **income elasticity of demand**.

18. *College Tuition (from the GRE economics test)*   A time-series study of the demand for higher education, using tuition charges as a price variable, yields the following result:
$$\frac{dq}{dp} \cdot \frac{p}{q} = -0.4,$$
where p is tuition and q is the quantity of higher education. Which of the following is suggested by the result?
  (a) As tuition rises, students want to buy a greater quantity of education.
  (b) As a determinant of the demand for higher education, income is more important than price.
  (c) If colleges lowered tuition slightly, their total tuition receipts would increase.
  (d) If colleges raised tuition slightly, their total tuition receipts would increase.
  (e) Colleges cannot increase enrollments by offering larger scholarships.

COMMUNICATION AND REASONING EXERCISES

19. Complete the following sentence. If demand is inelastic, then revenue will decrease if _____ .

20. Complete the following sentence. If demand has unit elasticity, the revenue will decrease if _____ .

21. Your calculus study group is discussing elasticity of demand, and a member of the group asks the following question. "Since elasticity of demand measures the response of demand to change in unit price, what is the difference between elasticity of demand and the quantity $-dq/dp$?" How would you respond?

22. Another member of your study group claims that unit elasticity of demand need not always correspond to maximum revenue. Is he correct? Explain your answer.

▶ ___ You're the Expert

**PRODUCTION
LOT SIZE
MANAGEMENT**

Your publishing company, Knockem Dead Paperbacks, Inc., is planning the production of its latest best-seller, *Henrietta's Heaving Heart* by Celestine A. Lafleur. Sales are projected at 100,000 books per month in the next year. Your job is to coordinate print runs of the book in order to meet the anticipated demand and also minimize total costs to Knockem Dead, Inc.

Each print run has a setup cost of $5,000, each book costs $1 to produce, and monthly storage costs for books awaiting shipment average 1¢ per book. What are you to do?

First you test some scenarios to decide on your strategy. If you decide to print all 1,200,000 books (the estimated demand for the year: 100,000 per month for 12 months) in a single run at the start of the year and sales run as predicted, then the number of books in stock begins at 1,200,000 and decreases to zero by the end of the year, as shown in Figure 1.

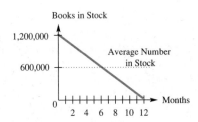

FIGURE 1

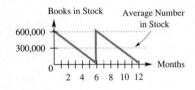

FIGURE 2

On average, you will be storing 600,000 books for 12 months at 1¢ per book, giving a total storage cost of $600,000 \times 12 \times .01 = \$72,000$, while the setup cost for the single print run will be $5,000. When you add to this the total cost of producing 1,200,000 books at $1 per book, your total cost will come out to $1,277,000.

If, on the other hand, you decide to cut down on storage costs by printing the book in two runs of 600,000 each, you will get the picture shown in Figure 2.

As shown in the figure, the storage cost is cut in half, because on average there are only 300,000 books in stock. Thus, the total storage cost is $36,000, while the setup cost has doubled to $10,000 (there are now two runs). The production costs are the same: 1,200,000 books at $1 per book. The total cost is now reduced to $1,246,000, a savings of $31,000 compared to your first scenario.

"Aha!" you say to yourself, "Why not drastically cut costs by setting up a run every month?" You calculate that the setup costs alone would be $12 \times \$5,000 = \$60,000$, which is already more than the setup plus storage costs for two runs. Perhaps, then, you should investigate three runs, four runs, and so on, until you reach the lowest cost. This seems a laborious process, especially because you will have to repeat it again when planning for Lafleur's sequel, *Lorenzo's Longing Lips,* due to be released next year. Realizing that this is an optimization problem, you decide to use some calculus to help you come up with a *formula* that you can use for all future plans. So you get to work.

Instead of working with the number 1,200,000, you use the letter N so that you can be as flexible as possible. (What if sales suddenly drop halfway through the year and you have to redo your calculations from scratch?) Thus, you have a total of N books to be produced for the year. You now calculate the total cost of producing them in x production runs (in the first scenario, $x = 1$, while in the second, $x = 2$). Because you are to produce a total of N books in x production runs, you will have to produce N/x books in each of the x runs. N/x is called the **lot size**. As you can see from the diagrams above, the average number of books in storage will be half that amount, $N/(2x)$.

Now you can calculate the total cost for a year. Write P for the setup cost of a single print run ($P = \$5,000$ in your case) and c for the *annual* cost of storing a book (to convert all of the time measurements to years; $c = \$0.12$ here). Finally, write b for the cost of producing a single book ($b = \$1$ here). The cost breakdown is now as follows.

Setup Costs: x print runs @ P dollars per run $\qquad\qquad Px$

Storage Costs: $N/(2x)$ books stored @ c dollars per year $\quad cN/(2x)$

Production Costs: N books @ b dollars per book $\qquad\underline{\qquad Nb\qquad}$

$$\textbf{Total Cost:}\quad Px + \frac{cN}{2x} + Nb$$

Remember that P, N, c, and b are all constants, while x is the only variable: the number of print runs. Thus, you have the cost function

$$C(x) = Px + \frac{cN}{2x} + Nb,$$

and you are trying to find the value of x that will minimize this total cost. But that's easy! All you need to do is find the local extrema and select the absolute minimum (if any).

First, the domain of the function is $(0, +\infty)$ because there is an x in the denominator, and x can't be negative. Next, you locate the extrema.

Set $C'(x) = 0$ and solve for x:

$$P - \frac{cN}{2x^2} = 0,$$

giving

$$2x^2 = \frac{cN}{P},$$

so

$$x = \sqrt{\frac{cN}{2P}}.$$

You have found only one stationary point. There are no singular points or endpoints. In order to decide whether this represents a local maximum or minimum, you try the second-derivative test.

$$C''(x) = \frac{cN}{x^3}.$$

Because all the numbers (including x) are positive, so is $C''(x)$, so you have a minimum. This also tells you that the whole curve is concave up, and hence that you have an *absolute* minimum.

So now you are practically done! You are absolutely certain that the value of x that gives the lowest total cost is $\sqrt{cN/(2P)}$. You now substitute the numbers to see what this says about *Henrietta's Heaving Heart*.

$$x = \sqrt{\frac{cN}{2P}} = \sqrt{\frac{(0.12)(1,200,000)}{2(5,000)}} \approx 3.79$$

Don't be disappointed that the answer is not a whole number (whole numbers are rarely found in real scenarios). What the answer does indicate is that either 3 or 4 print runs will cost the least money. If you take $x = 3$, you get a total cost of

$$C(3) = (5,000)(3) + \frac{(0.12)(1,200,000)}{(2)(3)} + (1,200,000)(1)$$

$$= \$1,239,000,$$

while if you take $x = 4$, you get a total cost of

$$C(4) = (5,000)(4) + \frac{(0.12)(1,200,000)}{(2)(4)} + (1,200,000)(1)$$

$$= \$1,238,000.$$

So, four print runs will allow you to minimize your total costs.

Exercises

1. *Lorenzo's Longing Lips* will sell 2,000,000 copies in a year. The remaining costs are the same. How many print runs should you now use?

2. In general, what happens to the number of runs that minimizes cost if both the setup cost and the total number of books are doubled?

3. In general, what happens to the number of runs that minimizes cost if the setup cost increases by a factor of 4?

4. Assuming that the total number of copies and storage costs are as originally stated, find the setup cost that would necessitate a single print run.

5. Assuming that the total number of copies and setup cost are as originally stated, find the storage cost that would necessitate a print run each month.

6. If you look at Figure 2, you will notice that we assumed that all the books in each run were manufactured in a very short time; otherwise the figure might have looked more like Figure 3, which shows the inventory assuming a slower rate of production. How would this affect the answer?

7. Referring to the general situation discussed above, find the cost function (cost as a function of total number of books produced) and average cost function, assuming that the number of runs is chosen to minimize total cost.

8. Let $\overline{C}$ represent the average cost function, calculate $\lim_{N \to +\infty} \overline{C}(N)$, and interpret the result.

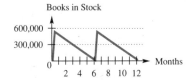

Books in Stock

600,000

300,000

0   2 4 6 8 10 12   Months

FIGURE 3

▶ ▬▬ **Review Exercises**

In Exercises 1–16, find all of the local and absolute extrema of the given functions on the given domain (if supplied) or on the largest possible domain (if no domain is supplied).

1. $f(x) = 2x^2 - 2x - 1$ on $[0, 3]$

2. $f(x) = -x^2 - 2x + 3$ on $[0, 3]$

3. $g(x) = 2x^3 - 6x + 1$ on $[-2, \infty)$

4. $g(x) = x^3 - 12x + 1$ on $[-2, \infty)$

5. $g(t) = \frac{1}{4}t^4 + t^3 + t^2$ on $(-\infty, \infty)$

6. $g(t) = 3t^4 + 28t^3 + 72t^2 + 1$ on $(-\infty, \infty)$

7. $f(t) = \frac{t + 1}{(t - 1)^2}$, $-2 \le t \le 2$

8. $f(t) = \frac{t - 1}{(t + 1)^2}$ on $[-2, 2]$

9. $f(t) = (t - 1)^{2/3}$

10. $f(t) = (t^2 + 1)^{2/3}$

11. $g(x) = x - 3x^{1/3}$

12. $g(x) = \frac{1}{x} + \frac{1}{x^2}$

13. $f(r) = \frac{1}{2}r^2 - \ln r$ on $(0, +\infty)$

14. $f(r) = r^2 + \ln r$ on $(0, +\infty)$

15. $g(t) = e^{t^2} + 1$

16. $g(t) = t + e^{-t}$

Exercises 17–32: For these exercises, carefully sketch the graphs of the functions in Exercises 1–16.

17. See Exercise 1.

18. See Exercise 2.

19. See Exercise 3.

20. See Exercise 4.

21. See Exercise 5.

22. See Exercise 6.

23. See Exercise 7.

24. See Exercise 8.

25. See Exercise 9.

26. See Exercise 10.

27. See Exercise 11.

28. See Exercise 12.

29. See Exercise 13.

30. See Exercise 14.

31. See Exercise 15.

32. See Exercise 16.

33. Maximize $P = xy^2$
with $x \geq 0$, $y \geq 0$ and $x^2 + y^2 = 75$.

34. Minimize $P = (1/2)x^2 + y^2$
with $x > 0$, $y > 0$, and $xy^2 = 125$.

35. Minimize $S = 3x + y + z$
with $xy = yz = 9$ and x, y and $z > 0$.

36. Maximize $S = xyz$
with $x^2 + y^2 = x^2 + z^2 = 1$ and x, y, $z > 0$.

37. What are the dimensions of the rectangle with largest area that can be inscribed in the first quadrant of the xy-plane under the curve $y = 1 - x^2$?

38. What are the dimensions of the rectangle with largest area that can be inscribed above the x-axis and under the curve $y = (1 - x^2)^{1/2}$?

APPLICATIONS

39. *Gas Mileage*   My Chevy's gas mileage (in miles per gallon) is given as the following function of speed:

$$M(x) = (x/1000 + 1/x)^{-1}$$

(x is the speed in miles per hour, $M(x)$ the gas mileage in miles per gallon). At what speed would you recommend I drive my Chevy in order to maximize fuel economy?

40. *Fast Cars*   My Zazna's gas mileage (in miles per gallon) is given as the following function of speed:

$$M(x) = (x/1000 + 4/x)^{-1}$$

(x is the speed in miles per hour, $M(x)$ the gas mileage in miles per gallon). At what speed would you recommend I drive it in order to maximize fuel economy?

41. *Wooden Beams*   The strength of a rectangular wooden beam is given by the formula $S = cwt^2$, where c is a constant (depending on the units chosen and the length of the beam), w is its cross-sectional width, and t is its thickness (cross-sectional height). Find the ratio of thickness to width of the strongest beam that can be cut from a circular log. [The equation of a circle of radius r is $x^2 + y^2 = r^2$.]

42. *Wooden Beams*   The stiffness of a rectangular wooden beam is given by the formula $S = cwt^3$, where c is a constant (depending on the units chosen and the length of the beam), w is its cross-sectional width, and t is its thickness (cross-sectional height). Find the ratio of thickness to width of the stiffest beam that can be cut from a circular log. [The equation of a circle of radius r is $x^2 + y^2 = r^2$.]

43. *Revenue*   The Fancy French Perfume company is deciding on the price of its latest men's cologne, "Deadly." After extensive market research, the company has come up with the demand equation $q = 1,000 - 200p^2 + 20,000p$, where p is the price

(in dollars) per 10-oz bottle and q is the number of 10-oz bottles it can sell to a leading department store. What price will bring in the largest revenue?

44. *Profit*   The Fancy French Perfume company is also deciding on the price of the after-shave version of its new cologne, "Deadly." Again, at $\$p$ per 10-oz bottle, it can sell $q = 1,000 - 200p^2 + 20,000p$ bottles, but each bottle costs the company $\$10$ to make. What should the price be to bring the largest profit?

45. *Average Cost*   The cost to Fullcourt Press of printing q books is

$$C(q) = 10,000 + 20q + \frac{1}{100}q^2.$$

At what production level is the average cost per book the lowest, and what is the least average cost per book?

46. *Average Profit*   The demand for Fullcourt Press's latest book is given by

$$q = 100,000 - 2,500p,$$

where q is the number of books it can sell per week at a price of p dollars per book. Using the production cost given in the preceding exercise, find the price that will maximize the average weekly profit.

47. *Hotel Rooms*   You have noticed that the occupancy of your 200-room hotel is very sensitive to price increases. If you charge $\$60$ per day for a room, your hotel is usually fully booked, and each $\$10$ increase in the daily fee results in 20 additional vacant rooms. How much should you charge per day in order to maximize your revenue? How many rooms will be vacant at that pricing strategy?

48. *Parking Garage Fees*   You manage a parking lot with a 400-car capacity in the theater district and have noticed that you can operate at full capacity if you charge $\$5$ per hour during peak theater hours. Raising the price tends to drive occupancy down, with a loss of 10 cars for each additional $1 per hour parking fee. How much should you charge to maximize your total revenue, and how many cars will frequent your lot at that price?

49. *Demand*   The demand equation for roses at Flower Emporium is

$$9p^2 + 25q^2 = 22,500,$$

where q is the number of dozen roses that can be sold per week at $$p$ per dozen. If the price is now $30, and Flower Emporium is decreasing the price by $2 per week, how are its sales changing?

50. *Demand*   Repeat the preceding exercise, with a demand equation of

$$9p^2 + 16q^2 = 14,400.$$

51. *Electronics*   Two variable resistances R_1 and R_2 connected in parallel produce a combined resistance R given by the equation

$$\frac{1}{R} = \frac{1}{R_1} + \frac{1}{R_2}.$$

At a certain moment in time, $R_1 = 6$ ohms and is increasing at 2 ohms/s, while $R_2 = 1$ ohm and is decreasing at 1 ohm/s. What is happening to the overall resistance?

52. *Relativity*   Einstein's theory of special relativity predicts that the mass m of a particle moving with velocity v is given by

$$m = \frac{m_0}{\sqrt{1 - \dfrac{v^2}{c^2}}},$$

where m_0 is its rest mass and c is the speed of light ($c \approx 3 \times 10^8$ m./s). If a particle with a rest mass of 100 g is moving at 30% the speed of light and accelerating at 1,000 m/s, how fast is its mass increasing at that instant?

53. *Shadows*   A 6-ft-tall man is walking away from a street lamp at 3 ft/s. The street lamp is 15 ft above the ground. How fast is the length of his shadow increasing when he is 10 ft away from the lamp?

54. *Punch Bowls*   A hemispherical punch bowl of radius 20 cm is being filled with fruit punch so that its depth is increasing at a rate of 2 cm/s. How fast is its volume increasing at the instant when the depth has reached 10 cm? [The volume of punch at level h is given by $V = \frac{1}{3}\pi h^2(60 - h)$ cm³.]

55. *Elasticity of Demand*   Calculate the elasticity of demand for the demand equation given in Exercise 43 when the price is set at $60 per bottle. Then use the elasticity of demand to show that revenue is maximized when $p = 66.67 per bottle.

56. *Elasticity of Demand*   Use elasticity of demand to calculate how items should be priced in order to maximize revenue if the demand equation is given by $q = 300 - p$.

57. *Elasticity of Demand*   The demand equation for roses at Flower Emporium is

$$9p^2 + 25q^2 = 22,500,$$

where q is the number of dozen roses that can be sold per week at $$p$ per dozen. Use the elasticity of demand to find the price of roses that maximizes weekly revenue.

58. *Elasticity of Demand*   Repeat the preceding exercise, with a demand equation of

$$9p^2 + 16q^2 = 14,400.$$

Source: Courtesy Circuit City.

The Integral

APPLICATION ▶ Sunny Electronics Company's chief competitor, Cyberspace Electronics, Inc. has just launched a new home entertainment system that competes directly with Sunny's World Entertainment System. Cyberspace is offering a 2-year limited warranty on its product, so Sunny is thinking of offering a 20-year pro-rated warranty. According to this warranty, if anything goes wrong, Sunny will refund the original $1,000 cost of the system depreciated continuously at an annual rate of 12%. What will it cost Sunny to provide this warranty?

INTRODUCTION ▸Roughly speaking, calculus is divided into two parts: **differential calculus** (the calculus of derivatives) and **integral calculus,** which is the subject of this chapter and the next. Integral calculus is concerned with problems that are the "reverse" of the problems seen in differential calculus. We begin by studying **anti-derivatives,** functions whose derivatives are a given function. The ability to compute antiderivatives allows us to solve many problems in economics, physics, and geometry, including the not-so-obviously re-lated problem of computing areas of complicated regions.

6.1 THE INDEFINITE INTEGRAL

Having studied differentiation in the previous three chapters, we now discuss how to *reverse* the process.

▼ **EXAMPLE 1**

Given that the derivative of $F(x)$ is $2x$, what is $F(x)$?

SOLUTION   After a moment's thought, we recall that the derivative of x^2 is $2x$. Thus, the original function might well have been $F(x) = x^2$. On the other hand, it could have been $F(x) = x^2 + 7$, since the derivative of $x^2 + 7$ is also $2x$. In fact, it could have been $F(x) = x^2 + C$, where C is any constant whatsoever, since the derivative of $x^2 + C$ is $2x$ no matter what the value of C is. Thus, there are *infinitely many answers,* one for each choice of a constant C.

This motivates the following definition.

ANTIDERIVATIVE

By an **antiderivative** of a function $f(x)$, we mean a function $F(x)$ whose derivative is $f(x)$.

▼ **EXAMPLE 2**

Find an antiderivative of $f(x) = 4x^3 + 1$.

SOLUTION   Another way of phrasing the question is this:

$$4x^3 + 1 \ \textit{is the derivative of what function?}$$

Searching our memories once again, we recall that $4x^3$ by itself is the derivative of x^4, while 1 by itself is the derivative of x. Also, if we add x^4 and x, the derivative of the sum is the sum of the derivatives. So $4x^3 + 1$ is the derivative of $x^4 + x$. In other words, an antiderivative of $4x^3 + 1$ is $F(x) = x^4 + x$.

Before we go on... Notice once again that we could add any constant to this answer and get another answer. For example, $F(x) = x^4 + x - 78.2$ is also an answer. So, any function of the form $F(x) = x^4 + x + C$, where C stands for an arbitrary constant (positive or negative), would be an antiderivative of $4x^3 + 1$.

▼ **EXAMPLE 3**

What possible antiderivatives are there of $f(x) = x$?

SOLUTION   The derivative of x^2 is $2x$, which is twice as big as $f(x)$. So we try $\frac{1}{2}x^2$, which has derivative x, as desired. Thus, all the functions represented by $F(x) = \frac{1}{2}x^2 + C$ are antiderivatives of $f(x)$.

Before we go on... In fact, these are *all* the possible antiderivatives of $f(x)$. We'll say more about this in a little while.

We have seen that we get infinitely many antiderivatives of a given function $f(x)$ by adding on an arbitrary constant. We refer to this *collection* of antiderivatives as the **indefinite integral of $f(x)$ with respect to x**. (As with derivatives, the phrase "with respect to x" simply reminds us that the variable is x, and not some other letter.) The term "indefinite" reminds us that there is no single definite answer, but instead a whole *family* of answers, one for each choice of a constant C.

Thus, Example 1 tells us that

the indefinite integral of $2x$ with respect
to x is $x^2 + C$.

C will always stand for an arbitrary constant and is usually called the **constant of integration.** Now it would be convenient to use symbols in place of the words "the indefinite integral of $f(x)$ with respect to x" (symbols are both more concise and more precise), and mathematicians have developed such a shorthand: We write the indefinite integral of $f(x)$ with respect to x as $\int f(x)dx$. Here, the symbol $\int$ stands for "the indefinite integral of," and dx stands for "with respect to x."

$$\int \qquad f(x) \qquad dx$$

(the indefinite integral of) $f(x)$ (with respect to x)

The function $f(x)$ (of which we are taking the indefinite integral) is called the **integrand.** If this strikes you as a peculiar notation to choose, we shall give some explanation of it in Section 5.

▼   **EXAMPLE 4**

The indefinite integral of $2x$ with respect to x is $x^2 + C$. In symbols,

$$\int 2x \, dx = x^2 + C.$$

Before we go on... In order to check the truth of this statement, remember that it claims that the derivative of $x^2 + C$ is $2x$. In fact,

$$\frac{d}{dx}[x^2 + C] = 2x. \; ✔$$

▼   **EXAMPLE 5**

$$\int [2e^{2x} + 8(x - 1)] \, dx = e^{2x} + 4(x - 1)^2 + C$$

Before we go on... Again, to check this we compute the derivative of the right-hand side.

$$\frac{d}{dx}[e^{2x} + 4(x - 1)^2 + C] = 2e^{2x} + 8(x - 1) \; ✔$$

How did we find the antiderivative in the first place? Read on.

Now we would like to make the process of finding indefinite integrals more mechanical. For example, it would be very nice to have a power rule for indefinite integrals similar to the one we already have for derivatives. Let us look at a few powers of x and see if we can notice a pattern.

In Example 3 we saw that the indefinite integral of x is $\frac{1}{2}x^2 + C$, or $x^2/2 + C$. In other words,

$$\int x \, dx = \frac{x^2}{2} + C.$$

What about the indefinite integral of x^2? Let us try a "trial and error" approach. Remember that we are trying to find a function whose derivative is x^2. For a first guess, let us try x^3. Its derivative is $3x^2$, which is three times the answer we want. Thus, we try one-third of that, $\frac{1}{3}x^3$, or $x^3/3$. The derivative of that is exactly x^2, so we have what we wanted,

$$\int x^2 \, dx = \frac{x^3}{3} + C.$$

Now we begin to see a pattern, and we make the following guess.

$$\int x^3 \, dx = \frac{x^4}{4} + C$$

To check that this guess is correct, we take the derivative of the right-hand side, and indeed, we get x^3. Thus, we speculate as follows.

$$\int x^n \, dx = \frac{x^{n+1}}{n+1} + C \quad (n \neq -1)$$

Q Why the restriction on n?

A If we put $n = -1$, the right-hand side of the formula would not make sense because there would be a zero in the denominator.

We'll look more closely at this exception in a moment. First, let us check this formula by taking the derivative of the right-hand side.

$$\frac{d}{dx}\left(\frac{x^{n+1}}{n+1} + C\right) = \frac{(n+1)x^{n+1-1}}{n+1} = x^n$$

Thus, $\frac{x^{n+1}}{n+1} + C$ *is* an antiderivative of x^n, and so the formula works.

Now let us turn to the exceptional case $n = -1$. In this case, we are seeking the indefinite integral of x^{-1}, or $1/x$. Prodding our memories a little, we recall that there is a function whose derivative is $1/x$, namely, $\ln x$. We also know that the derivative of $\ln|x|$ is $1/x$. (This permits x to be negative as well. Put another way, the domain of $\ln|x|$ is the same as the domain of $1/x$, which is not true of $\ln x$.) Thus,

$$\int x^{-1} \, dx = \int \frac{1}{x} \, dx = \ln|x| + C.$$

Summarizing, we have the following.

POWER RULE FOR THE INDEFINITE INTEGRAL

$$\int x^n \, dx = \frac{x^{n+1}}{n+1} + C \quad (n \neq -1)$$

$$\int x^{-1} \, dx = \int \frac{1}{x} \, dx = \ln|x| + C$$

In words, the first formula tells us that *to find the indefinite integral of x^n (if $n \neq -1$) add 1 to the exponent n and then divide by the new exponent.*

Thus, for example,

$$\int \frac{1}{x^{57}} \, dx = \int x^{-57} \, dx = \frac{x^{-56}}{-56} + C = -\frac{x^{-56}}{56} + C$$

$$= -\frac{1}{56x^{56}} + C,$$

since adding one to -57 gives -56 (not -58!).

As another example of this rule, consider the integral $\int 1 dx$, which is usually written simply as $\int dx$. Since $1 = x^0$, the power rule tells us that

$$\int 1\, dx = \int x^0\, dx = \frac{x^1}{1} + C = x + C.$$

If this seems peculiar, just check that the derivative of $x + C$ is indeed 1.

Here is another antiderivative that is easy to calculate. Since e^x is its own derivative, it is also its own antiderivative. This gives the following rule.

INDEFINITE INTEGRAL OF e^x

$$\int e^x\, dx = e^x + C$$

Q What about more complicated functions, such as $2x^3 + 6x^5 - 1$?

A We need analogs of the theorems on derivatives of sums, differences, and constant multiples.

RULES FOR THE INDEFINITE INTEGRAL

(a) $\displaystyle\int [f(x) \pm g(x)]\, dx = \int f(x)\, dx \pm \int g(x)\, dx$

(b) $\displaystyle\int kf(x)\, dx = k\int f(x)\, dx$ (k constant)

The first rule says what we already saw in Example 2: that the integral of the sum of two functions is the sum of the individual integrals. The same goes for the difference of two functions. In other words, to take the integral of a sum, first take the integrals separately, and then add the answers. The second rule says that to take the integral of a constant times a function, take the integral of the function by itself, and then multiply the answer by that constant. Why are these rules true? For the reason we noticed in Example 2: the derivative of a sum is the sum of the derivatives, and so on.

▼   **EXAMPLE 6**

Find $\int (x^3 + x^5 - 1)\, dx$.

SOLUTION   Applying the addition rule (a), we get

$$\int (x^3 + x^5 - 1)\, dx = \int x^3\, dx + \int x^5\, dx - \int 1\, dx$$

$$= \frac{x^4}{4} + \frac{x^6}{6} - x + C.$$

Before we go on... Let us check the answer.

$$\frac{d}{dx}\left(\frac{x^4}{4} + \frac{x^6}{6} - x + C\right) = x^3 + x^5 - 1 \quad ✔$$

Q Why is there only a single arbitrary constant C?

A We could have written the answer as $\frac{1}{4}x^4 + D + \frac{1}{6}x^6 + E - x + F$, where D, E, and F are all arbitrary constants. Now suppose that, for example, we set $D = 1$, $E = -2$ and $F = 6$. Then the particular antiderivative we get is $\frac{1}{4}x^4 + \frac{1}{6}x^6 - x + 5$, which has the form $\frac{1}{4}x^4 + \frac{1}{6}x^6 - x + C$. Thus, we could have chosen the single constant C to be 5 and we would have obtained the same answer. In other words, the answer $\frac{1}{4}x^4 + \frac{1}{6}x^6 - x + C$ is just as general as the answer $\frac{1}{4}x^4 + D + \frac{1}{6}x^6 + E - x + F$.

In practice, we don't bother rewriting the integral as several separate integrals but do this in our heads. To put this another way, we just integrate term-by-term, in the same way that we differentiate term-by-term.

▼ **EXAMPLE 7**

Find $\int(4x^3 + 6x^4 - 1)\, dx$.

SOLUTION   Applying the addition rule first, we get

$$\int (4x^3 + 6x^4 - 1)\, dx = \int 4x^3\, dx + \int 6x^4\, dx - \int 1\, dx.$$

We now apply the constant multiple rule (b) to the first two terms.

$$= 4\int x^3\, dx + 6\int x^4\, dx - \int 1\, dx$$

$$= 4\frac{x^4}{4} + 6\frac{x^5}{5} - x + C$$

$$= x^4 + \frac{6x^5}{5} - x + C$$

Before we go on... Let us check our answer.

$$\frac{d}{dx}\left(x^4 + \frac{6x^5}{5} - x + C\right) = 4x^3 + 6x^4 - 1 \quad ✔$$

Again, we would usually skip the middle steps and write

$$\int (4x^3 + 6x^4 - 1)\, dx = 4\frac{x^4}{4} + 6\frac{x^5}{5} - x + C.$$

It is helpful to think of the constant factors 4 and 6 as "going along for the ride" just as they do when we take derivatives.

▼ **EXAMPLE 8**

Find $\displaystyle\int \left(\frac{x^{3.2}}{3} + \frac{1}{x^2} - \frac{6}{x} + 7.3e^x\right) dx$.

SOLUTION   The integration rules tell us that we can take the indefinite integrals of each term in turn, and add or subtract the answers, as the case may be. First, though, we should rewrite the integrand in exponent form so that we can apply the power rule to the appropriate terms.

$$\int \left(\frac{x^{3.2}}{3} + \frac{1}{x^2} - \frac{6}{x} + 7.3e^x\right) dx = \int \left(\frac{1}{3}x^{3.2} + x^{-2} - 6x^{-1} + 7.3e^x\right) dx$$

$$= \frac{1}{3}\left(\frac{x^{4.2}}{4.2}\right) + \frac{x^{-1}}{(-1)} - 6\ln|x| + 7.3e^x + C$$

$$= \frac{x^{4.2}}{12.6} - \frac{1}{x} - 6\ln|x| + 7.3e^x + C$$

Before we go on... Notice again how the constant factors $\frac{1}{3}$, 6, and 7.3 "went along for the ride."

   As usual, let us check our answer.

$$\frac{d}{dx}\left(\frac{x^{4.2}}{12.6} - \frac{1}{x} - 6\ln|x| + 7.3e^x + C\right)$$

$$= \frac{x^{3.2}}{3} + \frac{1}{x^2} - \frac{6}{x} + 7.3e^x \quad ✔$$

▼ **EXAMPLE 9**

Find $\displaystyle\int \left(\frac{u^{3.2}}{3} + \frac{1}{u^2} - \frac{6}{u} + 7.3e^u\right) du$.

SOLUTION   This looks very similar to the last example, except for one thing: the letter x has been replaced by the letter u. We are asking for a function *of u* whose derivative *with respect to u* is $u^{3.2}/3 + 1/u^2 - 6/u + 7.3e^u$. There is nothing special about the letter x—we can use u instead if we like. Thus, we get

$$\int \left(\frac{u^{3.2}}{3} + \frac{1}{u^2} - \frac{6}{u} + 7.3e^u\right) du$$

$$= \frac{u^{4.2}}{12.6} - \frac{1}{u} - 6\ln|u| + 7.3e^u + C.$$

Before we go on... In checking our answer, we take the derivative *with respect to u.*

$$\frac{d}{du}\left(\frac{u^{4.2}}{12.6} - \frac{1}{u} - 6\ln|u| + 7.3e^u + C\right)$$

$$= \frac{u^{3.2}}{3} + \frac{1}{u^2} - \frac{6}{u} + 7.3e^u$$

▼ **EXAMPLE 10**   Cost and Marginal Cost

The marginal cost (in dollars) to produce baseball caps at a production level of x caps is $3.20 - 0.001x$, and it is found that the cost of producing 50 caps is $200. Find the cost function.

SOLUTION   We are asked to find the cost function $C(x)$, given that the *marginal* cost function is $3.20 - 0.001x$. Recalling that the marginal cost function is the derivative of the cost function, we have

$$C'(x) = 3.2 - 0.001x,$$

and must find $C(x)$. Now $C(x)$ must be an antiderivative of $C'(x)$, so we write

$$C(x) = \int (3.20 - 0.001x)\, dx$$

$$= 3.20x - 0.001\frac{x^2}{2} + K \qquad \text{\small K is the constant of integration.}$$

$$= 3.20x + 0.0005x^2 + K.$$

(Why did we use K and not C for the constant of integration?) Now, unless we know a value for K, we don't really know what the cost function is. However, there is a further piece of information we have ignored: the cost of producing 50 baseball caps is $200. In other words,

$$C(50) = 200.$$

Substituting in our formula for $C(x)$,

$$C(50) = 3.20(50) + 0.0005(50)^2 + K = 200$$

i.e.,

$$161.25 + K = 200$$

so

$$K = 38.75.$$

Now that we know what K is, we can write down the cost function.

$$C(x) = 3.20x + 0.0005x^2 + 38.75$$

Before we go on...

Q What is the significance of the term 38.75?

A If we take $x = 0$, we obtain

$$C(0) = 3.20(0) + 0.0005(0)^2 + 38.75,$$

or

$$C(0) = 38.75.$$

Thus 38.75 is the cost of producing zero items: in other words, the **fixed cost.**

We end this section with a few important ideas.

Let us return for a moment to the beginning of all this, when we said that *all* antiderivatives of $2x$ are of the form $x^2 + C$, for some constant C or the other. If you are as skeptical as we sincerely hope you are, then there should be a nagging doubt at the back of your mind: could there be some mysterious function we've possibly never heard of, *other than $x^2 + C$,* with the property that its derivative is also $2x$? Just because we haven't encountered any such function, it doesn't mean that there isn't one lurking around somewhere. (We might conceivably detect some signal from the Andromeda galaxy telling us about such a function.) Thus, we ought to experience a little discomfort when we claim that *the* indefinite integral of $2x$ is $x^2 + C$.

Well, let us put our minds at ease. We claim that, if $F(x)$ and $G(x)$ are both antiderivatives of $f(x)$, then F and G can differ by at most a constant. In other words, if $F'(x) = G'(x) = f(x)$, then $G(x) = F(x) + C$ for some constant C. To see why, think about what the equation $F'(x) = G'(x)$ says (see Figure 1).

If $F'(x) = G'(x)$ for all x, then F and G have the *same slope* at each point. This means that their graphs must be *parallel,* and remain exactly the same vertical distance apart. But that is the same as saying that the functions differ by a constant.*

In summary, if you know one antiderivative $F(x)$ of $f(x)$, then you know them all, since any other antiderivative of f is just $F(x)$ plus some constant. Thus, if we write $F(x) + C$, we can be confident that we have written down all possible antiderivatives of f.

Let us now explore the notation we've developed. Since $\int f(x)\,dx$ means the indefinite integral of $f(x)$, its derivative must be $f(x)$. In other words,

$$\frac{d}{dx} \int f(x)\,dx = f(x).$$

In words, this complicated-looking formula is proclaiming nothing more than the fact that the derivative of a function whose derivative is $f(x)$, is $f(x)$! In other words, *the derivative of the indefinite integral of a function is the function we started with.* For example, if we start with $f(x) = x^3$, take the indefinite integral, which is $\frac{1}{4}x^4 + C$, and then take the derivative, we wind up with what we started with, x^3. This is, after all, how we have been checking our work.

Now, this sounds like we are saying that differentiation and integration are inverse processes: if you start with a function, take its indefinite integral, and then take the derivative of the answer, you will get the function you started with. To be truly inverse, the same thing should happen in reverse, so let us try it. Start with a function $F(x)$, take its derivative, and *then* take the indefinite integral of the answer. For example, if we start with

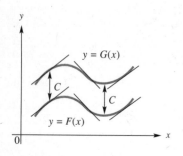

y

$y = G(x)$

C

C

$y = F(x)$

0

x

FIGURE 1

▼ * This argument can be turned into a more rigorous proof—that is, a proof that does not rely on geometric concepts such as "parallel graphs."

$F(x) = x^2 + 1$, and first take the derivative, we get $2x$. Taking the indefinite integral now gives $x^2 + C$, where we don't know what the constant is anymore. Thus, we lose information by going this route. We don't get the original function $f(x)$ back again, but we get $f(x) + C$. Thus, what we *can* say is:

$$\int \left(\frac{d}{dx} f(x) \right) dx = f(x) + C.$$

This loss of information really takes place when we first take the derivative. For example, the derivative of $x^2 + 1$ is the same as the derivative of $x^2 + 3$. To summarize:

RELATIONSHIP BETWEEN THE INDEFINITE INTEGRAL AND THE DERIVATIVE

$$\frac{d}{dx} \int f(x) \, dx = f(x)$$

$$\int \left(\frac{d}{dx} f(x) \right) dx = f(x) + C$$

▶ **6.1 EXERCISES**

Evaluate the integrals in Exercises 1–10 mentally.

1. $\int x^5 \, dx$

2. $\int x^7 \, dx$

3. $\int 6 \, dx$

4. $\int (-5) \, dx$

5. $\int x \, dx$

6. $\int (-x) \, dx$

7. $\int (x^2 - x) \, dx$

8. $\int (x + x^3) \, dx$

9. $\int (1 + x) \, dx$

10. $\int (4 - x) \, dx$

Evaluate the integrals in Exercises 11–32.

11. $\int x^{-5} \, dx$

12. $\int x^{-7} \, dx$

13. $\int \left(u^2 - \frac{1}{u} \right) du$

14. $\int \left(v^{-2} + \frac{2}{v} \right) dv$

15. $\int (3x^4 - 2x^{-2} + x^{-5} + 4) \, dx$

16. $\int (4x^7 - x^{-3} + 1) \, dx$

17. $\int \left(2e^x + \frac{5}{x} \right) dx$

18. $\int \left(2e^x + \frac{5}{x} + x^{-2} \right) dx$

19. $\int \left(\frac{1}{x} + \frac{2}{x^2} - \frac{1}{x^3} \right) dx$

20. $\int \left(\frac{3}{x} - \frac{1}{x^5} + \frac{1}{x^7} \right) dx$

21. $\int (3x^{0.1} - x^{4.3}) \, dx$

22. $\int \left(\frac{x^{2.1}}{2} - 2 \right) dx$

23. $\int \left(\frac{3}{x^{0.1}} - \frac{4}{x^{1.1}} \right) dx$

24. $\int \left(\frac{1}{x^{1.1}} - \frac{1}{x} \right) dx$

25. $\int \dfrac{x+2}{x^3}\, dx$

26. $\int \dfrac{x^2-2}{x}\, dx$

27. $\int \sqrt{x}\, dx$

28. $\int \sqrt[3]{x}\, dx$

29. $\int \left(2\sqrt[3]{x} - \dfrac{1}{2\sqrt{x}}\right) dx$

30. $\int \left(2\sqrt{x} - \dfrac{1}{2\sqrt{x}}\right) dx$

31. $\int \dfrac{x - 2\sqrt{x}}{x^2}\, dx$

32. $\int \dfrac{x^2 - 2\sqrt{x}}{x^3}\, dx$

33. Find $f(x)$ if $f(0) = 1$ and the tangent line at $(x, f(x))$ has slope x.

34. Find $f(x)$ if $f(1) = 1$ and the tangent line at $(x, f(x))$ has slope $\frac{1}{x}$.

35. Find $f(x)$ if $f(0) = 0$ and the tangent line at $(x, f(x))$ has slope $e^x - 1$.

36. Find $f(x)$ if $f(1) = -1$ and the tangent line at $(x, f(x))$ has slope $2e^x + 1$.

APPLICATIONS

37. *Marginal Cost*   The marginal cost of producing the xth box of light bulbs is $5 - (x/10{,}000)$, and the fixed cost is $20{,}000$. Find the total cost function $C(x)$.

38. *Marginal Cost*   The marginal cost of producing the xth box of computer disks is $10 + (x^2/100{,}000)$, and the fixed cost is $100{,}000$. Find the total cost function $C(x)$.

39. *Marginal Cost*   The marginal cost of producing the xth roll of film is given by $5 + 2x + \frac{1}{x}$. The total cost to produce one roll is $1{,}000$. Find the total cost function $C(x)$.

40. *Marginal Cost*   The marginal cost of producing the xth box of videotape is given by $10 + x + 1/x^2$. The total cost to produce one hundred boxes is $10{,}000$. Find the total cost function $C(x)$.

For Exercises 41 and 42, recall that the velocity of a particle moving in a straight line is given by $v = ds/dt$.

41. *Motion in a Straight Line*   The velocity of a particle moving in a straight line is given by $v(t) = t^2 + 1$.
   (a) Find an expression for the position s after a time t.
   (b) Given that $s = 1$ at time $t = 0$, find the constant of integration C, and hence find an expression for s in terms of t without any unknown constants.

42. *Motion in a Straight Line*   The velocity of a particle moving in a straight line is given by $v = 3e^t + t$.
   (a) Find an expression for the position s after a time t.
   (b) Given that $s = 3$ at time $t = 0$, find the constant of integration C, and hence find an expression for s in terms of t without any unknown constants.

COMMUNICATION AND REASONING EXERCISES

43. Give an argument for the rule that the integral of a sum is the sum of the integrals.

44. Is it true that $\int \dfrac{1}{x^3}\, dx = \ln (x^3) + C$? Give a reason for your answer.

45. Give an example to show that the integral of a product is not the product of the integrals.

46. Give an example to show that the integral of a quotient is not the quotient of the integrals.

47. If x represents the number of items manufactured and $f(x)$ represents the marginal cost per item, what does $\int f(x)\, dx$ represent? In general, how are the units of $f(x)$ and the units of $\int f(x)\, dx$ related?

48. Complete the following: $-1/x$ is an _____ of $1/x^2$, whereas $\ln x^2$ is not. $-1/x + C$ is the _____ of $1/x^2$, because the _____ of $-1/x + C$ is _____.

49. Complete the following sentence. If you take the _____ of the _____ of $f(x)$, you obtain $f(x)$ back. On the other hand, if you take the _____ of the _____ of $f(x)$, you obtain $f(x) + C$.

50. If a Martian told you that the Institute of Alien Mathematics, after a long and difficult search, has announced the discovery of a new antiderivative of $x - 1$ called $M(x)$ (the formula for $M(x)$ is classified information, and so cannot be revealed here) how would you respond?

▶ ═══ **6.2** SUBSTITUTION

The chain rule for derivatives gives us an extremely useful technique for finding antiderivatives. This technique is called **substitution** or **change of variables.** We'll start with an example to illustrate the mathematics behind the technique of substitution, then discuss how the technique is used in practice.

▼ **EXAMPLE 1**

Find $\int 2x(x^2 + 1)^{1/2} \, dx$.

SOLUTION   The answer is $\frac{2}{3}(x^2 + 1)^{3/2} + C$, and let us see why this is so. We simply need to check the derivative.

$$\frac{d}{dx}\left(\frac{2}{3}(x^2 + 1)^{3/2} + C\right) = \frac{2}{3} \cdot \frac{3}{2} \, (x^2 + 1)^{1/2} \cdot (2x)$$

$$= 2x(x^2 + 1)^{1/2} \quad ✔$$

Before we go on... Notice how we had to use the chain rule to take the derivative above. Let us start again and attempt to compute the integral from scratch. We want to find

$$\int 2x(x^2 + 1)^{1/2} \, dx.$$

Inspired by the chain rule, we let u be the quantity that is raised to the power, $u = x^2 + 1$. Then $du/dx = 2x$. Substituting these in the integral, we get

$$\int \left(\frac{du}{dx} u^{1/2}\right) dx, \text{ or } \int \left(u^{1/2} \frac{du}{dx}\right) dx.$$

Although we are tempted to cancel the dx's, we need to remember that dx is not a real number. The quantity in parentheses reminds us of the chain rule, and in fact

$$\frac{d}{dx}\left(\frac{2}{3} u^{3/2}\right) = u^{1/2} \frac{du}{dx}.$$

This is the same as saying that

$$\int \left(u^{1/2} \frac{du}{dx}\right) dx = \frac{2}{3} u^{3/2} + C.$$

If we now substitute $u = x^2 + 1$ into $\frac{2}{3} u^{3/2} + C$, we get the answer that we gave originally.

Let us reexamine the calculation. We saw that

$$\int \left(u^{1/2} \frac{du}{dx}\right) dx = \frac{2}{3} u^{3/2} + C.$$

On the other hand, we know from the power rule that

$$\int u^{1/2} \, du = \frac{2}{3} u^{3/2} + C$$

also. Therefore,

$$\int \left(u^{1/2} \frac{du}{dx} \right) dx = \int u^{1/2} \, du.$$

There is nothing special about $u^{1/2}$ here. For any function f we can say that

$$\int \left(f(u) \frac{du}{dx} \right) dx = \int f(u) \, du.$$

The point is that $\int f(u) \, du$ may be simpler than the original integral, just as $\int u^{1/2} \, du$ is simpler than $\int 2x(x^2 + 1)^{1/2} \, dx$. Comparing the above pairs of integrals suggests that*

$$\frac{du}{dx} \, dx = du.$$

We can also write this equation as

$$dx = \frac{1}{du/dx} \, du.$$

Now let us see how the technique works in practice.

▼ **EXAMPLE 2**

Calculate $\int x(x^2 + 1)^2 \, dx$.

SOLUTION   First, we decide what we are going to take as u. Although there is no rule that always works, the following often works.

Take u to be an expression that is being raised to a power.

In this example, $x^2 + 1$ is being raised to the second power, so let's try setting $u = x^2 + 1$.

The next step is to take the derivative of u with respect to x.

$$\frac{du}{dx} = 2x$$

We can now write the differential equation $du = \dfrac{du}{dx} \, dx$, which here is

$$du = 2x \, dx.$$

▼ * In fact, we saw this equation earlier in the section on linear approximation, where we called du and dx **differentials**.

Now we divide both sides by $2x$ to get

$$dx = \frac{1}{2x}\, du.$$

Let's summarize what we just did.

$u = x^2 + 1$	Decide what to take as u.
$\dfrac{du}{dx} = 2x$	Take the derivative with respect to x.
$du = 2x\, dx$	Write the equation $du = \dfrac{du}{dx}\, dx$.
$dx = \dfrac{1}{2x}\, du$	Solve for dx.

Now we are ready for the next step.

> *Substitute the expression for u in the original integral,
> and also substitute for dx.*

Thus, we get

$$\int x(x^2 + 1)^2\, dx = \int x\, u^2\, \frac{1}{2x}\, du.$$

Next, we do the following.

> *Simplify the integrand, leaving an integral in u only.*

That's easy: the x's cancel.* So we have

$$\int u^2\, \frac{1}{2}\, du = \frac{1}{2} \int u^2\, du.$$

Now we have an integral that is easy to calculate:

$$\frac{1}{2} \int u^2\, du = \frac{1}{2}\left(\frac{u^3}{3}\right) + C = \frac{u^3}{6} + C,$$

and we are almost done. We are looking for a function of x, so we substitute $u = x^2 + 1$. Thus, the answer is

$$\frac{(x^2 + 1)^3}{6} + C.$$

▼ * We'll tell you later what to do if they don't.

Before we go on... Here is the solution as we would usually write it.

$$u = x^2 + 1$$

$$\frac{du}{dx} = 2x$$

$$du = 2x \, dx$$

$$dx = \frac{1}{2x} \, du$$

$$\int x(x^2 + 1)^2 \, dx = \int x \, u^2 \, \frac{1}{2x} \, du \qquad \text{Substitute.}$$

$$= \int u^2 \, \frac{1}{2} \, du \qquad \text{Eliminate } x\text{'s.}$$

$$= \frac{1}{2} \int u^2 \, du \qquad \text{Simplify integral.}$$

$$= \frac{1}{2} \left(\frac{u^3}{3} \right) + C \qquad \text{Evaluate integral.}$$

$$= \frac{u^3}{6} + C$$

$$= \frac{(x^2 + 1)^3}{6} + C \qquad \text{Substitute.}$$

We should, as always, check our answer.

$$\frac{d}{dx} \left(\frac{(x^2 + 1)^3}{6} + C \right) = \frac{3(x^2 + 1)^2 \cdot 2x}{6} = x(x^2 + 1)^2 \quad \checkmark$$

▶ CAUTION

1. It is important to follow the following steps *in order*:
   First, eliminate *all the x*'s.
   Next, take the antiderivative with respect to *u*.
   Finally, substitute back for *u*.
2. If you can't eliminate *x* or you wind up with a mess, you may have made a bad choice for *u*, so try something else. (See Example 6 in this section for another idea.) ◀

▼ **EXAMPLE 3**

Evaluate $\int 3xe^{x^2} \, dx$.

SOLUTION   In the integrand, we have *e* raised to an expression. This is another place where *u*-substitution often works. We take $u = x^2$, the expression to which *e* is being raised.

$$u = x^2$$

$$\frac{du}{dx} = 2x$$

$$du = 2x\,dx$$

$$dx = \frac{1}{2x}\,du$$

$$\int 3xe^{x^2}\,dx = \int 3x\,e^u\,\frac{1}{2x}\,dx$$

$$= \int 3e^u\,\frac{1}{2}\,du$$

$$= \frac{3}{2}\int e^u\,du$$

$$= \frac{3}{2}\,e^u + C$$

$$= \frac{3}{2}\,e^{x^2} + C$$

Before we go on... We check our answer.

$$\frac{d}{dx}\left(\frac{3}{2}\,e^{x^2} + C\right) = \frac{3}{2}\,2x\,e^{x^2} = 3xe^{x^2} \quad ✔$$

▼ **EXAMPLE 4**

Evaluate $\displaystyle\int (3x + 8)^{-1}\,dx$.

SOLUTION   In the integrand, we have the expression $3x + 8$ raised to a power, so we take $u = 3x + 8$.

$$u = 3x + 8$$

$$\frac{du}{dx} = 3$$

$$du = 3\,dx$$

$$dx = \frac{1}{3}\,du$$

$$\int (3x + 8)^{-1} dx = \int u^{-1} \frac{1}{3} \, du$$

$$= \frac{1}{3} \int u^{-1} \, du$$

$$= \frac{1}{3} \ln|u| + C$$

$$= \frac{1}{3} \ln|3x + 8| + C$$

Before we go on... First, we check the answer.

$$\frac{d}{dx}\left(\frac{1}{3} \ln|3x + 8| + C\right) = \frac{1}{3} \cdot \frac{1}{3x + 8} \cdot 3$$

$$= (3x + 8)^{-1} \checkmark$$

Notice something interesting: the answer suggests that

$$\int (ax + b)^{-1} \, dx = \frac{1}{a} \ln|ax + b| + C$$

for any constants a and b (with $a \neq 0$). In fact, this is easy to see if we simply redo the calculation with $3x + 8$ replaced by $ax + b$. This is a useful result to remember, and we will collect this and similar results at the end of this section.

▼ EXAMPLE 5

Evaluate $\displaystyle\int \frac{2x + 1}{(3x^2 + 3x - 5)^{1/3}} \, dx.$

SOLUTION   The quantity being raised to a power is in the denominator of the integrand. Thus, we try putting $u = 3x^2 + 3x - 5$. (Putting $u = 2x + 1$ won't work. Try it. If there are two competing candidates for u, it is sometimes best to put u equal to the more complicated of the two.)

$$u = 3x^2 + 3x - 5$$

$$\frac{du}{dx} = 6x + 3$$

$$du = (6x + 3) \, dx$$

$$dx = \frac{1}{6x + 3} \, du$$

$$\int \frac{2x + 1}{(3x^2 + 3x - 5)^{1/3}} \, dx = \int \frac{2x + 1}{u^{1/3}(6x + 3)} \, du$$

Now we would normally cancel x's, but the x's don't seem to want to cancel! If we look at what we have for a moment, though, we notice something: $6x + 3 = 3(2x + 1)$, and the quantity $2x + 1$ cancels.

$$= \int \frac{2x + 1}{u^{1/3}3(2x + 1)} \, du$$

$$= \int \frac{1}{3u^{1/3}} \, du$$

$$= \frac{1}{3} \int \frac{1}{u^{1/3}} \, du$$

$$= \frac{1}{3} \int u^{-1/3} \, du \qquad \text{Remember to convert to exponential form.}$$

$$= \frac{1}{3} \cdot \frac{u^{2/3}}{(2/3)} + C$$

$$= \frac{1}{3} \cdot \frac{3}{2} u^{2/3} + C = \frac{1}{2} u^{2/3} + C$$

$$= \frac{1}{2} (3x^2 + 3x - 5)^{2/3} + C$$

Before we go on... We check our answer.

$$\frac{d}{dx} \left(\frac{1}{2} (3x^2 + 3x - 5)^{2/3} + C \right) = \frac{1}{3} (3x^2 + 3x - 5)^{-1/3}(6x + 3)$$

$$= \frac{6x + 3}{3(3x^2 + 3x - 5)^{1/3}}$$

$$= \frac{2x + 1}{(3x^2 + 3x - 5)^{1/3}} \quad \checkmark$$

▼ EXAMPLE 6

Evaluate $\int \dfrac{2x}{(x - 5)^2} \, dx$.

SOLUTION   We try putting $u = x - 5$.

$$u = x - 5$$
$$\frac{du}{dx} = 1$$
$$du = dx$$
$$dx = du$$

$$\int \frac{2x}{(x - 5)^2} \, dx = \int \frac{2x}{u^2} \, du$$

We have arrived at the step where the x's should cancel, and they don't. This is where we need to do something new.

> *If there are x's left over after cancellation, go back to the equation relating x and u, solve for x, and substitute in the integrand.*

In this case, the equation relating x and u is the first equation in the box: $u = x - 5$. We solve for x, getting $x = u + 5$. We now substitute this expression for x into the integral.

$$\int \frac{2x}{u^2}\, du = \int \frac{2(u + 5)}{u^2}\, du$$

$$= 2 \int \frac{u + 5}{u^2}\, du$$

Now that we have gotten rid of the x's, how do we take the antiderivative? There are two possible ways of doing this. The first method is to break the integrand into two fractions u/u^2 and $5/u^2$ and integrate each term separately. The second method—which really amounts to the same thing—is to move the denominator up to the numerator by changing the sign of the exponent, and then use the distributive law.

$$= 2 \int (u + 5)u^{-2}\, du$$

$$= 2 \int (u^{-1} + 5u^{-2})\, du$$

$$= 2(\ln |u| - 5u^{-1}) + C$$

$$= 2(\ln |x - 5| - 5(x - 5)^{-1}) + C$$

Before we go on... We check our answer.

$$\frac{d}{dx}(2(\ln |x - 5| - 5(x - 5)^{-1}) + C) = 2\left(\frac{1}{x - 5} + 5(x - 5)^{-2}\right)$$

$$= 2\,\frac{x - 5 + 5}{(x - 5)^2}$$

$$= \frac{2x}{(x - 5)^2} \quad \checkmark$$

SHORTCUTS

If a and b are constants with $a \neq 0$, then we have the following formulas. (You have already seen one of them in Example 4. All of them can be obtained using the substitution $u = ax + b$. They will appear in the exercises.)

SHORTCUTS: INTEGRALS OF EXPRESSIONS INVOLVING $(ax + b)$

Rule	Example

$$\int (ax + b)^n \, dx = \frac{(ax + b)^{n+1}}{a(n + 1)} + C \text{ (if } n \neq -1)$$

$$\int (3x - 1)^2 \, dx = \frac{(3x - 1)^3}{3(3)} + C$$

$$= \frac{(3x - 1)^3}{9} + C$$

$$\int (ax + b)^{-1} \, dx = \frac{1}{a} \ln|ax + b| + C$$

$$\int (3 - 2x)^{-1} dx = \frac{1}{(-2)} \ln|3 - 2x| + C$$

$$= -\frac{1}{2} \ln|3 - 2x| + C$$

$$\int e^{ax+b} \, dx = \frac{1}{a} e^{ax+b} + C$$

$$\int e^{-x+4} \, dx = \frac{1}{(-1)} e^{-x+4} + C$$

$$= -e^{-x+4} + C$$

We end this section with a little advice.

HINTS AND GENERAL GUIDELINES

1. Always remember not to substitute back for u until the end of the calculation, after you have taken the antiderivative.
2. If an x does not cancel in the integrand, do one of the following.
   (a) Try another substitution.
   (b) Go back to the formula for u as a function of x, solve the equation for x as a function of u, and substitute for x.
3. Don't ever bother using the substitution $u = x$. All this does is replace the letter x with the letter u throughout, giving you the same integral you started with!
4. To check your answer, simply take its derivative. You should get the integrand you started with.

▶ **6.2 EXERCISES**

In Exercises 1–30, evaluate the given integrals.

1. $\int (3x + 1)^5 \, dx$

2. $\int (-x - 1)^7 \, dx$

3. $\int (-2x + 2)^{-2} \, dx$

4. $\int (2x)^{-1} \, dx$

5. $\int x(3x^2 + 3)^3 \, dx$

6. $\int x(-x^2 - 1)^3 \, dx$

7. $\int 2x\sqrt{3x^2 - 1} \, dx$

8. $\int 3x\sqrt{-x^2 + 1} \, dx$

9. $\int xe^{-x^2+1} \, dx$

10. $\int xe^{2x^2-1}dx$

11. $\int (x + 1)e^{-(x^2+2x)} dx$

12. $\int (2x - 1)e^{2x^2-2x} dx$

13. $\int \dfrac{-2x - 1}{(x^2 + x + 1)^3} dx$

14. $\int \dfrac{x^3 - x^2}{3x^4 - 4x^3} dx$

15. $\int \dfrac{x^2 + x^5}{\sqrt{2x^3 + x^6 - 5}} dx$

16. $\int \dfrac{2(x^3 - x^4)}{(5x^4 - 4x^5)^5} dx$

17. $\int 2x\sqrt{x + 1}\, dx$

18. $\int \dfrac{x}{\sqrt{x + 1}} dx$

19. $\int x(x - 2)^5 dx$

20. $\int x(\sqrt[3]{x} - 2) dx$

21. $\int \dfrac{3e^{-1/x}}{x^2} dx$

22. $\int \dfrac{2e^{2/x}}{x^2} dx$

23. $\int x(x^2 + 1)^{1.3} dx$

24. $\int \dfrac{x}{(3x^2 - 1)^{0.4}} dx$

25. $\int (1 + 9.3e^{3.1x-2}) dx$

26. $\int (3.2 - 4e^{1.2x-3}) dx$

27. $\int \dfrac{e^{-0.05x}}{1 - e^{-0.05x}} dx$

28. $\int \dfrac{3e^{1.2x}}{2 + e^{1.2x}} dx$

29. $\int ((2x - 1)e^{2x^2-2x} + xe^{x^2}) dx$

30. $\int (xe^{-x^2+1} + e^{2x}) dx$

In Exercises 31–34, derive the equations, where a and b are constants with a ≠ 0.

31. $\int (ax + b)^n dx = \dfrac{(ax + b)^{n+1}}{a(n + 1)} + C \quad (n \neq -1)$

32. $\int (ax + b)^{-1} dx = \dfrac{1}{a} \ln |ax + b| + C$

33. $\int e^{ax+b} dx = \dfrac{1}{a} e^{ax+b} + C$

34. $\int \dfrac{1}{(ax + b)^n} dx = -\dfrac{1}{a(n - 1)(ax + b)^{n-1}} + C \quad (n \neq 1)$

Use the formulas in Exercises 31 through 34 to calculate the integrals in Exercises 35–44 mentally.

35. $\int e^{-x} dx$

36. $\int e^{x-1} dx$

37. $\int e^{2x-1} dx$

38. $\int e^{-3x} dx$

39. $\int (2x + 4)^2 dx$

40. $\int (3x - 2)^4 dx$

41. $\int \dfrac{1}{5x - 1} dx$

42. $\int (x - 1)^{-1} dx$

43. $\int (1.5x)^3 dx$

44. $\int e^{2.1x} dx$

45. Find $f(x)$ if $f(0) = 0$ and the tangent line at $(x, f(x))$ has slope $x(x^2 + 1)^3$.

46. Find $f(x)$ if $f(1) = 0$ and the tangent line at $(x, f(x))$ has slope $\dfrac{x}{x^2 + 1}$.

47. Find $f(x)$ if $f(0) = \dfrac{1}{2e}$ and the tangent line at $(x, f(x))$ has slope xe^{x^2-1}.

48. Find $f(x)$ if $f(1) = -1 + \dfrac{1}{e}$ and the tangent line at x has slope $(x - 1)e^{x^2-2x}$.

APPLICATIONS

49. *Motion in a Straight Line*   The velocity of a particle moving in a straight line is given by $v = t(t^2 + 1)^4 + t$.
(a) Find an expression for the position s after a time t.
(b) Given that $s = 1$ at time $t = 0$, find the constant of integration C, and hence find an expression for s in terms of t without any unknown constants.

50. *Motion in a Straight Line*   The velocity of a particle moving in a straight line is given by $v = 3te^{t^2}t$.
(a) Find an expression for the position s after a time t.
(b) Given that $s = 3$ at time $t = 0$, find the constant of integration C, and hence find an expression for s in terms of t without any unknown constants.

51. *Cost*   The marginal cost of producing the xth box of light bulbs is $5 + \sqrt{x + 1}$, and the fixed cost is $20,000. Find the total cost function $C(x)$.

52. *Cost*   The marginal cost of producing the xth box of computer disks is $(10 + x\sqrt{x^2 + 1})/100,000$, and the fixed cost is $100,000. Find the total cost function $C(x)$.

53. *Cost*   The marginal cost of producing the xth roll of film is given by $5 + 1/(x + 1)$. The total cost to produce one roll is $1,000. Find the total cost function $C(x)$.

54. *Cost*   The marginal cost of producing the xth box of videotape is given by $10 - \dfrac{x}{(x^2 + 1)^2}$. The total cost to produce one-hundred boxes is $10,000. Find the total cost function $C(x)$.

COMMUNICATION AND REASONING EXERCISES

55. Are there any circumstances under which one should use the substitution $u = x$? Illustrate your answer by means of an example.

56. Give an example of an integral that can be calculated by using the substitution $u = x^2 + 1$. Justify your claim by carrying out the calculation.

57. Give an example of an integral that can be calculated by using the power rule for antiderivatives, and also by using the substitution $u = x^2 + x$. Justify your claim by carrying out the calculations.

58. At what stage of a calculation using a u-substitution should one substitute back for u in terms of x: before or after taking the antiderivative?

59. You are asked to calculate $\displaystyle\int \frac{u}{u^2 + 1}\, du$. What is wrong with the substitution $u = u^2 + 1$?

60. What is wrong with the following "calculation" of $\displaystyle\int \frac{1}{x^2 - 1}\, dx$?

$$\int \frac{1}{x^2 - 1} = \int \frac{1}{u} \qquad \text{(using the substitution } u = x^2 - 1)$$

$$= \ln |u| + C$$

$$= \ln |x^2 - 1| + C$$

▶ ═════ **6.3** APPLICATIONS OF THE INDEFINITE INTEGRAL

APPLICATIONS TO BUSINESS

We have already seen some simple applications of the indefinite integral to business (Example 10 of Section 1 and exercises in Sections 1 and 2). We now look at several other scenarios in which the indefinite integral is useful.

▼ **EXAMPLE 1**   Volume Discount

A software company offers volume discounts to buyers of its program. If a customer buys a number of copies, then the xth copy will cost the customer $500/\sqrt{x + 1}$. What is the total cost to buy x copies?

SOLUTION   Let $C(x)$ be the total cost to buy x copies. What we are told is $C'(x)$, the marginal cost.

$$C'(x) = \frac{500}{\sqrt{x + 1}}$$

Thus,

$$C(x) = \int C'(x) \, dx$$

$$= \int \frac{500}{\sqrt{x + 1}} \, dx.$$

We can compute this using a simple substitution.

$$u = x + 1$$

$$du = dx$$

So

$$C(x) = \int \frac{500}{\sqrt{u}} \, du$$

$$= \int 500 u^{-1/2} \, du$$

$$= \frac{500 u^{1/2}}{1/2} + K \qquad (K \text{ is the constant of integration.})$$

$$= 1{,}000 \, (x + 1)^{1/2} + K.$$

We now need to find K. If a customer buys no copies of the program, the customer pays nothing, so $C(0) = 0$, which gives

$$0 = 1{,}000 \, (0 + 1)^{1/2} + K$$

$$= 1{,}000 + K,$$

so

$$K = -1{,}000.$$

Therefore,

$$C(x) = 1{,}000 \, (x + 1)^{1/2} - 1{,}000.$$

Before we go on... This is in fact a scheme that some software publishers and copy centers do use for volume discounts.

You might object that we should compute the total cost by adding up the selling prices of each copy, so

$$C(x) = \frac{500}{\sqrt{2}} + \frac{500}{\sqrt{3}} + \ldots + \frac{500}{\sqrt{x + 1}}.$$

You would be right. However, just as marginal cost is really only an estimate of the cost of an item, so the answer we obtained here by integration is only an estimate of the total revenue. We shall return to the relationship between integration and summation in Section 5.

Q Is there a "cue" to tell us when we need to take the integral? (For instance, how did you know to take the integral to obtain the cost of x items in the above example?)

A Luckily, there is an easy way to tell us whether an integral is needed. First look at the units of measurement of the function we are given and the one we are asked to find. In the above example, we were given cost per item and were asked to calculate cost. The function we were given is measured in dollars per item, while the cost function we want is measured in dollars. We now write an equation relating these units of measurement.

$$\text{Dollars} = \frac{\text{Dollars}}{\text{Item}} \times \text{Items}$$

$$\uparrow \qquad \uparrow \qquad \uparrow$$
$$\text{Cost} = \int \text{Cost per item } dx$$

(The second line is the integral equation we got in the example.) This is the cue we need: *multiplying units of measurement corresponds to taking the integral* (just as dividing units of measurement corresponds to taking the derivative).

▼ | **EXAMPLE 2** | Total Sales

My publisher tells me that I can expect monthly sales of my epic novel to be given by $1{,}000(1 - e^{-0.5t})$ copies, where t is the time in months after its publication.

(a) Find a formula for the number of copies $S(t)$ my novel will sell in the first t months.

(b) If there were 20,000 pre-publication orders for my novel, find a formula for the total number of copies $T(t)$ that my novel will have sold t months after publication. How many copies will sell in the first year?

SOLUTION **(a)** The formula $1{,}000(1 - e^{-0.5t})$ gives the number of copies sold per month. In other words, the accumulated sales are going up at a rate of $1{,}000(1 - e^{-0.5t})$ copies per month, or

$$S'(t) = 1{,}000(1 - e^{-0.5t}).$$

Thus,

$$S(t) = \int S'(t)\, dt$$

$$= \int 1{,}000\, (1 - e^{0.5t})\, dt$$

$$= 1{,}000 \int (1 - e^{-0.5t})\, dt$$

$$= 1{,}000 \left(t + \frac{e^{-0.5t}}{0.5} \right) + C$$

$$= 1{,}000\, (t + 2e^{-0.5t}) + C,$$

and it remains to find the constant C. To do this, note that $S(t)$ represents total sales starting at publication ($t = 0$) and ending at a future time t. Since we are counting only sales after publication,

$$S(0) = 0.$$

Substituting gives

$$S(0) = 1,000(0 + 2e^{-0.5(0)}) + C = 0,$$

or

$$2,000 + C = 0,$$

so

$$C = -2,000.$$

Thus, the total sales are given by

$$S(t) = 1,000(t + 2e^{-0.5t}) - 2,000 \text{ copies.}$$

(b) Note that the solution to (a) gives the number of copies that will be sold *starting now*. To obtain the total sales T, we must add the 20,000 copies already sold, so

$$T(t) = 1,000(t + 2e^{-0.5t}) - 2,000 + 20,000$$
$$= 1,000(t + 2e^{-0.5t}) + 18,000.$$

In one year's time, the total sales will therefore amount to

$$T(12) = 1,000 (12 + 2e^{-(0.5)(12)}) + 18,000$$
$$\approx 30,002 \text{ copies.}$$

Before we go on... How did we recognize that an integral was called for in the first place? We recognized that we were given a rate of change (copies per month). Alternatively, we could have used an analysis of units of measurement: we were given monthly sales, measured in copies per month, and we wanted total sales, measured in copies. So we write

$$\text{Copies} = \frac{\text{Copies}}{\text{Month}} \times \text{Months}$$

$$\text{Total Sales} = \int \text{Monthly sales } dt.$$

Note also that $S(t)$ and $T(t)$ differ by a constant, and so they have the same derivative: $S'(t) = T'(t) = 1,000(1 - e^{-0.5t})$.

How reasonable is the model? Figure 1 shows graphing calculator plots of $T'(t)$ and $T(t)$.

The graph of $T'(t)$ on the left shows a reasonable sales pattern for a new product: a fast initial increase, and then a gradual leveling off, in this

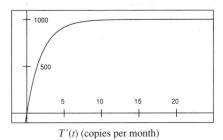

$T'(t)$ (copies per month)

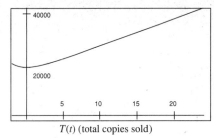

$T(t)$ (total copies sold)

FIGURE 1

case at 1,000 copies per month. The accumulated sales graph on the right thus becomes approximately linear with slope 1,000 for large values of t.

▼   **EXAMPLE 3**   Total Demand with Varying Price

Enormous State University (ESU) charges $2,000 per year for tuition and currently has an enrollment of 24,000 undergraduates. It finds that its net annual gain in enrollment can be predicted by the formula $q = 2,400 - 0.25p$, where q is the net annual gain (in students per year) and p is the tuition fee it charges. ESU's financial planning calls for its tuition to be $p = 2,000 + 500t$, t years from now. Express the enrollment as a function of t and predict the enrollment 10 years from now.

SOLUTION   The problem asks for enrollment as a function of time, which we denote by $E(t)$. We are given an equation for the net gain in enrollment per year: $q = 2,400 - 0.25p$. Its units, students per year, tell us that this is the derivative, $E'(t)$, so we have

$$E'(t) = 2,400 - 0.25p.$$

The problem now is that $E'(t)$ is given as a function of p rather than t, but we can rectify this by substituting the formula for p, $p = 2,000 + 500t$. This gives

$$E'(t) = 2,400 - 0.25(2,000 + 500t)$$
$$= 1,900 - 125t.$$

Thus,

$$E(t) = \int (1,900 - 125t) \, dt$$
$$= 1,900t - 62.5t^2 + C,$$

and it remains to calculate the constant C. For this, we use the additional piece of information that, at time $t = 0$, the enrollment is 24,000. Thus,

$E(0) = 24,000$, and so

$$24,000 = 1,900(0) - 62.5(0)^2 + C,$$

giving

$$C = 24,000.$$

Thus,

$$E(t) = 1,900t - 62.5t^2 + 24,000.$$

In 10 years' time, the enrollment will be

$$E(10) = 1,900(10) - 62.5(10)^2 + 24,000$$
$$= 36,750 \text{ students.}$$

Before we go on... Note that we did not use directly the information that ESU currently charges $2,000 in annual tuition. This information is implied by the equation $p = 2,000 + 500t$. Notice also that the unit analysis here is similar to that of the previous example: Students = (Students/Year) × Years.

 Figure 2 shows a graphing calculator plot of the graph of $E(t)$ for $t \le 50$.

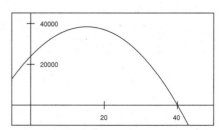

FIGURE 2

The graph predicts that enrollment will peak in about 15 years and that ESU cannot continue to raise fees indefinitely. If it does, it will need to close its doors in 40 years' time! A word of caution: We used a linear demand equation, and we have seen that linear models tend to be accurate only for a small range of values. (We have also not taken into account such factors as inflation and the fees charged by other universities.)

MOTION

Calculus is the language of the physical sciences. In the remainder of this section we discuss in some detail the application of calculus to motion, as an example of the intertwining of mathematics and physics that is an important part of both.

We begin by bringing together some facts scattered through the last several chapters having to do with an object moving in a straight line.

POSITION, VELOCITY, AND ACCELERATION

1. If $s = s(t)$ is the position of an object at time t, then its velocity is given by the derivative

$$v = \frac{ds}{dt}.$$

In short, *velocity is the derivative of position.*

2. The acceleration of an object is given by the derivative

$$a = \frac{dv}{dt}.$$

In short, *acceleration is the derivative of velocity.*

3. On the planet Earth, a freely falling body experiencing no air resistance accelerates at approximately 32 feet per second per second, or 32 ft/s^2.

Our first goal is to answer the following question. *Suppose that at time $t = 0$, I throw a ball up at a specified velocity v_0 from a specified position s_0. What is its position after t seconds?*

We are going to "work backwards" through the three points above. We shall restate them in reverse order, and also convert the derivative formulas into integral formulas. First, we make the convention that height above the ground is a positive number s, so that if s is increasing, velocity (its derivative) is positive. This means that since the acceleration due to gravity acts downward it is negative.

ACCELERATION, VELOCITY, AND POSITION: INTEGRAL FORM

I: $a(t) = -32$ ft/s.^{2}
II: $v(t) = \int a(t)\, dt$
III: $s(t) = \int v(t)\, dt$

Now we work from acceleration to position.

Step 1:   Substitute a in II and solve for v:

$$v(t) = \int a(t)\, dt = \int (-32)\, dt = -32t + C,$$

so $$v(t) = -32t + C.$$

Now we have to deal with the constant C. Looking back at the question we wish to answer, we see that at time $t = 0$, the velocity is v_0 (we call this the **initial velocity**). In other words, when $t = 0$, $v = v_0$. Substituting this into the above equation, we get

$$v_0 = -32(0) + C,$$

so $C = v_0$. Thus,

$$v(t) = v_0 - 32t.$$

Step 2   Substitute the formula for v into III and solve for s:

$$s(t) = \int v(t)\, dt = \int (v_0 - 32t)\, dt = v_0 t - 16t^2 + C,$$

so

$$s(t) = v_0 t - 16t^2 + C.$$

Now we have to deal once again with the constant C. Looking back at the question again, we see that at time $t = 0$, the position is specified as s_0 (we call this the **initial position**). In other words, when $t = 0$, $s = s_0$. Substituting this into the above equation, we get

$$s_0 = v_0(0) - 16(0) + C,$$

so that $C = s_0$. Thus,

$$s(t) = s_0 + v_0 t - 16t^2.$$

This formula tells us exactly where the ball is at time t, and answers our original question.

VELOCITY AND POSITION OF AN OBJECT MOVING VERTICALLY UNDER GRAVITY

The following equations describe the vertical motion of an object under gravity without air resistance.

Velocity Formula   **Position Formula**

$$v(t) = v_0 - 32t \qquad\qquad s(t) = s_0 + v_0 t - 16t^2$$

v is velocity in feet per second, s is position (height) in feet, and t is time in seconds.

Remember our convention that up is positive. Thus, v_0 refers to the initial *upward* velocity and s to the height above the ground.

These formulas allow us to answer a number of questions.

▼ EXAMPLE 4

A stone is tossed upward at 30 ft/s.

(a) How fast and in what direction is it going after 5 seconds?
(b) Where will it be after 5 seconds?

SOLUTION   **(a)** The question here asks for velocity, so we use the velocity formula

$$v = v_0 - 32t.$$

The first sentence tell us that the initial velocity is 30 ft/s upwards. Thus, $v_0 = 30$. The question asks us what happens at $t = 5$. We substitute these values into the equation and obtain

$$v = 30 - 32(5) = -130 \text{ ft/s}.$$

Thus, at time $t = 5$, the stone is *falling* at a speed of 130 ft/s.
  (b) Here, we're looking for the position at $t = 5$, so we use the position formula

$$s = s_0 + v_0 t - 16^2.$$

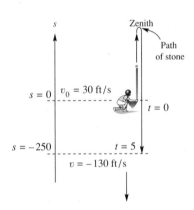

FIGURE 3

To use this formula, we need the values of all the constants on the right-hand side. Now we already know that $t = 5$ and $v_0 = 30$. What about s_0? This is the initial position, the position of the stone at time $t = 0$. Since we are not told this, we are free to set up our own convention and take the initial position as zero. In other words, we take $s_0 = 0$. Now we have all the values we need, and

$$s = 0 + 30(5) - 16(25)$$
$$= -250 \text{ ft.}$$

Thus, the stone is at a location 250 feet *below* its position when tossed. In other words, it has reached its **zenith** (highest point) and then dropped past its initial location, as shown in Figure 3.

Before we go on... You might be curious as to where and when the stone reaches its zenith and starts to fall. The next example shows how we can find this.

▼ EXAMPLE 5

A bullet is fired upward at 1,000 ft/s. How high does it get?

SOLUTION . We are asked for the position of the bullet at its zenith, so we turn to the position formula

$$s = s_0 + v_0 t - 16t^2.$$

As in the last example, we take $s_0 = 0$, and we are given $v_0 = 1,000$. What about the value of t? We haven't been told what t is, so it appears we are stuck.

When we run into a roadblock, a good strategy is to go back and check whether there is any information that we haven't yet used. The question asks about the highest point reached by the projectile. Although we don't know the value of t at that instant, we *do* know something: the zenith is the point of *maximum* height, so the bullet has velocity zero then (in other words, it has ceased going up and is about to come down). Thus, we look at the *velocity* equation

$$v = v_0 - 32t.$$

Now we know that $v_0 = 1,000$, and we want to find the t that will make $v = 0$:

$$0 = 1,000 - 32t,$$

so

$$t = 1,000/32 = 31.25 \text{ seconds.}$$

Now we can go back to the position equation and substitute for all the quantities on the right-hand side.

$$s = 0 + 1,000(31.25) - 16(31.25)^2 = 15,625 \text{ ft}$$

The equations we have been using ignore air resistance. The next example shows one way of modeling the effect of air resistance on a falling object: we assume that the acceleration decays exponentially.

▼ **EXAMPLE 6**   Skydiving

A skydiver experiences an acceleration of

$$a(t) = -32e^{-0.2t} \text{ ft/s}^2$$

t seconds after jumping from a plane. If the plane is flying at a height of 15,000 feet when the skydiver jumps, find his height as a function of time. If his parachute fails to deploy, when will he hit the ground, and how fast will he be going at the time?

SOLUTION   As before, we need to work backwards from acceleration to velocity and then to position. We start with

$$v(t) = \int a(t) \, dt = \int (-32e^{0.2t}) \, dt$$

$$= -32 \int e^{-0.2t} \, dt$$

$$= -32 \frac{e^{-0.2t}}{-0.2} + C$$

$$= 160e^{-0.2t} + C.$$

Now, to find C, we think about the initial velocity. As the skydiver jumps out of the plane, his initial velocity will be 0; that is, $v(0) = 0$, so

$$160e^0 + C = 0$$

and

$$160 + C = 0,$$

giving

$$C = -160.$$

Therefore,

$$v(t) = 160e^{-0.2t} - 160.$$

To find the height, we use

$$s(t) = \int v(t)\, dt,$$

so

$$s(t) = \int (160e^{-0.2t} - 160)\, dt$$

$$= 160\, \frac{e^{-0.2t}}{-0.2} - 160t + C$$

$$= -800e^{0.2t} - 160t + C.$$

The initial height is 15,000 feet, so $s(0) = 15,000$, giving

$$-800e^0 - 160(0) + C = 15,000,$$

and hence

$$C = 15,800.$$

We can now write

$$s(t) = -800e^{-0.2t} - 160t + 15,800.$$

To determine when he will hit the ground, we have to solve $s(t) = 0$, or

$$-800e^{-0.2t} - 160t + 15,800 = 0.$$

Unfortunately, this cannot be solved analytically, but we can approximate the answer using a graphing calculator or computer. Figure 4 shows the graph of s.

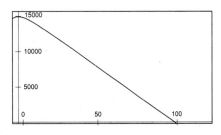

FIGURE 4

The graph shows that the time of impact will be around 100 seconds. Zooming in gives the more accurate estimate of $t = 98.75$ seconds. The velocity at impact is then

$$v(98.75) = 160e^{-0.2(98.75)} - 160 \approx 160 \text{ ft/s.}$$

Before we go on... The graph of velocity is shown in Figure 5.

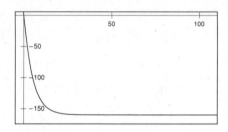

FIGURE 5

Because acceleration decays towards 0, the velocity approaches a limit, in this case -160 ft/s. You can think of the effect of wind resistance as limiting the velocity, keeping it from getting arbitrarily large. The limiting velocity is called the **terminal velocity.**

► **6.3 EXERCISES**

APPLICATIONS

1. *Cost* If your fixed cost to manufacture light bulbs is $10,000 and your marginal cost is $100 + 0.01x$ for the xth box of bulbs, find your total cost for x boxes.

2. *Cost* If your fixed cost to manufacture computer disks is $15,000 and your marginal cost is $200 - 0.05x$ for the xth box of disks, find your total cost for x boxes.

3. *Cost* If your fixed cost to manufacture film is $10,000 and your marginal cost is $100 + 20e^{-0.01x}$ for the xth box of film, find your total cost for x boxes.

4. *Cost* If your fixed cost to manufacture videotape is $15,000 and your marginal cost is $200 + 50e^{-0.05x}$ for the xth box of tape, find your total cost for x boxes.

5. *Volume Discounts* Your office furniture company offers a volume discount on chairs: the first chair costs $200, and you reduce the cost by $2 per chair for each

additional chair up to 100 chairs. Find the marginal cost function, and then determine how much you charge for x chairs.

6. *Volume Discounts* Your office furniture company offers a volume discount on tables: the first table costs $1,000, and you reduce the cost by $50 per table for each additional table up to 20 tables. Find the marginal cost function, and then determine how much you charge for x tables.

7. *Volume Discounts* Your software company offers a volume discount on its latest program: you charge $5,000/(x + 10)$ for the xth copy purchased by a customer. How much do you charge for a total of x copies?

8. *Volume Discounts* Your software company plans to offer the following volume discount on its next release: you will charge $50,000/(x + 10)^2$ for the xth

copy bought by a customer. How much will you charge for x copies?

9. *Average Cost* If your fixed cost to manufacture light bulbs is $10,000 and your marginal cost is $200 + 0.02x$ for the xth box of bulbs, find the production level that will minimize your *average* cost per box, and find the minimum average cost.

10. *Average Cost* If your fixed cost to manufacture computer disks is $9,000 and your marginal cost is $50 + 0.2x$ for the xth box of disks, find the production level that will minimize your *average* cost per box, and find the minimum average cost.

11. *Average Revenue* A sporting goods manufacturer offers a volume discount to stores on boogie boards: it charges $20 + 20e^{-0.5x}$ for the xth boogie board. What is the average revenue per boogie board if a store buys x boogie boards?

12. *Average Revenue* A sporting goods manufacturer offers a volume discount to stores on skateboards: it charges $10 + 15e^{-0.01x}$ for the xth skateboard. What is the average revenue per skateboard if a store buys x skateboards?

13. *Total Sales* Daily sales of your T-shirts over the next seven days are predicted by the equation

$$q = 10 - 0.2t^2,$$

where q represents the number of T-shirts sold per day t days from now.
(a) Give an expression for the total sales $s(t)$ after t days, assuming your total sales now ($t = 0$) are zero.
(b) How many T-shirts can you expect to sell in the next seven days?

14. *Declining Sales* Five years ago, your rock group's just-released CD, "Galactic Explosion," was selling 10,000 copies per year, but since that time, sales have declined by 2,000 each year, so that annual sales can be modeled by the linear equation

$$q = 10,000 - 2,000t,$$

where q represents annual sales of your CD, and t is the time in years ($t = 0$ corresponds to this date 5 years ago.)
(a) Give an expression for the total sales $s(t)$ at time t.
(b) What is the total number of CDs sold to date?

15. *Foreign Investments* 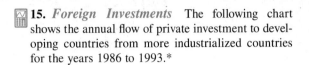 The following chart shows the annual flow of private investment to developing countries from more industrialized countries for the years 1986 to 1993.*

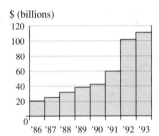

$ (billions)

(a) Let $q(t)$ represent the annual quantity of private investments (in billions of dollars), where t is the number of years since the start of 1986. Which of the following models best fits the data shown? (Feel free to use a graphing calculator.)
(A) $q(t) = 10 + 12.5t$
(B) $q(t) = 10 + 35.5\sqrt{t}$
(C) $q(t) = 10 + 1.56t^2$

(b) Use the best-fit model from part (a) to obtain an equation for the total flow $P(t)$ of private investment from January 1, 1986 until t years later.
(c) Use the model to predict how much money will have been sent to developing countries from January 1, 1986 through January 1, 1996.

16. *Foreign Investments* 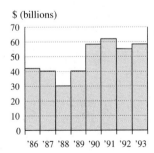 The following chart shows the annual flow of government loans and grants to developing countries from more industrialized countries.*

$ (billions)

▼ * Source: World Bank (*The New York Times*, Dec. 17, 1993, p. D1.)

(a) Let $q(t)$ represent the annual quantity of loans (in billions of dollars), where t is the number of years since the start of 1986. Which of the following models best fits the data shown? (Feel free to use a graphing calculator.)

(A) $q(t) = 38 - 0.6(t - 3)^2$
(B) $q(t) = 38 + 0.6(t - 3)^2$
(C) $q(t) = 20 + e^{0.54t}$

(b) Use the best-fit model from part (a) to obtain an equation for the total flow $G(t)$ of government loans and grants from January 1, 1986 until t years later.

(c) Use the model to predict how much money will have been sent to developing countries from January 1, 1986 through January 1, 1996.

17. *Exports*   Based on figures released by the U.S. Agriculture Department, the value of U.S. pork exports from 1985 through 1993 can be approximated by the equation

$$q = \frac{460e^{(t-4)}}{1 + e^{(t-4)}},$$

where q represents annual exports (in millions of dollars) and t the time in years, with $t = 0$ corresponding to January 1, 1985.* The figure shows the actual data with the graph of the equation superimposed.

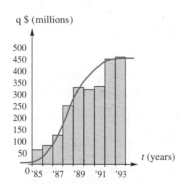

q $ (millions)

(a) Use the model to obtain an equation for total U.S. pork exports E in billions of dollars as a function of t. (Assume zero total exports as of January 1, 1985; that is, $E(0) = 0$.)

(b) Use your equation to predict the value of total U.S. pork exports to the nearest million dollars from January 1, 1985 to January 1, 2000.

18. *Sales*   The weekly demand for your company's Lo-Cal Chocolate Mousse can be modeled by the equation

$$q = \frac{50e^{2t-1}}{1 + e^{2t-1}},$$

where q is the number of gallons sold per week and t is the time in weeks. At present ($t = 0$) total sales of Lo-Cal mousse amount to 100 gallons.

(a) Express the total sales $s(t)$ of Lo-Cal mousse as a function of time t.

(b) Estimate the total sales of Lo-Cal mousse one year from now.

19. *Production*   The production of cellular phones at your electronics plant has a Cobb-Douglas production function given by

$$P = 10x^{0.3}y^{0.7},$$

where P is the number of cellular phones it produces per year, x is the number of employees, and y is the daily operating budget (in dollars). You have decided to increase the daily operating budget linearly from the present level of $6,000 to $10,000 over the next two years. Your plant employs 100 workers.

(a) Determine the productivity P as a function of t, where t is measured in years from now.

(b) Estimate the total production of cellular phones over the next two years. (Give your answer to the nearest cellular phone.)

20. *Production*   Referring to the Cobb-Douglas production formula in the previous exercise, assume that you wish to maintain a daily operating budget of $6,000 but plan to increase the work force linearly from the present 100 employees to 200 over the coming two years.

(a) Determine the productivity P as a function of t, where t is measured in years from now.

(b) Estimate the total production of cellular phones over the next two years. (Give your answer to the nearest cellular phone.)

Using the Logistic Equation to Predict Sales[†]   The logistic equation has the form

$$y(x) = \frac{NP_0}{P_0 + (N - P_0)e^{-kx}},$$

where N, P_0 and k are positive constants. (See the accompanying graph.)

---

*This is the authors' model based on figures published by the U.S. Agriculture Department quoting annual sales from 1985 through 1993. (*The New York Times*, Dec. 12, 1993, p. D1.)

[†]See "You're the Expert" at the end of the chapter on logarithmic and exponential functions for a discussion of the use of the logistic equation in modeling epidemics.

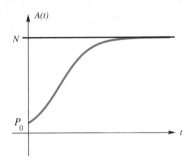

This equation is useful in modeling and predicting the demand for a new product, where

y = *demand in annual sales*

x = *number of years since the introduction of the product to the market*

P_0 = *initial demand (demand at time x = 0)*

N = *total potential demand*

k = *approximate initial rate of growth of demand (a good approximation if N is large compared with P_0).*

In Exercises 21–26, we use the logistic curve to model the growing demand for several therapeutic drugs.

21. Show that the logistic equation can be rewritten in the form

$$y(x) = \frac{NP_0 e^{kx}}{P_0 e^{kx} + (N - P_0)}.$$

22. Using the result in the preceding exercise and a suitable substitution, show that

$$\int y(x)\, dx = \frac{N}{k} \ln \left(P_0 e^{kx} + (N - P_0) \right) + C.$$

23. *Eli Lilly Corp.*'s human growth hormone Humatrope has a potential market estimated at roughly N = $160 million per year, and its total Humatrope

sales can be approximated by the logistic function with k = 2.7 and P_0 = 0.0025. $y(x)$ represents annual sales in millions of dollars, and x represents the number of years since the drug's approval by the FDA in 1987.* Use the model to give an estimate (to the nearest $10 million) of the value of total sales of Humatrope over the ten-year period beginning in 1987. (Hint: Use the result from Exercise 22.)

24. *Genentech*'s human growth hormone Protropin has a potential market estimated at roughly N = $300 million per year, and its total Protropin sales can be approximated by the logistic function with k = 1.7 and P_0 = 0.20. $y(x)$ represents annual sales in millions of dollars, and x represents the number of years since the drug's approval by the FDA in 1986.* Use the model to give an estimate (to the nearest $10 million) of the value of total sales of Protropin over the ten-year period beginning in 1986. (Hint: Use the result from Exercise 22.)

25. *(Graphing calculator or computer required)* *Genzyme*'s drug against Gaucher's disease, Ceredase, cost the company $30 million to develop and has a potential market of $900 million per year. At the time of FDA approval (1991) annual sales were estimated at $50 million. Assuming an initial rate of growth specified by k = 0.5 (that is, approximately a 50% initial growth rate), use a graphing calculator to estimate how long (to the nearest year) it will take *Genzyme* to earn 10 times development costs from sales of Ceredase.[†] (Hint: Use the result from Exercise 22.)

26. *(Graphing calculator or computer required)* Repeat the previous exercise using an initial growth rate specified by k = 0.3 (that is, approximately a 30% initial growth rate). (Hint: Use the result from Exercise 22.)

▼ * The model is a very crude one, based on 1991 sales data, total sales data through May, 1992, and very rough estimates of the potential market and selling price. Source: Senate Judiciary Committee; Subcommittee on Antitrust and Monopoly/*The New York Times*, May 14, 1992, p. D1.

† Initial sales figure based on sales for 9 months of 1991 immediately after FDA approval. Source: ibid.

Vertical Motion   *In Exercises 27–38, neglect the effects of air resistance.*

27. If a stone is dropped from a rest position above the ground, how fast, and in what direction, will it be traveling after 10 seconds?

28. If a stone is thrown upward at 10 feet per second, how fast, and in what direction, will it be traveling after 10 seconds?

29. Show that if a projectile is thrown upward with a velocity of $v_0/32$ ft/s, then it will reach its highest point after $v_0/32$ seconds.

30. Use the result of the preceding exercise to show that if a projectile is thrown upward with a velocity of v_0 ft/s, then its highest point will be $v_0^2/64$ feet above the starting point.

In Exercises 31–38, use the results of the previous two exercises.

31. I threw a ball up in the air to a height of 20 feet. How fast was the ball traveling when it left my hand?

32. I threw a ball up in the air to a height of 40 feet. How fast was the ball traveling when it left my hand?

33. I threw a ball up in the air to a height of 20 feet. Where was the ball after 4 seconds?

34. I threw a ball up in the air to a height of 40 feet. Where was the ball after 5 seconds?

35. A piece of chalk is tossed vertically upward by Professor Schwartzenegger and hits the ceiling 100 feet above with a *BANG*.
    (a) What is the minimum speed the piece of chalk must have been traveling to enable it to hit the ceiling?
    (b) Assuming that Professor Schwartzenegger in fact tossed the piece of chalk up at 100 ft/s, how fast would it have been moving when it struck the ceiling?

COMMUNICATION AND REASONING EXERCISES

43. Why might it be a bad idea to offer a volume discount in which the marginal cost is $100e^{-0.01x}$?

44. Would it be a bad idea to offer a volume discount in which the marginal cost is $10,000/(x + 10)^2$?

45. Complete the following sentence. A unit analysis tells us that if $P(s)$ and $Q(s)$ are functions with the property

(c) Assuming that Professor Schwartzenegger tossed the chalk up at 100 ft/s, and that it recoils from the ceiling with the same speed it had at the instant it hit, how long will it take the chalk to make the return journey and hit the ground?

36. A projectile is fired vertically upwards from ground level at 16,000 ft/s.
    (a) How high does it go?
    (b) How long does it take to reach its zenith (highest point)?
    (c) How fast is it traveling when it hits the ground?

37. *Strength*   Professor Strong can throw a 10-pound, dumbbell twice as high as Professor Weak can. How much faster can Professor Strong throw it?

38. *Weakness*   Professor Weak can throw a computer disk three times as high as Professor Strong can. How much faster can Professor Weak throw it?

39. *Varying Acceleration*   A particle in a nuclear accelerator undergoes an acceleration given by

$$a = 3t^2 - 4t + 5 \text{ ft/s}^2.$$

Assuming it started from rest, how far has it traveled in 120 seconds?

40. *Varying Acceleration*   The Starship Galactica undergoes an acceleration given by

$$a = e^{2(t-2)} + 5 \text{ ft/s}^2.$$

Assuming it started from rest, how far has it traveled in 60 seconds?

41. *Fast Cars*   Professor Hare's Gran Turismo can do 0 to 60 mph in 3 seconds. Assuming constant acceleration, how fast is it accelerating? [60 mph = 88 ft/s]

42. *Slow Cars*   Professor Turtle's Slomobile can only do 0 to 60 mph in 23 seconds. Assuming constant acceleration, how fast is it accelerating? [60 mph = 88 ft/s]

that units of $P(s) \times$ *units of* $s =$ *units of* $Q(s)$, then _____ .

46. Why are the units of measurement of $\int f(x)\,dx$ equal to the units of measurement of $f(x)$ times the units of measurement of x?

▶ ▬▬▬▬ **6.4** GEOMETRIC DEFINITION OF THE DEFINITE INTEGRAL

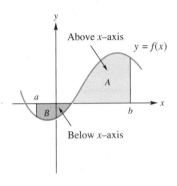

FIGURE 1

This section and the next are in many ways the intellectual climax of the course so far. We are about to see that the antiderivative provides a solution to what appears to be a completely unrelated problem. We call Newton and Leibniz the inventors of calculus largely because they made this connection.

Up to this point we have been talking about the *indefinite* integral. This suggests that there may be such a thing as the *definite* integral. Indeed there is. To introduce it, we are now going to talk—of all things—about *area*. Consider the graph of a function f, as in Figure 1.

We have chosen two values a and b of x and have shaded the region enclosed by the vertical lines $x = a$ and $x = b$, the x-axis, and the curve $y = f(x)$. We would like to find the area of the shaded region, but for several reasons, we shall *subtract* any area that lies below the x-axis (the grey area in the figure). We say that the **definite integral of f from a to b** is the total "net" area: $A - B$. For example, if $A = 5$ square units, and $B = 2$ square units, then the definite integral is equal to $5 - 2 = 3$. We write the definite integral as $\int_a^b f(x)\, dx$. This is read as "the integral from a to b of $f(x)$, with respect to x," or "the integral from $x = a$ to $x = b$ of $f(x)$, with respect to x." We summarize this with the following definition.

GEOMETRIC DEFINITION OF THE DEFINITE INTEGRAL

If f is a function whose domain contains the closed interval $[a, b]$, then the **definite integral of $f(x)$ from $x = a$ to $x = b$** is defined as

(Area between the vertical lines $x = a$ and $x = b$ that is *above* the x-axis and below the graph of $f(x)$) −

(Area between the vertical lines $x = a$ and $x = b$ that is *below* the x-axis and above the graph of $f(x)$)

assuming that these areas exist and are finite. We denote the definite integral of $f(x)$ from a to b by $\int_a^b f(x)\, dx$, where we read the symbols as follows.

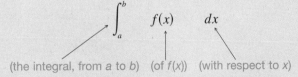

$$\int_a^b \quad f(x) \quad dx$$

(the integral, from a to b)   (of f(x))   (with respect to x)

The numbers a and b are called the **limits of integration.** The number a is called the **lower** limit and b the **upper** limit.

▶ NOTES

1. The reason we call this the *geometric* definition of the definite integral is that we have defined it using the geometric notion of area. In the next section we shall give an algebraic definition.

2. The stipulation that the areas in question exist and are finite puts restrictions on the functions f that we can integrate. Although we shall not go into the fine details of the theory, examples of functions whose integrals we can find are the functions that are continuous on $[a, b]$ and the functions that are *bounded* and have only finitely many discontinuities in $[a, b]$.* ◄

Q The notation looks suspiciously like the *indefinite* integral with some new decorations. What on earth has this to do with antiderivatives?

A Therein lies the surprise at the heart of calculus, but we'll keep you in suspense a moment longer. In the meantime, remember this:

$\int f(x) \, dx$ means the *indefinite* integral of $f(x)$ with respect to x—in other words, the general form of the antiderivative of $f(x)$.

$\int_b^a f(x) \, dx$ stands for the *definite* integral of $f(x)$ from a to b—in other words, the area as discussed above.

▼ **EXAMPLE 1**

Find $\int_0^1 1 \, dx$.

SOLUTION   This is the area of the region shaded in Figure 2.
   Since this is a square with sides equal to 1, its area is 1, so

$$\int_0^1 1 \, dx = 1.$$

Before we go on... What, then, is $\int_0^2 1 \, dx$? $\int_0^1 2 \, dx$? $\int_0^1 (-1) \, dx$? (Answers: 2, 2, and -1.)

▼ **EXAMPLE 2**

Calculate $\int_0^1 x \, dx$.

SOLUTION   This is the area of the region shaded in Figure 3.
   This is a triangle with height 1 and base 1, so it has area $\frac{1}{2}$. Thus,

$$\int_0^1 x \, dx = \frac{1}{2}.$$

Before we go on... What, then, is $\int_0^2 x \, dx$? $\int_0^1 2x \, dx$? $\int_0^1 (-x) \, dx$? (Answers: 2, 1, and $-\frac{1}{2}$.)

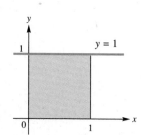

FIGURE 2

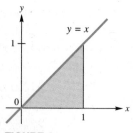

FIGURE 3

▼ *$f(x)$ is bounded on $[a, b]$ if the part of its graph corresponding to $a \le x \le b$ lies entirely between two horizontal lines. For example, the function given by

$$f(x) = \begin{cases} \dfrac{1}{x} & \text{if } x > 0 \\ 0 & \text{if } x = 0 \end{cases}$$

is not bounded on $[0, 1]$, since its graph goes up to infinity near $x = 0$. Thus we shall not (yet) try to compute $\int_0^1 f(x) \, dx$. (See the section on improper integrals in the next chapter for more on this example.)

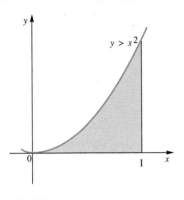

FIGURE 4

Q What about $\int_0^1 x^2 \, dx$?

A This is the area of the region under part of a parabola, as shown in Figure 4. Unfortunately, this is a very difficult area to calculate geometrically. In fact, the culmination of one part of Greek geometry was a calculation by Archimedes of (what amounted to) this area.

So as not to keep you in suspense any longer, we shall tell you how to calculate areas such as $\int_0^1 x^2 \, dx$, and then in Section 6 we'll see just why the calculation works. In order to express the answer in the simplest possible way, we first introduce some notation.

If F is any function, by $[F(x)]_a^b$ we shall mean the difference $F(b) - F(a)$.

Here are some quick examples.

$$[x^2 + 1]_{-1}^1 = [(1^2 + 1) - ((-1)^2 + 1)]$$
$$= [1 + 1 - (1 + 1)] = 0$$

$$[x + e^x]_0^1 = [(1 + e^1) - (0 + e^0)]$$
$$= [1 + e - 0 - 1] = e$$

$$[3(x^3 - x)]_{-2}^2 = 3[(x^3 - x)]_{-2}^2$$
$$= 3[(2^3 - 2) - ((-2)^3 - (-2))]$$
$$= 3[8 - 2 + 8 - 2] = 36$$

Notice the care you must take to get all of the signs right.

Here is one of the most important results in calculus.

THE FUNDAMENTAL THEOREM OF CALCULUS (FTC)

If $F(x)$ is any antiderivative of $f(x)$, and $f(x)$ is continuous on the interval $[a, b]$, then

$$\int_a^b f(x) \, dx = [F(x)]_a^b = F(b) - F(a).$$

Actually, this is only half of the FTC. In Section 6 we shall see the whole thing and also discuss why it is true. For the moment, just appreciate what it says: in order to calculate areas, all we need to be able to do is calculate antiderivatives. But this is not hard, as we've seen. This result has been called the biggest accident in all of mathematics. Who would have thought that the problems of finding areas and finding rates of change should be at all related?

▼ **EXAMPLE 3**

Calculate $\int_0^1 x^2 \, dx$.

SOLUTION To calculate this integral, we follow the FTC, which says: first find an antiderivative, then evaluate at the limits of integration, and then subtract.

$$\int_0^1 x^2\, dx = \left[\frac{x^3}{3}\right]_0^1 = \frac{1}{3}[x^3]_0^1 = \frac{1}{3}[1^3 - 0^3] = \frac{1}{3}.$$

Thus, the area in Figure 4 is exactly $\frac{1}{3}$ square units.

Before we go on...

Q What happened to the constant of integration C?

A You can leave it out, since the FTC says that you can use *any* antiderivative you like, and so we chose the simplest-looking antiderivative, which is $x^3/3$. Here is what happens if we use another antiderivative, $(x^3/3) + 2$.

$$\int_0^1 x^3\, dx = \left[\frac{x^3}{x} + 2\right]_0^1 = \left(\frac{1^3}{3} + 2\right) - \left(\frac{0^3}{3} + 2\right)$$

$$= \frac{1}{3} + 2 - 2 = \frac{1}{3}.$$

Thus, we get the same answer as before. Notice how the constant 2 canceled. This is what would happen to C if we left it in, so we leave it out.

AVOIDING ERRORS IN SIGNS

When evaluating the definite integral, you can avoid common errors in signs as follows. When you have found the antiderivative and obtained, say, $\left[x - \frac{x^3}{3}\right]_{-1}^1$, here is what you can do.

1. Make the template

$$[(\quad) - (\quad)],$$

leaving the insides of the parentheses blank.

2. Fill in the value of the antiderivative at the upper limit of integration in the first blank and its value at the lower limit of integration in the second blank, getting, for example,

$$\left[\left(1 - \frac{1^3}{3}\right) - \left((-1) - \frac{(-1)^3}{3}\right)\right].$$

Now we have all the needed parentheses in place, and we will not forget to distribute minus signs correctly.

▼ **EXAMPLE 4**

Calculate the total area enclosed by the x-axis, the vertical lines $x = -2$ and $x = 2$, and the curve $y = x^2 - 1$.

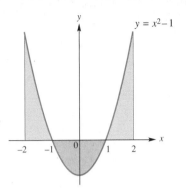

FIGURE 5

SOLUTION   We first make a quick sketch of the curve, which we recognize as a parabola with x-intercepts -1 and 1. The graph is shown in Figure 5.

If we were to simply calculate the definite integral $\int_{-2}^{2}(x^2 - 1)\,dx$, we would wind up with the *net* area, the sum of the two pieces on either side *minus* the piece in the middle below the x-axis. This is not what the question is asking for: we are trying to calculate the *total* area, and this means *adding* the areas of the three pieces. To do this, we must calculate the area of each piece separately.

$$\text{Right-hand piece: } \int_{1}^{2}(x^2 - 1)\,dx = \left[\frac{x^3}{3} - x\right]_{1}^{2}$$

$$= \left[\left(\frac{2^3}{3} - 2\right) - \left(\frac{1^3}{3} - 1\right)\right]$$

$$= \frac{8}{3} - 2 - \frac{1}{3} + 1$$

$$= \frac{4}{3}$$

$$\text{Middle piece: } \int_{-1}^{1}(x^2 - 1)\,dx = \left[\frac{x^3}{3} - x\right]_{-1}^{1}$$

$$= \left[\left(\frac{1^3}{3} - 1\right) - \left(\frac{(-1)^3}{3} - (-1)\right)\right]$$

$$= \frac{1}{3} - 1 + \frac{1}{3} - 1$$

$$= -\frac{4}{3}$$

This integral is negative because the area lies below the x-axis. The actual area of the middle piece must be $(+)\frac{4}{3}$ square units.

Left-hand piece: By the symmetry of the curve, this has the same area as the right-hand piece, which is $\frac{4}{3}$.

Thus, the total area is $\frac{4}{3} + \frac{4}{3} + \frac{4}{3} = 4$.

Before we go on... If we had simply calculated $\int_{-2}^{2}(x^2 - 1)\,dx$, we would have obtained

$$\int_{-2}^{2}(x^2 - 1)\,dx = \left[\frac{x^3}{3} - x\right]_{-2}^{2} = \left[\left(\frac{2^3}{3} - 2\right) - \left(\frac{(-2)^3}{3} - (-2)\right)\right]$$

$$= \frac{8}{3} - 2 + \frac{8}{3} - 2 = \frac{4}{3}.$$

Notice two things. First, this is the wrong answer. Second, it is the sum of the areas of the two pieces above the x-axis, minus the area below the x-axis. That is, it is $\frac{4}{3} + \frac{4}{3} - \frac{4}{3}$.

▼ | **EXAMPLE 5**

Calculate $\int_1^2 (2x - 1)e^{2x^2-2x}\, dx$.

SOLUTION This is one of those integrals that requires a u-substitution. We have two ways we can proceed. We could first make the substitution, then take the antiderivative, then substitute for u, and finally evaluate at $x = 2$ and $x = 1$ and subtract.

We prefer the following procedure. When we make the u-substitution, we calculate the u-values corresponding to $x = 2$ and $x = 1$ and change the limits of integration. This will save us having to put everything back in terms of x before substituting the limits. Here we go.

$$u = 2x^2 - 2x$$

$$\frac{du}{dx} = 4x - 2$$

$$dx = \frac{1}{4x - 2}\, du$$

When $x = 2$, $u = 4$ (substituting $x = 2$ in the formula
$$u = 2x^2 - 2x)$$
When $x = 1$, $u = 0$ (substituting $x = 1$ in the formula
$$u = 2x^2 - 2x)$$

Now, in the substitution step, we change the limits $x = 1$ and $x = 2$ to the corresponding limits $u = 0$ and $u = 4$.

$$
\underset{\substack{\nearrow \\ x\text{-value}}}{\overset{\substack{x\text{-value} \\ \searrow}}{\int_1^2}} (2x - 1)e^{2x^2-2x}\, dx = \underset{\substack{\nearrow \\ u\text{-value}}}{\overset{\substack{u\text{-value} \\ \searrow}}{\int_0^4}} (2x - 1)e^u \frac{1}{4x - 2}\, du
$$

$$= \int_0^4 (2x - 1)e^u \frac{1}{2(2x - 1)}\, du$$

$$= \int_0^4 e^u \frac{1}{2}\, du$$

$$= \frac{1}{2} \int_0^4 e^u\, du$$

$$= \frac{1}{2} \left[e^u \right]_0^4$$

$$= \frac{1}{2}(e^4 - e^0) = \frac{1}{2}(e^4 - 1)$$

Before we go on... The alternative process outlined in the first paragraph is this:

$$\int (2x - 1)\, e^{2x^2 - 2x}\, dx = \int e^u \frac{1}{2}\, du \quad \text{(after substitution)}$$

$$= \frac{1}{2} e^u + C$$

$$= \frac{1}{2} e^{2x^2 - 2x} + C,$$

and so

$$\int_1^2 (2x - 1) e^{2x^2 - 2x}\, dx = \left[\frac{1}{2} e^{2x^2 - 2x}\right]_1^2 = \frac{1}{2}(e^4 - 1).$$

▼ **EXAMPLE 6**

Calculate $\int_{-1}^1 x(3x^2 + 3)^3\, dx$.

SOLUTION   We use substitution.

$$u = 3x^2 + 3$$

$$\frac{du}{dx} = 6x$$

$$dx = \frac{1}{6x}\, du$$

When $x = 1$, $u = 6$.
When $x = -1$, $u = 6$.

Substituting,

$$\int_{-1}^1 x(3x^2 + 3)^3\, dx = \int_6^6 xu^3 \frac{1}{6x}\, du$$

$$= \frac{1}{6} \int_6^6 u^3\, du$$

$$= \frac{1}{6}\left[\frac{u^4}{4}\right]_6^6 = \frac{1}{24}(6^4 - 6^4) = 0.$$

Before we go on... Looking back at the calculation, notice that we could have saved ourselves a lot of work if we had paused to think after getting

$$\frac{1}{6} \int_6^6 u^3\, du.$$

This is the definite integral of u^3 from $u = 6$ to $u = 6$ and is the area of a region with 0 width. No wonder the answer is 0!

▼ **EXAMPLE 7**   Displacement

A car traveling down a road has velocity $v(t) = 30 + 2t$ mph at time t hours. How far does it travel between times $t = 1$ and $t = 5$?

SOLUTION   The most straightforward way to do this would be to find the position $s(t)$ and then calculate the difference $s(5) - s(1)$ to find the difference in the car's positions at the first hour and the fifth. (Notice that the car never turns around to double back on itself, so $s(5) - s(1)$ *is* the total distance traveled.) There is a small problem with carrying out this calculation, however: we do not know where the car starts, so we can determine s only up to a constant of integration. This is only a small problem, since the constant will cancel out in the difference. On the other hand, since $s(t)$ is an antiderivative of $v(t)$, the difference $s(5) - s(1)$ is the value of $\int_1^5 v(t)\, dt$, so we can instead calculate this definite integral.

$$\int_1^5 v(t)\, dt = \int_1^5 (30 + 2t)\, dt = \left[30t + t^2 \right]_1^5$$
$$= (150 + 25) - (30 + 1)$$
$$= 144 \text{ miles}$$

The way in which we solved the last example gives another useful interpretation of the integral. Remember that $f'(x)$ gives the rate of change of $f(x)$. If we integrate the rate of change over the closed interval $[a, b]$, that is, compute the definite integral $\int_a^b f'(x)\, dx$, we get the **total change** of $f(x)$ between $x = a$ and $x = b$, which is $f(b) - f(a)$. In the case of motion, the total change of position from one time to another is also known as the **displacement.** So, if $v(t)$ is velocity, $\int_a^b v(t)\, dt$ gives the displacement from time a to time b.

THE DEFINITE INTEGRAL AS TOTAL CHANGE

Given the rate of change $f'(x)$ of a quantity $f(x)$, the total change of $f(x)$ between $x = a$ and $x = b$ is given by:

$$\text{Total change in } f(x) = f(b) - f(a) = \int_a^b f'(x)\, dx.$$

▼ **EXAMPLE 8**   Total Revenue

According to data published in *The New York Times*, the annual revenue of *United Airlines* increased more-or-less linearly from January 1988 through December 1992 and could be modeled by the equation

$$r(t) = 8.50 + 0.95t,$$

where $r(t)$ is *United Airlines'* annual revenue in billions of dollars, and t is time in years since January 1988.*

(a) What was the total revenue earned by *United Airlines* over the reported period?

(b) Obtain an equation giving the total revenue earned by *United Airlines* since January 1, 1988 as a function of t, where t is the time in years since January 1, 1988.

(c) Assuming the trend continues, use the model to predict the total revenue earned by *United Airlines* from January 1988 through December 1994.

SOLUTION **(a)** Let $R(t)$ be the total revenue earned by *United Airlines*, up to time t. The given function $r(t)$ is revenue per year, and is thus the derivative of $R(t)$. Hence, the total change in $R(t)$ from $t = 0$ to $t = 5$ is

$$R(5) - R(0) = \int_0^5 r(t)\, dt$$

$$= \int_0^5 (8.50 + 0.95t)\, dt$$

$$= [8.50t + 0.475t^2]_0^5$$

$$= (8.50(5) + 0.475(5)^2) - (8.50(0) + 0.475(0)^2)$$

$$= \$54.375 \text{ billion.}$$

(b) We are asked for $R(t) - R(0)$. We have already done most of the work in part (a).

$$R(t) - R(0) = [8.50t + 0.475t^2]_0^t$$

$$= (8.50(t) + 0.475(t)^2) - (8.50(0) + 0.475(0)^2)$$

$$= 8.50t + 0.475t^2$$

(c) We are asked for $R(7) - R(0)$. We can use the answer to part (b) with $t = 7$.

$$R(7) - R(0) = 8.50(7) + 0.475(7)^2$$

$$= \$82.775 \text{ billion}$$

Before we go on... Figure 6 shows a facsimile of the revenue chart that actually appeared in the cited *New York Times* article, with the graph of our linear model superimposed.

Notice something interesting: each rectangle in the shaded region has width one unit and height equal to the revenue earned by *United Airlines* during the specified year. Thus, the area of, say, the rectangle corresponding to 1991 is $1 \times 11.7 = 11.7$, the total revenue for that year. In other words,

---

▼ * The model is the authors', based on data published in *The New York Times*, Dec. 24, 1993, p. D1. (Source: Company Reports.)

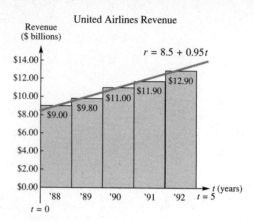

FIGURE 6

the total revenue earned by United Airlines from t = 0 to t = 5 is equal to the total shaded area. But this area is closely approximated by the integral we calculated in part (a), since the integral gives the area under the line r = 8.5 + 0.95t between the vertical lines t = 0 and t = 5.

    We can check our answer against the actual total revenue. Adding the areas of the rectangles gives

$$8.0 + 9.8 + 11.0 + 11.7 + 12.9 = \$53.4 \text{ billion}.$$

Our linear model gave us

$$\int_0^5 8.5 + 0.95t \, dt = \$54.375 \text{ billion},$$

which is close to the actual revenue.

Q Why did we bother with the linear model at all, since the actual data are readily available?

A The usefulness of the linear model is that it allows us to *predict*, as we did in part (c) of the example. Further, we have already seen numerous examples in the text of how mathematical models can assist in financial planning. Having an equation to work with is far more useful than having a collection of numbers.

Q How did you obtain the linear model?

A If you look at the revenue chart, you will notice that annual revenue appears to increase linearly with time. We noticed this and proceeded to find the equation of the line passing through two selected points. You should try to come up with your own linear model and perhaps improve on ours.*

▼ *We shall see in the chapter on functions of several variables how to find the *best* linear approximation of a collection of data such as the data above. In fact, this very example will come up again in that context.

▶ **6.4 EXERCISES**

Evaluate the definite integrals in Exercises 1–26.

1. $\displaystyle\int_{-1}^{1} (x^2 + 2)\, dx$ **2.** $\displaystyle\int_{-2}^{1} (x - 2)\, dx$ **3.** $\displaystyle\int_{-2}^{2} (x^3 - 2x)\, dx$ **4.** $\displaystyle\int_{-1}^{1} (2x^3 + x)\, dx$

5. $\displaystyle\int_{1}^{3} \left(\frac{2}{x^2} + 3x\right) dx$ **6.** $\displaystyle\int_{2}^{3} \left(x + \frac{1}{x}\right) dx$ **7.** $\displaystyle\int_{0}^{1} (2.1x - 4.3x^{1.2})\, dx$ **8.** $\displaystyle\int_{-1}^{0} (4.3x^2 - 1)\, dx$

9. $\displaystyle\int_{0}^{1} 2e^x\, dx$ **10.** $\displaystyle\int_{-1}^{0} 3e^x\, dx$ **11.** $\displaystyle\int_{-1}^{1} e^{2x-1}\, dx$ **12.** $\displaystyle\int_{0}^{2} e^{-x+1}\, dx$

13. $\displaystyle\int_{0}^{50} e^{-0.02x-1}\, dx$ **14.** $\displaystyle\int_{-20}^{0} 3e^{2.2x}\, dx$ **15.** $\displaystyle\int_{0}^{1} \sqrt{x}\, dx$ **16.** $\displaystyle\int_{-1}^{1} \sqrt[3]{x}\, dx$

17. $\displaystyle\int_{-1.1}^{1.1} e^{x+1}\, dx$ **18.** $\displaystyle\int_{0}^{\sqrt{2}} x\sqrt{2x^2 + 1}\, dx$ **19.** $\displaystyle\int_{-\sqrt{2}}^{\sqrt{2}} 3x\sqrt{2x^2 + 1}\, dx$ **20.** $\displaystyle\int_{-1.2}^{1.2} e^{-x-1}\, dx$

21. $\displaystyle\int_{0}^{1} 5xe^{x^2+2}\, dx$ **22.** $\displaystyle\int_{0}^{2} \frac{3x}{x^2 + 2}\, dx$ **23.** $\displaystyle\int_{0}^{1} x\sqrt{2x + 1}\, dx$ **24.** $\displaystyle\int_{-1}^{0} 2x\sqrt{x + 1}\, dx$

25. $\displaystyle\int_{2}^{3} \frac{x^2}{(x^3 - 1)}\, dx$ **26.** $\displaystyle\int_{2}^{3} \frac{x}{(2x^2 - 5)}\, dx$

Calculate the total areas in Exercises 27–34.

27. Bounded by the line $y = x$, the x-axis, and the lines $x = 0$ and $x = 1$

28. Bounded by the line $y = 2x$, the x-axis, and the lines $x = 1$ and $x = 2$

29. Bounded by the curve $y = \sqrt{x}$, the x-axis, and the lines $x = 0$ and $x = 4$

30. Bounded by the curve $y = 2\sqrt{x}$, the x-axis, and the lines $x = 0$ and $x = 16$

31. Bounded by the curve $y = x^2 - 1$, the x-axis, and the lines $x = 0$ and $x = 4$

32. Bounded by the curve $y = 1 - x^2$, the x-axis, and the lines $x = -1$ and $x = 2$

33. Bounded by the x-axis, the curve $y = xe^{x^2}$, and the lines $x = 0$ and $x = (\ln 2)^{1/2}$

34. Bounded by the x-axis, the curve $y = xe^{x^2-1}$ and the lines $x = 0$ and $x = (1 + \ln 3)^{1/2}$

APPLICATIONS

35. *Displacement*   A car traveling down a road has a velocity of $v(t) = 60 - e^{-t/10}$ mph at time t hours. Find the total distance it travels from time $t = 1$ hour to time $t = 6$.

36. *Displacement*   A ball thrown in the air has a velocity of $v(t) = 100 - 32t$ ft/s at time t seconds. Find the total displacement of the ball between times $t = 1$ second and $t = 7$ seconds, and interpret your answer.

37. *Cost*   The marginal cost of producing the xth box of light bulbs is $5 + x^2/1{,}000$. Determine how much would be added to the total cost by a change in production from $x = 10$ to $x = 100$ boxes.

38. *Revenue*   The marginal revenue of the xth box of computer disks sold is $100e^{-x/1{,}000}$. Find the revenue generated by selling items 100 through 1,000.

39. *Revenue* The *EMC Corporation,* which specializes in the manufacture of computer data storage systems, underwent rapid growth during the period 1989–1994, and its annual revenue (in billions of dollars) was approximately

$$r(t) = 0.13e^{0.44t} \quad (0 \le t \le 5)$$

where t is the number of years since June, 1989.* Use this model to estimate the total revenue (to the nearest $0.1 billion) earned by *EMC* from June 1989 to June 1994.

40. *Revenue* *Carnival Corporation*'s airline services grew rapidly during the period 1989–1994, and its resulting annual revenues (in millions of dollars) can be approximated by

$$r(t) = 0.33t^3 + 0.929t^2 + 15.02t + 6.43$$
$$(0 \le t \le 5),$$

where t is the number of years since June 1989.* Use this model to estimate the total revenue (to the nearest $1 million) earned by Carnival from its airline service from June 1990 to June 1994.

41. *Oil Exploration Costs* Annual spending by U.S. oil companies on domestic oil exploration dropped from $9.2 billion in 1981 to $0.8 billion in 1992.[†]
(a) Assuming a linear decrease in annual spending over the given period, give the annual amount $c(t)$ (in billions of dollars) spent on domestic oil exploration as a function of time t in years since 1981.
(b) Use your linear model to estimate the total amount spent on domestic oil exploration over the given period.

42. *Oil Exploration Costs* Annual spending by U.S. oil companies on oil exploration in foreign countries dropped from $4.5 billion in 1981 to $4.0 billion in 1992.[†]
(a) Assuming a linear decrease in annual spending over the given period, give the annual amount $c(t)$ (in billions of dollars) spent on foreign oil exploration as a function of time t in years since 1981.

(b) Use your linear model to estimate the total amount spent on foreign oil exploration over the given period.

43. *Embryo Development* The oxygen consumption of a bird embryo increases from the time the egg is laid through the time the chick hatches. In the case of a typical galliform bird, the oxygen consumption (in milliliters per hour) can be approximated by

$$c(t) = -0.00271t^3 + 0.137t^2 - 0.892t + 0.149$$
$$(8 \le t \le 30),$$

where t is the time (in days) since the egg was laid.[‡] (An egg will typically hatch at around $t = 28$.) Find the total amount of oxygen consumed during the ninth and tenth days ($t = 8$ to $t = 10$). (Warning: be careful with the units of time.)

44. *Embryo Development* The oxygen consumption of a turkey embryo increases from the time the egg is laid through the time the chick hatches. In the case of a brush turkey, the oxygen consumption (in milliliters per hour) can be approximated by

$$c(t) = -0.00118t^3 + 0.119t^2 - 1.83t + 3.972$$
$$(20 \le t \le 50),$$

where t is the time (in days) since the egg was laid.[‡] (An egg will typically hatch at around $t = 50$.) Find the total amount of oxygen consumed during the twenty-first and twenty-second days ($t = 20$ to $t = 22$). (Warning: be careful with the units of time.)

Exercises 45–48 are almost identical to exercises from the previous section, but you should now use definite integrals to answer the questions.

45. *Exports* Based on figures released by the U.S. Agriculture Department, the value of U.S. pork exports from 1985 through 1993 can be approximated by the equation

$$q = \frac{460e^{(t-4)}}{1 + e^{(t-4)}},$$

▼ * The model is the authors'. Sources for data: Company Reports/IDC/*The New York Times,* November 9, 1994, p. D8; October 23, 1994, p. F5.

† Source: American Petroleum Institute (*The New York Times,* November 8, 1993, p. D3.)

‡ The model approximates graphical data published in the article "The Brush Turkey" by Roger S. Seymour, *Scientific American,* December 1991, pp. 108–14.

where q represents annual exports (in millions of dollars) and t the time in years with $t = 0$ corresponding to January 1, 1985.* Use the model to predict the total value of U.S. pork exports to the nearest million dollars from January 1, 1985 to January 1, 2000.

46. Sales   The weekly demand for your company's Lo-Cal Chocolate Mousse can be modeled by the equation

$$q = \frac{50e^{2t-1}}{1 + e^{2t-1}},$$

where q is the number of gallons sold per week and t is the time in weeks. Estimate the total sales of Lo-Cal mousse for the coming year.

47. Production   The production of cellular phones at your electronics plant has a Cobb-Douglas production function given by

$$p = 10x^{0.3}y^{0.7},$$

where p is the number of cellular phones it produces per year, x is the number of employees, and y is the daily operating budget (in dollars). You have decided to increase the daily operating budget linearly from the present level of $6,000 to $10,000 over the next two years. Your plant employs 100 workers. Estimate the total production of cellular phones over the next two years. (Give your answer to the nearest cellular phone.)

48. Production   Referring to the Cobb-Douglas production formula in the previous exercise, assume that you wish to maintain a daily operating budget of $6,000 but plan to increase the work force linearly from the present 100 employees to 200 over the coming two years. Estimate the total production of cellular phones over the next two years. (Give your answer to the nearest cellular phone.)

49. Fuel Consumption   The way Professor Waner drives, he burns gas at the rate of $1 - e^{-t}$ gallons each hour, t hours after a fill-up. Find the number of gallons of gas he burns in the first 10 hours after a fill-up.

50. Fuel Consumption   The way Professor Costenoble drives, he burns gas at the rate of $1/(t + 1)$ gallons each hour, t hours after a fill-up. Find the number of gallons of gas he burns in the first 10 hours after a fill-up.

51. Health Spending   The following chart shows approximate figures for annual spending on health care in the United States from 1981 to 1994, in billions of dollars.[†]

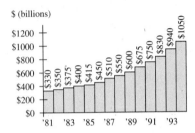

$ (billions)

(a) Take t to be time in years since January 1981, and use the figures for 1981 ($t = 0$) and 1993 ($t = 12$) to construct an exponential model of the form $s(t) = Pe^{kt}$ for annual spending, where $s(t)$ is the annual spending (in billions of dollars) at time t.

(b) Use your model to estimate the total amount spent on health care in the U.S. to the nearest $1 billion for the period shown, and compare it with the result given by the data in the chart. (Hint: The period shown extends to January 1995.)

(c) Use your model to predict total spending on health care in the U.S. from January 1995 to January 2000 to the nearest $1 billion.

52. Health Spending   Repeat the previous exercise, but this time base your model on the figures for 1982 ($t = 1$) and 1994 ($t = 13$).

53. Sales   Weekly sales of your Jurassic Park T-shirts have been falling continuously at a rate of 5% per week. Assuming you are now selling 50 T-shirts per week, how many shirts will you sell during the coming year? (Round your answer to the nearest shirt.)

54. Sales   Annual sales of fountain pens in Littleville are presently 4,000 per year and are continuously increasing at a rate of 10% per year. How many fountain pens will be sold over the next five years?

▼ * This is the authors' model based on figures published by the U.S. Agriculture Department quoting annual sales from 1985 through 1993. (*The New York Times*, Dec. 12, 1993, p. D1.)

[†] Source: Department of Health and Human Services, Department of Commerce (*The New York Times*, December 29, 1993, p. A12.) The figures are estimates based on the graph.

55. *Kinetic Energy* The work done in accelerating an object from velocity v_0 to velocity v_1 is given by

$$W = \int_{v_0}^{v_1} v \frac{d}{dv}(p)\, dv,$$

where p is its momentum, given by $p = mv$ (m = mass).

Assuming that m is a constant, show that

$$W = \frac{1}{2} mv_1{}^2 - \frac{1}{2} mv_0{}^2.$$

The quantity $\frac{1}{2} mv^2$ is referred to as the **kinetic energy** of the object, so the work required to accelerate an object is given by its change in kinetic energy.

56. *Einstein's Energy Equation* According to the special theory of relativity, the apparent mass of an object depends on its velocity according to the formula

$$m = \frac{m_0}{\left(1 - \dfrac{v^2}{c^2}\right)^{1/2}},$$

where v is its velocity, m_0 is the "rest mass" of the object (that is, its mass when $v = 0$), and c is the velocity of light: approximately 3×10^8 meters per second.

(a) Show that if $p = mv$ is the momentum,

$$\frac{d}{dv}(p) = \frac{m}{\left(1 - \dfrac{v^2}{c^2}\right)^{3/2}}.$$

(b) Use the integral formula for W in the preceding exercise, together with the result in part (a), to show that the work required to accelerate an object from a velocity of v_0 to v_1 is given by

$$W = \frac{m_0 c^2}{\sqrt{1 - \dfrac{v_1{}^2}{c^2}}} - \frac{m_0 c^2}{\sqrt{1 - \dfrac{v_0{}^2}{c^2}}}.$$

We call the quantity $\dfrac{m_0 c^2}{\sqrt{1 - \dfrac{v^2}{c^2}}}$ the **total relativistic energy** of an object moving at velocity v. Thus, the work to accelerate an object from one velocity to another is given by the change in its total relativistic energy.

(c) Deduce (as Albert Einstein did) that the total relativistic energy of a body at rest with rest mass m is given by the famous equation

$$E = mc^2.$$

57. *Total Cost* Use the Fundamental Theorem of Calculus to show that if $m(x)$ is the marginal cost at a production level of x items, then the cost function $C(x)$ is given by

$$C(x) = C(0) + \int_0^x m(t)\, dt.$$

What term do we use for $C(0)$?

58. *Total Sales* The total cost of producing x items is given by

$$C(x) = 246.76 + \int_0^x 5t\, dt.$$

Find the fixed cost and the marginal cost of producing the tenth item.

COMMUNICATION AND REASONING EXERCISES

59. Explain how the indefinite integral and the definite integral are related.

60. Give an example of a nonzero function whose definite integral is zero.

61. Give a nontrivial example of a velocity function that will produce a displacement of 0 from time $t = 0$ to time $t = 10$.

62. Complete the following sentence. The total sales from time a to time b are obtained from the marginal sales by taking its ———— ———— from ———— to ————.

63. Give an example of a decreasing function $f(x)$ with the property that $\int_a^b f(x)\, dx$ is positive for every choice of a and $b > a$.

64. Explain why, in computing the total of a quantity from its rate of change, it is useful to have the definite integral subtract area below the x-axis.

▶ ═══ **6.5** ALGEBRAIC DEFINITION OF THE DEFINITE INTEGRAL

When we studied the derivative, we had two ways of thinking about it: *geometrically*, as the slope of the tangent line, and *algebraically*, as the limit of a difference quotient. It was the algebraic approach that we used to accurately define the derivative, as the geometric idea of the tangent line is too vague by itself.

So far, we have only given a geometric definition of the definite integral, in terms of area. The trouble is that we have been assuming that we know what "area" is, without attempting to define it precisely.* Thus, we really need to replace the intuitive notion of area with a precise algebraic definition. On a more practical note, we shall see that the algebraic formulation of the definite integral enables us to approximate definite integrals quickly, easily, and accurately using technology when there is no closed form formula for the antiderivative.

To do this we use an idea that dates back at least to the Greeks.[†] We try to compute the area of a complicated region by filling it with shapes whose areas we know. The shape whose area we can agree on is the rectangle—its area is its height times its width.

Suppose that we have a continuous function $f(x)$ defined on the interval $[a, b]$, and we wish to know the area under the graph of f: that is, the definite integral of f from a to b. For now we shall assume that f is positive. We approximate the area under the graph of f using rectangular "tiles," as shown in Figure 1.

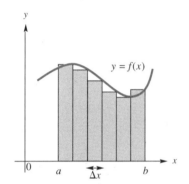

FIGURE 1

Of course, this will not give us the exact area, as you can see, but it will give us an approximation. This is the approach we've used throughout calculus: find an approximation that we can calculate, and then pass to a limit to get the exact answer.

In order to write down the sum of the areas of these rectangles, our approximation to the area under the graph, we need to know the dimensions of the rectangles. First, we take all the rectangles in Figure 1 to have the same width, Δx. Although there happen to be six of them in the figure, we will consider the general case of n rectangles of equal width. Since the interval $[a, b]$ has width $b - a$, the width of each rectangle should be one nth of this amount, or

$$\Delta x = \frac{b - a}{n}.$$

▼ * If you think that you already know exactly what we mean by "area," we challenge you to try to come up with explanation that would be meaningful to someone who had never heard of area before.

[†] This idea shows up in the Greeks' "method of exhaustion," but this would be too tiring for us to get into.

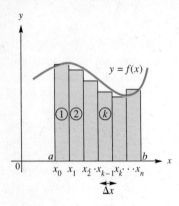

FIGURE 2

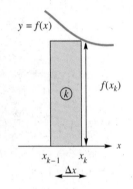

FIGURE 3

Before looking at the heights of the rectangles, we first give names to all the values of x that mark the boundaries between successive rectangles (see Figure 2). The first rectangle has its left edge at $x_0 = a$ and its right edge Δx units to the right, at x_1. Thus, $x_1 = a + \Delta x$. The second rectangle has its left edge at x_1 and its right edge another Δx units to the right, at x_2, with $x_2 = a + 2\Delta x$. In general, the kth rectangle has its left edge at x_{k-1} and its right edge Δx units to the right, at x_k, with $x_k = a + k\Delta x$. The last rectangle (the nth one) has its left edge at x_{n-1} and its right edge at $x_n = b$.

We now turn to the heights of the rectangles. If you look closely at Figures 1 and 2, you will notice that the right edge of each rectangle touches the graph of $y = f(x)$. Thus, the first rectangle has a height of $f(x_1)$, since its right edge has x-coordinate x_1 and the height of the graph at $x = x_1$ is given by $f(x_1)$. Similarly, the kth rectangle has height $f(x_k)$, and the last (nth) rectangle has height $f(x_n)$. Figure 3 shows an enlargement of the kth rectangle.

Its area is $height \times width = f(x_k) \, \Delta x$. Now we add together the areas of all the rectangles, and obtain

$$\text{Approximate value of } \int_a^b f(x) \, dx = f(x_1) \, \Delta x + f(x_2) \, \Delta x$$
$$+ \cdots + f(x_k) \, \Delta x$$
$$+ \cdots + f(x_n) \, \Delta x \quad .$$

Since sums such as these are often used in mathematics, a shorthand has been developed for them. We write

$$\sum_{k=1}^{n} f(x_k) \, \Delta x = f(x_1) \, \Delta x + f(x_2) \, \Delta x + \cdots + f(x_k) \, \Delta x$$
$$+ \cdots + f(x_n) \, \Delta x.$$

The symbol "Σ" is the Greek letter sigma, and stands for **summation.** The letter k here is called the index of summation, and we can think of it as counting off the rectangles. We read the notation as "the sum from $k = 1$ to n of the quantities $f(x_k) \, \Delta x$." Think of it as a set of instructions:

Set $k = 1$, and obtain $f(x_1) \, \Delta x$.

Set $k = 2$, and obtain $f(x_2) \, \Delta x$.

. . .

Set $k = n$, and obtain $f(x_n) \, \Delta x_n$.

Now sum all the above quantities.

Q Since this is an approximation to the definite integral $\int_a^b f(x) \, dx$, how do we get the *exact* area?

A By our usual technique of passing to a limit. We can obtain more and more accurate approximations to the area by taking smaller values of Δx (or equivalently, taking larger and larger values for n). This is suggested by Figure 4, which shows the approximations for $n = 6$ and $n = 12$. Thus, to obtain the *exact* area or definite integral, we pass to the limit as $\Delta x \to 0$. In other words, we make the following definition.

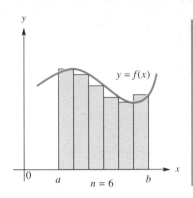

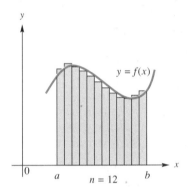

FIGURE 4

ALGEBRAIC DEFINITION OF THE DEFINITE INTEGRAL

If $f(x)$ is any continuous function, then the **definite integral of f from a to b** is defined to be

$$\int_a^b f(x)\,dx = \lim_{\Delta x \to 0} \sum_{k=1}^n f(x_k)\,\Delta x,$$

where $\Delta x = \dfrac{b-a}{n}$, and $x_k = a + k\,\Delta x$. The limit is obtained by letting $\Delta x \to 0$, or equivalently, by letting $n \to +\infty$.

(Actually, this is one of several definitions, all equivalent for continuous f but different for pathological functions.) By the way, you can see now where the notation for the integral came from: the "$\int$" is an elongated "S," the Roman equivalent of the Greek Σ, $f(x_k)$ becomes $f(x)$, and Δx becomes dx in the limit.

The sum

$$\sum_{k=1}^n f(x_k)\,\Delta x$$

by itself is called the **nth right-hand sum** of f (because we have taken the height of each rectangle to be the value of the function at the x-coordinate of its *right* edge). We get the **nth left-hand sum** by replacing $f(x_k)$ by $f(x_{k-1})$ in the above formula. The limit is the same as for the right-hand sum. We can get other kinds of "in-between" sums by using $f(x_k^*)$ instead of $f(x_k)$, where x_k^* is any value of x between x_{k-1} and x_k. All these choices lead to the same result in the limit. All these different kinds of sums are referred to as **Riemann sums,** after the nineteenth-century mathematician Bernhard Riemann.

▶ NOTES
1.   The fact that the limit exists, and is the same for the left- and right-hand sums, is not easy to prove. It depends on the fact that we assumed that $f(x)$ was continuous on the interval $[a, b]$.
2.   Our assumption that the function is positive can be dropped. The formula for the nth sum automatically subtracts area below the x-axis, since if the graph lies below the x-axis at $x = x_k$, $f(x_k)$ will be negative.
3.   The subdivision of the interval $[a, b]$ into n equal parts is usually referred to as a **partition** of $[a, b]$.   ◀

▼   **EXAMPLE 1**

Find the left- and right-hand sums of $f(x) = x^2$ on the interval $[0, 1]$ with $n = 5$. Then calculate the average of the two sums, and compare the answer with the actual integral. How accurate is it?

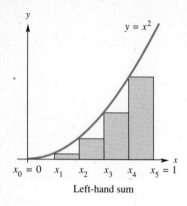

Left-hand sum

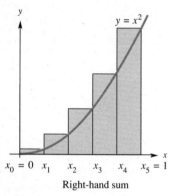

Right-hand sum

FIGURE 5

SOLUTION   Before we do the calculations, look at Figure 5, which shows the rectangles for these sums.

Looking at the figure, we expect the left-hand sum be smaller than the actual area and the right-hand sum to be bigger. We now turn to the calculation.

Step 1:   First calculate $\Delta x = \dfrac{b - a}{n}$.

Here, $a = 0$, $b = 1$ and $n = 5$, so $\Delta x = (1 - 0)/5 = 0.2$.

Step 2:   Calculate the values of x_k and $f(x_k)$ for $0 \le k \le n$. Since $n = 5$, we must calculate 6 values of $f(x_k)$. These are shown in the following table.

k	0	1	2	3	4	5
$x_k = a + k\Delta x$	0	0.2	0.4	0.6	0.8	1.0
$f(x_k)$	0	0.04	0.16	0.36	0.64	1.0

Step 3:   For the *left-hand* sum, add the values of $f(x_k)$ starting with the leftmost ($k = 0$) and ending with $k = n - 1 = 4$, and then multiply the sum by Δx. For the *right-hand sum,* add the values of $f(x_k)$, starting with the second ($k = 1$) and ending at $k = n = 5$, and then multiply the sum by Δx.

Left-hand sum: $(0 + 0.04 + 0.16 + 0.36 + 0.64)(0.2) = 0.24$

Right-hand sum: $(0.04 + 0.16 + 0.36 + 0.64 + 1.0)(0.2) = 0.44$

Both sums are approximations of the definite integral $\int_0^1 x^2\, dx$. The Fundamental Theorem of Calculus tells us that the exact answer is

$$\int_0^1 x^2\, dx = \left[\frac{x^3}{3}\right]_0^1 = \frac{1}{3} = 0.3333 \ldots$$

Although neither the left-hand sum nor the right-hand sum gives a good approximation, we are asked to also look at the average of the two sums.

$$\text{average sum} = \frac{\text{Left-hand sum} + \text{Right-hand sum}}{2}$$

$$= \frac{0.24 + 0.44}{2} = 0.34,$$

which is a far better approximation. For reasons that will become clearer in Section 7, this average is referred to as the **trapezoidal sum.**

Now we can complete the answer to the question: we compare the trapezoidal sum with the actual sum by taking the magnitude of their difference,

$$\left|0.34 - 0.3333 \ldots\right| = 0.00666 \ldots$$

This is the **error term.** Now round this error to a single significant digit.

$$0.00666 \ldots \approx 0.007$$

We say our answer is accurate to within $\pm\, 0.007$.

Before we go on... If we use Riemann sums to approximate an integral, we will usually not be able to determine the exact error (otherwise we would know the exact answer, so there would be little point in computing an approximation!). However, there are ways of computing the maximum possible error. We shall return to this topic in Section 7.

GRAPHING CALCULATORS

Calculating left- and right-hand sums can be a tedious process, especially for large values of n. (If you chose $n = 100$, for example, you would be kept busy filling in values in a table with 101 columns!) Luckily, we have programmable graphing calculators and computers to do the chore for us. In the appendix you will find a graphing calculator program called SUMS that computes both the left and right Riemann sums. (It works on the TI-82 and higher numbered models. You should read through the section on programming in your calculator instruction booklet before attempting to enter this program.)

    The next example uses the TI-82 program.

▼   **EXAMPLE 2**

Use a graphing calculator to calculate the left- and right-hand sums for the integral

$$\int_{-1}^{1} \sqrt{1 - x^2}\; dx,$$

with $n = 100$, 200, and 500.

 SOLUTION   In the "Y =" window, we enter the function:

$$Y_1 = (1 - X^2)^{\wedge}(0.5)$$

We then hit $\boxed{\text{PRGM}}$, select the program SUMS (the name we have given to the program) and enter 100 for N, -1 for the left endpoint, and 1 for the right endpoint. We obtain the following answers.

             Left-hand sum: 1.56913 . . .

             Right-hand sum: 1.56913 . . .

With $N = 200$, we obtain

             Left-hand sum: 1.57020 . . .

             Right-hand sum: 1.57020 . . .

With $N = 500$, we obtain

             Left-hand sum: 1.57064 . . .

             Right-hand sum: 1.57064 . . .

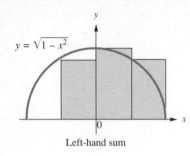

Left-hand sum

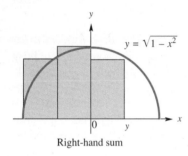

Right-hand sum

FIGURE 6

Before we go on...

Q Why are the two sums always the same?

A To see why, we need to take a look at the graph (which you can readily do by hitting GRAPH on your calculator). Figure 6 shows the graph with the left- and right-hand sums drawn for $n = 4$.

From the diagram, we see that the symmetry of the semicircle and the fact that we used an even number for n accounts for the fact that the sums were equal. Neither is likely to be the exact answer.

In fact, the exact area is half the area of a disc of radius 1. Since a disc of radius r has an area of πr^2, the area we approximated is $\frac{1}{2}\pi(1)^2 = \pi/2 = (3.141592 \ldots)/2 = 1.570796 \ldots$, which is indeed different from our approximations. The accuracy of our approximation with $N = 100$ is

$$\text{Error} = |1.570796\ldots - 1.56913\ldots| = 0.0016\ldots$$

Rounding to one significant digit gives 0.002. Thus, our answer is accurate to within ± 0.002. With $N = 200$, our accuracy improves to

$$|1.5707\ldots - 1.57020\ldots| = 0.0005\ldots$$

With $N = 500$, our accuracy is

$$|1.5707\ldots - 1.57064\ldots| = 0.0002\ldots$$

Thus, our approximation is improving as N gets larger, albeit slowly.

We are not in a position to calculate the integral using the Fundamental Theorem of Calculus, since the antiderivative of $\sqrt{1 - x^2}$ involves a function we have not encountered—the inverse sine function. Thus, we must make do with our numerical approach. This is one of the ways in which π itself can be computed (though not the most efficient way, by any means).

The following example appeared as an exercise in Section 3. We can now look at it in a new light.

▼ **EXAMPLE 3**   Exports

Based on figures released by the U.S. Agriculture Department, the value of U.S. pork exports from 1985 through 1993 can be approximated by the equation

$$q = \frac{460e^{(t-4)}}{1 + e^{(t-4)}},$$

where q represents annual exports (in millions of dollars) and t the time in years, with $t = 0$ corresponding to January 1, 1985.* Figure 7 shows the actual data with the graph of the equation superimposed.

▼ * This is the authors' model based on figures published by the U.S. Agriculture Department quoting annual sales from 1985 through 1993. (*The New York Times*, Dec. 12, 1993, p. D1.) (The figures shown on the graph are estimates based on the published graph.)

FIGURE 7

Notice that the chart looks suspiciously like the approximation of the definite integral by a sum of areas of rectangles, though it is not exactly a left-hand, right-hand, or any other form of Riemann sum (since our mathematical curve does not pass through the tops of all the rectangles). In other words, when we calculate the total U.S. pork exports over the given period by summing the individual figures, we are really adding the areas of the rectangles in the figure (since each rectangle has width 1 unit), and so we are finding a Riemann sum, more or less.

If we add up the areas of the rectangles shown in the diagram, we get

$$(60 + 80 + 125 + 250 + 330 + 320 + 335 + 450 + 460)(1)$$
$$= \$2,410 \text{ billion.}$$

If we now calculate the left- and right-hand sums of $q(t)$ with $n = 9$ (corresponding to the divisions shown in the graph), and round the answers to the nearest $1 billion, we obtain

Left-hand sum: $2,070 billion

Right-hand sum: $2,519 billion.

Their average is the trapezoidal sum $\dfrac{2,070 + 2,519}{2} = \$2,294.5$ billion, so we're not far off the mark with our model! (We can hardly expect the actual pork export figures to coincide with an abstract mathematical function. But as we have seen, the model does allow us to predict future exports as well as to perform other kinds of analysis. It's good to have such a model around.)

▶ ## 6.5 EXERCISES

Calculate the left-hand, right-hand, and trapezoidal (average) sums for the integrals in Exercises 1–10. What is the accuracy of the trapezoidal sum?

1. $\int_0^2 (4x - 1) \, dx$, $n = 4$

2. $\int_{-1}^1 (1 - 3x) \, dx$, $n = 4$

3. $\int_{-2}^2 x^2 \, dx$, $n = 4$

4. $\int_1^5 x^2 \, dx$, $n = 4$

5. $\int_{-1}^1 x^3 \, dx$, $n = 5$

6. $\int_0^1 x^3 \, dx$, $n = 5$

7. $\int_0^{10} e^{-x} \, dx$, $n = 5$

8. $\int_{-5}^5 e^{-x} \, dx$, $n = 5$

9. $\int_0^1 \dfrac{1}{1 + x} \, dx$, $n = 5$

10. $\int_0^1 \dfrac{x}{1 + x^2} \, dx$, $n = 5$

Calculate the trapezoidal sum for each of the integrals in Exercises 11–16.

11. $\int_0^1 4 \sqrt{1 - x^2} \, dx$, $n = 5$

12. $\int_0^1 \dfrac{4}{1 + x^2} \, dx$, $n = 5$

13. $\int_0^{10} e^{-x^2} \, dx$, $n = 10$

14. $\int_0^{100} e^{-x^2} \, dx$, $n = 10$

15. $\int_0^1 \ln(x^2 + 1) \, dx$, $n = 10$

16. $\int_0^{10} e^{-x} \, dx$, $n = 10$

Exercises 17–22 require the use of a graphing calculator with the program described in the text, or a similar computer program. In each case, calculate the left-hand, right-hand and trapezoidal sums for the given integrals with (a) $n = 100$ and (b) $n = 200$. Round all answers to four decimal places. [Your graphing calculator should have sin *and* π *buttons.]*

17. $\int_0^1 4\sqrt{1 - x^2}\, dx$

18. $\int_0^1 \dfrac{4}{1 + x^2}\, dx$

19. $\int_2^3 \dfrac{2x^{1.2}}{1 + 3.5x^{4.7}}\, dx$

20. $\int_3^4 3xe^{1.3x}\, dx$

21. $\int_0^\pi \sin(x)\, dx$

22. $\int_0^\pi 2(\sin(x))^2\, dx$

APPLICATIONS

23. *Sales*   Estimated sales of *IBM*'s mainframe computers and related peripherals showed the following pattern of decline over the years 1990 through 1994.*

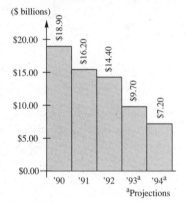

($ billions)

($18.90, $16.20, $14.40, $9.70, $7.20)

'90 '91 '92 '93ᵃ '94ᵃ
ᵃProjections

(a) Using the figures for 1990 and 1994, modexl *IBM*'s annual mainframe sales using a linear function $s(t) = mt + b$, where $s(t)$ is the annual sales in billions of dollars and t is time in years since the start of 1990. (Take $t = 0.5$ for the 1990 figure and $t = 4.5$ for the 1994 figure.)

(b) Approximate $\int_0^5 s(t)\, dt$ by a trapezoidal sum with $n = 5$. How accurately does your model estimate *IBM*'s total mainframe sales over the given period? How do you account for this discrepancy?

24. *Total Profit*   The profitability of *United Airlines* for the years 1988 through 1992 showed the following decline.†

(a) Using the figures for 1988 and 1992, model *United*'s annual profits using a linear function $p(t) = mt + b$, where $p(t)$ is the annual profit in millions of dollars and t is time in years since the

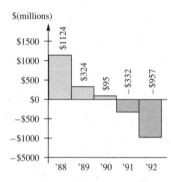

$(millions)

($1124, $324, $95, –$332, –$957)

'88 '89 '90 '91 '92

start of 1988. (Take $t = 0.5$ for the 1988 figure and $t = 4.5$ for the 1992 figure.)

(b) Approximate $\int_0^5 p(t)\, dt$ by a trapezoidal sum with $n = 5$. How accurately does your model estimate *United*'s aggregate profit or loss over the given period? Explain any discrepancy you find.

25. *Cost*   The marginal cost function for the manufacture of portable CD players is given by

$$C'(x) = 20 - \frac{x}{5,000},$$

where x is the number of CD players manufactured. Use the trapezoidal sum with $n = 1$ to estimate the cost of producing CD players 11 through 100. Compare your answer with the answer given by the Fundamental Theorem of Calculus.

26. *Cost*   Repeat the previous exercise using the marginal cost function

$$C'(x) = 25 - \frac{x}{50}.$$

▼ * Source: Salomon Brothers (*The New York Times*, October 26, 1993, p. D1.)
   † Source: Company Reports (*The New York Times*, December 24, 1993, p. D1.)

27. *Surveying* My uncle intends to build a kidney-shaped swimming pool in his small yard, and the town zoning board will approve the project only if the total area of the pool does not exceed 500 square feet. The accompanying figure shows a diagram of the planned swimming pool, with measurements of its width at the indicated points. Will my uncle's plans be approved? (Use left- and right-hand sum estimates.)

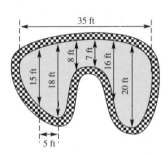

28. *Pollution* An aerial photograph of an ocean oil spill shows the pattern in the accompanying diagram. Assuming that the oil slick has a uniform depth of 0.01 meters, how many cubic meters of oil would you estimate to be in the spill? (Volume = Area × Thickness)

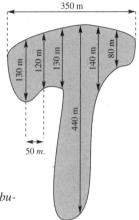

The Normal Curve   *The normal distribution curve, which models the distributions of data in a wide range of applications, is given by the function*

$$p(x) = \frac{1}{\sqrt{2\pi}\,\sigma}e^{-(x-\mu)^2/2\sigma^2},$$

where $\pi = 3.14159265\ldots$ and σ and μ are constants called the standard deviation and the mean. Exercises 29 and 30 illustrate its use and require a graphing calculator programmed to calculate trapezoidal sums.

29. *Test Scores* Enormous State University's Calculus I test scores are modeled by the normal distribution with $\mu = 72.6$ and $\sigma = 5.2$. The percentage of students who obtained scores between a and b on the test is given by

$$\int_a^b p(x)\,dx.$$

(a) Use a trapezoidal Riemann sum with $n = 40$ to estimate the percentage of students who obtained between 60 and 100 on the test.

(b) What percentage of students scores less than 30?

30. *Consumer Satisfaction* In a survey, consumers were asked to rate a new toothpaste on a scale of

1–10. The resulting data are modeled by a normal distribution with $\mu = 4.5$ and $\sigma = 1.0$. The percentage of consumers who rated the toothpaste with a score between a and b on the test is given by

$$\int_a^b p(x)\,dx.$$

(a) Use a trapezoidal Riemann sum with $n = 10$ to estimate the percentage of customers who rated the toothpaste 5 or above. (Use the range 4.5 to 10.5)

(b) What percentage of customers rated the toothpaste 0 or 1? (Use the range -0.5 to 1.5)

COMMUNICATION AND REASONING EXERCISES

31. Let $f(x) = mx + c$, where m and c are constants. Why does the trapezoidal approximation to $\int_a^b f(x)\,dx$ gives the exact answer with $n = 1$? Convince yourself of this by sketching the left-hand and right-hand sums for $\int_0^1 (1 - x)\,dx$.

32. Let $f(x)$ be a decreasing function on the interval $[a, b]$. (That is, $f'(x) \leq 0$ on the interval.) Demonstrate by means of a sketch that the left-hand sum is always greater than or equal to the right-hand sum. What can you say about *increasing* functions?

33. Let $f(x)$ be an increasing function. Draw a picture and demonstrate that the difference between the right- and left-hand sums for $\int_a^b f(x)\, dx$ is $(f(b) - f(a))\, \Delta x$. Conclude that the difference goes to 0 as $n \to +\infty$

34. Another approximation of the integral is the **midpoint** approximation, in which we compute the sum

$$\sum_{i=1}^{n} f(\bar{x}_k)\, \Delta x,$$

where $\bar{x}_k = (x_{k-1} + x_k)/2$ is the point midway between the left and right endpoints of the interval $[x_{k-1}, x_k]$. Why is it true that the midpoint approximation is exact if f is linear (compare Exercise 31)?

▶ === **6.6** THE FUNDAMENTAL THEOREM OF CALCULUS

Although we have been using the Fundamental Theorem of Calculus rather heavily, we have not explained where it comes from. Now that we are familiar with the definite integral as an area or limit of a sum, the proof of the Fundamental Theorem will seem less mysterious. Here is the complete statement of the theorem.

> **THEOREM (THE FUNDAMENTAL THEOREM OF CALCULUS, OR FTC)**
>
> Let f be any continuous function defined on the interval $[a, b]$. Then
>
> 1. $f(x)$ has an antiderivative, namely, $\int_a^x f(t)\, dt$, and
> 2. if $F(x)$ is any antiderivative of $f(x)$, then
>
> $$\int_a^b f(x)\, dx = F(b) - F(a).$$

We shall outline the proof of these two statement over the next several pages. By the way, what we are about to do was first done by the English mathematician Isaac Barrow (1630–1677), who was Newton's teacher. It was Newton, however, who recognized its importance.

For convenience, we shall take f as a function of t rather than x (see Figure 1). Our figures will assume that f is always positive, but this is not necessary.

Concentrating on the second statement of the FTC, we consider the question: How do we calculate the definite integral $\int_a^b f(t)\, dt$? We start by considering a harder question. We shall try to calculate, for every x between a and b, the area

$$A(x) = \int_a^x f(t)\, dt.$$

This is the area under the curve from $t = a$ to $t = x$, as shown in Figure 2. Observe that varying x means moving the vertical line on the right of the shaded area. This changes the value of $A(x)$. In other words, $A(x)$ *depends on x*. In fact, it is a *rule* that gives us a number—namely, the area—for any specified value of x. Thus, it is a perfectly respectable function

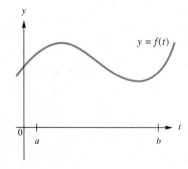

FIGURE 1

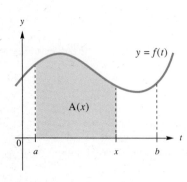

FIGURE 2

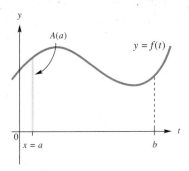
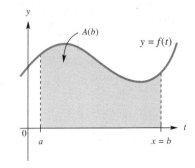

FIGURE 3

of x, even if we do not have a formula for its values. The domain of $A(x)$ is $[a, b]$. Figure 3 shows two specific values of $A(x)$.

From the figure, we see that $A(a) = 0$, and $A(b) = \int_a^b f(t)\, dt$, which is what we are trying to calculate.

Q Now what do we do with this new function $A(x)$?

A Since we are doing calculus, why not take its derivative?

Q But how do you take the derivative of such a strange function?

A The same way you take the derivative of *any* new function: by using the definition

$$A'(x) = \lim_{h \to 0} \frac{A(x + h) - A(x)}{h}.$$

This may not be as easy as it looks. We do not have an algebraic formula for $A(x)$, so how do we deal with the difference quotient? All we know about $A(x + h)$ and $A(x)$ is that they represent areas, so we sketch them in Figure 4.

The figure shows us that $A(x + h) - A(x)$ is equal to the bigger area minus the smaller area, giving the area of the "sliver" on the right.

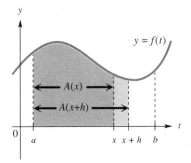
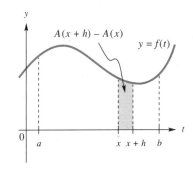

FIGURE 4

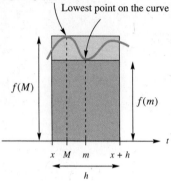

Highest point on the curve

Lowest point on the curve

$f(M)$

$f(m)$

x M m $x + h$

h

FIGURE 5

Thus, we write

$$A'(x) = \lim_{h \to 0} \frac{A(x + h) - A(x)}{h} = \lim_{h \to 0} \frac{\text{Area of sliver}}{h}.$$

Now we still need to say something about the area of the sliver. To do this, we are going to look at a magnified picture of the sliver, and we'll also allow for the fact that the curve might have been more wobbly than we originally drew it (see Figure 5).

Now what on earth have we done to it? First, notice that our magnified view shows the wobbles we promised. Second, we have chosen two new t-values: M and m. M is the value of t between x and $x + h$ that gives the maximum value of f, while m is the t-value that gives the minimum value of f. These give the highest and lowest points of the curve restricted to the interval $[x, x + h]$. We have also shaded in two rectangles. The taller one, of height $f(M)$ and width h, is clearly an overestimation of the area of the sliver, while the shorter one, of height $f(m)$ and width h, is an underestimation of the area of the sliver. In other words, the area of the sliver is somewhere *between* the areas of the two rectangles. We can represent this by the inequality

Area of short rectangle $\leq$ Area of sliver $\leq$ Area of tall rectangle,

or

$$h\, f(m) \leq \text{Area of sliver} \leq h\, f(M).$$

Now we divide the whole inequality by h to get

$$\frac{h\, f(m)}{h} \leq \frac{\text{Area of sliver}}{h} \leq \frac{h\, f(M)}{h}$$

(we are assuming that $h > 0$ as in the picture, but something similar can be done if $h < 0$) or, canceling the h's,

$$f(m) \leq \frac{\text{Area of sliver}}{h} \leq f(M).$$

Now remember that we are trying to find the limit of the middle term as h approaches zero. First, let us look at what is happening to the two end terms as $h \to 0$. Since both the points m and M are forced to lie between x and $x + h$, it follows that every time we choose a smaller value for h, both m and M must be selected once again, closer and closer to x (since $x + h$ is approaching x). In other words, both m and M approach x as $h \to 0$. Thus (because of the continuity of f), both $f(m)$ and $f(M)$ are approaching $f(x)$ as $h \to 0$. We have the following situation as $h \to 0$.

$$f(m) \leq \frac{\text{Area of sliver}}{h} \leq f(M)$$

approaching $f(x)$ approaching $f(x)$

Since the two end terms have limit $f(x)$ as $h \to 0$, and since the middle quantity is sandwiched between them, it too must be approaching $f(x)$ as $h \to 0$.

So we conclude

$$A'(x) = \lim_{h \to 0} \frac{A(x+h) - A(x)}{h} = \lim_{h \to 0} \frac{\text{Area of sliver}}{h} = f(x).$$

Thus, we have, with our bare hands, calculated the derivative of $A(x)$ and found that $A'(x) = f(x)$.

Q Good job! But what is the point of all of this?

A We have found that *the derivative of $A(x)$ is $f(x)$*. Rephrasing this,

the area function $A(x)$ is an antiderivative of $f(x)$.

So we are finally beginning to see a connection between area and antiderivatives. This is, in fact, the first statement of the FTC.

Now let us return to the question: How do we calculate the definite integral $\int_a^b f(t)\, dt$? We are asking for the value of $A(b)$, which we now know to be an antiderivative of $f(x)$, evaluated at $x = b$. Let us think for a moment about *any* antiderivative $F(x)$ of $f(x)$. Since both $F(x)$ and $A(x)$ are antiderivatives of $f(x)$, they differ by a constant, so we can write

$$F(x) = A(x) + C.$$

Now compare $F(b) - F(a)$ with $A(b) - A(a)$:

$$F(b) - F(a) = (A(b) + C) - (A(a) + C)$$
$$= A(b) - A(a).$$

But recall that $A(a) = 0$. Thus,

$$F(b) - F(a) = A(b) = \int_a^b f(t)\, dt.$$

This is the second statement of the FTC.

▶ **NOTES**
1. The first statement of the Fundamental Theorem of Calculus is more interesting than you might think at first. It says that if you have any continuous function whatsoever, then there is an antiderivative for it lurking around somewhere.
2. Notice that another way of saying that $\int_a^x f(t)\, dt$ is an antiderivative of $f(x)$ is to say that its derivative is $f(x)$. In other words, we have the following formula.

DERIVATIVE OF THE INTEGRAL

$$\frac{d}{dx} \int_a^x f(t)\, dt = f(x)$$

3. The second part of the theorem justifies the method of calculating areas we used in the last section, and it took mathematicians a long time to realize this! Prior to the discovery of the Fundamental Theorem, mathematicians calculated areas under curves by using Riemann sums. They sliced the area into thin rectangles to get an approximate area and then tried to take the limit as the rectangles got narrower and narrower. As a result, the calculation of even simple integrals such as $\int_b^a e^x \, dx$ became a formidable task, worthy of a published article in a mathematics journal! ◄

▼ **EXAMPLE 1**

Find the area function $A(x) = \int_1^x f(t) \, dt$ if $f(x) = 1/x^2$. Graph both $f(x)$ and $A(x)$.

SOLUTION   We compute $A(x)$ using the Fundamental Theorem of Calculus: we first find an antiderivative of f.

$$A(x) = \int_1^x \frac{1}{t^2} \, dt = \int_1^x t^{-2} \, dt = -t^{-1} \Big|_1^x = -x^{-1} - (-1) = 1 - \frac{1}{x}$$

The graphs of $f(x)$ and $A(x)$ are shown in Figure 6.

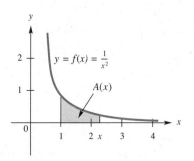
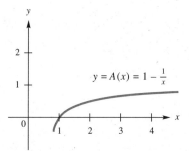

FIGURE 6

Before we go on... Notice that $A(x)$ is an antiderivative of $f(x)$. It is the particular antiderivative such that $A(1) = 0$, since $\int_1^1 1/t^2 \, dt$, the area from 1 to 1, must be 0. Notice also that $A(x)$ increases but never gets larger than 1. We shall see the significance of this in the next chapter when we discuss improper integrals.

▼ **EXAMPLE 2**

Find $\dfrac{d}{dx} \displaystyle\int_0^x e^{t^2} \, dt$.

SOLUTION   We have $\dfrac{d}{dx} \displaystyle\int_0^x e^{t^2} \, dt = e^{x^2}$ by the first statement in the Fundamental Theorem.

Before we go on... The remarkable thing about this example is that it is impossible to write an explicit formula for $\int_0^x e^{t^2}\, dt$ in terms of "elementary" functions (which include all those we have talked about and a few more). Nonetheless, we found an explicit formula for the *derivative* of this function.

▶ **6.6 EXERCISES**

1. Use geometry (not antiderivatives) to compute $A(x) = \int_0^x 1\, dt$, and verify that $A'(x) = 1$.

2. Use geometry (not antiderivatives) to compute $A(x) = \int_0^x t\, dt$, and verify that $A'(x) = x$.

3. Use geometry to compute $A(x) = \int_2^x (t - 2)\, dt$, and verify that $A'(x) = x - 2$.

4. Use geometry to compute $A(x) = \int_2^x 1\, dt$, and verify that $A'(x) = 1$.

5. Use geometry to compute $A(x) = \int_a^x c\, dt$, and verify that $A'(x) = c$. (c is a constant.)

6. Use geometry to compute $A(x) = \int_a^x (t - 2)\, dt$, and verify that $A'(x) = x - 2$.

For each of the functions in Exercises 7–16, use anti-derivatives to find the area function $A(x) = \int_a^x f(t)\, dt$. Sketch both $f(x)$ and $A(x)$.

7. $f(x) = x, \quad a = 0$

8. $f(x) = 2x, \quad a = 1$

9. $f(x) = x^2, \quad a = 0$

10. $f(x) = x^2 + 1, \quad a = 0$

11. $f(x) = e^x, \quad a = 0$

12. $f(x) = e^{-x}, \quad a = 0$

13. $f(x) = 1/x, \quad a = 1$

14. $f(x) = 1/(x + 1), \quad a = 0$

15. $f(x) = \begin{cases} x & \text{if } 0 \le x \le 1 \\ 2 & \text{if } x > 1 \end{cases}, \quad a = 0$

16. $f(x) = \begin{cases} 2x & \text{if } 0 \le x \le 2 \\ 1 & \text{if } x > 2 \end{cases}, \quad a = 0$

APPLICATIONS

17. *The Natural Logarithm Returns* Suppose that you had never heard of the natural logarithm function and were therefore stuck when you tried to find an antiderivative of $\frac{1}{x}$. Here is how you might proceed.
   (a) What does the Fundamental Theorem of Calculus give as an antiderivative of $\frac{1}{x}$?
   (b) Choosing $a = 1$, give this "new" function the name $M(x)$, and derive the following properties: $M(1) = 0;\ M'(x) = \frac{1}{x}$.
   (c) Use the chain rule for derivatives to show that if a is any positive constant, then

$$\frac{d}{dx}(M(ax)) = \frac{1}{x}.$$

   (d) Let $\quad F(x) = M(ax) - M(x)$. Deduce that $F'(x) = 0$, and hence that $F(x) = constant$.
   (e) By setting $x = 1$, find the value of this constant, and hence deduce that

$$M(ax) = M(a) + M(x).$$

18. *The Error Function* The **error function**, erf (x), is defined by

$$\text{erf}(x) = \frac{2}{\sqrt{\pi}} \int_0^x e^{-t^2}\, dt.$$

This function is very important in statistics.
   (a) Find erf $'(x)$.
   (b) Use a trapezoidal sum with $n = 5$ to approximate erf (1) and erf (2).
   (c) Find an antiderivative of e^{-x^2} in terms of erf (x).
   (d) Use the answers to parts (b) and (c) to approximate

$$\int_1^2 e^{-x^2}\, dx.$$

In the Exercises 19–22, use a graphing calculator or computer to obtain the graphs of the given function with the given ranges. Use "trace" or some other method to evaluate the function as requested. (Your answers should be accurate to two decimal places. The instructions after the exercises refer to the TI-82 and similar models.)

19. $f(x) = \int_0^x e^{t^2}\,dt$ $(-3 \le x \le 3, -1 \le y \le 2)$. Calculate $f(0)$, $f(0.5)$, and $f(1)$.
[On the TI-82, set $Y_1 = \text{fnInt }(e^\wedge\,(-T^\wedge\,2),\ T,\ 0,\ X)$ in the "Y=" menu and hit GRAPH. Have patience! The calculator must compute each y-coordinate by doing the integral numerically.]

20. $f(x) = \int_0^x t^2 e^t\,dt$ $(-3 \le x \le 3, -1 \le y \le 2)$. Calculate $f(-1)$, $f(0)$, and $f(1)$.
[$Y_1 = \text{fnInt}(T^\wedge\,2e^\wedge T,\ T,\ 0,\ X)$]

21. $f(x)$ is the antiderivative of $\sqrt{1 - x^2}$ with the property that $f(0) = 0$ $(-1 \le x \le 1, -1 \le y \le 1)$. Calculate $f(-0.5)$ and $f(0.5)$.

22. $f(x)$ is the antiderivative of $\dfrac{1}{\sqrt{1 - x^2}}$ with the property that $f(0) = 0$ $(-0.5 \le x \le 0.5, -2 \le y \le 2)$. Calculate $f(-0.5)$ and $f(0.5)$.

COMMUNICATION AND REASONING EXERCISES

23. What does the Fundamental Theorem of Calculus permit one to do?

24. Your friend has just told you that the function $f(x) = e^{-x^2}$ can't be integrated and hence has no antiderivative. Is your friend correct? Explain your answer.

25. Use the FTC to find an antiderivative F of

$$f(x) = \begin{cases} 0 & \text{if } x < 0 \\ 1 & \text{if } x \ge 0. \end{cases}$$

Is it true that $F'(x) = f(x)$ for *all x?*

26. According to the Fundamental Theorem of Calculus as stated in this text, which functions are guaranteed to have antiderivatives? What does the answer to the previous exercise tell you about the theorem as stated in this text?tx

---

▶ ═══ **6.7** NUMERICAL INTEGRATION

The Fundamental Theorem of Calculus gives us an exact formula for computing $\int_a^b f(x)\,dx$, *provided* we can find an antiderivative for f. This method of evaluating definite integrals is called the **analytic** method. However, as we have seen, there are times when this is difficult or impossible. In these cases, it is usually good enough to find an approximate, or **numerical** solution, and there are some very simple ways to do this. We have already been using the left-hand and right-hand Riemann sums, as well as their average, which we referred to, mysteriously, as the "trapezoidal" sum.

In this section, we shall take this a step further. First, we shall find out where the term "trapezoidal" comes from. Second, we shall develop a more efficient method of approximating the integral, called Simpson's rule. Finally, we shall discuss the accuracy of both the trapezoid rule and Simpson's rule, so that we can choose the number of partitions with more confidence when approximating the integral by a sum.

THE TRAPEZOID RULE AND SIMPSON'S RULE

The **trapezoid rule** starts by taking a simple partition of $[a, b]$ into n intervals of the same width, just as we did with the left- and right-hand sums. Since the total width must be $b - a$, each small interval must have width $\Delta x = (b - a)/n$. As we did with the Riemann sums, we take

$$x_0 = a$$

$$x_1 = a + \Delta x = a + \frac{b - a}{n}$$

$$x_2 = a + 2\Delta x = a + 2\frac{b - a}{n}$$

$$\ldots$$

$$x_k = a + k\Delta x = a + k\frac{b - a}{n}$$

$$\ldots$$

$$x_n = a + n\Delta x = b.$$

The trapezoid rule gives us the following approximation to the definite integral.

TRAPEZOID RULE

If $x_k = a + k\Delta x = a + k\dfrac{b - a}{n}$, then

$$\int_a^b f(x)\, dx \approx \frac{b - a}{2n}[f(x_0) + 2f(x_1) + 2f(x_2)$$
$$+ \ldots + 2f(x_{n-1}) + f(x_n)].$$

Q This doesn't look like a Riemann sum at all. Why is it an approximation to the integral?

A Recall that the left-hand and right-hand sums were given by

$$\text{Left-hand sum} = \Delta x[f(x_0) + f(x_1) + f(x_2) + \ldots + f(x_{n-1})]$$

and

$$\text{Right-hand sum} = \Delta x[f(x_1) + f(x_2) + \ldots + f(x_{n-1}) + f(x_n)].$$

The average—or what we referred to as the "trapezoidal" sum, is given by

$$\text{Average} = \frac{\text{Left-hand sum + Right-hand sum}}{2}$$

$$= \frac{1}{2}\Delta x[f(x_0) + f(x_1) + f(x_2) + \ldots + f(x_{n-1}) + f(x_1)$$
$$+ f(x_2) + \ldots + f(x_{n-1}) + f(x_n)]$$

$$= \frac{\Delta x}{2}[f(x_0) + 2f(x_1) + 2f(x_2) + \ldots + 2f(x_{n-1}) + f(x_n)]$$

$$= \frac{b-a}{2n}[f(x_0) + 2f(x_1) + 2f(x_2) + \ldots + 2f(x_{n-1}) + f(x_n)],$$

which is the trapezoid rule. In other words, the trapezoid rule just gives the trapezoidal sum, or the average of the left- and right-hand sums. Thus, although it is not really a Riemann sum, it is the average of two Riemann sums.

Q Where does the term "trapezoid" come from?

A Look at Figure 1.

Instead of approximating the area in each strip with the area of a rectangle, as we do for Riemann sums, the idea is to use a different shape. Between x_{k-1} and x_k we take the shaded region shown in Figure 2.

This is a **trapezoid,** a four-sided region with two opposite sides parallel. In this case it is the two vertical sides that are parallel. The area of a trapezoid is the average length of the parallel sides, times the distance between them, which in this case gives

$$\frac{f(x_{k-1}) + f(x_k)}{2} \cdot \frac{b-a}{n} = \frac{b-a}{2n}[f(x_{k-1}) + f(x_k)].$$

If we now add up the area of these trapezoids, each $f(x_k)$ will appear twice, except for $f(a)$ and $f(b)$, which appear once each. The sum is then the formula given as the trapezoid rule.

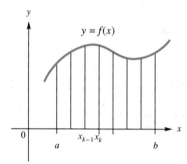

FIGURE 1

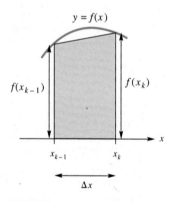

FIGURE 2

▼ | **EXAMPLE 1** |

Use the trapezoid rule to approximate $\int_0^1 x^2\, dx$, **(a)** using 5 intervals and **(b)** using 10 intervals.

SOLUTION   Instead of calculating the left- and right-hand sums separately and then taking the average, we can calculate the trapezoid rule in one step using the same table we used for the Riemann sums.

(a) $n = 5$, so $\Delta x = \dfrac{b - a}{n} = \dfrac{1 - 0}{5} = 0.2$.

k	0	1	2	3	4	5
$x_k = a + k\,\Delta x$	0	0.2	0.4	0.6	0.8	1.0
$f(x_k)$	0	0.04	0.16	0.36	0.64	1.0

By the trapezoid rule,

$$\int_0^1 x^2 \, dx \approx \frac{b - a}{2n} \left[f(x_0) + 2f(x_1) + 2f(x_2) + \ldots + 2f(x_{n-1}) + f(x_n) \right]$$
$$= 0.1[0 + 2(0.04) + 2(0.16) + 2(0.36) + 2(0.64) + 1.0]$$
$$= 0.34.$$

(b) $n = 10$, so $\Delta x = \dfrac{b - a}{n} = \dfrac{1 - 0}{10} = 0.1$.

x	0	1	2	3	4	5	6	7	8	9	10
$x_k = a + k\,\Delta x$	0	0.1	0.2	0.3	0.4	0.5	0.6	0.7	0.8	0.9	1.0
$f(x_k)$	0	0.01	0.04	0.09	0.16	0.25	0.36	0.49	0.64	0.81	1.0

By the trapezoid rule,

$$\int_0^1 x^2 \, dx \approx \frac{b - a}{2n} \left[f(x_0) + 2f(x_1) + 2f(x_2) + \ldots + 2f(x_{n-1}) + f(x_n) \right]$$
$$= 0.05[0 + 2(0.10) + 2(0.04) + 2(0.09) + 2(0.16) + 2(0.25)$$
$$+ 2(0.36) + 2(0.49) + 2(0.64) + 2(0.81) + 1.0]$$
$$= 0.335.$$

Before we go on... The exact answer, of course, is

$$\int_0^1 x^2 \, dx = \left[\frac{x^3}{3} \right]_0^1 = \frac{1}{3} = 0.3333\ldots$$

▼ EXAMPLE 2

Use the trapezoid rule with 6 intervals to approximate

$$\int_0^6 e^{-x^2} \, dx.$$

SOLUTION   This example is interesting because it has been shown that the antiderivative of e^{-x^2} cannot be written in terms of "elementary" functions (which include those we have talked about here, and a few more).* However, we can easily approximate this integral. Since we are using $n = 6$, we have $\Delta x = (b - a)/n = (6 - 0)/6 = 1$, and so

$$\int_0^6 e^{-x^2}\, dx \approx 0.5[e^0 + 2e^{-1} + 2e^{-4} + 2e^{-9} + 2e^{-16} + 2e^{-25} + e^{-36}]$$

$$\approx 0.886.$$

Before we go on... You would probably be surprised to learn that the exact value of the slightly larger integral, $\int_0^\infty e^{-x^2}\, dx$, is in fact $\sqrt{\pi}/4$. Since $\sqrt{\pi}/4 = 0.8862.\ldots$, this agrees with our answer to three decimal places. (The area $\int_6^\infty e^{-x^2}\, dx$ is exceptionally small.)

Simpson's rule is another approximation of the integral. Again, we start by partitioning $[a, b]$ into intervals all of the same width, but this time we must use an even number of intervals, so n will be even.

SIMPSON'S RULE

If n is even, and $x_k = a + k\, \Delta x = a + k\dfrac{b - a}{n}$, then

$$\int_a^b f(x)\, dx \approx \frac{b - a}{3n}\, [f(a) + 4f(x_1) + 2f(x_2) + 4f(x_3)$$

$$+ \ldots + 2f(x_{n-1}) + 4f(x_{n-1}) + f(b)].$$

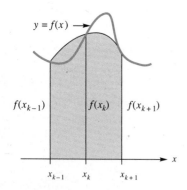

$y = f(x)$

$f(x_{k-1})$ $f(x_k)$ $f(x_{k+1})$

x_{k-1} x_k x_{k+1}

FIGURE 3

Q Why?

A As with the trapezoid rule, we want to approximate the areas in each strip by something more complicated than a rectangle. This time we take the strips in pairs (which is why we need an even number of them) and draw a *parabola* through the three points $(x_{k-1},\ f(x_{k-1}))$, $(x_k,\ f(x_k))$, and $(x_{k+1}, f(x_{k+1}))$, as shown in Figure 3.

It is then not too difficult to find the equation of this parabola (it has the form $y = Ax^2 + Bx + C$), and from that to find the area underneath by integrating. The remarkably simple answer is

$$\text{Area under parabola} = \frac{b - a}{3n}\, [f(x_{k-1}) + 4f(x_k) + f(x_{k+1})]$$

When we add the area under the parabola over the first two strips to the area under the parabola over the third and fourth strips, and so on, we get Simpson's rule.

▼ * In one of the exercises of the previous section, we saw that its antiderivative is a constant multiple of the **error function**, erf(x), and we calculated a few of its values.

▼ EXAMPLE 3

Use 4 intervals in Simpson's rule to approximate $\int_0^1 x^2 \, dx$.

SOLUTION   Since $n = 4$, we have $(b - a)/n = 1/4$, and Simpson's rule tells us that

$$\int_0^1 x^2 \, dx \approx \frac{1}{12}\left[0 + 4\frac{1}{16} + 2\frac{4}{16} + 4\frac{9}{16} + 1\right]$$

$$= \left(\frac{1}{12}\right)\left(\frac{64}{16}\right)$$

$$= \frac{1}{3}.$$

Before we go on... This is the *exact* answer. What is going on? Remember that Simpson's rule is based on approximating the graph by quadratic functions. If the function is already quadratic, as it is here, the approximation is exact.

▼ EXAMPLE 4

Use 6 intervals in Simpson's rule to approximate $\int_0^6 e^{-x^2} \, dx$.

SOLUTION   As in Example 2, $(b - a)/n = 1$, so Simpson's rule gives us

$$\int_0^6 e^{-x^2} \, dx \approx \frac{1}{3}[e^0 + 4e^{-1} + 2e^{-4} + 4e^{-9} + 2e^{-16} + 4e^{-25} + e^{-36}]$$

$$\approx 0.8362.$$

USING A GRAPHING CALCULATOR OR SPREADSHEET

We have already seen a graphing calculator program to compute the left- and right-hand Riemann approximations of $\int_a^b f(x) \, dx$. Further, we have seen that the trapezoid rule amounts to the average of the left- and right-hand sums, so instead of writing a new program for the trapezoid rule, all we need do is to add a few lines of code to the program SUMS to compute the average, and we obtain a program that shows all three sums. This is the program TRAP in the appendix. The program following it, SIMP, calculates the sum for Simpson's rule.

▶ NOTE   To run these programs, don't forget to first enter the function you are integrating under Y_1 in the "Y=" window. ◀

The appendix also contains instructions for using a computer spreadsheet to do these calculations.

▼ **EXAMPLE 5**

Use the trapezoid rule and Simpson's rule to approximate $\int_{-1}^{1} 2\sqrt{1 - x^2}\, dx$ with $n = 200$. How accurate are these approximations?

SOLUTION   This integral gives twice the area of a semicircle with radius 1, so it should be equal to π. Using the two rules, we get

$$\text{Trapezoid rule: } \int_{-1}^{1} 2\sqrt{1 - x^2}\, dx \approx 3.14041 \ldots$$
$$\text{Simpson's rule: } \int_{-1}^{1} 2\sqrt{1 - x^2}\, dx \approx 3.14113 \ldots$$

The actual value of π is $3.141592654 \ldots$, so to estimate the errors, we round the differences to one significant figure:

$$\text{Trapezoid rule error: } |3.14041 - 3.14159| = 0.00118 \approx 0.001$$
$$\text{Simpson's rule error: } |3.14113 - 3.14159| = 0.00046 \approx 0.0005.$$

Thus Simpson's rule is more accurate—to within $\pm\, 0.0005$—since it has the smallest error.

Before we go on... Your graphing calculator should also have its own (more sophisticated) algorithm for evaluating a definite integral numerically: On a TI-82, enter

$$\text{fnInt}(Y_1, X, -1, 1)$$

and it gives the more accurate answer of

$$3.141593074.$$

Its error is $|3.141593074 - 3.141592654| = 0.0000004204 \approx 0.0000004.$

ACCURACY

If you look at Example 1, you see that the trapezoid rule gets closer to the right answer for larger n. If you compare Examples 2 and 4, you see that we have two different approximations for the same integral. This raises some interesting and important questions: How large does n have to be to get "close enough" to the right answer? Which of the answers in Examples 2 and 4 is closest to the right answer? To answer these questions, we need to know something about the **error** in these two rules: that is, how far they are from the right answer.

$$\text{Error} = |\text{Approximation} - \text{Exact answer}|$$

Q Doesn't this raise a "Catch 22" situation? In order to know the error, we need to know the exact answer. But if we knew the exact answer, then we would hardly need to find a numerical approximation in the first place!

A We remedy this dilemma as follows: since we can't always calculate exactly what the error is, we look instead for a **bound** on the error. For instance, instead of trying to say "the error is exactly 0.001," we say instead, "the error is no larger than 0.001."

The following formulas give bounds on the errors for the rules we have been using.

> **THE ERRORS IN THE TRAPEZOID RULE AND SIMPSON'S RULE**
>
> If $f''(x)$ is continuous in $[a, b]$, then the error in the trapezoid rule is no larger than
>
> $$\frac{(b - a)^3}{12n^2}\,|\,f''(M)\,|,$$
>
> where $|\,f''(M)\,|$ is the largest value of $|\,f''(x)\,|$ in $[a, b]$.
>
> If $f^{(4)}(x)$ is continuous in $[a, b]$, then the error in Simpson's rule is no larger than
>
> $$\frac{(b - a)^5}{180n^4}\,|\,f^{(4)}(M)\,|,$$
>
> where $|\,f^{(4)}(M)\,|$ is the largest value of $|\,f^{(4)}(x)\,|$ in $[a, b]$.

We will not talk about where these facts come from, as their derivations are beyond the scope of this book. However, they are not hard to use.

▼ **EXAMPLE 6**

How accurate is the calculation in Example 2?

SOLUTION   In that example we used 6 intervals in the trapezoid rule to estimate

$$\int_0^5 e^{-x^2}\, dx.$$

In order to use the error estimate, we need to know the largest value of $f''(x)$ on the interval $[0, 6]$, for $f(x) = e^{-x^2}$. Calculating,

$$f'(x) = -2xe^{-x^2}, \text{ and}$$
$$f''(x) = 2(2x^2 - 1)e^{-x^2}.$$

Since we want to find the extreme values of f'', we calculate its derivative,

$$f'''(x) = 4x(3 - 2x^2)e^{-x^2}.$$

Now $f'''(x) = 0$ only when $x = 0$ or $3 - 2x^2 = 0$, so $x = 0$ or $\pm\sqrt{\frac{3}{2}} \approx \pm 1.225$. Checking values in the interval $[0, 6]$, we get the following.

x	0	1.225	6
$f''(x)$	-2	0.89	3×10^{-14}

The largest value of $|f''(x)|$ is therefore 2. This tells us that the error is no larger than

$$\frac{(6 - 0^3)}{12 \cdot 6^2} \cdot 2 = 1.$$

This tells us that all we really know from the calculation we did is that

$$-0.114 \le \int_0^5 e^{-x^2}\, dx \le 1.886.$$

This does not give us much confidence in our calculation.

Before we go on... In fact, chances are that the calculation is more accurate than that, but the mathematics we've just done gives us no reason to believe this.

▼ **EXAMPLE 7**

How accurate is the calculation in Example 4?

SOLUTION   That example used 6 intervals in Simpson's rule to approximate

$$\int_0^5 e^{-x^2}\, dx.$$

In order to estimate the error, we need to find the largest value of $|f^{(4)}(x)|$, for x in $[0, 6]$, with $f(x) = e^{-x^2}$ again. Calculating derivatives beyond those we know from above,

$$f^{(4)}(x) = 4(4x^4 - 12x^2 + 3)e^{-x^2}, \text{ and}$$
$$f^{(5)}(x) = -8x(4x^4 - 15x^2 + 15)e^{-x^2}.$$

To find the extreme values of $f^{(4)}(x)$, we need to know where $f^{(5)}(x) = 0$, but the only place this happens is when $x = 0$. (This can easily be checked graphically. Alternatively, notice that, if $f^{(5)}(x) = 0$, either $x = 0$, or $4x^4 - 15x^2 + 15 = 0$. But this equation has no solutions, since the similar quadratic $4u^2 - 15u + 15 = 0$ has none.) Checking values, we get the following.

x	0	6
$f^{(4)}(x)$	12	4×10^{-12}

The largest value of $|f^{(4)}(x)|$ is 12, so the error in Simpson's rule is no larger than

$$\frac{(6 - 0)^5}{180 \cdot 6^4} \cdot 12 = 0.4.$$

This means that we can say that

$$0.4362 \leq \int_0^5 e^{-x^2}\, dx \leq 1.2362.$$

Again, we would not be surprised to find out that our calculation was more accurate than that, but we have no real reason to believe so.

▼   **EXAMPLE 8**

Going back to Example 1, how large must n be to approximate the answer to 5 decimal places?

SOLUTION   We are now asking that the error be no larger than 0.000001. To get the formula for the error, we need to find the largest value of $\left| f''(x) \right|$ on $[0, 1]$, but this is easy since $f''(x) = 2$ is constant: its largest value is 2. The error if we use n intervals is then

$$\frac{(1 - 0)^3}{12n^2} \cdot 2 = \frac{1}{6n^2}.$$

Our question is: How large must n be so that $1/(6n^2)$ is smaller than 0.000001? If we set $1/(6n^2) = 0.000001$ and solve for n, we get $n = 408.2$, meaning that we need $n \geq 409$ to be guaranteed 5 decimal places of accuracy.

Before we go on... This large a sum is much too tedious to compute by hand but easy enough to do with a programmable calculator or computer.

▼   **EXAMPLE 9**

How large should n be in Example 4 to approximate the answer to 5 decimal places?

SOLUTION   Our error estimate is the one we found in Example 6: the error is no larger than $6^5 \cdot 12/(180n^4) = 518.4/n^4$. If we set this equal to 0.000001, we find that $n = 150.9$, so we need n to be at least 151 to be guaranteed 5 digits of accuracy.

USING A GRAPHING CALCULATOR TO FIND
THE ERROR ESTIMATES

We have included two further graphing calculator programs in the appendix—called TRAPERR and SIMPERR—that do the calculation of the error estimates for you. The next example demonstrates their use on the TI-82.

▼ **EXAMPLE 10**

Use a graphing calculator to approximate $\int_{0.5}^{5} x \ln x \, dx$ with both the trapezoid rule and Simpson's rule using 250 partitions. How reliable are these answers?

SOLUTION   We must first take care of the "Y=" window, so we need all derivatives up to the fourth.

$$f(x) = x \ln x \qquad\qquad f'(x) = \ln x + 1$$

$$f''(x) = \frac{1}{x} \qquad\qquad f'''(x) = -\frac{1}{x^2}$$

$$f^{(4)}(x) = \frac{2}{x^3}$$

We enter these in the "Y=" window as follows.

$$Y_1 = X*\ln(X)$$
$$Y_2 = abs(1/X)$$
$$Y_4 = abs(2/(X^3))$$

We then run TRAP and SIMP, with $N = 250$, $A = 0.5$, $B = 5$, and obtain

$$\text{Trapezoid rule:} \int_{0.5}^{5} x \ln x \, dx \approx 14.01717947$$

$$\text{Simpson's rule:} \int_{0.5}^{1} x \ln x \, dx \approx 14.01711731.$$

To obtain the error estimates, we run TRAPERR and SIMPERR with the above values and find

Trapezoid error estimate: $0.000243 \approx 0.0002$

Simpson error estimate: $0.0000000419904 \approx 0.00000004$

Clearly, the approximation given by Simpson's rule is far more accurate, so we conclude

$$\int_{0.5}^{5} x \ln x \, dx = 14.01711731 \pm 0.00000004.$$

In other words, this estimate is accurate to 7 decimal places!

Before we go on... Let us compare this with the TI-82 graphing calculator's built-in integration algorithm. Enter

$$\text{fnInt(XlnX, X, Ø.5,5)}$$

and press ⌈ENTER⌉. The TI-82 gives the answer 14.0171173, which has one less significant digit than we obtained with Simpson's rule. In other words, Simpson's rule with $n = 250$ gives slightly more accurate information.

▶ NOTE   You may suspect that the trapezoid rule is always less accurate, but that is not the case. It is possible to construct functions for which the trapezoid rule gives a far more accurate approximation than Simpson's rule (in fact, there is an example in the text above). Further, calculating the error for the trapezoid rule is easier, since taking the second derivative involves less work than taking the fourth derivative. (Think about $f(x) = 1/(1 + x^2)$, for example.) ◀

In this section we have just scratched the surface of a large subject. For just about any problem in which it might be difficult to find an exact answer analytically, people have worked out approximations. These problems include finding solutions to equations, computing integrals, and many more. This field of **numerical methods** has been around for as long as people have wanted to do computations. It became particularly important when high-speed computers and calculators were developed, so that the routine calculations could be done much more quickly.* Today, numerical methods underlie the large-scale computer calculations used for economic forecasting, weather forecasting, and, much more successfully, research in physics. A key to using any approximation is knowing something about how bad an approximation it is, so all approximations have error estimates similar to those we gave for the trapezoid rule and Simpson's rule.

▶ ___6.7 EXERCISES___

In each of Exercises 1–6, use the trapezoid rule with 10 intervals to approximate the integral. Use the error estimate to tell how accurate your answer is. If possible, compute the integral exactly using the Fundamental Theorem of Calculus, and compare the exact answer to your approximation.

1. $\int_0^2 2x\, dx$ **2.** $\int_{-1}^1 (x - 1)\, dx$ **3.** $\int_0^3 x^3\, dx$ **4.** $\int_1^3 (x^3 + 1)\, dx$

5. $\int_1^5 \ln x\, dx$ **6.** $\int_2^6 \ln x\, dx$

7–12. Repeat Exercises 1–6, but use Simpson's rule with 10 intervals.

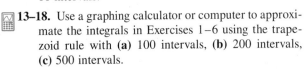 **13–18.** Use a graphing calculator or computer to approximate the integrals in Exercises 1–6 using the trapezoid rule with (a) 100 intervals, (b) 200 intervals, (c) 500 intervals.

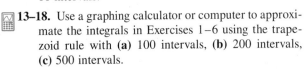 **19–24.** Use a graphing calculator or computer to approximate the integrals in Exercises 1–6 using Simpson's rule with (a) 100 intervals, (b) 200 intervals, (c) 500 intervals.

25. Consider $\int_0^2 x^4\, dx$. How large should n be in order to compute this integral to 3 decimal places using the trapezoid rule?

26. Consider $\int_0^2 x^5\, dx$. How large should n be in order to compute this integral to 3 decimal places using the trapezoid rule?

▼ * The Manhattan Project to develop the atomic bomb, during World War II, used a room full of calculators. This is what they called the women hired to do the lengthy calculations on the mechanical contraptions that were available at the time. The first electronic computers were developed during this war to speed up these and other wartime calculations.

27. Repeat Exercise 25 using Simpson's rule.

28. Repeat Exercise 26 using Simpson's rule.

29. Consider $\int_0^{10} xe^{-x}\,dx$. How large should n be in order to compute this integral to 3 decimal places using the trapezoid rule?

30. Consider $\int_0^{10} x^2e^{-x}\,dx$. How large should n be in order to compute this integral to 3 decimal places using the trapezoid rule?

31. Repeat Exercise 29 using Simpson's rule.

32. Repeat Exercise 30 using Simpson's rule.

Computing the Length of a Graph   *The length of the graph of the function f from x = a to x = b (see the figure) can be shown to be*

$$\text{Length} = \int_a^b \sqrt{1 + [f'(x)]^2}\,dx$$

Use this formula to approximate the lengths of the graphs of the functions in Exercises 33–36. In each

case, use Simpson's rule with **(a)** *6 partitions (if you are doing it by hand)* **(b)** *200 partitions (if you are using technology).*

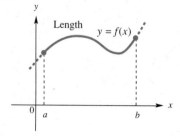

33. $f(x) = x^2$, x in $[0, 1]$

34. $f(x) = x^2/2$, x in $[0, 2]$

35. $f(x) = e^x$, x in $[0, 1]$

36. $f(x) = e^{-x}$, x in $[5, 10]$

COMMUNICATION AND REASONING EXERCISES

37. Looking at the error estimate for the trapezoid rule, by how much will the error shrink if you increase n by a factor of 10? What does this say about the increase in the number of digits of accuracy of the estimate given by the rule?

38. Looking at the error estimate for Simpson's rule, by how much will the error shrink if you increase n by a factor of 10? What does this say about the increase in the number of digits of accuracy of the estimate given by the rule?

▶ ▓▓▓▓**You're the Expert** ▓▓▓▓▓▓▓▓▓▓▓▓▓▓▓▓▓▓▓▓▓▓▓▓▓▓

THE COST OF ISSUING A WARRANTY*

You are a consultant to Sunny Electronics Company. Sunny's chief competitor, Cyberspace Electronics, Inc., has just launched a new home entertainment system that competes directly with Sunny's World Entertainment System (WES). Cyberspace is offering a 2-year limited warranty on its product, so Sunny is thinking of offering a 20-year pro-rated warranty. According to this warranty, if anything goes wrong, Sunny will refund the original $1,000 cost of the system depreciated continuously at an annual rate of 12%. Sunny has asked you to estimate what it will cost the company to provide this warranty.

Using the formula for continuous depreciation, you calculate that it will cost Sunny

$$C(t) = 1,000e^{-0.12t}$$

▼ * This situation was inspired by the article "Determination of Warranty Reserves" by W.W. Menke, *Management Sciences*, Vol. 15, No. 10, June, 1969. (The Institute of Management Sciences)

to replace a system that is t years old. You realize that some information is missing—you would like to know how the failure rate of the World Entertainment System (WES) increases with time. For instance, what percentage of products will fail after 5 years? 10 years? After searching the literature for a while, you find a reasonable model that predicts the percentage of items that will fail after time t:

$$p(t) = 1 - e^{-t/N},$$

where t is the time in years and N is the average lifetime of the product. For the WES, the average lifetime happens to be 15 years. So you use $p(t) = 1 - e^{-t/15}$, whose graph is shown in Figure 1.

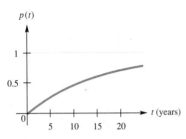

This is the total percentage that fail in t years. Thus, for example, the percentage of systems that will fail by the end of the first year is $p(1) = 1 - e^{-1/15} \approx 0.0645$, or 6.45%, while the percentage that will fail by the end of 5 years is $p(5) = 1 - e^{-5/15} \approx 0.283$, or 28.3%.

You start by trying to estimate the cost to the company for small time periods, say once a quarter, and you realize that it would be more helpful to know the *failure rate*—the percentage of entertainment systems that will fail per year. Recalling the ideas of calculus, you remember that the failure rate is given by the derivative,

$$p'(t) = \frac{1}{15}e^{-t/15}.$$

Thus, for example, the percentage of sets that fail in the fifth year is approximately $p'(5) \approx 0.048$, or 4.8%.

For the sake of definiteness, you suppose that the company has sold 100 sets, and you calculate the costs one quarter at a time. During the first quarter ($t = 0.25$ years), the percentage of the 100 sets that will fail is approximately

$$\text{Annual failure rate} \times \frac{1}{4} \approx p'(0.25) \cdot (0.25).$$

Thus, the cost to the company for the 100 sets is

$$\text{Number of failures} \times \text{Cost per failure}$$
$$\approx 100p'(0.25) \cdot (0.25) \times 1000e^{-0.12(0.25)}$$
$$= 100{,}000p'(0.25)e^{-0.12(0.25)}(0.25)$$

For the second quarter, the cost is approximately

$$100{,}000p'(0.50)e^{-0.12(0.50)}(0.25),$$

since now $t = 0.5$ years. Adding all these costs through the end of the eightieth quarter gives

$$100{,}000p'(0.25)e^{-0.12(0.25)}(0.25)$$
$$+ \ 100{,}000p'(0.50)e^{-0.12(0.50)}(0.25)$$
$$+100{,}000p'(0.75)e^{-0.12(0.75)}(0.25)$$
$$+ \ldots + 100{,}000p'(80)e^{-0.12(80)}(0.25).$$

FIGURE 1

Just as you are about to calculate this awful-looking sum with 80 terms, you notice that it looks suspiciously like a Riemann sum. In fact, you can write it as

$$\sum_{k=1}^{80} 100{,}000 \, p'(t_k) e^{-0.12 t_k} \, \Delta t,$$

where Δt is the time period (one quarter $= 0.25$ years) and $t_k = k \, \Delta t$. Further, you realize that you will get more accurate estimates by taking smaller and smaller time periods: in other words, by taking the limit of the sum as $\Delta t \to 0$. The limit is the definite integral

$$\lim_{\Delta t \to 0} \sum_{k=1}^{n} 100{,}000 \, p'(t_k) e^{-0.12 t_k} \, \Delta t,$$

$$= \int_0^{20} 100{,}000 \, p'(t) e^{-0.12 t} \, dt.$$

(The limits of integration must be in units of t, that is, in years.) Substituting the formula for $p'(t)$ gives

$$\begin{aligned}
\text{Total cost} &= \int_0^{20} 100{,}000 \cdot \frac{1}{15} e^{-t/15} e^{-0.12 t} \, dt \\
&= \frac{100{,}000}{15} \int_0^{20} e^{-t/15} e^{-0.12 t} \, dt \\
&= \frac{100{,}000}{15} \int_0^{20} e^{-14 t/75} \, dt \\
&= \frac{100{,}000}{15} \left(-\frac{75}{14} \right) \left[e^{-14 t/75} \right]_0^{20} \\
&= \$34{,}860.25
\end{aligned}$$

This is the cost of providing warranty coverage for 100 sets. Thus, for one set the cost will be $348.60, more than a third of the cost of a new set.

Exercises

1. Find the cost per system if Sunny decides to limit the warranty to 10 years.

2. How many years should the warranty last to cost Sunny no more than $100 per set?

3. If Sunny kept the 20-year warranty plan but increased the discount rate from 12% to 30%, how would this affect the warranty cost per set?

4. The sales manager at Sunny tells you that the company is prepared to spend no more than $100 per set on warranty costs but would like to keep the 20-year pro-rated warranty. What could you recommend?

5. A junior executive in the company suggests that, under the original plan, extending the warranty to 30 years will make very little difference. How accurate is her assessment?

6. Suppose that Sunny used linear depreciation rather than exponential depreciation. In particular, suppose that Sunny will pay $C(t) = 1{,}000 - 50t$ if a system

fails after t years (up to 20 years). How much will it cost Sunny to provide this warranty? (You will have to use numerical integration. The integral can also be evaluated using integration by parts, a technique that will be discussed in the next chapter.)

7. Suppose that Sunny used linear depreciation over 10 years rather than 20 (so that $C(10) = 0$). How much would it cost to provide this warranty?

▶ ## Review Exercises

Evaluate the integrals in Exercises 1–34.

1. $\int 10x^{10}\, dx$

2. $\int 2x^7\, dx$

3. $\int \frac{3}{x^4}\, dx$

4. $\int \frac{4}{x^5}\, dx$

5. $\int \left(\sqrt{x} + \frac{1}{x}\right) dx$

6. $\int \left(\sqrt[3]{x} - \frac{2}{x}\right) dx$

7. $\int \left(2e^x + 3x - \frac{4}{x}\right) dx$

8. $\int \left(5e^x - 2x^2 + \frac{3}{x}\right) dx$

9. $\int (x + 2)^{10}\, dx$

10. $\int (2x - 3)^5\, dx$

11. $\int x\sqrt[3]{1 + x^2}\, dx$

12. $\int x^2\sqrt{x^3 + 2}\, dx$

13. $\int \frac{x}{x^2 + 1}\, dx$

14. $\int \frac{x^2}{x^3 + 2}\, dx$

15. $\int 5e^{-2x}\, dx$

16. $\int 4e^{2x+1}\, dx$

17. $\int (xe^{x^2} - 3x)\, dx$

18. $\int \left(xe^{x^2+1} - \frac{1}{x + 2}\right) dx$

19. $\int \frac{4.7x^{0.2}}{x^{1.2} - 4}\, dx$

20. $\int \frac{2.3x^{0.1} - 4}{x^{0.9}}\, dx$

21. $\int \frac{e^{0.3t}}{1 + 2e^{0.3t}}\, dt$

22. $\int \frac{e^{0.3t} - 1}{e^{0.3t}}\, dt$

23. $\int_0^1 (x^3 + 2x)\, dx$

24. $\int_{-1}^1 (x^3 + x^2)\, dx$

25. $\int_1^2 \frac{3}{x^2}\, dx$

26. $\int_{-2}^{-1} \frac{2}{x}\, dx$

27. $\int_0^1 (e^{-x} + x)\, dx$

28. $\int_0^{10} xe^{-x^2}\, dx$

29. $\int_0^2 x^2\sqrt{x^3 + 1}\, dx$

30. $\int_1^3 x^3\sqrt{x^2 - 1}\, dx$

31. $\int_0^1 (4 + x^{0.2}(2 - x^{1.2})^4)\, dx$

32. $\int_0^1 \left(3 - \frac{x^{0.2}}{1 + x^{1.2}}\right) dx$

33. $\int_{-1}^1 xe^{-0.3x^2}\, dx$

34. $\int_0^4 \frac{e^{0.3t}}{1 + e^{0.3t}}\, dt$

Find the areas of the regions described in Exercises 35–46.

35. Bounded by $y = 1 - x^2$, the x-axis, and the lines $x = -1$ and $x = 1$

36. Bounded by $y = 1 - x^4$, the x-axis, and the lines $x = -1$ and $x = 1$

37. Bounded by $y = 1/x$, the x-axis, and the lines $x = 1$ and $x = 10$

38. Bounded by $y = 2/x$, the x-axis, and the lines $x = 2$ and $x = 10$

39. Bounded by $y = xe^{-x^2}$, the x-axis, and the lines $x = 0$ and $x = 5$

40. Bounded by $y = 1 - e^{3x-4}$, the x-axis, and the lines $x = 0$ and $x = 5$

41. Bounded by $y = 1 - 2x^2$ and the x-axis

42. Bounded by $y = 1 - 2x^4$ and the x-axis

43. Bounded by $y = x^4 - x^2$ and the x-axis

44. Bounded by $y = x^4 - 5x^2 + 4$ and the x-axis

45. Bounded by $y = x^2 - x^3$, the x-axis, and the line $x = -1$

46. Bounded by $y = e^x - e^{-x}$, the x-axis, and the lines $x = -1$ and $x = 1$

Evaluate the left- and right-hand Riemann approximations and the trapezoid approximation for the integrals in Exercises 47–50 using the stated number of partitions. (Round your answers to two decimal places.)

47. $\displaystyle\int_{-1}^{1} e^{-x^2}\, dx, \quad n = 4$

48. $\displaystyle\int_{0}^{5} e^{-x^2}\, dx, \quad n = 5$

49. $\displaystyle\int_{0.5}^{2} \ln x\, dx, \quad n = 3$

50. $\displaystyle\int_{0}^{1} \sqrt{1 + \sqrt{x}}\, dx, \quad n = 2$

Use a computer or graphing calculator to evaluate the left- and right-hand Riemann approximations for the integrals in Exercises 51–54 using the stated number of partitions. (Round your answers to four decimal places.)

51. $\displaystyle\int_{-1}^{1} e^{x^2}\, dx, \quad n = 100, 200, 500$

52. $\displaystyle\int_{0}^{1} e^{x^2}\, dx, \quad n = 100, 200, 500$

53. $\displaystyle\int_{0.5}^{2} \frac{dx}{4 + \ln x}, \quad n = 100, 200, 500$

54. $\displaystyle\int_{0}^{1} \sqrt{1 + x^2}\, dx, \quad n = 100, 200, 500$

55–62. Using a graphing calculator or computer, approximate the integrals in Exercises 47–54 using Simpson's rule with $n = 50, 100,$ and 500.

63–70. For each of the integrals given in 47–54, determine how large n must be to approximate the integral to within ± 0.001 using the trapezoid rule.

71–78. For each of the integrals given in 47–54, determine how large n must be to approximate the integral to within ± 0.001 using Simpson's rule.

APPLICATIONS

79. *Magazine Sales* The accompanying chart shows annual newsstand sales of all magazines in the U.S. (in millions of magazines).*
  (a) Model the annual sales with a linear function $s(t)$ using the data from 1982 ($t = 0$) and 1992 ($t = 10$).
  (b) Compare the actual sales over the given time period with the total sales predicted by your model.
  (c) Use your model to estimate the number of magazines sold in the U.S. from the start of 1980 ($t = -2$) to the start of 1990.

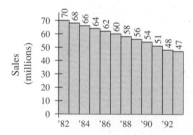

▼ *Source: Vos, Gruppo & Capell Inc./*The New York Times*, Dec. 8, 1993, p. D6.

80. *Oversupply*   The following chart shows the number of unsold newsstand magazines in the U.S.*

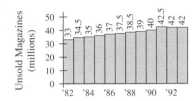

(a) Model the annual surplus with a linear function $s(t)$ using the data from 1982 ($t = 0$) and 1990 ($t = 8$).

(b) Compare the total number of unsold magazines for the given time period with the figure predicted by your model.

(c) Use your model to estimate the total number of unsold magazines in the U.S. from the start of 1980 ($t = -2$) to the start of 1990.

81. *Acceleration*   On the surface of the moon, the downward acceleration due to gravity is 5.3 ft/s², and there is no air resistance. If a ball is thrown upward from the ground at a velocity of 100 ft/s, how high will it get?

82. *Acceleration*   If a ball is dropped from a height of 1,000 ft above the surface of the moon (see the preceding exercise), how long will it take to drop to the ground? How fast will it be falling when it hits the ground?

83. *Acceleration*   A rocket accelerates at a rate of $320 - 320e^{-t/60}$ ft/s² at t seconds after liftoff. How far will it rise in the first minute?

84. *Acceleration*   A race car accelerates from a stop at a rate of $e^{-t/120}$ ft/s² after t seconds. How far will it travel in the second minute?

85. *Cost*   The marginal cost of Better Baby Buggies Inc.'s Turbo model is $50 + 40/(x + 40)$ for the xth buggy made in a week, and the company's fixed costs amount to $50,000 each week. Find the total cost to make x Turbo buggies in one week.

86. *Cost*   The marginal cost of Better Baby Buggies Inc.'s Gran Turismo model is $75 + 500/(x + 10)^2$ for the xth buggy made in a week, and the company's

for the xth buggy made in a week, and the company's fixed costs amount to $20,000 each week. Find the total cost to make x Gran Turismo buggies in one week.

87. *Revenue*   Due to volume discounts, Better Baby Buggies receives $75 + 25e^{-x}$ for the xth Turbo buggy it sells. Find the total revenue the company receives by selling x Turbo buggies.

88. *Revenue*   Due to volume discounts and other effects, Better Baby Buggies receives $100 + 50xe^{-x^2}$ for the xth Gran Turismo buggy it sells. Find the total revenue the company receives by selling x Gran Turismo buggies.

89. *Demand for Solar Cells*   Based on data from 1978 through 1993, the demand for solar energy can be modeled by

$$q(p) = \frac{268.227}{p^{1.1990}},$$

where p is the cost of solar cells for each watt of capacity (in constant 1973 dollars) and q is the annual sales of solar cells in millions of watts (or megawatts) of capacity shipped.[†] At the time of this writing, the *Enron Corporation* is planning to build a 100-megawatt solar power plant in the southern Nevada desert.

(a) Use the demand equation to estimate what the solar cells may cost Enron (to the nearest $1 per watt of capacity).

(b) Assume that the (rounded) cost to Enron is the lowest that is currently (January 1995) technologically feasible, but that this cost will continue to decrease exponentially, halving every 4.342 years.[‡] Obtain the price p per watt as a function of time t.

(c) Use the demand equation together with the result from part (b) to estimate the total sales of solar cells (in megawatts of capacity) from January 1995 through January 1998.

90. *Demand for Solar Cells*   Repeat the preceding exercise using the rougher demand equation

$$q(p) = \frac{300}{p}.$$

▼ *Source: Vos, Gruppo & Capell Inc./*The New York Times*, Dec. 8, 1993, p. D6.

[†] The model is a power regression based on data gleaned from published graphs. Source: PV News/Worldwatch Institute/*The New York Times*, November 15, 1994, p. D2.

[‡] Based on an exponential regression of the above data.

CHAPTER 7

Tapping the Rich May Prove Tricky

By SYLVIA NASAR

As Bill Clinton's new economic team considers how best to turn his campaign promises on taxes into legislative proposals, it may feel the ground shifting.

His campaign tossed around a host of proposals on government financing, from spending cuts to getting more tax revenue from corporations and individuals, all aimed at getting the resources to accomplish goals ranging from middle-class tax relief to public works to halving the deficit. But the consensus among experts was that many of the proposals were unrealistic, and that the proposal most likely to yield significant new money was tax increases on the wealthiest taxpayers.

Getting the really rich—the people who reaped an outsize share of the economic gains of the 1980s—to pay more was a major plank of the Democratic campaign. For Clinton tax purposes, a couple with income of $200,000 and a single taxpayer with $150,000 count as rich.

LIKELY TO FALL SHORT

The problem is that while higher tax rates on high incomes are likely to provide a good deal of new money, they are not likely, barring an unexpectedly buoyant economy, to generate the $92 billion over four years that the Clinton camp has claimed.

Many experts had been skeptical of that claim from the start. Congressional Budget Office estimates put the added revenue at a maximum of $80 billion over four years and estimates from Treasury

officials and the Republican side of the Joint Economic Committee of Congress are even lower.

But even as long-term deficit projections are looking gloomier, so also is the outlook for collecting as much from the rich as had been projected even in the lower estimates.

Perhaps the biggest consideration is one raised in a new study of how very rich taxpayers react to higher taxes. The study published by the National Bureau of Economic Research strongly suggests that extremely rich people—the top slice of the top 1 percent of taxpayers—have considerable flexibility to expose less of their income to taxation. . . .

"If there's a significant change in rates, say from 31 percent to 41 percent, people will change their behavior to take advantage of ways to defer income," Professor Poterba said. "It would set into motion a return to pre-1986 tax shelters."

Another reason for thinking that the rich will yield less revenue than Clinton tax planners had hoped is that the rich may not possess as many riches as they used to. The latest Internal Revenue Service summary of tax returns, for 1989, released this year, shows that many high fliers—real estate empire builders, retailers and newly redundant executives—had their wings clipped during the last four years of economic drift.

The number of taxpayers reporting pretax income of $1 million or more dropped from 62,000 in 1988 to 58,000 in 1989, and their share of total income shrank from 5.5 percent to 4.7 percent. . . .

Source: From Sylvia Nasar, "Tapping the Rich May Prove Tricky," *The New York Times*, December 12, 1992, p. 39.

Further Integration Techniques and Applications of the Integral

APPLICATION ▶ You have just been hired by the incoming administration to coordinate national tax policy, and the so-called experts on your staff can't seem to agree on which of three tax proposals will result in the highest revenue for the government. The data you have are the two income tax proposals (graphs of tax vs. income) and the distribution of incomes in the country. How do you use this information to decide which tax policy will result in the most revenue?

| INTRODUCTION | ▶ We have seen that integration is a powerful tool for the measurement of area and the computation of total sales, revenues, and profits. In this chapter, we shall first look at some further techniques for the computation of integrals and then at some more applications of the integral. We shall then see how to extend the definition of the definite integral to include integrals over infinite intervals, and show how such integrals can be used for long-term forecasting. Finally, we shall introduce the beautiful theory of differential equations and their numerous applications. |

▶ 7.1 INTEGRATION BY PARTS

In this section we shall discuss **integration by parts,** an integration technique that comes from the product rule for derivatives. The tabular method we present here has been around for some time and makes integration by parts quite simple, particularly in problems where it has to be iterated. The particular version we present was developed and taught to us by Professor Dan Rosen.

To start with, we introduce a little notation in order to simplify things while we introduce integration by parts (we shall use this notation only in the next few pages). If f is any function, then we shall denote its derivative by $D(f)$ and an antiderivative by $I(f)$. Thus, for example, if $f(x) = 2x^2$, then

$$D(f) = 4x,$$

while

$$I(f) = \frac{2x^3}{3}.$$

[If we wished, we could instead take $I(f) = 2x^3/3 + 46$.]

INTEGRATION BY PARTS FORMULA

$$\int (f \cdot g)\, dx = f \cdot I(g) - \int [D(f) \cdot I(g)]\, dx$$

In words, this says:

The integral of a product of two functions is the first times the integral of the second minus the **integral of** (the derivative of the first times the integral of the second).

If we don't like having to find the integral of $f \cdot g$, integration by parts allows us to consider instead the (possibly) easier integral of $D(f) \cdot I(g)$.

Q Where does the integration by parts formula come from?

A We start by applying the product rule to the function $f \cdot I(g)$:

$$D(f \cdot I(g)) = D(f) \cdot I(g) + f \cdot D(I(g))$$
$$= D(f) \cdot I(g) + f \cdot g$$

since $D(I(g))$ is the derivative of the antiderivative of g. Integrating both sides gives

$$f \cdot I(g) = \int [D(f) \cdot I(g)] \, dx + \int (f \cdot g) \, dx.$$

If we now subtract $\int[D(f) \cdot I(g)] \, dx$ from both sides, we get the integration by parts formula.

It is convenient to arrange the calculation of the integration by parts formula in tabular form:

D	**I**
$+ f$	g
$- D(f) \longrightarrow$	$I(g)$

The column marked D is the differentiation column, and the column marked I is the integration column. We place f in the D column and g in the I column to indicate that we will be differentiating f and integrating g (as in the formula). The slanted arrow reminds us that we are to multiply the two functions it links, f and $I(g)$, and the color plus sign on the left reminds us that $(+)f \cdot I(g)$ appears in the answer. The horizontal arrow linking $D(f)$ and $I(g)$ reminds us that we are to take the product of $D(f)$ and $I(g)$, *and integrate.* The color minus sign on the left reminds us that it is $-\int D(f) \cdot I(g) \, dx$ that appears in the answer. Combining the contributions of the two lines gives us $f \cdot I(g) - \int D(f) \cdot I(g)$, which is the integration by parts formula.

▼ **EXAMPLE 1**

Calculate $\int xe^x \, dx$.

SOLUTION First notice that none of the techniques of integration that we've talked about up to now will help us. In particular, we cannot simply find antiderivatives of x and e^x and multiply them together: you should check that $\frac{x^2}{2} e^x$ is *not* an antiderivative for xe^x. So, let's try integration by parts. We are required to find the integral of the *product* of x and e^x. Now, here we must make a decision: while using integration by parts, which of the two functions gets differentiated, the x or the e^x? Since the derivative of x is just 1, differ entiating makes it simpler, so let us try putting x in the D column. This means that e^x has to go in the I column. We then get the following table:

D	I
$+$ x	e^x
$-$ 1 $\longrightarrow$	e^x

We can now read the answer as

$$\int xe^x \, dx = xe^x - \int 1 \cdot e^x \, dx = xe^x - e^x + C.$$

Before we go on... Had we decided to put e^x in the D column instead, we would have had the following table:

D	I
$+$ e^x	x
$-$ e^x $\longrightarrow$	$\frac{1}{2}x^2$

From this table we get

$$\int xe^x \, dx = \frac{1}{2}x^2 e^x - \int \frac{1}{2}x^2 e^x \, dx.$$

To evaluate this requires computing a worse integral than the one we started with! How do we know beforehand which way to go? We don't. We have to be willing to do a little trial and error. We try it one way, and if it doesn't make things simpler, we try it another way. *Remember, though, that the function we put in the I column must be one that we can integrate.*

▼   **EXAMPLE 2**

Calculate $\int x^2 e^{-x} \, dx$.

SOLUTION   Again, we have a product—the integrand is the product of x^2 and e^x. Since differentiating x^2 makes it simpler, we put it in the D column and get

D	I
$+$ x^2	e^{-x}
$-$ $2x$ $\longrightarrow$	$-e^{-x}$

(Recall that the integral of e^{-x} is $-e^{-x}$.) Thus,

$$\int x^2 e^{-x}\,dx = -x^2 e^{-x} - \int 2x(-e^{-x})\,dx$$

The last integral is simpler than the one we started with, but still involves a product. It is a good candidate for another integration by parts. The table we would use would start with $2x$ in the D column and $-e^{-x}$ in the I column, but notice that this is exactly what we see in the last row of the table above. Therefore, we *continue the process,* elongating the table:

D	I
$+\ x^2$	e^{-x}
$-\ 2x$	$-e^{-x}$
$+\ 2$	e^{-x}

Notice how the signs on the left alternate, because we really need to compute $-\int 2x(-e^{-x})dx$. Now we would still have to compute an integral (the integral of the product of the functions in the bottom row) to complete the computation. But why stop here? Let us continue the process one more step:

D	I
$+\ x^2$	e^{-x}
$-\ 2x$	$-e^{-x}$
$+\ 2$	e^{-x}
$-\ 0$	$-e^{-x}$

In the bottom line we see that all that is left to integrate is $0 \cdot (-e^{-x}) = 0$. Since an indefinite integral of 0 is 0, we can simply read the answer as

$$\int x^2 e^{-x}\,dx = x^2(-e^{-x}) - 2x(e^{-x}) + 2(-e^{-x}) + C$$
$$= -x^2 e^{-x} - 2xe^{-x} - 2e^{-x} + C$$
$$= -e^{-x}(x^2 + 2x + 2) + C.$$

Before we go on... Since that took several steps, let's check our work.

$$\frac{d}{dx}[-e^{-x}(x^2 + 2x + 2) + C] = e^{-x}(x^2 + 2x + 2) - e^{-x}(2x + 2)$$
$$= x^2 e^{-x} \ \checkmark$$

In the last example we saw a technique that we can summarize as follows.

INTEGRATING A POLYNOMIAL TIMES A FUNCTION

If one of the factors in the integrand is a polynomial, and the other is a function that can be integrated repeatedly, put the polynomial in the D column and keep differentiating until you get zero. Then complete the I column to match, and read the answer.

◀ ─

Notice that we could have used this technique in Example 1 by continuing the table one more step. To show you how easy it is, let us try the following.

▼ **EXAMPLE 3**

Evaluate $\int (3x^3 - x^2)e^{2x}\, dx$.

SOLUTION   We put the polynomial $3x^3 - x^2$ in the D column and proceed as above.

D	I
$+\;3x^3 - x^2$	e^{2x}
$-\;9x^2 - 2x$	$\dfrac{e^{2x}}{2}$
$+\;18x - 2$	$\dfrac{e^{2x}}{4}$
$-\;\;\;18$	$\dfrac{e^{2x}}{8}$
$+\;\;\;0$	$\dfrac{e^{2x}}{16}$

We now read the answer, getting

$$\int (3x^3 - x^2)e^{2x}\, dx = (3x^3 - x^2)\frac{e^{2x}}{2} - (9x^2 - 2x)\frac{e^{2x}}{4} + (18x - 2)\frac{e^{2x}}{8} - 18\frac{e^{2x}}{16} + C$$

$$= e^{2x}\left[\frac{3x^3 - x^2}{2} - \frac{9x^2 - 2x}{4} + \frac{18x - 2}{8} - \frac{18}{16}\right] + C$$

$$= e^{2x}\left[\frac{3x^3}{2} - \frac{11x^2}{4} + \frac{11x}{4} - \frac{11}{8}\right] + C$$

Before we go on... As usual, we can check the answer by taking its derivative. You should do this.

It is not always the case that the integrand is a polynomial times something easy to integrate, so we can't always expect to end up with a zero in the *D* column. In that case we have to hope that at some point we will be able to integrate the product of the functions in the last row. Here is an example.

▼ **EXAMPLE 4**

Find $\int x \ln x \, dx$

SOLUTION   This is a product, so it is a good candidate for integration by parts. Our first impulse is to differentiate *x*, but that would mean integrating ln *x*, and we do not (yet) know how to do that. So we try it the other way around and hope for the best.

D	*I*
+ ln *x*	*x*
− $\dfrac{1}{x}$	$\dfrac{x^2}{2}$

Why did we stop? If we continued the table, both columns would get more complicated. However, if we stop here we get

$$\int x \ln x \, dx = (\ln x)\left(\frac{x^2}{2}\right) - \int \left(\frac{1}{x}\right)\left(\frac{x^2}{2}\right) dx$$

$$= \frac{x^2}{2} \ln x - \frac{1}{2} \int x \, dx$$

$$= \frac{x^2}{2} \ln x - \frac{x^2}{4} + C.$$

▼ **EXAMPLE 5**

Find $\int \ln x \, dx$.

SOLUTION   Now ln *x* as it stands does not look like a product. (It is the natural logarithm of *x*, *not* the "natural logarithm times *x*.") We can, however, *make* it into a product by thinking of it as 1 · ln *x*. Now you are probably very tempted to put the 1 in the *D* column, since its derivative is zero. But this will force you to put the ln *x* in the *I* column, and we don't yet know what the integral of ln x is—this is the problem we were given! So we do what we did in the last example and put ln x in the D column.

D	I
$+$ ln x	1
$-\dfrac{1}{x}$	x

Why did we stop there? The product of $\frac{1}{x}$ and x is just 1, and we certainly know how to integrate that. Thus, we have

$$\int \ln x \; dx = x \ln x - \int \left(\frac{1}{x}\right) x \; dx$$

$$= x \ln x - \int 1 \; dx$$

$$= x \ln x - x + C.$$

Now we know what the integral of ln x is!

The technique of integration by parts can also be used to evaluate definite integrals.

▼ **EXAMPLE 6**

Evaluate $\displaystyle\int_0^1 (x^2 + 1)e^{-2x+1} \; dx$.

SOLUTION According to the Fundamental Theorem of Calculus, we should find an antiderivative and then evaluate at the limits of integration.

D	I
$+\ x^2 + 1$	e^{-2x+1}
$-\ 2x$	$-\dfrac{e^{-2x+1}}{2}$
$+\ 2$	$\dfrac{e^{-2x+1}}{4}$
$-\ 0$	$-\dfrac{e^{-2x+1}}{8}$

Thus

$$\int_0^1 (x^2 + 1)e^{-2x+1}\, dx = -\left[e^{-2x+1}\left(\frac{x^2+1}{2} + \frac{2x}{4} + \frac{2}{8}\right)\right]_0^1$$

$$= -\left[e^{-2x+1}\left(\frac{x^2}{2} + \frac{x}{2} + \frac{3}{4}\right)\right]_0^1$$

$$= -\left[e^{-3}\left(\frac{1}{2} + \frac{1}{2} + \frac{3}{4}\right) - e\left(\frac{3}{4}\right)\right]$$

$$= \frac{3}{4}e - \frac{7}{4}e^{-3}$$

$$\approx 1.95.$$

HINTS AND GENERAL GUIDELINES FOR INTEGRATION BY PARTS

1. To integrate a product in which one factor is a polynomial and the other can be integrated several times, put the polynomial in the D column and the other factor in the I column. Then differentiate the polynomial until you get zero.

2. If one of the factors is a polynomial but the other factor cannot be integrated easily, put the polynomial in the I column and the other factor in the D column. Stop the table as soon as the product of the functions in the bottom row can be integrated.

3. If neither factor is a polynomial, put the factor that seems easiest to integrate in the I column and the other factor in the D column. Again, stop the table as soon as the product of the functions in the bottom row can be integrated.

4. If your method doesn't work, try switching the functions in the D and I columns. If there is still a problem, try some other method (substitution, for example).

▶ **7.1 EXERCISES**

Evaluate the integrals in Exercises 1–24.

1. $\displaystyle\int 2xe^x\, dx$

2. $\displaystyle\int 3xe^{-x}\, dx$

3. $\displaystyle\int (3x - 1)e^{-x}\, dx$

4. $\displaystyle\int (1 - x)e^x\, dx$

5. $\displaystyle\int (x^2 - 1)e^{2x}\, dx$

6. $\displaystyle\int (x^2 + 1)e^{-2x}\, dx$

7. $\displaystyle\int (x^2 + 1)e^{-2x+4}\, dx$

8. $\displaystyle\int (x^2 + 1)e^{3x+1}\, dx$

9. $\displaystyle\int \frac{x^2 - x}{e^x}\, dx$

10. $\displaystyle\int \frac{2x + 1}{e^{3x}}\, dx$

11. $\displaystyle\int x^3 \ln x\, dx$

12. $\displaystyle\int x^2 \ln x\, dx$

13. $\int (t^2 + 1)\ln(2t)\, dt$ **14.** $\int (t^2 - t)\ln(-t)\, dt$ **15.** $\int t^{1/3}\ln t\, dt$ **16.** $\int t^{-1/2}\ln t\, dt$

17. $\int_0^1 (x + 1)e^x\, dx$ **18.** $\int_{-1}^1 (x^2 + x)e^{-x}\, dx$ **19.** $\int_0^1 x^2 (x + 1)^{10}\, dx$ **20.** $\int_0^1 x^3 (x + 1)^{10}\, dx$

21. $\int_1^2 x \ln(2x)\, dx$ **22.** $\int_1^2 x^2 \ln(3x)\, dx$ **23.** $\int_0^1 x \ln(x + 1)\, dx$ **24.** $\int_0^1 x^2 \ln(x + 1)\, dx$

25. Find the area bounded by the curve $y = xe^{-x}$, the x-axis, and the lines $x = 0$ and $x = 10$.

26. Find the area bounded by the curve $y = x \ln x$, the x-axis, and the lines $x = 1$ and $x = e$.

27. Find the area bounded by the curve $y = (x + 1)\ln x$, the x-axis, and the lines $x = 1$ and $x = 2$.

28. Find the area bounded by the curve $y = (x - 1)e^x$, the x-axis, and the lines $x = 0$ and $x = 2$.

APPLICATIONS

29. *Displacement*   A rocket rising from the ground has a velocity of $2{,}000te^{-t/120}$ ft/s after t seconds. How far does it rise in the first two minutes?

30. *Sales*   Weekly sales of graphing calculators can be modeled by the equation

$$s(t) = 10 - te^{-t/20},$$

where s is the number of calculators sold per week after t weeks. How many graphing calculators (to the nearest unit) will be sold in the first 20 weeks?

31. *Total Cost*   The marginal cost of the xth box of light bulbs is $10 + [\ln(x + 1)]/(x + 1)^2$, while the fixed cost is \$5,000. Find the total cost to make x boxes of bulbs.

32. *Total Revenue*   The marginal revenue for selling the xth box of light bulbs is $10 + 0.001x^2e^{-x/100}$. Find the total revenue generated by selling 200 boxes of bulbs.

33. *Revenue*   You have been raising the price of your Jurassic Park T-shirts by 50¢ per week, and sales have been falling continuously at a rate of 2% per week. Assuming you are now selling 50 T-shirts per week and charging \$10 per T-shirt, how much revenue will you generate during the coming year? (Round your answer to the nearest dollar.)

34. *Revenue*   Luckily, sales of your "This is your brain on Calculus . . ." T-shirts are now 50 T-shirts per week and increasing continuously at a rate of 5% per week. You are now charging \$10 per T-shirt and are decreasing the price by 50¢ per week. How much revenue will you generate during the next six weeks?

35. *Revenue*   About 70.5 million magazines were sold in the U.S. in 1982, and this number has been falling by about 2.5 million per year since then.* Assume that the average price of magazines was \$2.00 in 1982 and that this price has been increasing continuously at 5% per year. How much revenue was generated by sales of magazines in the period 1982–1992? (Give your answer to the nearest \$1 million.) [*Hint:* Annual revenue after t years = Annual sales × Price per magazine]

36. *Lost Revenue*   In 1982, about 35 million magazines placed on newsstands in the U.S. were not sold. This number has increased by about 0.6 million per year.* Assume that the average price of magazines was \$2.00 in 1982 and that this price has been increasing continuously at 5% per year. How much revenue was lost through unsold magazines in the period 1982–1992? (Give your answer to the nearest \$1 million.) [*Hint:* Annual revenue after t years = Number of magazines × Price per magazine]

▼   * These figures were for sales of 123 leading magazines at newsstands. Source: *The New York Times*, Dec. 6, 1993, p. D6.

► ═══════ **7.2** INTEGRATION USING TABLES

Calculating antiderivatives can sometimes be a time-consuming and difficult process, but fortunately there are a number of published tables that give formulas for commonly occurring integrals. Such tables are called "standard integral" tables.* A short version of such a table appears in Appendix D, and we shall be using this table in the examples that follow.

▼ **EXAMPLE 1**

Evaluate $\int \dfrac{1}{9 - x^2} dx$.

SOLUTION   The first thing to notice is that all of the integrals in the table use the variable u instead of x. Since the variable is only a "dummy" variable, this does not matter. You can freely replace x with u or vice versa.

Now we would like to see a heading in the table that says "Integrals Containing $a - u^2$." The closest thing is one saying "Integrals Containing $a^2 \pm u^2$ $(a > 0)$," and under this heading we see

$$\int \frac{du}{a^2 - u^2} = \frac{1}{2a} \ln \left| \frac{a + u}{a - u} \right| + C.$$

For $a^2 - u^2$ to match $9 - x^2$, we must take $a = 3$ (as the heading says, we must take $a > 0$). Substituting x for u and 3 for a, we obtain

$$\int \frac{1}{9 - x^2} dx = \frac{1}{3} \ln \left| \frac{3 + x}{3 - x} \right| + C.$$

Before we go on... How was this integral calculated in the first place? Someone noticed that

$$\frac{1}{a^2 - u^2} = \frac{1}{2a} \left(\frac{1}{a + u} + \frac{1}{a - u} \right),$$

which can be integrated fairly easily. This observation has been generalized into a technique called "partial fractions" which we shall not study in this book.[†]

Often you have to manipulate an integral to make it look like one in the table.

─────────

▼ * One commonly used published source is *CRC Standard Mathematical Tables and Formulae* (Cleveland: CRC Press, Inc., 1991).
[†] By the way, why would the substitution $u = 9 - x^2$ *not* have worked in this example?

▼ **EXAMPLE 2**

Evaluate $\displaystyle\int \frac{1}{\sqrt{4-9x^2}}\, dx$.

SOLUTION The closest thing we can find in the table comes under the heading "Integrals Containing $\sqrt{a^2-u^2}$ $(a>0)$."

$$\int \frac{du}{\sqrt{a^2-u^2}} = \frac{1}{a} \sin^{-1}\frac{u}{a} + C$$

(The function $\sin^{-1}(x)$ is called the **inverse sine function.** Although we have not talked about it, it can be easily calculated on any scientific or graphing calculator. Just be sure your calculator is set to "radians" mode rather than degrees.) Now we can set $a = 2$, but we want $u^2 = 9x^2$, which suggests the following substitution.

$$u = 3x$$
$$\frac{du}{dx} = 3$$
$$dx = \frac{1}{3}\, du$$

This gives

$$\int \frac{1}{\sqrt{4-9x^2}}\, dx = \frac{1}{3}\int \frac{du}{\sqrt{4-u^2}}$$
$$= \frac{1}{3}\cdot\frac{1}{2}\sin^{-1}\frac{u}{2} + C$$
$$= \frac{1}{6}\sin^{-1}\frac{3x}{2} + C.$$

Before we go on... Now you see why the table uses u instead of x. We often need to make a substitution before using the table.

▼ **EXAMPLE 3**

Evaluate $\displaystyle\int_{-3}^{0} \frac{dx}{x^2 + 6x + 12}$.

SOLUTION The only entries in the table with quadratics in the denominator have either $a^2 + u^2$ or $a^2 - u^2$, not a general quadratic. However, recall the technique of "completing the square."

$$x^2 + 6x + 12 = (x^2 + 6x) + 12$$
$$= (x^2 + 6x + 9) + 12 - 9 \qquad \text{Add and subtract the}$$
$$\text{square of half of 6.}$$
$$= (x + 3)^2 + 3$$

So we can rewrite the integral as follows:

$$\int_{-3}^{0} \frac{dx}{x^2 + 6x + 12} = \int_{-3}^{0} \frac{dx}{(x + 3)^2 + 3}.$$

This is beginning to look like $\int \frac{du}{a^2 + u^2}$. We want $a^2 = 3$, so $a = \sqrt{3}$, and $u^2 = (x + 3)^2$, so we substitute:

$$u = x + 3$$
$$\frac{du}{dx} = 1$$
$$dx = du$$
When $x = -3$, $u = 0$.
When $x = 0$, $u = 3$.

So

$$\int_{-3}^{0} \frac{dx}{(x + 3)^2 + 3} = \int_{0}^{3} \frac{du}{3 + u^2}$$

$$= \frac{1}{\sqrt{3}} \tan^{-1} \frac{u}{\sqrt{3}} \Big|_{0}^{3} \qquad \text{according to the table}$$

$$= \frac{1}{\sqrt{3}} (\tan^{-1}\sqrt{3} - \tan^{-1}0)$$

$$\approx 0.6046.$$

HINTS AND GENERAL GUIDELINES FOR USING INTEGRATION TABLES

1. Try a substitution to make your integral look like one in the table.
2. If you have a quadratic expression, complete the square.
3. Use a bigger table if one is available.

In real life, tables of integrals are being replaced by computer software that can do symbolic integration, and computer programs that essentially encode large tables of integrals are in common use today. This is the tool that many people who actually work with integrals would reach for first.

▶ **7.2 EXERCISES**

Evaluate the following integrals using the table on the inside cover of this textbook.

1. $\displaystyle\int \frac{x}{1 + x}\, dx$ **2.** $\displaystyle\int \frac{dx}{x(1 + x)}$ **3.** $\displaystyle\int x\sqrt{1 + 2x}\, dx$ **4.** $\displaystyle\int x^2\sqrt{2 + x}\, dx$

5. $\displaystyle\int \sqrt{x^2 + 4}\, dx$ **6.** $\displaystyle\int \sqrt{x^2 - 9}\, dx$ **7.** $\displaystyle\int \frac{dx}{1 + x^2}\, dx$ **8.** $\displaystyle\int \frac{dx}{\sqrt{1 - x^2}}$

9. $\displaystyle\int \frac{dx}{\sqrt{3 + x^2}}$ **10.** $\displaystyle\int \frac{dx}{5 - x^2}$ **11.** $\displaystyle\int \frac{dx}{1 - 4x^2}$ **12.** $\displaystyle\int \frac{dx}{1 + 4x^2}$

13. $\displaystyle\int 2x\sqrt{2x^2 - 1}\, dx$ **14.** $\displaystyle\int (9x^2 + 2)^{3/2}\, dx$ **15.** $\displaystyle\int \frac{dx}{3x^2 - 1}$ **16.** $\displaystyle\int \frac{dx}{1 - 3x^2}$

17. $\displaystyle\int \frac{dx}{x^2 + 6x + 10}$ **18.** $\displaystyle\int \frac{dx}{x^2 + 2x - 3}$ **19.** $\displaystyle\int \sqrt{x^2 - x + 3}\, dx$ **20.** $\displaystyle\int \sqrt{x^2 + x + 1}\, dx$

21. $\displaystyle\int \frac{dx}{3x^2(2 + 4x)}$ **22.** $\displaystyle\int \frac{3x}{(2 + 4x^2)}\, dx$ **23.** $\displaystyle\int \frac{e^x}{9 - e^{2x}}\, dx$ **24.** $\displaystyle\int \frac{e^x}{e^{2x} + 4}\, dx$

25. $\displaystyle\int \frac{1}{x(1 + (\ln x)^2)}\, dx$ **26.** $\displaystyle\int \frac{\ln x}{x(1 + \ln x)}\, dx$

APPLICATIONS

27. *Cost*   If a company's fixed cost to manufacture light bulbs is $10,000 and its marginal cost is $105x^2\sqrt{1 + x}$ for the xth crate of bulbs, find the total cost for x crates.

28. *Cost*   If a company's fixed cost to manufacture computer disks is $15,000 and its marginal cost is $15x\sqrt{2 + x}$ for the xth box of disks, find the total cost for x boxes.

29. *Cost*   If a company's fixed cost to manufacture rolls of film is $10,000 and its marginal cost is $30/\sqrt{x^2 + 9}$ for the xth roll, find the total cost for x rolls.

30. *Cost*   If a company's fixed cost to manufacture video tapes is $15,000 and its marginal cost is $50e^x/\sqrt{9 + e^{2x}}$ for the xth box of tapes, find the total cost for x boxes.

31. *Volume Discounts*   Your software company offers a volume discount on its latest program: you charge $5,000/\sqrt{x^2 + 1}$ for the xth copy purchased by a customer. How much do you charge for a total of x copies?

32. *Volume Discounts*   Your software company plans to offer the following volume discount on its next release: you will charge $50,000/\sqrt{x^2 + 50}$ for the xth copy bought by a customer. How much will you charge for x copies?

33. *Average Cost*   If a company's fixed cost to manufacture light bulbs is $10,000 and its marginal cost is $10/\sqrt{4 + x^2}$ for the xth box of bulbs, find its *average* cost function (per box).

34. *Average Cost*   If a company's fixed cost to manufacture computer disks is $9,000 and its marginal cost is $10/(5 + 3x)$ for the xth box of disks, find its *average* cost function (per box).

35. *Varying Acceleration*   A particle in a nuclear accelerator undergoes an acceleration of

$$a(t) = \frac{1,000}{(1 + t^2)^{3/2}}\ \text{ft/s}^2.$$

Assuming the particle started from rest, how far has it traveled in 120 seconds?

36. *Varying Acceleration*   The Starship Galactica undergoes an acceleration given by

$$a(t) = \frac{t}{\sqrt{1 + t}}\ \text{ft/s}^2.$$

Assuming the ship started from rest, how far has it traveled in 60 seconds?

37. Locate at least one integral on the table that can be calculated using a substitution. Calculate it using the substitution, and check that your answer agrees with that in the table.

38. Enlarge the table of integrals by adding at least one new integral to the category "Integrals Containing $\sqrt{a^2 - u^2}$."

39. Your friend claims that there can be no such thing as a complete table of integrals. Discuss the merits of this claim.

40. Another friend claims that there should be a single algebraic formula for all integrals of the form $\int u^n \sqrt{u^2 \pm a^2}\, du$ $(n = 0, 1, 2, \ldots)$. What evidence is there to refute this claim?

▶ ═══════ **7.3** AREA BETWEEN TWO CURVES AND APPLICATIONS

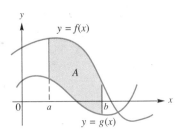

FIGURE 1

With the definite integral, we can calculate the area between the graph of a function and the x-axis. With a little more work, we can calculate the area between two curves. Figure 1 shows the graphs of two functions, $f(x)$ and $g(x)$, with $f(x) \geq g(x)$ for every x in the interval $[a, b]$.

Q How can we find the area between the graphs of two functions?

A Pretty easily, actually, by using the following formula.

AREA BETWEEN TWO GRAPHS

If $f(x) \geq g(x)$ for all x in $[a, b]$, then the area A of the region between the graphs of f and g and between $x = a$ and $x = b$ is

$$A = \int_a^b [f(x) - g(x)]\, dx.$$

This formula is best justified by looking at some examples.

▼ **EXAMPLE 1**

Find the area of the region between the graphs of $f(x) = x$ and $g(x) = x^2$, for x in $[0, 1]$.

SOLUTION   The area in question is shown in Figure 2.

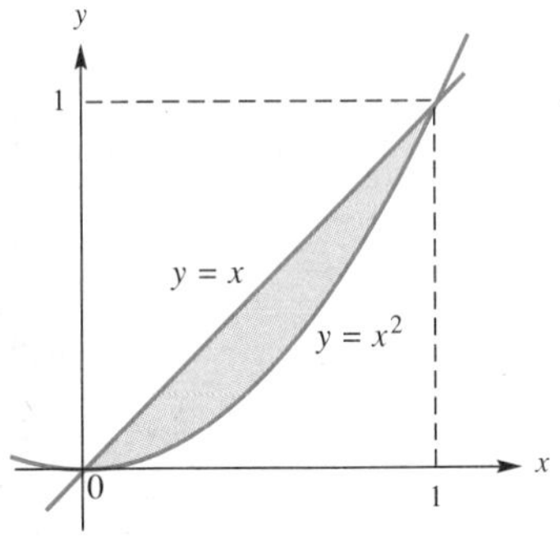

From the picture it is easy to see that $f(x) \geq g(x)$ for x in $[0, 1]$, so the area is

$$\int_0^1 [x - x^2]\, dx = \left[\frac{x^2}{2} - \frac{x^3}{3}\right]_0^1 = \left(\frac{1}{2} - \frac{1}{3}\right) - (0 - 0) = \frac{1}{6}.$$

FIGURE 2

Before we go on... Why is this the correct calculation? Let us go back to what we already know: the area of the region between the graph of f and the x-axis is given by $\int_a^b f(x)\,dx$ when f is positive. Shown in Figure 3 are two such regions: under $y = x$ and under $y = x^2$.

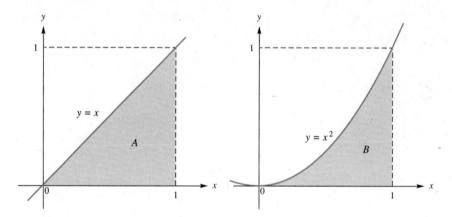

FIGURE 3

The area A, under $y = x$, is

$$\int_0^1 x\,dx = \left[\frac{x^2}{2}\right]_0^1 = \frac{1}{2}$$

while the area B, under $y = x^2$, is

$$\int_0^1 x^2\,dx = \left[\frac{x^3}{3}\right]_0^1 = \frac{1}{3}.$$

Now, the area we are asked to find is the *difference* between these two areas, since it is the area of the part of the first region that is not also in the second region. This means that the area that we want is $A - B = \frac{1}{2} - \frac{1}{3} = \frac{1}{6}$. We can write this using integrals as

$$A - B = \int_0^1 x\,dx - \int_0^1 x^2\,dx = \int_0^1 [x - x^2]\,dx = \int_0^1 [f(x) - g(x)]\,dx,$$

which is the formula we used in the first place.

▼   **EXAMPLE 2**

Find the area of the region between the graphs of $f(x) = 1 - x^2$ and $g(x) = x^2 - 1$, for x in $[-1, 1]$.

SOLUTION   The region is shown in Figure 4.

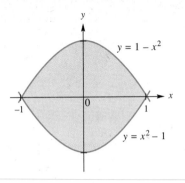

FIGURE 4

Since $1 - x^2 \geq x^2 - 1$ for x in $[-1, 1]$, the area is

$$\int_{-1}^{1} [(1 - x^2) - (x^2 - 1)] \, dx = \int_{-1}^{1} (2 - 2x^2) \, dx$$

$$= 2 \left[x - \frac{x^3}{3} \right]_{-1}^{1}$$

$$= 2 \left[\left(1 - \frac{1}{3} \right) - \left(-1 + \frac{1}{3} \right) \right] = \frac{8}{3}.$$

Before we go on... To see why the formula works in this case, look at Figure 5, which shows the area broken into two pieces.

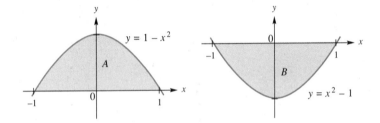

FIGURE 5

We know how to calculate the areas of these two regions separately. The area A is

$$\int_{-1}^{1} (1 - x^2) \, dx = \left[x - \frac{x^3}{3} \right]_{-1}^{1} = \frac{4}{3}.$$

To calculate the area B, we compute

$$\int_{-1}^{1} (x^2 - 1) \, dx = \left[\frac{x^3}{3} - x \right]_{-1}^{1} = -\frac{4}{3}.$$

Since this is negative, the area B is actually

$$B = -\int_{-1}^{1} (x^2 - 1) \, dx = -\left(-\frac{4}{3} \right) = \frac{4}{3}.$$

Thus, the total area that the problem asks us to find is $A + B = \frac{8}{3}$. In terms of integrals, this is

$$A + B = \int_{-1}^{1} (1 - x^2) \, dx - \int_{-1}^{1} (x^2 - 1) \, dx = \int_{-1}^{1} [(1 - x^2) - (x^2 - 1)] \, dx$$

$$= \int_{-1}^{1} [f(x) - g(x)] \, dx.$$

To convince ourselves that the formula will always work, we should look at one more case, the one in which both graphs are below the x-axis. We shall leave it to you to make up such an example. The next example looks at another question: What if the two graphs cross?

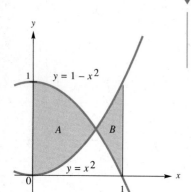

FIGURE 6

▼ **EXAMPLE 3**

Find the area of the region between $y = x^2$ and $y = 1 - x^2$ for x in $[0, 1]$.

SOLUTION   The area is shown in Figure 6. From the figure, we can see that neither graph lies above the other throughout the whole interval. To remedy this, we break the area into the two pieces on either side of the point at which the graphs cross, then compute the size of each area separately. In order to do this, we need to know exactly where that crossing point is. The crossing point is where $x^2 = 1 - x^2$, so we solve for x.

$$x^2 = 1 - x^2$$
$$2x^2 = 1$$
$$x^2 = \frac{1}{2}$$

$$x = \pm\frac{1}{\sqrt{2}} = \pm0.707\ldots$$

Since we are only interested in the interval $[0, 1]$, the crossing point we're interested in is at $x = 1/\sqrt{2} \approx 0.707$.

Now, to compute the areas A and B we need to know one more thing: which graph is on top in each of these areas? We can see that from the figure, but what if the functions were more complicated and we could not easily draw the graphs? We need not worry. If we make the wrong choice for the top function, then the integral will simply come out to be the negative of the area (why?) so we can simply take the absolute value of the integral to get the area of the region in question. We have

$$A = \int_0^{1/\sqrt{2}} [(1 - x^2) - x^2]\, dx = \int_0^{1/\sqrt{2}} (1 - 2x^2)\, dx$$

$$= \left[x - \frac{2x^3}{3} \right]_0^{1/\sqrt{2}}$$

$$= \left(\frac{1}{\sqrt{2}} - \frac{1}{3\sqrt{2}} \right) - (0 - 0) = \frac{\sqrt{2}}{3},$$

and

$$B = \int_{1/\sqrt{2}}^{1} [x^2 - (1 - x^2)]\, dx = \int_{1/\sqrt{2}}^{1} (2x^2 - 1)\, dx$$

$$= \left[\frac{2x^2}{3} - x \right]_{1/\sqrt{2}}^{1}$$

$$= \left(\frac{2}{3} - 1 \right) - \left(\frac{1}{3\sqrt{2}} - \frac{1}{\sqrt{2}} \right)$$

$$= \frac{\sqrt{2} - 1}{3}.$$

This gives a total area of $A + B = \dfrac{2\sqrt{2} - 1}{3} \approx 0.6095$.

Before we go on...

Q What would have happened if we had not calculated the two areas separately, but tried to do the whole calculation with one integral?

A We would have obtained the wrong answer:

$$\int_0^1 [(1 - x^2) - x^2]\, dx = \int_0^1 [1 - 2x^2]\, dx$$

$$= \left[x - \frac{2x^3}{3}\right]_0^1$$

$$= \frac{1}{3},$$

which is not even close to the right answer. What this integral calculated was actually $A - B$, rather than $A + B$. Why?

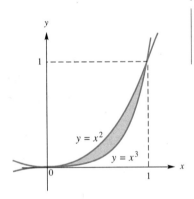

FIGURE 7

▼   **EXAMPLE 4**

Find the area of the region enclosed by $y = x^2$ and $y = x^3$.

SOLUTION   This example has a new wrinkle: we are not told what interval to use for x. However, if we look at the graph in Figure 7, we see that the question can really have only one meaning.
    We are being asked to find the area of the shaded sliver, which is the only part of the picture that is actually *enclosed* by the two graphs. This sliver is bounded on either side by the two points where the graphs cross, so our first task is to find those points. They are the points where $x^2 = x^3$, so we solve

$$x^2 = x^3$$
$$x^3 - x^2 = 0$$
$$x^2(x - 1) = 0$$
$$x = 0 \quad \text{or} \quad x = 1.$$

Thus, we must integrate over the interval $[0, 1]$. Although we see from the diagram that the graph of $y = x^2$ is above that of $y = x^3$, let us pretend that we didn't notice that, and calculate

$$\int_0^1 (x^3 - x^2)\, dx = \left[\frac{x^4}{4} - \frac{x^3}{3}\right]_0^1$$

$$= \left(\frac{1}{4} - \frac{1}{3}\right) - (0 - 0) = -\frac{1}{12}.$$

This tells us that the required area is $\frac{1}{12}$ square units, and also that we had our integral reversed. Had we calculated $\int_0^1 (x^2 - x^3)\, dx$ instead, we would have obtained the correct answer, $\frac{1}{12}$.

> ## FINDING THE AREA BETWEEN THE GRAPHS OF $f(x)$ AND $g(x)$
>
> 1. Find all points of intersection by solving $f(x) = g(x)$ for x. This either determines the interval over which you will integrate, or breaks up a given interval into regions between the intersection points.
> 2. Find the area of each region between intersection points by integrating the difference of the larger and the smaller function. (If you accidentally take the smaller minus the larger, the integral will calculate the negative of the area, so just take the absolute value.)
> 3. Add together the areas you found in Step 2 to get the total area.

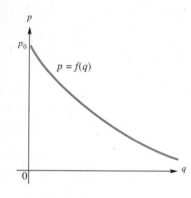

FIGURE 8

We now turn to some interesting applications to business and finance.

CONSUMERS' SURPLUS

Consider a general demand curve presented, as is traditional in economics, as $p = f(q)$, where p is unit price and q is demand measured, say, in annual sales (Figure 8).

We can interpret f as follows:

> $f(q)$ is the price at which the demand will be q units per year.

The price p_0 shown on the graph is the highest price that customers are willing to pay.

Now suppose that a manufacturer sets the unit price at $\bar{p} < p_0$. Then the corresponding demand in units per year is $\bar{q}$, where $f(\bar{q}) = \bar{p}$ (Figure 9). There are some consumers who are willing to pay a higher price than $\bar{p}$, and these consumers save money by paying the lower price. This raises the following question.

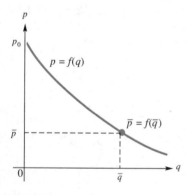

FIGURE 9

Q If the price is set at $\bar{p}$, how much will be saved per year by consumers who were willing to pay a higher price?

A We can calculate this from the graph of the demand function as follows. First partition the interval $[0, \bar{q}]$ into n subintervals of equal length, as we did when we calculated Riemann sums. Figure 10 shows a typical subinterval.

The interval $[q_{k-1}, q_k]$ represents units q_{k-1} through q_k, and the price consumers would have been willing to pay for these units is approximately $f(q_k)$. But they only paid $\bar{p}$ for each of these units, so they saved a total of

$$\text{Savings in price} \times \text{Number of units} = (f(q_k) - \bar{p})\Delta q.$$

This is the area of the shaded rectangle in Figure 10. Now, these savings apply only to the units between q_{k-1} and q_k. To obtain the total savings, we

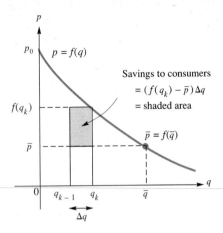

FIGURE 10

must add together the portions corresponding to the entire interval* $[0, \bar{q}]$, and so

$$\text{Total savings} \approx (f(q_1) - \bar{p})\Delta q + (f(q_2) - \bar{p})\Delta q$$
$$+ \ldots + (f(q_n) - \bar{p})\Delta q.$$

But this is the right-hand Riemann sum for the function $f(q) - \bar{p}$ and so, passing to the limit, we obtain the exact savings as

$$\text{Total savings} = \int_0^{\bar{q}} (f(q) - \bar{p})dq.$$

This total savings is called the **consumers' surplus** and is represented by the area enclosed by the graphs of $p = f(q)$ and $p = \bar{p}$ in Figure 11.

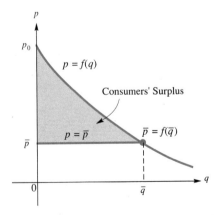

FIGURE 11

▼ * We stop at $\bar{q}$ because if the price is set at $\bar{p}$, only $\bar{q}$ units are sold, and we are only interested in the savings on units actually sold.

To summarize:

CONSUMERS' SURPLUS

The **consumers' surplus** is the total amount saved by consumers who were willing to pay more than the selling price of $\bar{p}$ per unit, and is given by

$$CS = \int_0^{\bar{q}} (f(q) - \bar{p})\, dq,$$

where $f(\bar{q}) = \bar{p}$. Geometrically, it is the area of the region enclosed by the graphs of $p = f(q)$ and $p = \bar{p}$ for $0 \le q \le \bar{p}$.

▼ **EXAMPLE 5**   Consumers' Surplus

Your used CD store has an exponential demand equation of the form

$$p = 15e^{-0.01q},$$

where q represents daily sales of CDs and p is the price you charge per used CD. Calculate the daily consumers' surplus if you sell your used CDs at \$5 each.

SOLUTION   We are given $f(q) = 15e^{-0.01q}$ and $\bar{p} = 5$. We still need $\bar{q}$. By definition,

$$f(\bar{q}) = \bar{p};$$

that is,

$$15e^{-0.01\bar{q}} = 5,$$

so we must solve for $\bar{q}$:

$$e^{-0.01\bar{q}} = \frac{1}{3},$$

so

$$-0.01\bar{q} = \ln\left(\frac{1}{3}\right) = -\ln 3.$$

Thus,

$$\bar{q} = \frac{\ln 3}{0.01} \approx 109.8612.$$

We now have

$$CS = \int_0^{\bar{q}} (f(q) - \bar{p}) \, dq$$

$$= \int_0^{109.8612} (15e^{-0.01q} - 5) \, dq$$

$$= \left[\frac{15}{-0.01} e^{-0.01q} - 5q \right]_0^{109.8612}$$

$$= (-500 - 549.306) - (-1,500 - 0)$$

$$= \$450.69 \text{ per day.}$$

PRODUCERS' SURPLUS

We can also consider extra income earned by producers. Consider a supply equation of the form $p = f(q)$, where now $f(q)$ is the price at which a supplier is willing to supply q items (per time period). Because a producer is generally willing to supply more units at a larger price per unit, a supply curve usually has a positive slope, as shown in Figure 12.

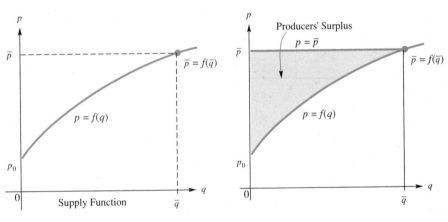

FIGURE 12   **FIGURE 13**

The price p_0 is the lowest price that a producer is willing to charge. Now suppose an item is priced at $\bar{p}$. Then a producer who is prepared to charge less than that will earn "extra" money as a result of the higher selling price $\bar{p}$. The total extra earnings by such producers is called the **producers' surplus.** Arguing as before, we can say that this is given by the area shown in Figure 13.

PRODUCERS' SURPLUS

The **producers' surplus** is the extra amount earned by producers who were willing to charge less than the selling price of $\bar{p}$ per unit, and is given by

$$PS = \int_0^{\bar{q}} (\bar{p} - f(q))\, dq,$$

where $f(\bar{q}) = \bar{p}$. Geometrically, it is the area of the region enclosed by the graphs of $p = f(q)$ and $p = \bar{p}$ for $0 \le q \le \bar{q}$.

▼   **EXAMPLE 6**   Producers' Surplus

My dorm-room tie-dye T-shirt enterprise has grown to the extent that I am now able to produce them in bulk, and several sororities have begun placing orders. I have informed one sorority that I am prepared to supply $20\sqrt{p - 4}$ T-shirts at a price of p dollars per shirt. What is the total surplus to my enterprise if I sell them to the sorority at \$8 each?

SOLUTION   We need to calculate the producers' surplus when $\bar{p} = 8$. The supply equation is

$$q = 20\sqrt{p - 4},$$

but in order to use the formula for producers' surplus we need to express p as a function of q. Therefore, we solve the supply equation for q. First, square both sides to remove the radical sign:

$$q^2 = 400(p - 4)$$

so

$$p - 4 = \frac{q^2}{400},$$

giving

$$p = f(q) = \frac{q^2}{400} + 4.$$

We now need the value of $\bar{q}$ corresponding to $\bar{p} = 8$. The easiest way to find it is to substitute $p = 8$ in the original equation, which gives

$$\bar{q} = 20\sqrt{8 - 4} = 20\sqrt{4} = 40.$$

Thus,

$$PS = \int_0^{\bar{q}} (\bar{p} - f(q)) \, dq$$

$$= \int_0^{40} \left(8 - \left(\frac{q^2}{400} + 4 \right) \right) dq$$

$$= \int_0^{40} \left(4 - \frac{q^2}{400} \right) dq$$

$$= \left[4q - \frac{q^3}{1,200} \right]_0^{40} = \$142.22.$$

Before we go on...

Q How do we interpret this?

A First notice that, since $\bar{q} = 40$, I will be selling the sorority 40 T-shirts at \$8 per shirt. On the other hand, had the sorority been more savvy, it could have said the following: "For the nth T-shirt that you make, charge the lowest price at which you are willing to supply n T-shirts." That request would not violate my supply policy, but would result in a lower price for the total of 40 T-shirts (since the supply curve dictates that smaller amounts correspond to lower prices per unit). The \$142.22 represents the extra amount I will get by charging \$8 for all 40 T-shirts.

▼ **EXAMPLE 7**   Equilibrium Price

Referring to Example 6, a representative informs me that the sorority is prepared to order only $\sqrt{200(16 - p)}$ T-shirts at p dollars each. At the same time, I wish to maintain my policy of supplying $20\sqrt{p - 4}$ T-shirts at p dollars each. I would like to produce as many T-shirts for them as possible in order to avoid being left with unsold T-shirts. What price should I charge per T-shirt, and what then are the consumers' and producers' surpluses?

SOLUTION   The price that guarantees neither a shortage nor a surplus of T-shirts is the equilibrium price, the price where supply equals demand. We have

$$\text{Supply:} \quad q = 20\sqrt{p - 4}$$
$$\text{Demand:} \quad q = \sqrt{200(16 - p)}.$$

Equating these gives

$$20\sqrt{p - 4} = \sqrt{200(16 - p)}$$

so

$$400(p - 4) = 200(16 - p),$$

giving

$$400p - 1{,}600 = 3{,}200 - 200p$$
$$200p = 1{,}600,$$

or

$$p = \$8 \text{ per T-shirt.}$$

We therefore take $\bar{p} = 8$. The corresponding value for q is obtained by substituting $p = 8$ into either the demand or supply equation.

$$\bar{q} = 20\sqrt{8 - 4} = 40$$

Thus, $\bar{p} = 8$ and $\bar{q} = 40$. We must now calculate the consumers' surplus and the producers' surplus. We have already calculated the producers' surplus for $\bar{p} = 8$ in Example 6, getting

$$PS = \$142.22.$$

For the consumers' surplus, we must first express p as a function of q for the demand equation. Thus, we solve the demand equation for p as we did for the supply equation in Example 6, and we obtain

$$\text{Demand:}\quad f(q) = 16 - \frac{q^2}{200}.$$

Therefore,

$$CS = \int_0^{\bar{q}} (f(q) - \bar{p})\, dq$$

$$= \int_0^{40} \left[\left(16 - \frac{q^2}{200} \right) - 8 \right] dq$$

$$= \int_0^{40} \left(8 - \frac{q^2}{200} \right) dq$$

$$= \left[8q - \frac{q^3}{600} \right]_0^{40} = \$213.33.$$

Before we go on... Figure 14 shows both the consumers' and producers' surpluses on the same graph.

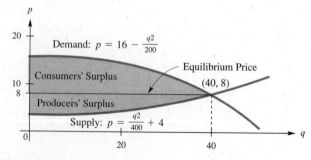

FIGURE 14

Q How do we interpret this result?

A We saw that I earned an extra $142.22 (producers' surplus) due to the fact that the sorority was not as savvy as I had expected. On the other hand, *I* could also have been more savvy and said: "I will charge you as much for the *n*th T-shirt as you are prepared to pay per shirt for *n* T-shirts," and I would have earned $213.33 more (the consumers' surplus). As it turns out, neither I nor the sorority were savvy enough, and it seems that I lost more as a result than they did!

▶ **7.3 EXERCISES**

Find the areas of the indicated regions in Exercises 1–24 (a graphing calculator may be useful for Exercises 21–24).

1. Between $y = x^2$ and $y = -1$ for x in $[-1, 1]$.

2. Between $y = x^3$ and $y = -1$ for x in $[-1, 1]$.

3. Between $y = -x$ and $y = x$ for x in $[0, 2]$.

4. Between $y = -x$ and $y = x/2$ for x in $[0, 2]$.

5. Between $y = x$ and $y = x^2$ for x in $[-1, 1]$.

6. Between $y = x$ and $y = x^3$ for x in $[-1, 1]$.

7. Between $y = e^x$ and $y = x$ for x in $[0, 1]$.

8. Between $y = e^{-x}$ and $y = -x$ for x in $[0, 1]$.

9. Between $y = (x - 1)^2$ and $y = -(x - 1)^2$ for x in $[0, 1]$.

10. Between $y = x^2(x^3 + 1)^{10}$ and $y = -x(x^2 + 1)^{10}$ for x in $[0, 1]$.

11. Enclosed by $y = x$ and $y = x^4$.

12. Enclosed by $y = x$ and $y = -x^4$.

13. Enclosed by $y = x^3$ and $y = x^4$.

14. Enclosed by $y = x$ and $y = x^3$.

15. Enclosed by $y = x^2$ and $y = x^4$.

16. Enclosed by $y = x^4 - x^2$ and $y = x^2 - x^4$.

17. Enclosed by $y = e^x$, $y = 2$, and the y-axis.

18. Enclosed by $y = e^{-x}$, $y = 3$, and the y-axis.

19. Enclosed by $y = \ln x$, $y = 2 - \ln x$, and $x = 4$.

20. Enclosed by $y = \ln x$, $y = 1 - \ln x$, and $x = 4$.

21. Enclosed by $y = e^x$ and $y = 2x + 1$.

22. Enclosed by $y = e^x$ and $y = x + 2$.

23. Enclosed by $y = \ln x$ and $y = \frac{x}{2} - \frac{1}{2}$.

24. Enclosed by $y = \ln x$ and $y = x - 2$.

Calculate the consumers' surplus for each of the demand equations in Exercises 25–36 at the indicated unit price $\bar{p}$.

25. $p = 10 - 2q, \quad \bar{p} = 5$

26. $p = 100 - q, \quad \bar{p} = 20$

27. $p = 100 - 3\sqrt{q}, \quad \bar{p} = 76$

28. $p = 10 - 2q^{1/3}, \quad \bar{p} = 6$

29. $p = 500e^{-2q}, \quad \bar{p} = 100$

30. $p = 100 - e^{0.1q}, \quad \bar{p} = 50$

31. $q = 100 - 2p, \quad \bar{p} = 20$

32. $q = 50 - 3p, \quad \bar{p} = 10$

33. $q = 100 - 0.25p^2, \quad \bar{p} = 10$

34. $q = 20 - 0.05p^2, \quad \bar{p} = 5$

35. $q = 500e^{-0.5p} - 50, \quad \bar{p} = 1$

36. $q = 100 - e^{0.1p}, \quad \bar{p} = 20$

Calculate the producers' surplus for each of the supply equations in Exercises 37–48 at the indicated unit price $\bar{p}$.

37. $p = 10 + 2q, \quad \bar{p} = 20$

38. $p = 100 + q, \quad \bar{p} = 200$

39. $p = 10 + 2q^{1/3}, \quad \bar{p} = 12$

40. $p = 100 + 3\sqrt{q}, \quad \bar{p} = 124$

41. $p = 500e^{0.5q}, \quad \bar{p} = 1,000$

42. $p = 100 + e^{0.01q}, \quad \bar{p} = 120$

43. $q = 2p - 50, \quad \bar{p} = 40$

44. $q = 4p - 1,000, \quad \bar{p} = 1,000$

45. $q = 0.25p^2 - 10, \quad \bar{p} = 10$

46. $q = 0.05p^2 - 20, \quad \bar{p} = 50$

47. $q = 500e^{0.05p} - 50, \quad \bar{p} = 10$

48. $q = 10(e^{0.1p} - 1), \quad \bar{p} = 5$

APPLICATIONS

49. *College Tuition*   A study of about 1800 colleges and universities in the U.S. resulted in the demand equation $q = 9859.39 - 2.17p$, where q is the enrollment at a college or university and p is the average annual tuition (plus fees) it charges.* Officials at Enormous State University have developed a policy whereby the number of students it will accept per year at a tuition level of p dollars is given by $q = 100 + 0.5p$. Find the equilibrium tuition price $\bar{p}$ and the consumers' and producers' surplus at this tuition level. How much less would the university be earning per year if it decided to charge the nth student it admits an annual tuition of $2(n - 100)$? (Express all answers to the nearest dollar.)

50. *Fast Food*   A fast-food outlet finds that the demand equation for its new side-dish, "Sweetdough Tidbit" is given by

$$p = \frac{128}{(q + 1)^2},$$

where p is the price (in cents) per serving, and q is the number of servings that can be sold per hour at this price. At the same time, the franchise is prepared to sell $q = 0.5p - 1$ servings per hour at a price of p cents. Find the equilibrium price $\bar{p}$ and the consumers' and producers' surplus at this price level. How much less would the franchise be earning per hour if it sold the nth serving per hour of Sweetdough Tidbits for $2(n + 1)$ cents?

51. *Sales*   The following graph shows the demand and supply curves for your "$E = mc^2$" T-shirts. Assuming that you sell your T-shirts at the equilibrium price, who has the larger surplus—you or the consumer?

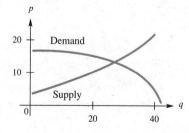

▼ * Based on a study by A.L. Ostrosky, Jr., and J. V. Koch, as cited in their book, *Introduction to Mathematical Economics* (Prospect Heights, IL: Waveland Press, 1979), p. 133.

52. *Sales* Repeat the preceding exercise using the following curves:

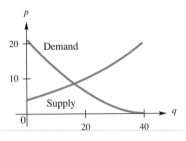

53. *Linear Demand* Given a linear demand equation of the form $q = -mp + b$ $(m > 0)$, find a formula for the consumers' surplus at a price level of $\bar{p}$ per unit.

54. *Linear Supply* Given a linear supply equation of the form $q = mp + b$ $(m > 0)$, find a formula for the producers' surplus at a price level of $\bar{p}$ per unit.

55. *Cost* *Snapple Beverage Corp.*'s annual revenue R and profit P in millions of dollars for the period 1989 through 1993 can be approximated by
$$R = 16.15e^{0.87t}$$
and
$$P = 3.93e^{t},$$
where t is the number of years since 1989.*
(a) Use these models to estimate *Snapple*'s accumulated costs for this period.
(b) How does this relate to the area between two curves?
(c) What does a comparison of the exponents in the formulas for R and P tell you about *Snapple*?

56. *Cost* *Microsoft Corp.*'s annual revenue R and profit P in billions of dollars for the period 1986 through 1994 can be approximated by
$$R = 26.27e^{0.37t}$$
and
$$P = 0.6256e^{0.37t},$$
where t is the number of years since 1986.[†]
(a) Use these models to estimate its accumulated costs for this period.
(b) How does this relate to the area between two curves?
(c) What does a comparison of the exponents in the formulas for R and P tell you about *Microsoft*?

57. *Subsidizing Emission Control* The marginal cost to the utilities industry of reducing sulfur emissions at several levels of reduction is shown in the following table.[‡]

Reduction (millions of tons)	8	10	12
Marginal Cost (dollars per ton)	270	360	780

(a) Show by substitution that this data can be modeled by
$$C'(q) = 1{,}000{,}000(41.25q^2 - 697.5q + 3{,}210),$$
where q is the reduction in millions of tons of sulfur and $C'(q)$ is the marginal cost of reducing emissions (in dollars per million tons of reduction).[§]
(b) If the government subsidizes sulfur emissions at a rate of $400 per ton, find the total net cost to the utilities industry of removing 12 million tons of sulfur.
(c) The answer to part (b) is represented by the area between two graphs. What are the equations of these graphs?

▼ * Obtained using exponential regression on graphical data published in *The New York Times,* July 10, 1994, Section 13, p. 1. (Raw data were estimated from a graph. Source: Snapple Beverage Corp./Datastream/Nielsen North America.)

† These equations were obtained using exponential regression on graphical data published in *The New York Times,* July 18, 1994, p. D1. (Raw data were estimated from a graph. Source: Computer Intelligence Infocorp./Microsoft company records.)

‡ See "You're the Expert" in the chapter "Introduction to the Derivative." These figures were produced in a computerized study of reducing sulfur emissions from the 1980 level by the given amounts. Source: Congress of the United States, Congressional Budget Office, *Curbing Acid Rain: Cost, Budget and Coal Market Effects* (Washington, D.C.: Government Printing Office, 1986): xx, xxii, 23, 80.

§ You can obtain this equation directly from the data (as we did) by assuming that $C'(q) = aq^2 + bq + c$, substituting the values of $C'(q)$ for the given values of q, and then solving for a, b, and c.

58. *Variable Emissions Subsidies*   Referring to the marginal cost equation in Exercise 57, assume that the government subsidy for sulfur emissions is given by the formula

$$S'(q) = 500,000,000q,$$

where q is the level of reduction in millions of tons and $S'(q)$ is the marginal subsidy in dollars per million tons.

(a) Find the total net cost to the utilities industry of removing 12 million tons of sulfur. Interpret your answer.

(b) The answer to part (a) is represented by the net area between two graphs. What are the equations of these graphs?

COMMUNICATION AND REASONING EXERCISES

59. *Foreign Trade*   The following graph shows Canada's monthly exports and imports for the year ending February 1, 1994.* What does the area between the export and import curves represent?

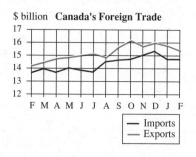

$ billion   **Canada's Foreign Trade**

— Imports
— Exports

60. *Foreign Trade*   The following graph shows a fictitious country's monthly exports and imports for

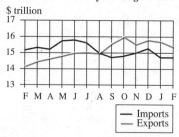

Fictitious Country's Foreign Trade
$ trillion

— Imports
— Exports

the year ending February 1, 2003. What does the total area enclosed by the export and import curves represent, and what does the definite integral of the difference, Exports − Imports, represent?

61. What is wrong with the following claim: "I own 100 units of *Abbott Laboratories, Inc.* stocks that originally cost me $22 per share. My net income from this investment over the period March 5–April 27, 1993 is represented by the area between the stock price curve and the purchase price curve as shown on the following graph."[†]

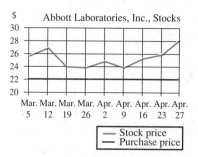

$ Abbott Laboratories, Inc., Stocks

— Stock price
— Purchase price

62. Is it always true that a company's total profit over a one-year period is represented by the area between the daily revenue and daily cost curves? Illustrate your answer by means of an example.

▼ * Source: Statistics Canada/Globe and Mail, April 20, 1994, p. B3.
  † Source: News Reports/*Chicago Tribune*, April 28, 1993, Section 3, p. 6.

► ═══ **7.4** AVERAGES AND MOVING AVERAGES

AVERAGES

You probably already know how to find the average of a collection of numbers. If you want to find the average of, say, 20 numbers, you simply add them up and divide by 20. More generally, if you want to find the **average,** or **mean** of the n numbers $y_1, y_2, y_3, \ldots y_n$, you add them up and divide by n.

AVERAGE OR MEAN OF A SET OF VALUES

$$\bar{y} = \frac{y_1 + y_2 + \ldots + y_n}{n}.$$

Let us revisit an example from the last chapter.

▼ **EXAMPLE 1**   Pork Exports

According to figures released by the U.S. Agriculture Department, the yearly value of U.S. pork exports were as shown in Figure 1.*

The average value of pork exports for the 9-year period shown is

$$\bar{y} = \frac{y_1 + y_2 + \cdots + y_n}{n}$$

$$= \frac{60 + 80 + 125 + 250 + 330 + 320 + 325 + 450 + 460}{9}$$

$$= \frac{2,400}{9} \approx \$266.67 \text{ million per year.}$$

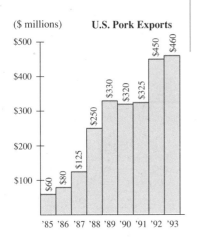

($ millions)   **U.S. Pork Exports**

$500, $400, $300, $200, $100
$60 $80 $125 $250 $330 $320 $325 $450 $460
'85 '86 '87 '88 '89 '90 '91 '92 '93

FIGURE 1

Before we go on... The height of each rectangle in the bar graph is equal to the value of exports for that year; the first rectangle has height 60, the second 80, and so on. Since the rectangles have width 1 unit (1 year), the *area* of the first rectangle is 60, the area of the second is 80, and so on. Thus, the sum of the numbers $60 + 80 + \ldots$ is also the sum of the areas of these rectangles. The average is obtained by dividing this number by 9, which happens to be the total width of the bar graph. In other words,

$$\text{Average height of bar graph} = \frac{\text{Area of bar graph}}{\text{Width of bar graph}}.$$

▼ * Source: U.S. Agriculture Department. (*The New York Times,* Dec. 12, 1993, p. D1.) (The figures shown on the graph are estimates based on the published graph.)

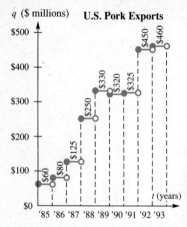

Graph of step function with domain [0, 9]

FIGURE 2

We can also think of the area of the bar graph in Example 1 as the area under a step function (Figure 2). If we call this step function f and take its domain as $[a, b]$, we therefore have

$$\text{Average value of } f = \frac{\text{Area under graph}}{b - a}.$$

Now the area under the graph is calculated by the definite integral, so we make the following definition.

AVERAGE VALUE OF A FUNCTION

The **average**, or **mean**, of a function $f(x)$ on the interval $[a, b]$ is given by

$$\bar{f} = \frac{1}{b - a} \int_a^b f(x)\, dx.$$

This formula is useful in situations where we need to average a continuous function instead of a collection of numbers. Such a situation would arise, for example, if we use a mathematical model to approximate a large collection of data.

▼   **EXAMPLE 2**

Find the average of $f(x) = x$ on the interval $[0, 4]$.

SOLUTION

$$\bar{f} = \frac{1}{b - a} \int_a^b f(x)\, dx$$

$$= \frac{1}{4 - 0} \int_0^4 x\, dx$$

$$= \frac{1}{4} \left[\frac{x^2}{2} \right]_0^4 = \frac{1}{8}(16 - 0) = 2$$

Before we go on... Look at the graph of this function together with the average value 2 (Figure 3).

The area under the graph of f is the area of the triangle shown in the figure, which is $(\frac{1}{2})(4)(4) = 8$. We can construct a rectangle with the same area and the same base $[0, 4]$ by taking the height of the rectangle to be 2, the average value of f. Thus, we can think of the average as the height of a rectangle whose area is the same as the area under the graph of f. This equality of areas follows from the equation

$$(b - a)\bar{f} = \int_a^b f(x)\, dx.$$

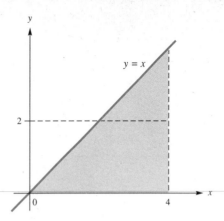

FIGURE 3

▼ **EXAMPLE 3**   More on Pork Exports

In the last chapter we used the following function to model the pork export figures shown in Example 1:

$$q(t) = \frac{460e^{(t-4)}}{1 + e^{(t-4)}},$$

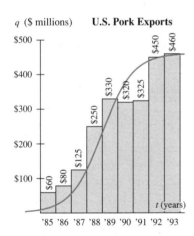

q ($ millions)   **U.S. Pork Exports**

FIGURE 4

where *q* represents annual sales (in millions of dollars) and *t* is the time in years, with *t* = 0 corresponding to Jan. 1, 1985. Figure 4 shows the actual sales data with the graph of the equation superimposed. Calculate $\bar{q}$ for the period Jan. 1, 1985 to Jan. 1, 1994.

SOLUTION   We use the formula

$$\bar{q} = \frac{1}{b - a} \int_a^b q(t)\, dt$$

$$= \frac{1}{9 - 0} \int_0^9 \frac{460e^{(t-4)}}{1 + e^{(t-4)}}\, dt$$

$$= \frac{1}{9} \left[460 \ln\left(1 + e^{(t-4)}\right)\right]_0^9$$

$$= \frac{460}{9}\left[\left(\ln\left(1 + e^5\right)\right) - \left(\ln\left(1 + e^{-4}\right)\right)\right]$$

$$\approx \$254.97 \text{ million per year.}$$

Before we go on... We saw in Example 1 that the average of the actual sales was $266.67 million. Although the accuracy of our model leaves something to be desired, we can use our model to estimate the average value of pork exports for periods into the future for which actual data are not yet available. Also, a more carefully constructed model might yield more accurate estimates.

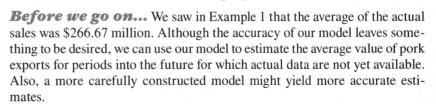

▼ █ **EXAMPLE 4**   Average Balance

The People's Credit Union pays 3% interest, compounded continuously, and at the end of the year pays a bonus of 1% of the average amount in each account during the year. If you deposit $10,000 at the beginning of the year, how much interest and how large a bonus will you get?

SOLUTION   Using the continuous compound interest formula, the amount of money you have in the account at time t is

$$A(t) = 10,000e^{0.03t},$$

where t is measured in years. For a period of one year, t runs from 0 to 1. At the end of the year the account will have

$$A(1) = \$10,304.55,$$

so you will have earned $304.55 interest. To compute the bonus, we first need to find the average amount in the account. But that means we need to find the average of $A(t)$ over the interval $[0, 1]$. Thus,

$$\bar{A} = \frac{1}{b - a} \int_a^b A(t) \, dt$$

$$= \frac{1}{1 - 0} \int_0^1 10,000e^{0.03t} \, dt$$

$$= \frac{10,000}{0.03} \left[e^{0.03t} \right]_0^1$$

$$\approx 333,333.33 \left[e^{0.03} - 1 \right] \approx \$10,151.51.$$

The bonus is 1% of this, or $101.52.

Before we go on... The 1% bonus was one-third of the total interest. Think about why this happened. What fraction of the total interest would the bonus be if the interest rate was 4%, 5%, or 10%?

MOVING AVERAGES

Suppose that you follow the performance of a company's stock by recording the daily closing price. These numbers may jump around quite a bit. In order to see any trends, you would like a way to "smooth out" this data. The **moving average** is one common way to do that.

▼ **EXAMPLE 5**   Stock Prices

The following table shows Colossal Conglomerate Corp.'s closing stock prices for a 20-day period:

Day	1	2	3	4	5	6	7	8	9	10
Price	20	22	21	24	24	23	25	26	20	24
Day	11	12	13	14	15	16	17	18	19	20
Price	26	26	25	27	28	27	29	27	25	24

Plot these prices and also the five-day moving average.

SOLUTION

The five-day moving average is the average of each day's price together with the preceding four days. We can compute this starting on the fifth day. The numbers we get are these.

Day	1	2	3	4	5	6	7	8	9	10
Moving Average					22.2	22.8	23.4	24.4	23.6	23.6
Day	11	12	13	14	15	16	17	18	19	20
Moving Average	24.2	24.4	24.2	25.6	26.4	26.6	27.2	27.6	27.2	26.4

Colossal Conglomerate Corp.

FIGURE 5

The closing stock prices and moving averages are plotted in Figure 5.

As you can see, the moving average is less volatile. Since it averages the stock's performance over five days, a single day's fluctuation is smoothed out. Look at day 9 in particular. The moving average also tends to lag behind the actual performance, since it takes into account past history. Look at the downturns at days 6 and 18 in particular.

Before we go on... The period of 5 days is arbitrary. Using a longer period of time would smooth the data more, but increase the lag. For data used as economic indicators, such as housing prices or retail sales, it is common to compute the 1-year moving average to smooth out seasonal variations.

You can program your graphing calculator to compute the moving averages of a list of data and to plot the scatter graph shown in the figure. We have included instructions and a program for the TI-82 in Appendix B. Computer spreadsheets are also quite useful, and Appendix C includes an example of this as well.

If we use a mathematical model to describe a large collection of data, we may want to compute the moving average of a continuous function. This we can do using the definite integral.

▼   **EXAMPLE 6**

Find the 2-unit moving average of $f(x) = x^2$.

SOLUTION   We are looking for the function $\bar{f}(x) = $ average of f on $[x - 2, x]$. We compute this using the integral formula we gave at the beginning of this section.

$$
\begin{aligned}
\bar{f}(x) &= \frac{1}{x - (x - 2)} \int_{x-2}^{x} t^2 \, dt \\
&= \frac{1}{2} \left[\frac{t^3}{3} \right]_{x-2}^{x} = \frac{1}{6}[x^3 - (x - 2)^3] \\
&= \frac{1}{6}(x^3 - (x^3 - 6x^2 + 12x - 8)) = x^2 - 2x + \frac{4}{3}
\end{aligned}
$$

The graphs are shown in Figure 6. Notice again that the moving average does not reach the extreme minimum that the original function does, and lags behind it in the sense that its minimum occurs later than that of $f(x)$.

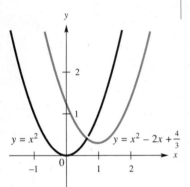

$y = x^2$

$y = x^2 - 2x + \frac{4}{3}$

FIGURE 6

If you look at the first and second steps in the calculation we did in the above example, you will see that the n-unit moving average of a function can be written as follows.

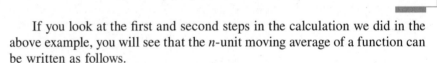

n-UNIT MOVING AVERAGE OF A FUNCTION

The n-unit moving average of a function is given by

$$
\bar{f}(x) = \frac{1}{n} \int_{x-n}^{x} f(t) \, dt.
$$

USING A GRAPHING CALCULATOR OR COMPUTER

The following example shows how one can use a graphing calculator to calculate and plot the moving average of a function. (See the appendix for a program to compute and plot moving averages of lists of data.)

▼   **EXAMPLE 7**

Use a graphing calculator to plot the 3-unit moving average of

$$
f(x) = \frac{x}{1 + |x|}, \quad -5 \le x \le 5.
$$

SOLUTION   This function is a little tricky to integrate analytically, since $|x|$ is defined differently for positive and negative values of x, so we let our graphing calculator approximate the integral for us. (The format that follows is for the TI-82 calculator. Other graphing calculators should be similar.)
   Enter

$$Y_1 = X/(1 + \text{abs }(X))$$

$$Y_2 = (1/3) \text{ fnInt } (T/(1 + \text{abs}(T)), T, X - 3, X)$$

The Y_2 entry is a numerical approximation of the 3-unit moving average of $f(x)$,

$$\bar{f}(x) = \frac{1}{3} \int_{x-3}^{x} \frac{t}{1 + |t|}\, dt.$$

We set the viewing window ranges to $-5 \le x \le 5$ and $-1 \le y \le 1$, and plot these curves. (You will need to wait a while for the plot of the moving average—the calculator has to do a numerical integration to obtain each point on the graph.) The result is shown in Figure 7.

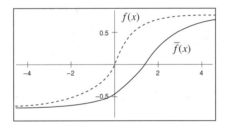

FIGURE 7

Before we go on... We can also use the calculator to evaluate the moving average at any value of x. For instance, to calculate $\bar{f}(1.2)$ on a TI-82 calculator, we enter

$$1.2 \rightarrow X \ \boxed{\text{ENTER}}$$
$$Y_2 \qquad \boxed{\text{ENTER}}$$

This has the effect of setting $x = 1.2$ and then evaluating the function Y_2. The answer we obtain is

$$\bar{f}(1.2) \approx -0.11960.$$

▶   **7.4 EXERCISES**

Find the average value of the indicated quantity in Exercises 1 through 4.

1.   **Total Assets of Utilities Funds**

Source: Morningstar Inc./
New York Times, Jan. 2, 1994

2.   **Number of Personal Computers in the
U.S. with CD–ROM Drives**

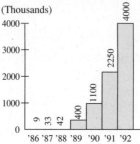

Source: Info Tech/New York Times,
Dec. 28, 1993

3.   **Value of Announced Mergers
and Corporate Transactions**

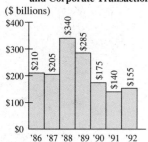

Source: Securities Data Company/
New York Times, Jan. 3, 1994

4.   **Annual Rate of Increase in Medical Costs in
New York City Metropolitan Area**

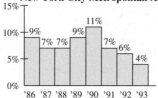

Source: Federal Bureau of
Labor Statistics/New York Times,
Dec. 25, 1993

*Find the averages of the functions in Exercises 5–10 over the given intervals. Plot
each function and its average on the same graph, as in Example 2.*

5. $f(x) = x^3$ over $[0, 2]$

6. $f(x) = x^3$ over $[-1, 1]$

7. $f(x) = x^3 - x$ over $[0, 2]$

8. $f(x) = x^3 - x$ over $[0, 1]$

9. $f(x) = e^{-x}$ over $[0, 2]$

10. $f(x) = e^x$ over $[-1, 1]$

*Plot the sequences in Exercises 11 and 12 and their 5-unit moving averages, as in
Example 5.*

11. 1, 2, 3, 4, 3, 2, 3, 4, 5, 6, 5, 6, 7, 8, 9, 8, 7, 8, 9, 10

12. 1, 2, 3, 4, 5, 6, 7, 6, 5, 4, 3, 2, 3, 4, 5, 6, 7, 8, 9, 10

*Calculate the 5-unit moving average of each of the functions in Exercises 13–20.
Plot each function and its moving average on the same graph, as in Example 6.
(You may use a graphing calculator or a computer for these plots, but you should
compute the moving averages analytically.)*

13. $f(x) = x^3$

14. $f(x) = x^3 - x$

15. $f(x) = x^{2/3}$

16. $f(x) = x^{2/3} + x$

17. $f(x) = e^{0.5x}$

18. $f(x) = e^{-0.02x}$

19. $f(x) = \sqrt{x}$

20. $f(x) = x^{1/3}$

APPLICATIONS

21. *Investments*   If you invest $10,000 at 8% interest compounded continuously, what is the average amount in your account over one year?

22. *Investments*   If you invest $10,000 at 12% interest compounded continuously, what is the average amount in your account over one year?

23. *Average Balance*   Suppose that you have an account (paying no interest) into which you deposit $3,000 at the beginning of each month. You withdraw money so that the amount in the account decreases linearly to 0 by the end of the month. Find the average

amount in the account over a period of several months. (Assume that the account starts at $0 at $t = 0$ months.)

24. *Average Balance*   Suppose that you have an account (paying no interest) into which you deposit $4,000 at the beginning of each month. You withdraw $3,000 during the course of each month, in such a way that the amount decreases linearly. Find the average amount in the account in the first two months. (Assume that the account starts at $0 at $t = 0$ months.)

In Exercises 25 through 28, use a graphing calculator or graphing computer software to plot the given functions together with their three-unit moving averages.

25. $f(x) = \dfrac{10x}{1 + 5|x|}$ **26.** $f(x) = \dfrac{1}{1 + e^x}$ **27.** $f(x) = \ln(1 + x^2)$ **28.** $f(x) = e^{1-x^2}$

29. *Expansion of Fast-Food Outlets*   The following chart shows the approximate number of *McDonald's*® restaurants in the United States at year-end, in thousands.*

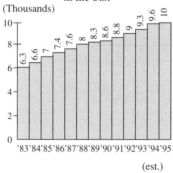

Number of McDonald's Restaurants in the U.S.
(Thousands)

'83 '84 '85 '86 '87 '88 '89 '90 '91 '92 '93 '94 '95
(est.)

6.3 6.6 7 7.4 7.6 8 8.3 8.6 8.8 9 9.3 9.6 10

(a) Use the 1985 and 1995 figures to model these data with a linear function.
(b) Find and plot the 4-year moving average of your model.
(c) What can you say about the slope of the moving average?
(d) (For programmed calculator or computer) Use the graphing calculator or spreadsheet program described in Appendices B and C to plot the actual data and 4-year moving averages.

30. *Fast-Food Customers*   The following graph shows the declining number of Americans (year-end figures in thousands) per fast-food outlet in the U.S.[†]

▼ * Source: Technomics/U.S. Department of Commerce/*The New York Times,* Jan. 9, 1994, p. F5.
† Source: Technomics/*The New York Times,* Jan. 9, 1994, p. F5.

Americans per Store (thousands)

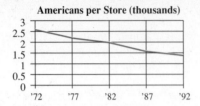

(a) Use the 1972 and 1992 figures to model these data with a linear function. (Give constants correct to two decimal places.)

(b) Find and plot the 5-year moving average of your model.

(c) What can you say about the slope of the moving average?

(d) (For programmed calculator or computer) Use the graphing calculator or spreadsheet program described in Appendices B and C to plot the actual data and 5-year moving averages.

31. *Moving Average of a Linear Function*   Find a formula for the a-unit moving average of a general linear function $f(x) = mx + b$.

32. *Moving Average of an Exponential Function*   Find a formula for the a-unit moving average of a general exponential function $f(x) = Ae^{kx}$.

33. *Market Share*   The following graph shows the market share of *McDonald's*® restaurants in the fast-food business in the U.S.*

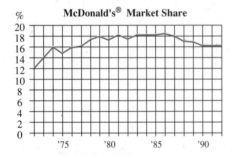

(a) With $t = 0$ representing 1972, obtain a quadratic model

$$p(t) = at^2 + bt + c$$

that approximates the percentage market share enjoyed by *McDonald's*. (Use the data for $t = 0$, $t = 10$ and $t = 20$, rounded to the nearest percentage.)

(b) Use your model to estimate *McDonald's* average market share over the period shown in the graph.

(c) Use your model to predict *McDonald's* average market share for the three years ending in the year 2,000.

▼ *Source: Technomics/U.S. Department of Commerce/*The New York Times*, Jan. 9, 1994, p. F5.

34. *Growth in Prozac Sales*   The following graph shows the annual growth in new prescriptions for Prozac® since its introduction in 1988.*

% Growth in New Prescriptions for Prozac®

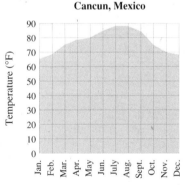

(a) With $t = 0$ representing 1989, obtain a quadratic model

$$p(t) = at^2 + bt + c$$

that approximates the percentage growth in Prozac prescriptions. (Use the data for $t = 0$, $t = 2$ and $t = 4$, rounded to the nearest 10%.)

(b) Use your model to estimate the average annual rate of growth in Prozac prescriptions over the period shown in the graph.

(c) Use your model to predict the average rate of growth of Prozac prescriptions for the three years ending in 1996.

35. *Fair Weather*[†]   The Cancun Royal Hotel's advertising brochure features the following chart showing the year-round temperature.

Year-round Temperatures in Cancun, Mexico

(a) Estimate and plot the two- and three-month moving averages. (Use a graphing calculator program, if available.)

(b) What can you say about the 24-month moving average?

(c) Comment on the limitations of a quadratic model for this data.

▼ [*] Sources: Mehta and Isaly, Smith Barney Shearson; Lehman Brothers/*The New York Times,* Jan. 9, 1994, p. F7.

[†] Inspired by an exercise in the Harvard Consortium Calculus project (p. 76). Temperatures are fictitious.

36. *Foul Weather*   Repeat Exercise 35 using the following data from the Tough Traveler Lodge in Frigidville.

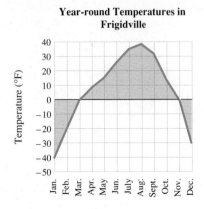

37. *Average Rate of Increase*   In Exercise 4 you computed the average annual rate of increase in medical costs in New York City, but you did it in a somewhat naive way. Here is a better way:
  (a) Taking into account compounding, find the total percent increase in medical costs from 1986 through 1993 (using the data in Exercise 4).
  (b) Find the annual rate of increase that (taking into account compounding) would have produced the same total percent increase over that period. How does this answer compare with your answer to Exercise 4?

38. *Average Rate of Inflation*   Suppose that the government's monthly inflation figures for a year are as follows:

Jan.	Feb.	Mar.	Apr.	May	Jun.	Jul.	Aug.	Sept.	Oct.	Nov.	Dec.
0.5%	1%	0.7%	1.3%	0.8%	0.6%	0.5%	0.2%	0.3%	0.4%	0.1%	0.2%

  (a) Taking into account compounding, find the inflation rate for the year.
  (b) Find the monthly rate that would have produced the same inflation for the year. How does this compare with the average of the monthly rates?

Modeling with the Cosine Function   *Exercises 39–42 require the use of a graphing calculator or computer. (Note: Make sure that your calculator is set to* radians *mode before you begin.) The accompanying figure shows a sketch of the graph of the function*

$$f(x) = C + A\cos\left(\frac{2\pi}{k}(x - d)\right)\quad (C, A, k, d \text{ constants}).$$

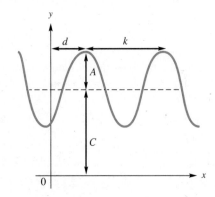

39. Use a graphing calculator or computer to plot the graph of $y = 5 + 4\cos(x - 0.1)$ and its 2-unit moving average.

40. Use a graphing calculator or computer to plot the graph of $y = 2 + 0.5\cos(2(x - 1))$ and its 3-unit moving average.

41. *Electrical Current* The typical voltage V supplied by an electrical outlet in the U.S. is given by

$$V(t) = 165\cos(120\pi t),$$

where t is time in seconds.

(a) Use a graphing calculator to find the average voltage over the interval $[0, \frac{1}{6}]$. How many times does the voltage reach a maximum in one second? (This is referred to as the number of **cycles per second.**)

(b) Use a graphing calculator to plot the function $S(t) = (V(t))^2$ over the interval $[0, \frac{1}{6}]$.

The **root mean square** voltage is given by the formula

$$V_{rms} = \sqrt{\overline{S}},$$

where $\overline{S}$ is the average value of $S(t)$ over one cycle. Calculate V_{rms}.

42. *Tides* The depth of water at my favorite surfing spot varies from 5 ft to 15 ft, depending on the tide. Last Sunday, high tide occurred at 5:00 AM, and the next high tide occurred at 6:30 PM. Use the cosine model to describe the depth of water as a function of time t in hours since midnight on Sunday morning. What was the average depth of the water between 10:00 AM and 2:00 PM?

COMMUNICATION AND REASONING EXERCISES

43. What property does a (non-constant) function have if its average value over an interval is zero? Sketch a graph of such a function.

44. Can the average value of a function f on an interval be larger than its value at any point in that interval? Explain.

45. Explain why it is sometimes more useful to consider the moving average of a stock price rather than the stock price itself.

46. Your monthly salary has been steadily increasing for the past year, and your average monthly salary was x dollars. Would you have earned more money if you were paid x dollars per month? Explain your answer.

47. Criticize the following claim: "the average value of a function on an interval is midway between its highest and lowest value."

48. Your manager tells you that 12-month moving averages gives at least as much information as shorter-term moving averages, and very often more. How would you argue that she is wrong?

49. Which of the following most closely approximates the original function: **(a)** its 10-unit moving average **(b)** its 1-unit moving average **(c)** its 0.8-unit moving average. Explain your answer.

50. Is an increasing function larger or smaller than its one-unit moving average? Explain.

▶ ══ **7.5** IMPROPER INTEGRALS AND APPLICATIONS

All the definite integrals we have seen have the form $\int_a^b f(x)\,dx$, where a and b are finite and $f(x)$ is piecewise-continuous on the closed interval $[a, b]$. There are occasions when we would like to relax these requirements, and when we do so we obtain what are called **improper integrals.** There are various types of improper integrals.

INTEGRALS IN WHICH A LIMIT OF INTEGRATION IS INFINITE

These are integrals of the form

$$\int_{a}^{+\infty} f(x)\ dx, \quad \int_{-\infty}^{b} f(x)\ dx, \quad \text{or} \quad \int_{-\infty}^{+\infty} f(x)\ dx.$$

Let us concentrate for a moment on the first form, $\int_{a}^{+\infty} f(x)\ dx$. If, instead of $+\infty$, we took a very large positive number M, then the integral would be just the very ordinary-looking definite integral $\int_{a}^{M} f(x)\ dx$. We might be tempted to say that $\int_{a}^{+\infty} f(x)\ dx$ *means* $\int_{a}^{M} f(x)\ dx$, where M is some very large number. But how large should M be? By now the answer is probably suggesting itself to you: we take the limit, as M approaches $+\infty$, of the integral $\int_{a}^{M} f(x)\ dx$.

IMPROPER INTEGRAL WITH AN INFINITE LIMIT OF INTEGRATION

We define

$$\int_{a}^{+\infty} f(x)\ dx = \lim_{M \to +\infty} \int_{a}^{M} f(x)\ dx,$$

provided the limit exists. If the limit exists, we say that $\int_{a}^{+\infty} f(x)\ dx$ **converges.** Otherwise, we say that $\int_{a}^{+\infty} f(x)\ dx$ **diverges.** Similarly, we define

$$\int_{-\infty}^{b} f(x)\ dx = \lim_{M \to -\infty} \int_{M}^{b} f(x)\ dx,$$

provided the limit exists. Finally, we define

$$\int_{-\infty}^{+\infty} f(x)\ dx = \int_{-\infty}^{a} f(x)\ dx + \int_{a}^{+\infty} f(x)\ dx$$

for some convenient a, provided *both* integrals on the right converge.

Q We know that the integral can be used to calculate the area under the curve. Is this also true for improper integrals?

A Yes. The best way to illustrate this is by looking at several examples.

▼ **EXAMPLE 1**

Calculate $\displaystyle\int_{1}^{+\infty} \frac{1}{x^2}\ dx$ and $\displaystyle\int_{-\infty}^{-1} \frac{1}{x^2}\ dx$.

SOLUTION   We do the first one, and leave the second—an almost identical calculation—to you. The definition tells us that we must first evaluate $\int_{1}^{M} (1/x^2)\ dx$ and then take the limit as $M \to +\infty$. Now

$$\int_{1}^{M} \frac{1}{x^2}\ dx = \int_{1}^{M} x^{-2}\ dx = -[x^{-1}]_{1}^{M} = 1 - \frac{1}{M}.$$

Thus,
$$\int_1^{+\infty} \frac{1}{x^2}\, dx = \lim_{M \to +\infty} \int_1^M \frac{1}{x^2}\, dx$$
$$= \lim_{M \to +\infty} \left(1 - \frac{1}{M}\right) = 1.$$

In other words, the integral converges to 1.

Before we go on... How do we interpret this integral geometrically? The integral $\int_1^M (1/x^2)\, dx$ calculates the area shaded in Figure 1(a).

Geometrically, letting M approach $+\infty$ corresponds to moving the right-hand boundary in Figure 1(a) farther and farther to the right, resulting in Figure 1(b). In other words, the improper integral $\int_1^{+\infty} (1/x^2)\, dx$ calculates the area of an *infinitely long* region.

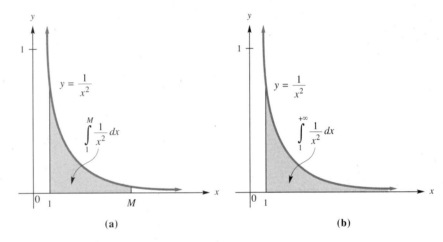

(a)　　　　　　　　　**(b)**

FIGURE 1

Q You said $\int_1^{+\infty} (1/x^2)\, dx = 1$. The infinitely long region shown in Figure 1(b) has an area of only 1 square unit?!

A Exactly! Think of it this way: suppose you had enough paint to cover exactly 1 square unit of area. If you paint the region in Figure 1(a) and then keep painting, extending the right edge farther and farther to the right, you would never run out of paint. This is one of the places where mathematics seems to contradict common sense. But common sense is notoriously unreliable when dealing with infinities.

▼　**EXAMPLE 2**

Compute $\displaystyle\int_1^{+\infty} \frac{1}{x}\, dx$.

SOLUTION Proceeding as before, we first calculate

$$\int_1^M \frac{1}{x}\, dx = \left[\ln|x|\right]_1^M = \ln M - \ln 1 = \ln M.$$

Thus,

$$\int_1^{+\infty} \frac{1}{x}\,dx = \lim_{M\to+\infty} (\ln M) = +\infty.$$

The integral diverges to $+\infty$.

Before we go on... Figure 2 shows the area in question, compared with $\int_1^{+\infty} (1/x^2)\,dx$.

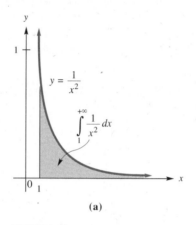

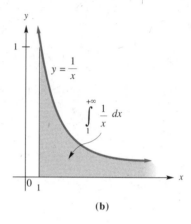

(a)   **(b)**

FIGURE 2

Although the infinitely long area in Figure 2(a) is exactly 1 square unit, the infinitely long area in Figure 2(b) is infinitely large! This means that, no matter how much paint you were given (even a million gallons) you would eventually run out of paint if you tried to paint the region on the right.

▼ **EXAMPLE 3**   Future Sales of Freon

It is estimated that from the year 2000, sales of freon will decrease continuously at a rate of about 15% per year. Further, sales of freon at the start of the year 2000 are predicted to be 35 million pounds per year.* What will be the total future sales of freon starting in the year 2000?

SOLUTION   We use the continuous decay formula to predict annual sales of freon beginning in the year 2000.

$$s(t) = Ae^{-kt}$$
$$= 35e^{-0.15t} \quad (t \text{ is the number of years since 2000})$$

▼ *These figures are approximations based on published data. (Source: The Automobile Consulting Group/*The New York Times,* December 26, 1993, p. F23.)

We are asked for *total* sales beginning at $t = 0$. Recalling that total sales can be computed as the definite integral of annual sales, we obtain

$$\text{Total sales beginning in 2000} = \int_{0}^{+\infty} 35e^{-0.15t}\,dt = \lim_{M \to +\infty} \int_{0}^{M} 35e^{-0.15t}\,dt$$

$$= \lim_{M \to +\infty} 35 \int_{0}^{M} e^{-0.15t}\,dt$$

$$= -35 \lim_{M \to +\infty} \left[\frac{e^{-0.15t}}{0.15} \right]_{0}^{M}$$

$$= \frac{35}{0.15} \lim_{M \to +\infty} (e^{-0.15M} - e^{0})$$

$$= -\frac{35}{0.15}(-1) \approx 233.33 \text{ million pounds},$$

(since $\lim_{M \to +\infty} e^{-0.15M} = 0$).

 Although it is not easy to program a graphing calculator to calculate a limit, we can approximate the above result numerically by making a table of values for $\int_{0}^{M} 35e^{-0.15t}\,dt$ for larger and larger values of M. The following table shows the result of applying Simpson's rule using the graphing calculator program discussed in the appendix. N is the number of subdivisions. (Why are we using more subdivisions for increasing values of M?)

M	20	40	100
N	40	80	200
$\int_{0}^{M} 35e^{-0.15t}\,dt$	221.7	232.8	233.3

Alternatively, we could have used the built-in integration feature ("fnInt" on the TI-82 and TI-85) and let the calculator worry about the value of N.

INTEGRALS IN WHICH THE INTEGRAND BECOMES INFINITE AT AN ENDPOINT

These are integrals of the form

$$\int_{a}^{b} f(x)\,dx,$$

where $f(x) \to \pm\infty$ as either $x \to a$ or $x \to b$. We sometimes say that $f(x)$ **becomes infinite** at either $x = a$ or $x = b$.

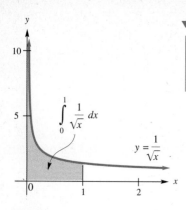

FIGURE 3

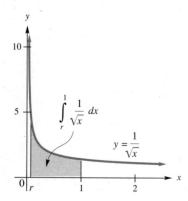

FIGURE 4

▼ **EXAMPLE 4**

Calculate the integral

$$\int_0^1 \frac{1}{\sqrt{x}}\, dx.$$

SOLUTION Notice that $f(x)$ becomes infinite at $x = 0$, and so the integral is improper. Looking at Figure 3, we see that the corresponding region is infinitely long in the vertical direction rather than the horizontal direction.

Since the problem occurs at $x = 0$ (where the integrand is not defined) we approximate the integral by starting just to the right of $x = 0$ at, say $x = r$, and then letting r approach zero from the right. Figure 4 shows the region corresponding to such an approximation.

We then take

$$\int_0^1 \frac{1}{\sqrt{x}}\, dx = \lim_{r \to 0^+} \int_r^1 \frac{1}{\sqrt{x}}\, dx$$

$$= \lim_{r \to 0^+} \int_r^1 x^{-1/2}\, dx$$

$$= \lim_{r \to 0^+} [2x^{1/2}]_r^1 = \lim_{r \to 0+} (2 - 2r^{1/2}) = 2.$$

Thus, as in Example 1 we have an infinitely long region with finite area.

In general, we make the following definition.

IMPROPER INTEGRAL WHERE THE INTEGRAND BECOMES INFINITE

If $f(x)$ is defined for all x with $a < x \le b$ but becomes infinite at $x = a$, we define

$$\int_a^b f(x)\, dx = \lim_{r \to a^+} \int_r^b f(x)\, dx,$$

provided the limit exists. Similarly, if $f(x)$ is defined for all x with $a \le x < b$ but becomes infinite at $x = b$, we define

$$\int_a^b f(x)\, dx = \lim_{r \to b^-} \int_a^r f(x)\, dx.$$

provided the limit exists. In either case, if the limit exists, we say that $\int_a^b f(x)\, dx$ **converges.** Otherwise, we say that $\int_a^b f(x)\, dx$ **diverges.**

▶ **CAUTION**

If the integrand becomes infinite at both endpoints or becomes infinite at some point strictly between a and b, then the above definition does not apply. We shall see how to deal with such cases in the examples. ◀

▼ **EXAMPLE 5**

Investigate the convergence of the integral $\displaystyle\int_{-1}^{3} \frac{x}{x^2 - 9}\, dx$.

SOLUTION   First, we check to see whether the integrand becomes infinite for any value of x in the range of integration. To do this, we set the denominator equal to zero and solve for x.

$$x^2 - 9 = 0$$

gives

$$(x - 3)(x + 3) = 0,$$

so

$$x = -3 \quad \text{or} \quad x = 3.$$

The first of these solutions is outside the range of integration, so we ignore it. The second, $x = 3$, happens to be one of the endpoints of the range of integration, so we conclude that we do have an improper integral, and use the definition above.

$$\int_{-1}^{3} \frac{x}{x^2 - 9}\, dx = \lim_{r \to 3^-} \int_{-1}^{r} \frac{x}{x^2 - 9}\, dx$$

The integral on the right can be calculated using the substitution $u = x^2 - 9$.

$$u = x^2 - 9$$
$$\frac{du}{dx} = 2x$$
$$dx = \frac{1}{2x}\, du$$
When $x = r$, $u = r^2 - 9$.
When $x = -1$, $u = -8$.

Thus,

$$\int_{-1}^{r} \frac{x}{x^2 - 9}\, dx = \frac{1}{2} \int_{-8}^{r^2-9} \frac{1}{u}\, du$$
$$= \frac{1}{2}\Big[\ln |u| \Big]_{-8}^{r^2-9}$$
$$= \frac{1}{2}[\ln |r^2 - 9| - \ln 8].$$

Taking the limit,

$$\int_{-1}^{3} \frac{x}{x^2 - 9} \, dx = \lim_{r \to 3^-} \int_{-1}^{r} \frac{x}{x^2 - 9} \, dx$$

$$= \lim_{r \to 3^-} \left(\frac{1}{2} [\ln |r^2 - 9| - 8] \right) = -\infty.$$

Thus, the integral diverges to $-\infty$.

Before we go on... Figure 5 shows the area in question. The figure makes it clear why the integral diverged to $-\infty$ instead of $+\infty$.

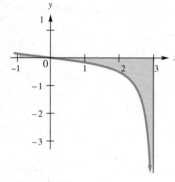

FIGURE 5

In the next example the integrand becomes infinite at a value strictly between the limits of integration.

▼   **EXAMPLE 6**

Investigate the convergence of the integral $\int_{-3}^{2} \frac{1}{x^2} \, dx$.

SOLUTION   We first check to see whether the integrand becomes infinite at any value of x in the range of integration, and we notice that $x = 0$ is the only such value of x. Since 0 lies between -3 and 2, we conclude that we have an improper integral. But $x = 0$ is not an endpoint of the range of integration, so the above definition does not apply.

Q   If the integrand becomes infinite at some point c *inside* the interval $[a, b]$ of integration, what do we do?

A   Break up the interval into two intervals $[a, c]$ and $[c, b]$. Then deal with each of these intervals separately.

In our case, the integrand is not defined at $x = 0$, so we rewrite the integral as

$$\int_{-3}^{2} \frac{1}{x^2} \, dx = \int_{-3}^{0} \frac{1}{x^2} \, dx + \int_{0}^{2} \frac{1}{x^2} \, dx.$$

Each integral on the right is an improper integral that we know how to handle. The first is

$$\int_{-3}^{0} \frac{1}{x^2} \, dx = \lim_{r \to 0^-} \int_{-3}^{r} \frac{1}{x^2} \, dx$$

$$= \lim_{r \to 0^-} \left(\frac{1}{3} - \frac{1}{r} \right) = +\infty,$$

so this integral diverges to $+\infty$. Since this piece of the integral diverges, there is no hope of the original integral converging, so we can stop here and conclude that the given integral diverges.

Before we go on... If we had been sloppy and not bothered to check whether we had an improper integral to begin with, we would have obtained the incorrect answer,

$$\int_{-3}^{2} \frac{1}{x^2}\, dx = \left[-\frac{1}{x}\right]_{-3}^{2} = \frac{1}{6}. \quad \text{WRONG}$$

The moral is that even the most innocent-looking definite integral may be an improper integral in disguise! We must remember to check to see if the integrand becomes infinite at some point between a and b.

▶ NOTE   Suppose that an improper integral I can be expressed as a sum of two or more improper integrals,

$$I = J + K + \ldots$$

Then (1) if one or more of the integrals $J, K, \ldots$ diverges, so does I, and (2) if all of the integrals $J, K, \ldots$ converge, then so does I. Moreover, I converges to the sum of the individual integrals.   ◀

▶ NOTE ON GRAPHING CALCULATORS AND COMPUTERS

Q What would happen if we tried to calculate the integral $\int_{-3}^{2} \frac{1}{x^2}\, dx$ using a graphing calculator or computer?

A The best way to answer this question is to try it on a graphing calculator. For example, you can use the TI-82 to evaluate this integral numerically by entering

fnInt (1/X^2, X, −3, 2)

and pressing ENTER . We tried it, waited a while, and then the calculator said that there was an error. (Apparently its algorithm detected that something was amiss.*)

Not willing to give up so easily, we next tried the trapezoid rule with $N = 200$. This time, the calculator gave a "division by zero" error. We quickly realized that the reason for this was that $x = 0$ was one of the points in the subdivision, and the error resulted from the attempt to evaluate $f(0)$. To remedy that, we used $N = 201$ instead (so that 0 would not be a subdivision point) and obtained the answer 437.81. We then tried again with $N = 501$, and obtained the answer 1,092.51. In other words, the larger the choice for N, the larger the value of the integral, suggesting once again that the integral diverges to $+\infty$.

---

* Most graphing calculator and computer integration algorithms work by using smaller and smaller subdivisions until the answers converge to a fixed real number. If, as is the case here, the integral diverges, then the answer fails to converge, the algorithm will fail, and an error will be returned.

However, do not feel confident that you can use numerical integration to determine the convergence or divergence of an improper integral. For instance, if you evaluate $\int_{-1}^{1} \frac{1}{x}\, dx$ with the trapezoid rule, you will obtain the following answers:

N	201	301	413	665
$\int_{-1}^{1} \frac{1}{x}\, dx$ by trap. rule	5×10^{-12}	1×10^{-11}	4×10^{-11}	-2×10^{-11}

These values are quite small, but getting no closer to zero (this is probably round-off error). It is not true that the value of this integral is zero. If you use the method of Example 6 to investigate this integral, you will find that it diverges: $\int_{-1}^{0} \frac{1}{x}\, dx$ diverges to $-\infty$ and $\int_{0}^{1} \frac{1}{x}\, dx$ diverges to $+\infty$. Why are the trapezoid sums so close to zero, then? ◄

We end this section with an example where the integral is improper in more than one way.

▼ **EXAMPLE 7**

Investigate the convergence of $\displaystyle\int_{0}^{+\infty} \frac{1}{x^{0.6}}\, dx$.

SOLUTION   The given integral is improper in two ways: one of the endpoints is infinite, and the integrand becomes infinite at the other endpoint. As in Example 6, we must first break the integral into two improper integrals,

$$\int_{0}^{1} \frac{1}{x^{0.6}}\, dx \text{ and } \int_{1}^{+\infty} \frac{1}{x^{0.6}}\, dx.$$

(We could have broken it at any positive value, but we chose 1 for convenience.) In the first integral the integrand becomes infinite at one of the endpoints, and in the second one of the endpoints is infinite. We now investigate these integrals separately.

$$\int_{0}^{1} x^{-0.6}\, dx = \lim_{r \to 0^+} \int_{r}^{1} x^{-0.6}\, dx$$

$$= \lim_{r \to 0^+} \left[\frac{x^{0.4}}{0.4} \right]_{r}^{1}$$

$$= \lim_{r \to 0^+} \left(\frac{1}{0.4} - \frac{r}{0.4} \right) = \frac{1}{0.4} = \frac{5}{2}$$

$$\int_{1}^{+\infty} x^{-0.6}\, dx = \lim_{M \to +\infty} \int_{1}^{M} x^{-0.6}\, dx$$

$$= \lim_{M \to +\infty} \left[\frac{x^{0.4}}{0.4} \right]_{1}^{M} = +\infty$$

Since the second integral diverges, we conclude that the original integral diverges as well.

▶ __7.5 EXERCISES__

Decide whether each of the integrals in Exercises 1 through 26 converges. If the integral converges, compute its value.

1. $\displaystyle\int_{1}^{+\infty} x\, dx$

2. $\displaystyle\int_{0}^{+\infty} e^{-x}\, dx$

3. $\displaystyle\int_{-2}^{+\infty} e^{-0.5x}\, dx$

4. $\displaystyle\int_{1}^{+\infty} \frac{1}{x^{1.5}}\, dx$

5. $\displaystyle\int_{-\infty}^{2} e^{x}\, dx$

6. $\displaystyle\int_{-\infty}^{-1} \frac{1}{x^{1/3}}\, dx$

7. $\displaystyle\int_{-\infty}^{-2} \frac{1}{x^{2}}\, dx$

8. $\displaystyle\int_{-\infty}^{0} e^{-x}\, dx$

9. $\displaystyle\int_{0}^{+\infty} x^{2} e^{-6x}\, dx$

10. $\displaystyle\int_{0}^{+\infty} (2x - 4)e^{-x}\, dx$

11. $\displaystyle\int_{0}^{5} \frac{2}{x^{1/3}}\, dx$

12. $\displaystyle\int_{0}^{2} \frac{1}{x^{2}}\, dx$

13. $\displaystyle\int_{-1}^{2} \frac{3}{(x + 1)^{2}}\, dx$

14. $\displaystyle\int_{-1}^{2} \frac{3}{(x + 1)^{1/2}}\, dx$

15. $\displaystyle\int_{-1}^{2} \frac{3x}{x^{2} - 1}\, dx$

16. $\displaystyle\int_{-1}^{2} \frac{3}{x^{1/3}}\, dx$

17. $\displaystyle\int_{-2}^{2} \frac{1}{(x + 1)^{1/5}}\, dx$

18. $\displaystyle\int_{-2}^{2} \frac{2x}{\sqrt{4 - x^{2}}}\, dx$

19. $\displaystyle\int_{-1}^{1} \frac{2x}{x^{2} - 1}\, dx$

20. $\displaystyle\int_{-1}^{2} \frac{2x}{x^{2} - 1}\, dx$

21. $\displaystyle\int_{-\infty}^{+\infty} xe^{-x^{2}}\, dx$

22. $\displaystyle\int_{-\infty}^{\infty} xe^{1 - x^{2}}\, dx$

23. $\displaystyle\int_{0}^{+\infty} \frac{1}{x \ln x}\, dx$

24. $\displaystyle\int_{0}^{+\infty} \ln x\, dx$

25. $\displaystyle\int_{0}^{+\infty} \frac{2x}{x^{2} - 1}\, dx$

26. $\displaystyle\int_{-\infty}^{0} \frac{2x}{x^{2} - 1}\, dx$

APPLICATIONS

27. *Sales*   My financial adviser has predicted that annual sales of BATMAN® T-shirts will continue to decline continuously at a rate of 10% each year. At the moment, I have 3,200 of the shirts in stock and am selling them at a rate of 200 per year. Will I ever sell them all?

28. *Revenue*   Alarmed about the sales prospects for my BATMAN T-shirts (see Exercise 27) I will try to make up lost revenues by increasing the price by $1 each year. I now charge $10 per shirt. What is the total amount of revenue I can expect to earn from sales of my T-shirts, assuming the sales levels described in the previous exercise?

29. *Advertising*   Spending on cigarette advertising in the United States declined from close to $600 million per year at the start of 1991 to half that amount three years later.* Use a continuous decay model to forecast the total revenue that will be spent on cigarette advertising beginning in January 1991.

30. *Sales*   Sales of the text *I Love Calculus* have been declining steadily by 5% per year. Assuming that *I Love Calculus* currently sells 5,000 copies per year and that sales will continue this pattern of decline, calculate total future sales of the text.

31. *Variable Sales*   The value of your Chateau Petit Mont Blanc 1963 vintage burgundy is continuously increasing at 40% per year, and you have a supply of 1,000 liter bottles worth $85 each at today's prices. In order to ensure a steady income, you have decided to sell your wine at a diminishing rate—starting at 500 bottles per year and then decreasing this figure exponentially at a fractional rate of 100% per year. How much income (to the nearest dollar) can you expect to generate by this scheme?

32. *Panic Sales*   Unfortunately, your large supply of Chateau Petit Mont Blanc is continuously turning to vinegar at a fractional rate of 60% per year! You have thus decided to sell off your Petit Mont Blanc at $50

▼ *Sources: Advertising Age/Competitive Media Reporting (*The New York Times,* March 3, 1994, p. D1.)

per bottle, but the market is a little thin, and you can only sell 400 bottles per year. Since you have no way of knowing which bottles now contain vinegar until they are opened, you shall have to give refunds for all the bottles of vinegar. What will your net income be before all the wine turns to vinegar?

33. *Foreign Investments*   According to data published by the World Bank, the annual flow of private investment to developing countries from more developed countries is approximately

$$q(t) = 10 + 1.56t^2,$$

where $q(t)$ is the annual investment in billions of dollars, and t is time in years since the start of 1986.* Assuming a world-wide inflation rate of 5% per year, find the value of all private aid to developing countries from 1986 on in constant dollars. (The constant dollar value of $q(t)$ dollars t years from now is given by $q(t)e^{-rt}$, where r is the fractional rate of inflation.)

34. *Foreign Investments*   Repeat the previous exercise using the following model for government loans and grants to developing countries.*

$$q(t) = 38 + 0.6(t - 3)^2$$

35. *Pork Exports*   According to figures released by the U.S. Department of Agriculture, the value of U.S. pork exports can be modeled by the equation

$$q(t) = \frac{460e^{(t - 4)}}{1 + e^{(t - 4)}},$$

where t is time in years since the start of 1985, and $q(t)$ is annual exports in millions of dollars. Investigate the integrals $\int_0^{+\infty} q(t)\, dt$ and $\int_{-\infty}^0 q(t)\, dt$ and interpret your answers.

36. *Sales*   The weekly demand for your company's LoCal Mousse is modeled by the equation

$$q(t) = \frac{50e^{2t - 1}}{1 + e^{2t - 1}},$$

where t is time from now in weeks and $q(t)$ is the number of gallons sold per week. Investigate the integrals $\int_0^{+\infty} q(t)\, dt$ and $\int_{-\infty}^0 q(t)\, dt$ and interpret your answers.

37. *Meteor Impacts*   The frequency of meteor impacts on earth can be modeled by

$$n(k) = \frac{1}{5.6997k^{1.081}},$$

where $n(k) = N'(k)$, and $N(k)$ is the average number of meteors of energy less than or equal to k megatons that will hit the earth in one year.[†] (A small nuclear bomb releases on the order of one megaton of energy.)
(a) How many meteors of energy at least $k = 0.2$ hit the earth each year?
(b) Investigate and interpret the integral

$$\int_0^1 n(k)\, dk.$$

38. *Meteor Impacts*   (continuing the previous exercise)
(a) Explain why the integral

$$\int_a^b kn(k)\, dk$$

computes the total energy released each year by meteors with energies between a and b megatons.
(b) Compute and interpret

$$\int_0^1 kn(k)\, dk.$$

(c) Compute and interpret

$$\int_1^{+\infty} kn(k)\, dk.$$

39. *The Gamma Function*   The gamma function is defined by the formula

$$\Gamma(x) = \int_0^{+\infty} t^{x - 1}e^{-t}\, dt.$$

(a) Find $\Gamma(1)$ and $\Gamma(2)$ (assume that $\lim_{M \to +\infty} Me^{-M} = 0$).
(b) Use integration by parts to show that, for a positive integer n, $\Gamma(n + 1) = n\Gamma(n)$.
(c) Deduce that $\Gamma(n) = (n - 1)!$ for every positive integer n.

▼ * The authors' approximation, based on data published by the World Bank/*The New York Times*, Dec. 17, 1993, p. D1.
† This is the authors' model, based on data published by NASA International Near-Earth Object Detection Workshop. (*The New York Times*, Jan. 25, 1994, p. C1.)

40. *Laplace Transforms*   The Laplace Transform $F(x)$ of a function $f(t)$ is given by the formula

$$F(x) = \int_0^{+\infty} f(t)e^{-xt}\, dt.$$

(a) Find $F(x)$ for $f(t) = 1$ and for $f(t) = t$.
(b) Find a formula for $F(x)$ if $f(t) = t^n$ ($n = 1, 2, 3, \ldots$).
(c) Find a formula for $F(x)$ if $f(t) = e^{at}$ (a constant).

The Normal Curve   *The following exercises require the use of a graphing calculator or computer programmed to do numerical integration. The* normal distribution *curve, which models the distributions of data in a wide range of applications, is given by the function*

$$p(x) = \frac{1}{\sqrt{2\pi}\,\sigma}\, e^{-(x-\mu)^2/2\sigma^2},$$

where $\pi = 3.14159265\ldots$, *and* σ *and* μ *are constants.*

41. With $\sigma = 4$ and $\mu = 1$, use a graphing calculator to approximate $\int_{-\infty}^{+\infty} p(x)\, dx$.

42. With $\sigma = 1$ and $\mu = 0$, find $\int_0^{+\infty} p(x)\, dx$.

43. With $\sigma = 1$ and $\mu = 0$, find $\int_1^{+\infty} p(x)\, dx$.

44. With $\sigma = 1$ and $\mu = 0$, find $\int_{-\infty}^1 p(x)\, dx$.

COMMUNICATION AND REASONING EXERCISES

45. It sometimes happens that the Fundamental Theorem of Calculus gives the correct answer for an improper integral. Explain why the FTC gives the correct answer for improper integrals of the form

$$\int_{-a}^a \frac{1}{x^r}\, dx$$

if $r < 1$.

46. Why can't the Fundamental Theorem of Calculus be used to evaluate $\displaystyle\int_{-1}^1 \frac{1}{x}\, dx$?

47. Why can't the Fundamental Theorem of Calculus be used to evaluate $\displaystyle\int_1^{+\infty} \frac{1}{x^2}\, dx$?

48. How can you use a graphing calculator to approximate improper integrals? (Your discussion should refer to each type of improper integral.)

49. Make up an interesting application problem whose solution is

$$\int_{10}^{+\infty} 100te^{-0.2t}\, dt = \$1,015.01.$$

50. Make up an interesting application problem whose solution is $\displaystyle\int_{100}^{+\infty} \frac{1}{r^2}\, dr = 0.01.$

▶ ═══ **7.6** DIFFERENTIAL EQUATIONS AND APPLICATIONS

A **differential equation** is an equation that involves the derivative of an unknown function. To **solve** a differential equation means to find that unknown function. The field of differential equations is a very active area of study in mathematics, and we shall see only a small part of it in this section.

▼ **EXAMPLE 1**

Solve the differential equation $\dfrac{dy}{dx} = x^2$.

SOLUTION   To "solve" the differential equation means to find the unknown function, y. But since

$$\frac{dy}{dx} = x^2,$$

y must be an antiderivative of x^2, so

$$y = \int x^2 \, dx = \frac{x^3}{3} + C.$$

Before we go on... The differential equation we solved is called a **first order** differential equation, because it involves only the first derivative.

The solution $y = (x^3/3) + C$ is called the **general solution** of the differential equation $dy/dx = x^2$ because it actually represents infinitely many different solutions, one for each choice of the constant C. We can get **particular solutions** by choosing particular values for C. For instance, $y = (x^3/3) - 4$ is the particular solution corresponding to the choice $C = -4$.

We did not have to work hard to solve the equation in Example 1. In fact, any differential equation of the form

$$\frac{dy}{dx} = f(x)$$

can be solved by integrating. The solution is

$$y = \int f(x) \, dx.$$

We shall call a differential equation of this form **elementary.**

ELEMENTARY DIFFERENTIAL EQUATIONS
An **elementary** differential equation has the form

$$\frac{dy}{dx} = f(x).$$

Its solution is

$$y = \int f(x) \, dx.$$

Not all differential equations are elementary, as the next example shows.

▼ **EXAMPLE 2**

Given the differential equation $\dfrac{dy}{dx} = \dfrac{x}{y^2}$,

(a) find the general solution, and express y as a function of x, and

(b) find the particular solution that satisfies $y(0) = 2$.

SOLUTION

(a) First notice that we do not have an elementary differential equation because we have a function of both x and y on the right-hand side. We cannot simply integrate the right-hand side with respect to x because the variable y appears there, and y is some as-yet-unknown function of x.

Instead of trying to integrate immediately, we use a method called "separation of variables," and proceed as follows.

Step 1: Separate the xs from the ys algebraically.

$$\frac{dy}{dx} = \frac{x}{y^2}$$

gives

$$y^2 \, dy = x \, dx.$$

Step 2: Integrate both sides:

$$\int y^2 \, dy = \int x \, dx,$$

so that

$$\frac{y^3}{3} = \frac{x^2}{2} + C.$$

We now have y as an implicit function of x. Since we want an *explicit* function of x (if possible), we solve the equation for y.

$$y^3 = \frac{3x^2}{2} + 3C = \frac{3x^2}{2} + D,$$

where we have renamed the arbitrary constant $3C$ as D. Thus,

$$y = \left(\frac{3x^2}{2} + D\right)^{1/3} \quad (D \text{ arbitrary})$$

is the general solution.

(b) Recall that we get particular solutions by choosing values for the arbitrary constant (D in this case). We want the particular solution that gives $y(0) = 2$, that is, $y = 2$ when $x = 0$. Substituting these values in our general solution gives

$$2 = (0 + D)^{1/3},$$

so that

$$D^{1/3} = 2$$

giving

$$D = 2^3 = 8.$$

Thus, the particular solution we want is

$$y = \left(\frac{3x^2}{2} + 8\right)^{1/3}.$$

Before we go on... We can check our general solution as follows.

$$\frac{dy}{dx} = \frac{1}{3}\left(\frac{3x^2}{2} + D\right)^{-2/3}(3x) = x\left(\frac{3x^2}{2} + D\right)^{-2/3}$$

$$\frac{x}{y^2} = \frac{x}{\left(\left(\frac{3x^2}{2} + D\right)^{1/3}\right)^2} = x\left(\frac{3x^2}{2} + D\right)^{-2/3}$$

Therefore, $\frac{dy}{dx} = \frac{x}{y^2}$ as desired.

Notice that we were able to separate the xs from the ys in the above example precisely because the right-hand side, x/y^2, was a product of a function of x (namely, x itself) and a function of y (namely, $1/y^2$). We can formalize this result as follows.

> **SEPARABLE DIFFERENTIAL EQUATIONS**
>
> A **separable** differential equation has the form
>
> $$\frac{dy}{dx} = f(x)g(y).$$
>
> We solve a separable differential equation by separating the xs and the ys algebraically and then integrating.

▼ **EXAMPLE 3**   Rising Medical Costs

The cost of medical care in the New York City metropolitan area was going up continuously at an average percentage rate of 7.3% in the 1980s and early 1990s.* Find a formula for medical cost y as a function of time t.

▼ *Source: Federal Bureau of Labor Statistics/*The New York Times,* Dec. 25, 1993.

SOLUTION   This is the kind of continuous growth problem we studied in the chapter on exponential and logarithmic functions. But bear with us: we shall now derive the formula we used in that chapter. First, note that we can rephrase the first sentence as

the rate of increase of y equals 7.3% of y,

or

$$\frac{dy}{dt} = 0.073y.$$

Here the variables are y and t, and we separate them to obtain

$$\frac{1}{y}\, dy = 0.073\; dt.$$

Integrating, we get

$$\int \frac{1}{y}\, dy = \int 0.073\; dt,$$

giving

$$\ln |y| = 0.073t + C.$$

Since we are trying to find y as a function of t, we solve this equation for y. We can rewrite the equation in exponential form as

$$|y| = e^{0.073t} + C$$
$$= e^{C}\, e^{0.073t}.$$

Thus,

$$y = \pm\, e^{C}\, e^{0.073t}.$$

Since C is an arbitrary constant, the quantity $\pm\, e^{C}$ is also an arbitrary constant, so we can write

$$y = Ae^{0.073t},$$

which is the formula we have used before for continuous growth. The constant A is the value of y at time $t = 0$. In other words, A is the cost of medical care at time 0.

▼ **EXAMPLE 4**   Newton's Law of Cooling

Newton's Law of Cooling states that a hot object cools at a rate proportional to the difference between its temperature and the temperature of the surrounding environment. Obtain a formula showing how the temperature of an object changes with time t.

SOLUTION   The unknown function here is the temperature, $H(t)$, where t is time. Newton's Law of Cooling tells us that $H(t)$ *decreases* at a rate proportional to the difference between $H(t)$ and E, the temperature of the surrounding environment (E is called the *ambient temperature*). In other words,

$$\frac{dH}{dt} = -k(H - E),$$

where k is a positive constant.* Since E is a constant, we can separate the variables as follows.

$$\frac{dH}{H - E} = -k\,dt.$$

Integrating,

$$\int \frac{dH}{H - E} = \int (-k)\,dt,$$

or

$$\ln|H - E| = -kt + C.$$

To solve for H, first rewrite the expression in exponential form.

$$H - E = e^{-kt + C}$$

(Note that $H - E$ is positive, so we don't need absolute values.)

$$= e^C e^{-kt}$$

$$= A e^{-kt}$$

So

$$H(t) = E + A e^{-kt}.$$

Before we go on...

Q What is the significance of the constant A?

A Consider the temperature at time $t = 0$:

$$H(0) = E + A e^0 = E + A.$$

In words, at time $t = 0$ the temperature of the object equals the temperature of the environment plus A. In other words, A is the initial difference between the ambient temperature and the object. For example, it you place a red-hot ingot at 300°F in a bucket of 65°F water, the initial difference in temperature is 235°F, so the temperature of the ingot will follow the curve

$$H = 65 + 235 e^{-kt}.$$

▼ * When we say that quantity Q is *proportional* to quantity R, we mean that $Q = kR$ for some constant k. The constant k is referred to as the **constant of proportionality.**

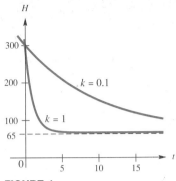

FIGURE 1

The constant k depends on the object that is cooling and the unit of measurement of time t. Figure 1 shows the graph of this function for $k = 1$ and $k = 0.1$.

Note that $\lim\limits_{t \to +\infty} H(t) = \lim\limits_{t \to +\infty} (65 + 235e^{-kt}) = 65$, as illustrated on the graph.

▼   **EXAMPLE 5**   Determining a Demand Equation

Doreen has noticed that the elasticity of demand for her "I ❤ Calculus" T-shirts can be modeled by the linear equation

$$E = 0.5p - 1.5,$$

where p is the price per T-shirt. Also, she can sell 20 T-shirts per week at a price of \$5 per T-shirt. Find a demand equation for her T-shirts by expressing weekly sales q as a function of unit price p.

SOLUTION   First, recall from the chapter on applications of the derivative that the elasticity of demand is given by

$$E = -\frac{dq}{dp} \cdot \frac{p}{q}.$$

The information we are given is that

$$-\frac{dq}{dp} \cdot \frac{p}{q} = 0.5p - 1.5.$$

This is a differential equation in which we can separate the variables by multiplying both sides of the equation by the quantity dp/p.

$$-\frac{1}{q}\,dq = \frac{0.5p - 1.5}{p}\,dp$$

or

$$-\frac{1}{q}\,dq = 0.5 - \frac{1.5}{p}.$$

Integrating,

$$-\int \frac{1}{q}\,dq = \int \left(0.5 - \frac{1.5}{p}\right) dp,$$

or

$$-\ln |q| = 0.5p - 1.5 \ln |p| + C.$$

Because the quantities p and q must be positive, we may drop the absolute value signs and solve for q.

$$\ln q = -0.5p + 1.5 \ln p - C$$

so

$$q = e^{-0.5p + 1.5 \ln p - C}$$
$$= e^{-0.5p} e^{1.5 \ln p} e^{-C}$$
$$= Ae^{-0.5p} e^{\ln(p^{1.5})} \qquad \text{By the rules for logarithms}$$
$$= Ap^{1.5} e^{-0.5p}. \qquad \text{Since } e^{\ln x} = x$$

We now have the demand equation, except for the constant A. But there is one piece of information we haven't used yet: $q = 20$ when $p = 5$. If we substitute this into our demand equation, we get

$$20 = A(5^{1.5}) e^{-0.5(5)}.$$

Thus,

$$A = \frac{20}{5^{1.5} e^{-2.5}} \approx 21.7927.$$

Now that we know the value of A, we can write our demand equation as

$$q = 21.7927 p^{1.5} e^{-0.5p}.$$

(This is the particular solution that fits the given data.)

Before we go on... A linear model for elasticity of demand has led to an exponential model for the demand equation. Is this a plausible equation? Figure 2 shows a graphing calculator plot of q as a function of p.

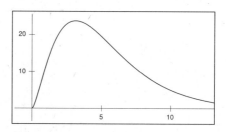

FIGURE 2

The curve behaves reasonably for $p \geq \$3$. On the other hand, the curve seems to indicate that at prices less than \$3 per T-shirt, demand will increase with increasing price! This is one of the shortcomings of relying on a linear model for elasticity of demand. Recall from Chapter 1 that linear models are often useful only for a small range of values. We therefore view the part of the demand curve with $p \leq 3$ with some suspicion.

► __7.6 EXERCISES__

Find the general solutions of the differential equations in Exercises 1 through 10.
Where possible, solve for y as a function of x.

1. $\dfrac{dy}{dx} = x^2 + \sqrt{x}$

2. $\dfrac{dy}{dx} = \dfrac{1}{x} + 3$

3. $\dfrac{dy}{dx} = \dfrac{x}{y}$

4. $\dfrac{dy}{dx} = \dfrac{y}{x}$

5. $\dfrac{dy}{dx} = xy$

6. $\dfrac{dy}{dx} = x^2 y$

7. $\dfrac{dy}{dx} = (x + 1)y^2$

8. $\dfrac{dy}{dx} = \dfrac{1}{(x + 1)y^2}$

9. $x\dfrac{dy}{dx} = \dfrac{1}{y}\ln x$

10. $\dfrac{1}{x}\dfrac{dy}{dx} = \dfrac{1}{y}\ln x$

For each differential equation in Exercises 11 through 20, find the particular so-
lution indicated.

11. $\dfrac{dy}{dx} = x^3 - 2x, \quad y = 1$ when $x = 0$

12. $\dfrac{dy}{dx} = 2 - e^{-x}, \quad y = 0$ when $x = 0$

13. $\dfrac{dy}{dx} = \dfrac{x^2}{y^2}, \quad y = 2$ when $x = 0$

14. $\dfrac{dy}{dx} = \dfrac{y^2}{x^2}, \quad y = \dfrac{1}{2}$ when $x = 1$

15. $x\dfrac{dy}{dx} = y, \quad y(1) = 2$

16. $x^2 \dfrac{dy}{dx} = y, \quad y(1) = 1$

17. $\dfrac{dy}{dx} = x(y + 1), \quad y(0) = 0$

18. $\dfrac{dy}{dx} = \dfrac{y + 1}{x}, \quad y(0) = 2$

19. $\dfrac{1}{x}\dfrac{dy}{dx} = \dfrac{y^2}{x^2 + 1}, \quad y(0) = -1$

20. $\dfrac{1}{x}\dfrac{dy}{dx} = \dfrac{y}{(x^2 + 1)^2}, \quad y(0) = 1$

APPLICATIONS

21. *Sales*   Your monthly sales of Tofu Ice Cream are falling 5% per month. If you currently sell 1,000 quarts per month, find the differential equation describing your change in sales, and then solve to predict your monthly sales.

22. *Profit*   Your monthly profit on sales of Avocado Ice Cream are rising 10% per month. If you currently make a profit of $15,000 per month, find the differential equation describing your change in profit, and solve to predict your monthly profits.

23. *Market Saturation*   You have just introduced a new computer to the market. You predict that you will eventually sell 100,000 computers and that your monthly rate of sales will be 10% of the difference between the saturation value and the total number you have sold up to that point. Find a differential equation for your total sales (as a function of the month) and solve. (What are your total sales at the moment when you first introduce the computer?)

24. *Market Saturation*   Repeat Exercise 23, assuming that monthly sales will be 5% of the difference be-

tween the saturation value and the total sales so that point, and assuming that you sell 5,000 computers to customers just before placing the computer on the market.

25. *Logistic Equation*   There are many examples of growth in which the rate of growth is slow at first, becomes faster, and then slows again as a limit is reached. This can be described by the differential equation

$$\frac{dy}{dt} = ay(L - y),$$

where a is a constant and L is the limit of y. Show by substitution that

$$y = \frac{CL}{e^{-aLt} + C}$$

is a solution to this equation, where C is an arbitrary constant.

26. *Logistic Equation*   Using separation of variables and integration with a table of integrals, solve the differential equation in the preceding exercise to derive the solution given there.

Exercises 27–30 require the use of a graphing calculator or computer.

 27. *Market Saturation*   You have just introduced a new model of TV. You predict that the market will saturate at 2,000,000 TVs and that your total sales will be governed by the equation

$$\frac{dS}{dt} = \frac{1}{4}S(2 - S),$$

where S is the total sales in millions of TVs and t is measured in months. If you give away 1,000 TV sets when you first introduce the TV, what will S be? Sketch the graph of S as a function of t. About how long will it take to saturate the market? (See Exercise 25.)

 28. *Epidemics*   A certain epidemic of influenza is predicted to follow the function defined by

$$\frac{dA}{dt} = \frac{1}{10}A(20 - A),$$

where A is the number of people infected (in millions) and t is the number of months after the epidemic starts. If 20,000 cases are reported initially, find $A(t)$

and sketch its graph. When is A growing fastest? How many people will eventually be affected? (See Exercise 25.)

 29. *Growth of Tumors*   The growth of tumors in animals can be modeled by the Gompertz equation

$$\frac{dy}{dt} = -ay \ln\left(\frac{y}{b}\right),$$

where y is the size of a tumor, t is time, and a and b are constants that depend on the type of tumor and the units of measurement.

(a) Solve for y as a function of t.

(b) If $a = 1$, $b = 10$, and $y(0) = 5$ cm^3 (with t measured in days), find the specific solution and graph it.

 30. *Growth of Tumors*   Continuing the preceding exercise, suppose that $a = 1$, $b = 10$, and $y(0) = 15$ m^3. Find the specific solution and graph it. Comparing its graph to that obtained in the previous exercise, what can you say about tumor growth in these instances?

COMMUNICATION AND REASONING EXERCISES

31. What is the difference between a particular solution and the general solution of a differential equation? How do we get a particular solution from the general solution?

32. Why is there always an arbitrary constant in the general solution of a differential equation? Why are there not two or more arbitrary constants in a first-order differential equation?

33. Show by example that a **second-order** differential equation, one involving the second derivative y'', usually has two arbitrary constants in its general solution.

34. Find a differential equation that is not separable.

35. Find a differential equation whose general solution is $y = 4e^{-x} + 3x + C$.

36. Explain how, knowing the elasticity of demand as a function of either price or demand, you may find the demand equation.

▶ ▃▃ **You're the Expert**

ESTIMATING TAX
REVENUES

You have just been hired by the incoming administration of a certain country as chief consultant for national tax policy and have been getting conflicting advice from the so-called finance experts on your staff. Several of them have come up with plausible suggestions for new tax structures, and your job is to choose the plan that results in the most revenue for the government.

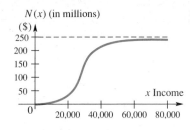

FIGURE 1

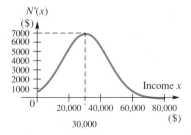

FIGURE 2

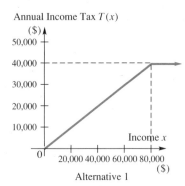

FIGURE 3

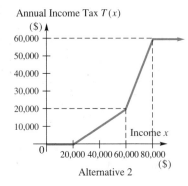

FIGURE 4

Before you can evaluate their plans, you realize that it is essential to know the income distribution: that is, how many people earn how much money. One might think that the most useful way of specifying income distribution would be to use a function that gives the exact number $f(x)$ of people earning a given salary x. This would necessarily be a **discrete** function—it only makes sense if x happens to be a whole number of cents. There is, after all, no one earning a salary of exactly \$12,000.142567! Further, this function would behave rather erratically, since there are, for example, probably many more people making a salary of exactly \$30,000 than exactly \$30,000.01. It is far more convenient to start with the function defined by

$$N(x) = \text{total number of people earning between 0 and } x \text{ dollars.}$$

The graph of $N(x)$ might look like the one shown in Figure 1.

Notice that the curve climbs most steeply at around $x = \$30,000$, which suggests that most people are earning around that amount. If we take the *derivative* of $N(x)$, we get an **income distribution function.** Its graph might look like the one shown in Figure 2.

The derivative $N'(x)$ peaks at \$30,000. Since the derivative measures the rate of change, its value at x is the additional number of taxpayers per \$1 increase in salary. Thus, its value at, say, $x = 20,000$ tells us the approximate number of people earning between \$20,000 and \$20,001, so the fact that $N'(20,000) = 5,000$ tells us that approximately 5,000 people are earning a salary of \$20,000. In other words, N' shows the distribution of incomes among the population, hence the name "distribution function."*

You thus send a memo to your experts requesting the income distribution function for the nation. After much collection of data, they tell you that the income distribution function is

$$N'(x) = 7,000 \, e^{-(x-30,000)^2/400,000,000}.$$

This is in fact the function whose graph is shown in Figure 2, and it is an example of a **normal** distribution. Notice that the curve is symmetric around the median income of \$30,000 and that about 7,000 people are earning between \$30,000 and \$30,001 annually.[†]

Given this income distribution, your financial experts have come up with two possible tax policies (Figures 3 and 4).

In the first alternative, all taxpayers pay half of their income in taxes, except that no one pays more than \$40,000 in taxes. In the second alternative, there are four **tax brackets,** described by the following table.

▼ * We shall look at a very similar idea later in the chapter on probability and calculus.

[†] You might find it odd that you weren't given the original function N, but it will turn out that you don't need it. How would you compute it? Is it possible to write down a formula for N?

Income	Marginal tax rate
$0–20,000	0%
$20,000–60,000	50%
$60,000–80,000	200%
Above $80,000	0%

Now you must determine which alternative will generate the largest annual tax revenue.

Each of Figures 3 and 4 is the graph of a function, T. Before calculating the formulas for these functions, you try to work with the general situation. You have an income distribution function N' and a tax function T, both functions of annual income. You need to find a formula for total tax revenues. You begin by trying subdivision. First, you decide to work only with incomes in some finite bracket $[0, M]$ by using a cutoff—say, $M = \$10$ million. (You shall eventually let M approach $+\infty$.) Next, you subdivide the interval $[0, M]$ into a large number of intervals of width Δx. If $[x_{k-1}, x_k]$ is a typical such interval, you need to calculate the approximate tax revenue from people whose total incomes lie between x_{k-1} and x_k. You will then sum over k to get the total revenue.

You need to know how many people are making incomes between x_{k-1} and x_k. Because $N(x_k)$ people are making incomes *up to* x_k and $N(x_{k-1})$ people are making incomes up to x_{k-1}, the number of people making incomes between x_{k-1} and x_k is $N(x_k) - N(x_{k-1})$. Because x_k is very close to x_{k-1}, the incomes of these people are all approximately equal to x_k dollars, so each of these taxpayers is paying an annual tax of about $T(x_k)$. This gives a tax revenue of

$$[N(x_k) - N(x_{k-1})]T(x_k).$$

Now you do a clever thing. You write $x_k - x_{k-1} = \Delta x$ and replace the quantity $N(x_k) - N(x_{k-1})$ by

$$\frac{N(x_k) - N(x_{k-1})}{\Delta x}\, \Delta x.$$

This gives you a tax revenue of about

$$\frac{N(x_k) - N(x_{k-1})}{\Delta x}\, T(x_k)\, \Delta x.$$

from wage-earners in the bracket $[x_{k-1}, x_k]$. Summing over k gives an approximate total revenue of

$$\sum_{k=1}^{n} \frac{N(x_k) - N(x_{k-1})}{\Delta x}\, T(x_k)\, \Delta x.$$

The larger n is, the more accurate your estimate will be, so you take the limit of the sum as $n \to \infty$. When you do this, two things happen. First, the quantity

$$\frac{N(x_k) - N(x_{k-1})}{\Delta x}$$

approaches the derivative, $N'(x_k)$. Second, the sum, which you recognize as a Riemann sum, approaches the integral

$$\int_0^M N'(x)T(x) \, dx.$$

You now take the limit as $M \rightarrow +\infty$ and obtain

$$\text{Total tax revenue} = \int_0^{+\infty} N'(x)T(x) \, dx.$$

This indefinite integral is fine in theory, but the actual calculation will have to be done numerically, so you stick with the upper limit of $10 million for now, and you will have to check that it is reasonable at the end (notice that, by the graph of N', it appears that extremely few, if any, people earn that much). Now you already have a formula for $N'(x)$, but you still need to write formulas for the tax functions $T(x)$ for both alternatives.

Alternative 1   The graph in Figure 3 rises linearly from 0 to 40,000 as x ranges from 0 to 80,000, and then stays constant at 40,000. Thus, the slope of the first part is $\frac{40,000}{80,000} = \frac{1}{2}$. The taxation function is therefore

$$T(x) = \begin{cases} \frac{x}{2} & \text{if } 0 \leq x \leq 80,000 \\ 40,000 & \text{if } x \geq 80,000. \end{cases}$$

To perform the integration, you need to break the integral into two pieces, the first from 0 to 80,000 and the second from 80,000 to 10,000,000. In other words,

$$R_1 = \int_0^{80,000} (7,000e^{-(x-30,000)^2/400,000,000}) \frac{x}{2} \, dx +$$

$$\int_{80,000}^{10,000,000} (7,000e^{-(x-30,000)^2/400,000,000})40,000 \, dx.$$

You decide not to attempt this by hand*! You use numerical integration software to obtain a grand total of $R_1 = \$3,732,760,000,000$, or \$3.73276 trillion.[†]

Alternative 2   The graph in Figure 4 rises linearly from 0 to 20,000 as x ranges from 20,000 to 60,000, rises from 20,000 to 60,000 as x ranges from

---

▼ * The first integral requires a substitution and integration by parts. The second cannot be done in elementary terms at all.

[†] Rounded to six significant digits.

60,000 to 80,000, and then stays constant at 60,000. Thus, the slope of the first incline is $\frac{1}{2}$ and the slope of the second incline is 2 (this is why the *marginal* tax rates are 50% and 200% respectively). The taxation function is therefore

$$T(x) = \begin{cases} 0 & \text{if } 0 \leq x \leq 20{,}000 \\ \dfrac{x - 20{,}000}{2} & \text{if } 20{,}000 \leq x \leq 60{,}000 \\ 20{,}000 + 2(x - 60{,}000) & \text{if } 60{,}000 \leq x \leq 80{,}000 \\ 60{,}000 & \text{if } x \geq 80{,}000 \end{cases}$$

Values of x between 0 and 20,000 do not contribute to the integral, and so

$$R_2 = \int_{20{,}000}^{60{,}000} (7{,}000e^{-(x-30{,}000)^2/400{,}000{,}000}) \left(\frac{x - 20{,}000}{2} \right) dx$$

$$+ \int_{60{,}000}^{80{,}000} (7{,}000e^{-(x-30{,}000)^2/400{,}000{,}000})(20{,}000 + 2(x - 60{,}000))\, dx$$

$$+ \int_{80{,}000}^{10{,}000{,}000} (7{,}000e^{-(x-30{,}000)^2/400{,}000{,}000})60{,}000\, dx.$$

This gives $R_2 = \$1.52016$ trillion—considerably less than Alternative 1. Thus, even though this alternative taxes the wealthy more heavily, it yields less total revenue.

Now what about the cutoff at $10,000,000 annual income? If you try either integral again with an upper limit of $100 million, you will see no change in either one to 6 significant digits. There simply are not enough taxpayers earning an income above $10,000,000 to make a difference. You conclude that your answers are sufficiently accurate and that the first alternative provides the most tax revenue.

Exercises

(Exercises 1–6 require the use of a graphing calculator or computer.)

In Exercises 1–6, calculate the total tax revenue for a country with the given income distribution and tax policies.

1. $N'(x) = 3{,}000e^{-(x-10{,}000)^2/10{,}000}$, 25% tax on all income

2. $N'(x) = 3{,}000e^{-(x-10{,}000)^2/10{,}000}$, 45% tax on all income

3. $N'(x) = 5{,}000e^{-(x-30{,}000)^2/100{,}000}$, no tax on an income below $30,000, $10,000 tax on any income of $30,000 or above.

4. $N'(x) = 5{,}000e^{-(x-30{,}000)^2/100{,}000}$, no tax on an income below $50,000, $20,000 tax on any income of $50,000 or above.

$$\frac{N(x_k) - N(x_{k-1})}{\Delta x}$$

approaches the derivative, $N'(x_k)$. Second, the sum, which you recognize as a Riemann sum, approaches the integral

$$\int_0^M N'(x)T(x) \, dx.$$

You now take the limit as $M \to +\infty$ and obtain

$$\text{Total tax revenue} = \int_0^{+\infty} N'(x)T(x) \, dx.$$

This indefinite integral is fine in theory, but the actual calculation will have to be done numerically, so you stick with the upper limit of $10 million for now, and you will have to check that it is reasonable at the end (notice that, by the graph of N', it appears that extremely few, if any, people earn that much). Now you already have a formula for $N'(x)$, but you still need to write formulas for the tax functions $T(x)$ for both alternatives.

Alternative 1   The graph in Figure 3 rises linearly from 0 to 40,000 as x ranges from 0 to 80,000, and then stays constant at 40,000. Thus, the slope of the first part is $\frac{40,000}{80,000} = \frac{1}{2}$. The taxation function is therefore

$$T(x) = \begin{cases} \frac{x}{2} & \text{if } 0 \le x \le 80{,}000 \\ 40{,}000 & \text{if } x \ge 80{,}000. \end{cases}$$

To perform the integration, you need to break the integral into two pieces, the first from 0 to 80,000 and the second from 80,000 to 10,000,000. In other words,

$$R_1 = \int_0^{80{,}000} (7{,}000e^{-(x-30{,}000)^2/400{,}000{,}000}) \frac{x}{2} \, dx +$$

$$\int_{80{,}000}^{10{,}000{,}000} (7{,}000e^{-(x-30{,}000)^2/400{,}000{,}000})40{,}000 \, dx.$$

You decide not to attempt this by hand*! You use numerical integration software to obtain a grand total of $R_1 = \$3{,}732{,}760{,}000{,}000$, or $3.73276 trillion.†

Alternative 2   The graph in Figure 4 rises linearly from 0 to 20,000 as x ranges from 20,000 to 60,000, rises from 20,000 to 60,000 as x ranges from

▼ * The first integral requires a substitution and integration by parts. The second cannot be done in elementary terms at all.

† Rounded to six significant digits.

60,000 to 80,000, and then stays constant at 60,000. Thus, the slope of the first incline is $\frac{1}{2}$ and the slope of the second incline is 2 (this is why the *marginal* tax rates are 50% and 200% respectively). The taxation function is therefore

$$T(x) = \begin{cases} 0 & \text{if } 0 \le x \le 20{,}000 \\ \dfrac{x - 20{,}000}{2} & \text{if } 20{,}000 \le x \le 60{,}000 \\ 20{,}000 + 2(x - 60{,}000) & \text{if } 60{,}000 \le x \le 80{,}000 \\ 60{,}000 & \text{if } x \ge 80{,}000 \end{cases}$$

Values of x between 0 and 20,000 do not contribute to the integral, and so

$$R_2 = \int_{20{,}000}^{60{,}000} (7{,}000e^{-(x-30{,}000)^2/400{,}000{,}000}) \left(\frac{x - 20{,}000}{2} \right) dx$$

$$+ \int_{60{,}000}^{80{,}000} (7{,}000e^{-(x-30{,}000)^2/400{,}000{,}000})(20{,}000 + 2(x - 60{,}000)) \, dx$$

$$+ \int_{80{,}000}^{10{,}000{,}000} (7{,}000e^{-(x-30{,}000)^2/400{,}000{,}000}60{,}000 \, dx.$$

This gives $R_2 = \$1.52016$ trillion—considerably less than Alternative 1. Thus, even though this alternative taxes the wealthy more heavily, it yields less total revenue.

Now what about the cutoff at \$10,000,000 annual income? If you try either integral again with an upper limit of \$100 million, you will see no change in either one to 6 significant digits. There simply are not enough taxpayers earning an income above \$10,000,000 to make a difference. You conclude that your answers are sufficiently accurate and that the first alternative provides the most tax revenue.

Exercises

(Exercises 1–6 require the use of a graphing calculator or computer.)

In Exercises 1–6, calculate the total tax revenue for a country with the given income distribution and tax policies.

1. $N'(x) = 3{,}000e^{-(x-10{,}000)^2/10{,}000}$, 25% tax on all income

2. $N'(x) = 3{,}000e^{-(x-10{,}000)^2/10{,}000}$, 45% tax on all income

3. $N'(x) = 5{,}000e^{-(x-30{,}000)^2/100{,}000}$, no tax on an income below \$30,000, \$10,000 tax on any income of \$30,000 or above.

4. $N'(x) = 5{,}000e^{-(x-30{,}000)^2/100{,}000}$, no tax on an income below \$50,000, \$20,000 tax on any income of \$50,000 or above.

5. $N'(x) = 7,000e^{-(x-30,000)^2/400,000,000}$, $T(x)$ with the following graph:

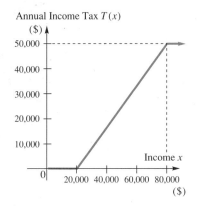

Annual Income Tax $T(x)$
($)

6. $N'(x) = 7,000e^{-(x-30,000)^2/400,000,000}$, $T(x)$ with the following graph:

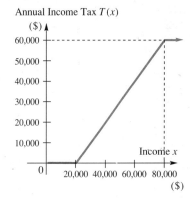

Annual Income Tax $T(x)$
($)

▶ ### Review Exercises

Evaluate the integrals in Exercises 1–30.

1. $\displaystyle\int (x^2 + 2)e^x \, dx$

2. $\displaystyle\int (x^2 + x)e^{2x} \, dx$

3. $\displaystyle\int x^2 \ln 2x \, dx$

4. $\displaystyle\int \sqrt{x} \ln \sqrt{x} \, dx$

5. $\displaystyle\int_0^1 x^2 e^x \, dx$

6. $\displaystyle\int_{-2}^2 (x^3 + 1)e^{-x} \, dx$

7. $\displaystyle\int_1^e x^2 \ln x \, dx$

8. $\displaystyle\int_1^{2e} (x + 1)\ln x \, dx$

9. $\displaystyle\int \frac{dx}{9 + 4x^2}$

10. $\displaystyle\int \frac{dx}{9 - 4x^2}$

11. $\displaystyle\int \frac{dx}{\sqrt{9 + 4x^2}}$

12. $\displaystyle\int \frac{dx}{\sqrt{9 - 4x^2}}$

13. $\displaystyle\int \sqrt{x^2 + 2x + 2} \, dx$

14. $\displaystyle\int \sqrt{x^2 + 2x - 2} \, dx$

15. $\displaystyle\int \frac{e^{2x}}{1 + e^{4x}} \, dx$

16. $\displaystyle\int \frac{e^{2x}}{e^{4x} - 1} \, dx$

17. $\displaystyle\int_0^1 \frac{e^{2x}}{\sqrt{9 + 4e^{4x}}} \, dx$

18. $\displaystyle\int_{-1}^0 \frac{e^{2x}}{\sqrt{9 - 4e^{4x}}} \, dx$

19. $\displaystyle\int_1^e \frac{\sqrt{(\ln x)^2 + 1}}{x} \, dx$

20. $\displaystyle\int_1^e \frac{\ln x \sqrt{(\ln x)^2 + 1}}{x} \, dx$

21. $\displaystyle\int_1^\infty \frac{1}{x^5} \, dx$

22. $\displaystyle\int_1^\infty \frac{1}{x^{0.6}} \, dx$

23. $\displaystyle\int_{-\infty}^1 e^{x/2} \, dx$

24. $\displaystyle\int_{-\infty}^0 xe^{x/2} \, dx$

25. $\displaystyle\int_{-\infty}^\infty \frac{1}{x^2} \, dx$

26. $\displaystyle\int_{-\infty}^\infty e^{-x} \, dx$

27. $\displaystyle\int_0^1 \frac{1}{(x - 1)^2} \, dx$

28. $\displaystyle\int_0^1 \frac{1}{x^2 - 1} \, dx$

29. $\displaystyle\int_0^1 \frac{1}{\sqrt{1 - x}} \, dx$

30. $\displaystyle\int_0^1 \frac{1/\sqrt{x}}{1 - \sqrt{x}} \, dx$

Find the areas indicated in Exercises 31–40.

31. Between $y = x^3$ and $y = 1 - x^3$ for x in $[0, 1]$.

32. Between $y = x^3$ and $y = 1 - x^3$ for x in $[-1, 1]$.

33. Between $y = e^x$ and $y = 2e^{-x}$ for x in $[0, 2]$.

34. Between $y = e^x$ and $y = e^{-x}$ for x in $[0, 2]$.

35. Between $y = 1 - x^2$ and $y = x^2$.

36. Between $y = 1 - x^4$ and $y = x^4$.

37. Between $y = x^3 - x$ and $y = x - x^3$.

38. Between $y = x^4 - x^2 + 1$ and $y = x^4$.

39. Between $y = \dfrac{1}{x^2 + 1}$ and $y = \dfrac{1}{x^2 - 1}$ for x in $[0, \frac{1}{2}]$.

40. Between $y = \dfrac{1}{x^2 + 4}$ and $y = \dfrac{1}{2x^2 + 3}$ for x in $[0, 1]$.

41–44. Find the averages of the functions in Exercises 5–8.   **45–48.** Find the averages of the functions in Exercises 17–20.

In Exercises 49–52, find the 2-unit moving averages of the given functions. If you are using a graphing calculator or computer software, plot the original function and its moving average on the same graph.

49. $f(x) = x^{2/3}$

50. $f(x) = x^{4/3}$

51. $f(x) = \ln x$

52. $f(x) = x \ln x$

53. $\dfrac{dy}{dx} = x^2 y^2$

54. $\dfrac{dy}{dx} = x^3 y^2$

55. $\dfrac{dy}{dx} = xy + x + y + 1$

56. $\dfrac{dy}{dx} = xy + 2x$

57. $\dfrac{dy}{dx} = \dfrac{1}{y}$

58. $\dfrac{dy}{dx} = \dfrac{1}{y^2}$

59. $xy \dfrac{dy}{dx} = 1$, $y(1) = 1$

60. $x^2 y \dfrac{dy}{dx} = 1$, $y(1) = 1$

61. $y(x^2 + 1) \dfrac{dy}{dx} = xy^2$, $y(0) = 2$

62. $xy^2 \dfrac{dy}{dx} = y(x^2 + 1)$, $y(1) = 10$

APPLICATIONS

63. *Dangerous Weapons*   The following graph shows the number of deaths per year in the U.S. resulting from injuries related to motor vehicles and firearms.*

(a) Estimate the definite integral of the function represented by the top graph from '73 to '92. What does the answer tell you?

(b) Estimate the area between the graphs from '73 to '92. What does the answer tell you?

Deaths Per Year in U.S.

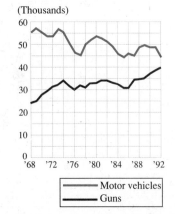

64. *Dangerous Weapons*   Referring to the graph in the previous exercise, estimate the average difference between the two graphs from '68 to '88. What does the answer tell you?

65. *Cost*   Your factory buys nuts, currently priced at $2 per box, but the price is rising 3% per month continuously due to high inflation. You currently buy 1,000 boxes per month, but your purchases are increasing by 100 boxes per month. How much will you spend on nuts in the next 12 months?

66. *Cost*   Your factory buys bolts, currently priced at $1.50 per box, but the price is rising 2.5% per month continuously due to high inflation. You currently buy 5,000 boxes per month, but your purchases are declining by 200 boxes per month. How much will you spend on bolts in the next 12 months?

67. *Supply and Demand*   The supply equation for fertilizer is

$$p = (q - 20)^{0.5} \qquad (q \geq 20),$$

where p is the price (in dollars) per 20-lb bag and q is the weekly supply. The demand equation is

$$p = (70 - q)^{0.5} \qquad (q \leq 70).$$

Find the equilibrium price and the consumers' and producers' surpluses at that price.

▼ *Source: National Center for Health Statistics/The New York Times, January 26, 1994, p. A12.*

68. *Supply and Demand* The supply equation for compost is

$$p = (q - 10)^{0.7} \qquad (q \geq 10),$$

where p is the price (in dollars) per 20-lb bag and q is the weekly supply. The demand equation is

$$p = (60 - q)^{0.7} \qquad (q \leq 60).$$

Find the equilibrium price and the consumers' and producers' surpluses at that price.

69. *Motion* A cannonball launched into the air from ground level had an upward velocity of $v(t) = 320 - 32t$ ft/s after t seconds.
   (a) Find the area under the graph of v for $0 \leq t \leq 10$. What does this area represent?
   (b) Find the time T at which the ball returns to the ground, and then find the area under the graph for $0 \leq t \leq T$ without further calculation.

70. *Motion* Two cars pass the same point on the New York Thruway at time $t = 0$ hours. Car A maintains a speed of $v_A = 55$ mph, while car B has velocity $v_B(t) = 30 + 20t$ for $0 \leq t \leq 2$ and then $v_B(t) = 70 - 20t$ for $2 \leq t \leq 4$.
   (a) Find the area between the graphs of v_A and v_B for $0 \leq t \leq 1$. What does this area represent?
   (b) When will the two cars meet for the second time? Express your answer both as a time and in terms of the areas between the graphs of v_A and v_B.

71. *Pedestrian Safety* The following chart shows the number of pedestrians killed by vehicles in New York City each year over the given period.*
   (a) Find the average number of pedestrians killed in New York per year over the given period.
   (b) Give a rough sketch of the 2-year moving average without doing any actual computation, and compare your sketch with that of the actual 2-year moving average.

**Pedestrians Killed by Vehicles
in New York City**

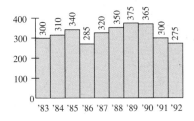

72. *Growth of Medical Costs* Repeat the previous exercise using the following chart, which shows the annual rate of increase in medical costs in the New York City Metropolitan Center.[†]

**Rate of Increase in Medical Costs in the
New York City Metropolitan Center**

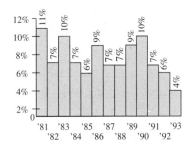

73. *Investments* If your investment of $10,000 in Tarnished Teak Enterprises is continuously depreciating at 6% per year, what is the average amount in your account over one year?

74. *Bacterial Growth* If a culture of bacteria is doubling every hour, what is the average population over the first two hours, assuming that the culture contains a million organisms at the start?

75. *Investments* You are offered an investment whose instantaneous rate of growth (in dollars per year) is always 10% of the square root of the amount in the account. If you invest $10,000 initially, find the amount present as a function of time.

76. *Loans* A junior executive at a bank came up with the following idea: The bank should offer an account accumulating interest at a rate (in dollars per year) equal to 1% of the square of the amount in the account at any time. Looking at what would happen if someone were to deposit $100 in such an account, explain why the executive was fired on the spot.

▼ *Source: Department of Transportation/*The New York Times,* January 21, 1994, p. B1.
   [†] Source: Federal Bureau of Labor Statistics/*The New York Times,* December 25, 1993, p. L 36.

CHAPTER **8**

"Royal Flush"

After some lean years, luxury homes are starting to sell again on Long Island

A Long Island house that looks like a million bucks may cost a good deal less now. And buyers are starting to notice.

The market for luxury houses, battered by the recession, is experiencing a resurgence. Most of the buyers are Long Island residents who have been able to sell their middle-market houses and take advantage of the not-so-sky-high price of living in luxury.

Last year was the Daniel Gale Agency's best year ever for selling luxury houses—those priced at $400,000 or more—said Bonnie Devendorf, associate broker and manager of the firm's Locust Valley office. She said the firm, which covers the North Shore from Stony Brook to Port Washington, is on its way this year to beat the record, despite a $9 million home sale in 1993.

"Year over year, the market above $400,000 is certainly up from last year," said Jim Linane, regional manager of Chase Manhattan Personal Financial Services, which specializes in mortgages of more than $300,000. He said such areas as Port Washington, Manhasset and Jericho are experiencing a lot of activity. "They're fairly convenient to the railroad and are good commuter towns," he said. "The South Shore Suffolk area is a little bit slow this year. But anything's that's a decent commute into the city seems to be pretty hot."

An easy commute into New York was a major consideration for Steve and Carol Sterneck, who recently moved from Manhattan into an East Hills home. "It was a combination of factors that led to the decision to move out of the city and into a home," Steve Sterneck said. . . .

Source: From Jacqueline Henry, "Royal Flush," *New York Newsday*, 1994, p. D8.

Functions of Several Variables

APPLICATION ▶ Your market research of real estate investments reveals the following sales figures for new homes of different prices over the past year.

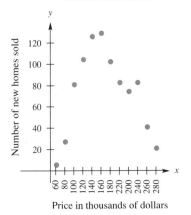

Sales of New Homes

As the chief economist for Sturdy Homes, Inc., you would like to come up with a demand equation for home buying trends.

589

INTRODUCTION ▶ We have studied functions of a single variable quite extensively now. However, not every function that arises is a function of a single variable. For example, suppose you run a company that produces two types of floppy disks, low-density and high-density, and you wish to come up with a cost equation. If your factory produces x low-density disks and y high-density disks, your cost equation might look something like this:

$$C(x, y) = 0.2x + 0.4y + 0.1\sqrt{xy}.$$

This is a function of two variables, since it *depends on* both x and y. As we shall see, we can extend the techniques of calculus to such functions, allowing us to examine marginal costs, for example. One of the main applications we shall look at is optimization: finding the maximum or minimum of a function of two or more variables.

▶ 8.1 FUNCTIONS OF TWO OR MORE VARIABLES

Recall that a function of one variable is a rule for manufacturing a new number $f(x)$ from a single number x. In the same vein, a **function, f, of x and y** is a rule for manufacturing a new number $f(x, y)$ from a *pair* of numbers (x, y). For instance, the cost function $C(x, y) = 0.2x + 0.4y + 0.1\sqrt{xy}$ mentioned in the introduction is such a function. The function $f(x, y) = x^2 + y^2$ is another example. Figure 1 illustrates this concept—in goes a pair of numbers and out comes a single number.

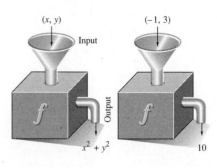

$$f(x, y) = x^2 + y^2$$

FIGURE 1

For example, with $f(x, y) = x^2 + y^2$, $f(-1, 3) = (-1)^2 + 3^2 = 10$. In other words, we evaluated $f(-1, 3)$ by substituting the quantity (-1) for x and quantity 3 for y.

▼ **EXAMPLE 1**

Let $f(x, y) = x^2 - 2xy + y - 1$. Find $f(-1, 1), f(0, 0), f(a, 1), f(a, b)$, and $f(a + h, b)$.

SOLUTION   To calculate $f(-1, 1)$ we substitute the quantity (-1) for x and 1 for y. Thus,

$$f(-1, 1) = (-1)^2 - 2(-1)(1) + (1) - 1 = 3.$$

Similarly,

$$f(0, 0) = 0^2 - 2(0)(0) + (0) - 1 = -1,$$
$$f(a, 1) = a^2 - 2a(1) + (1) - 1 = a^2 - 2a,$$
$$f(a, b) = a^2 - 2ab + b - 1,$$
$$f(a + h, b) = (a + h)^2 - 2(a + h)b + b - 1.$$

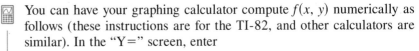 You can have your graphing calculator compute $f(x, y)$ numerically as follows (these instructions are for the TI-82, and other calculators are similar). In the "Y=" screen, enter

$$Y_1 = X^\wedge 2 - 2XY + Y - 1$$

Then, to evaluate, say, $f(-4, 3)$, enter

$$-4 \rightarrow X$$
$$3 \rightarrow Y$$
$$Y_1$$

and the calculator will evaluate the function and give the answer, $f(-4, 3) = 42$.

This procedure can be laborious if we want to calculate $f(x, y)$ for several different values of x and y, so we have supplied a program to remedy this. (See Appendix B.)

▼ **EXAMPLE 2**   Cost Functions

Suppose that you own a company making two models of stereo speakers: the Ultra Mini and the Big Stack. Your total monthly cost (in dollars) to make x Ultra Minis and y Big Stacks is given by

$$C(x, y) = \$10,000 + 20x + 40y.$$

What is the significance of each term in this formula?

SOLUTION   The terms have meanings similar to those we saw for cost functions of a single variable. The constant term 10,000 is the **fixed cost,** the amount it costs even if you make no speakers, because

$$C(0, 0) = 10,000.$$

The term $20x$ indicates that each Ultra Mini adds $20 to the total cost. We say that $20 is the **marginal cost** of each Ultra Mini. The term $40y$ indicates that each Big Stack adds $40 to the total cost. The marginal cost of each Big Stack is $40.

Before we go on... This is an example of a **linear** function of two variables. The coefficients of x and y play roles similar to that of the slope of a line. In particular, they give the rates of change of the function as each variable increases while the other stays constant (think about it).

▼   **EXAMPLE 3**   Cost Functions

Another possibility for the cost function in the previous example is

$$C(x, y) = \$10,000 + 20x + 40y + 30\sqrt{x + y}.$$

We shall see later that it is still possible to compute the marginal cost of each model of speaker, and the extra term has the effect that the marginal costs decrease. This might be the result of economies produced by buying common materials in bulk. Notice that the extra term is really a function of the total number of speakers produced, $x + y$.

Before we go on... What values of x and y may we substitute into $C(x, y)$? Certainly we must have $x \geq 0$ and $y \geq 0$, because it makes no sense to speak of manufacturing a negative number of speakers. There is certainly also some upper bound to the number of speakers that can be made in a month. This might take one of several forms. It might be that $x \leq 100$ and $y \leq 75$. The inequalities $0 \leq x \leq 100$ and $0 \leq y \leq 75$ describe the region in the plane shaded in Figure 2.

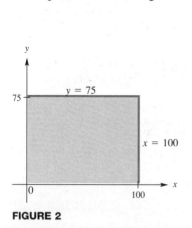

FIGURE 2

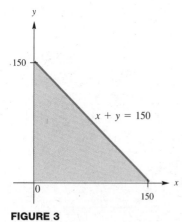

FIGURE 3

Another possibility would be that the *total* number of speakers is bounded—say, $x + y \leq 150$. This, together with $x \geq 0$ and $y \geq 0$, describes the region shaded in Figure 3.

In either case, the region shown represents the pairs (x, y) for which $C(x, y)$ is defined. Just as in the case of a function of one variable, we call this region the **domain** of the function. As before, when the domain is not given explicitly we agree to take the largest possible domain.

▼ **EXAMPLE 4**   Revenue

Revenue is given by the equation

$$R = pq,$$

where p is the price per item and q is the number of items sold. We can now think of this as a function of two variables and write it as $R(p, q)$.

Before we go on... In examples earlier in this book we reduced R to a function of one variable by writing, for example, q as a function of p. This gives

$$R(p) = R(p, q(p)).$$

We used similar methods in, for example, optimization problems to reduce what were really functions of two or more variables to functions of one variable alone. In this chapter we shall see how to apply calculus directly to functions of several variables.

The next example reintroduces the Cobb-Douglas production formula, this time viewed as a function of two variables.

▼ **EXAMPLE 5**   Cobb–Douglas Production Function

Productivity usually depends on both labor and capital. For example, we can measure the productivity of an automobile manufacturing plant by counting the number of automobiles it produces per year. As a measure of labor, we can use the number of employees, and we can use the annual operating budget total as our measure of capital. The Cobb-Douglas production function then has the form

$$P(x, y) = Kx^a y^{1-a},$$

where P stands for the number of automobiles produced per year, x is the number of employees, and y is the annual operating budget. The numbers K and a are constants that depend on the particular factory we are looking at, with a between 0 and 1. For instance, we might have $P(x, y) = 10x^{0.3}y^{0.7}$ for our particular automobile factory. Values for these constants can be determined using actual production data. (We shall be doing this in the exercises.)

Before we go on... Note that if we have either $x = 0$ or $y = 0$, then $P(x, y) = 0$. This is consistent with common sense—we can hardly expect anything to be produced without labor or capital.

Functions of several variables can arise in geometry as well. For example, the area of a rectangle with height h and width w is

$$A(h, w) = hw.$$

Here is another useful example.

▼   **EXAMPLE 6**   Distance

Express the distance of the point (x, y) to the point $(3, 2)$ as a function of the two variables x and y. Use your function to find the distance from $(-1, 1)$ to $(3, 2)$.

SOLUTION   If we plot (x, y) and $(3, 2)$, we might get the picture shown in Figure 4. We need to calculate the distance d.

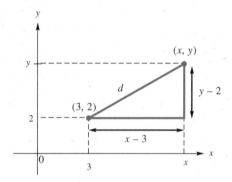

FIGURE 4

By the Pythagorean theorem applied to the right triangle shown, we get

$$d^2 = (x - 3)^2 + (y - 2)^2.$$

Taking square roots (d is supposed to be a distance, so we take the positive square root), we get

$$d = \sqrt{(x - 3)^2 + (y - 2)^2}.$$

Thus, the distance is given by the function

$$d(x, y) = \sqrt{(x - 3)^2 + (y - 2)^2}.$$

Now, to find the distance from $(-1, 1)$ to $(3, 2)$, all we need to do is compute $d(-1, 1)$.

$$d(-1, 1) = \sqrt{(-1 - 3)^2 + (1 - 2)^2} = \sqrt{17}$$

There is nothing special about having chosen the particular point $(3, 2)$ in the above example. If we replace it with the more general point (a, b), we get the following useful formula for the distance between two points.

DISTANCE BETWEEN TWO POINTS IN THE PLANE

The distance between the points (a, b) and (x, y) is

$$d = \sqrt{(x - a)^2 + (y - b)^2}.$$

If (a, b) happens to be the origin, so that $a = b = 0$, we get the following special case.

DISTANCE TO THE ORIGIN IN THE PLANE

The distance from the point (x, y) to the origin is

$$d = \sqrt{x^2 + y^2}.$$

This gives the following equation for the circle centered at the origin with radius r.

$$\sqrt{x^2 + y^2} = r$$

Squaring both sides gives the following, which we shall need in later sections.

EQUATION OF THE CIRCLE OF RADIUS r CENTERED AT THE ORIGIN

$$x^2 + y^2 = r^2$$

We looked briefly at Newton's Law of Gravity in the chapter on functions and their graphs. Recall that according to Newton's Law, the gravitational force exerted on a particle with mass m by another particle with mass M is given by the following function of distance.

$$F(r) = G\frac{Mm}{r^2}$$

Here, r is the distance between the two particles in meters, the masses M and m are given in kilograms, $G \approx 6.67 \times 10^{-11}$, and the resulting force is measured in newtons.*

▼ **EXAMPLE 7**   Newton's Law of Gravity Revisited

Find the gravitational force exerted on a particle with mass m situated at the point (x, y) by another particle with mass M situated at the point (a, b). Express the answer as a function of the coordinates of the particle with mass m.

SOLUTION   We already have the formula we need—almost. The formula for gravitational force is expressed as a function of the distance, r, between the

_____

▼ *A newton is the force that will cause a 1-kilogram mass to accelerate at 1 meter/sec².

two particles. Since we are given the coordinates of the two particles, we can express r in terms of these coordinates by using the formula for distance.

$$r = \sqrt{(x - a)^2 + (y - b)^2}$$

Substituting for r, we get

$$F(x, y) = G\frac{Mm}{(x - a)^2 + (y - b)^2}.$$

Before we go on... Notice that $F(a, b)$ is not defined, since substituting $x = a$ and $y = b$ would make the denominator equal 0. Thus, the largest possible domain of F excludes the point (a, b). Since (a, b) is the only value of (x, y) for which F is not defined, we can deduce that *the domain of F consists of all points (x, y) except for (a, b).* In other words, the domain of F is the whole xy-plane with the single point (a, b) missing.*

Q What is the difference between the function $F(r)$ and the new function we just obtained, $F(x, y)$?

A The old function required us to calculate the distance between the two particles before we could find the force, as it was a function of distance r. The new function gives the force as soon as we know the coordinates of the particles.

Q Why have we expressed F as a function of x and y only, and not also as a function of a and b?

A It is a matter of interpretation. When we write F as a function of x and y we are thinking of a and b as *constants*. For instance, (a, b) could be taken to be the coordinates of the sun—which we often assume to be fixed in space—while (x, y) could be thought of as the coordinates of the earth—which is moving around the sun. In that case it is most natural to think of x and y as variable and a and b as constant.

In case you have the impression that the only interesting functions are those of one or two variables, consider the following example of a function of *three* variables.

▼ **EXAMPLE 8**   Faculty Salaries

David Katz came up with the following function for the faculty salary of a professor with 10 years of teaching experience in a large university.

$$S(x, y, z) = 13{,}005 + 230x + 18y + 102z$$

Here, S is the salary in 1969–1970 in dollars per year, x is the number of books the professor has published, y is the number of articles published, and

▼ **Mathematicians often refer to this as a "punctured plane."*

z is the number of "excellent" articles published.* What salary would you expect a professor with 10 years' experience to have earned in 1969–1970 if she has published two books, 20 articles, and 3 "excellent" articles?

SOLUTION   All we need to do is calculate

$$S(2, 20, 3) = 13,005 + 230(2) + 18(20) + 102(3)$$
$$= \$14,131.$$

Before we go on... This is an example of a **linear** function of three variables. Katz came up with his model by surveying a large number of faculty members and then finding the linear function "best" fitting the data. Such models are called **linear regression** models. We shall see, in the section on **least-squares fit,** how to find the coefficients in such a model given the data obtained in the survey.

What does this model say about the value of a single book or a single article? If a book takes 15 times as long to write as an article, how would you recommend a professor spend her writing time?

The general form of a linear function of several variables is this:

LINEAR FUNCTION OF SEVERAL VARIABLES

A **linear function of the variables $x, y, z, \ldots$** is a function of the form

$$f(x, y) = a + bx + cy + dz + \ldots \quad (a, b, c, d, \ldots \text{ constants}).$$

▶ ___ **8.1 EXERCISES**

For each of the functions in Exercises 1–4, evaluate (**a**) $f(0, 0)$; (**b**) $f(1, 0)$; (**c**) $f(0, -1)$; (**d**) $f(a, 2)$; (**e**) $f(y, x)$; (**f**) $f(x + h, y + k)$.

1. $f(x, y) = x^2 + y^2 - x + 1$

2. $f(x, y) = x^2 - y - xy + 1$

3. $f(x, y) = \sqrt{(x - 1)^2 + (y - 2)^2}$

4. $f(x, y) = \sqrt{(x + 1)^2 + (y - 1)^2}$

For each of the functions in Exercises 5–8, evaluate (**a**) $g(0, 0, 0)$; (**b**) $g(1, 0, 0)$; (**c**) $g(0, 1, 0)$; (**d**) $g(z, x, y)$; (**e**) $g(x + h, y + k, z + l)$, *provided such a value exists.*

5. $g(x, y, z) = e^{x+y+z}$

6. $g(x, y, z) = \ln(x + y + z)$

7. $g(x, y, z) = \dfrac{xyz}{x^2 + y^2 + z^2}$

8. $g(x, y, z) = \dfrac{e^{xyz}}{x + y + z}$

▼ * David A. Katz, "Faculty Salaries, Promotions and Productivity at a Large University," *American Economic Review,* June 1973, pp. 469–477. Prof. Katz's equation actually included other variables, such as the number of dissertations supervised, so our equation assumes that all of these are zero.

In each of Exercises 9–12, find the distance between the given pairs of points.

9. $(1, -1)$ and $(2, -2)$

10. $(1, 0)$ and $(6, 1)$

11. $(a, 0)$ and $(0, b)$

12. (a, a) and (b, b)

13. Find k so that $(1, k)$ is equidistant from $(0, 0)$ and $(2, 1)$.

14. Find k so that (k, k) is equidistant from $(-1, 0)$ and $(0, 2)$.

15. Describe the set of points (x, y) such that $(x - 2)^2 + (y + 1)^2 = 9$.

16. Describe the set of points (x, y) such that $(x + 3)^2 + (y - 1)^2 = 4$.

17. Describe the set of points (x, y) such that $x^2 + 6x + y^2 + 4y + 7 = 0$.

18. Describe the set of points (x, y) such that $x^2 - 4x + y^2 + 2y - 3 = 0$.

APPLICATIONS

Exercises 19 through 22 involve the Cobb-Douglas production function (see Example 5). Recall that it has the form

$$P(x, y) = Kx^a y^{1-a}$$

where P stands for the number of items produced per year, x is the number of employees, and y is the annual operating budget. (The numbers K and a are constants that may depend on the particular situation we are looking at, with $0 \le a \le 1$.)

19. *Productivity*   How many items will be produced per year by a company with 100 employees and an annual operating budget of $500,000, if $K = 1,000$, $a = 0.5$?

20. *Productivity*   How many items will be produced per year by a company with 50 employees and an annual operating budget of $1,000,000, if $K = 1,000$, $a = 0.5$?

21. *Production Modeling with Cobb-Douglas*   Two years ago, my piano manufacturing plant employed 1,000 workers, had an operating budget of $1 million, and turned out 100 pianos. Last year, I slashed the operating budget to $10,000, and production dropped to 10 pianos.
  (a) Use the data for each of the two years and the Cobb-Douglas formula to obtain two equations in K and a.
  (b) Take logs of both sides in each equation and obtain two linear equations in a and $\log K$.
  (c) Solve these equations to obtain values for a and K.

  (d) Use these values in the Cobb-Douglas formula to predict production if I increase the operating budget back to $1 million but lay off half the work force.

22. *Production Modeling with Cobb-Douglas* Repeat Exercise 21 using the following data: Two years ago: 1,000 employees, $1 million operating budget, 100 pianos. Last year: 1,000 employees, $100,000 operating budget, 10 pianos.

23. *Modeling Spending with a Linear Function* The following table shows total U.S. personal income and consumer spending in trillions of dollars for three months in 1992.*

	April	July	Nov.
Income	5.05	5.05	5.1
Spending	4	4.05	4.15

Model monthly spending as a function of monthly income and time using a linear function of the form

$$s(i, t) = ai + bt + c \qquad (a, b, c \text{ constants}),$$

where s represents monthly consumer spending, i represents monthly income (in trillions of dollars) and t represents time in months since April 1992. (Round the constants to two decimal places.)

24. *Modeling International Investments with a Linear Function*   The following table shows the annual flow of private investment and government loans

---

*Source: National Association of Purchasing Management, Department of Commerce (*The New York Times,* June 2, 1993, Section 3, p. 1.) We have rounded all figures to the nearest $0.05 trillion.

(in billions of dollars) to developing countries for the indicated years.*

	1986	1990	1992
Private	20	40	100
Government	40	60	60

Model annual private investment as a function of annual government investment and time using a linear function of the form

$$p(g, t) = ag + bt + c \qquad (a, b, c \text{ constants}),$$

where p represents annual private investment, g represents annual government investment, and t represents the time in years since 1986.

25. *Demand for Beer*   Economist Richard Stone obtained a demand function of the following form for beer in pre-World-War-II Great Britain.

$$Q(y, p, r) = Ky^{-0.023}p^{-1.040}r^{0.939}$$

Here, Q is the value of total annual sales of beer, y is the total real income in Great Britain, p is the average retail price of beer, and r is the average retail price of all other commodities. K is a positive constant that depends on the units of beer and currency.[†]
(a) Does the demand for beer increase or decrease with increasing values of r?
(b) If $K = 200$, find $Q(2 \times 10^8, 0.5, 500)$ to the nearest whole number, and interpret your answer.

26. *The Logistic Function*   One form of the logistic equation[‡] is

$$f(r, a, t) = \frac{K}{1 + e^{-r(t-a)}} + L,$$

where $K > 0$ and $L \geq 0$ are constants.
(a) Does the value of f increase or decrease with increasing t?
(b) Does the value of f increase or decrease with increasing a?

(c) Assume that $K = 1, L = a = 0$. Use a graphing calculator (or some other method) to determine the effect of increasing r on the graph of f versus t.

27. *Utility*   Suppose your newspaper is trying to decide between two competing desktop publishing software packages, "Macro Publish" and "Turbo Publish." You estimate that if you purchase x copies of Macro Publish and y copies of Turbo Publish, your company's daily productivity will be given by the function

$$U(x, y) = 6x^{0.8}y^{0.2} + x.$$

$U(x, y)$ is measured in pages per day (U is called a **utility function**). If $x = y = 10$, calculate the effect of increasing x by one unit, and interpret the result.

28. *Housing Costs*[§]   The cost C (in dollars) of building a house is related to the number k of carpenters used and the number e of electricians used by

$$C(k, e) = 15,000 + 50k^2 + 60e^2.$$

If $k = e = 10$, compare the effects of increasing k by one unit and increasing e by one unit. Interpret the result.

29. The volume of an ellipsoid with cross-sectional radii a, b, and c is given by the formula

$$V(a, b, c) = \tfrac{4}{3}\pi abc.$$

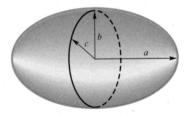

(a) Find at least two sets of values for a, b, and c so that $V(a, b, c) = 1$.
(b) Find the value of a such that $V(a, a, a) = 1$, and sketch the resulting ellipsoid.

▼

* Source: World Bank (*The New York Times*, Dec. 17, 1993, p. D1.)
† Source: Richard Stone, "The Analysis of Market Demand," *Journal of the Royal Statistical Society* 108 (1945); 286–382.
‡ See "You're the Expert" in the chapter on logarithmic and exponential functions for a discussion of the logistic equation.
§ Based on an Exercise in *Introduction to Mathematical Economics* by A.L. Ostrosky Jr. and J.V. Koch (Prospect Heights, IL: Waveland Press, 1979.)

30. The volume of a right elliptical cone with height h and radii a and b of its base is given by the formula

$$V(a, b, h) = \tfrac{1}{3}\pi abh.$$

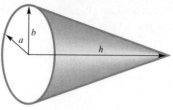

(a) Find at least two sets of values for a, b, and h so that $V(a, b, h) = 1$.

(b) Find the value of a such that $V(a, a, a) = 1$, and sketch the resulting cone.

Complete the tables in Exercises 31–34.

The use of a graphing calculator or computer is suggested for Exercises 33 through 38.

31.

x	y	$f(x, y) = x^2\sqrt{1 + xy}$
-1	-1	
1	12	
0.3	0.5	
41	42	

32.

x	y	$f(x, y) = x^2 e^y$
0	2	
-1	5	
1.4	2.5	
11	9	

33.

x	y	$f(x, y) = x\ln(x^2 + y^2)$
3	1	
1.4	-1	
e	0	
0	e	

34.

x	y	$f(x, y) = \dfrac{x}{x^2 - y^2}$
-1	2	
0	0.2	
0.4	2.5	
10	0	

35. *Level Curves*   The height of each point in a hilly region is given as a function of its coordinates by the formula

$$f(x, y) = y^2 - x^2.$$

(a) Use a graphing calculator to plot the curves on which the height is 0, 1, and 2 on the same set of axes. These are called **level curves of** f.

(b) *Without* using a graphing calculator, sketch the level curve $f(x, y) = 3$.

(c) *Without* using a graphing calculator, sketch the curves $f(y, x) = 1$ and $f(y, x) = 2$.

36. *Isotherms*   The temperature (in degrees Fahrenheit) at each point in a region is given as a function of the coordinates by the formula

$$T(x, y) = 60.5(x - y^2).$$

(a) Use a graphing calculator to sketch the curves on which the temperature is $0°$, $30°$, and $90°$. These curves are called **isotherms**.

(b) Without using a graphing calculator, sketch the isotherms corresponding to $20°$, $50°$ and $100°$.

(c) What do the isotherms corresponding to negative temperatures look like?

37. *Pollution*   The burden of man-made aerosol sulfate in the earth's atmosphere, in grams per square meter, is given by

$$B(x, n) = \frac{xn}{A},$$

where x is the total weight of aerosol sulfate emitted into the atmosphere per year and n is the number of years it remains in the atmosphere. A is the surface area of the earth, approximately 5.1×10^{14} square meters.[*]

(a) Calculate the burden, given the current (1995) estimated values of $x = 1.5 \times 10^{14}$ grams per year, and $n = 5$ days (0.014 years).

(b) If x and n are as above, what does the function $W(x, n) = xn$ measure?

▼ [*] Source: Robert J. Charlson and Tom M. L. Wigley, "Sulfate Aerosol and Climatic Change," *Scientific American*, February, 1994, pp. 48–57.

38. *Pollution* The amount of aerosol sulfate (in grams) was approximately 45×10^{12} grams in 1940, and has been increasing exponentially ever since, with a doubling time of approximately 20 years.[*] Use the model from the previous exercise to give a formula for the atmospheric burden of aerosol sulfate as a function of the time t in years since 1940 and the number of years n it remains in the atmosphere.

39. *Alien Intelligence* Frank Drake, an astronomer at the University of California at Santa Cruz, devised the following equation to estimate the number of planet-based civilizations in our Milky Way galaxy willing and able to communicate with Earth.[†]

$$N(R, f_p, n_e, f_l, f_i, f_c, L) = R\, f_p\, n_e\, f_l\, f_i\, f_c\, L$$

R = the number of new stars formed in our galaxy each year
f_p = the fraction of those stars that have planetary systems
n_e = the average number of planets in each such system that can support life
f_l = the fraction of such planets on which life actually evolves
f_i = the fraction of life-sustaining planets on which intelligent life evolves
f_c = the fraction of intelligent-life-bearing planets on which the intelligent beings develop the means and the will to communicate over interstellar distances
L = the average lifetime of such technological civilizations (in years)

(a) What would be the effect on N if any one of the variables were doubled?
(b) How would you modify the formula if you were interested only in the number of intelligent-life-bearing planets in the galaxy?

(c) How could one convert this function into a linear function?
(d) (For discussion) Try to come up with an estimate of N.

40. *More Alien Intelligence* The formula given in the previous exercise restricts attention to planet-based civilizations in our galaxy. Give a formula that includes intelligent planet-based aliens from the galaxy Andromeda. (Assume that all the variables used in the formula for the Milky Way have the same values for Andromeda.)

41. *Minivan Sales* Chrysler's percentage share of the U.S. minivan market in the period 1993–1994 could be approximated by the linear function

$$c(x, y, z) = 72.3 - 0.8x - 0.2y - 0.7z,$$

where x is the percentage of the market held by foreign manufacturers, y is *General Motors'* percentage share, and z is *Ford's* percentage share.[‡]
(a) Results for the third quarter of 1994 showed *Chrysaler's* share as 38.8%, *GM's* share as 20.1%, and *Ford's* share as 32.9%. According to the model, what was the share held by foreign manufacturers?
(b) Which of the three competitors would you regard as representing the greatest potential harm to *Chrysler's* minivan sales?

42. *Minivan Sales* Referring to the model in the previous exercise, use the fact that the variables x, y, z, and c together account for 100% of all minivan sales in the U.S. to obtain c as a function of y and z only. What does your model say about *Chrysler's* domestic competitors?

COMMUNICATION AND REASONING EXERCISES

43. Illustrate by means of an example how a real-valued function of the two variables x and y gives different real-valued functions of one variable when we restrict y to be different constants.

44. Give an example of a function of the two variables x and y with the property that interchanging x and y has no effect.

▼ [*] Source: Robert J. Charlson and Tom M. L. Wigley, "Sulfate Aerosol and Climatic Change," *Scientific American,* February, 1994, pp. 48–57.
[†] Source: "First Contact" (Plume Books/Penguin Group)/*The New York Times,* October 6, 1992, p. C1.
[‡] The model is the authors'. Source for raw data: Ford Motor Company/*New York Times,* November 9, 1994, p. D5.

45. Give an example of a function f of the two variables x and y with the property that $f(x, y) = -f(y, x)$.

46. Suppose that $C(x, y)$ represents the cost of x CDs and y cassettes. If $C(x, y + 1) < C(x + 1, y)$ for every $x \geq 0$ and $y \geq 0$, what does this tell you about the cost of CDs and cassettes?

47. Brand Z's annual sales are affected by the sales of related products X and Y, as follows. Each $1-million increase in sales of brand X causes a $2.1-million decline in sales of brand Z, whereas each $1-million increase in sales of brand Y results in an increase of $0.4 million in sales of brand Z. Currently, brands X, Y, and Z are each selling $6 million per year. Model the sales of brand Z using a linear function.

48. Let $f(x, y, z) = 43.2 - 2.3x + 11.3y - 4.5z$. Complete the following sentence. An increase of 1 in the value of y causes the value of f to _____ by _____, whereas increasing the value of x by 1 and _____ the value of z by _____ causes a decrease of 11.3 in the value of f.

▶ **8.2** THREE-DIMENSIONAL SPACE AND THE GRAPH OF A FUNCTION OF TWO VARIABLES

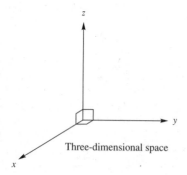

Two-dimensional space

Three-dimensional space

FIGURE 1

Just as functions of a single variable have graphs, so do functions of two or more variables. Let us return for a moment to the graph of a function of a single variable, such as $f(x) = x^2 + 1$. The procedure we used was the following.

1. Replace $f(x)$ with a new variable y, getting $y = x^2 + 1$.
2. Graph the resulting equation. (This is the standard parabola $y = x^2$ raised one unit above the y-axis.)

Now let us try this procedure with a function of two variables, say, $f(x, y) = x^2 + y^2$.

1. Replace $f(x, y)$ with a new variable z (we are already using x and y), getting $z = x^2 + y^2$.
2. Graph the resulting equation. How?

In order to graph this equation, we shall need *three* axes: the x-, y-, and z-axes. In other words, our graph will live in **three-dimensional space,** or **3-space.** *

Just as we had two mutually perpendicular axes in two-dimensional space (the "xy-plane"), so we have three mutually perpendicular axes in three-dimensional space (Figure 1).

▼ * If we were dealing instead with a function of *three* variables, then we would need to go to *four-dimensional* space. Here we run into visualization problems (to say the least!) so we won't discuss the graphs of functions of three or more variables in this text.

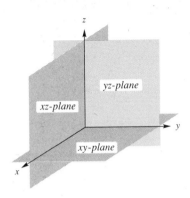

FIGURE 2

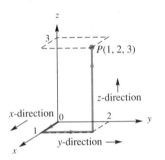

FIGURE 3

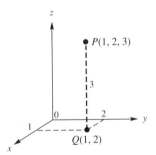

FIGURE 4

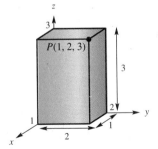

FIGURE 5

In the three-dimensional picture, the x-axis is meant to be coming straight out of the page at you. (Another way to visualize it is to imagine that you are looking at the bottom left-hand corner of your classroom. The corner itself is the origin, while the blackboard is on the "yz-plane.") Notice that, in both 2-space and 3-space, the axis labeled with the last letter always goes up. The z-direction is the "up" direction in 3-space, rather than the y-direction.

There are three important planes associated with these axes: the xy-plane, the yz-plane, and the xz-plane. These planes are shown in Figure 2.

Note that any two of these planes intersect in one of the axes (for example, the xy- and xz-planes intersect in the x-axis) while all three meet at the origin. Also notice that the xy-plane consists of all points with z-coordinate zero, the xz-plane consists of all points with $y = 0$, and the yz-plane consists of all points with $x = 0$.

Now we describe a way of assigning coordinates to points in 3-space, much as we did in 2-space. In 3-space, each point has *three* coordinates, as you might expect: the x-coordinate, the y-coordinate, and the z-coordinate. To see how this works, look at the following examples.

▼ **EXAMPLE 1**

Locate the point $(1, 2, 3)$ in 3-space.

SOLUTION   Our procedure is similar to the one we used in 2-space: start at the origin, proceed 1 unit in the x direction (towards you along the x-axis), then proceed 2 units in the y direction (to the right), and finally, proceed three units in the z direction (straight up). We wind up at the point P shown in Figure 3.

Before we go on... There are several other ways of thinking about the location of P. First, the fact that the x-coordinate of P is 1 means that P is 1 unit toward us from the back plane (the yz-plane). The fact that its y-coordinate is 2 means that it is two units to the right of the xz-plane. The fact that its z-coordinate is 3 means that it is three units above the "ground" (the xy-plane).

Here is another, extremely useful, way of thinking about the location of P. First, look at the x- and y-coordinates, obtaining the point $Q(1, 2)$ in the xy-plane. The point we want is then three units vertically above the point Q, since the z-coordinate of a point is just its height. This strategy is shown in Figure 4.

Here is yet another approach. Imagine a rectangular box situated with one of its corners at the origin, as shown in Figure 5. The corner opposite the origin is our point P.

▼ **EXAMPLE 2**

Locate the points $P(0, -1, 2)$, $Q(-2, 3, 1)$, and $R(-1, -2, 0)$.

SOLUTION   As in 2-space, negative coordinates indicate movement in directions opposite to positive coordinates. For the point P, we go 0 units in the x

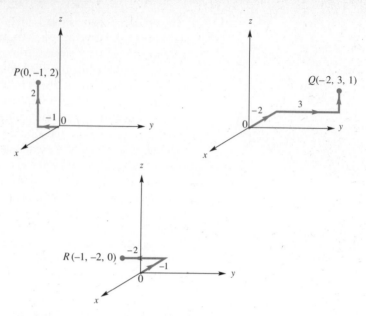

FIGURE 6

direction, -1 units in the y direction (that is, 1 unit to the left) and 2 units up. We locate the other two points in a similar way, as shown in Figure 6.

Always remember this:
The z-coordinate of a point is its height above the xy-plane.

Our next task is to describe the graph of a function $f(x, y)$ of two variables. Recall that the graph is the set of all the points (x, y, z) with $z = f(x, y)$. In other words, for *every* point (x, y) in the domain of f, the z-coordinate is given by evaluating the function at (x, y). Thus, there will be a point on the graph above *every* point in the domain of f, so that the graph will in fact be a *curved surface* of some sort.

▼ **EXAMPLE 3**

Describe the graph of $f(x, y) = x^2 + y^2$.

SOLUTION Your first thought might be to make a table of values. You could choose some values for x and y, and then, for each such pair, calculate $z = x^2 + y^2$. For example, you might get the following table.

(x, y)	$(0, 0)$	$(1, 0)$	$(0, 1)$	$(1, 1)$	$(-1, 0)$	$(0, -1)$	$(-1, -1)$
$z = x^2 + y^2$	0	1	1	2	1	1	2

This gives the following seven points on the graph of f: $(0, 0, 0)$, $(1, 0, 1)$, $(0, 1, 1)$, $(1, 1, 2)$, $(-1, 0, 1)$, $(0, -1, 1)$, $(-1, -1, 2)$. These points are shown in Figure 7.

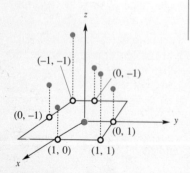

FIGURE 7

· The points on the xy-plane we chose for our table are marked with hollow circles, while the corresponding points on the graph are marked with solid dots. The problem is that this small number of points hardly tells us what the surface looks like, and even if we plotted more points it is not clear that we would get anything more than a mass of circles on the page.

What can we do then? There are several alternatives. One place to start is to let a graphing calculator or computer draw the graph. The graph in Figure 8 was generated by a computer.

The computer software produces this drawing by calculating some values of $f(x, y)$, and then joining adjacent points with lines to create the appearance of a surface with a grid drawn on it. Most calculators or computers allow you to rotate the drawing to look at it from different sides and get a better idea what the surface looks like. This particular surface is called a **paraboloid.**

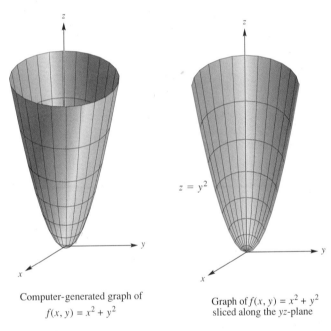

Computer-generated graph of
$f(x, y) = x^2 + y^2$

FIGURE 8

Graph of $f(x, y) = x^2 + y^2$
sliced along the yz-plane

FIGURE 9

Q The name "paraboloid" suggests "parabola," and the surface is reminiscent of a parabola. Why is this?

A In answering this question we shall see a useful analytic technique for understanding a graph. The technique is to slice through the surface with various planes. This will take a little while to explain.

If we slice vertically through this surface along the yz-plane, we get the picture in Figure 9.

The front edge, where we cut, looks like a parabola, and it is one. To see why, notice that the yz-plane is the set of points where $x = 0$. To get the intersection of $x = 0$ and $z = x^2 + y^2$, we substitute $x = 0$ in the second equation, getting

$$z = y^2.$$

This is the equation of a parabola in the yz-plane.

Similarly, we can slice through the surface with the xz-plane by setting $y = 0$. This gives the parabola $z = x^2$ in the xz-plane (Figure 10).

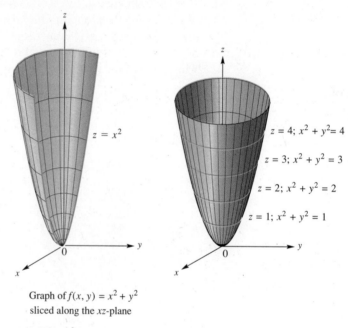

Graph of $f(x, y) = x^2 + y^2$
sliced along the xz-plane

FIGURE 10   **FIGURE 11**

We can also look at horizontal slices through the surface: that is, slices by planes parallel to the xy-plane. These are given by setting $z = c$ for various numbers c. For example, if we set $z = 1$, we will see only the points with height 1. Substituting in the equation $z = x^2 + y^2$ gives the equation

$$1 = x^2 + y^2,$$

which is the equation of a circle of radius 1. If we set $z = 4$, we get the equation of a circle of radius 2.

$$4 = x^2 + y^2$$

In general, if we slice through the surface at height $z = c$, we get a circle (of radius $\sqrt{c}$). Figure 11 shows several of these circles.

Looking at these circular slices, it becomes clear that this surface is the one we get by taking the parabola $z = x^2$ and spinning it around the z-axis. This is an example of what is known as a **surface of revolution.**

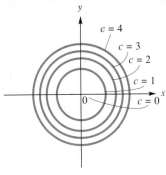

Level Curves of the Paraboloid
$z = x^2 + y^2$

FIGURE 12

Before we go on... Notice that each horizontal slice through the surface was obtained by putting $z = constant$. This gave us an equation in x and y, describing a curve. These curves are called the **level curves** of the surface $z = f(x, y)$. In this example, the equations are of the form $x^2 + y^2 = constant$, and so the level curves are circles. Figure 12 shows the level curves for $c = 0, 1, 2, 3,$ and 4.

The level curves give a contour map of the surface. Each curve shows you all of the points on the surface at a particular height c. You can use this contour map to visualize the shape of the surface. In your mind's eye, move the contour at $c = 1$ to a height of 1 unit above the xy-plane, the contour at $c = 2$ to a height of 2 units above the xy-plane, and so on. You will end up with something like Figure 11.

Graphing many level curves by hand is laborious, and this is one place that graphing calculators come in handy.

▼ **EXAMPLE 4**

Describe the graph of $g(x, y) = \frac{1}{2}x + \frac{1}{3}y - 1$.

SOLUTION   Notice first that g is a linear function of x and y. Figure 13 shows a portion of the graph.

In fact, the graph is an infinite flat surface. In general, the graph of any linear function of two variables will be a plane. To better picture exactly what plane this is, we can find the three **intercepts,** the places where the plane crosses the coordinate axes, as shown in the figure.

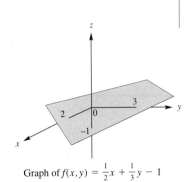

Graph of $f(x, y) = \frac{1}{2}x + \frac{1}{3}y - 1$

FIGURE 13

x-intercept   This is where the plane crosses the x-axis. Since points on this axis have both their y- and z-coordinates 0, we set $y = 0$ and $z = 0$, which gives

$$0 = \frac{1}{2}x - 1.$$

Hence, $x = 2$ is the place where the plane crosses the x-axis.

y-intercept   Set $x = 0$ and $z = 0$, getting

$$0 = \frac{1}{3}y - 1.$$

This gives $y = 3$ as the place where the plane crosses the y-axis.

z-intercept   Set $x = 0$ and $y = 0$ to get $z = -1$ as the place where the plane crosses the z-axis.

Three points are enough to define a plane, so we can say that the plane is the one passing through the three points $(2, 0, 0)$, $(0, 3, 0)$, and $(0, 0, -1)$.

Before we go on... What do the level curves of a linear function of two variables look like?

EXAMPLE 5

Describe the graph of $f(x, y) = \sqrt{4 - x^2 - y^2}$.

SOLUTION   Before doing any drawing or analyzing, we need to think about the domain of f. We cannot take the square root of a negative number, so we must have

$$4 - x^2 - y^2 \geq 0.$$

If we write this inequality as

$$x^2 + y^2 \leq 4,$$

we see that it describes the region inside the circle of radius 2 centered at the origin. Thus, the domain is the disc shown in Figure 14.

Therefore, the graph of f will be a surface lying entirely above this disc.

Figure 15 shows a computer-generated drawing of this graph.

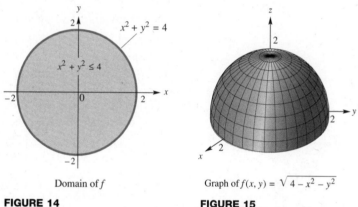

Domain of f	Graph of $f(x, y) = \sqrt{4 - x^2 - y^2}$
FIGURE 14	**FIGURE 15**

(Actually, we have deliberately generated the graph in an unconventional way to emphasize its "circularity." If we were to draw the graph in the conventional way, we would get something like Figure 16. Why does the bottom look ragged in this drawing?)

The surface in Figure 15 resembles a hemisphere. Let us see if slices confirm this.

Slice by the yz-plane   Set $x = 0$ to get the following equations.

$$z = \sqrt{4 - y^2}$$
$$z^2 = 4 - y^2$$
$$y^2 + z^2 = 4$$

The last one is the equation of a circle of radius 2 centered at the origin in the yz-plane. Actually, we see only the top half of this circle, since the equation $z = \sqrt{4 - y^2}$ tells us that we shall take only positive z. (See Figure 17.)

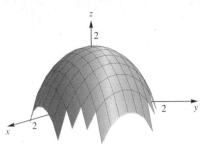

Graph of $f(x, y) = \sqrt{4 - x^2 - y^2}$

FIGURE 16

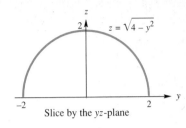

Slice by the yz-plane

FIGURE 17

Slice by the xz-plane This is similar. If we set $y = 0$, we get

$$x^2 + z^2 = 4.$$

This is the circle of radius 2 centered at the origin in the xz-plane. Again, we really see only the top half of the circle.

Let us now look at some level curves.

Slice by z = 0 Setting $z = 0$ gives the following equations.

$$0 = \sqrt{4 - x^2 - y^2}$$
$$0 = 4 - x^2 - y^2$$
$$x^2 + y^2 = 4$$

This is the circle of radius 2 in the xy-plane. This time we see the whole circle. In fact, this is the boundary of the domain of f.

Slice by z = 1 Now we get these equations.

$$1 = \sqrt{4 - x^2 - y^2}$$
$$1 = 4 - x^2 - y^2$$
$$x^2 + y^2 = 3$$

This is the equation of the circle of radius $\sqrt{3} \approx 1.7$.

Slice by z = 2 This gives $x^2 + y^2 = 0$, which says that both x and y must be 0. Thus, at a height of 2 there is only the single point $(0, 0, 2)$ on the surface.

If we slice by a plane $z = c$ for any height c between 0 and 2, we will get a circle.

However, if $c > 2$, then we will end up with an equation with $x^2 + y^2$ equal to a negative number, which is impossible. This means that there are no points on the surface with height greater than 2. Figure 18 shows the level curves $z = c$ with $c = 0, 1, 2$.

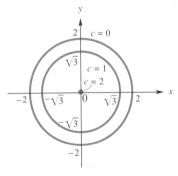

Level curves (slices by $z = c$)

FIGURE 18

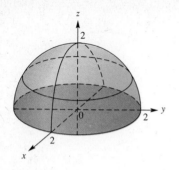

FIGURE 19

Putting these slices together in 3-space gives us Figure 19, confirming that the surface is indeed a hemisphere of radius 2 centered at the origin.

Put another way, this is the surface of revolution obtained by taking the semicircle $z = \sqrt{4 - x^2}$ and spinning it around the z-axis.

Before we go on... If we start with the equation $z = \sqrt{4 - x^2 - y^2}$, squaring both sides and rearranging terms gives

$$x^2 + y^2 + z^2 = 4.$$

By reasoning similar to that in the previous section, this is the equation of a sphere of radius 2 centered at the origin (see the exercises). The original equation tells us, however, that z must be taken to be positive, so the surface is just the upper hemisphere.

▼ **EXAMPLE 6**

Describe the graph of $f(x, y) = xy$.

SOLUTION   The graph is shown in Figure 20.

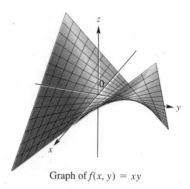

Graph of $f(x, y) = xy$

FIGURE 20

This is an example of a "saddle point" at the origin (we shall return to this in a later section). Slices in various directions show interesting features. Let us begin with the level curves.

Slice by $z = c$ This gives $xy = c$, which is a hyperbola. The level curves for various values of c are shown in Figure 21.

The case $c = 0$ is interesting: $xy = 0$ has as its graph the union of the x-axis and the y-axis (why?).

Slice by $x = c$ This gives $z = cy$, the line with slope c through the origin. It is surprising that even though the surface clearly curves, there are straight lines that can be drawn on it. In fact, all of the grid lines drawn in Figure 20 are straight.

Looking at Figure 20, we see that it might be interesting to look at the slices by vertical planes at 45° angles to the xz- and yz-planes.

Slice by $y = x$ This gives $z = x^2$, the parabola. This is the upward curve in the saddle.

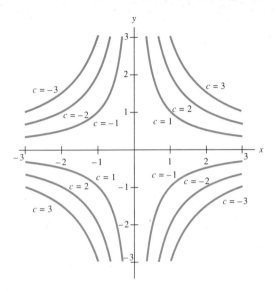

FIGURE 21

Slice by y = −x   This gives $z = -x^2$, a downward curving parabola. This is the downward curve in the saddle.

Before we go on... The point of this example is that slices in different directions can show very different aspects of a surface. Try planes parallel to the coordinate planes, but also be willing to look at planes suggested by the surface itself.

▶   **8.2 EXERCISES**

1. Sketch the cube with vertices $(0, 0, 0)$, $(1, 0, 0)$, $(0, 1, 0)$, $(0, 0, 1)$, $(1, 1, 0)$, $(1, 0, 1)$, $(0, 1, 1)$, and $(1, 1, 1)$.

2. Sketch the cube with vertices $(-1, -1, -1)$, $(1, -1, -1)$, $(-1, 1, -1)$, $(-1, -1, 1)$, $(1, 1, -1)$, $(1, -1, 1)$, $(-1, 1, 1)$, and $(1, 1, 1)$.

3. Sketch the pyramid with vertices $(1, 1, 0)$, $(1, -1, 0)$, $(-1, 1, 0)$, $(-1, -1, 0)$, and $(0, 0, 2)$.

4. Sketch the solid with vertices $(1, 1, 0)$, $(1, -1, 0)$, $(-1, 1, 0)$, $(-1, -1, 0)$, $(0, 0, -1)$, and $(0, 0, 1)$.

Sketch the planes in Exercises 5–10.

5. $z = -2$

6. $z = 4$

7. $y = 2$

8. $y = -3$

9. $x = -3$

10. $x = 2$

Match the graphs with the equations in Exercises 11–18.

11. $f(x, y) = 1 - 3x + 2y$

12. $f(x, y) = 1 - \sqrt{x^2 + y^2}$

13. $f(x, y) = 1 - (x^2 + y^2)$

14. $f(x, y) = y^2 - x^2$

15. $f(x, y) = -\sqrt{1 - (x^2 + y^2)}$

16. $f(x, y) = 1 + (x^2 + y^2)$

17. $f(x, y) = \dfrac{1}{x^2 + y^2}$

18. $f(x, y) = 3x - 2y + 1$

A

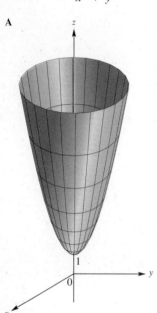

B

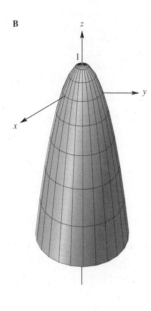

C

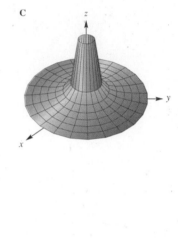

D

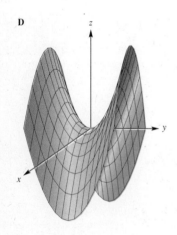

E

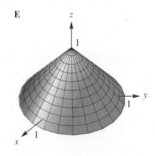

F

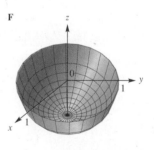

G

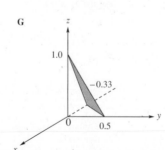

H

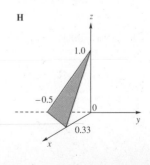

Sketch the graphs of the functions in Exercises 19–40.

19. $f(x, y) = 1 - x - y$

20. $f(x, y) = x + y - 2$

21. $g(x, y) = 2x + y - 2$

22. $g(x, y) = 3 - x + 2y$

23. $h(x, y) = x + 2$

24. $h(x, y) = 3 - y$

25. $r(x, y) = x + y$

26. $r(x, y) = x - y$

27. $s(x, y) = 2x^2 + 2y^2$. Show cross sections at $z = 1$ and $z = 2$.

28. $s(x, y) = -(x^2 + y^2)$. Show cross sections at $z = -1$ and $z = -2$.

29. $t(x, y) = x^2 + 2y^2$. Show cross sections at $x = 0$ and $z = 1$.

30. $t(x, y) = \frac{1}{2}x^2 + y^2$. Show cross sections at $x = 0$ and $z = 1$.

31. $f(x, y) = 2 + \sqrt{x^2 + y^2}$. Show cross sections at $z = 3$ and $y = 0$.

32. $f(x, y) = 2 - \sqrt{x^2 + y^2}$. Show cross sections at $z = 0$ and $y = 0$.

33. $f(x, y) = -2\sqrt{x^2 + y^2}$. Show cross sections at $z = -4$ and $y = 1$.

34. $f(x, y) = 2 + 2\sqrt{x^2 + y^2}$. Show cross sections at $z = 4$ and $y = 1$.

35. $f(x, y) = y^2$

36. $g(x, y) = x^2$

37. $h(x, y) = \dfrac{1}{y}$

38. $k(x, y) = e^y$

39. $f(x, y) = e^{-(x^2 + y^2)}$

40. $g(x, y) = \dfrac{1}{\sqrt{x^2 + y^2}}$

APPLICATIONS

A graphing calculator is suggested for Exercises 41 through 50.

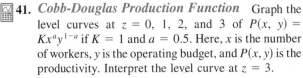

41. Cobb-Douglas Production Function   Graph the level curves at $z = 0$, 1, 2, and 3 of $P(x, y) = Kx^a y^{1-a}$ if $K = 1$ and $a = 0.5$. Here, x is the number of workers, y is the operating budget, and $P(x, y)$ is the productivity. Interpret the level curve at $z = 3$.

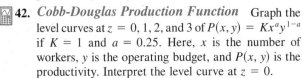

42. Cobb-Douglas Production Function   Graph the level curves at $z = 0, 1, 2$, and 3 of $P(x, y) = Kx^a y^{1-a}$ if $K = 1$ and $a = 0.25$. Here, x is the number of workers, y is the operating budget, and $P(x, y)$ is the productivity. Interpret the level curve at $z = 0$.

43. Utility   Suppose that your newspaper is trying to decide between two competing desktop publishing software packages, "Macro Publish" and "Turbo Publish." You estimate that if you purchase x copies of Macro Publish and y copies of Turbo Publish, your company's daily productivity will be given by

$$U(x, y) = 6x^{0.8}y^{0.2} + x,$$

where $U(x, y)$ is measured in pages per day. (U is called a *utility function*.) Graph the level curves at $z = 0$, 10, 20, and 30. What does the level curve at $z = 0$ tell you?

44. Utility   Suppose that your small publishing company is trying to decide between two competing desktop publishing software packages, "Macro Publish" and "Turbo Publish." You estimate that if you purchase x copies of Macro Publish and y copies of Turbo Publish, your company's daily productivity will be given by

$$U(x, y) = 5x^{0.2}y^{0.8} + x,$$

where $U(x, y)$ is measured in pages per day. Graph the level curves at $z = 0$, 10, 20, and 30. Give a formula for the level curve at $z = 30$ specifying y as a function of x. What does this curve tell you?

45. *Housing Costs* *   The cost C of building a house is related to the number k of carpenters used and the number e of electricians used by

$$C(k, e) = 15,000 + 50k^2 + 60e^2.$$

Describe the level curves $C = 30,000$ and $C = 40,000$. What do these level curves represent?

46. *Housing Costs* *   The cost C of building a house (in a different area from that in the previous exercise) is related to the number k of carpenters used and the number e of electricians used by

$$C(k, e) = 15,000 + 70k^2 + 40e^2.$$

Describe the slices by the planes $k = 2$ and $e = 2$. What do these slices represent?

47. *Area*   The area of a rectangle of height h and width w is $A(h, w) = hw$. Sketch a few level curves of A. Looking at your graph, if the perimeter $h + w$ of the rectangle is constant, what h and w give the largest

area? (We suggest you draw in the line $h + w = c$ for several values of c.)

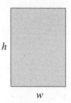

48. *Area*   The area of an ellipse with semimajor axis a and semiminor axis b is $A(a, b) = \pi ab$. Sketch the graph of A. If $a^2 + b^2$ is constant, what a and b give the largest area?

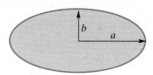

COMMUNICATION AND REASONING EXERCISES

49. Show that the distance between the points (x, y, z) and (a, b, c) is given by the following **three-dimensional distance formula.**

$$d = \sqrt{(x - a)^2 + (y - b)^2 + (z - c)^2}.$$

The following diagram should be of assistance.

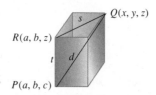

50. Use the result of the previous exercise to show that the sphere of radius r centered at the origin is given by the following equation.

$$x^2 + y^2 + z^2 = r^2$$

51. Why is three-dimensional space used to represent the graph of a function of two variables?

52. Why is it that we can sketch the graphs of functions of two variables on the two-dimensional flat surfaces of these pages?

▶ ══════ **8.3** PARTIAL DERIVATIVES

Recall that if f is a function of x, then df/dx measures how fast f changes as x increases. If f is a function of two or more variables we can ask how fast f changes as each variable increases (and the others remain fixed). These rates of change are called the "partial derivatives of f," and measure how each variable contributes to the change in f. Here is a more precise definition.

▼   * Based on an exercise in *Introduction to Mathematical Economics* by A. L. Ostrosky, Jr., and J. V. Koch (Prospect Heights, IL: Waveland Press, 1979.)

> **PARTIAL DERIVATIVES**
>
> The **partial derivative of f with respect to x** is the derivative of f with respect to x, treating all other variables as constant; the **partial derivative of f with respect to y** is the derivative of f with respect to y, treating all other variables as constant; and so on for other variables.

The partial derivatives are written as $\partial f/\partial x$, $\partial f/\partial y$, and so on. The symbol "∂" is used (instead of "d") to remind us that there is more than one variable and that we are holding the other variables fixed.

▼ **EXAMPLE 1**

Let $f(x, y) = x^2y + y^2x - xy + y$. Find $\partial f/\partial x$ and $\partial f/\partial y$.

SOLUTION   To find $\partial f/\partial x$, we take the derivative with respect to x, thinking of y as a constant.

$$\frac{\partial f}{\partial x} = 2xy + y^2 - y$$

Notice that $\partial/\partial x[x^2y]$ is the derivative of x^2 times a *constant,* since we are treating y as constant. Therefore, it is $2x$ times that constant—that is, $2xy$. Similarly, $\partial/\partial x[y] = 0$, since the derivative of a constant is 0.

We obtain $\partial f/\partial y$ as the derivative of f with respect to y, regarding x as constant.

$$\frac{\partial f}{\partial y} = x^2 + 2xy - x + 1$$

▼ **EXAMPLE 2**   Productivity

The Cobb-Douglas production function for the Handy Gadget Company is

$$P = 100x^{0.3}y^{0.7},$$

where P is the number of gadgets it turns out per month, x is the number of employees at the company, and y is the monthly operating budget. Handy Gadgets employs 50 people. How fast is the number of gadgets produced going up as the operating budget increases from a level of $10,000 per month?

SOLUTION   We are looking for the rate of increase of production P with respect to the operating budget y, with x held fixed at 50. Thus, we take the partial derivative $\partial P/\partial y$.

$$\frac{\partial P}{\partial y} = 70x^{0.3}y^{-0.3}$$

We must now evaluate this derivative at $x = 50$ and $y = 10,000$. We obtain

$$\left.\frac{\partial P}{\partial y}\right|_{(50,\,10,000)} = 70(50)^{0.3}(10,000)^{-0.3} \approx 14.28.$$

How do we interpret this? First, $\partial P/\partial y$ is measured in gadgets/dollar (per month), so the answer tells us that productivity goes up by about 14 gadgets per month for every additional dollar added to the monthly operating budget. Not bad!

Before we go on... It might be illuminating to calculate the actual productivity at these figures. This we do by evaluating the original function P at the given values of x and y.

$$P(50, 10,000) = 100(50)^{0.3}(10,000)^{0.7} \approx 204,028 \text{ gadgets per month}$$

Thus, the increase of 14 in the number of gadgets is not so wonderful after all. A \$1 increase in the budget represents a 0.01% increase in operating costs, but this results in an increase of about only 0.0067% in productivity.

GEOMETRIC INTERPRETATION OF PARTIAL DERIVATIVES

Q We know that if f is a function of one variable x, the derivative df/dx gives the slopes of the tangent lines to its graph. What about the *partial* derivatives of a function of several variables?

A Suppose that f is a function of x and y. By definition, $\partial f/\partial x$ is the derivative of the function of x you get by holding y fixed. If you evaluate this derivative at the point (a, b), you are holding y fixed at the value b, taking the ordinary derivative of the resulting function of x, and evaluating this

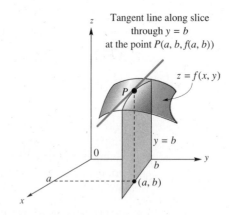

$\left.\dfrac{\partial f}{\partial x}\right|_{(a,\,b)}$ **is the slope of the tangent line at the point**

$P(a, b, f(a, b))$ **along the slice through** $y = b$.

FIGURE 1

at $x = a$. Now, holding y fixed at b amounts to slicing through the graph of f along the plane $y = b$, resulting in a curve. Thus, the partial derivative is the slope of the tangent line to this curve at the point where $x = a$ and $y = b$, along the plane $y = b$ (Figure 1).

Note that this fits with our interpretation of $\partial f/\partial x$ as the rate of increase of f with increasing x when y is held fixed at b.

The other partial derivative, $\partial f/\partial y|_{(a, b)}$ is, similarly, the slope of the tangent line at the same point $P(a, b, f(a, b))$, but along the slice by the plane $x = a$. You should to draw the corresponding picture for this on your own.

▼ **EXAMPLE 3**

Calculate $\partial f/\partial x$, $\partial f/\partial y$, and $\partial f/\partial z$ if $f(x, y, z) = xy^2z^3 - xy$.

SOLUTION   Although we now have three variables, the calculation remains the same: $\partial f/\partial x$ is the derivative of f with respect to x, holding *both* other variables, y and z, constant.

$$\frac{\partial f}{\partial x} = y^2z^3 - y$$

Similarly, $\partial f/\partial y$ is the derivative of f with respect to y, with both x and z held constant.

$$\frac{\partial f}{\partial y} = 2xyz^3 - x$$

Finally, to find $\partial f/\partial z$, we hold both x and y constant and take the derivative with respect to z.

$$\frac{\partial f}{\partial z} = 3xy^2z^2$$

Before we go on... The procedure is the same for any number of variables: to get the partial derivative with respect to any one variable, we treat all the others as constants.

▼ **EXAMPLE 4**   Gravity in 3-Space

According to Newton's Law of Gravity, the gravitational force exerted on a particle with mass m situated at the point (x, y, z) by another particle with mass M situated at the point (a, b, c) is given by

$$F(x, y, z) = G\frac{Mm}{(x - a)^2 + (y - b)^2 + (z - c)^2}.$$

(All units of distance are in meters, the masses M and m are in kilograms, $G \approx 6.67 \times 10^{-11}$, and the resulting force is measured in newtons.) Suppose that a 1,000-kg mass is situated at the origin and that your 100-kg space module is at the point (10, 100, 1,000) and traveling in the x-direction at one meter per second. How fast is the gravitational force your module experiences decreasing?

SOLUTION   First, substitute for the constants M, m, a, b, and c.

$$F(x, y, z) = G \frac{100,000}{x^2 + y^2 + z^2}$$

We'll substitute for G later. Also, since y and z are not changing, we *could* substitute for them now, but we'll leave them as variables for the time being and substitute later. Now the question we have to answer is really this: How fast is F decreasing per unit of increase in x? In other words, we need to find the partial derivative, $\partial F / \partial x$, and evaluate it at (10, 100, 1,000). To do this, we hold y and z constant and differentiate with respect to x.

$$\frac{\partial F}{\partial x} = -G \frac{200,000x}{(x^2 + y^2 + z^2)^2}$$

Evaluating at (10, 100, 1,000) now gives

$$\left.\frac{\partial F}{\partial x}\right|_{(10, 100, 1,000)} = -G \frac{2,000,000}{(1,010,100)^2} \approx -1.31 \times 10^{-16}.$$

Thus, the gravitational force is *decreasing* at 1.31×10^{-16} newtons per second.

Before we go on... Here are some variations on this problem. If we were traveling at 2 meters per second instead of one meter per second, the rate of decrease would be twice that amount. More generally, if we were traveling at v meters per second, the rate would be v times that amount.* If we were traveling in the y-direction instead of the x-direction, we would have taken the partial derivative with respect to y instead of x, and similarly for the z-direction.

▼   **EXAMPLE 5**   Cost Functions

Returning to an example from Section 1, suppose that you own a company making two models of stereo speakers, the Ultra Mini and the Big Stack. Your total monthly cost (in dollars) to make x Ultra Minis and y Big Stacks is given by

$$C(x, y) = \$10,000 + 20x + 40y.$$

What is the significance of $\partial C / \partial x$ and $\partial C / \partial y$?

▼ * Can you see how this is really a related rates problem in disguise?

SOLUTION   First, we compute these partial derivatives.

$$\frac{\partial C}{\partial x} = 20$$

$$\frac{\partial C}{\partial y} = 40$$

These are the amounts that each additional Ultra Mini or Big Stack will add to the total cost, respectively. These are, in other words, the **marginal costs** of each model of speaker.

Before we go on... How much does the cost rise if you increase x by Δx and y by Δy? In this example, the change in cost is given by

$$\Delta C = 20\,\Delta x + 40\,\Delta y = \frac{\partial C}{\partial x}\,\Delta x + \frac{\partial C}{\partial y}\,\Delta y.$$

This leads to the **chain rule for several variables.** Part of this rule says that if x and y are both functions of t, then C is a function of t through them, and the rate of change of C with respect to t can be calculated as

$$\frac{dC}{dt} = \frac{\partial C}{\partial x} \cdot \frac{dx}{dt} + \frac{\partial C}{\partial y} \cdot \frac{dy}{dt}.$$

We shall not have a chance to use this interesting result further in this book.

▼ **EXAMPLE 6**   Cost Functions

Another possibility for the cost function in the previous example is

$$C(x, y) = \$10{,}000 + 20x + 40y + 30\sqrt{x + y}.$$

Now what are the marginal costs of the two models of speakers?

SOLUTION   We compute the partial derivatives.

$$\frac{\partial C}{\partial x} = 20 + \frac{15}{\sqrt{x + y}}$$

$$\frac{\partial C}{\partial y} = 40 + \frac{15}{\sqrt{x + y}}$$

Now the marginal costs depend on the total production level $x + y$. As production increases, the marginal costs decline. For example, if $x + y = 9$, then it will cost an additional \$25 for each Ultra Mini and \$45 for each Big Stack. On the other hand, if $x + y = 25$, it will cost only \$23 for each additional Ultra Mini and only \$43 for each Big Stack.

Just as for functions of a single variable, we can calculate second derivatives. Suppose, for example, that we have a function of x and y, say, $f(x, y) =$

$x^2 - x^2y^2$. We know that

$$\frac{\partial f}{\partial x} = 2x - 2xy^2.$$

The symbol $\partial/\partial x$ means "the partial derivative with respect to x," just as d/dx stood for "the derivative with respect to x" in the chapter on derivatives. If we now want to take the partial derivative with respect to x once again, there is nothing to stop us.

$$\frac{\partial}{\partial x}\left(\frac{\partial f}{\partial x}\right) = 2 - 2y^2$$

This is called the **second partial derivative $\partial^2 f/\partial x^2$.**

We get the following derivatives similarly.

$$\frac{\partial f}{\partial y} = -2x^2y$$

$$\frac{\partial^2 f}{\partial y^2} = -2x^2$$

Now what if we instead take the partial derivative with respect to y of $\partial f/\partial x$?

$$\frac{\partial^2 f}{\partial y\,\partial x} = \frac{\partial}{\partial y}\left(\frac{df}{dx}\right) = \frac{\partial}{\partial y}(2x - 2xy^2) = -4xy$$

Here, $\partial^2 f/\partial y\,\partial x$ means "first take the partial derivative with respect to x, and then with respect to y," and is called a **mixed partial derivative.** If we differentiate in the opposite order, we get

$$\frac{\partial^2 f}{\partial x\,\partial y} = \frac{\partial}{\partial x}\left(\frac{\partial f}{\partial y}\right) = \frac{\partial}{\partial x}(-2x^2y) = -4xy,$$

the same as $\partial^2 f/\partial y\,\partial x$. This is no coincidence—the mixed partial derivatives $\partial^2 f/\partial x\,\partial y$ and $\partial^2 f/\partial y\,\partial x$ will always agree as long as the first partial derivatives are both differentiable functions of x and y and the mixed partial derivatives are continuous. Since all the functions we shall use are of this type, we can take the derivatives in any order we like when calculating mixed derivatives.

Before we go to the exercises, here is another notation for partial derivatives.

$$f_x \text{ means } \frac{\partial f}{\partial x}.$$

$$f_y \text{ means } \frac{\partial f}{\partial x}.$$

$$f_{xy} \text{ means } (f_x)_y = \frac{\partial^2 f}{\partial y\,\partial x}. \qquad \text{Note the order in which the derivatives are taken.}$$

$$f_{yx} \text{ means } (f_y)_x = \frac{\partial^2 f}{\partial x\,\partial y}.$$

We shall sometimes use this more convenient notation, especially for second-order partial derivatives.

▶ **8.3 EXERCISES**

In each of Exercises 1–16, find $\partial f/\partial x$, $\partial f/\partial y$, $\partial^2 f/\partial x^2$, $\partial^2 f/\partial y^2$, $\partial^2 f/\partial x\,\partial y$, and $\partial^2 f/\partial y\,\partial x$, and then evaluate them at the point $(1, -1)$ if possible.

1. $f(x, y) = 3x^2 - y^3 + x - 1$

2. $f(x, y) = x^{1/2} - 2y^4 + y + 6$

3. $f(x, y) = 3x^2 y$

4. $f(x, y) = x^4 y^2 - x$

5. $f(x, y) = x^2 y^3 - x^3 y^2 - xy$

6. $f(x, y) = x^{-1}y^2 + xy^2 + xy$

7. $f(x, y) = (2xy + 1)^3$

8. $f(x, y) = 1/(xy + 1)^2$

9. $f(x, y) = e^{x + y}$

10. $f(x, y) = e^{2x + y}$

11. $f(x, y) = 5x^{0.6}y^{0.4}$

12. $f(x, y) = -2x^{0.1}y^{0.9}$

13. $f(x, y) = 4.1x^{1.2}e^{-0.2y}$

14. $f(x, y) = x + \dfrac{e^{0.1y}}{x^2}$

15. $f(x, y) = e^{0.2xy}$

16. $f(x, y) = xe^{xy}$

In each of Exercises 17–28, find $\partial f/\partial x$, $\partial f/\partial y$ and $\partial f/\partial z$, and then evaluate them at the point $(0, -1, 1)$ if possible.

17. $f(x, y, z) = xyz$

18. $f(x, y, z) = xy + xz - yz$

19. $f(x, y, z) = -\dfrac{4}{x + y + z^2}$

20. $f(x, y, z) = \dfrac{6}{x^2 + y^2 + z^2}$

21. $f(x, y, z) = xe^{yz} + ye^{xz}$

22. $f(x, y, z) = xye^z + xe^{yz} + e^{xyz}$

23. $f(x, y, z) = x^{0.1}y^{0.4}z^{0.5}$

24. $f(x, y, z) = 2x^{0.2}y^{0.8} + z^2$

25. $f(x, y, z) = e^{xyz}$

26. $f(x, y, z) = \ln(x + y + z)$

27. $f(x, y, z) = \dfrac{2,000z}{1 + y^{0.3}}$

28. $f(x, y, z) = \dfrac{e^{0.2x}}{1 + e^{-0.1y}}$

APPLICATIONS

29. *Minivan Sales*   *Chrysler's* percentage share of the U.S. minivan market in the period 1993–1994 could be approximated by the linear function

$$c(x, y, z) = 72.3 - 0.8x - 0.2y - 0.7z,$$

where x is the percentage share of the market held by foreign manufacturers, y is *General Motors'* percentage share, and z is *Ford's* percentage share.* Calculate the partial derivatives $\partial c/\partial x$, $\partial c/\partial y$, and $\partial c/\partial z$, and interpret the results.

30. *Minivan Sales*   In the previous exercise, if we take into account the fact that the variables x, y, z, and c together account for 100% of all minivan sales in the U.S., we obtain

$$c = -38.5 + 3.0y + 0.5z,$$

where y is *General Motors'* percentage share of the market, and z is *Ford's* percentage share. Calculate the partial derivatives $\partial c/\partial y$ and $\partial c/\partial z$ and interpret the results, explaining why the coefficients of y and z are positive.

▼ *The model is the authors'. Source for raw data: Ford Motor Company/The New York Times, November 9, 1994, p. D5.*

31. *Marginal Cost* Your weekly cost to manufacture x cars and y trucks is given by

$$C(x, y) = \$200{,}000 + 6{,}000x$$
$$+ 4{,}000y - 100{,}000e^{-0.01(x+y)}.$$

What is the marginal cost of cars? Of trucks? How do these marginal costs behave as total production rises?

32. *Marginal Cost* Your weekly cost to manufacture x bicycles and y tricycles is given by

$$C(x, y) = \$20{,}000 + 60x + 20y + 50\sqrt{xy}.$$

What is the marginal cost of bicycles? Of tricycles? How do these marginal costs behave as x and y increase?

33. *Average Cost* If you average your costs over your total production, you get the **average cost,** written $\bar{C}$.

$$\bar{C}(x, y) = \frac{C(x, y)}{x + y}$$

Find the average cost for the cost function in Exercise 31. Then find the marginal average cost of a car and the marginal average cost of a truck at a production level of 50 cars and 50 trucks. Round your answer to two decimal places. Interpret your answers.

34. *Average Cost* Find the average cost for the cost function in Exercise 32 (see the previous exercise). Then find the marginal average cost of a bicycle and the marginal average cost of a tricycle at a production level of 5 bicycles and 5 tricycles. Round your answer to two decimal places. Interpret your answers.

35. *Marginal Revenue* As manager of an auto dealership, you offer a car rental company the following deal. You will charge $\$15{,}000$ per car and $\$10{,}000$ per truck, but you will then give the company a discount of $\$5{,}000$ times the square root of the total number of vehicles it buys from you. Looking at your marginal revenue, is this a good deal for the rental company?

36. *Marginal Revenue* As marketing director for a bicycle manufacturer, you come up with the following scheme. You will offer to sell a dealer x bicycles and y tricycles for

$$R(x, y) = \$3{,}500 - 3{,}500e^{-0.02x-0.01y}.$$

Find your marginal revenue for bicycles and for tricycles. Are you likely to be fired for your suggestion?

37. *Research Productivity* Here we apply a variant of the Cobb-Douglas function to the modeling of research productivity. A mathematical model of research productivity at a particular physics laboratory is given by

$$P = 0.04x^{0.4}y^{0.2}z^{0.4},$$

where P is the annual number of ground-breaking research papers produced by the staff, x is the number of physicists on the research team, y is the laboratory's annual research budget, and z is the annual National Science Foundation subsidy to the laboratory. Find the rate of increase of research papers per government-subsidy-dollar at a subsidy level of $\$1{,}000{,}000$ per year and a staff level of 10 physicists if the annual budget is $\$100{,}000$.

38. *Research Productivity* A major drug company estimates that the annual number P of patents for new drugs developed by its research team is best modeled by the formula

$$P = 0.3x^{0.3}y^{0.4}z^{0.3},$$

where x is the number of research biochemists on the payroll, y is the annual research budget, and z is size of the bonus awarded to discoverers of new drugs. Assuming that the company has 12 biochemists on the staff, an annual research budget of $\$500{,}000$ and pays $\$40{,}000$ bonuses to developers of new drugs, calculate the rate of growth in the annual number of patents per new research staff member.

39. *Utility* Your newspaper is trying to decide between two competing desktop publishing software packages, "Macro Publish" and "Turbo Publish." You estimate that if you purchase x copies of Macro Publish and y copies of Turbo Publish, your company's daily productivity will be given by

$$U(x, y) = 6x^{0.8}y^{0.2} + x.$$

$U(x, y)$ is measured in pages per day. (Recall that U is called a *utility function*.)

(a) Calculate $\partial U/\partial x|_{(10,\ 5)}$ and $\partial U/\partial y|_{(10,\ 5)}$ to two decimal places, and interpret the results.

(b) What does the ratio $\partial U/\partial x|_{(10,\ 5)}/\partial U/\partial y|_{(10,\ 5)}$ tell you about the usefulness of these products?

40. *Grades*[*] A production formula for a student's performance on a difficult English examination is

---

▼ [*] Based on an exercise in *Introduction to Mathematical Economics* by A. L. Ostrosky, Jr., and J. V. Koch (Prospect Heights, IL: Waveland Press, 1979).

given by

$$g(t, x) = 4tx - 0.2t^2 - x^2.$$

Here g is the grade the student can expect to obtain, t is the number of hours of study for the examination, and x is the student's grade point average.
(a) Calculate $\partial g/\partial t|_{(10, 3)}$ and $\partial g/\partial x|_{(10, 3)}$ and interpret the results.
(b) What does the ratio $\partial g/\partial t|_{(10, 3)}/\partial g/\partial x|_{(10, 3)}$ tell you about the relative merits of study and grade point average?

41. *Electrostatic Repulsion* If positive electric charges of Q and q coulombs are situated at positions (a, b, c) and (x, y, z) respectively, then the force of repulsion they experience is given by

$$F = K\frac{Qq}{(x - a)^2 + (y - b)^2 + (z - c)^2},$$

where $K \approx 9 \times 10^9$, F is given in newtons, and all positions are measured in meters. Assume that a charge of 10 coulombs is situated at the origin, and that a second charge of 5 coulombs is situated at $(2, 3, 3)$ and moving in the y-direction at one meter per second. How fast is the electrostatic force it experiences decreasing?

42. *Electrostatic Repulsion* Repeat the preceding exercise, assuming that a charge of 10 coulombs is situated at the origin and that a second charge of 5 coulombs is situated at $(2, 3, 3)$ and moving in the negative z-direction at one meter per second.

43. *Investments* Recall that the compound interest formula for annual compounding is

$$A(P, r, t) = P(1 + r)^t,$$

where A is the future value of an investment of P dollars after t years at an interest rate of r.
(a) Calculate $\partial A/\partial P$, $\partial A/\partial r$, and $\partial A/\partial t$, all evaluated at $(100, 0.10, 10)$. (Round answers to two decimal places.) Interpret your answers.
(b) What does the function $\partial A/\partial P|_{(100, 0.10, t)}$ of t tell you about your investment?

44. *Investments* Repeat the preceding exercise using the formula for continuous compounding,

$$A(P, r, t) = Pe^{rt}.$$

45. *Modeling with the Cobb-Douglas Production Formula* Assume you are given a production formula of the form

$$P(x, y) = Kx^a y^b \qquad (a + b = 1).$$

(a) Obtain formulas for $\partial P/\partial x$ and $\partial P/\partial y$, and show that $\partial P/\partial x = \partial P/\partial y$ precisely when $x/y = a/b$.
(b) Let x be the number of workers a firm employs and let y be its monthly operating budget in thousands of dollars. Assume that the firm currently employs 100 workers and has a monthly operating budget of $200,000. If each additional worker contributes as much to productivity as each additional $1,000 per month, find values of a and b that would model the firm's productivity.

46. *Housing Costs** The cost C of building a house is related to the number k of carpenters used and the number e of electricians used by

$$C(k, e) = 15{,}000 + 50k^2 + 60e^2.$$

If 3 electricians are currently employed in building your new house, and the marginal cost per additional electrician is the same as the marginal cost per additional carpenter, how many carpenters are being used? (Round your answer to the nearest carpenter.)

47. *Nutrient Diffusion* Suppose that one cubic centimeter of nutrient is placed at the center of a circular petri dish filled with water. We might wonder how it is distributed after a time of t seconds. According to the classical theory of diffusion, the concentration of nutrient (in parts of nutrient per part of water) after a time t is given by

$$u(r, t) = \frac{1}{4\pi Dt}e^{-r^2/(4Dt)}.$$

Here D is the *diffusivity*, which we will take to be 1, and r is the distance from the center in centimeters. How fast is the concentration increasing at a distance of 1 cm from the center three seconds after the nutrient is introduced?

48. *Nutrient Diffusion* Referring to the previous exercise, how fast is the concentration increasing at a distance of 4 cm from the center four seconds after the nutrient is introduced?

▼ * Based on an exercise in *Introduction to Mathematical Economics* by A. L. Ostrosky, Jr., and J. V. Koch (Prospect Heights, IL: Waveland Press, 1979).

COMMUNICATION AND REASONING EXERCISES

49. Given that $f(a, b) = r$, $f_x(a, b) = s$, and $f_y(a, b) = t$, complete the following sentence: _____ is increasing at a rate of _____ units per unit of x, _____ is increasing at a rate of _____ units per unit of y, and the value of _____ is _____ when $x = $ _____ and $y = $ _____.

50. A firm's productivity depends on two variables, x and y. Currently, $x = a$ and $y = b$, and the firm's productivity is 4,000 units. Productivity is increasing at a rate of 400 units per unit *decrease* in x, and is decreasing at a rate of 300 units per unit increase in y. What does all of this information tell you about the firm's productivity function $g(x, y)$?

51. Give an example of a function $f(x, y)$ with $f(1, 1) = 10$, $f_x(1, 1) = -2$, and $f_y(1, 1) = 3$.

52. Give an example of a function $f(x, y, z)$, all of whose partial derivatives are nonzero constants.

53. The graph of $z = b + mx + ny$ (b, m, and n constants) is a plane.
   (a) Explain the geometric significance of the numbers b, m, and n.
   (b) Show that the equation of the plane passing through (h, k, l) with slope m in the x-direction (in the sense of $\partial/\partial x$ and slope n in the y-direction is
$$z - l = m(x - h) + n(y - k).$$

54. The **tangent plane** to the graph of $f(x, y)$ at $P(a, b, f(a, b))$ is the plane containing the lines tangent to the slice through the graph by $y = b$ (as in Figure 1 in the text) and the slice through the graph by $x = a$. Use the result of the preceding exercise to show that the equation of the tangent plane is
$$z = f(a, b) + f_x(a, b)(x - a) + f_y(a, b)(y - b).$$

▶ ══════ **8.4** MAXIMA AND MINIMA

In the chapter on applications of the derivative we saw how to locate local extrema of a function of a single variable. In this section, we extend our methods to functions of two variables. Similar techniques work for functions of three or more variables.

Figure 1 shows a portion of the graph of the function $f(x, y) = 2(x^2 + y^2) - (x^4 + y^4) + 3$.

The graph resembles a "fling carpet" and there are several interesting points, marked a, b, c, and d on the graph. The point a has coordinates $(0, 0, f(0, 0))$, is directly above the origin $P(0, 0)$, and is the lowest point on the portion of the graph shown. Thus, we say that f has a **local minimum** at $(0, 0)$, since $f(0, 0)$ is smaller than $f(x, y)$ for any (x, y) near the point P. Similarly, the point b is higher than any point in the vicinity, $f(1, 1) \geq f(x, y)$ for any (x, y) near Q. Thus, we say that f has a **local maximum** at $(1, 1)$. The points c and d represent a new phenomenon and are called **saddle points.** They are neither local maxima nor local minima, but seem to be a little of both. To see more clearly what features a saddle point has, look at Figure 2, which shows a portion of the graph near the point c.

It is easy to see from this picture where the term "saddle point" comes from. Now, if we slice through the graph along $y = 1$, we get a curve on which c is the lowest point. Thus, c looks like a local minimum along this slice. On the other hand, if we slice through the graph along $x = 0$, we get another curve at right angles to the first one, and on which c is the *highest* point, so that c looks like a local maximum along this slice. This kind of

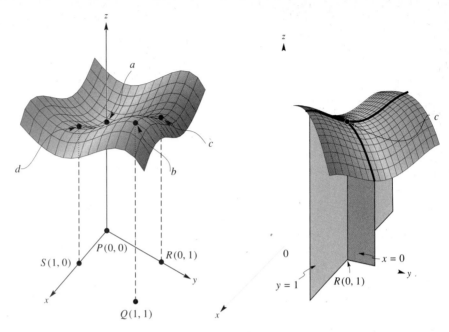

FIGURE 1 **FIGURE 2**

behavior characterizes a saddle point: f has a **saddle point** at (r, s) if f has a local minimum at (r, s) along some slices through that point and a local maximum along other slices through that point. If you look at the other saddle point, d, in Figure 1, you see the same behavior.

We also notice the following Figure 1.

1. The points P, Q, R, and S are all in the **interior** of the domain of f. That is, none of them lies on the boundary of the domain. Said another way, we can move some distance in any direction from any of these points without leaving the domain of f.

2. The tangent lines along the slices through these points parallel to the x- and y-axes are *horizontal*. Thus, the partial derivatives, $\partial f/\partial x$ and $\partial f/\partial y$, will be zero when evaluated at any of the points P, Q, R, and S. This gives us a way of locating candidates for local extrema and saddle points.

LOCATING CANDIDATES FOR LOCAL EXTREMA AND SADDLE POINTS IN THE INTERIOR OF THE DOMAIN OF f

Set $\partial f/\partial x = 0$ and $\partial f/\partial y = 0$ simultaneously, and solve for x and y. Then check that the resulting points (x, y) are in the interior of the domain of f.

Points at which all the partial derivatives are zero are called **critical points.** Thus, it is the critical points that are the only candidates for local extrema and saddle points in the interior of the domain of f.* Let's apply this principle to the function whose graph was shown in Figure 1.

▼   EXAMPLE 1

Locate all the critical points of the function

$$f(x, y) = 2(x^2 + y^2) - (x^4 + y^4) + 3.$$

SOLUTION   According to the principle above, we must set $\partial f/\partial x = 0$ and $\partial f/\partial y = 0$ and solve for x and y.

$$\frac{\partial f}{\partial x} = 4x - 4x^3 = 4x(1 - x^2)$$

$$\frac{\partial f}{\partial y} = 4y - 4y^3 = 4y(1 - y^2)$$

Setting these equal to zero gives us the following simultaneous equations.

$$4x(1 - x^2) = 0$$
$$4y(1 - y^2) = 0$$

The first equation has solutions $x = 0$, 1, or -1, while the second has solutions $y = 0$, 1, or -1. Since these are simultaneous equations, we need values of x and y that satisfy *both* equations. We can do this by choosing any one of the values for x that satisfies the first and any one of the values of y that satisfies the second. This gives us a total of *nine* critical points:

$$(0, 0), (0, 1), (0, -1), (1, 0), (1, 1), (1, -1), (-1, 0),$$
$$(-1, 1), \text{ and } (-1, -1).$$

Four of these are the points $P, Q, R,$ and S on the graph in Figure 1. The other five are also points at which f has extreme or saddle points, but Figure 1 shows only a small portion of the actual graph, and the remaining five points are out of range. (We'll see more of the graph of f a little later on.)

To get the points on the graph corresponding to these x- and y-values, we take the z-coordinate to be $f(x, y) = 2(x^2 + y^2) - (x^4 + y^4) + 3$, getting the nine points

$$(0, 0, 3), (0, 1, 4), (0, -1, 4), (1, 0, 4), (1, 1, 5),$$
$$(1, -1, 5), (-1, 0, 4), (-1, 1, 5), \text{ and } (-1, -1, 5).$$

▼ * We'll be looking at extrema on the *boundary* of the domain of a function in the next section. What we are calling critical points correspond to the *stationary* points of a function of one variable. We shall not consider the analogs of the singular points.

Before we go on...

Q How we can tell whether each of these candidates actually gives a saddle point or a local extremum?

A There is an analog of the second-derivative test that is useful in determining which candidates are extrema and which are saddle points. We shall discuss it after the next example.

▼ **EXAMPLE 2**

Locate all possible extrema and saddle points on the graph of

$$f(x, y) = e^{-(x^2+y^2)}.$$

SOLUTION   The partial derivatives of f are

$$\frac{\partial f}{\partial x} = -2xe^{-(x^2+y^2)}$$

$$\frac{\partial f}{\partial y} = -2ye^{-(x^2+y^2)}.$$

Setting these equal to zero gives the following equations.

$$-2xe^{-(x^2+y^2)} = 0$$
$$-2ye^{-(x^2+y^2)} = 0$$

The first equation implies that $x = 0$,* and the second implies that $y = 0$. Thus, the only critical point is $(0, 0)$, and the corresponding point on the graph is $(0, 0, 1)$, since $f(0, 0) = 1$.

Before we go on...

Q Since we haven't drawn the graph of f, how do we know whether f has an extremum or saddle point there?

A If you look at the function $f(x, y) = e^{-(x^2+y^2)}$, you will notice that the exponent is always zero or negative. If you raise e to a negative number, the result is less than 1, so 1 must be the maximum value the function can take. In other words, $(0, 0, 1)$ is an **absolute maximum** of f. Figure 3 shows the graph of this function.

Notice that this surface is radially symmetric about the z-axis (why?) and levels off toward $z = 0$ as x or y becomes large.

The following test gives us a way of deciding whether a critical point gives a local maximum, minimum, or saddle point.

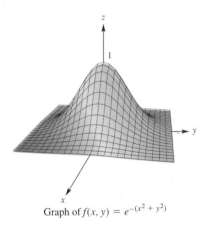

Graph of $f(x, y) = e^{-(x^2 + y^2)}$

FIGURE 3

▼ * Recall that if a product of two numbers is zero, one or the other must be zero. In our case, the number $e^{-(x^2+y^2)}$ can't be zero (since e^u is never zero), giving the result claimed.

SECOND-DERIVATIVE TEST FOR FUNCTIONS OF TWO VARIABLES

Suppose $f(x, y)$ is a function of two variables and that (a, b) is a critical point in the interior of the domain of f (so that $f_x(a, b) = 0$ and $f_y(a, b) = 0$). Let H be the quantity

$$f_{xx}(a, b)f_{yy}(a, b) - [f_{xy}(a, b)]^2.$$

Then

   f has a local minimum at (a, b) if $H > 0$ and $f_{xx}(a, b) > 0$,

   f has a local maximum at (a, b) if $H > 0$ and $f_{xx}(a, b) < 0$, and

   f has a saddle point at (a, b) if $H < 0$.

If $H = 0$, the test tells us nothing, so we need to look at the graph to see what is going on.

▶ **NOTE**   There is a second-derivative test for functions of three or more variables, but it is considerably more complicated. We shall stick with functions of two variables for the most part in this book. The justification of the second-derivative test is beyond the scope of this book.   ◀

Let us use the second-derivative test to continue Example 1.

▼ **EXAMPLE 3**

Use the second-derivative test to classify all the critical points of the function $f(x, y) = 2(x^2 + y^2) - (x^4 + y^4) + 3$.

SOLUTION   We found the following nine critical points in Example 1:

$$(0, 0), (0, 1), (0, -1), (1, 0), (1, 1), (1, -1), (-1, 0),$$
$$(-1, 1), \text{ and } (-1, -1).$$

We now need to apply the second-derivative test to each point in turn. First, we calculate all the second derivatives we shall need:

$$f_x = 4x - 4x^3, \quad \text{so} \quad f_{xx} = 4 - 12x^2, \quad \text{and} \quad f_{xy} = 0,$$

and

$$f_y = 4y - 4y^3, \quad \text{so} \quad f_{yy} = 4 - 12y^2.$$

   The point $(0, 0)$: $f_{xx}(0, 0) = 4$, $f_{yy}(0, 0) = 4$, and $f_{xy}(0, 0) = 0$. Thus, $H = 4(4) - 0^2 = 16$. Since $H > 0$ and $f_{xx}(0, 0) > 0$, we have a local minimum at $(0, 0, 3)$.

   The point $(0, 1)$: $f_{xx}(0, 1) = 4$, $f_{yy}(0, 1) = -8$, and $f_{xy}(0, 1) = 0$. Thus, $H = 4(-8) - 0^2 = -32$. Since $H < 0$, we have a saddle point at $(0, 1, 4)$.

   The point $(0, -1)$: $f_{xx}(0, -1) = 4$, $f_{yy}(0, -1) = -8$ and $f_{xy}(0, -1) = 0$. These are the same values we got for the last point, so we again have a saddle point at $(0, -1, 4)$.

The point $(1, 0)$: $f_{xx}(1, 0) = -8, f_{yy}(1, 0) = 4$, and $f_{xy}(1, 0) = 0$. Thus, $H = (-8)4 - 0^2 = -32$. Since $H < 0$, we have a saddle point at $(1, 0, 4)$.

The point $(1, 1)$: $f_{xx}(1, 1) = -8, f_{yy}(1, 1) = -8$, and $f_{xy}(1, 1) = 0$. Thus, $H = (-8)^2 - 0^2 = 64$. Since $H > 0$ and $f_{xx}(1, 1) < 0$, we have a local maximum at $(1, 1, 5)$.

The point $(1, -1)$: These are the same values we got for the last point, so we again have a local maximum at $(1, -1, 5)$.

The point $(-1, 0)$: These are the same values we got for the point $(1, 0)$, so we have a saddle point at $(-1, 0, 4)$.

The point $(-1, 1)$: These are the same values we got for the point $(1, 1)$, so we have a local maximum at $(-1, 1, 5)$.

The point $(-1, -1)$: These are again the same values we got for the point $(1, 1)$, so we have a local maximum at $(-1, -1, 5)$.

Before we go on... Figure 4 shows a larger portion of the graph of $f(x, y)$ including all nine critical points. We leave it to you to spot their locations on the graph.

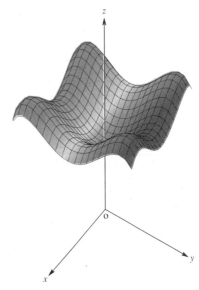

Graph of $f(x, y) = 2(x^2 + y^2) - (x^4 + y^4) + 3$

FIGURE 4

▼ **EXAMPLE 4**

Locate and classify all critical points of

$$f(x, y) = x^2 + 2y^2 + 2xy + 4x.$$

SOLUTION   We first calculate the first-order partial derivatives.

$$f_x = 2x + 2y + 4$$
$$f_y = 2x + 4y$$

Setting these equal to zero gives a system of two linear equations in two unknowns.

$$x + y = -2$$
$$x + 2y = 0$$

This system has solution $(-4, 2)$, so this is our only critical point. The z-coordinate of this point is $f(-4, 2) = -8$. The second partial derivatives are

$$f_{xx} = 2, \quad f_{xy} = 2, \quad f_{yy} = 4.$$

Notice that all these derivatives are constants—they have the same value at every point (a, b). Thus, $H = 2 \cdot 4 - 2^2 = 4$. Since $H > 0$ and $f_{xx} > 0$, the second-derivative test tells us that we have a local minimum at $(-4, 2, -8)$.

▼   **EXAMPLE 5**

Locate and classify all critical points of $f(x, y) = x^2y - x^2 - y^2$.

SOLUTION

$$f_x = 2xy - 2x = 2x(y - 1)$$
$$f_y = x^2 - 2y$$

Setting these equal to zero gives

$$x = 0 \text{ or } y = 1$$
$$x^2 = 2y.$$

We get a solution by choosing either $x = 0$ or $y = 1$ and substituting into $x^2 = 2y$, giving two possibilities: $0 = 2y$ and $x^2 = 2$. The first, $0 = 2y$, gives $(0, 0)$. The second, $x^2 = 2$, gives $x = \pm\sqrt{2}$ and hence the points $(\sqrt{2}, 1)$ and $(-\sqrt{2}, 1)$.

Thus, we have three critical points. To apply the second-derivative test, we calculate the second derivatives.

$$f_{xx} = 2y - 2$$
$$f_{xy} = 2x$$
$$f_{yy} = -2$$

We now look at each critical point in turn.

The point $(0, 0)$: $f_{xx}(0, 0) = -2$, $f_{xy}(0, 0) = 0$, $f_{yy}(0, 0) = -2$, so $H = 4$. Since $H > 0$ and $f_{xx}(0, 0) < 0$, the second-derivative test tells us that f has a local maximum at $(0, 0, 0)$.

The point $(\sqrt{2}, 1)$: $f_{xx}(\sqrt{2}, 1) = 0$, $f_{xy}(\sqrt{2}, 1) = 2\sqrt{2}$, and $f_{yy}(\sqrt{2}, 1) = -2$, so $H = -8$. Since $H < 0$, we know that f has a saddle point at $(\sqrt{2}, 1, -1)$.

The point $(-\sqrt{2}, 1)$**:** $f_{xx}(-\sqrt{2}, 1) = 0$, $f_{xy}(-\sqrt{2}, 1) = -2\sqrt{2}$, and $f_{yy}(\sqrt{2}, 1) = -2$, so once again $H = -8$, and the point $(-\sqrt{2}, 1, -1)$ is a saddle point.

Before we go on... Figure 5 shows the graph of f. See if you can spot the three critical points.

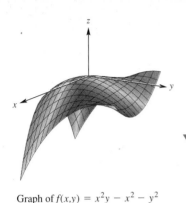

Graph of $f(x,y) = x^2y - x^2 - y^2$

FIGURE 5

▼ EXAMPLE 6

Locate and classify all the critical points of $f(x, y) = 2x^4 - 6x^2y + y^4$.

SOLUTION

$$f_x = 8x^3 - 12xy = 4x(2x^2 - 3y)$$
$$f_y = 4y^3 - 6x^2 = 2(2y^3 - 3x^2)$$
$$f_{xx} = 24x^2 - 12y$$
$$f_{xy} = -12x$$
$$f_{yy} = 12y^2$$

To obtain the critical points, we set f_x and f_y equal to zero and solve. Setting $f_x = 0$ gives

$$x = 0 \text{ or } 3y = 2x^2.$$

Setting $f_y = 0$ gives

$$2y^3 = 3x^2.$$

To solve these equations simultaneously, we choose either $x = 0$ or $3y = 2x^2$ and combine it with $2y^3 = 3x^2$.

The combination $x = 0$ and $2y^3 = 3x^2$ gives the solution $(0, 0)$.

The combination $3y = 2x^2$ and $2y^3 = 3x^2$ is more tricky, but we can solve each of these equations for x^2 and equate the two answers:

$$x^2 = \tfrac{3}{2}y \text{ and } x^2 = \tfrac{2}{3}y^3$$
$$\tfrac{3}{2}y = \tfrac{2}{3}y^3$$

Thus, $y(\tfrac{3}{2} - \tfrac{2}{3}y^2) = 0$, giving $y = 0$ or $y = \pm\tfrac{3}{2}$. If $y = 0$, then $x = 0$, and we have the point $(0, 0)$ once again. If $y = \tfrac{3}{2}$, then $x^2 = (\tfrac{3}{2})^2$, so $x = \pm\tfrac{3}{2}$. If $y = -\tfrac{3}{2}$, then $x^2 = -(\tfrac{3}{2})^2$, giving no real solution for x. Thus the only solutions are the following.

$$(0, 0), (\tfrac{3}{2}, \tfrac{3}{2}) \text{ and } (-\tfrac{3}{2}, \tfrac{3}{2})$$

The point $(0, 0)$**:** All the second derivatives are zero when evaluated at $(0, 0)$, so the second-derivative test fails. This means that we must look at the function and try to decide whether there is a local extremum at $(0, 0)$. Now $f(0, 0) = 0$. If we slice the graph along $y = 0$, we get $f(x, 0) = 2x^4$, which we know possesses a minimum at $x = 0$. Thus, our critical point is the lowest

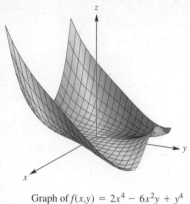

Graph of $f(x,y) = 2x^4 - 6x^2y + y^4$

FIGURE 6

point on the graph for this slice. If we slice along $x = 0$, we get $f(0, y) = y^4$, so there is again a minimum in this slice. Thus, it's beginning to look as though $(0, 0)$ is a local minimum. Now might be a good time to look at the graph, as drawn by computer (see Figure 6).

The graph appears to get lower in the direction of $x = y$, so let us try taking a slice along this plane. If we set $x = y$, we get $f(x, x) = 2x^4 - 6x^3 + x^4 = 3x^4 - 6x^3 = 3x^3(x - 2)$. Now this is negative for all values of x between 0 and 2, showing that the surface dips below the xy-plane along this slice. In other words, $(0, 0)$ can't be a local minimum, since the surface is lower than $z = 0$ at points nearby. We conclude that $(0, 0, 0)$ is neither a local maximum nor a local minimum. Moreover, it isn't a saddle point either. (Why?)

The point $(\frac{3}{2},\frac{3}{2})$: $f_{xx}(\frac{3}{2},\frac{3}{2}) = 36$, $f_{xy}(\frac{3}{2},\frac{3}{2}) = -18$, $f_{yy}(\frac{3}{2},\frac{3}{2}) = 27$. Thus $H = 648 > 0$, and we have a local minimum at $(\frac{3}{2}, \frac{3}{2}, -\frac{81}{16})$.

The point $(-\frac{3}{2},\frac{3}{2})$: $f_{xx}(-\frac{3}{2},\frac{3}{2}) = 36$, $f_{xy}(-\frac{3}{2},\frac{3}{2}) = -18$, $f_{yy}(-\frac{3}{2},\frac{3}{2}) = 27$. Thus again $H > 0$, and we have another local minimum at $(-\frac{3}{2}, \frac{3}{2}, -\frac{81}{16})$.

▶ **8.4 EXERCISES**

In Exercises 1 through 6, classify each labeled point on the graph as either:
  (a) *a local maximum;* **(b)** *a local minimum;* **(c)** *a saddle point;*
  (d) *a critical point, but neither a local extremum nor a saddle point;*
  (e) *none of the above.*

1.

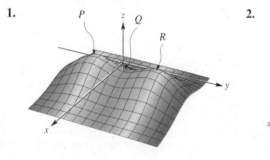

2.

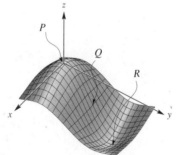

3.

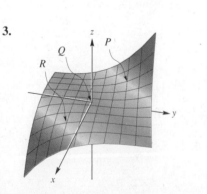

4.

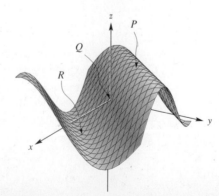

5.

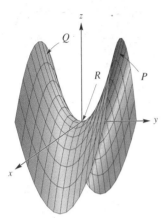

6.

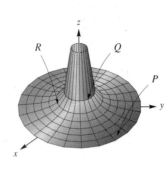

7. Sketch the graph of a function that has one local extremum and no saddle points.

8. Sketch the graph of a function that has one saddle point and one local extremum.

Locate and classify all critical points of each function in Exercises 9–26.

9. $f(x, y) = x^2 + y^2 + 1$

10. $f(x, y) = 4 - (x^2 + y^2)$

11. $g(x, y) = 1 - x^2 - x - y^2 + y$

12. $g(x, y) = x^2 + x + y^2 - y - 1$

13. $h(x, y) = x^2 y - 2x^2 - 4y^2$

14. $h(x, y) = x^2 + y^2 - y^2 x - 4$

15. $s(x, y) = e^{x^2 + y^2}$

16. $s(x, y) = e^{-(x^2 + y^2)}$

17. $t(x, y) = x^4 + 8xy^2 + 2y^4$

18. $t(x, y) = x^3 - 3xy + y^3$

19. $f(x, y) = x^2 + y - e^y$

20. $f(x, y) = xe^y$

21. $f(x, y) = e^{-(x^2 + y^2 + 2x)}$

22. $f(x, y) = e^{-(x^2 + y^2 - 2x)}$

23. $f(x, y) = xy + \dfrac{2}{x} + \dfrac{2}{y}$

24. $f(x, y) = xy + \dfrac{4}{x} + \dfrac{2}{y}$

25. $g(x, y) = x^2 + y^2 + \dfrac{2}{xy}$

26. $g(x, y) = x^3 + y^3 + \dfrac{3}{xy}$

APPLICATIONS

27. *Average Cost* Your bicycle factory makes two models, 5-speeds and 10-speeds. Each week, the total cost to make x 5-speeds and y 10-speeds is given by

$$C(x, y) = \$10,000 + 50x + 70y + 0.0125xy.$$

At what production levels is your average cost least? (Remember that the average cost is given by $\overline{C}(x, y) = C(x, y)/(x + y)$.)

28. *Average Cost* Your bicycle factory makes two models, 5-speeds and 10-speeds. Each week, the total cost to make x 5-speeds and y 10-speeds is given by

$$C(x, y) = \$140,000 + 20x + 90y + 0.07xy.$$

At what production levels is your average cost least?

29. *Average Cost* Let $C(x, y)$ be any cost function. Show that when the average cost is minimized, the marginal costs C_x and C_y both equal the average cost. Explain why this is reasonable.

30. *Average Profit* Let $P(x, y)$ be any profit function. Show that when the average profit is maximized, the marginal profits P_x and P_y both equal the average profit. Explain why this is reasonable.

31. *Revenue* Your company manufactures two models of stereo speakers, the Ultra Mini and the Big Stack. Demand for each depends partly on the price of the other. If one is expensive, more people will buy the other. If p_1 is the price per pair of the Ultra Mini, and

p_2 is the price of the Big Stack, demand for the Ultra Mini is given by

$$q_1(p_1, p_2) = 100{,}000 - 100p_1 + 10p_2,$$

where q_1 represents the number of pairs of Ultra Minis that will be sold in a year. The demand for the Big Stack is given by

$$q_2(p_1, p_2) = 150{,}000 + 10p_1 - 100p_2.$$

COMMUNICATION AND REASONING EXERCISES

33. Let $H = f_{xx}(a, b)f_{yy}(a, b) - f_{xy}(a, b)^2$. What condition on H guarantees that f has a local extremum at the point (a, b)?

34. Let H be as in the previous exercise. Give an example to show that it is possible to have $H = 0$ and a local minimum at (a, b).

35. Suppose that when the graph of $f(x, y)$ is sliced by a vertical plane through (a, b) parallel to either the xz-plane or the yz-plane, the resulting curve has a local maximum at (a, b). Does this mean that f has a local maximum at (a, b)? Explain your answer.

Find the prices for the Ultra Mini and the Big Stack that will maximize your total revenue.

32. *Revenue*   Repeat the previous exercise, with the following demand functions.

$$q_1(p_1, p_2) = 100{,}000 - 100p_1 + p_2$$
$$q_2(p_1, p_2) = 150{,}000 + p_1 - 100p_2$$

36. Suppose that f has a local maximum at (a, b). Does it follow that when the graph of f is sliced by a vertical plane parallel to either the xz-plane or the yz-plane, the resulting curve has a local maximum at (a, b)? Explain your answer.

37. The tangent plane to a graph was introduced in the exercises in the previous section. Explain why the tangent plane is parallel to the xy-plane at a local maximum or minimum of $f(x, y)$.

38. Explain why the tangent plane is parallel to the xy-plane at a saddle point of $f(x, y)$.

▶ ══ 8.5 CONSTRAINED MAXIMA AND MINIMA AND APPLICATIONS

So far we have looked only at the local extrema of f that lie in the interior of the domain of f. There may also be local extrema on the boundary of the domain (just as, for a function of one variable, the endpoints of the domain may be local extrema). This situation arises, for example, in optimization problems with constraints, similar to those we saw in the chapter on applications of the derivative. Here is a typical example.

$$\text{Maximize } S = xy + 2xz + 2yz$$
$$\text{subject to } xyz = 4$$
$$\text{and } x \geq 0,\ y \geq 0,\ z \geq 0.$$

There are two kinds of constraints in this example: equations and inequalities. The inequalities specify a restriction on the domain of S. Our strategy for solving such problems is essentially the same as in the chapter on applications of the derivative. First, we use any equality constraints to eliminate variables. In the examples in this section we will be able to reduce to a function of only two variables. The inequality constraints then help define the domain of this function. Next, we locate any critical points in the interior of the domain. Finally, we look at the boundary of the domain. When there is no boundary to worry about, another method, called the *method of Lagrange multipliers*, comes in handy.

We first look at functions of two variables with restricted domains, so we can see how to handle the boundaries.

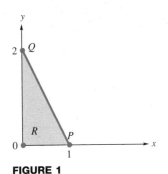

FIGURE 1

▼ **EXAMPLE 1**

Find the maximum and minimum value of $f(x, y) = xy - x - 2y$ on the triangular region R with vertices $(0, 0)$, $(1, 0)$, and $(0, 2)$.

SOLUTION   The *domain* of f is the region R, which is shown in Figure 1. *Locate critical points in the interior of the domain.*   We have

$$f_x = y - 1$$
$$f_y = x - 2$$
$$f_{xx} = 0$$
$$f_{xy} = 1$$
$$f_{yy} = 0.$$

The only critical point is thus $(2, 1)$. Since this lies outside the domain (the region R) we ignore it. Thus, there are no critical points in the interior of the domain of f.

Locate local extrema on the boundary of the domain.   The boundary of the domain consists of three line segments, OP, OQ, and PQ. We deal with these one at a time.

Segment OP   This line segment has equation $y = 0, 0 \le x \le 1$. Along this segment we see $f(x, 0) = -x$. We find the local extrema of this function of one variable by the methods we used in the chapter on applications of the derivative. There are no critical points, and there are two endpoints, $x = 0$ and $x = 1$. Since $y = 0$, this gives us the following two candidates for local extrema: $(0, 0, 0)$ and $(1, 0, -1)$.

Segment OQ   This line segment has equation $x = 0, 0 \le y \le 2$. Along this segment we see $f(0, y) = -2y$. We now locate the local extrema of this function of one variable. Once again, there are only the endpoints, $y = 0$ and $y = 2$. Since $x = 0$, this gives us the two candidates $(0, 0, 0)$ and $(0, 2, -4)$.

Segment PQ   This line segment has equation $y = -2x + 2$ with $0 \le x \le 1$. Along this segment we see

$$f(x, -2x + 2) = x(-2x + 2) - x - 2(-2x + 2)$$
$$= -2x^2 + 5x - 4.$$

This function of x (whose graph is an upside-down parabola) has a stationary maximum when its derivative, $-4x + 5$ is 0, or $x = \frac{5}{4}$. Since this is bigger than 1, it lies outside the domain $0 \le x \le 1$. Thus, we reject it. There are no other critical points, and the endpoints are $x = 0$ and $x = 1$. When $x = 0$, $y = -2(0) + 2 = 2$, giving the point $(0, 2, -4)$. When $x = 1$, $y = -2(1) + 2 = 0$, giving the point $(1, 0, -1)$.

Thus, the candidates for maxima and minima are $(0, 0, 0)$, $(1, 0, -1)$, and $(0, 2, -4)$ (which happen to lie over the corner points of the domain R). Since their z-coordinates give the value of f, we see that f has an absolute maximum of 0 at the point $(0, 0)$ and an absolute minimum of -4 at the point $(0, 2)$.

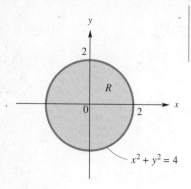

FIGURE 2

▼ **EXAMPLE 2**

Find the maximum and minimum value of $f(x, y) = x^2 + 2y^2 - x$ on the circular region specified by $x^2 + y^2 \leq 4$.

SOLUTION   The domain R consists of all points inside the circle $x^2 + y^2 = 4$, which is the circle of radius 2 centered at the origin, as shown in Figure 2.

Locate critical points in the interior of the domain.   We have the following.

$$f_x = 2x - 1$$
$$f_y = 4y$$
$$f_{xx} = 2$$
$$f_{xy} = 0$$
$$f_{yy} = 4$$

To get the critical points, we set f_x and f_y equal to zero and solve for x and y, getting $x = \frac{1}{2}$ and $y = 0$. Thus, the only critical point is $(\frac{1}{2}, 0)$. This lies in the interior of the domain R, so we check to see whether it is a local extremum. Since $H = 8 > 0$ and f_{xx} is positive, we have a local minimum at $(\frac{1}{2}, 0, -\frac{1}{4})$.

Locate local extrema on the boundary of the domain.   The boundary of the domain is the circle of radius 2 and has equation $x^2 + y^2 = 4$, or $y^2 = 4 - x^2$. We can substitute for y^2 in the equation $z = x^2 + 2y^2 - x$ and obtain the curve $z = x^2 + 2(4 - x^2) - x = -x^2 - x + 8$. This has a stationary maximum when $-2x = 1$ or $x = -\frac{1}{2}$. When $x = -\frac{1}{2}$, $y^2 = 4 - (-\frac{1}{2})^2 = \frac{15}{4}$, so $y = \pm\sqrt{15}/2 \approx 1.94$. Thus, we have two candidates: $(-\frac{1}{2}, \sqrt{15}/2, \frac{33}{4})$ and $(-\frac{1}{2}, -\sqrt{15}/2, \frac{33}{4})$.

   Thus, our three candidates for extrema are $(\frac{1}{2}, 0, -\frac{1}{4})$, $(-\frac{1}{2}, \sqrt{15}/2, \frac{33}{4})$ and $(-\frac{1}{2}, -\sqrt{15}/2, \frac{33}{4})$. Looking at the z-coordinates, we see that f has an absolute maximum of $\frac{33}{4}$ at $(-\frac{1}{2}, \sqrt{15}/2)$ and $(-\frac{1}{2}, -\sqrt{15}/2)$, and an absolute minimum of $-\frac{1}{4}$ at $(\frac{1}{2}, 0)$.

We now look at an example with a constraint equation.

▼ **EXAMPLE 3**   Minimizing Area

Find the dimensions of a rectangular box with no top having a volume of 4 cubic feet and the least possible surface area.

SOLUTION   Our first task is to rephrase this as a mathematical optimization problem. Figure 3 shows a picture of the box with dimensions x, y, and z. We want to minimize the total surface area, which is given by

$$S = xy + 2xz + 2yz.$$

This is our **objective function.** We can't simply choose x, y, and z to all be zero, since the enclosed volume must be 4 cubic feet. In other words,

$$xyz = 4.$$

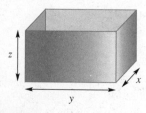

FIGURE 3

This is our constraint equation. There are further unstated constraints: $x \geq 0$, $y \geq 0$, and $z \geq 0$. We now restate the problem as follows.

$$Minimize\ S = xy + 2xz + 2yz$$
$$subject\ to\ xyz = 4,$$
$$x \geq 0, y \geq 0, z \geq 0.$$

As suggested in the discussion before Example 1, we will proceed as follows:

Solve the constraint equation for one of the variables, and substitute in the objective function.   Solving the constraint equation for z gives

$$z = \frac{4}{xy}.$$

Substituting this into the objective function gives a function of only two variables.

$$S(x, y) = xy + \frac{8}{y} + \frac{8}{x}.$$

Minimize the resulting function of two variables.   A word about the domain of S: Since we said that x and y must be nonnegative, we already know that (x, y) is restricted to the first quadrant. Moreover, since S is not defined if either x or y is zero, we exclude the x- and y-axes from the domain of S and are left with the interior of the first quadrant as our domain. Since this has no boundary, we look only for critical points in the interior.

$$S_x = y - \frac{8}{x^2}$$

$$S_y = x - \frac{8}{y^2}$$

$$S_{xx} = \frac{16}{x^3}$$

$$S_{xy} = 1$$

$$S_{yy} = \frac{16}{y^3}$$

We now equate the first partial derivatives to zero and solve for x and y.

$$y = \frac{8}{x^2}$$

$$x = \frac{8}{y^2}$$

Substituting the first of these in the second gives the following.

$$x = \frac{x^4}{8}$$

$$x^4 - 8x = 0$$

$$x(x^3 - 8) = 0$$

The two solutions are $x = 0$, which we reject (since x cannot be zero), and $x = 2$. Substituting this in $y = 8/x^2$ gives $y = 2$ also. Thus, the only critical point is $(2, 2)$. To apply the second-derivative test, we compute $S_{xx}(2, 2) = 2$, $S_{xy}(2, 2) = 1$, $S_{yy}(2, 2) = 2$, so that $H > 0$, and we have a local minimum at $(2, 2)$.

Q Is this local minimum an absolute minimum?

A Yes. There must be a least surface area among all boxes that hold 4 cubic feet (why?). Since this would give a local minimum of S, and since the only possible local minimum of S occurs at $(2, 2)$, this is the absolute minimum.

To finish the example, we get the value of z by substituting the values of x and y into the constraint equation $z = 4/xy$, which gives $z = 1$. Thus, the required dimensions of the box are

$$x = 2 \text{ ft}$$
$$y = 2 \text{ ft}$$
$$z = 1 \text{ ft.}$$

▼ **EXAMPLE 4**   Maximizing Profit

You own a company making two models of stereo speakers, the Ultra Mini and the Big Stack. Each Ultra Mini requires 1 square foot of fabric and 3 feet of wire, while each Big Stack requires 5 square feet of fabric and 9 feet of wire. You have each week 100 square feet of fabric and 270 feet of wire to use. Your profit function is

$$P(x, y) = 10x + 60y + 40\sqrt{x + y}.$$

Here x is the number of Ultra Minis, y is the number of Big Stacks, and P is your profit in dollars. Find the number of each model you should make each week in order to maximize your profit.

SOLUTION   The constraints in this problem come from the limited amount of fabric and wire available. The constraint on fabric is

$$x + 5y \leq 100.$$

The constraint on wire is

$$3x + 9y \leq 270.$$

There are also the constraints $x \geq 0$ and $y \geq 0$ that are often present in applications. Our problem is then the following.

$$\textit{Maximize } P = 10x + 60y + 40\sqrt{x + y}$$
$$\textit{subject to } x + 5y \leq 100$$
$$3x + 9y \leq 270$$
$$x \geq 0 \textit{ and } y \geq 0.$$

If we graph the domain R, we get Figure 4.

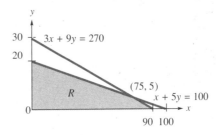

FIGURE 4

(The point of intersection $(75, 5)$ is found by solving the system of linear equations given by the equations of the two lines.) Since we have no equality constraints, we proceed as follows.

Locate critical points in the interior of the domain.

$$\frac{\partial P}{\partial x} = 10 + \frac{20}{\sqrt{x + y}}$$

$$\frac{\partial P}{\partial y} = 60 + \frac{20}{\sqrt{x + y}}$$

These can never equal 0, since the radical is always positive. Therefore, there are no critical points in the interior of the domain.

Locate local extrema on the boundary of the domain.   The boundary of the domain consists of four line segments, which we consider one at a time.

$$y = 0, \quad 0 \leq x \leq 90$$

Here, $P = 10x + 40\sqrt{x}$, and $P' = 10 + 20/\sqrt{x}$ is never 0. Thus, the only points we need to look at are the endpoints $x = 0$ and $x = 90$, which give $(0, 0, 0)$ and $(90, 0, 904.2)$.

$$x = 0, \quad 0 \leq y \leq 20$$

Here, $P = 60y + 40\sqrt{y}$. Again, there are no critical points, only the endpoints $y = 0$ and $y = 20$. This gives one more point, $(0, 20, 1378.9)$.

$$x' = 100 - 5y, \quad 5 \leq y \leq 20$$

This gives

$$P = 10(100 - 5y) + 60y + 40\sqrt{100 - 4y}$$
$$= 1{,}000 + 10y + 40\sqrt{100 - 4y}$$
$$P' = 10 - \frac{80}{\sqrt{100 - 4y}}.$$

Setting the derivative equal to 0 and solving, we get $y = 9$. Substituting this in $x = 100 - 5y$, we get $x = 55$. Now we check the second derivative.

$$P'' = -\frac{160}{(100 - 4y)^{3/2}} < 0$$

This tells us that we have found a local maximum. In fact, it must be the maximum value on this line segment. The point on the graph is $(55, 9, 1410)$.

$$x = 90 - 3y, \quad 0 \leq y \leq 5$$

This gives

$$P = 10(90 - 3y) + 60y + 40\sqrt{90 - 2y}$$
$$= 900 + 30y + 40\sqrt{90 - 2y}$$
$$P' = 30 - \frac{40}{\sqrt{90 - 2y}}.$$

Setting this equal to 0 and solving for y gives $y = 44\frac{1}{9}$, which is outside the range $0 \leq y \leq 5$. Therefore, there are no critical points on this line segment. Its endpoints are $y = 0$ and $y = 5$, giving $(90, 0, 904.2)$ and $(75, 5, 1407.8)$.

Looking at the heights of all of the points we have found, the highest is $(55, 9, 1410)$, so this gives you the largest profit. In other words, you should make 55 Ultra Minis and 9 Big Stacks each week, giving you the largest possible profit of $1410 per week.

Before we go on... If you have studied linear programming, this should remind you of the problems you solved by that technique. However, since the objective function is not linear, the techniques of linear programming fail to solve this problem. In particular, notice that the solution we found is *not* at a corner of the feasible region, but in the middle of one edge. It is also quite possible that the solution could be in the interior of the region.

THE METHOD OF LAGRANGE MULTIPLIERS

Suppose we have an optimization problem in which it is difficult or impossible to solve a constraint equation for one of the variables. Then we can use the method of **Lagrange multipliers** to avoid this difficulty. We shall restrict attention to the case of a single constraint equation, although the method generalizes to any number of constraint equations.

LOCATING LOCAL EXTREMA USING THE METHOD OF LAGRANGE
MULTIPLIERS

To locate the candidates for local extrema of a function $f(x, y, \ldots)$
subject to the constraint $g(x, y, \ldots) = 0$, solve the following system
of equations for $x, y, \ldots$, and λ.

$$f_x = \lambda g_x$$
$$f_y = \lambda g_y$$
$$\ldots$$
$$g = 0.$$

The unknown λ is called a **Lagrange multiplier.** The points
$(x, y, \ldots)$ that occur in solutions are then the candidates for the
local extrema of f subject to $g = 0$.

▼   **EXAMPLE 5**

Use the method of Lagrange multipliers to find the maximum value of
$f(x, y) = 2xy$ subject to $x^2 + 4y^2 = 32$.

SOLUTION   We start by rewriting the problem in standard form.

$$\text{Maximize } f(x, y) = 2xy \text{ subject to } x^2 + 4y^2 - 32 = 0.$$

Here, $g(x, y) = x^2 + 4y^2 - 32$, and the system of equations we need to
solve is

$$f_x = \lambda g_x, \quad \text{or} \quad 2y = 2\lambda x,$$
$$f_y = \lambda g_y, \quad \text{or} \quad 2x = 8\lambda y,$$
$$g = 0, \quad \text{or} \quad x^2 + 4y^2 - 32 = 0.$$

A convenient way to solve such a system is to solve one of the equations for
λ and then substitute in the remaining equations. Thus, we start by solving the
first equation to obtain

$$\lambda = \frac{y}{x}.$$

(A word of caution: since we divided by x, we made the implicit assumption
that $x \neq 0$, so before continuing, we should check what happens if $x = 0$. But
if $x = 0$, then the first equation, $2y = 2\lambda x$, tells us that $y = 0$ as well, and
this contradicts the third equation: $x^2 + 4y^2 - 32 = 0$. Thus, we can rule
out the possibility that $x = 0$.)

Substituting in the remaining equations gives

$$x = \frac{4y^2}{x}, \quad \text{or} \quad x^2 = 4y^2,$$

$$\text{and} \quad x^2 + 4y^2 - 32 = 0.$$

Notice how we have reduced the number of unknowns and also the number of equations by one. We can now substitute $x^2 = 4y^2$ in the last equation, obtaining

$$4y^2 + 4y^2 - 32 = 0,$$

or

$$8y^2 = 32,$$

giving

$$y = \pm 2.$$

We now substitute back to obtain

$$x^2 = 4y^2 = 16,$$

or

$$x = \pm 4.$$

We don't need the value of λ, so we won't solve for it. Thus, the candidates for local extrema are given by $x = \pm 4$ and $y = \pm 2$, giving the four points $(-4, -2)$, $(-4, 2)$, $(4, -2)$, and $(4, 2)$.

Recall that we are seeking the values of x and y that give the maximum value for $f(x, y) = 2xy$. Since we now have only four points to choose from, we compare the values of f at these four points and conclude that the maximum value of f occurs when $(x, y) = (-4, -2)$ or $(4, 2)$.

Before we go on...

Q Something is suspicious here. We didn't check to see whether these candidates were local extrema to begin with, let alone absolute extrema! How do we justify this omission?

A One of the difficulties with using the method of Lagrange multipliers is that it does not provide us with a test analogous to the second-derivative test for functions of several variables. However, if you grant that the function in question does have an absolute maximum, then we require no test, since one of the candidates must give this maximum.

Q But how do we know that the given function has an absolute maximum?

A The best way to see this is by giving a geometric interpretation. The constraint $x^2 + 4y^2 = 32$ tells us that the point (x, y) must lie on the ellipse shown in Figure 5. The function $f(x, y) = 2xy$ gives the area of the rectangle shaded in Figure 5.

Since there is a largest possible rectangle of this type, the function f must have an absolute maximum for at least one pair of coordinates (x, y).

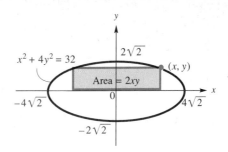

FIGURE 5

Q When can I use the method of Lagrange multipliers? When *should* I use it?

A We have only discussed the method when there is a single equality constraint. There is a generalization, which we shall not discuss, that works when there are more equality constraints (as a hint, we need to introduce one multiplier for each constraint). So, if you have a problem with more than one equality constraint, or with any inequality constraints, you must use the method we discussed earlier in this section. On the other hand, if you have one equality constraint, and it would be difficult to solve it for one of the variables, then you should use Lagrange multipliers. We have noted in the exercises where you might use Lagrange multipliers.

Q Why does the method of Lagrange multipliers work?

A An adequate answer is beyond the scope of this book.

▶ **8.5 EXERCISES**

Find the maximum and minimum values, and the points at which they occur, for each function in Exercises 1–16.

1. $f(x, y) = x^2 + y^2, \quad 0 \le x \le 2, \quad 0 \le y \le 2$

2. $g(x, y) = \sqrt{x^2 + y^2}, \quad 1 \le x \le 2, \quad 1 \le y \le 2$

3. $h(x, y) = (x - 1)^2 + y^2, \quad x^2 + y^2 \le 4$

4. $k(x, y) = x^2 + (y - 1)^2, \quad x^2 + y^2 \le 9$

5. $f(x, y) = e^{x^2+y^2}, \quad 4x^2 + y^2 \le 4$

6. $g(x, y) = e^{-(x^2+y^2)}, \quad x^2 + 4y^2 \le 4$

7. $h(x, y) = e^{4x^2+y^2}, \quad x^2 + y^2 \le 1$

8. $k(x, y) = e^{-(x^2+4y^2)}, \quad x^2 + y^2 \le 4$

9. $f(x, y) = x + y + 1/(xy), \quad x \ge \frac{1}{2}, \quad y \ge \frac{1}{2}, \quad x + y \le 3$

10. $g(x, y) = x + y + 8/(xy), \quad x \ge 1, \quad y \ge 1, \quad x + y \le 6$

11. $h(x, y) = xy + 8/x + 8/y, \quad x \ge 1, \quad y \ge 1, \quad xy \le 9$

12. $k(x, y) = xy + 1/x + 8/y$, $\quad x \geq \frac{1}{4}$, $\quad y \geq \frac{1}{4}$, $\quad xy \leq 9$

13. $f(x, y) = x^2 + 2x + y^2$, on the region in the figure.

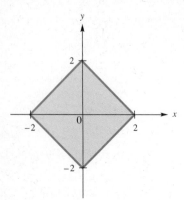

14. $g(x, y) = x^2 + y^2$, on the region in the figure.

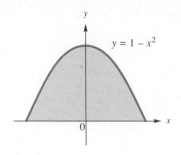

$y = 1 - x^2$

15. $h(x, y) = x^3 + y^3$, on the region in the figure.

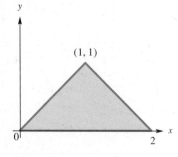

(1, 1)

16. $k(x, y) = x^3 + 2y^3$, on the region in the figure.

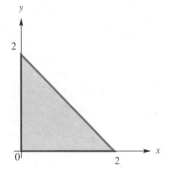

17. At what points on the sphere $x^2 + y^2 + z^2 = 1$ is the product xyz a maximum? (The method of Lagrange multipliers can be used here.)

18. At what point on the surface $z = (x^2 + x + y^2 + 4)^{1/2}$ is the quantity $x^2 + y^2 + z^2$ a minimum? (The method of Lagrange multipliers can be used here.)

APPLICATIONS

19. *Cost*   Your bicycle factory makes two models, 5-speeds and 10-speeds. Each week, your total cost to make x 5-speeds and y 10-speeds is given by

$$C(x, y) = \$10,000 + 50x + 70y - 0.5xy.$$

You want to make between 100 and 150 5-speeds, and between 80 and 120 10-speeds. What combination will cost you the least? What combination will cost you the most?

20. *Cost*   Your bicycle factory makes two models, 5-speeds and 10-speeds. Each week, your total cost to

make x 5-speeds and y 10-speeds is given by

$$C(x, y) = \$10,000 + 50x + 70y - 0.46xy.$$

You want to make between 100 and 150 5-speeds, and between 80 and 120 10-speeds. What combination will cost you the least? What combination will cost you the most?

21. *Profit*   Your software company sells two programs, Walls and Doors. Your profit from selling x copies of Walls and y copies of Doors is given by

$$P(x, y) = 20x + 40y - 0.1(x^2 + y^2).$$

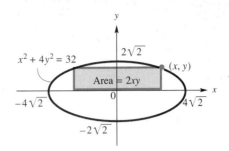

FIGURE 5

Q When can I use the method of Lagrange multipliers? When *should* I use it?

A We have only discussed the method when there is a single equality constraint. There is a generalization, which we shall not discuss, that works when there are more equality constraints (as a hint, we need to introduce one multiplier for each constraint). So, if you have a problem with more than one equality constraint, or with any inequality constraints, you must use the method we discussed earlier in this section. On the other hand, if you have one equality constraint, and it would be difficult to solve it for one of the variables, then you should use Lagrange multipliers. We have noted in the exercises where you might use Lagrange multipliers.

Q Why does the method of Lagrange multipliers work?

A An adequate answer is beyond the scope of this book.

▶ **8.5 EXERCISES**

Find the maximum and minimum values, and the points at which they occur, for each function in Exercises 1–16.

1. $f(x, y) = x^2 + y^2, \quad 0 \le x \le 2, \quad 0 \le y \le 2$

2. $g(x, y) = \sqrt{x^2 + y^2}, \quad 1 \le x \le 2, \quad 1 \le y \le 2$

3. $h(x, y) = (x - 1)^2 + y^2, \quad x^2 + y^2 \le 4$

4. $k(x, y) = x^2 + (y - 1)^2, \quad x^2 + y^2 \le 9$

5. $f(x, y) = e^{x^2+y^2}, \quad 4x^2 + y^2 \le 4$

6. $g(x, y) = e^{-(x^2+y^2)}, \quad x^2 + 4y^2 \le 4$

7. $h(x, y) = e^{4x^2+y^2}, \quad x^2 + y^2 < 1$

8. $k(x, y) = e^{-(x^2+4y^2)}, \quad x^2 + y^2 \le 4$

9. $f(x, y) = x + y + 1/(xy), \quad x \ge \frac{1}{2}, \quad y \ge \frac{1}{2}, \quad x + y \le 3$

10. $g(x, y) = x + y + 8/(xy), \quad x \ge 1, \quad y \ge 1, \quad x + y \le 6$

11. $h(x, y) = xy + 8/x + 8/y, \quad x \ge 1, \quad y \ge 1, \quad xy \le 9$

12. $k(x, y) = xy + 1/x + 8/y, \quad x \geq \frac{1}{4}, \quad y \geq \frac{1}{4}, \quad xy \leq 9$

13. $f(x, y) = x^2 + 2x + y^2$, on the region in the figure.

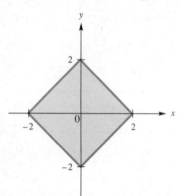

14. $g(x, y) = x^2 + y^2$, on the region in the figure.

$y = 1 - x^2$

15. $h(x, y) = x^3 + y^3$, on the region in the figure.

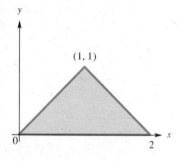

(1, 1)

16. $k(x, y) = x^3 + 2y^3$, on the region in the figure.

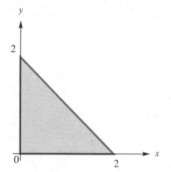

17. At what points on the sphere $x^2 + y^2 + z^2 = 1$ is the product xyz a maximum? (The method of Lagrange multipliers can be used here.)

18. At what point on the surface $z = (x^2 + x + y^2 + 4)^{1/2}$ is the quantity $x^2 + y^2 + z^2$ a minimum? (The method of Lagrange multipliers can be used here.)

APPLICATIONS

19. *Cost*   Your bicycle factory makes two models, 5-speeds and 10-speeds. Each week, your total cost to make x 5-speeds and y 10-speeds is given by

$$C(x, y) = \$10,000 + 50x + 70y - 0.5xy.$$

You want to make between 100 and 150 5-speeds, and between 80 and 120 10-speeds. What combination will cost you the least? What combination will cost you the most?

20. *Cost*   Your bicycle factory makes two models, 5-speeds and 10-speeds. Each week, your total cost to

make x 5-speeds and y 10-speeds is given by

$$C(x, y) = \$10,000 + 50x + 70y - 0.46xy.$$

You want to make between 100 and 150 5-speeds, and between 80 and 120 10-speeds. What combination will cost you the least? What combination will cost you the most?

21. *Profit*   Your software company sells two programs, Walls and Doors. Your profit from selling x copies of Walls and y copies of Doors is given by

$$P(x, y) = 20x + 40y - 0.1(x^2 + y^2).$$

If you can sell a maximum of 200 copies of the two programs together, what combination will bring you the largest profit?

22. **Profit** Your software company sells two programs, Walls and Doors. Your profit from selling x copies of Walls and y copies of Doors is given by

$$P(x, y) = 20x + 40y - 0.1(x^2 + y^2).$$

If you can sell a maximum of 400 copies of the two programs together, what combination will bring you the largest profit?

23. **Temperature** The temperature at the point (x, y) on the square with vertices $(0, 0)$, $(0, 1)$, $(1, 0)$, and $(1, 1)$ is given by $T(x, y) = x^2 + 2y^2$. Find the hottest and coldest points on the square.

24. **Temperature** The temperature at the point (x, y) on the square with vertices $(0, 0)$, $(0, 1)$, $(1, 0)$, and $(1, 1)$ is given by $T(x, y) = x^2 + 2y^2 - x$. Find the hottest and coldest points on the square.

25. **Temperature** The temperature at the point (x, y) on the disc $\{(x, y) \mid x^2 + y^2 \le 1\}$ is given by $T(x, y) = x^2 + 2y^2 - x$. Find the hottest and coldest points on the disc.

26. **Temperature** The temperature at the point (x, y) on the disc $\{(x, y) \mid x^2 + y^2 \le 1\}$ is given by $T(x, y) = 2x^2 + y^2$. Find the hottest and coldest points on the disc.

The method of Lagrange multipliers can be used for Exercises 27–38.

27. **Geometry** Find the point on the plane $-2x + 2y + z - 5 = 0$ closest to the point $(-1, 1, 3)$.

28. **Geometry** Find the point on the plane $2x - 2y - z + 1 = 0$ closest to the point $(1, 1, 0)$.

29. **Geometry** What point on the surface $z = x^2 + y - 1$ is closest to the origin?

30. **Geometry** What point on the surface $z = x + y^2 - 3$ is closest to the origin?

31. **Construction Cost** A closed rectangular box is made with two kinds of materials. The top and bottom are made with heavy-duty cardboard costing 20¢ per square foot, while the sides are made with lightweight cardboard costing 10¢ per square foot. Given that the box is to have a capacity of 2 cubic feet, what should its dimensions be if the cost is to be minimized?

32. **Construction Cost** Repeat the previous exercise if the heavy-duty cardboard costs 30¢ per square foot

and the lightweight cardboard costs 5¢ per square foot.

33. **Construction Cost** Referring to Exercise 31, my company wishes to manufacture boxes with a capacity of 2 cubic feet as cheaply as possible, but unfortunately, the company that manufactures the cardboard is unable to give me price quotes for the heavy-duty and lightweight cardboard as yet. Find formulas for the dimensions of the box in terms of the price per square foot of heavy-duty and lightweight cardboard.

34. **Construction Cost** Repeat the previous exercise, assuming that only the bottoms of the boxes are to made using heavy-duty cardboard.

35. **Package Dimensions** The *U.S. Postal Service* (USPS) will accept only packages with a length plus girth no more than 108 inches. (See figure.)
What is the largest-volume package that the USPS will accept?

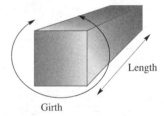

Length

Girth

36. **Package Dimensions** The *United Parcel Service* (UPS) will accept only packages with a length no more than 108 inches and length plus girth no more than 130 inches (See figure above.) What is the largest-volume package that UPS will accept?

37. **Geometry** Find the dimensions of the rectangular box with least volume that can be inscribed above the xy-plane and under the paraboloid $z = 1 - (x^2 + y^2)$.

38. **Geometry** Find the dimensions of the rectangular box with least volume that can be inscribed above the xy-plane and under the paraboloid $z = 2 - (2x^2 + y^2)$.

39. **Resource Allocation** You manage an ice cream factory that makes two flavors: Creamy Vanilla and Continental Mocha. Into each quart of Creamy Vanilla go two eggs and three cups of cream. Into each quart of Continental Mocha go one egg and three cups of cream. You have in stock 500 eggs and 900 cups of cream. Your profit on x quarts of vanilla and y quarts

of mocha are $P(x, y) = 3x + 2y - 0.01(x^2 + y^2)$. How many quarts of each flavor should you produce in order to make the largest profit?

40. *Resource Allocation*   Repeat the preceding exercise using the profit function $P(x, y) = 3x + 2y - 0.005(x^2 + y^2)$.

41. *Resource Allocation*   Urban Institute of Technology's Math Department offers two courses: Finite Math and Calculus. Each section of Finite Math has 60 students, while each section of Calculus has 50. The department is allowed to offer a total of up to 110 sections. Further, there are no more than 6,000 students who would like to take a math course. The university's profit on x sections of Finite Math and y sections of Calculus is

$$P(x, y) = \$5,000,000(1 - e^{-0.02x - 0.01y})$$

(the profit being the difference between what the students are charged and what the professors are paid). How many sections of each course should the department offer in order to make the largest profit?

42. *Resource Allocation*   Repeat the preceding exercise using the profit function

$$P(x, y) = \$5,000,000(1 - e^{-0.01x - 0.02y}).$$

43. *Nutrition*   *Gerber* Mixed Cereal for Baby costs 10¢ per serving. *Gerber* Mango Tropical Fruit Dessert

costs 53¢ per serving. If you want the product of the number of servings of each to be at least 10 per day, how can you do so at the least cost?

44. *Nutrition*   Repeat the preceding exercise if instead you want the product of the number of servings of cereal and the square of the number of servings of dessert to be at least 10.

45. *Purchasing*   The ESU Business School is buying computers. It has two models to choose from, the Pomegranate and the Ami. Each Pomegranate comes with 4 MB of memory and 80 MB of disk space, while each Ami has 3 MB of memory and 100 MB of disk space. For reasons related to its accreditation, the school would like to be able to say that it has a total of at least 480 MB of memory and at least 12,800 MB of disk space. Because of complicated volume pricing, the cost to the school of x Pomegranates and y Amis is

$$C(x, y) = x^2 + y^2 - 200x - 200y + 220,000.$$

How many of each kind of computer should the school buy in order to minimize the average cost per computer?

46. *Purchasing*   Repeat the preceding exercise assuming that the cost is

$$C(x, y) = x^2 + y^2 - 300x - 200y + 332,500.$$

COMMUNICATION AND REASONING EXERCISES

47. If the partial derivatives of a function of several variables are never zero, is it possible for the function to have local extrema on some domain? Explain your answer.

48. Suppose we know that $f(x, y)$ has an absolute maximum somewhere in the domain D, and that (a, b) is the only point in D such that $f_x(a, b) = f_y(a, b) = 0$. Must it be the case that f has an absolute maximum at (a, b)? Explain.

49. Under what circumstances would it be necessary to use the method of Lagrange multipliers?

50. Under what circumstances would the method of Lagrange multipliers not apply?

51. A **linear programming problem in two variables** is a problem of the following form.

Maximize (or minimize) $f(x, y)$

subject to constraints $C(x, y) \geq 0$ or $C(x, y) \leq 0$.

Here, the objective function f and the constraints C are linear functions. There may be several such linear constraints in one problem. Explain why the solution cannot occur in the interior of the domain of f.

52. Continuing the preceding exercise, explain why the solution will actually be at a corner of the domain of f (where two or more of the line segments making up the boundary meet). This result—or rather a slight generalization of it—is known as the Fundamental Theorem of Linear Programming.

▶ ━━━━ **8.6** LEAST-SQUARES FIT

Throughout this book we have used functions to model relationships between variables, for example the relationship between price and demand. Often these functions were linear. In this section, we discuss how to come up with such a model.

In the simplest case, we have two data points and we only need to find the equation of the line passing through them. However, it often happens that we have many data points that don't quite lie on one line. The problem then is to find the line coming closest to passing through all of the points.

Suppose, for example, that your market research of real estate investments reveals the following sales figures for new homes of different prices over the past year (this is some of the data referred to at the beginning of the chapter).

Price (Thousands of $)	Sales of New Homes This Year
$150–$169	126
$170–$189	103
$190–$209	82
$210–$229	75
$230–$249	82
$250–$269	40
$270–$289	20

If we simplify the situation by replacing each of the price ranges by a single price in the middle of the range, we get the following table:

Price (Thousands of $)	Sales of New Homes This Year
$160	126
$180	103
$200	82
$220	75
$240	82
$260	40
$280	20

We would like to use these data to construct a demand function for the real estate market. (Recall that a demand function gives demand, q—measured here by annual sales—as a function of unit price, p.) Figure 8.6.1 shows a plot of q versus p.

The data definitely suggest a straight line more-or-less, and hence an approximately linear relationship between p and q. Figure 8.6.2 shows several possible "straight-line fits."

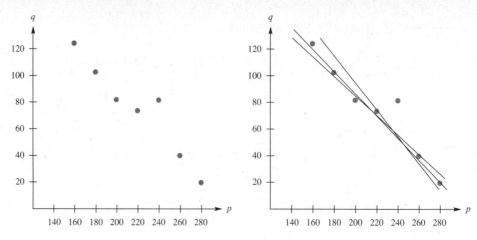

FIGURE 1   **FIGURE 2**

The question now is, "What line best fits the data?"

To answer this question, we need to be clear as to just what we mean by "best fits." Call the linear function whose graph is the desired straight line $l(p)$, and let $q(p)$ be the actual sales at price p. Since we want $l(p)$ to be as good an approximation to $q(p)$ as possible, what would seem best is that $l(p)$ be as close as possible to the actual sales $q(p)$ for each price p in our table. The distances between the **predicted values** $l(p)$ and the **observed values** $q(p)$ appear as the vertical distances shown in Figure 8.6.3, and are calculated as $|q(p) - l(p)|$ for each value of p for which we have a data point.

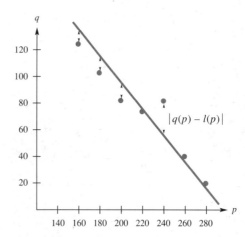

FIGURE 3

$\boldsymbol{Q}$ Since we want the vertical distances to be as small as possible, why can't we set them all to zero and solve?

$\boldsymbol{A}$ If this were possible, then there would be a straight line that passes through all the data points. A look at the graph shows that this is not the case.

Q Then why not find the line that minimizes *all* the vertical distances $|q(p) - l(p)|$?

A This is not possible either. The line that minimizes the first two distances is the line that passes through the first two data points, since it makes the distances 0. But this line certainly does not minimize the distance to the third point. In other words, there is a trade-off: making some distances smaller makes others larger.

Q So what do we do?

A Since we cannot minimize *all* of the distances, we minimize some reasonable combination of them.

Now, one reasonable combination of the distances would be their *sum,* but that turns out the be difficult to work with (because of the absolute values). Instead, we use the sum of the *squares* of the distances. In other words, if we denote the various values of *p* by $p_1, p_2, \ldots, p_n$, then we shall minimize the quantity

$$S = (l(p_1) - q(p_1))^2 + (l(p_2) - q(p_2))^2$$
$$+ \ldots + (l(p_n) - q(p_n))^2.$$

The line that minimizes *S* is called the **least-squares line** or **best-fit line** associated with the given data.

Q Is there a good reason for looking at the sum of the squares of the distances?

A There are several. As we mentioned above, it is technically easier to use the sum of the squares than to use the sum of the distances themselves. On a more theoretical level, we can view the observed values $(q(p_1), q(p_2), \ldots, q(p_n))$ and predicted values $(l(p_1), l(p_2), \ldots, l(p_n))$ as being two points in *n*-dimensional space, and we can view our goal as the minimization of the distance between these two points. The distance between these points is given by a generalization of the formula for the distance between two points in 2-space or 3-space:

$$d = \sqrt{(l(p_1) - q(p_1))^2 + (l(p_2) - q(p_2))^2 + \cdots + (l(p_n) - q(p_n))^2}.$$

The sum of squares, *S*, is therefore the square of the distance between two points in *n*-space, and minimizing *S* minimizes the distance between these points.

To see how to minimize *S* we begin with a simpler example.

▼ **EXAMPLE 1**

Find the least-squares line associated with the following data:

p	1	2	3	4
q	1.5	1.6	2.1	3.0

SOLUTION   Since we are looking for a linear approximation of q as a function of p, we take l to have the form

$$l(p) = mp + b.$$

Our job is to determine the constants m and b. We must now minimize the quantity

$$S = (l(1) - q(1))^2 + (l(2) - q(2))^2 + (l(3) - q(3))^2$$
$$+ (l(4) - q(4))^2$$
$$= (m + b - 1.5)^2 + (2m + b - 1.6)^2$$
$$+ (3m + b - 2.1)^2 + (4m + b - 3.0)^2.$$

The variables here are m and b, so we have a function of two variables to minimize, and we proceed as in Section 4. We begin by taking partial derivatives.

$$S_m = 2(m + b - 1.5) + 4(2m + b - 1.6)$$
$$+ 6(3m + b - 2.1) + 8(4m + b - 3.0)$$
$$S_b = 2(m + b - 1.5) + 2(2m + b - 1.6)$$
$$+ 2(3m + b - 2.1) + 2(4m + b - 3.0)$$

We must now set these equal to zero and solve for the two unknowns m and b. In order to make this easier, we first gather terms.

$$S_m = 60m + 20b - 46 = 0$$
$$S_b = 20m + 8b - 16.4 = 0$$

This is a system of two linear equations in the two unknowns m and b, and has solution $m = 0.5$, $b = 0.8$ (you can check by solving for yourself!). Thus, our least-squares line is

$$l(p) = 0.5p + 0.8.$$

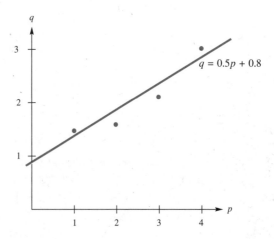

$q = 0.5p + 0.8$

FIGURE 4

Before we go on... We didn't check that these values of m and b gave an absolute minimum of S. But we can argue as follows: There must be a minimum value of S somewhere, it must show up as a critical point, and since we have located the only possible critical point, this must be it!

Figure 8.6.4 shows the data points and the least-squares line.

Notice that the line doesn't pass through even one of the original points, and yet it is the straight line that best approximates them.

Before going on to another example, let us return to the calculation we just did and see if we can come up with a *formula* for m and b. To do this, we consider the general situation with a table of given data.

p	p_1	p_2	$\cdots$	p_n
q	q_1	q_2	$\cdots$	q_n

The linear function we are after has the form $l(p) = mp + b$, and we want to minimize the quantity

$$S = (l(p_1) - q(p_1))^2 + (l(p_2) - q(p_2))^2$$
$$+ \ldots + (l(p_n) - q(p_n))^2$$
$$= (mp_1 + b - q_1)^2 + (mp_2 + b - q_2)^2$$
$$+ \ldots + (mp_n + b - q_n)^2.$$

We set the partial derivatives equal to 0.

$$S_m = 2p_1(mp_1 + b - q_1) + 2p_2(mp_2 + b - q_2)$$
$$+ \ldots + 2p_n(mp_n + b - q_n) = 0$$
$$S_b = 2(mp_1 + b - q_1) + 2(mp_2 + b - q_2)$$
$$+ \ldots + 2(mp_n + b - q_n) = 0$$

Gathering terms and dividing by 2, we get

$$(p_1^2 + p_2^2 + \ldots + p_n^2)m + (p_1 + p_2 + \ldots + p_n)b$$
$$= (p_1q_1 + p_2q_2 + \ldots + p_nq_n)$$
$$(p_1 + p_2 + \ldots + p_n)m + nb = (q_1 + q_2 + \ldots + q_n).$$

We can rewrite these equations more neatly using Σ-notation.

$$\left(\sum p_i^2\right)m + \left(\sum p_i\right)b = \sum p_iq_i$$
$$\left(\sum p_i\right)m + nb = \sum q_i$$

We conclude the following.

LEAST-SQUARES LINE

The **least-squares line** line through $(p_1, q_1), (p_2, q_2), \ldots, (p_n, q_n)$ has the form

$$q = mp + b,$$

where the constants m and b are the solutions to the system of equations

$$\left(\sum p_i^2 \right) m + \left(\sum p_i \right) b = \sum p_i q_i$$

$$\left(\sum p_i \right) m + nb = \sum q_i.$$

Let us now return to the data on demand for real estate with which we began this section.

▼   **EXAMPLE 2**   Demand

Find a linear demand equation that best fits the following data, and use it to predict annual sales of homes priced at $140,000.

Price (Thousands of $)	Sales of New Homes This Year
$160	126
$180	103
$200	82
$220	75
$240	82
$260	40
$280	20

SOLUTION   We first calculate the following quantities needed in the formula.

$$\sum p_i^2 = \text{sum of squares of } p\text{-values} = 350{,}000$$

$$\sum p_i = \text{sum of } p\text{-values} = 1{,}540$$

$$\sum q_i = \text{sum of } q\text{-values} = 528$$

$$\sum p_i q_i = \text{sum of products} = 107{,}280$$

Thus, the system of equations we must solve is

$$350,000m + 1,540b = 107,280$$
$$1,540m + 7b = 528.$$

Solving this system by hand or with the aid of a calculator or computer, we obtain the following solution (rounded to four significant digits).

$$m \approx -0.7929, \, b \approx 249.9$$

Thus, our least-squares line is

$$q = -0.7929p + 249.9.$$

We can now use this equation to predict the annual sales of homes priced at $140,000, as we were asked to do. Remembering that p is the price in thousands of dollars, we set $p = 140$ and solve for q, getting $q \approx 139$. Thus, our model predicts that approximately 139 homes will have been sold in the range $140,000–$159,000.

Before we go on... We must remember that these figures were for sales in a *range* of prices. For instance, it would be extremely unlikely that 139 homes would have been sold at exactly $140,000. On the other hand, it does predict that, were we to place 139 homes on the market at $140,000, we could expect to sell them all.*

Figure 8.6.5 shows the original data, together with the least-squares line.

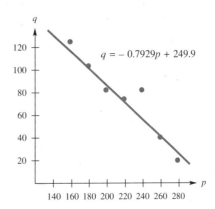

FIGURE 5

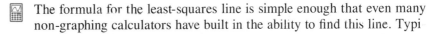
The formula for the least-squares line is simple enough that even many non-graphing calculators have built in the ability to find this line. Typi

---

▼ * This is really a debatable point; we clearly wouldn't sell *any* if they were not perceived by prospective buyers are being worth $140,000 or if there were already a glut of $140,000 homes on the market. Thus, real-life planning is often more complicated than our simple model permits. Still, it remains a powerful predictive tool.

cally, you enter the data points one by one. As you do this, the calculator computes the sums Σp_i, Σq_i, Σp_i^2, and $\Sigma p_i q_i$. One press of a button then calculates m and b. Spreadsheets are also handy for calculating these coefficients and generally can calculate them automatically (under the name **linear regression**).

▶ NOTE   In the above example, we used a linear demand model based on data in the range $160,000–$280,000 to predict demand at $140,000. This process is called **extrapolation,** since we have chosen a price *outside* the specified range. However, we have already seen in Chapter 1 the errors that arise when we try to extrapolate a linear model—the further we extrapolate the model, the less reliable it becomes. (This does not apply only to linear models, but to any mathematical model.) For instance, if in the previous example we had used the model to predict the demand for $350,000 homes, our linear model would predict a negative demand. Thus, extrapolation of any mathematical model beyond the range for which data are available must be done with caution. We are on firmer ground with **interpolation,** whereby we predict a value *inside* our range of data.   ◀

▼ **EXAMPLE 3**   Population Growth

The U.S. population, according to the U.S. Bureau of the Census, was as follows in the beginning of this century (all figures in thousands of people).

Year	1900	1910	1920	1930	1940	1950
Population	76,212	92,228	106,022	123,203	132,165	151,326

Fit a function of the form $A = Pe^{kt}$ to this data, where A is the population and t is the number of years after 1900. Use the model to predict the population in 1990.

SOLUTION   The equation we are asked to fit is not linear, but there is a clever way to apply the techniques of this section anyway. If we take the natural logarithm of both sides of the equation, we get

$$\ln A = \ln P + kt$$

This equation is linear *in t and* $\ln A$. We can fit a least-squares line to the data we obtain by replacing the population figures with their natural logarithms.

t	0	10	20	30	40	50
$\ln A$	11.24127	11.43202	11.57140	11.72159	11.79181	11.92719

Thus, the system of equations we must solve is

$$350{,}000m + 1{,}540b = 107{,}280$$
$$1{,}540m + 7b = 528.$$

Solving this system by hand or with the aid of a calculator or computer, we obtain the following solution (rounded to four significant digits).

$$m \approx -0.7929, \ b \approx 249.9$$

Thus, our least-squares line is

$$q = -0.7929p + 249.9.$$

We can now use this equation to predict the annual sales of homes priced at $140,000, as we were asked to do. Remembering that p is the price in thousands of dollars, we set $p = 140$ and solve for q, getting $q \approx 139$. Thus, our model predicts that approximately 139 homes will have been sold in the range $140,000–$159,000.

Before we go on... We must remember that these figures were for sales in a *range* of prices. For instance, it would be extremely unlikely that 139 homes would have been sold at exactly $140,000. On the other hand, it does predict that, were we to place 139 homes on the market at $140,000, we could expect to sell them all.*

Figure 8.6.5 shows the original data, together with the least-squares line.

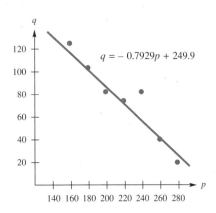

FIGURE 5

The formula for the least-squares line is simple enough that even many non-graphing calculators have built in the ability to find this line. Typi-

▼ * This is really a debatable point; we clearly wouldn't sell *any* if they were not perceived by prospective buyers are being worth $140,000 or if there were already a glut of $140,000 homes on the market. Thus, real-life planning is often more complicated than our simple model permits. Still, it remains a powerful predictive tool.

cally, you enter the data points one by one. As you do this, the calculator computes the sums Σp_i, Σq_i, Σp_1^2, and $\Sigma p_i q_i$. One press of a button then calculates m and b. Spreadsheets are also handy for calculating these coefficients and generally can calculate them automatically (under the name **linear regression**).

▶ NOTE   In the above example, we used a linear demand model based on data in the range $160,000–$280,000 to predict demand at $140,000. This process is called **extrapolation,** since we have chosen a price *outside* the specified range. However, we have already seen in Chapter 1 the errors that arise when we try to extrapolate a linear model—the further we extrapolate the model, the less reliable it becomes. (This does not apply only to linear models, but to any mathematical model.) For instance, if in the previous example we had used the model to predict the demand for $350,000 homes, our linear model would predict a negative demand. Thus, extrapolation of any mathematical model beyond the range for which data are available must be done with caution. We are on firmer ground with **interpolation,** whereby we predict a value *inside* our range of data.   ◀

▼ EXAMPLE 3   Population Growth

The U.S. population, according to the U.S. Bureau of the Census, was as follows in the beginning of this century (all figures in thousands of people).

Year	1900	1910	1920	1930	1940	1950
Population	76,212	92,228	106,022	123,203	132,165	151,326

Fit a function of the form $A = Pe^{kt}$ to this data, where A is the population and t is the number of years after 1900. Use the model to predict the population in 1990.

SOLUTION   The equation we are asked to fit is not linear, but there is a clever way to apply the techniques of this section anyway. If we take the natural logarithm of both sides of the equation, we get

$$\ln A = \ln P + kt$$

This equation is linear *in t and* $\ln A$. We can fit a least-squares line to the data we obtain by replacing the population figures with their natural logarithms.

t	0	10	20	30	40	50
ln A	11.24127	11.43202	11.57140	11.72159	11.79181	11.92719

We then compute (rounding to four significant digits)

$$\sum t_i^2 = 5{,}500$$

$$\sum t_i = 150$$

$$\sum t_i \ln A_i = 1{,}765$$

$$\sum \ln A_i = 69.69.$$

This gives us the following system of equations.

$$5{,}500m + 150b = 1{,}765$$
$$150m + 6b = 69.69$$

The solution to this system is $m = 0.013$ and $b = 11.3$, so our least-squares line is

$$\ln A = 11.3 + 0.013t.$$

This means that $k = 0.013$, and $\ln P = 11.3$, so $P = 80{,}800$. Our exponential curve is then

$$A = 80{,}800e^{0.013t} \text{ (population in thousands } t \text{ years after 1900).}$$

Figure 8.6.6 shows the actual population figures and the curve we just found.

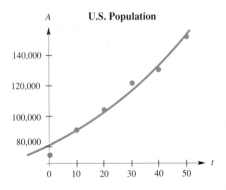

FIGURE 6

Our equation predicts the following population in 1990.

$$A = 80{,}800e^{0.013(90)} \approx 260{,}000 \text{ thousand}$$

The actual population, according to the 1990 census, was 250,000 thousand people.

Before we go on... Although we extrapolated well beyond the given range of data, the answer was fairly accurate. Is there a moral to this? We doubt it.

▶ ## 8.6 EXERCISES

Find the least-squares line through each set of points in Exercises 1–4. Graph the data and the least-squares line.

1. (1, 1), (2, 2), (3, 4)

2. (0, 1), (1, 2), (2, 2)

3. (0, −1), (1, 3), (4, 6), (5, 0)

4. (2, 4), (6, 8), (8, 12), (10, 0)

APPLICATIONS

The use of a graphing calculator or computer spreadsheet is recommended for Exercises 5–24.

 5. *Pollution Control*   According to recent surveys, the percentage of new plant and equipment expenditures on pollution control by U.S. manufacturing companies is as shown: *

1975	1980	1981	1984	1987
9.3	4.8	4.3	3.3	4.3

Use a least-squares line to estimate the figure for 1985.

 6. *Pollution Control*   The percentage of new plant and equipment expenditures on pollution control by U.S. public utility companies is as shown:*

1975	1980	1981	1984	1987
8.4	8.1	7.3	6.8	5.1

Use a least-squares line to estimate the figure for 1985.

 7. *Oil Recovery*   In 1978, Congress conducted a study of the amount of additional oil that can be extracted from existing oil wells by "enhanced recovery techniques" (such as injecting solvents or steam into an oil well to lower the density of the oil). As the price of oil increases, the amount of oil that can be recovered economically in this manner also increases. The following table gives the study's estimates of recoverable oil based on the price per barrel: [†]

Price per Barrel	$12	$14	$22	$30
Recovery (billions of barrels)	21.2	29.4	41.6	49.2

▼ [†] Survey of Current Business **58, 62, 66, 68**

*Source: U.S. Congress, Office of Technology Assessment, *Enhanced Oil Recovery Potential in the U.S.* (Washington, CD: OTA, 1978): 7. The recovery figures are based on a 10% minimum rate of return, and the prices are in constant 1976 dollars rounded to the nearest $1.

Use a least-squares line to estimate the additional amount of oil that can be economically recovered if the price of oil were to drop to $10 per barrel.

8. *Depletion of Natural Gas Reserves*   The following table shows the estimated U.S. natural gas reserves in trillions of cubic feet:*

1980	1982	1984	1986	1988
194	200	197	190	183

Use a least-squares linear model to estimate the rate at which natural gas was depleted during the given period.

9. *Life Expectancy*   Life expectancy at birth in the U.S. for people born in various years is given in the following table.

1920	1930	1940	1950	1960	1970	1980	1990
54.1	59.7	62.9	68.2	69.7	70.8	73.7	75.4

Using a least-squares line, estimate the rate at which life expectancy was changing during the given period.

10. *Infant Mortality*   The infant mortality rates (deaths per 1,000 live births) in the U.S. for various years are given in the following table.

1960	1970	1980	1985	1986	1987	1988
26	20	12.6	10.6	10.4	10.1	10

Use a least-squares line to predict the infant mortality rate in the year 2010. Is this model realistic? What might be a better model?

11. *Profit*   The accompanying chart shows the net income (profit) of the *Walt Disney Company* for the years 1984–1992.[†]

▼   *Source: U.S. Energy Information Administration, International Energy Annual (Washington, DC: Government Printing Office): annual.

[†]Profits estimated to the nearest $5 million. Source: Company Reports/*The New York Times*, December, 1992, p. D1.

Net Income of the Walt Disney Company

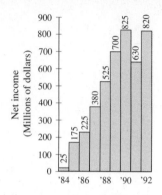

Find a least-squares linear model for this data. (Find profit p as a function of the year t, with $t = 0$ corresponding to 1980.) Use your model to estimate *Disney*'s profit (to the nearest million dollars) in 1993.

12. *Stock Prices*   Repeat the previous exercise, but this time model *Walt Disney*'s stock price. (See the chart.)*

Stock Price of the Walt Disney Company

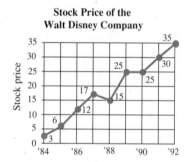

13. *Gross National Product*   The GNP (in billions of dollars, to the nearest billion dollars) of the U.S. for the years 1960–1990 is given in the following table.[†]

1960	1970	1975	1980	1989	1990
515	1,016	1,598	2,732	5,201	5,465

Fit a least-squares line to this data, and graph both it and the data. Predict the GNP in the year 2000. (Answer to the nearest billion dollars).

14. *Gross National Product*   Fit an exponential function $A = Pe^{kt}$ to the data in the preceding exercise. Take t to be the year since 1960, and round all constants in your answer to four significant digits. Which of these two exercises gives a better fit? (Compare their graphs with the data.)

▼ * Stock prices are rough estimates. Source: Ibid.

† Source: Bureau of Economic Analysis, U.S. Department of Commerce.

15. *Population*   The population of the U.S., in thousands, is given by decade in the following table.*

1790	1800	1810	1820	1830	1840	1850
3,929	5,308	7,240	9,638	12,861	17,063	23,192

1860	1870	1880	1890	1900	1910	1920
31,443	38,558	50,189	62,980	76,212	92,228	106,022

1930	1940	1950	1960	1970	1980	1990
123,203	132,165	151,326	179,323	203,302	226,542	248,710

Fit an exponential equation $A = Pe^{kt}$ to this data, taking t to be the time since 1790, and graph both it and the data. Predict the population in the year 2000.

16. *Population*   Repeat the preceding exercise, fitting a least-squares line. Which of these two exercises provides the better fit?

17. *Stock Market*   The yearly highs of the Dow Jones Industrial Average (to the nearest point) for the decade of the 1980s are given in the following table.

1980	1981	1982	1983	1984	1985
1000	1024	1071	1287	1287	1553

	1986	1987	1988	1989	1990
	1956	2722	2184	2791	3000

Graph these data, judge whether a line or an exponential function provides the best fit, and then find a function of that kind to fit the data. What would you predict for the high in the year 2000?

18. *Stock Market*   The yearly lows of the Dow Jones Industrial Average (to the nearest point) for the decade of the 1980s are given in the following table.

1980	1981	1982	1983	1984	1985
759	824	776	1027	1087	1185

	1986	1987	1988	1989	1990
	1502	1739	1879	2145	2365

▼ *Source: Bureau of the Census, U.S. Department of Commerce.

Graph these data, judge whether a line or an exponential function provides the best fit, and then find a function of that kind to fit the data. What ould you predict for the low in the year 2000?

19. *Inflation*   The consumer price index (CPI) for various years from 1970 to 1990 is given in the following table* (1982–4 = 100).

1970	1975	1980	1985	1987	1988	1989	1990
38.8	53.8	82.4	107.6	113.6	118.3	124.0	130.7

Fit an exponential curve to these data. Predict the CPI for the year 2000.

20. *Inflation*   The buying power of a dollar for various years from 1970 to 1990 is given in the following table (calculated by taking 100/CPI, 1982–4 = 1.00).

1970	1975	1980	1985	1987	1988	1989	1990
2.58	1.86	1.21	0.929	0.880	0.845	0.806	0.765

Fit an exponential curve to these data. Predict the buying power of a dollar in the year 2000. How is your answer here related to the answer to the previous exercise?

Best-Fit Power Function   *Exercises 21 and 22 are based on modeling using a "power" function of the form*

$$y(x) = ax^b,$$

where a and b are constants to be determined. If we take the natural logarithm of both sides of this equation, we get

$$\ln y = \ln (ax^b) = \ln a + b\ln x,$$

showing a linear relationship between $\ln y$ *and* $\ln x$*. Thus, given a sequence of data points* (x_i, y_i)*, we can therefore fit a power curve* $y = ax^b$ *to it by finding the best-fit straight line for the points* $(\ln x_i, \ln y_i)$*. The slope is then b, and the y-intercept is* $\ln a$*.*

21. *Best-Fit Demand Curve*   You have the following data showing the weekly sales at various prices for your fluorescent blue shower curtains.

Price, p	$10	$15	$20	$25	$30
Weekly Sales, q	30	15	10	5	3

Find the best-fit power function, and graph it together with the given data points.

22. *Best-Fit Demand Curve*   Repeat the preceding exercise using the following data for sales of your fluorescent pink shower curtains:

---

* Source: Bureau of Labor Statistics, U.S. Department of Labor.

Price, p	$10	$15	$20	$25	$30
Weekly Sales, q	500	400	300	250	220

 23. Cobb-Douglas Production Function   Recall that the Cobb-Douglas production function has the form

$$P(x, y) = Kx^a y^{1-a},$$

where P stands for the number of items produced per unit time, x is the number of employees, and y is the operating budget for that time. The numbers K and a are constants that depend on the particular situation we are looking at, with a between 0 and 1. Show how, by taking logs of both sides, you can obtain a linear relationship between $\ln(P/y)$ and $\ln(x/y)$. Explain how this relationship allows you to estimate K and a given a number of data points (x_i, y_i, P_i).

24. Cobb-Douglas Production Function   Suppose that you monitor production levels in your factory, and over a period of seven days you get the following data for daily production.

Workers, x	100	110	90	100	95	105	110
Budget, $y	10,000	9,000	9,000	12,000	11,000	9,500	10,000
Production, P	400	410	350	430	400	405	425

Fit a Cobb-Douglas production function of the form $P = Kx^a y^{1-a}$ to this data, and predict your daily production if you use 120 workers and a budget of $12,000. (See Exercise 23.)

COMMUNICATION AND REASONING EXERCISES

25. If the points $(x_1, y_1), (x_2, y_2), \ldots, (x_n, y_n)$ lie on a straight line, what can you say about the least-squares line associated with these points?

26. What can you say about the least-squares line associated with two data points?

27. If all but one of the points $(x_1, y_1), (x_2, y_2), \ldots, (x_n, y_n)$ lie on a straight line, does the least-squares line associated with these points pass through all but one of these points?

28. Must the least-squares line pass through at least one of the data points? Illustrate your answer with an example.

29. Why must care be taken when using mathematical models to extrapolate?

30. Model your mathematics test scores so far using a least-squares line, and use it to predict your score on the next test.

▶══ **8.7** DOUBLE INTEGRALS

When discussing functions of one variable, we computed the area under a graph by integration. The analog for the graph of a function of two variables would be the *volume* under the graph, as in Figure 1.

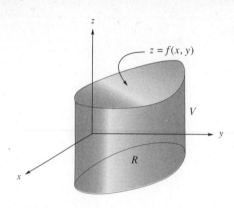

FIGURE 1

By analogy with the definite integral of a function of one variable, we make the following definition.

> ## GEOMETRIC DEFINITION OF THE DOUBLE INTEGRAL
>
> The **double integral of $f(x, y)$ over the region R in the xy-plane** is defined as
>
> (volume *above* the region R and under the graph of f) $-$
>   (volume *below* the region R and above the graph of f).
>
> We denote the double integral of $f(x, y)$ over the region R by
>
> $$\iint\limits_{R} f(x, y) \, dx \, dy.$$

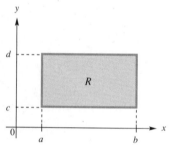

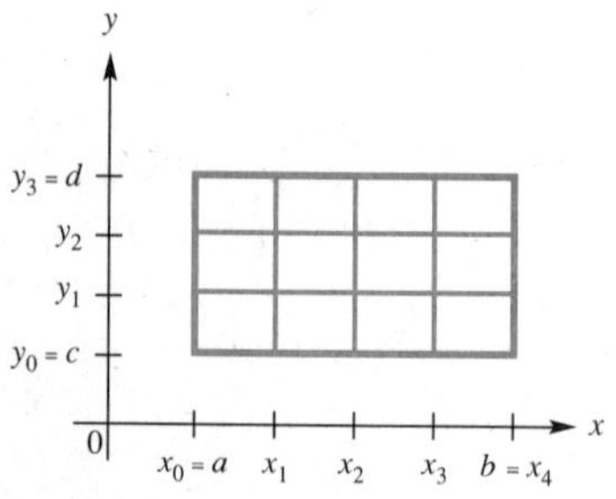

FIGURE 2

As we saw in the case of the definite integral of a function of one variable, we also desire an *algebraic,* or *numerical,* definition for two reasons: to make the mathematical definition more precise—so as not to rely on the notion of "volume"—and for direct computation of the integral using technology.

We start with the simplest case, when the region R is a rectangle $a \leq x \leq b$ and $c \leq y \leq d$. (See Figure 2.)

In order to compute the volume over R, we mimic what we did to find the area under the graph of a function of one variable. We break up the interval $[a, b]$ into m intervals all of width $\Delta x = (b - a)/m$, and we break up $[c, d]$ into n intervals all of width $\Delta y = (d - c)/n$. Figure 3 shows an example with $m = 4$ and $n = 3$.

This gives us mn rectangles defined by $x_{i-1} \leq x \leq x_i$ and $y_{j-1} \leq y \leq y_j$. Over one of these rectangles, f is approximately equal to its value at one corner, say, $f(x_i, y_j)$. The volume under f over this small rectangle is then approximately the volume of the rectangular brick shown in Figure 4.

FIGURE 3

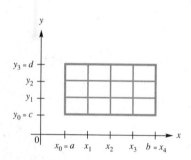

This brick has height $f(x_i, y_j)$, and its base is Δx by Δy. Its volume is $f(x_i, y_j) \, \Delta x \, \Delta y$. Adding together the volumes of all of the bricks over the small rectangles in R, we get

$$\iint\limits_{R} f(x, y) \, dx \, dy \approx \sum_{j=1}^{n} \sum_{i=1}^{m} f(x_i, y_j) \, \Delta x \, \Delta y.$$

This double sum is called a **double Riemann sum**. Algebraically, we define the double integral to be the limit of the Riemann sums as m and n go to infinity.

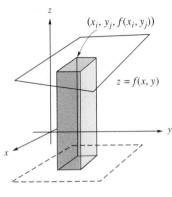

FIGURE 4

ALGEBRAIC DEFINITION OF THE DOUBLE INTEGRAL

$$\iint\limits_{R} f(x, y) \, dx \, dy \approx \lim_{m \to \infty} \lim_{n \to \infty} \sum_{j=1}^{n} \sum_{i=1}^{m} f(x_i, y_j) \, \Delta x \, \Delta y$$

▶ **NOTE**   This definition is adequate (the limit exists) when f is continuous. More elaborate definitions are needed for badly behaved functions.

This definition also gives us a clue about how to compute a double integral. The innermost sum is $\sum_{i=1}^{m} f(x_i, y_j) \, \Delta x$, which is a Riemann sum for $\int_a^b f(x, y_j) \, dx$. The innermost limit is therefore

$$\lim_{m \to \infty} \sum_{i=1}^{m} f(x_i, y_j) \, \Delta x = \int_a^b f(x, y_j) \, dx.$$

The outermost limit is then also a Riemann sum, and we get the following way of calculating double integrals. ◀

COMPUTING THE DOUBLE INTEGRAL OVER A RECTANGLE

If R is the rectangle $a \le x \le b$ and $c \le y \le d$, then

$$\iint\limits_{R} f(x, y) \, dx \, dy = \int_c^d \left(\int_a^b f(x, y) \, dx \right) dy = \int_a^b \left(\int_c^d f(x, y) \, dy \right) dx.$$

▶ **NOTE**   The second formula comes from switching the order of summation in the double sum. ◀

▼ **EXAMPLE 1**

Let $f(x, y) = xy$, and let R be the rectangle $0 \le x \le 1$ and $0 \le y \le 1$. Compute

$$\iint\limits_{R} xy \, dx \, dy.$$

SOLUTION

$$\iint\limits_R xy \, dx \, dy = \int_0^1 \int_0^1 xy \, dx \, dy$$

(We usually drop the parentheses like this.) We compute this **iterated integral** from the inside out. First, we compute

$$\int_0^1 xy \, dx.$$

To do this computation, we do as we did when finding partial derivatives: we treat y as a constant. This gives

$$\int_0^1 xy \, dx = \left[\frac{x^2}{2} \cdot y\right]_{x=0}^1 = \frac{1}{2}y - 0 = \frac{1}{2}y.$$

We can now calculate the outer integral.

$$\int_0^1 \int_0^1 xy \, dx \, dy = \int_0^1 \frac{1}{2}y \, dy = \left[\frac{1}{4}y^2\right]_0^1 = \frac{1}{4}$$

Before we go on... We could also reverse the order of integration.

$$\int_0^1 \int_0^1 xy \, dy \, dx = \int_0^1 \left[x\frac{y^2}{2}\right]_{y=0}^1 dx = \int_0^1 \frac{1}{2}x \, dx = \left[\frac{1}{4}x^2\right]_0^1 = \frac{1}{4}$$

▼ **EXAMPLE 2**

If R is the rectangle $0 \le x \le 2$, $1 \le y \le 3$, compute

$$\iint\limits_R (e^x + y) \, dx \, dy.$$

SOLUTION

$$\iint\limits_R (e^x + y) \, dx \, dy = \int_1^3 \int_0^2 (e^x + y) \, dx \, dy$$

$$= \int_1^3 \left[e^x + xy\right]_{x=0}^2 dy$$

$$= \int_1^3 [(e^2 + 2y) - (1 + 0)] \, dy$$

$$= \int_1^2 (2y + e^2 - 1) \, dy$$

$$= \left[y^2 + e^2y - y\right]_1^2$$

$$= 2 + e^2$$

Before we go on... Again, we could have reversed the order of integration. You should do this for practice.

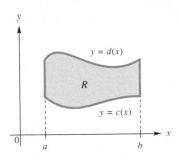

FIGURE 5

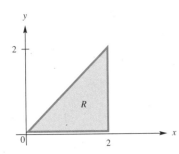

FIGURE 6

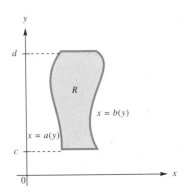

FIGURE 7

Often, we need to integrate over regions R that are not rectangular. There are two cases that come up. First, we may have a region like the one shown in Figure 5.

In this region, the bottom and top sides are defined by functions $y = c(x)$ and $y = d(x)$ respectively, so that the whole region can be described by the inequalities $a \le x \le b$ and $c(x) \le y \le d(x)$. To evaluate a double integral over such a region, we have the following.

COMPUTING THE DOUBLE INTEGRAL OVER A NONRECTANGULAR REGION

If R is the region $a \le x \le b$ and $c(x) \le y \le d(x)$, then we integrate over R according to the following equation.

$$\iint_R f(x, y)\, dx\, dy = \int_a^b \int_{c(x)}^{d(x)} f(x, y)\, dy\, dx$$

▼ **EXAMPLE 3**

If R is the triangle shown in Figure 6, compute

$$\iint_R x\, dx\, dy.$$

SOLUTION R is the region described by $0 \le x \le 2, 0 \le y \le x$.

$$\iint_R x\, dx\, dy = \int_0^2 \int_0^x x\, dy\, dx$$

$$= \int_0^2 \left[xy \right]_{y=0}^x dx$$

$$= \int_0^2 x^2\, dx$$

$$= \left[\frac{x^3}{3} \right]_0^2$$

$$= \frac{8}{3}$$

Another type of region is shown in Figure 7. This is the region described by $c \le y \le d$ and $a(y) \le x \le b(y)$. To evaluate a double integral over such a region, we have the following.

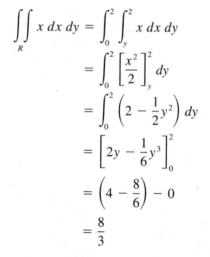

> **INTEGRATION OVER A NONRECTANGULAR REGION (CONT.)**
>
> If R is the region $c \leq y \leq d$ and $a(y) \leq x \leq b(y)$, then we integrate over R according to the following equation.
>
> $$\iint_R f(x, y) \, dx \, dy = \int_c^d \int_{a(y)}^{b(y)} f(x, y) \, dx \, dy$$

▼ **EXAMPLE 4**

Redo Example 3, integrating in the opposite order.

SOLUTION We can do this if we can describe the region in Figure 6 in the way shown in Figure 7. In fact, it is the region $0 \leq y \leq 2$ and $y \leq x \leq 2$. A good way of seeing this description is to draw a horizontal line through the region, as in Figure 8.

The possible heights for such a line are $0 \leq y \leq 2$. The line extends from $x = y$ on the left to $x = 2$ on the right, so $y \leq x \leq 2$. We can now compute the integral.

$$\iint_R x \, dx \, dy = \int_0^2 \int_y^2 x \, dx \, dy$$

$$= \int_0^2 \left[\frac{x^2}{2} \right]_y^2 dy$$

$$= \int_0^2 \left(2 - \frac{1}{2}y^2 \right) dy$$

$$= \left[2y - \frac{1}{6}y^3 \right]_0^2$$

$$= \left(4 - \frac{8}{6} \right) - 0$$

$$= \frac{8}{3}$$

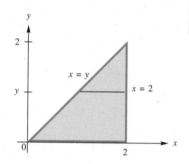

FIGURE 8

Before we go on... Many regions can be described in two different ways. Sometimes one description will be much easier to work with than the other, so it pays to try both.

There are many applications of double integrals besides finding volumes. We can also use them to find *averages*. Remember that the average of $f(x)$ on $[a, b]$ is given by $\int_a^b f(x) \, dx$ divided by $b - a$, the length of the interval.

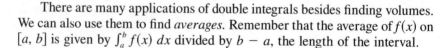

AVERAGE OF A FUNCTION OF TWO VARIABLES

The **average of $f(x, y)$ on the region R** is

$$\bar{f} = \frac{1}{A} \iint_R f(x, y) \, dx \, dy.$$

Here, A is the area of R.

▼ **EXAMPLE 5**

Find the average value of $f(x, y) = x + y$ over the region R shown in Figure 9.

SOLUTION   It looks easiest to describe R as the region $-1 \leq x \leq 1$ and $0 \leq y \leq 1 - x^2$. We first compute the area of R, which is the area under the graph of $y = 1 - x^2$.

$$A = \int_{-1}^{1} (1 - x^2) \, dx = \frac{4}{3}$$

Now we compute the double integral.

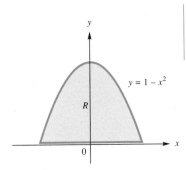

y

$y = 1 - x^2$

R

0   x

FIGURE 9

$$\iint_R (x + y) \, dx \, dy = \int_{-1}^{1} \int_0^{1-x^2} (x + y) \, dy \, dx$$

$$= \int_{-1}^{1} \left[xy + \frac{y^2}{2} \right]_{y=0}^{1-x^2} dx$$

$$= \int_{-1}^{1} \left(x(1 - x^2) + \frac{1}{2}(1 - x^2)^2 \right) dx$$

$$= \int_{-1}^{1} \left(\frac{1}{2} + x - x^2 - x^3 + \frac{1}{2}x^4 \right) dx$$

$$= \left[\frac{x}{2} + \frac{x^2}{2} - \frac{x^3}{3} - \frac{x^4}{4} + \frac{x^5}{10} \right]_{-1}^{1}$$

$$= \frac{8}{15}$$

We can now compute the average.

$$\bar{f} = \frac{1}{(4/3)} \cdot \frac{8}{15} = \frac{2}{5}$$

Before we go on... If we wanted to describe the region with the order of integration reversed, it would be $0 \leq y \leq 1$ and $-\sqrt{1 - y} \leq x \leq \sqrt{1 - y}$. This leads to a more difficult integral to evaluate.

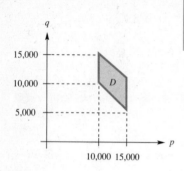

FIGURE 10

▼ **EXAMPLE 6**   Average Revenue

Your marketing department estimates that if you price your new line of cars at p dollars per car, you will be able to sell between $q = 20,000 - p$ and $q = 25,000 - p$ cars in the first year. If you price the cars somewhere between $10,000 and $15,000, what is the average of all the possible revenues you could have in a year?

SOLUTION   Revenue is given by $R = pq$ as usual, and we are told that $10,000 \le p \le 15,000$ and $20,000 - p \le q \le 25,000 - p$. This domain D of prices and demands is shown in Figure 10.

To average the revenue R over the domain D we need to compute the area A of D. Using either calculus or geometry, we get $A = 25,000,000$. We then need to integrate R over D.

$$\iint_D pq \, dp \, dq = \int_{10,000}^{15,000} \int_{20,000-p}^{25,000-p} pq \, dq \, dp$$

$$= \int_{10,000}^{15,000} \left[\frac{1}{2}pq^2\right]_{q=20,000-p}^{25,000-p} dp$$

$$= \frac{1}{2} \int_{10,000}^{15,000} \left[p(25,000 - p)^2 - p(20,000 - p)^2\right] dp$$

$$= \frac{1}{2} \int_{10,000}^{15,000} (225,000,000p - 10,000p^2) \, dp$$

$$= 3,072,916,666,666,667$$

We get the following average.

$$\bar{R} = \frac{3,072,916,666,666,667}{25,000,000}$$

$$\approx \$122,900,000 \text{ per year}$$

Before we go on... To check that this is a reasonable answer, notice that the revenues at the corners of the domain are $100,000,000 per year, $150,000,000 per year (at two corners), and $75,000,000 per year. Some of these are smaller than the average and some larger, as we would expect. You should also check that the maximum possible revenue is $156,250,000 per year (what is the minimum possible revenue?)

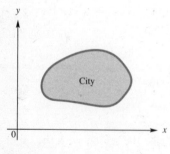

FIGURE 11

Another useful application comes about when we consider density. For example, suppose that $P(x, y)$ represents the population density (in people per square mile, say) in the city shown in Figure 11.

If we break the city up into small rectangles (for example, city blocks), then the population in the small rectangle $x_{i-1} \le x \le x_i$ and $y_{j-1} \le y \le y_j$ will be approximately $P(x_i, y_j) \, \Delta x \, \Delta y$. Adding up all of these population estimates, we get

$$\text{Total population} \approx \sum_{i=1}^{m} \sum_{j=1}^{n} P(x_i, y_j) \, \Delta x \, \Delta y.$$

Since this is a double Riemann sum, we get the following calculation of the population of the city when we take the limit as m and n go to infinity.

$$\text{Total population} = \iint_{\text{City}} P(x, y) \, dx \, dy$$

▼ **EXAMPLE 7**   Population

Squaresville is a city in the shape of a square 5 miles on a side. The population density at a distance of x miles east and y miles north of the southwest corner is $P(x, y) = x^2 + y^2$ thousand people per square mile. Find the total population of Squaresville.

SOLUTION   Squaresville is pictured in Figure 12, in which we put the origin in the southwest corner of the city.

To compute the total population, we integrate the population density over the city S.

$$\begin{aligned}
\text{Population} &= \iint_{S} P(x, y) \, dx \, dy \\
&= \int_{0}^{5} \int_{0}^{5} (x^2 + y^2) \, dx \, dy \\
&= \int_{0}^{5} \left[\frac{x^3}{3} + xy^2 \right]_{x=0}^{5} dy \\
&= \int_{0}^{5} \left(\frac{125}{3} + 5y^2 \right) dy \\
&= \left[\frac{125}{3}y + \frac{5}{3}y^3 \right]_{0}^{5} \\
&= \frac{1,250}{3} \approx 417 \text{ thousand people}
\end{aligned}$$

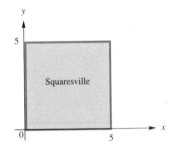

y

5

Squaresville

0   5 x

FIGURE 12

Before we go on... Note that the average population density is the total population divided by the area of the city, which is about 17 thousand people per square mile. Compare this calculation with the calculations of averages in the previous two examples.

► **8.7 EXERCISES**

Computed the integrals in Exercises 1–16.

1. $\int_0^1 \int_0^1 (x - 2y)\, dx\, dy$

2. $\int_{-1}^1 \int_0^2 (2x + 3y)\, dx\, dy$

3. $\int_0^1 \int_0^2 (ye^x - x - y)\, dx\, dy$

4. $\int_1^2 \int_2^3 \left(\frac{1}{x} + \frac{1}{y}\right) dx\, dy$

5. $\int_0^3 \int_0^2 e^{x+y}\, dx\, dy$

6. $\int_0^1 \int_0^1 e^{x-y}\, dx\, dy$

7. $\int_0^1 \int_0^{2-y} x\, dx\, dy$

8. $\int_0^1 \int_0^{2-y} y\, dx\, dy$

9. $\int_{-1}^1 \int_{y-1}^{y+1} e^{x+y}\, dx\, dy$

10. $\int_0^1 \int_y^{y+2} \frac{1}{\sqrt{x+y}}\, dx\, dy$

11. $\int_0^1 \int_{-x^2}^{x^2} x\, dy\, dx$

12. $\int_1^4 \int_{-\sqrt{x}}^{\sqrt{x}} \frac{1}{x}\, dy\, dx$

13. $\int_0^1 \int_0^x e^{x^2}\, dy\, dx$

14. $\int_0^1 \int_0^{x^2} e^{x^3+1}\, dy\, dx$

15. $\int_1^3 \int_{1-x}^{8-x} \sqrt[3]{x+y}\, dy\, dx$

16. $\int_1^2 \int_{1-2x}^{x^2} \frac{x+1}{(2x+y)^2}\, dy\, dx$

In each of Exercises 17–24, integrate the given function over the indicated domain. (Remember that you often have a choice as to the order of integration.)

17. $f(x, y) = 2$

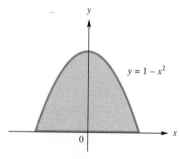
$y = 1 - x^2$

18. $f(x, y) = x$

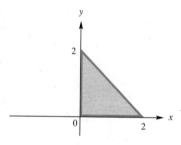

19. $f(x, y) = 1 + y$

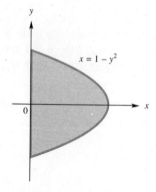
$x = 1 - y^2$

20. $f(x, y) = e^{x+y}$

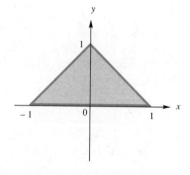

21. $f(x, y) = xy^2$

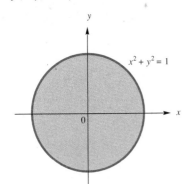

$x^2 + y^2 = 1$

22. $f(x, y) = xy^2$

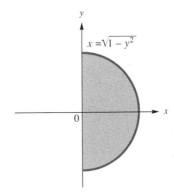

$x = \sqrt{1 - y^2}$

23. $f(x, y) = x^2 + y^2$

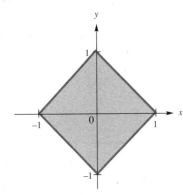

24. $f(x, y) = x^2$

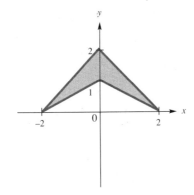

25–32. Find the average value of each function in Exercises 17–24.

In each of Exercises 33–40, sketch the region over which you are integrating, then write down the integral with the order of integration reversed (changing the limits of integration as necessary).

33. $\displaystyle\int_0^1 \int_0^{1-y} f(x, y)\, dx\, dy$

34. $\displaystyle\int_{-1}^1 \int_0^{1+y} f(x, y)\, dx\, dy$

35. $\displaystyle\int_{-1}^1 \int_0^{\sqrt{1+y}} f(x, y)\, dx\, dy$

36. $\displaystyle\int_{-1}^1 \int_0^{\sqrt{1-y^2}} f(x, y)\, dx\, dy$

37. $\displaystyle\int_0^2 \int_1^{4/x^2} f(x, y)\, dy\, dx$

38. $\displaystyle\int_1^{10} \int_0^{\ln x} f(x, y)\, dy\, dx$

39. $\displaystyle\int_0^2 \int_{2x}^4 f(x, y)\, dy\, dx$

40. $\displaystyle\int_{-2}^2 \int_{-\frac{1}{2}\sqrt{4-x^2}}^{\frac{1}{2}\sqrt{4-x^2}} f(x, y)\, dy\, dx$

41. Find the volume under the graph $z = 1 - x^2$ over the region $0 \le x \le 1$ and $0 \le y \le 2$.

42. Find the volume under the graph of $z = 1 - x^2$ over the triangle $0 \le x \le 1$, $0 \le y \le 1-x$.

43. Find the volume of the tetrahedron shown in the figure (its corners are $(0, 0, 0)$, $(1, 0, 0)$, $(0, 1, 0)$, and $(0, 0, 1)$).

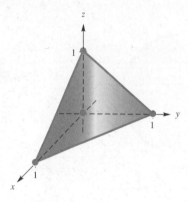

44. Find the volume of the tetrahedron with corners at $(0, 0, 0)$, $(a, 0, 0)$, $(0, b, 0)$, and $(0, 0, c)$.

APPLICATIONS

45. *Cobb-Douglas Production Function* The Cobb-Douglas production function for the Handy Gadget Company is given by

$$P = 10,000x^{0.3}y^{0.7},$$

where P is the number of gadgets it turns out per month, x is the number of employees at the company, and y is the monthly operating budget in thousands of dollars. Because the company hires part-time workers, it uses anywhere between 45 and 55 workers each month, and its operating budget varies from $8,000 to $12,000 per month. What is the average number of gadgets it can turn out per month? (Answer to the nearest gadget.)

46. *Cobb-Douglas Production Function* Repeat the preceding exercise using the production function

$$P = 10,000x^{0.7}y^{0.3}.$$

47. *Revenue* Your latest CD-ROM of clip-art is expected to sell between $q = 8,000 - p^2$ and $q = 10,000 - p^2$ copies if priced at p dollars. You plan to set the price between $40 and $50. What are the maximum and minimum possible revenues you can make? What is the average of all the possible revenues that you can make?

48. *Revenue* Your latest CD-ROM drive is expected to sell between $q = 180,000 - p^2$ and $q = 200,000 - p^2$ units if priced at p dollars. You plan to set the price between $300 and $400. What are the maximum and minimum possible revenues you can make? What is the average of all the possible revenues that you can make?

49. *Revenue* Your self-published novel has demand curves between $p = 15,000/q$ and $p = 20,000/q$. You expect to sell between 500 and 1,000 copies. What are the maximum and minimum possible revenues you can make? What is the average of all the possible revenues that you can make?

50. *Revenue* Your self-published book of poetry has demand curves between $p = 80,000/q^2$ and $p = 100,000/q^2$. You expect to sell between 50 and 100 copies. What are the maximum and minimum possible revenues you can make? What is the average of all the possible revenues that you can make?

51. *Population Density* The town of West Podunk is shaped like a rectangle 20 miles wide and 30 miles tall (see the figure). It has a population density of $P(x, y) = e^{-0.1(x+y)}$ hundred people x miles east and y

miles north of the southwest corner of town. What is the total population of the town?

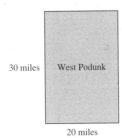

20 miles

52. *Population Density* The town of East Podunk is shaped like a triangle with a base of 20 miles and a height of 30 miles (see the figure). It has a population density of $P(x, y) = e^{-0.1(x+y)}$ hundred people x miles east and y miles north of the southwest corner of town. What is the total population of the town?

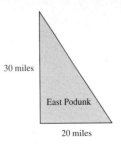

53. *Temperature* The temperature at the point (x, y) on the square with vertices $(0, 0)$, $(0, 1)$, $(1, 0)$, and $(1, 1)$ is given by $T(x, y) = x^2 + 2y^2$. Find the average temperature on the square.

54. *Temperature* The temperature at the point (x, y) on the square with vertices $(0, 0)$, $(0, 1)$, $(1, 0)$, and $(1, 1)$ is given by $T(x, y) = x^2 + 2y^2 - x$. Find the average temperature on the square.

COMMUNICATION AND REASONING EXERCISES

55. Explain how double integrals can be used to compute **(a)** the area between two curves in the xy-plane; **(b)** the volume of solids in 3-space.

56. Complete the following sentence: The first step in calculating an integral of the form $\int_a^b \int_{c/x}^{d(x)} f(x, y) \, dy \, dx$ is to evaluate the integral _____, obtained by holding _____ constant and integrating with respect to _____.

57. Show that if a, b, c, and d are constant, then

$$\int_a^b \int_c^d f(x)g(y) \, dx \, dy = \int_c^d f(x) \, dx \int_a^b g(y) \, dy.$$

Test this result on the integral $\int_0^1 \int_1^2 ye^x \, dx \, dy$.

58. If the units of $f(x, y)$ are bootlags per square meter, and x and y are given in meters, what are the units of $\int_a^b \int_{c/x}^{d(x)} f(x, y) \, dx \, dy$?

▶ **You're the Expert**

CONSTRUCTING A
BEST-FIT DEMAND
CURVE

You are a consultant for a new real estate development company that has purchased a large tract of land in Monmouth County, New Jersey, and intends to build a number of homes there. The company would like to target a single price bracket, and it has hired you to determine what its most profitable course of action might be. It typically sells new homes at 30% over cost, and it plans to build as many homes as the market will bear. Further, the company's market research team has come up with the following sales figures for new homes in Monmouth County over the past year.

Price (Thousands of $)	Sales of New Homes This Year
$50–$69	9
$70–$89	25
$90–$109	80
$110–$129	112
$130–$149	125
$150–$169	126
$170–$189	103
$190–$209	82
$210–$229	75
$230–$249	82
$250–$269	40
$270–$289	20

What would you advise the company to do?

You decide that your best strategy is to first find a demand equation based on these data. You first recast the data by replacing each price range with a single price in the middle of the range, coming up with the following table (in which p is in thousands of dollars) and graph (Figure 1).

p	60	80	100	120	140	160	180	200	220	240	260	280
q	9	25	80	112	125	126	103	82	75	82	40	20

Sales of New Homes

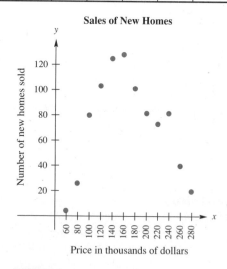

Price in thousands of dollars

FIGURE 1

You decide at once that fitting a least-squares line is not the way to go. The plotted data reminds you of a parabola, not a straight line. So you decide instead to try a demand equation of the form $q = f(p) = ap^2 + bp + c$.

You now need to find values for a, b, and c that give the parabola that best fits the data.

As a first step, you decide to look at the general situation. You have a collection of data points (p_1, q_1), (p_2, q_2), $\ldots$, (p_n, q_n), and desire the least-squares fit quadratic function $f(p) = ap^2 + bp + c$. Thus, you need to minimize the quantity

$$
\begin{aligned}
S &= (f(p_1) - q(p_1))^2 + (f(p_2) - q(p_2))^2 \\
&\quad + \ldots + (f(p_n) - q(p_n))^2 \\
&= (ap_1^2 + bp_1 + c - q_1)^2 + (ap_2^2 + bp_2 + c - q_2)^2 \\
&\quad + \ldots + (ap_n^2 + bp_n + c - q_n)^2.
\end{aligned}
$$

Noting that the variables are a, b, and c, you take partial derivatives.

$$
\begin{aligned}
S_a &= 2p_1^2(ap_1^2 + bp_1 + c - q_1) + 2p_2^2(ap_2^2 + bp_2 + c - q_2) \\
&\quad + \ldots + 2p_n^2(ap_n^2 + bp_n + c - q_n) \\
S_b &= 2p_1(ap_1^2 + bp_1 + c - q_1) + 2p_2(ap_2^2 + bp_2 + c - q_2) \\
&\quad + \ldots + 2p_n(ap_n^2 + bp_n + c - q_n) \\
S_c &= 2(ap_1^2 + bp_1 + c - q_1) + 2(ap_2^2 + bp_2 + c - q_2) \\
&\quad + \ldots + 2(ap_n^2 + bp_n + c - q_n)
\end{aligned}
$$

To locate the absolute minimum, you must set these equal to zero and solve for a, b, and c. You notice right away that these equations are linear in a, b, and c, so you group terms, divide by 2, and get

$$
\begin{aligned}
(p_1^4 + p_2^4 + \ldots + p_n^4)a &+ (p_1^3 + p_2^3 + \ldots + p_n^3)b \\
+ (p_1^2 + p_2^2 + \ldots + p_n^2)c & \\
&= (p_1^2 q_1 + p_2^2 q_2 + \ldots + p_n^2 q_n) \\
(p_1^3 + p_2^3 + \ldots + p_n^3)a &+ (p_1^2 + p_2^2 + \ldots + p_n^2)b \\
+ (p_1 + p_2 + \ldots + p_n)c & \\
&= (p_1 q_2 + p_2 q_2 + \ldots + p_n q_n) \\
(p_1^2 + p_2^2 + \ldots + p_n^2)a &+ (p_1 + p_2 + \ldots + p_n)b + nc \\
&= (q_1 + q_2 + \ldots + q_n).
\end{aligned}
$$

Using Σ-notation, these equations read more simply:

$$
\begin{aligned}
\left(\sum p_i^4 \right)a + \left(\sum p_i^3 \right)b + \left(\sum p_i^2 \right)c &= \sum p_i^2 q_i \\
\left(\sum p_i^3 \right)a + \left(\sum p_i^2 \right)b + \left(\sum p_i \right)c &= \sum p_i q_i \\
\left(\sum p_i^2 \right)a + \left(\sum p_i \right)b + nc &= \sum q_i.
\end{aligned}
$$

Now all you need to do is evaluate the coefficients and solve. Going back to the data, we have the following calculations.

$$\sum p_i^4 \approx 20{,}427{,}200{,}000 \qquad \sum p_i^3 = 88{,}128{,}000$$

$$\sum p_i^2 = 404{,}000 \qquad \sum p_i = 2{,}040$$

$$\sum p_i^2 q_i = 27{,}523{,}200 \qquad \sum p_i q_i = 148{,}760$$

$$\sum q_i = 879$$

Thus, your system of equations is

$$20{,}427{,}200{,}000a + 88{,}128{,}000b + 404{,}000c = 27{,}523{,}200$$
$$88{,}128{,}000a + 404{,}000b + 2{,}040c = 148{,}760$$
$$404{,}000a + 2{,}040b + 12c = 879.$$

Using your calculator, you obtain

$$a \approx -0.008626, \quad b \approx 2.921, \quad c \approx -132.9.$$

Thus, your demand equation is

$$q = -0.008626p^2 + 2.921p - 132.9.$$

Before going any further, you have your graphing calculator draw the graph of q superimposed on the data with which you started (Figure 2).

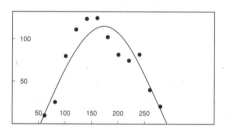

FIGURE 2

Not a bad fit! Now you get to work maximizing the profit. In order to do this, you need to find a formula giving the profit P as a function of the unit price p. Now, you know that the company can sell q homes at a unit price of p, where q is given by the formula $q = -0.008634p^2 + 2.924p - 133.1$. What you must compute is the *profit* it will obtain from the sale of these homes. Since it sells them at 30% over cost, it makes a profit of \$30 on every \$130 it charges. Since its total revenue will be

$$R = pq = p(-0.008626p^2 + 2.921p - 132.9),$$

its profit is $\frac{30}{130}$ of the revenue.

$$P = \frac{30}{130}(-0.008626p^3 + 2.921p^2 - 132.9p)$$

To maximize the profit, we set the derivative of profit equal to zero, and solve.

$$\frac{30}{130}(-0.025878p^2 + 5.842p - 132.9) = 0$$

Using the quadratic formula, we get two stationary points, $p = 200.1$ and $p = 25.67$. You then check that the lower value gives a local minimum, and it is the higher price that gives a local maximum. Thus, you report that according to your expert opinion, the company should target the $170,000–189,000 market range. Further, you inform the company that it can sell $q(200.1) \approx 106$ homes (assuming there are no competitors in Monmouth County!)

Exercises

 1. Find the best *cubic* approximation $q = ap^3 + bp^2 + cp + d$ to the data in this section. Sketch its graph together with the data points. If you used this approximation, would you change your advice to the development company?

 2. *Investments in South Africa* It is 1994, you are the CEO of a large chain of fast-food stores, and you are considering expanding to South Africa. You have decided that the most prudent policy would be to wait until the total number of U.S. companies with direct investment in South Africa is at least 500. The following chart shows the number of U.S. companies with direct investment in South Africa for the years 1986 through 1994.*

(a) If you use a best-fit quadratic approximation of the number of companies as a function of the number of years since 1986, when should you invest in South Africa?

(b) Comment on the long-term implications of your model.

_____

▼ *Source: Investor Responsibility Research Center/*New York Times*, January 31, 1994, p. D1.

► Review Exercises

Evaluate $g(0, 0, 0)$, $g(1, 0, 0)$, $g(0, 1, 0)$, $g(x, x, x)$, and $g(x, y + k, z)$ in each of Exercises 1–4, provided such a value exists.

1. $g(x, y, z) = xy(x + y - z) + x^2$

2. $g(x, y, z) = x \ln(xyz)$

3. $g(x, y, z) = xe^{xy+z}$

4. $g(x, y, z) = yz/(x^2 + y^2)$

Find the distance between the pairs of points given in Exercises 5–8.

5. $(1, 2)$ and $(3, 0)$
6. $(1, -1)$ and $(-2, 1)$
7. (a, b) and (a, c).
8. (a, b) and (b, a)

Exercises 9 and 10 involve the Cobb-Douglas production function. Recall that it has the form

$$P(x, y) = Kx^a y^{1-a},$$

where P stands for the number of items produced a year, x is the number of employees, and y is the annual operating budget.

9. How many items will be produced per year by a company with 1,000 employees and an annual operating budget of \$100,000 if $K = 100$, $a = 0.3$?

10. How many items will be produced per year by a company with 500 employees and an annual operating budget of \$100,000 if $K = 1,000$, $a = 0.3$?

In each of Exercises 11 and 12, find f_x, f_y, f_z, f_{xy}, f_{xz}, and f_{zz}.

11. $f(x, y, z) = x^2 + xyz$

12. $f(x, y, z) = 4x^2 y + z^2 - y$

Find f_x, f_y, f_z, and $f_x (0, 1, 0)$ in Exercises 13 and 14.

13. $f(x, y, z) = x/(x^2 + y^2 + z^2)$

14. $f(x, y, z) = (x^2 + y^2 + z^2)^{1/2}$

Find $f_{xx} + f_{yy} + f_{zz}$ in each of Exercises 15 and 16.

15. $f(x, y, z) = (x^2 + y^2 + z^2)^{-1/2}$

16. $f(x, y, z) = (x^2 + y^2 + z^2)^{-1}$

17. *Gravity in 3-Space* The gravitational force exerted on a particle with mass m situated at the point (x, y, z) by another particle with mass M situated at the point (a, b, c) is given by

$$F(x, y, z) = G\frac{Mm}{(x - a)^2 + (y - b)^2 + (z - c)^2}.$$

(All units of distance are in meters, the masses M and m are given in kilograms, $G \approx 6.67 \times 10^{-11}$, and the resulting force is given in newtons.) Suppose that a 3,000-kg mass is situated at the point $(1, 1, 1)$ and that a 10-kg object is at the point $(10, 100, 100)$ and traveling in the z-direction at one meter per second. How fast is the gravitational force it experiences decreasing?

18. *Gravity* Referring to the preceeding exercise, suppose that a 3,000-kg mass is situated at the point $(1, 0, -1)$ and that a 10-kg object is at the point $(100, 100, 100)$ and traveling in the negative x-direction at one meter per second. How fast is the gravitational force it experiences increasing?

In each of Exercises 19–28, locate and classify all critical points.

19. $f(x, y) = (x - 1)^2 + (2x - 3)^2$

20. $f(x, y) = x^2 - 2(y - 1)^2$

21. $g(x, y) = (x - 1)^2 - 3y^2 + 9$

22. $g(x, y) = 3(x + 1)^2 + (4y - 2)^2$

23. $h(x, y) = e^{xy}$

24. $h(x, y) = e^{x+y}$

25. $j(x, y) = xy + x^2$

26. $j(x, y) = xy + x^2 + y^2$

27. $f(x, y) = \ln(x^2 + y^2) - (x^2 + y^2)$

28. $f(x, y) = \ln\sqrt{x^2 + y^2} - \sqrt{x^2 + y^2}$

29. Find the point on the surface $z = \sqrt{x^2 + 2(y-3)^2}$ closest to the origin.

30. Find the point on the paraboloid $z = (x - 1)^2 + (y - 3)^2$ closest to the point $(1, 3, -2)$.

31. What point on the surface $z = 1/x$ is closest to the origin?

32. What point on the surface $z = 1/x$ is closest to the point $(0, 1, 0)$?

APPLICATIONS

33. *Temperature*   The temperature at the point (x, y) on the triangle with vertices $(0, 1)$, $(3, 0)$, and $(0, 0)$ is given by $T(x, y) = (x - 1)^2 + y^2$. Find the hottest and coldest points on the triangle.

34. *Temperature*   The temperature at the point (x, y) on the triangle with vertices $(0, -1)$, $(0, 1)$, and $(1, 0)$ is given by $T(x, y) = (x + 1)^2 - 3y^2$. Find the hottest and coldest points on the triangle.

35. *Cost*   To make x pencils and y pens costs you

$$C(x, y) = 0.001x^2 + 0.002y^2 - 2x - 4.8y + 10,880.$$

If you make between 900 and 1,100 pencils and between 1,000 and 1,500 pens, what combination will cost you the least, and what combination will cost you the most?

36. *Cost*   To make x nuts and y bolts costs you

$$C(x, y) = 0.01x + 0.02y + \frac{12,100}{x} + \frac{57,600}{y}.$$

If you make between 1,000 and 1,300 nuts and between 1,000 and 1,500 bolts, what combination will cost you the least, and what combination will cost you the most?

Find the least-squares line through each of the sets of points in Exercises 37–40.

37. $(1, 2)$, $(3, -4)$, $(5, 0)$

38. $(0, -1)$, $(-1, 2)$, $(2, -2)$

39. $(0, 1)$, $(1, 2)$, $(4, 2)$, $(5, 0)$

40. $(2, 4)$, $(6, 6)$, $(8, 0)$, $(10, 0)$

41. *Population*   The population of the world in various years is given in the following table (populations in billions).

Year	1	1650	1850	1930	1975	1991
Population	0.2	0.5	1	2	4	5.384

Fit an exponential curve to these data, and predict the world population in 2000 and in 2050. How good a fit is this?

42. *Money Supply*   The per capita money in circulation in the U.S. in various years is given in the following table.*

1910	1950	1970	1980	1985	1990
34.07	179.03	265.39	558.28	778.58	1028.71

Which would fit these data better, a line or an exponential curve? Fit curves of both types to these data, and compare their predictions of the per capita money in circulation in 2010.

▼ * Source: Financial Management Service, U.S. Department of the Treasury.

Compute the double integrals in Exercises 43–48.

43. $\displaystyle\int_{1}^{2}\int_{1}^{3}\sqrt{x+y}\,dx\,dy$

44. $\displaystyle\int_{1}^{2}\int_{0}^{2}xye^{x+y}\,dx\,dy$

45. $\displaystyle\int_{0}^{2}\int_{x}^{2x}\frac{1}{x^{2}+1}\,dy\,dx$

46. $\displaystyle\int_{0}^{1}\int_{0}^{x^{2}}xe^{-y}\,dy\,dx$

47. $\displaystyle\iint_{R}xy\,dx\,dy$, where R is the region shown in the figure.

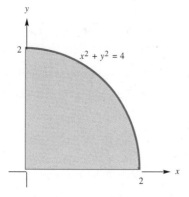

48. $\displaystyle\iint_{R}(x^{2}-y^{2})dx\,dy$, where R is the region shown in the figure.

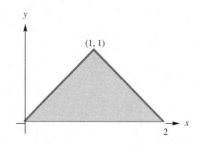

49–54. Compute the averages of the functions given in Exercises 43–48 over the given domains.

55. Find the volume under the graph of $z=4-x^{2}-y^{2}$ over the rectangle $-1\le x\le1$ and $-1\le y\le1$.

56. Find the volume under the graph of $z=1-y$ over the region in the xy-plane between the parabola $y=1-x^{2}$ and the x-axis.

57. *Profit*   If you make and sell x pencils and y pens, you will make a profit of

$$P(x,y)=0.03x+0.10y.$$

If you sell between 1,200 and 1,500 pencils and be-tween 1,800 and 2,000 pens, what is the average of all your possible profits?

58. *Profit*   If you make and sell x nuts and y bolts, you will make a profit of

$$P(x,y)=0.01x+0.05y.$$

If you sell between 1,000 and 1,200 nuts and between 1,500 and 1,600 bolts, what is the average of all your possible profits?

APPENDIX

Algebra Review

INTRODUCTION ▶ In this appendix we review a number of things from algebra that you need to know to get the most out of this book. You can work through this appendix, treating it as a refresher course in the basics of real numbers and algebraic expressions, or you can refer to parts of it as you need them.

Remember this crucial fact: The letters that we use in algebraic expressions stand for numbers. This means that what we say about numbers is true also about algebraic expressions. It also means that if you are not sure whether some algebraic manipulation you are about to do is legitimate, you can try it with numbers to see if it really works.

▶ **A.1** OPERATIONS ON THE REAL NUMBERS

There are five important operations on the set of real numbers: addition, subtraction, multiplication, division, and exponentiation. "Exponentiation" means the raising of a real number to a power; for instance, $3^2 = 3 \cdot 3 = 9$; $2^3 = 2 \cdot 2 \cdot 2 = 8$.

When we write an expression involving two or more of these operations, such as

$$2(3 - 5) + 4 \cdot 5,$$

or

$$\frac{2 \cdot 3^2 - 5}{4 - (-1)},$$

we agree to use the following rules to decide on the order in which we do the operations.

STANDARD ORDER OF OPERATIONS

1. **Parentheses and Fraction Bars**   Calculate the values of all expressions inside parentheses or brackets first, working from the innermost parentheses out. When dealing with a fraction bar, calculate the numerator and denominator separately and then do the division.
   Examples:

   $$6(2 + [3 - 5] - 4) = 6(2 + (-2) - 4) = 6(-4) = -24$$

   $$\frac{(4 - 2)}{3(-2 + 1)} = \frac{2}{3(-1)} = \frac{2}{-3} = -\frac{2}{3}$$

2. **Exponents**   Next, raise all numbers to the indicated powers.
   Examples:

   $$(2 + 4)^2 = 6^2 = 36$$

   $$\left.\begin{array}{l} (2 \cdot 3)^2 = 6^2 = 36 \\ 2 \cdot 3^2 = 2 \times 9 = 18 \end{array}\right\} \quad \text{Note the distinction here.}$$

   $$2\left(\frac{3}{4 - 5}\right)^2 = 2\left(\frac{3}{-1}\right)^2 = 2(-3)^2 = 2 \times 9 = 18$$

3. **Multiplication and Division**   Next, do all the multiplications and divisions from left to right.
   Examples:

   $$6(2[3 - 5] \div 4 \cdot (-1)) \div 2 = 6(2(-2) \div 4 \cdot (-1)) \div 2$$
   $$= 6(-4 \div 4 \cdot (-1)) \div 2$$
   $$= 6((-1) \cdot (-1)) \div 2$$
   $$= 6 \cdot 1 \div 2 = 3$$

$$4\left(\frac{2(4-2)}{3(-2\cdot5)}\right) = 4\left(\frac{2(2)}{3(-10)}\right) = 4\left(\frac{4}{-30}\right) = \frac{16}{-30} = -\frac{8}{15}$$

$$4\left(\frac{2(4^2\cdot2)}{3(-2+5)}\right) = 4\left(\frac{2(16\cdot2)}{3(3)}\right) = 4\left(\frac{64}{9}\right) = \frac{256}{9}$$

4. **Addition and Subtraction**   Last, do the remaining additions and subtractions from left to right.
   Examples:

$$2(3-5)^2 + 6 - 1 = 2(-2)^2 + 6 - 1$$
$$= 2(4) + 6 - 1 = 8 + 6 - 1 = 13$$

$$\left(\frac{1}{2}\right)^2 - (-1)^2 + 4 = \frac{1}{4} - 1 + 4 = -\frac{3}{4} + 4 = \frac{13}{4}$$

CALCULATORS, GRAPHING CALCULATORS, AND COMPUTERS

Any good calculator will respect the standard order of operations. However, we must be careful with division and exponentiation and often must use parentheses. The following table gives some examples of simple mathematical expressions and their calculator equivalents in the functional format used in most graphing calculators and computer programs. (These expressions would be entered in a scientific calculator in essentially the same way.)

Here is one more fact about calculators (and calculations in general): A calculation can never give you an answer more accurate than the numbers you start with. As a general rule of thumb, if you have numbers measuring something in the real world (time, length, or gross domestic product, for example) and these numbers are accurate only to a certain number of digits, then any calculations you do with them will be accurate only to that many digits (at best). For example, if someone tells you that a rectangle has sides of length 2.2 ft and 4.3 ft, you can say that the area is (approximately) 9.5 sq ft, rounding to two significant digits. If you report that the area is 9.46 sq ft, as your calculator will tell you, the third digit is probably suspect. We shall usually use data that are accurate to three or four digits, and round our answers to that size.

Mathematical Expression	Calculator Equivalent	Comments
$\dfrac{2}{3-5}$	2/(3−5)	Note the use of parentheses instead of the fraction bar. If we omit the parentheses, we get the expression shown next.
$\dfrac{2}{3}-5$	2/3−5	The calculator automatically follows the usual order of operations.
$\dfrac{2}{3 \times 5}$	2/(3*5)	Putting the denominator in parentheses ensures that the multiplication is carried out first. The symbol "*" is usually used for multiplication in graphing calculators and computers.
$\dfrac{2}{3} \times 5$	(2/3)*5	Putting the fraction in parentheses ensures that it is calculated first. Some calculators will interpret 2/3*5 as $\frac{2}{3 \times 5}$ but 2/3(5) as $\frac{2}{3} \times 5$.
$\dfrac{2-3}{4+5}$	(2−3)/(4+5)	Note once again the use of parentheses in place of the fraction bar.
2^3	2^3	The caret "^" is commonly used to denote exponentiation.
2^{3-4}	2^(3−4)	Be careful to use parentheses to tell the calculator where the exponent ends. Enclose the *entire exponent* in parentheses.
$2^3 - 4$	2^3−4	Without parentheses, the calculator will follow the usual order of operations: exponentiation first, then subtraction.
3×2^{-4}	3*2^(−4)	The shorter minus sign stands for the negation sign, which on some calculators is a separate key.
$2^{-4 \times 3} \times 5$	2^(−4*3)*5	Note once again how parentheses enclose the entire exponent.
$\dfrac{2^{3-2} \times 5}{2-7}$	2^(3 − 2)*5/(2 − 7) or (2^(3 − 2)*5)/(2 − 7)	Notice again the use of parentheses to hold the denominator together. We could also have enclosed the numerator in parentheses, although this is optional (why?).
$\dfrac{2^{3-6}+1}{2-4^3}$	(2^(3 − 6)+1)/(2 − 4^3)	Here, it is necessary to enclose each of the numerator and denominator in parentheses.
$2^{3-6} + \dfrac{1}{2} - 4^3$	2^(3 − 6)+1/2 − 4^3	This is the effect of leaving out the parentheses around the numerator and denominator in the previous expression.

▶ **A-1 EXERCISES**

Calculate each of the expressions in Exercises 1–14, expressing your answer as a whole number or a fraction in lowest terms:

1. $2(4 + (-1))(2 \cdot -4)$

2. $3 + ([4 - 2] \cdot 9)$

3. $\dfrac{3 + ([3 + (-5)])}{3 - 2 \times 2}$

4. $\dfrac{12 - (1 - 4)}{2(5 - 1) \cdot 2 - 1}$

5. $2 \cdot (-1)^2 \div 2$

6. $2 + 4 \cdot 3^2$

7. $2 \cdot 4^2 + 1$

8. $1 - 3 \cdot (-2)^2 \times 2$

9. $\dfrac{3 - 2(-3)^2}{-6(4 - 1)^2}$

10. $\dfrac{1 - 2(1 - 4)^2}{2(5 - 1)^2 \cdot 2}$

11. $3\left(\dfrac{-2 \cdot 3^2}{-(4 - 1)^2}\right)$

12. $-\left(\dfrac{8(1 - 4)^2}{-9(5 - 1)^2}\right)$

13. $3\left(1 - \left(-\dfrac{1}{2}\right)^2\right)^2 + 1$

14. $3\left(\dfrac{1}{9} - \left(\dfrac{2}{3}\right)^2\right)^2 + 1$

Convert each of the expressions in Exercises 15–30 into its calculator equivalent as in the table in the text.

15. $3 \times (2 - 5)$

16. $4 + \dfrac{5}{9}$

17. $\dfrac{3}{2 - 5}$

18. $\dfrac{4 - 1}{3}$

19. $\dfrac{3 - 1}{8 + 6}$

20. $3 + \dfrac{3}{2 - 9}$

21. $3 - \dfrac{4 + 7}{8}$

22. $\dfrac{4 \times 2}{\left(\dfrac{2}{3}\right)}$

23. $\dfrac{\left(\dfrac{2}{3}\right)}{5}$

24. $\dfrac{2}{\left(\dfrac{3}{5}\right)}$

25. $3^{4-5} \times 6$

26. $\dfrac{2}{3 + 5^{7-9}}$

27. $3\left(1 + \dfrac{4}{100}\right)^{-3}$

28. $3\left(\dfrac{1 + 4}{100}\right)^{-3}$

29. $3\left(1 - \left(-\dfrac{1}{2}\right)^2\right)^2 + 1$

30. $3\left(\dfrac{1}{9} - \left(\dfrac{2}{3}\right)^2\right)^2 + 1$

▶ ═══ **A.2** INTEGER EXPONENTS

POSITIVE EXPONENTS

If a is any real number and n is any *positive integer*, then by a^n we mean the quantity $a \cdot a \cdots a$ (n times); thus, $a^1 = a$, $a^2 = a \cdot a$,

$a^5 = a \cdot a \cdot a \cdot a \cdot a$. Here are some examples with actual numbers:

$$3^2 = 9, \quad 2^3 = 8, \quad 0^{34} = 0, \quad (-1)^5 = -1.$$

In the expression a^n, the number n is called the *exponent*, and the number a is called the *base*. The following rules show how to combine such expressions.

EXPONENT IDENTITIES

(a) $a^m a^n = a^{m+n}$ *Example:* $2^3 2^2 = 2^5 = 32$

(b) $\dfrac{a^m}{a^n} = a^{m-n}$ if $m > n$ and $a \neq 0$

Example: $\dfrac{4^3}{4^2} = 4^{3-2} = 4^1 = 4$

(c) $(a^n)^m = a^{nm}$ *Example:* $(3^2)^2 = 3^4 = 81$

(d) $(ab)^n = a^n b^n$ *Example:* $(4 \cdot 2)^2 = 4^2 2^2 = 64$

(e) $\left(\dfrac{a}{b}\right)^n = \dfrac{a^n}{b^n}$ if $b \neq 0$ *Example:* $\left(\dfrac{4}{3}\right)^2 = \dfrac{4^2}{3^2} = \dfrac{16}{9}$

▶ CAUTION

(a)   In identities (a) and (b), the bases of the expressions must be the same. For example, rule (a) gives $3^2 3^4 = 3^6$ but does *not* apply to $3^2 4^2$.

(b)   People sometimes invent their own identities, such as $a^m + a^n = a^{m+n}$, which is wrong! (If you don't believe this, try it with $a = m = n = 1$.) If you wind up with something like $2^3 + 2^4$, you are stuck with it—there are no identities around to simplify it further. ◀

▼ **EXAMPLE 1**   Positive Exponents

a. $10^2 10^3 = 10^{2+3}$ By (a)

$\qquad\quad = 10^5 = 100,000$

b. $\dfrac{4^6}{4^3} = 4^{6-3}$ By (a)

$\qquad = 4^3 = 64$

c. $\dfrac{(x^2)^3}{x^3} = \dfrac{x^6}{x^3}$ By (c)

$\qquad\qquad = x^{6-3}$ By (b)

$\qquad\qquad = x^3$

d. $\dfrac{(x^4 y)^3}{y} = \dfrac{(x^4)^3 y^3}{y}$ By (d)

$\qquad\qquad = \dfrac{x^{12} y^3}{y}$ By (c)

$\qquad\qquad = x^{12} y^{3-1}$ By (b)

$\qquad\qquad = x^{12} y^2$

$\pmb{Q}$ Just where do these identities come from?

$\pmb{A}$ Let us look at them one at a time.

(a) $a^m a^n = a^{m+n}$
We can see why this works by looking at an example:* Let's take $a = 3$, $m = 2$, and $n = 3$. Then the left-hand side is

$$3^2 3^3 = (3 \cdot 3)(3 \cdot 3 \cdot 3) = 3 \cdot 3 \cdot 3 \cdot 3 \cdot 3,$$

while the right-hand side is

$$3^{2+3} = 3^5 = 3 \cdot 3 \cdot 3 \cdot 3 \cdot 3,$$

the same thing.

(b) $\dfrac{a^m}{a^n} = a^{m-n}$
We check this by example again, this time with $a = 3$, $m = 5$, and $n = 2$. Then the left-hand side is

$$\frac{3^5}{3^2} = \frac{3 \cdot 3 \cdot 3 \cdot \cancel{3} \cdot \cancel{3}}{\cancel{3} \cdot \cancel{3}} = 3 \cdot 3 \cdot 3,$$

while the right-hand side is

$$3^{5-2} = 3^3 = 3 \cdot 3 \cdot 3,$$

the same thing.

(c) $(a^n)^m = a^{nm}$
Let's take $a = 3$, $m = 2$, and $n = 3$. Then the left-hand side is

$$(3^2)^3 = (3 \cdot 3)^3 = (3 \cdot 3)(3 \cdot 3)(3 \cdot 3),$$

while the right-hand side is

$$3^6 = 3 \cdot 3 \cdot 3 \cdot 3 \cdot 3 \cdot 3,$$

▼ *Although we are not giving a formal proof, examples are where the ideas for formal proofs usually come from, and it is the ideas we're interested in here.

while the right-hand side is

$$3^6 = 3 \cdot 3 \cdot 3 \cdot 3 \cdot 3 \cdot 3,$$

the same thing.

(d) $(ab)^n = a^n b^n$ and $\left(\dfrac{a}{b}\right)^n = \dfrac{a^n}{b^n}$

We'll check the first by means of an example, and leave the second for you. Let's take $a = 3$, $b = 5$, and $n = 4$. The left-hand side is then

$$(3 \cdot 5)^4 = (3 \cdot 5)(3 \cdot 5)(3 \cdot 5)(3 \cdot 5),$$

while the right-hand side is

$$3^4 5^4 = (3 \cdot 3 \cdot 3 \cdot 3)(5 \cdot 5 \cdot 5 \cdot 5),$$

which is the same thing.

Thus, these identities are really just observations about how multiplication works.

NEGATIVE AND ZERO EXPONENTS

It turns out to be very useful to allow ourselves to use exponents that are not positive integers. These are dealt with by the following definition.

NEGATIVE EXPONENTS

If a is any real number other than zero and n is any positive integer, then we define

$$a^{-n} = \frac{1}{a^n} = \frac{1}{a \cdot a \cdot \cdots a} \quad (n \text{ times}).$$

ZERO EXPONENTS

If a is any real number other than zero, then we define

$$a^0 = 1.$$

Thus, for instance,

$$2^{-3} = \frac{1}{2^3} = \frac{1}{8} \quad \text{and} \quad 3^0 = 1.$$

▼ **EXAMPLE 2**   Negative and Zero Exponents

(a) $3^0 = 1$

(b) $1{,}000{,}000^0 = 1$

(c) $3^{-3} = \dfrac{1}{3^3} = \dfrac{1}{27}$

(d) $x^{-1} = \dfrac{1}{x^1} = \dfrac{1}{x}$

(e) $(-3)^{-2} = \dfrac{1}{(-3)^2} = \dfrac{1}{9}$

(f) $1^{-27} = \dfrac{1}{1^{27}} = 1$

(g) $y^7 y^{-2} = y^7 \dfrac{1}{y^2} = y^5$

Q Where do these strange definitions come from?

A From the desire to keep the exponent identities true even if we allow 0 or negative exponents. For example, if the first exponent identity, $a^m a^n = a^{m+n}$, is to remain true, we must be able to say

$$a^0 a^n = a^{0+n} = a^n.$$

Dividing both sides by a^n gives $a^0 = 1$. So this *forces* us to say that $a^0 = 1$. On the other hand, the same identity tells us that

$$a^{-n} a^n = a^{-n+n} = a^0 = 1.$$

Dividing both sides by a^n gives $a^{-n} = 1/a^n$. Again, this forces us to say that this is what we mean by a^{-n}.

It turns out that once we make these agreements, all of the exponent identities remain true as stated, without any restrictions on the exponents m and n, as long as we make sure that $a \neq 0$ where having $a = 0$ would not make sense, namely, where 0^0 or 0 raised to a negative power would appear.*

▼ * Trying to raise 0 to a negative power would involve division by 0, so this is not defined. But what about raising 0 to the 0 power? If 0^0 had a definite value, then this value should be extremely close to the value of, say, $0^{0.000001}$. But 0 raised to any power is 0. Thus, 0^0 should be 0. On the other hand it should also be extremely close to $(0.000001)^0$, which is 1, since any nonzero number raised to 0 is 1. How can it be close to both 0 and 1?

▼ **EXAMPLE 3**

Simplify $\dfrac{x^4y^{-3}}{x^5y^2}$, and express the answer using no negative exponents.

SOLUTION

$$\frac{x^4y^{-3}}{x^5y^2} = x^{4-5}y^{-3-2} = x^{-1}y^{-5} = \frac{1}{xy^5}$$

▼ **EXAMPLE 4**

Simplify $\left(\dfrac{x^{-1}}{x^2y}\right)^5$, and express the answer using no negative exponents.

SOLUTION

$$\left(\frac{x^{-1}}{x^2y}\right)^5 = \frac{(x^{-1})^5}{(x^2y)^5} = \frac{x^{-5}}{x^{10}y^5} = \frac{1}{x^{15}y^5}$$

▶ ___ **A.2 EXERCISES**

Evaluate the expressions in Exercises 1–16.

1. 3^3
    2. $(-2)^3$
    3. $-(2 \cdot 3)^2$
    4. $(4 \cdot 2)^2$

5. $\left(\frac{-2}{3}\right)^2$
    6. $\left(\frac{3}{2}\right)^3$
    7. $(-2)^{-3}$
    8. -2^{-3}

9. $\left(\frac{1}{4}\right)^{-2}$
    10. $\left(\frac{-2}{3}\right)^{-2}$
    11. $2 \cdot 3^0$
    12. $3 \cdot (-2)^0$

13. $2^3 2^2$
    14. $3^2 3$
    15. $2^2 2^{-1} 2^4 2^{-4}$
    16. $5^2 5^{-3} 5^2 5^{-2}$

Simplify each expression in Exercises 17–30, expressing your answer with no negative exponents.

17. x^3x^2
    18. x^4x^{-1}
    19. $-x^2x^{-3}y$
    20. $-xy^{-1}x^{-1}$

21. $\dfrac{x^3}{x^4}$
    22. $\dfrac{y^5}{y^3}$
    23. $\dfrac{x^2y^2}{x^{-1}y}$
    24. $\dfrac{x^{-1}y}{x^2y^2}$

25. $\dfrac{(xy^{-1}z^3)^2}{x^2yz^2}$
    26. $\dfrac{x^2yz^2}{(xyz^{-1})^{-1}}$
    27. $\left(\dfrac{xy^{-2}z}{x^{-1}z}\right)^3$
    28. $\left(\dfrac{x^2y^{-1}z^0}{xyz}\right)^2$

29. $\left(\dfrac{x^{-1}y^{-2}z^2}{xy}\right)^{-2}$
    30. $\left(\dfrac{xy^{-2}}{x^2y^{-1}z}\right)^{-3}$

▶══ **A.3** RADICALS AND RATIONAL EXPONENTS

If a is any nonnegative real number, then its **square root** is the nonnegative number whose square is a. For example, the square root of 16 is 4, since $4^2 = 16$. We write the square root of n as $\sqrt{n}$. It is important to remember that $\sqrt{n}$ is never negative. Thus, for instance, $\sqrt{9}$ is 3, and not -3, even though $(-3)^2 = 9$. If we want to speak of the "negative square root" of 9, we write it as $-\sqrt{9} = -3$. If we wanted to write both square roots at once, we would write $\pm\sqrt{9} = \pm 3$.

The **cube root** of a real number a is the number whose cube is a. The cube root of a is written $\sqrt[3]{a}$ so that, for example, $\sqrt[3]{8} = 2$ (since $2^3 = 8$). Note that we can take the cube root of any number, whether positive, negative, or zero. For instance, the cube root of -8 is $\sqrt[3]{-8} = -2$, since $(-2)^3 = -8$. Unlike square roots, the cube root of a number may be negative. In fact, the cube root of a always has the same sign as a.

Higher roots are defined similarly. The **fourth root** of the *nonnegative* number a is defined as the nonnegative number whose fourth power is a, written $\sqrt[4]{a}$. The **fifth root** of any number a is the number whose fifth power is a, and so on.

▶ NOTE

We cannot take an even-numbered root of a negative number, but we can take an odd-numbered root of any number. Even roots are always positive, while odd roots have the same sign as the number we start with. ◀

▼ **EXAMPLE 1**

(a) $\sqrt{4} = 2$, since $2^2 = 4$.

(b) $\sqrt{16} = 4$, since $4^2 = 16$.

(c) $\sqrt{1} = 1$, since $1^2 = 1$.

(d) $\sqrt{2}$ is not a whole number, but is approximately equal to 1.414213562.

(e) $\sqrt{1 + 1} = \sqrt{2} \approx 1.414213562$. Note that we first added the quantities under the square root sign, and then took the square root. In general, $\sqrt{a + b}$ means the square root of the *quantity* $(a + b)$. The radical sign acts like a pair of parentheses or a fraction bar, telling us to evaluate what is inside before taking the root.

(f) $\sqrt[3]{27} = 3$, since $3^3 = 27$.

(g) $\sqrt[3]{-64} = -4$, since $(-4)^3 = -64$.

(h) $\sqrt[4]{16} = 2$, since $2^4 = 16$, but $\sqrt[4]{-16}$ is not defined.

(i) $\sqrt[5]{-1} = -1$, since $(-1)^5 = -1$. Similarly, $\sqrt[n]{-1} = -1$ if n is any odd number.

▶ CAUTION

It is important to remember that $\sqrt{a + b}$ is *not* equal to $\sqrt{a} + \sqrt{b}$ (consider $a = b = 1$, for example). Equating these expressions is a common error, so be careful! ◀

What *is* true is the following.

RADICALS OF PRODUCTS AND QUOTIENTS

If a and b are any real numbers (nonnegative in the case of even-numbered roots), then

(1) $\sqrt[n]{ab} = \sqrt[n]{a}\sqrt[n]{b}$;

(2) $\sqrt[n]{\dfrac{a}{b}} = \dfrac{\sqrt[n]{a}}{\sqrt[n]{b}}$ if $b \neq 0$.

Examples: $\sqrt{9 \cdot 4} = \sqrt{9}\sqrt{4} = 3 \times 2 = 6$.
(We could also have calculated this as follows:
$$\sqrt{9 \cdot 4} = \sqrt{36} = 6)$$
$$\sqrt{\frac{9}{4}} = \frac{\sqrt{9}}{\sqrt{4}} = \frac{3}{2}.$$

Rule 1 is similar to the rule $(a \cdot b)^2 = a^2 \cdot b^2$ for the square of a product, and Rule 2 is similar to the rule $(a/b)^2 = a^2/b^2$ for the square of a quotient.

▼ **EXAMPLE 2**

(a) $\sqrt{9 + 16} = \sqrt{25} = 5$ (It is *not* equal to $\sqrt{9} + \sqrt{16}$, which is $3 + 4 = 7$.)

(b) $\sqrt{4 \cdot 25} = \sqrt{100} = 10$. Also, $\sqrt{4 \cdot 25} = \sqrt{4}\sqrt{25} = 2 \times 5 = 10$.

(c) $2\sqrt{3 - 1} = 2\sqrt{2} \approx 2.828427$

(d) $\sqrt{4(3 + 13)} = \sqrt{4(16)} = \sqrt{4}\sqrt{16} = 2 \cdot 4 = 8$.

(e) If x is nonnegative, then $\sqrt{x^2} = x$.

(f) If x is negative, then $\sqrt{x^2} = -x$. For instance, $\sqrt{(-3)^2} = \sqrt{9} = 3 = -(-3)$.

(g) In general, $\sqrt{x^2} = |x|$, the **absolute value of x,** which is the nonnegative number with the same value as x. For instance, $|-3| = 3$, $|3| = 3$, and $|0| = 0$.

(h) In general, $\sqrt[n]{x^n} = x$ if n is odd, and $\sqrt[n]{x^n} = |x|$ if n is even.

(i) Whether or not x is negative, $\sqrt{x^4} = x^2$. (Why?)

(j) If x is nonnegative, then
$$\sqrt{x^3} = \sqrt{x^2 \cdot x} = \sqrt{x^2}\sqrt{x} = x\sqrt{x},$$

and
$$\sqrt{x^5} = \sqrt{x^4 \cdot x} = \sqrt{x^4}\sqrt{x} = x^2\sqrt{x}.$$

(k) $\sqrt{\dfrac{x^2 + y^2}{z^2}} = \dfrac{\sqrt{x^2 + y^2}}{\sqrt{z^2}} = \dfrac{\sqrt{x^2 + y^2}}{z}$, if z is positive. (We can't simplify the numerator any further.)

(l) $\sqrt[3]{-216} = \sqrt[3]{(-27)8} = \sqrt[3]{-27}\sqrt[3]{8} = (-3)2 = -6$

▼ **EXAMPLE 3**   Solving Equations

(a) Solve the equation $x^2 - 4 = 0$.

Adding 4 to both sides gives $x^2 = 4$. Since this says that the square of x is 4, the number x is either 2 or -2 since squaring either one gives 4. Thus, the solution is $x = \pm 2$. (We'll see later that the solution can also be obtained by factoring.)

(b) Solve for x: $x^2 - \frac{1}{2} = 0$.

Adding $\frac{1}{2}$ to both sides gives $x^2 = \frac{1}{2}$. Thus,

$$x = \pm \sqrt{\frac{1}{2}} = \pm \frac{1}{\sqrt{2}}.$$

(c) Solve the equation $x^3 + 64 = 0$.

Subtracting 64 from both sides of the equation gives $x^3 = -64$. Thus, $x = \sqrt[3]{-64} = -4$.

RATIONAL EXPONENTS

We already know what we mean by expressions such as x^4 and a^{-6}. The next step is to make sense of non-integral exponents as in $a^{1/2}$, 3^π, and similar beasts. First we look at rational exponents—that is, exponents of the form p/q with p and q integers.

Q What should we mean by $a^{1/2}$?

A The overriding concern here is that the exponent identities should remain true. In this case the identity to look at is the one that says that $(a^m)^n = a^{mn}$. This identity tells us that

$$(a^{1/2})^2 = a^1 = a.$$

That is, $a^{1/2}$, when squared, gives us a. But that must mean that $a^{1/2}$ is the *square root* of a—that is,

$$a^{1/2} = \sqrt{a}.$$

Q If q is a positive integer, what should we mean by $a^{1/q}$?

A Do as we did above, replacing 2 with q:

$$(a^{1/q})^q = a^1 = a,$$

so

$$a^{1/q} = \sqrt[q]{a}, \text{ the } q\text{th root of } a.$$

Notice that if a is negative, this makes sense only for q odd. To avoid this problem we usually stick to positive a.

Q If p and q are integers (q being positive), what should we mean by $a^{p/q}$?

A By the exponent identities

$$a^{p/q} = (a^p)^{1/q} = (a^{1/q})^p,$$

which gives us the following.

CONVERSION BETWEEN RATIONAL EXPONENTS AND RADICALS

If a is any nonnegative number, then

$$a^{p/q} = \sqrt[q]{a^p} = (\sqrt[q]{a})^p.$$

In particular,

$$a^{1/q} = \sqrt[q]{a}, \text{ the } q\text{th root of } a.$$

From left to right, we say that we are rewriting the expression in **radical form.** From right to left, we are rewriting the expression in **exponential form.** Again, if a is negative all of this makes sense only if q is odd. Also, the exponent is called a *rational exponent* since it is a rational number, p/q.

▼ **EXAMPLE 4**

$$4^{3/2} = (\sqrt{4})^3 = 2^3 = 8$$

$$8^{2/3} = (\sqrt[3]{8})^2 = 2^2 = 4$$

$$9^{-3/2} = \frac{1}{9^{3/2}} = \frac{1}{(\sqrt{9})^3} = \frac{1}{3^3} = \frac{1}{27}$$

All of the exponent identities continue to work when we allow rational exponents. In other words, we are free to use all the exponent identities even if the exponents are not integers.

▼ **EXAMPLE 5**

Simplify $2^2 2^{7/2}$.

SOLUTION

$$\begin{aligned} 2^2 2^{7/2} &= 2^{2+\frac{7}{2}} \\ &= 2^{\frac{11}{2}} \\ &= 2^{5\frac{1}{2}} \\ &= 2^{5+\frac{1}{2}} \\ &= 2^5 \cdot 2^{\frac{1}{2}} \\ &= 2^5 \sqrt{2} \end{aligned}$$

▼ **EXAMPLE 6**

Simplify $\dfrac{(x^3)^{5/3}}{x^3}$.

SOLUTION $\dfrac{(x^3)^{5/3}}{x^3} = \dfrac{x^5}{x^3} = x^2$

▼ **EXAMPLE 7**

Simplify $\dfrac{\sqrt{3}}{\sqrt[3]{3}}$.

SOLUTION $\dfrac{\sqrt{3}}{\sqrt[3]{3}} = \dfrac{3^{1/2}}{3^{1/3}} = 3^{\frac{1}{2}-\frac{1}{3}} = 3^{\frac{1}{6}} = \sqrt[6]{3}$

▼ **EXAMPLE 8**

Simplify $\sqrt[4]{a^6}$.

SOLUTION $\sqrt[4]{a^6} = a^{6/4} = a^{3/2} = a \cdot a^{1/2} = a\sqrt{a}$

▶ ___A.3 EXERCISES___

Evaluate the expressions in Exercises 1–16, rounding your answer to four significant digits where necessary.

1. $\sqrt{4}$ **2.** $\sqrt{5}$ **3.** $\sqrt{\dfrac{1}{4}}$ **4.** $\sqrt{\dfrac{1}{9}}$

5. $\sqrt{\dfrac{16}{9}}$ **6.** $\sqrt{\dfrac{9}{4}}$ **7.** $\dfrac{\sqrt{4}}{5}$ **8.** $\dfrac{6}{\sqrt{25}}$

9. $\sqrt{4 + 9}$ **10.** $\sqrt[4]{16 - 9}$ **11.** $\sqrt[3]{27 \div 8}$ **12.** $\sqrt[3]{8 \times 64}$

13. $\sqrt{(-2)^2}$ **14.** $\sqrt{(-1)^2}$ **15.** $\sqrt{\dfrac{1}{4}(1 + 5)}$ **16.** $\sqrt{\dfrac{1}{9}(3 + 33)}$

Simplify the expressions in Exercises 17–24, given that x, y, z, a, b, and c are positive real numbers.

17. $\sqrt{a^2 b^2}$ **18.** $\sqrt{\dfrac{a^2}{b^2}}$ **19.** $\sqrt{(x + 9)^2}$ **20.** $(\sqrt{x} + 9)^2$

21. $\sqrt[3]{x^3(a^3 + b^3)}$ **22.** $\sqrt[4]{\dfrac{x^4}{a^4 b^4}}$ **23.** $\sqrt{\dfrac{4xy^3}{x^2 y}}$ **24.** $\sqrt{\dfrac{4(x^2 + y^2)}{c^2}}$

Rewrite the expressions in Exercises 25–32 in exponential form.

25. $\sqrt{3}$ **26.** $\sqrt{8}$ **27.** $\sqrt{x^3}$ **28.** $\sqrt[3]{x^2}$

29. $\sqrt[3]{xy^2}$ **30.** $\sqrt{x^2 y}$ **31.** $\dfrac{x^2}{\sqrt{x}}$ **32.** $\dfrac{x}{\sqrt{x}}$

Rewrite the expressions in Exercises 33–38 in radical form.

33. $2^{2/3}$ **34.** $3^{4/5}$ **35.** $x^{4/3}$ **36.** $y^{7/4}$

37. $(x^{1/2} y^{1/3})^{1/5}$ **38.** $x^{-1/3} y^{3/2}$

Simplify the expressions in Exercises 39–48.

39. $4^{-1/2} 4^{7/2}$ **40.** $2^{1/a}/2^{2/a}$ **41.** $3^{2/3} 3^{-1/6} 3^{4/6} 3^{-4/6}$ **42.** $2^{1/3} 2^{-1} 2^{2/3} 2^{-1/3}$

43. $\dfrac{x^{3/2}}{x^{5/2}}$ **44.** $\dfrac{y^{5/4}}{y^{3/4}}$ **45.** $\dfrac{x^{1/2} y^2}{x^{-1/2} y}$ **46.** $\dfrac{x^{-1/2} y}{x^2 y^{3/2}}$

47. $\left(\dfrac{x}{y}\right)^{1/3} \left(\dfrac{y}{x}\right)^{2/3}$ **48.** $\left(\dfrac{x}{y}\right)^{-1/3} \left(\dfrac{y}{x}\right)^{1/3}$

Solve each equation in Exercises 49–56 for x, rounding your answer to four significant digits where necessary:

49. $x^2 - 16 = 0$ **50.** $x^2 - 1 = 0$ **51.** $x^2 - \frac{4}{9} = 0$

52. $x^2 - \frac{1}{10} = 0$ **53.** $x^2 - (1 + 2x)^2 = 0$ **54.** $x^2 - (2 - 3x)^2 = 0$

55. $x^5 + 32 = 0$ **56.** $x^4 - 81 = 0$

▶ ═══ **A.4** THE DISTRIBUTIVE LAW: MULTIPLYING ALGEBRAIC EXPRESSIONS

DISTRIBUTIVE LAW

The **distributive law** for real numbers states that

$$a(b \pm c) = ab \pm ac;$$
$$(a \pm b)c = ac \pm bc.$$

for any real numbers a, b, and c.

▼ **EXAMPLE 1**

(a) $2(x - \frac{1}{4})$ is *not* equal to $2x - \frac{1}{4}$, but is equal to $2x - 2(\frac{1}{4}) = 2x - \frac{1}{2}$.

(b) $x(x + 1) = x^2 + x$

(c) $2x(3x - 4) = 6x^2 - 8x$

(d) $(x - 4)x^2 = x^3 - 4x^2$

We can also use the distributive law to rewrite more complicated expressions.

▼ **EXAMPLE 2**

(a) $(x + 2)(x + 3) = (x + 2)x + (x + 2)3 = (x^2 + 2x) + (3x + 6)$
$$= x^2 + 5x + 6$$

(b) $(x + 2)(x - 3) = (x + 2)x - (x + 2)3 = (x^2 + 2x) - (3x + 6)$
$$= x^2 - x - 6$$

There is quicker way of expanding expressions such as this, called the "FOIL" method (First, Outer, Inner, Last). Consider, for instance, the expression $(x + 1)(x - 2)$. The FOIL method says: take the product of the *first* terms: $x \cdot x = x^2$, the product of the *outer* terms: $x \cdot (-2) = -2x$, the product of the *inner* terms: $1 \cdot x = x$, the product of the *last* terms: $1 \cdot (-2) = -2$, and then add them all up, getting $x^2 - 2x + x - 2 = x^2 - x - 2$.

▼ **EXAMPLE 3**

(a) $(x - 2)(2x + 5) = 2x^2 + 5x - 4x - 10 = 2x^2 + x - 10$

First   Outer   Inner   Last

(b) $(x^2 + 1)(x - 4) = x^3 - 4x^2 + x - 4$

(c) $(a - b)(a + b) = a^2 + ab - ab - b^2 = a^2 - b^2$

(d) $(a + b)^2 = (a + b)(a + b) = a^2 + ab + ab + b^2$
$$= a^2 + 2ab + b^2$$

(e) $(a - b)^2 = (a - b)(a - b) = a^2 - ab - ab + b^2$
$$= a^2 - 2ab + b^2$$

The last three are particularly important and are worth memorizing.

SPECIAL FORMULAS

$(a - b)(a + b) = a^2 - b^2$ **Difference of two squares**

$(a + b)^2 = a^2 + 2ab + b^2$ **Square of a sum**

$(a - b)^2 = a^2 - 2ab + b^2$ **Square of a difference**

▼ | **EXAMPLE 4**

(a) $(2 - x)(2 + x) = 4 - x^2$

(b) $(1 + a)(1 - a) = 1 - a^2$

(c) $(x + 3)^2 = x^2 + 6x + 9$

(d) $(4 - x)^2 = 16 - 8x + x^2$

Here are some longer examples.

▼ | **EXAMPLE 5**

(a) $(x + 1)(x^2 + 3x - 4) = (x + 1)x^2 + (x + 1)3x - (x + 1)4$

$$= (x^3 + x^2) + (3x^2 + 3x) - (4x + 4)$$

$$= x^3 + 4x^2 - x - 4$$

(b) $\left(x^2 - \dfrac{1}{x} + 1\right)(2x + 5) = \left(x^2 - \dfrac{1}{x} + 1\right)2x + \left(x^2 - \dfrac{1}{x} + 1\right)5$

$$= (2x^3 - 2 + 2x) + \left(5x^2 - \dfrac{5}{x} + 5\right)$$

$$= 2x^3 + 5x^2 + 2x + 3 - \dfrac{5}{x}$$

(c) $(x - y)(x - y)(x - y) = (x^2 - 2xy + y^2)(x - y)$

$$= (x^2 - 2xy + y^2)x - (x^2 - 2xy + y^2)y$$

$$= (x^3 - 2x^2y + xy^2) - (x^2y - 2xy^2 + y^3)$$

$$= x^3 - 3x^2y + 3xy^2 - y^3$$

▶ **A.4 EXERCISES**

Expand each expression in Exercises 1–22.

1. $x(4x + 6)$

2. $(4y - 2)y$

3. $(2x - y)y$

4. $x(3x + y)$

5. $(x + 1)(x - 3)$

6. $(y + 3)(y + 4)$

7. $(2y + 3)(y + 5)$

8. $(2x - 2)(3x - 4)$

9. $(2x - 3)^2$

10. $(3x + 1)^2$

11. $\left(x + \dfrac{1}{x}\right)^2$

12. $\left(y - \dfrac{1}{y}\right)^2$

13. $(2x - 3)(2x + 3)$

14. $(4 + 2x)(4 - 2x)$

15. $\left(y - \dfrac{1}{y}\right)\left(y + \dfrac{1}{y}\right)$

16. $(x - x^2)(x + x^2)$

17. $(x^2 + x - 1)(2x + 4)$

18. $(3x + 1)(2x^2 - x + 1)$

19. $(x^2 - 2x + 1)^2$

20. $(x + y - xy)^2$

21. $(y^3 + 2y^2 + y)(y^2 + 2y - 1)$

22. $(x^3 - 2x^2 + 4)(3x^2 - x + 2)$

▶ ▬▬▬ A.5 FACTORING ALGEBRAIC EXPRESSIONS

We can think of factoring as applying the distributive law in reverse. For example,

$$2x^2 + x = x(2x + 1),$$

which can be checked by using the distributive law. The first technique of factoring is to locate a **common factor**— that is, a term that occurs as a factor in each of the expressions being added or subtracted. For example, x is a common factor in $2x^2 + x$, since it is a factor of both $2x^2$ and x. On the other hand, x^2 is not a common factor, since it is not a factor of the second term, x.

Once we have located a common factor, we can "factor it out" by applying the distributive law.

▼ **EXAMPLE 1**

(a) $2x^3 - x^2 + x$ has x as a common factor, so

$$2x^3 - x^2 + x = x(2x^2 - x + 1).$$

(b) $2x^2 + 4x$ has $2x$ as a common factor, so

$$2x^2 + 4x = 2x(x + 2).$$

(c) $2x^2y + xy^2 - x^2y^2$ has xy as a common factor, so

$$2x^2y + xy^2 - x^2y^2 = xy(2x + y - xy).$$

(d) $(x^2 + 1)(x + 2) - (x^2 + 1)(x + 3)$ has $x^2 + 1$ as a common factor, so

$$\begin{aligned}
(x^2 + 1)(x + 2) - (x^2 + 1)(x + 3) &= (x^2 + 1)[(x + 2) - (x + 3)] \\
&= (x^2 + 1)(x + 2 - x - 3) \\
&= (x^2 + 1)(-1) \\
&= -(x^2 + 1) = -x^2 - 1.
\end{aligned}$$

(e) $12x(x^2 - 1)^5(x^3 + 1)^6 + 18x^2(x^2 - 1)^6(x^3 + 1)^5$ has $6x(x^2 - 1)^5(x^3 + 1)^5$ as a common factor, so

$$\begin{aligned}
12x(x^2 - 1)^5(x^3 + 1)^6 &+ 18x^2(x^2 - 1)^6(x^3 + 1)^5 \\
&= 6x(x^2 - 1)^5(x^3 + 1)^5[2(x^3 + 1) + 3x(x^2 - 1)] \\
&= 6x(x^2 - 1)^5(x^3 + 1)^5(2x^3 + 2 + 3x^3 - 3x) \\
&= 6x(x^2 - 1)^5(x^3 + 1)^5(5x^3 - 3x + 2).
\end{aligned}$$

We would also like to be able to reverse calculations such as $(x + 2)(2x - 5) = 2x^2 - x - 10$. That is, starting with the expression

$2x^2 - x - 10$, we would like to **factor** it and get back the original expression $(x + 2)(2x - 5)$. An expression of the form $ax^2 + bx + c$, where a, b, and c are real numbers, is called a **quadratic** expression in x. Thus, given a quadratic expression $ax^2 + bx + c$, we would like to write it in the form $(dx + e)(fx + g)$ for some real numbers d, e, f, and g. There are some quadratics, such as $x^2 + x + 1$, that cannot be factored in this form at all. Here, we shall consider only quadratics that do factor, and in such a way that the numbers d, e, f, and g are integers (whole numbers). (Other cases are fully discussed in Section 8.) The usual technique of factoring such quadratics is a "trial-and-error" approach, which we illustrate by means of examples.

▼   **EXAMPLE 2**

Factor $x^2 - 6x + 5$.

SOLUTION   Concentrate on the first and last terms:

$$x^2 \text{ has factors } x \text{ and } x \text{ (since } x \cdot x = x^2\text{);}$$
$$5 \text{ has factors } 5 \text{ and } 1.$$

Group them together and make an attempt.

$$(x + 5)(x + 1) = x^2 + 6x + 5$$

This is fine, except for the sign of the middle term. But notice that we can also get the 5 by multiplying (-5) and (-1). In other words, 5 also has factors (-5) and (-1). Using these instead gives

$$(x - 5)(x - 1) = x^2 - 6x + 5,$$

so we have found the correct factorization.

▼   **EXAMPLE 3**

Factor $x^2 - 4x - 12$.

SOLUTION   The first term, x^2, has factors x and x.
The last term, -12, factors in many ways—for example, as $(-3) \cdot 4$, $12 \cdot (-1)$ and $(-6) \cdot 2$, among others. Trying the first gives

$$(x - 3)(x + 4) = x^2 + x - 12. \qquad \text{No good.}$$

The second gives

$$(x + 12)(x - 1) = x^2 + 11x - 12. \qquad \text{No good.}$$

The third gives

$$(x - 6)(x + 2) = x^2 - 4x - 12. \qquad \text{Correct!}$$

Notice that in our third trial, we had two factors of the last term whose sum, -4, was the coefficient of x in the middle term. This is what we look for when factoring a quadratic with an initial term of x^2: *Find the numbers whose product is the constant term and whose sum is the coefficient of x.* In the last example, $(-6) \cdot 2 = -12$ and $-6 + 2 = -4$. When the coefficient of x^2 is *not* 1, we have to work a little harder.

▼ **EXAMPLE 4**

Factor the quadratic $4x^2 - 5x - 6$.

SOLUTION   Possible factorizations of $4x^2$ are $2x \cdot 2x$ or $4x \cdot x$. Possible factorizations of -6 are $(-3) \cdot 2$, $3 \cdot (-2)$, $(-6) \cdot 1$, or $6 \cdot (-1)$.

We now systematically try out all the possibilities until we come up with the correct one. For instance, the choices $2x \cdot 2x$ and $(-3) \cdot 2$ give

$$(2x - 3)(2x + 2) = 4x^2 - 2x - 6. \qquad \text{No good.}$$

The choices $4x \cdot x$ and $(-3) \cdot 2$ give

$$(4x - 3)(x + 2) = 4x^2 + 5x - 6.$$

This is *almost* correct, except for the sign of the middle term. We can fix this by switching signs.

$$(4x + 3)(x - 2) = 4x^2 - 5x - 6$$

▼ **EXAMPLE 5**

Factor $4x^2 - 25$.

SOLUTION   We recognize this as the difference of two squares.

$$4x^2 - 25 = (2x)^2 - 5^2$$
$$= (2x - 5)(2x + 5)$$

Not all quadratic expressions factor in this way. In Section 8 we look at a "test" that will tell us whether or not a given quadratic factors.

▼ **EXAMPLE 6**

Factor $x^4 - 5x^2 + 6$.

SOLUTION   This is not a quadratic, you say? Correct, it's a quartic (a fourth-degree expression). However, it looks rather like a quadratic. In fact, it is quadratic in x^2, meaning that it is

$$(x^2)^2 - 5(x^2) + 6 = y^2 - 5y + 6,$$

where $y = x^2$. The quadratic $y^2 - 5y + 6$ factors as

$$y^2 - 5y + 6 = (y - 3)(y - 2),$$

so

$$x^4 - 5x^2 + 6 = (x^2 - 3)(x^2 - 2).$$

This is sometimes a useful technique.

Our last example is here to remind you why we should want to factor polynomials in the first place. We shall return to this in Section A.7.

▼ **EXAMPLE 7**

Solve the equation $3x^2 + 4x - 4 = 0$.

SOLUTION   We first factor the left-hand side, getting

$$(3x - 2)(x + 2) = 0.$$

Thus, the product of the two quantities $(3x - 2)$ and $(x + 2)$ is zero. Now, if a product of two numbers is zero, it means that one or the other must be zero. In other words,

Either $3x - 2 = 0$, giving $x = \frac{2}{3}$,

or $x + 2 = 0$, giving $x = -2$.

Thus, there are two solutions: $x = \frac{2}{3}$, and $x = -2$.

▶ A.5 EXERCISES

In Exercises 1–18, factor the given expressions.

1. $2x + 3x^2$

2. $y^2 - 4y$

3. $6x^3 - 2x^2$

4. $3y^3 - 9y^2$

5. $x^2 - 8x + 7$

6. $y^2 + 6y + 8$

7. $x^2 + x - 12$

8. $y^2 + y - 6$

9. $2x^2 - 3x - 2$

10. $3y^2 - 8y - 3$

11. $6x^2 + 13x + 6$

12. $6y^2 + 17y + 12$

13. $12x^2 + x - 6$

14. $20y^2 + 7y - 3$

15. $x^2 + 4xy + 4y^2$

16. $4y^2 - 4xy + x^2$

17. $x^4 - 5x^2 + 4$

18. $y^4 + 2y^2 - 3$

19–36. *Set each of the expressions in Exercises 1–18 equal to zero and solve for the unknown (which is x in the odd-numbered exercises and y in the even-numbered exercises).*

In Exercises 37–44, factor each expression and simplify as much as possible.

37. $(x + 1)(x + 2) + (x + 1)(x + 3)$

38. $(x + 1)(x + 2)^2 + (x + 1)^2(x + 2)$

39. $(x^2 + 1)^5(x + 3)^4 + (x^2 + 1)^6(x + 3)^3$

40. $10x(x^2 + 1)^4(x^3 + 1)^5 + 15x^2(x^2 + 1)^5(x^3 + 1)^4$

41. $(x^3 + 1)\sqrt{x + 1} - (x^3 + 1)^2\sqrt{x + 1}$

42. $(x^2 + 1)\sqrt{x + 1} - \sqrt{(x + 1)^3}$

43. $\sqrt{(x + 1)^3} + \sqrt{(x + 1)^5}$

44. $(x^2 + 1)\sqrt[3]{(x + 1)^4} - \sqrt[3]{(x + 1)^7}$

▶ ══════ **A.6** RATIONAL EXPRESSIONS

A **rational expression** is an algebraic expression of the form P/Q, where P and Q are simpler expressions (usually polynomials), and the denominator Q is not zero.

▼ **EXAMPLE 1**

The following are rational expressions:

$$\frac{x^2 - 3x}{x}, \quad \frac{x + \dfrac{1}{x} + 1}{2x^2 - 1}, \quad \frac{2xy - y^2}{xy^2 - y}.$$

We can manipulate rational expressions in the same way that we manipulate fractions. Here are the basic rules.

ALGEBRA OF RATIONAL EXPRESSIONS

$$\frac{P}{Q} \cdot \frac{R}{S} = \frac{PR}{QS} \qquad \text{**Multiplication**}$$

$$\frac{P}{Q} + \frac{R}{S} = \frac{PS + RQ}{QS} \qquad \text{**Addition**}$$

$$\frac{P}{Q} - \frac{R}{S} = \frac{PS - RQ}{QS} \qquad \text{**Subtraction**}$$

$$\frac{1}{\left(\dfrac{P}{Q}\right)} = \frac{Q}{P} \qquad \text{**Reciprocals**}$$

$$\frac{\left(\dfrac{P}{Q}\right)}{\left(\dfrac{R}{S}\right)} = \frac{P}{Q} \cdot \frac{S}{R} = \frac{PS}{QR} \qquad \text{**Division**}$$

$$\frac{PR}{QR} = \frac{P}{Q} \qquad \text{**Cancellation**}$$

▼ **EXAMPLE 2**

(a) $\dfrac{x+1}{x} \cdot \dfrac{x-1}{2x+1} = \dfrac{(x+1)(x-1)}{x(2x+1)} = \dfrac{x^2-1}{2x^2+x}$

(b) $\dfrac{2x-1}{3x+2} + \dfrac{1}{x} = \dfrac{(2x-1)x + 1(3x+2)}{x(3x+2)} = \dfrac{2x^2+2x+2}{3x^2+2x}$

(c) $\dfrac{x}{3x+2} - \dfrac{x-4}{x} = \dfrac{x^2 - (x-4)(3x+2)}{x(3x+2)}$

$= \dfrac{x^2 - (3x^2 - 10x - 8)}{3x^2 + 2x}$

$= \dfrac{-2x^2 + 10x + 8}{3x^2 + 2x}$

(d) $\dfrac{1}{\left(\dfrac{2xy}{3x-1}\right)} = \dfrac{3x-1}{2xy}$

(e) $\dfrac{\left(\dfrac{x}{x-1}\right)}{\left(\dfrac{y-1}{y}\right)} = \dfrac{xy}{(x-1)(y-1)} = \dfrac{xy}{xy - x - y + 1}$

(f) $\dfrac{\left(\dfrac{1}{x+y} - \dfrac{1}{x}\right)}{y} = \dfrac{\left(\dfrac{x-(x+y)}{x(x+y)}\right)}{y} = \dfrac{\left(\dfrac{-y}{x(x+y)}\right)}{y}$

$= \dfrac{-y}{xy(x+y)} = -\dfrac{1}{x(x+y)}$

(g) $\dfrac{(x+1)(x+2)^2 - (x+1)^2(x+2)}{(x+2)^4}$

$= \dfrac{(x+1)(x+2)[(x+2) - (x+1)]}{(x+2)^4}$

$= \dfrac{(x+1)(x+2)(x+2-x-1)}{(x+2)^4} = \dfrac{(x+1)(x+2)}{(x+2)^4}$

$= \dfrac{x+1}{(x+2)^3}$

(h) $\dfrac{2x\sqrt{x+1} - x^2/\sqrt{x+1}}{x+1} = \dfrac{[2x(\sqrt{x+1})^2 - x^2]/\sqrt{x+1}}{x+1}$

$= \dfrac{2x(x+1) - x^2}{(x+1)\sqrt{x+1}}$

$= \dfrac{2x^2 + 2x - x^2}{(x+1)\sqrt{x+1}} = \dfrac{x^2 + 2x}{\sqrt{(x+1)^3}}$

$= \dfrac{x(x+2)}{\sqrt{(x+1)^3}}$

▶ A.6 EXERCISES

Rewrite each expression in Exercises 1–16 as a single rational expression, simplified as much as possible.

1. $\dfrac{x-4}{x+1} \cdot \dfrac{2x+1}{x-1}$

2. $\dfrac{2x-3}{x-2} \cdot \dfrac{x+3}{x+1}$

3. $\dfrac{x-4}{x+1} + \dfrac{2x+1}{x-1}$

4. $\dfrac{2x-3}{x-2} + \dfrac{x+3}{x+1}$

5. $\dfrac{x^2}{x+1} - \dfrac{x-1}{x+1}$

6. $\dfrac{x^2-1}{x-2} - \dfrac{1}{x-1}$

7. $\dfrac{1}{\left(\dfrac{x}{x-1}\right)} + x - 1$

8. $\dfrac{2}{\left(\dfrac{x-2}{x^2}\right)} - \dfrac{1}{x-2}$

9. $\dfrac{1}{x}\left(\dfrac{(x-3)}{xy} + \dfrac{1}{y}\right)$

10. $\dfrac{y^2}{x}\left(\dfrac{(2x-3)}{y} + \dfrac{x}{y}\right)$

11. $\dfrac{(x+1)^2(x+2)^3 - (x+1)^3(x+2)^2}{(x+2)^6}$

12. $\dfrac{6x(x^2+1)^2(x^3+2)^3 - 9x^2(x^2+1)^3(x^3+2)^2}{(x^3+2)^6}$

13. $\dfrac{\left((x^2-1)\sqrt{x^2+1} - \dfrac{x^4}{\sqrt{x^2+1}}\right)}{x^2+1}$

14. $\dfrac{\left(x\sqrt{x^3-1} - \dfrac{3x^4}{\sqrt{x^3-1}}\right)}{x^3-1}$

15. $\dfrac{\left(\dfrac{1}{(x+y)^2} - \dfrac{1}{x^2}\right)}{y}$

16. $\dfrac{\left(\dfrac{1}{(x+y)^3} - \dfrac{1}{x^3}\right)}{y}$

▶══ A.7 EQUATIONS

Here we shall pause to take an in-depth look at the concept of an *equation*. We have already solved various equations in the preceding sections, but have you ever stopped to think about just what is *meant* by an equation and the solution to an equation? After all, equations are what most people imagine mathematics is all about! Although we're sure you "know" what an equation is—you can recognize one when you see it—it is not a simple matter to say exactly what is meant by an equation. Here are some examples to illustrate this.

$$x + y = 2$$

is a very respectable equation, as are

$$x^2 + 3xy - \sqrt{z} = 11 \quad \text{and}$$
$$E = mc^2.$$

On the other hand,

$$x^2 + 1 = \sqrt{} \, ,$$
$$y + = 9, \quad \text{and}$$

The quick brown fox jumps over the lazy dog

are not equations.

An **equation** is a statement that two mathematical expressions are equal. In other words, it consists of two mathematical expressions separated by an equal sign.

Of course, this begs the question: What is a mathematical expression? Formally, a **mathematical expression** is a string of symbols including letters, numbers and some other characters $(+, -, \times, \div, \sqrt{},$ and so on) that obeys a lengthy list of "rules of syntax," such as the rule that says there should be expressions both before and after a plus sign. We shall not attempt to write down these rules—it would not be very illuminating. Remember this, though: The rules amount to saying that if you replaced all the letters in an expression with numbers, you would have a calculation that you can actually carry out.

Here is an important point to remember: The letters that occur in an equation signify numbers. Some stand for well-known numbers, such as π, c (the speed of light: 3×10^8m/s) or e (the base of natural logarithms: $2.71828 \ldots$). Some stand for **variables** or **unknowns.** Variables are quantities (such as length, height, or number of items) that can have many possible values, while unknowns are quantities whose values you may be asked to determine. The distinction between variables and unknowns is fuzzy, and mathematicians often use these terms interchangeably.

▼ **EXAMPLE 1**

$x + y = 7$ can be thought of as an equation in *two unknowns, x and y.* For example, x could stand for the number of days per week you attend math class and y for the number of days per week you don't attend math class. The equation $x + y = 7$ then amounts to the statement that there are a total of seven days in the week. If you knew the number x, you could find the remaining unknown, y.

We could also think of this as an equation in *two variables,* as the numbers x and y could vary depending on the week you're talking about.

Before we go on... It's interesting to notice that x and y do not vary randomly—again, if you know x then you know y. We can say that the value of y *depends on* the value of x. It's also common to say that y is a *function of x.*

Here is another important term: a **solution** to an equation in one or more unknowns is an assignment of numerical values to each of the unknowns, so that when these values are substituted for the unknowns, the equation becomes a *true statement about numbers.*

▼ **EXAMPLE 2**

The assignment $x = 2$ is a solution of $x^2 = 4$. To see this, substitute 2 for x. The given equation,

$$x^2 = 4$$

then becomes

$$2^2 = 4,$$

which is a true statement. By the same token, $x = -2$ is also a solution of $x^2 = 4$, since substituting -2 for x gives

$$(-2)^2 = 4,$$

which is another true statement.

▼ **EXAMPLE 3**

The assignments $x = 2$ and $y = \frac{1}{2}$ constitute a solution of $x^2 - 2y = 3$, since substitution yields the true statement $2^2 - 2(\frac{1}{2}) = 3$. Similarly, $x = \sqrt{3}$ and $y = 0$ is another solution. Try to see why without writing anything down.

Before we go on... There are in fact infinitely many possible solutions for this equation. You might try to come up with three or four more.

Now let's look at an important special case. An **equation in one unknown** has exactly one variable, and the symbol x is traditionally reserved for that purpose (like most traditions, it is not strictly followed). Here are a few equations in one unknown:

$$3x + 4 = 0$$
$$x^2 - 3x + 2 = 0$$
$$4x^4 + 11x^2 + 9 = 0$$
$$x^5 - 10x + 5 = 0$$
$$\sqrt{x - 2x^2} = 4^x.$$

On the other hand,

$$x^2 - 2y = 3$$

is an equation in two unknowns.

As pragmatic people, our first impulse upon seeing an equation in one unknown is to solve it.* We'll follow this impulse, and leave for later discussions the question of why we should want to do this.

▼ *Solving an equation means finding a solution for x.

There are two methods of solving an equation: analytical and numerical. To solve an equation **analytically** means to obtain exact solutions using algebraic rules. To solve it **numerically** means to use a computer or a graphing calculator to obtain solutions. In this appendix, we shall concentrate on the analytical approach.

First and foremost, we should point out that almost anything can happen when you try to solve an equation. Here are the possibilities, illustrated by examples.

SOLUTIONS OF EQUATIONS

1. Unique Solution
This means that the equation has one, and only one, solution.
Example: The equation $3x + 4 = 0$ has this property. Its only solution is $x = -\frac{4}{3}$.*
Sometimes the solution is not so easy to find, and often it cannot be found at all analytically. An example is $x^5 - 10x + 10 = 0$, whose unique real solution can only be found numerically.

2. Two or More Solutions
An equation can often have more than one solution.
Example: The equation $x^2 - 3x + 2 = 0$ has the two solutions $x = 1$ and $x = 2$. To see this, factor $x^2 - 3x + 2$ as $(x - 1)(x - 2)$. The equation then says that $(x - 1)(x - 2)$ must be zero. But if a product of two numbers is zero, at least one of those numbers is zero. In other words, either $(x - 1)$ or $(x - 2)$ is zero. We'll be discussing this method of solution in the next section. Just as in the case of a unique solution, multiple solutions may not be easy to find. An example is $x^5 - 10x + 5 = 0$, whose three real solutions can only be found numerically.

3. No Solutions
The equation $4x^4 + 11x^2 + 9 = 0$ has no real solutions whatsoever. Think for a while about why this should be the case before looking at the footnote.†

▼ * Try to verify this by solving the equation $3x + 4 = 0$ mentally. As you grow mathematically more sophisticated, you should try to solve more and more complicated equations mentally. Think of this as akin to technical exercises on a musical instrument.

† We first write the equation as $4x^4 + 11x^2 = -9$. If we substitute *any* value for x whatsoever, then the left-hand side, involving even powers of x, is nonnegative, whereas the right-hand side is manifestly negative, making the equation into a false statement. So there can't be a value for x that would make it true.

▶   ## A.7 EXERCISES

Mentally solve the equations in Exercises 1–12 for x. (That is, try to solve them by writing down as little as possible.)

1. $x + 1 = 0$

2. $x - 3 = 1$

3. $2x + 4 = 1$

4. $3x + 1 = x$

5. $x + 1 = 2x + 2$

6. $x + 1 = 3x + 1$

7. $\frac{x}{2} = x - \frac{1}{2}$

8. $\frac{1}{x} + 1 = 3$

9. $x + b = -x$

10. $ax = 1 \quad (a \neq 0)$

11. $ax + b = 0 \quad (a \neq 0)$

12. $ax + b = c \quad (a \neq 0)$

By factoring the left-hand side (or any other method) find all possible solutions for x in the equations in Exercises 13–26. Again, try to work mentally as far as possible. When you do decide to resort to pen and paper, be as meticulous as possible about the way you write mathematics, using as many steps as you like. This is the key to communicating your train of thought to the people assessing your work (they like that). Equally important, it helps you organize your thoughts more clearly. Also, see the footnotes for helpful hints in doing the last few.*

13. $x^2 - 1 = 0$

14. $x^2 - x = 0$

15. $3x^2 = 0$

16. $2x^2 = 2$

17. $2x^2 = 8$

18. $x^2 + 2x + 1 = 0$

19. $x^2 - 2x + 1 = 0$

20. $x^2 - 5x + 6 = 0$

21. $x^2 + 7x + 6 = 0$

22. $6x^2 + 7x + 1 = 0$

23. $x^2 + 1 = 0$

24. $x^3 + 1 = 0$

25. $x^4 + 1 = 0$

26. $x^4 - 1 = 0$

▶ ══════ ## A.8 SOLVING POLYNOMIAL EQUATIONS

We now restrict attention to special kinds of equations in one variable, called polynomial equations. (We already came across such equations in the previous section.)

Q Just what is a polynomial equation?

A Look at the equations in Exercises 13–26 in the previous section—they are all examples of polynomial equations. Notice that they have a similar form. For instance, the equation $6x^2 + 7x + 1 = 0$ has the form $ax^2 + bx + c = 0$, where $a = 6, b = 7$, and $c = 1$. Similarly, the equation $x^4 - 1 = 0$ has the form $ax^4 + bx^3 + cx^2 + dx + e = 0$, where $a = 1, b = 0, c = 0, d = 0$ and $e = -1$. Here is the general definition:

▼   * Recall that the difference of two squares is given by the formula:
$a^2 - b^2 = (a - b)(a + b)$.
You could also verify, by multiplying the whole thing out, that
$a^3 - b^3 = (a - b)(a^2 + ab + b^2)$.
Similarly,
$a^4 - b^4 = (a - b)(a^3 + a^2b + ab^2 + b^3)$.
Get the pattern yet?
One more:
$a^3 + b^3 = (a + b)(a^2 - ab + b^2)$.
This one generalizes only to odd powers.

> **POLYNOMIAL EQUATION**
>
> A **polynomial equation** is an equation that can be written in the form
>
> $$ax^n + bx^{n-1} + \ldots + rx + s = 0,$$
>
> where $a, b, \ldots, r,$ and s are constants.

We call the largest exponent of x appearing in a nonzero term of a polynomial the **degree** of that polynomial.

▼   **EXAMPLE 1**

(a) $x^2 - x - 1 = 0$ has degree 2, since the largest power of x that occurs is x^2. Degree 2 equations are also called **quadratic equations,** or just **quadratics.**

(b) $x^4 - x = 0$ has degree 4. It would be called a **quartic.**

(c) $3x + 1 = 0$ has degree 1, since the largest power of x that occurs is $x = x^1$. Degree 1 equations are called **linear equations.**

(d) $x^3 = 2x^2 + 1$ is a degree 3 polynomial (or **cubic**) in disguise. It can be rewritten as

$$x^3 - 2x^2 - 1 = 0,$$

which is in the standard form for a degree 3 equation.

Now comes the question: How do we solve these equations for x? This question was asked by mathematicians as early as 1600 BC. Let's look at these equations one degree at a time.

SOLUTION OF LINEAR EQUATIONS IN ONE UNKNOWN

By definition, a linear equation in one unknown can be written in the form

$$ax + b = 0 \quad \text{(with } a \text{ and } b \text{ being numbers and } a \neq 0\text{)}.$$

Solving this is a nice mental exercise: subtract b from both sides and then divide by a, getting $x = -b/a$. Don't bother memorizing this formula; just go ahead and solve linear equations as they arise. If you feel you need practice, see Exercises 1–12 in Section A.7.

SOLUTION OF QUADRATIC EQUATIONS IN ONE UNKNOWN

By definition, a quadratic equation in one unknown has the form

$$ax^2 + bx + c = 0 \quad \text{(with } a, b, \text{ and } c \text{ being numbers and } a \neq 0\text{).}^*$$

▼   * What happens if $a = 0$?

We're assuming that you saw these somewhere in high school but may be a little hazy as to the details of their solution. There are two ways of solving these equations—one works sometimes, and the other works every time.

SOLVING BY FACTORING

If we can factor a quadratic, we can solve the equation by setting each factor equal to 0.

This method is best illustrated by examples. (See the section on factoring for a review of how to factor quadratics.)

▼ **EXAMPLE 2**

Solve the quadratic equation

$$x^2 + 7x + 10 = 0.$$

SOLUTION   We factor the left-hand side, getting

$$(x + 5)(x + 2) = 0.$$

Since the product of the two quantities $(x + 5)$ and $(x + 2)$ is zero, it follows that one or the other of them must be zero. Thus,

$$\text{either } x + 5 = 0, \text{ in which case } x = -5,$$
$$\text{or } x + 2 = 0, \text{ in which case } x = -2.$$

This gives the two possible solutions: $x = -5$ and $x = -2$.

Before we go on... Instead of referring to $x = -5$ and $x = -2$ as the *solutions* of the equation $x^2 + 7x + 10 = 0$, we sometimes say that -5 and -2 are the **roots** of the quadratic $x^2 + 7x + 10$.

When solving equations it is very important to check your answers (both to catch mistakes and to screen out extraneous answers that creep in while solving certain more complicated equations). This is the easy part, though: just substitute your answers back into the equation to make sure they work. Here, we check that

$$(-5)^2 + 7(-5) + 10 = 25 - 35 + 10 = 0 \quad ✔$$

and

$$(-2)^2 + 7(-2) + 10 = 4 - 14 + 10 = 0. \quad ✔$$

▼ **EXAMPLE 3**

Solve the equation

$$2x^2 - 5x - 12 = 0.$$

SOLUTION   We again factor the left-hand side, getting

$$(2x + 3)(x - 4) = 0.$$

By the same reasoning as in the previous example, it follows that either $2x + 3 = 0$, giving $x = -\frac{3}{2}$, or $x - 4 = 0$, giving $x = 4$. Thus, the solutions are $x = -\frac{3}{2}$ and $x = 4$.

Before we go on... Again, we should check our answers.

$$2\left(-\frac{3}{2}\right)^2 - 5\left(-\frac{3}{2}\right) - 12 = \frac{9}{2} + \frac{15}{2} - 12 = 0 \quad ✔$$

and

$$2(4)^2 - 5(4) - 12 = 32 - 20 - 12 = 0. \quad ✔$$

Now, here's a nasty one.

▼ **EXAMPLE 4**

Solve the equation $x^2 + x + 1 = 0$.

SOLUTION   You can try to factor this one as long as you like, and we guarantee that you won't succeed.*

Here is a convenient test to see whether you should try to factor in the first place.

TEST FOR FACTORING

The quadratic $ax^2 + bx + c$, with a, b, and c being integers (whole numbers), factors into an expression of the form $(rx + s)(tx + u)$ with r, s, t, and u being integers precisely when the quantity $b^2 - 4ac$ is a perfect square (that is, it is the square of an integer). If this happens, we say that the quadratic **factors over the integers.**

◄

For example, the quadratic $2x^2 - 5x - 12$, from Example 3 above, has $a = 2, b = -5$, and $c = -12$, so $b^2 - 4ac = 121$. This is a perfect square, since $121 = 11^2$, and so we know that this quadratic has to factor over the integers. On the other hand, the quadratic $x^2 + 7x + 2$ has $a = 1, b = 7$,

---

▼ * That is, unless you're a hardened mathematician in the making, and decide to experiment with complex numbers (numbers that include the real numbers but also a square root of -1; see a footnote later in this section).

and $c = 2$, so $b^2 - 4ac = 41$, which is not a perfect square, showing us that this quadratic won't factor over the integers.

A more powerful method of solving quadratic equations, whether or not they factor over the integers, is to use the **quadratic formula.**

QUADRATIC FORMULA

The solutions of the general quadratic $ax^2 + bx + c = 0$ ($a \neq 0$) are given by

$$x = \frac{-b \pm \sqrt{b^2 - 4ac}}{2a}.$$

▼   **EXAMPLE 5**

Solve the equation $2x^2 - 5x - 12 = 0$ by using the quadratic formula. (This is the same equation we encountered in Example 3.)

SOLUTION   Here, we have $a = 2$, $b = -5$, and $c = -12$. According to the formula,

$$\begin{aligned}
x &= \frac{-b \pm \sqrt{b^2 - 4ac}}{2a} \\
&= \frac{5 \pm \sqrt{25 + 96}}{4} \\
&= \frac{5 \pm \sqrt{121}}{4} \\
&= \frac{5 \pm 11}{4} \\
&= \frac{16}{4} \text{ or } -\frac{6}{4} \\
&= 4 \text{ or } -\frac{3}{2},
\end{aligned}$$

as we saw before. (You can decide for yourself whether it was easier to factor or to use the quadratic formula.)

Before we continue with more examples, note that we got two different solutions from the formula in this example. A little thought will convince you that this happened because the quantity under the radical, $b^2 - 4ac$, was strictly positive, so that adding and subtracting its square root gave different answers. We call the quantity $\Delta = b^2 - 4ac$ the **discriminant** of the quadratic (Δ is the Greek letter delta) and we have the following general principle:

TEST FOR EXISTENCE OF REAL SOLUTIONS

1. If the discriminant $\Delta = b^2 - 4ac$ of a quadratic equation is positive, then there are two distinct real solutions.
2. If Δ is zero, then there is only one real solution.
3. If Δ is negative, then there are no real solutions.

▼ **EXAMPLE 6**

Solve $4x^2 = 12x - 9$.

SOLUTION   Before starting, we notice that the equation is not given in the standard form $ax^2 + bx + c = 0$. In order to obtain the values of a, b, and c, we first rewrite the equation in the standard form: $4x^2 - 12x + 9 = 0$. This gives $a = 4$, $b = -12$, and $c = 9$. Using the formula,

$$x = \frac{-b \pm \sqrt{b^2 - 4ac}}{2a}$$

$$= \frac{12 \pm \sqrt{144 - 144}}{8}$$

$$= \frac{12 \pm 0}{8} = \frac{12}{8} = \frac{3}{2}.$$

Note that in this example the discriminant was zero, so we had only one solution.

▼ **EXAMPLE 7**

Find the solution(s) of $x^2 + 2x - 1 = 0$.

SOLUTION   Using the formula yet again, we get

$$x = \frac{-b \pm \sqrt{b^2 - 4ac}}{2a}$$

$$= \frac{-2 \pm \sqrt{8}}{2}$$

$$= \frac{-2 \pm 2\sqrt{2}}{2}$$

$$= -1 \pm \sqrt{2}.$$

Thus, the two solutions are $x = -1 + \sqrt{2} = 0.414\ldots$ and $x = -1 - \sqrt{2} = -2.414\ldots$.

▼   **EXAMPLE 8**

Solve $x^2 + x + 1 = 0$.

SOLUTION   We again use the quadratic formula:

$$x = \frac{-b \pm \sqrt{b^2 - 4ac}}{2a}$$
$$= \frac{-1 \pm \sqrt{-3}}{2}.$$

This is an instance of a quadratic admitting *no* real solutions because its discriminant is negative.

Q   This is all very useful, but where on earth does the quadratic formula come from?

A   To see where it comes from, we solve another quadratic equation using a third method. Consider the equation $x^2 + 6x + 4 = 0$. You might recognize that $x^2 + 6x$ is the beginning of the perfect square $x^2 + 6x + 9 = (x + 3)^2$. Let us **complete the square** by rewriting our equation as

$$(x^2 + 6x + 9) - 9 + 4 = 0,$$

so that we can write it as

$$(x + 3)^2 - 5 = 0.$$

Then we can write

$$(x + 3)^2 = 5$$

and take the square root:

$$x + 3 = \pm\sqrt{5}.$$

Finally, we get

$$x = -3 \pm \sqrt{5} = -0.764 \text{ or } -5.236.$$

We get the quadratic formula by doing this again with the general equation. Here goes:

$$ax^2 + bx + c = 0.$$

First, divide out the nonzero number a, getting

$$x^2 + \frac{bx}{a} + \frac{c}{a} = 0.$$

Now add and subtract the quantity $\dfrac{b^2}{4a^2}$ to get

$$x^2 + \frac{bx}{a} + \frac{b^2}{4a^2} - \frac{b^2}{4a^2} + \frac{c}{a} = 0.$$

The reason we did this is to get the first three terms to factor as a perfect square:

$$\left(x + \frac{b}{2a}\right)^2 - \frac{b^2}{4a^2} + \frac{c}{a} = 0.$$

(Check this by multiplying out.) Adding $\dfrac{b^2}{4a^2} - \dfrac{c}{a}$ to both sides gives

$$\left(x + \frac{b}{2a}\right)^2 = \frac{b^2}{4a^2} - \frac{c}{a} = \frac{b^2 - 4ac}{4a^2}.$$

Taking square roots,

$$x + \frac{b}{2a} = \frac{\pm\sqrt{b^2 - 4ac}}{2a}.$$

Finally, adding $-\dfrac{b}{2a}$ to both sides yields the result,

$$x = -\frac{b}{2a} + \frac{\pm\sqrt{b^2 - 4ac}}{2a},$$

or

$$x = \frac{-b \pm \sqrt{b^2 - 4ac}}{2a}.$$

Done.

SOLUTION OF CUBIC EQUATIONS IN ONE UNKNOWN

By definition, a cubic equation in one unknown can be written in the form

$ax^3 + bx^2 + cx + d = 0$ (with a, b, c, and d being numbers and $a \neq 0$).

Now we get into something of a bind. While there is a perfectly respectable formula for the solutions, it is very complicated and involves the use of complex numbers rather heavily*. So we discuss instead a much simpler method that *sometimes* works nicely. Here is the method in a nutshell.

▼ * It was when this formula was discovered in the sixteenth century that complex numbers were first taken seriously. Although we would very much like to show you the formula, it is so involved that it would not fit well in this footnote.

SOLVING CUBICS BY FINDING ONE FACTOR

Step 1: By trial-and-error, find one solution $x = s$. If a, b, c, and d are integers, the only possible *rational* solutions are those of the form $s = \pm p/q$, where p is a factor of d and q is a factor of a.*

Step 2: It will now be possible to factor the cubic as

$$ax^3 + bx^2 + cx + d = (x - s)(ax^2 + ex + f) = 0.$$

To find $ax^2 + ex + f$, divide the cubic by $x - s$ using either long division or "synthetic division," a shortcut that would take us too far afield to describe.

Step 3: The factored equation says that either $x - s = 0$ or $ax^2 + ex + f = 0$. We already know that s is a solution, and now we see that the other solutions are the roots of the quadratic. Note that this quadratic may or may not have any real solutions, as usual.

Now let's do an example or two to illustrate this method.

▼ **EXAMPLE 9**

Solve the cubic $x^3 - x^2 + x - 1 = 0$.

SOLUTION   First, we find a single solution. Here, $a = 1$ and $d = -1$. Since the only factors of ± 1 are ± 1, the only possible rational solutions are $x = \pm 1$. By substitution, we see that $x = 1$ is a solution, since we get $1 - 1 + 1 - 1 = 0$. Thus, $(x - 1)$ is a factor. Dividing by $(x - 1)$ yields the quotient $(x^2 + 1)$. Here is the long division in all its gory detail:

$$
\begin{array}{r}
x^2 + 1 \\
x - 1 \overline{)\, x^3 - x^2 + x - 1} \\
\underline{x^3 - x^2 } \\
0 + x - 1 \\
\underline{x - 1} \\
0
\end{array}
$$

▼ *There may be *irrational* solutions, however; for example, $x^3 - 2 = 0$ has the single solution $x = \sqrt[3]{2}$.

Since the remainder is zero, we see that $(x - 1)$ is a factor of $x^3 - x^2 + x - 1$, as we claimed. The quotient $(x^2 + 1)$ appears on the top. Thus,

$$x^3 - x^2 + x - 1 = (x - 1)(x^2 + 1) = 0,$$

so that either $x - 1 = 0$ or $x^2 + 1 = 0$.

Since the discriminant of the quadratic $x^2 + 1$ is negative, we don't get any real solutions from $x^2 + 1 = 0$, so the only real solution is $x = 1$.

▼ **EXAMPLE 10**

Solve the cubic $2x^3 - 3x^2 - 17x + 30 = 0$.

SOLUTION   First we look for a single solution. Here, $a = 2$ and $d = 30$. The factors of a are ± 1 and ± 2, while the factors of d are ± 1, ± 2, ± 3, ± 5, ± 6, ± 15, and ± 30. This gives us a lare number of possible ratios: ± 1, ± 2, ± 3, ± 5, ± 6, ± 15, ± 30, $\pm \frac{1}{2}$, $\pm \frac{3}{2}$, $\pm \frac{5}{2}$, $\pm \frac{15}{2}$. Undaunted, we first try $x = 1$ and $x = -1$, getting nowhere. So we move on to $x = 2$, and we hit the jackpot, since substituting $x = 2$ gives $16 - 12 - 34 + 30 = 0$. Thus, $(x - 2)$ is a factor. Dividing yields the quotient $2x^2 + x - 15$. Here is the calculation.

$$
\begin{array}{r}
2x^2 + x - 15 \\
x - 2 \overline{)\,2x^3 - 3x^2 - 17x + 30} \\
\underline{2x^3 - 4x^2} \\
x^2 - 17x \\
\underline{x^2 - 2x} \\
-15x + 30 \\
\underline{-15x + 30} \\
0
\end{array}
$$

Thus,

$$2x^3 - 3x^2 - 17x + 30 = (x - 2)(2x^2 + x - 15) = 0.$$

Setting the factors equal to zero gives either $x - 2 = 0$ or $2x^2 + x - 15 = 0$.

We could solve the quadratic using the quadratic formula, but luckily, we notice that it factors as

$$2x^2 + x - 15 = (x + 3)(2x - 5).$$

Thus, the solutions are $x = 2$, $x = -3$, and $x = \frac{5}{2}$.

POSSIBLE OUTCOMES WHEN SOLVING A CUBIC EQUATION

If you consider all the cases, there are three possible outcomes when solving a cubic equation:

1. Three real solutions, as in Example 10.
2. One real solution, as in Example 9.
3. Two real solutions. Try, for example, $x^3 + x^2 - x - 1 = 0$, and see what happens.

◄—

SOLUTION OF HIGHER-ORDER POLYNOMIAL EQUATIONS IN ONE UNKNOWN

Logically speaking, our next step should be a discussion of quartics, then quintics, and so on forever. Well, we've got to stop somewhere, and cubics may be as good a place as any. On the other hand, since we've gotten so far, we ought to at least tell you what is known about higher-order polynomials.

Quartics

Just as in the case of cubics, there is a perfectly respectable formula to find the solutions of quartic equations.*

Quintics and Beyond

All good things must come to an end, we're afraid. It turns out that there is no "quintic formula." In other words, there is no single algebraic formula or collection of algebraic formulas that will give the solutions to all quintics. This question was settled by the Norwegian mathematician Niels Henrik Abel in 1824 after almost 300 years of controversy about this question. (In fact, several notable mathematicians had previously claimed to have devised formulas for solving the quintic, but these were all shot down by other mathematicians—this being one of the favorite pastimes of practitioners of our art.) The same negative answer applies to polynomial equations of degree 6 and higher. It's not that these equations don't have solutions, just that they can't be found using algebraic formulas.[†] However, there are certain special classes of polynomial equations that can be solved with algebraic methods. The way to identify such equations was discovered around 1829 by the French mathematician Évariste Galois.[‡]

▼ * See, for example, *First Course in the Theory of Equations* by L. E. Dickson (New York: J. Wiley & Sons) or *Modern Algebra* by B. L. van der Waerden (New York: Frederick Ungar Publishing Co.).

[†] What we mean by an "algebraic formula" is a formula in the coefficients using the operations of addition, multiplication, division, and the taking of radicals. Mathematicians call the use of such formulas in solving polynomial equations "solution by radicals." If you were a math major, you would eventually go on to study this under the heading of Galois Theory.

[‡] Both Abel and Galois died young. Abel died of tuberculosis at the age of 26, while Galois was killed in a duel at the age of 21.

▶ __A.8 EXERCISES__

Solve the equations in Exercises 1–6 for x (mentally, if possible).

1. $-x + 5 = 0$

2. $4x - 5 = 8$

3. $\frac{3}{4}x + 1 = 0$

4. $7x + 55 = 98$

5. $ax + b = c \quad (a \neq 0)$

6. $x - 1 = cx + d \quad (c \neq 1)$

By any method, determine all possible real solutions of each of the equations in Exercises 7–24. Check your answers by substitution.

7. $2x^2 + 7x - 4 = 0$

8. $x^2 + x + 1 = 0$

9. $x^2 - x + 1 = 0$

10. $2x^2 - 4x + 3 = 0$

11. $2x^2 - 5 = 0$

12. $3x^2 - 1 = 0$

13. $-x^2 - 2x - 1 = 0$

14. $2x^2 - x - 3 = 0$

15. $\frac{1}{2}x^2 - x - \frac{3}{2} = 0$

16. $-\frac{1}{2}x^2 - \frac{1}{2}x + 1 = 0$

17. $x^2 - x = 1$

18. $16x^2 = -24x - 9$

19. $x = 2 - \dfrac{1}{x}$

20. $x + 4 = \dfrac{1}{x - 2}$

21. $x^4 - 10x^2 + 9 = 0$

22. $x^4 - 2x^2 + 1 = 0$

23. $x^4 + x^2 - 1 = 0$

24. $x^3 + 2x^2 + x = 0$

Find all possible real solutions of each of the equations in Exercises 25–38.

25. $x^3 + 6x^2 + 11x + 6 = 0$

26. $x^3 - 6x^2 + 12x - 8 = 0$

27. $x^3 + 4x^2 + 4x + 3 = 0$

28. $y^3 + 64 = 0$

29. $x^3 - 1 = 0$

30. $x^3 - 27 = 0$

31. $y^3 + 3y^2 + 3y + 2 = 0$

32. $y^3 - 2y^2 - 2y - 3 = 0$

33. $x^3 - x^2 - 5x + 5 = 0$

34. $x^3 - x^2 - 3x + 3 = 0$

35. $2x^6 - x^4 - 2x^2 + 1 = 0$

36. $3x^6 - x^4 - 12x^2 + 4 = 0$

37. $(x^2 + 3x + 2)(x^2 - 5x + 6) = 0$

38. $(x^2 - 4x + 4)^2 (x^2 + 6x + 5)^3 = 0$

▶ ═══ **A.9** SOLVING MISCELLANEOUS EQUATIONS

There are equations that often arise in calculus that are not polynomial equations of low degree. Many of these complicated looking equations can be solved easily if you remember the following, which we used in the previous section.

SOLVING AN EQUATION OF THE FORM $P \cdot Q = 0$

If a product is equal to 0, then at least one of the factors must be 0. That is, if

$$P \cdot Q = 0,$$

then either $P = 0$ or $Q = 0$.

This is why we are interested in factoring in the first place. If you haven't already done so, you may want to go back and read the section on factoring. The equations we shall use as examples in this section are typical of many that turn up in calculus.

▼ **EXAMPLE 1**

Solve $x^5 - 4x^3 = 0$.

SOLUTION   We can factor out x^3 from the left-hand side to get

$$x^3(x^2 - 4) = 0.$$

Since this product equals 0, one of the factors must be 0. That means that either

$$x^3 = 0$$

or

$$x^2 - 4 = 0.$$

If $x^3 = 0$, then $x = 0$. On the other hand, if $x^2 - 4 = 0$, then

$$x^2 = 4,$$

and so $x = \pm 2$. This gives three solutions: $x = 0, \pm 2$.

Before we go on... We should always check our answers:

$$0^5 - 4 \cdot 0^3 = 0; \quad ✔$$
$$(-2)^5 - 4 \cdot (-2)^3 = -32 - 4(-8) = 0; \quad ✔$$
$$2^5 - 4 \cdot 2^3 = 32 - 4 \cdot 8 = 0. \quad ✔$$

▼ **EXAMPLE 2**

Solve $(x^2 - 1)(x + 2) + (x^2 - 1)(x + 4) = 0$.

SOLUTION   If we factor the left-hand side, we get

$$(x^2 - 1)[(x + 2) + (x + 4)] = (x^2 - 1)(2x + 6).$$

So we can rewrite the equation as

$$(x^2 - 1)(2x + 6) = 0.$$

Now, one of the factors must be 0. That means that either $x^2 - 1 = 0$ or $2x + 6 = 0$. If $x^2 - 1 = 0$, then $x = \pm 1$, and if $2x + 6 = 0$, then $x = -3$. So, there are three solutions: $x = -3, -1,$ or 1.

Before we go on... As usual, let us check our answers:

$$((-3)^2 - 1)(-3 + 2) + ((-3)^2 - 1)(-3 + 4)$$
$$= (9 - 1)(-1) + (9 - 1)(1) = -8 + 8 = 0 \quad ✔$$

$$((-1)^2 - 1)(-1 + 2) + ((-1)^2 - 1)(-1 + 4)$$
$$= (1 - 1)(1) + (1 - 1)(3) = 0 + 0 = 0; \; ✔$$
$$((1)^2 - 1)(1 + 2) + ((1)^2 - 1)(1 + 4)$$
$$= (1 - 1)(3) + (1 - 1)(5) = 0 + 0 = 0. \; ✔$$

▼ **EXAMPLE 3**

Solve $12x(x^2 - 1)^5(x^2 + 1)^6 + 12x(x^2 - 1)^6(x^2 + 1)^5 = 0$.

SOLUTION   Again, let us start by factoring the left-hand side:

$$12x(x^2 - 4)^5(x^2 + 2)^6 + 12x(x^2 - 4)^6(x^2 + 2)^5$$
$$= 12x(x^2 - 4)^5(x^2 + 2)^5[(x^2 + 2) + (x^2 - 4)]$$
$$= 12x(x^2 - 4)^5(x^2 + 2)^5(2x^2 - 2)$$
$$= 24x(x^2 - 4)^5(x^2 + 2)^5(x^2 - 1).$$

Setting this equal to 0, we get

$$24x(x^2 - 4)^5(x^2 + 2)^5(x^2 - 1) = 0,$$

which means that at least one of the factors of this product must be 0. Now it certainly cannot be the 24, but it could be the x: $x = 0$ is one solution. It could also be that

$$(x^2 - 4)^5 = 0,$$

or

$$x^2 - 4 = 0,$$

which has solutions $x = \pm 2$. Could it be that $(x^2 + 2)^5 = 0$? If so, then $x^2 + 2 = 0$, but this is impossible since $x^2 + 2 \geq 2$ no matter what x is. Finally, it could be that $x^2 - 1 = 0$, which has solutions $x = \pm 1$. This gives us five solutions to the original equation:

$$x = -2, -1, 0, 1, \text{ or } 2.$$

Before we go on... We should check our answers as usual, but we'll leave that to you.

▼ **EXAMPLE 4**

Solve $(x^2 - 1)(x^2 - 4) = 10$.

SOLUTION   Watch out! You may be tempted to say that $x^2 - 1 = 10$ or $x^2 - 4 = 10$, but this does not follow. If two numbers multiply to give you 10, what must they be? There are lots of possibilities: 2 and 5, 1 and 10, $-500,000$ and -0.00002 are just a few. The fact that the left-hand side is factored is nearly useless to us if we want to solve this equation. What we will have to do is multiply out, bring the 10 over to the left, and hope that we can factor what we get. Here goes:

$$x^4 - 5x^2 + 4 = 10$$
$$x^4 - 5x^2 - 6 = 0$$
$$(x^2 - 6)(x^2 + 1) = 0.$$

(Here we used a sometimes useful trick that we mentioned in Section 5: we treated x^2 like x and x^4 like x^2, so factoring $x^4 - 5x^2 - 6$ is essentially the same as factoring $x^2 - 5x - 6$.) *Now* we are allowed to say that one of the factors must be 0:

$$x^2 - 6 = 0 \text{ has solutions } x = \pm\sqrt{6} = \pm 2.449 \ldots$$

and

$$x^2 + 1 = 0 \text{ has no real solutions.}$$

Therefore we get exactly two solutions, $x = \pm\sqrt{6} = \pm 2.449 \ldots$.

Before we go on... Substituting into the original equation:

$$((\pm\sqrt{6})^2 - 1)((\pm\sqrt{6})^2 - 4) = (6 - 1)(6 - 4)$$
$$= 5 \cdot 2 = 10. \ ✔$$

In solving equations involving rational expressions, the following rule is very useful.

SOLVING AN EQUATION OF THE FORM $P/Q = 0$

$$\text{If } \frac{P}{Q} = 0, \text{ then } P = 0.$$

How else could a fraction equal 0? If that is not convincing, multiply both sides by Q (which cannot be 0 if the quotient is defined).

▼ **EXAMPLE 5**

Solve

$$\frac{(x + 1)(x + 2)^2 - (x + 1)^2(x + 2)}{(x + 2)^4} = 0.$$

SOLUTION   We can immediately set the numerator equal to 0:

$$(x + 1)(x + 2)^2 - (x + 1)^2(x + 2) = 0.$$

Factor:

$$(x + 1)(x + 2)[(x + 2) - (x + 1)] = 0$$

or

$$(x + 1)(x + 2)(1) = 0.$$

This gives us $x + 1 = 0$ or $x + 2 = 0$, so $x = -1$ or $x = -2$. But these are *not* both solutions to the original equation. Think about why for a moment before reading on.

The problem is that $x = -2$ does not make sense in the original equation: it makes the denominator 0. So it is not a solution, and $x = -1$ is the only solution.

Before we go on... Of course, we should check that $x = -1$ really is a solution:

$$\frac{(-1 + 1)(-1 + 2)^2 - (-1 + 1)^2(-1 + 2)}{(-1 + 2)^4} = 0 \quad ✔$$

One more comment: If we had simplified the left-hand side of the original equation before trying to solve it, we would have factored the top as we just did, but then we could have cancelled an $x + 2$ from the top and bottom, getting

$$\frac{(x + 1)}{(x + 2)^3} = 0.$$

Then, setting the top equal to 0 gives only $x + 1 = 0$, so $x = -1$ is the only answer we see. This involves about the same amount of work but has the advantage of not giving us that fake solution.

▼   **EXAMPLE 6**

Solve $1 - \dfrac{1}{x^2} = 0$.

SOLUTION   Write 1 as $\frac{1}{1}$, so that we now have a difference of two rational expressions,

$$\frac{1}{1} - \frac{1}{x^2} = 0.$$

To combine these we can put both over a common denominator of x^2, which gives

$$\frac{x^2 - 1}{x^2} = 0.$$

Now we can set the numerator, $x^2 - 1$, equal to zero. Thus,

$$x^2 - 1 = 0,$$

so

$$(x - 1)(x + 1) = 0,$$

giving $x = \pm 1$.

Before we go on... Check:

$$1 - 1/(\pm 1)^2 = 1 - 1 = 0. \quad ✔$$

This equation could also have been solved by writing

$$1 = \frac{1}{x^2}$$

and then multiplying both sides by x^2.

▼ **EXAMPLE 7**

Solve

$$\frac{2x - 1}{x} + \frac{3}{x - 2} = 0.$$

SOLUTION   We *could* first perform the subtraction on the left and then set the top equal to 0, but here is another approach. Subtracting the second expression from both sides gives

$$\frac{2x - 1}{x} = \frac{-3}{x - 2}.$$

Cross-multiplying (multiplying both sides by both denominators—that is, by $x(x - 2)$), now gives

$$(2x - 1)(x - 2) = -3x,$$

so

$$2x^2 - 5x + 2 = -3x.$$

Adding $3x$ to both sides gives the quadratic equation

$$2x^2 - 2x + 1 = 0.$$

The discriminant is $(-2)^2 - 4 \cdot 2 \cdot 1 = -4 < 0$, so we conclude that there is no real solution.

Before we go on... Notice that when we said $(2x - 1)(x - 2) = -3x$, we were *not* allowed to conclude that $2x - 1 = -3x$ or $x - 2 = -3x$.

▼ **EXAMPLE 8**

Solve

$$\frac{\left(2x\sqrt{x + 1} - \dfrac{x^2}{\sqrt{x + 1}}\right)}{x + 1} = 0.$$

SOLUTION   Setting the top equal to 0 gives

$$2x\sqrt{x+1} - \frac{x^2}{\sqrt{x+1}} = 0.$$

This still involves fractions. To get rid of the fractions, we could put everything over a common denominator ($\sqrt{x+1}$) and then set the top equal to 0, or we could multiply the whole equation by that common denominator in the first place to clear fractions. If we do the second, we get

$$2x(x+1) - x^2 = 0,$$

or

$$2x^2 + 2x - x^2 = 0,$$

or

$$x^2 + 2x = 0.$$

Factoring,

$$x(x+2) = 0,$$

so either $x = 0$ or $x + 2 = 0$, giving us $x = 0$ or $x = -2$. Again, one of these is not really a solution.

The problem is that $x = -2$ cannot be substituted into $\sqrt{x+1}$, since we would then have to take the square root of -1, and we are not allowing ourselves to do that. Therefore, $x = 0$ is the only solution.

Before we go on... As usual, we should check that $x = 0$ really works:

$$\frac{\left(2(0)\sqrt{0+1} - \frac{(0)^2}{\sqrt{0+1}}\right)}{0+1} = 0. \quad \checkmark$$

▶ **A.9 EXERCISES**

Solve the equations in Exercises 1–26.

1. $x^4 - 3x^3 = 0$

2. $x^6 - 9x^4 = 0$

3. $x^4 - 4x^2 = -4$

4. $x^4 - x^2 = 6$

5. $(x+1)(x+2) + (x+1)(x+3) = 0$

6. $(x+1)(x+2)^2 + (x+1)^2(x+2) = 0$

7. $(x^2+1)^5(x+3)^4 + (x^2+1)^6(x+3)^3 = 0$

8. $10x(x^2+1)^4(x^3+1)^5 - 10x^2(x^2+1)^5(x^3+1)^4 = 0$

9. $(x^3+1)\sqrt{x+1} - (x^3+1)^2\sqrt{x+1} = 0$

10. $(x^2+1)\sqrt{x+1} - \sqrt{(x+1)^3} = 0$

11. $\sqrt{(x+1)^3} + \sqrt{(x+1)^5} = 0$

12. $(x^2+1)\sqrt[3]{(x+1)^4} - \sqrt[3]{(x+1)^7} = 0$

Before we go on... Check:
$$1 - 1/(\pm 1)^2 = 1 - 1 = 0. \quad ✔$$

This equation could also have been solved by writing
$$1 = \frac{1}{x^2}$$

and then multiplying both sides by x^2.

▼   **EXAMPLE 7**

Solve
$$\frac{2x - 1}{x} + \frac{3}{x - 2} = 0.$$

SOLUTION   We *could* first perform the subtraction on the left and then set the top equal to 0, but here is another approach. Subtracting the second expression from both sides gives
$$\frac{2x - 1}{x} = \frac{-3}{x - 2}.$$

Cross-multiplying (multiplying both sides by both denominators—that is, by $x(x - 2)$), now gives
$$(2x - 1)(x - 2) = -3x,$$

so
$$2x^2 - 5x + 2 = -3x.$$

Adding $3x$ to both sides gives the quadratic equation
$$2x^2 - 2x + 1 = 0.$$

The discriminant is $(-2)^2 - 4 \cdot 2 \cdot 1 = -4 < 0$, so we conclude that there is no real solution.

Before we go on... Notice that when we said $(2x - 1)(x - 2) = -3x$, we were *not* allowed to conclude that $2x - 1 = -3x$ or $x - 2 = -3x$.

▼   **EXAMPLE 8**

Solve
$$\frac{\left(2x\sqrt{x + 1} - \dfrac{x^2}{\sqrt{x + 1}}\right)}{x + 1} = 0.$$

SOLUTION   Setting the top equal to 0 gives

$$2x\sqrt{x + 1} - \frac{x^2}{\sqrt{x + 1}} = 0.$$

This still involves fractions. To get rid of the fractions, we could put everything over a common denominator ($\sqrt{x + 1}$) and then set the top equal to 0, or we could multiply the whole equation by that common denominator in the first place to clear fractions. If we do the second, we get

$$2x(x + 1) - x^2 = 0,$$

or

$$2x^2 + 2x - x^2 = 0,$$

or

$$x^2 + 2x = 0.$$

Factoring,

$$x(x + 2) = 0,$$

so either $x = 0$ or $x + 2 = 0$, giving us $x = 0$ or $x = -2$. Again, one of these is not really a solution.

The problem is that $x = -2$ cannot be substituted into $\sqrt{x + 1}$, since we would then have to take the square root of -1, and we are not allowing ourselves to do that. Therefore, $x = 0$ is the only solution.

Before we go on... As usual, we should check that $x = 0$ really works:

$$\frac{\left(2(0)\sqrt{0 + 1} - \frac{(0)^2}{\sqrt{0 + 1}}\right)}{0 + 1} = 0. \quad ✔$$

► **A.9 EXERCISES**

Solve the equations in Exercises 1–26.

1. $x^4 - 3x^3 = 0$

2. $x^6 - 9x^4 = 0$

3. $x^4 - 4x^2 = -4$

4. $x^4 - x^2 = 6$

5. $(x + 1)(x + 2) + (x + 1)(x + 3) = 0$

6. $(x + 1)(x + 2)^2 + (x + 1)^2(x + 2) = 0$

7. $(x^2 + 1)^5(x + 3)^4 + (x^2 + 1)^6(x + 3)^3 = 0$

8. $10x(x^2 + 1)^4(x^3 + 1)^5 - 10x^2(x^2 + 1)^5(x^3 + 1)^4 = 0$

9. $(x^3 + 1)\sqrt{x + 1} - (x^3 + 1)^2\sqrt{x + 1} = 0$

10. $(x^2 + 1)\sqrt{x + 1} - \sqrt{(x + 1)^3} = 0$

11. $\sqrt{(x + 1)^3} + \sqrt{(x + 1)^5} = 0$

12. $(x^2 + 1)\sqrt[3]{(x + 1)^4} - \sqrt[3]{(x + 1)^7} = 0$

13. $(x + 1)^2(2x + 3) - (x + 1)(2x + 3)^2 = 0$

14. $(x^2 - 1)^2(x + 2)^3 - (x^2 - 1)^3(x + 2)^2 = 0$

15. $\dfrac{(x + 1)^2(x + 2)^3 - (x + 1)^3(x + 2)^2}{(x + 2)^6} = 0$

16. $\dfrac{6x(x^2 + 1)^2(x^2 + 2)^4 - 8x(x^2 + 1)^3(x^2 + 2)^3}{(x^2 + 2)^8} = 0$

17. $\dfrac{2(x^2 - 1)\sqrt{x^2 + 1} - \dfrac{x^4}{\sqrt{x^2 + 1}}}{x^2 + 1} = 0$

18. $\dfrac{4x\sqrt{x^3 - 1} - \dfrac{3x^4}{\sqrt{x^3 - 1}}}{x^3 - 1} = 0$

19. $x - \dfrac{1}{x} = 0$

20. $1 - \dfrac{4}{x^2} = 0$

21. $\dfrac{1}{x} - \dfrac{9}{x^3} = 0$

22. $\dfrac{1}{x} - \dfrac{1}{x + 1} = 0$

23. $\dfrac{x - 4}{x + 1} - \dfrac{x}{x - 1} = 0$

24. $\dfrac{2x - 3}{x - 1} - \dfrac{2x + 3}{x + 1} = 0$

25. $\dfrac{x + 4}{x + 1} + \dfrac{x + 4}{3x} = 0$

26. $\dfrac{2x - 3}{x} - \dfrac{2x - 3}{x + 1} = 0$

Using a Graphing Calculator

INTRODUCTION ▶ Graphing calculators can display graphs, perform matrix operations, perform computations on tables of statistical data, and be programmed for more specialized tasks. While earlier programmable calculators had some of these features, the main distinguishing feature of the graphing calculator is its graphics display, which is in effect a small computer monitor.

The variety of graphing calculators on the market is large and increasing rapidly. In this appendix, we shall use the Texas Instruments TI-82 an our example. If you have another model or brand you should be able to do everything we do here, but you may have to consult the manual for details on exactly which buttons to press when.

▶ B.1 GRAPHING FUNCTIONS

The graphing programs used by calculators work by plotting hundreds of points on the graph of a given function and then joining them, usually with straight line segments, creating the effect of a smooth curve.

Let us experiment by using the TI-82 to graph the two equations

$$x + 3y = 4,$$
$$y = -x^2 - 2x + 3.$$

Before we begin, we must first rewrite each equation in the form $y = $ *function of x*. The second is already in this form, and we solve the first for y to obtain

$$y = \frac{4 - x}{3}$$

To enter these functions of x, press $\boxed{Y=}$ to get the display "Y=" and enter the equations as they are written.

Keystrokes

$Y_1=(4-X)/3$ $\boxed{(}\ \boxed{4}\ \boxed{-}\ \boxed{X}\ \boxed{)}\ \boxed{\div}\ \boxed{3}\ \boxed{\text{ENTER}}$

$Y_2=-X^2-2X+3$ $\boxed{(\text{-})}\ \boxed{X}\ \boxed{X^2}\ \boxed{-}\ \boxed{2}\ \boxed{X}\ \boxed{+}\ \boxed{3}\ \boxed{\text{ENTER}}$

The parentheses are absolutely essential in the first equation. Also notice that this calculator distinguishes between "minus," $\boxed{-}$, and "negative," $\boxed{(\text{-})}$.

Before you go on to graph these functions, you should first set the **viewing window coordinates,** which determine the portion of the xy-plane you will view. The viewing window is determined by four numbers: Xmin, Xmax, Ymin, Ymax. Figure 1 shows the graphs of both equations with two viewing windows.

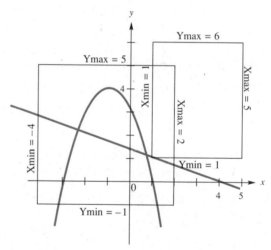

FIGURE 1

Notice that the viewing window specified by

Xmin=1, Xmax=5, Ymin=1, Ymax=6

contains no portion of either graph, so that if these were the settings you used, you would see a blank screen when you press $\boxed{\text{GRAPH}}$. Since this is not what we want, let us use the other window shown in Figure 1. To specify this window, press $\boxed{\text{WINDOW}}$ and enter the values shown.

`Xmin=-4`	ENTER	Remember to use negative, (-), rather than minus.
`Xmax=2`	ENTER	
`Xscl=1`	ENTER	It will place marks at 1-unit intervals on the x-axis.
`Ymin=-1`	ENTER	
`Ymax=5`	ENTER	
`Yscl=1`	ENTER	It will place marks at 1-unit intervals on the y-axis.

You are now ready to draw the graphs, so press [GRAPH]. Figure 2 shows what the display will look like.

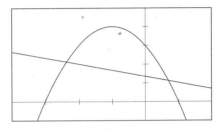

FIGURE 2

We shall now do two things with this graph.

FINDING THE COORDINATES OF THE VERTEX OF THE PARABOLA

(See the applications on maximizing revenue in Chapter 1.)

You can approximate the coordinates of the vertex by pressing [TRACE] and then using the left- and right-direction keys to move the cursor to a position as close to the vertex as possible. (See Figure 3, which shows the vertex at (1, 4).) The up- and down-arrow keys allow you to jump from one graph to the other.

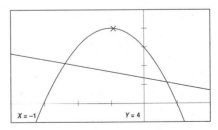

FIGURE 3

We were lucky to find the exact coordinates of the vertex in this example—usually we will not be able to do so. We should always "zoom in" for greater accuracy. To zoom in, use the trace feature to position the cursor at the point you want to examine more closely, press $\boxed{\text{ZOOM}}$, and select "2 : Zoom In" by pressing $\boxed{2}$ $\boxed{\text{ENTER}}$. The calculator then changes the viewing window to a small one centered at the cursor and redraws the graph as shown in the close-up view on the right in Figure 4.

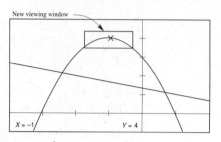
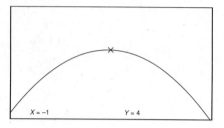

FIGURE 4

After zooming in, you should again press $\boxed{\text{TRACE}}$ to reposition the cursor on the curve and adjust its position for greater accuracy. You can zoom in again for even more accuracy.

LOCATING POINTS WHERE TWO GRAPHS CROSS

(See the material on locating break-even points in Chapter 1.)

To locate a point where two graphs cross, we can use the same technique we just used in locating the vertex: trace and zoom. We first use trace to position the curvor as close to an intersection point as possible, then use zoom followed by trace for more accuracy, repeating as many times as necessary.

SETTING THE *y*-COORDINATES OF THE VIEWING WINDOW AUTOMATICALLY

The following TI-82 program calculates the *y*-coordinates of the highest and lowest points on a graph and then sets the window coordinates so that the whole graph just fits in the window. (Read the section on programming in the manual before entering this program.)

Program to Set Window Coordinates (TI-82)

```
PROGRAM: WINDOW
:Input "ENTER XMIN",M
:Input "ENTER XMAX",N
:M→Xmin                      "Xmin" and "Xmax" are found
:N→Xmax                          under "Window" in VARS.
:(N−M)/95→D                  Increments of X (Graph window
                                 is 95 pixels wide.)

:M→X
:Y₁→L                        L is temporary Xmin.
```

```
Xmin=-4    [ENTER]
Xmax=2     [ENTER]
Xscl=1     [ENTER]
Ymin=-1    [ENTER]
Ymax=5     [ENTER]
Yscl=1     [ENTER]
```

```
:L→H                          H is temporary Xmax.
:For(X,M,N,D)
:If Y₁>H
:Then
:Y₁→H                         Increase H if a point on the graph is
                                  higher.

:End
:If Y₁<L
:Then
:Y₁→L                         Decrease L if a point of the graph is
                                  lower.

:End
:End                          End of loop
:L→Ymin
:H→Ymax
:Stop
```

Before running the program, the function you wish to graph should be entered as Y_1. The program will ask you for Xmin and Xmax, and it will then set the window coordinates automatically. If you press GRAPH after running the program, the calculator will draw the graph of Y_1 using the window determined by the program.

▶ ═══ **B.2** EVALUATING FUNCTIONS

The calculator can be used to evaluate functions in several different ways. Let us explore the function

$$f(x) = x^{1/3} + (x - 1)^{2/3}.$$

Although the TI-82 is quite happy taking the $\frac{1}{3}$ power of negative numbers, it does not evaluate the $\frac{2}{3}$ power of negative numbers. You can trick it into doing so by using the identity $(x - 1)^{2/3} = ((x - 1)^2)^{1/3}$:

$$Y_1 = X^{(1/3)} + ((X-1)^2)^{(1/3)}.$$

Figure 1 shows the graph of f using the viewing window defined by

$$\text{Xmin} = -3, \text{Xmax} = 2, \text{Ymin} = 0, \text{Ymax} = 2.$$

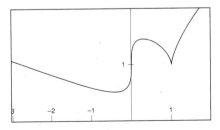

FIGURE 1

EVALUATING A FUNCTION ALGEBRAICALLY

Having entered the function in the "Y=" screen, you can repeatedly access it without having to retype the function. For instance, to evaluate $f(0.1532)$, enter (in the home screen)

$0.1532 \rightarrow$ X   [ENTER]   This sets $X = 0.1532$. The arrow is obtained by pressing [STO→]

Y_1 [ENTER]   This tells it to evaluate y as defined by Y_1 on the "Y="screen. To obtain it on the home screen, press

[2nd] [Y-VARS] [ENTER] [ENTER].

EVALUATING A FUNCTION GRAPHICALLY

Alternatively, you can use the graph to evaluate the function at a value of x and view the corresponding point on the graph at the same time. First graph the function using a suitable viewing window (as above). Although you could now use the trace feature to approximate, say, $f(0.1532)$ by searching for a point whose x-coordinate is as close to 0.1532 as possible, you can accomplish this more easily and accurately by using the "calc" feature, as follows.

With the graph displayed, press [2nd] [CALC], select "1:value" and press [ENTER]. You will then be asked for the value of X, so enter 0.1532 and press [ENTER]. The cursor will now be placed at the point on the graph where $x = 0.1532$ and the corresponding y-coordinate is $f(0.1532)$.

EVALUATING A FUNCTION USING A TABLE

Instead of using the graph, you can use a table to display the values of a function as follows. First, make sure that your function is entered in the "Y=" screen as above. Next, obtain the TABLE SETUP screen by pressing [2nd] [TblSet]. Under "Indpnt" select "Ask." (This gives you the option of entering values for the independent variable x.) Press [ENTER] and quit to return to the home screen. Now press [2nd] [TABLE], and you will obtain a table on which you can enter values of x, and immediately be shown the corresponding values for $f(x)$ in the Y_1 column.

Note that you can use this method to evaluate several functions at once by taking advantage of Y_2, Y_3, . . .

▶ ═══ B.3 CALCULATING DIFFERENCE QUOTIENTS

Recall that the **difference quotient** for the function f has the form

$$\frac{f(x + h) - f(x)}{h}.$$

In a typical situation, we might be given a value for x and be required to calculate the difference quotient for several values of h. As an example, we

```
:L→H
:For(X,M,N,D)
:If Y₁>H
:Then
:Y₁→H

:End
:If Y₁<L
:Then
:Y₁→L

:End
:End
:L→Ymin
:H→Ymax
:Stop
```

H is temporary Xmax.

Increase *H* if a point on the graph is
      higher.

Decrease *L* if a point of the graph is
      lower.

End of loop

Before running the program, the function you wish to graph should be entered as Y_1. The program will ask you for Xmin and Xmax, and it will then set the window coordinates automatically. If you press [GRAPH] after running the program, the calculator will draw the graph of Y_1 using the window determined by the program.

▶ ═══ B.2 EVALUATING FUNCTIONS

The calculator can be used to evaluate functions in several different ways. Let us explore the function

$$f(x) = x^{1/3} + (x - 1)^{2/3}.$$

Although the TI-82 is quite happy taking the $\frac{1}{3}$ power of negative numbers, it does not evaluate the $\frac{2}{3}$ power of negative numbers. You can trick it into doing so by using the identity $(x - 1)^{2/3} = ((x - 1)^2)^{1/3}$:

$$Y_1 = X^{\wedge}(1/3) + ((X-1)^2)^{\wedge}(1/3).$$

Figure 1 shows the graph of *f* using the viewing window defined by

$$\text{Xmin} = -3, \text{Xmax} = 2, \text{Ymin} = 0, \text{Ymax} = 2.$$

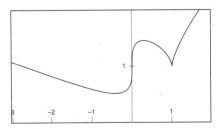

FIGURE 1

EVALUATING A FUNCTION ALGEBRAICALLY

Having entered the function in the "Y=" screen, you can repeatedly access it without having to retype the function. For instance, to evaluate $f(0.1532)$, enter (in the home screen)

0.1532 → X [ENTER]   This sets $X = 0.1532$. The arrow is obtained by pressing [STO→]

Y₁ [ENTER]   This tells it to evaluate y as defined by Y₁ on the "Y="screen. To obtain it on the home screen, press

[2nd] [Y-VARS] [ENTER] [ENTER].

EVALUATING A FUNCTION GRAPHICALLY

Alternatively, you can use the graph to evaluate the function at a value of x and view the corresponding point on the graph at the same time. First graph the function using a suitable viewing window (as above). Although you could now use the trace feature to approximate, say, $f(0.1532)$ by searching for a point whose x-coordinate is as close to 0.1532 as possible, you can accomplish this more easily and accurately by using the "calc" feature, as follows.

With the graph displayed, press [2nd] [CALC], select "1:value" and press [ENTER]. You will then be asked for the value of X, so enter 0.1532 and press [ENTER]. The cursor will now be placed at the point on the graph where $x = 0.1532$ and the corresponding y-coordinate is $f(0.1532)$.

EVALUATING A FUNCTION USING A TABLE

Instead of using the graph, you can use a table to display the values of a function as follows. First, make sure that your function is entered in the "Y=" screen as above. Next, obtain the TABLE SETUP screen by pressing [2nd] [TblSet]. Under "Indpnt" select "Ask." (This gives you the option of entering values for the independent variable x.) Press [ENTER] and quit to return to the home screen. Now press [2nd] [TABLE], and you will obtain a table on which you can enter values of x, and immediately be shown the corresponding values for $f(x)$ in the Y₁ column.

Note that you can use this method to evaluate several functions at once by taking advantage of Y₂, Y₃, . . .

▶ ══════ **B.3** CALCULATING DIFFERENCE QUOTIENTS

Recall that the **difference quotient** for the function f has the form

$$\frac{f(x + h) - f(x)}{h}.$$

In a typical situation, we might be given a value for x and be required to calculate the difference quotient for several values of h. As an example, we shall use the graphing calculator to find the difference quotient for $f(x) = \sqrt{x^2 - 1}$ at $x = 4$ and $h = \pm 0.1, \pm 0.01, \pm 0.001, \pm 0.0001$.

First, we evaluate the difference quotient for f at $x = 4$.

$$\frac{f(4 + h) - f(4)}{h} = \frac{\sqrt{(4 + h)^2 - 1} - \sqrt{4^2 - 1}}{h}$$

$$= \frac{\sqrt{(4 + h)^2 - 1} - \sqrt{15}}{h}$$

We wish to evaluate this function of h at the given values of h. To do this, we can employ the methods of the previous section: Go to the "Y=" screen and enter the above function with X playing the role of h as the independent variable.

$$\texttt{Y}_1\texttt{=(((4+X)\^{}2-1)\^{}0.5-15\^{}0.5)/X}$$

Evaluating for One Value of *h* at a Time

On the Home screen, enter

$\quad$ `0.1→X` $\boxed{\text{ENTER}}$

$\quad$ $\texttt{Y}_1$ $\qquad \boxed{\text{ENTER}}$

The value of the difference quotient will then be displayed as 1.031957242. Now repeat the procedure for all the other values of h listed above. If you wish to enter all the values of h on a table instead, proceed as follows.

Evaluating Using a Table

Follow the instructions in the last section on using a table to evaluate a function. If you then enter the values of h under the X column, the values of the difference quotient will appear on the Y_1 column.

X	Y_1
0.1	1.032
−0.1	1.0327
0.01	1.0327
−0.01	1.0328
0.001	1.0328
−0.001	1.0328
0.0001	1.0328

(The TI-82 permits only seven entries at a time in the table. To evaluate at other values, replace some of the values above with new ones.)

▶ ═══════ **B.4** PROGRAMS FOR LEFT- AND RIGHT-HAND SUMS, TRAPEZOID RULE, AND SIMPSON'S RULE

Below are listed two programs that you can use to calculate approximations to the definite integral. The first of these gives the left- and right-hand Riemann sums and trapezoidal sum of a given function, while the second gives the Simpson's rule approximation. To enter these programs, you should consult the section on programming in your user manual. (There is also a useful table of instructions and functions, together with the keystroke sequences needed to obtain them, at the back of the manual.)

Graphing Calculator Program to Compute Left- and Right-Hand Riemann Sums and Trapezoid Rule (TI-82)

```
PROGRAM: SUMS
:Input ("N? ",N)                        Prompts for the number of
                                           rectangles
:Input ("LEFT ENDPOINT? ",A)     Prompts for the left endpoint a
:Input ("RIGHT ENDPOINT? ",B)    Prompts for the right endpoint b
:(B-A)/N→D                              D is Δx = (b − a)/n.
:∅→R                                    R will eventually be the
                                           right-hand sum.
:∅→L                                    L will eventually be the left-hand
                                           sum.
:A→X                                    X is the current x-coordinate.
:For (I,1,N)                            Start of a loop—recall the sigma
                                           notation
:L+Y₁→L                                 Increment L by f(x_{i−1})
:A+I*D→X                                Corresponds to our formula
                                           x_i = a + iΔx
:R+Y₁→R                                 Increment R by f(x_i)
:End                                    End of loop
:L*D→L                                  Left sum
:R*D→R                                  Right sum
:(L+R)/2→T                              Trapezoidal sum
:Disp "LEFT SUM IS ",L
:Disp "RIGHT SUM IS ",R
:Disp "TRAP SUM IS ",T
:Stop
```

In the annotations above: D is $\Delta x = (b-a)/n$; increment L by $f(x_{i-1})$; $x_i = a + i\Delta x$; increment R by $f(x_i)$.

Graphing Calculator Program for Simpson's Rule (TI-82)

```
PROGRAM: SIMP
:Input ("EVEN NUMBER N≥2? ",N)   Make sure that N is even and
                                           at least 2.
:Input ("LEFT ENDPOINT? ",A)
:Input ("RIGHT ENDPOINT? ",B)
:(B−A)/N→D
:∅→S
```

```
:A→X
:For (I,1,N−1,2)
:S+Y₁→S
:A+I*D→X
:S+4*Y₁→S
:X+D→X
:S+Y₁→S
:End
:S*D/3→S
:Disp "SIMPSON SUM IS ",S
:Stop
```

The loop increments I in steps of 2.

The next four steps compute the sum

$f(x_{i-1}) + 4f(x_i) + f(x_{i+1})$.

Multiply by $\Delta x/3$.

Using the Programs

To use the above programs, first enter the function $f(x)$ whose sums you wish to compute as Y_1 in the "Y=" window. Then run the program. It will first ask for N, which is the number of partitions you wish to use. Enter the number (which must be at least 1) and then press ⌈ENTER⌉. (If you are running SIMP, make sure that the number you enter is even.) Next, it will ask for the values of the endpoints A and B in that order, so you enter them in the same way. Then wait a while (the larger N is, the longer the wait), and the answers will appear.

PROGRAMS TO ESTIMATE THE ERROR IN THE TRAPEZOID RULE AND SIMPSON'S RULE

The next two programs can be used to give upper bounds of the error in using either the trapeziod rule or Simpson's rule with a specified number of partitions.

Graphing Calculator Program to Compute the Error in the Trapezoid Rule (TI-82)

```
PROGRAM: TRAPERR
:Input ("EVEN N ",N)
:Input ("LEFT ENDPOINT? ",A)
:Input ("RIGHT ENDPOINT? ",B)
:fMax(Y₂,X,A,B)→X
:Y₂→R
:A→X
:Y₂→S
:B→X
:Y₂→T
:max({R,S,T})→M
:M*(B−A)∧3/(12*N∧2)→E
:Disp "ERROR BOUND IS ",E
:Stop
```

Number of partitions

Calculates maximum value of $|f''(x)|$ on $[a, b]$

The TI-82 sometimes ignores endpoints, so we look at their values as well and find the maximum of all the candidates.

Formula for error in trapezoid rule

Graphing Calculator Program to Compute the Error in Simpson's Rule (TI-82)

```
PROGRAM: SIMPERR
:Input ("EVEN N ",N)              'Number of partitions
:Input ("LEFT ENDPOINT? ",A)
:Input ("RIGHT ENDPOINT? ",B)
:fMax(Y₄,X,A,B)→X                 Calculates maximum
:Y₄→R                                value of | f⁽⁴⁾(x)| on [a, b]
:A→X                              The TI-82 sometimes ignores
:Y₄→S                                endpoints, so we look at their
:B→X                                 values as well and find the
:Y₄→T                                maximum of all the candidates.
:max({R,S,T})→M
:M*(B−A)^5/(180*N^4)→E            Formula for error in Simpson's
                                     rule
:Disp "ERROR BOUND IS ",E
:Stop
```

The `fMax(Y₄,X,A,B)→X` line calculates the maximum value of $|f^{(4)}(x)|$ on $[a, b]$.

The formula `:M*(B−A)^5/(180*N^4)→E` is the formula for error in Simpson's rule.

Using the Programs

To use TRAPERR, you must first enter the function $|f''(x)|$ as Y_2 in the "Y=" window. (We have used Y_2 since you may already have the original function $f(x)$ as Y_1.) To use SIMPERR, you must first enter the function $|f^{(4)}(x)|$ as Y_4 in the "Y=" window. This allows you to run TRAP, SIMP, TRAPERR, and SIMPERR without having to change anything in the "Y=" window.

▶ ══════ **B.5** CALCULATING AND PLOTTING AVERAGES AND MOVING AVERAGES

PLOTTING A LIST OF DATA

The following table, from an example in the chapter on applications of the integral, shows Colossal Conglomerate Corp.'s closing stock prices for a 20-day period:

Day	1	2	3	4	5	6	7	8	9	10
Price	20	22	21	24	24	23	25	26	20	24
Day	11	12	13	14	15	16	17	18	19	20
Price	26	26	25	27	28	27	29	27	25	24

To plot these data on a scatter graph, press [STAT], select EDIT, and press [ENTER] to obtain the list screen. If there is data on any of the lists shown, first clear it by selecting that list ($L_1, L_2, \ldots$) and pressing [CLEAR].

Next, enter the days 1, 2, 3, . . . in L_1 and the prices 20, 22, 21, 24, . . . in L_2. (Remember to press [ENTER] after each entry.)

To obtain a line graph of the stock prices, press [STAT PLOT], and select Plot1 and press [ENTER] to obtain the Plot1 menu. Next, select On, press [ENTER], and then select the icon representing the type of graph you want and again press [ENTER]. Next, press [ZOOM] [9] to set the *x*- and *y*-ranges to fit the data. The graph should then appear in the window.

▶ NOTE   If there are any active functions in the "Y=" screen, these too will be plotted if they are in range. To prevent this, either remove them or deactivate them (that is, prevent them from being plotted) by selecting the "=" signs in "Y_n= . . . " and pressing [ENTER]. You can reactivate them later by repeating this procedure. ◀

Graphing Calculator Program to Plot Moving Averages (TI-82)

```
PROGRAM: MOVINGAV
:Input ("NUMBER OF DATA POINTS ",N)
:Input ("PERIOD ",M)
:For (I,1,M−1,1)
:L₂(I) →L₃(I)

:End

:For (I,M,N,1)
:∅→S

:For(J,∅,M−1,1)
:S+L₂(I−J) →S
:End
:S/M→L₃(I)
:End
:Plot1(xyLine,L₁,L₂)
:Plot2(xyLine,L₁,L₃)
:FnOff
:PlotsOn 1,2
:ZoomStat
:Stop
```

Code	Comment
:Input ("PERIOD ",M)	For an *m*-unit moving average
:L₂(I) →L₃(I)	Set first $m − 1$ entries on L_3 to those of L_2;
:End	L_3 will be the moving averages list.
:∅→S	S will be the sum of data values.
:S/M→L₃(I)	Put *m*-unit average on list L_3
:FnOff	from "On/Off" in Y-VARS menu
:PlotsOn 1,2	from STATPLOT menu
:ZoomStat	"Zoom" menu—sets ranges

USING THE PROGRAM

To run the program, first enter the data on L_1 and L_2 as above. The program will display line graphs of both the values listed in L_1 and the moving averages in L_3. To access the moving averages directly, go to L_3.

Once you are done with statistical graphing and wish to reset the calculator to its normal mode (so that your statistical graphs will not be superimposed on every graph you draw), you should run the following little program. We have named it "AARESET" to ensure that it appears first on the list of programs, so that you can run it by simply pressing [PRGM], [ENTER], [ENTER].

Program to Reset the TI-82 to Normal Graphing Mode

```
PROGRAM: AARESET
:Plotsoff            From STATPLOT menu)
:FnOn                From "On/Off" in Y-VARS menu
:Stop
```

Be sure to run AARESET whenever you are done with your line graph plotting.

Alternatively, you can reset the calculator manually using the following sequence of keystrokes:

[2nd] [STAT PLOT], 4, [ENTER], [2nd], [Y-VARS], 5, 1, [ENTER].

▶ ══ B.6 EVALUATING A FUNCTION OF SEVERAL VARIABLES

The following little program will permit you to calculate $f(x, y)$ repeatedly with a minimum of keyboard work.

Graphing Calculator Program to Calculate $f(x, y)$ (TI-82)

```
PROGRAM:FXY
:1→A
:While A=1
:Input ("ENTER X ",X)
:Input ("ENTER Y ",Y)
:Disp "ANSWER",Y₈
:Input ("ENTER 1 TO CONTINUE ",A)
:End
:Stop
```

To run this program, first enter the function of two variables as Y_8 in the "Y=" screen. Then run the program. When prompted with "ENTER 1 TO CONTINUE," you can quit the program by entering any number other than 1. Entering 1 will cause it to ask for a new pair (x, y).

▶ NOTE   It is easy to adapt this program to calculate values of functions of three or more variables. We leave this to you.   ◀

▶ ══ B.7 CALCULATING PROBABILITIES ASSOCIATED WITH CONTINUOUS DISTRIBUTIONS

The following little program calculates the probability $P(A \leq X \leq B)$ associated with a normal distribution for a given mean and standard deviation. It can easily be adapted for use with either an exponential or beta distribution, and we leave that task to the interested reader.

Graphing Calculator Program to Calculate Probabilities Based on Normal Distribution (TI-82)

```
PROGRAM:NORMAL
:Input "ENTER MEDIAN MU ",M
:Input "ENTER ST DEV ",T
:Input "LEFT LIMIT ",A
:Input "RIGHT LIMIT ",B
:fnInt((1/T(2π)
^.5))e^(-(X-M)^2
/(2T²)),X,A,B)→P
```

Calculates $P(A \leq X \leq B)$

Value of median μ

Value of standard deviation σ

This is a single instruction line. The line breaks correspond to what should appear on your screen. "fnInt" can be obtained by pressing [MATH] [9]. The short minus sign is "negative" [(-)] and the longer one is "minus" [-].

```
:Disp "PROBABILITY IS ",P
:Stop
```

To Run the Program

When you run this little program, it asks for the mean, standard deviation, and the two endpoints A and B. It then goes ahead and calculates $P(A \leq X \leq B)$.

APPENDIX C

Using a Computer Spreadsheet

INTRODUCTION ▶Computer spreadsheets can be used to do many of the numerical calculations discussed in this book. In this appendix we show how to use a spreadsheet to calculate difference quotients, numerical approximations to integrals, and averages and moving averages.

For our examples we shall use the program Lotus 1-2-3, but other computer spreadsheets should be very similar. Consult the manual for your particular program for details on its commands.

C.1 CALCULATING DIFFERENCE QUOTIENTS

Recall that the **difference quotient** for the function f has the form

$$\frac{f(x + h) - f(x)}{h}.$$

In a typical situation, we might be given a value for x and be required to calculate the difference quotient for several values of h. As an example, we shall use the spreadsheet to find the difference quotient for $f(x) = \sqrt{x^2 - 1}$ at $x = 4$ and $h = \pm0.1, \pm0.01, \pm0.001, \pm0.0001$.

First, we evaluate the difference quotient for f at $x = 4$.

$$\frac{f(4 + h) - f(4)}{h} = \frac{\sqrt{(4 + h)^2 - 1} - \sqrt{4^2 - 1}}{h}$$

$$= \frac{\sqrt{(4 + h)^2 - 1} - \sqrt{15}}{h}$$

There are several ways of conveniently evaluating this at the eight values of h we want. Here is a nice way of doing it in Lotus 1-2-3 (other spreadsheets should be very similar). Start by entering the following in row 1:

	A	B	C
1		4	@SQRT(B1^2−1)

The 4 in position B1 is the value $x = 4$ that we are interested in, and the formula in C1 is $f(x)$. Now enter the following.

	A	B	C
1		4	@SQRT(B1^2−1)
2	−1	+$B$1 + A2	
3	−0.1		
4	−0.01		
5	−0.001		
6	−0.0001		
7	0.0001		
8	0.001		
9	0.01		
10	0.1		
11	1		

The values in the A column are the values of h that we want to use. The formula in B2 calculates $x + h$. If you now copy this formula and paste it into the cells B3 though B11, it will calculate in those cells the values $x + h$ for all the values of h you entered. (*Note:* Entering $B$1 instead of B1 makes Lotus 1-2-3 always take the value of x from B1. However, as you paste the formula into the other cells, the A2 in the formula will be changed to A3, A4, and so on, as appropriate, to pick up the correct value of h.) Now copy the formula in C1 into the cells C2 through C11, and the C column will contain the values $f(x + h)$ for various h. Finally, use the D column to calculate the difference quotients:

A	B	C	D
	4	@SQRT(B1^2−1)	
−1	+$B$1 + A2	@SQRT(B2^2−1)	+(C2−$C$1)/A2

The formula in D2 will calculate the difference quotient for the first value $h = -1$. If you now copy this formula into D3 through D11, you will calculate the remaining difference quotients. After calculation, the spreadsheet should look like this:

	A	B	C	D
1		4	3.8729833	
2	−1	3	2.8284271	1.0445562
3	−0.1	3.9	3.7696154	1.0336798
4	−0.01	3.99	3.8626545	1.0328819
5	−0.001	3.999	3.8719505	1.0328042
6	−0.0001	3.9999	3.8728801	1.0327964
7	0.0001	4.0001	3.8730866	1.0327947
8	0.001	4.001	3.8740161	1.032787
9	0.01	4.01	3.8833104	1.0327097
10	0.1	4.1	3.9761791	1.0319572
11	1	5	4.8989795	1.0259961

You can see from this that the derivative $f'(4)$ is approximately 1.03279.

▶ ═══════ **C.2** CALCULATING LEFT- AND RIGHT-HAND RIEMANN SUMS, THE TRAPEZOID RULE, AND SIMPSON'S RULE

In this example we shall calculate the left- and right-hand Riemann sum approximations to

$$\int_0^2 e^{-x^2}\, dx$$

for $n = 10$. We begin by computing the numbers x_k. Enter a in the top left cell of the spreadsheet, and in the cell below it calculate $x_1 = a + \Delta x = a + 2/10$.

	A	B	C	D
1	0			
2	+A1 + 2/10			

If we now copy the entry in cell A2 into cells A3 through A11, we get all the values of x_k.

	A	B	C	D
1	0			
2	0.2			
3	0.4			
4	0.6			
5	0.8			
6	1			
7	1.2			
8	1.4			
9	1.6			
10	1.8			
11	2			

In B1 we now enter the formula for $f(x) = e^{-x^2}$.

	A	B	C	D
1	0	@EXP(−A1^2)		
2	0.2			
3	0.4			
4	0.6			
5	0.8			
6	1			
7	1.2			
8	1.4			
9	1.6			
10	1.8			
11	2			

If we copy this formula into cells B2 through B11, column B will have the values $f(x_k)$.

	A	B	C	D
1	0	1		
2	0.2	0.9607894		
3	0.4	0.8521437		
4	0.6	0.6976763		
5	0.8	0.5272924		
6	1	0.3678794		
7	1.2	0.2369277		
8	1.4	0.1408584		
9	1.6	0.0773047		
10	1.8	0.0391638		
11	2	0.0183156		

We can now compute the left- and right-hand Riemann sums by entering their formulas in any convenient cells—say, C1 and C2.

	A	B	C	D
1	0	1	@SUM(B1..B10)*2/10	
2	0.2	0.9607894	@SUM(B2..B11)*2/10	
3	0.4	0.8521437		
4	0.6	0.6976763		
5	0.8	0.5272924		
6	1	0.3678794		
7	1.2	0.2369277		
8	1.4	0.1408584		
9	1.6	0.0773047		
10	1.8	0.0391638		
11	2	0.0183156		

After calculating the left- and right-hand Riemann sums, we can calculate the trapezoidal sum as their average.

	A	B	C	D
1	0	1	0.9800072	
2	0.2	0.9607894	0.7836703	
3	0.4	0.8521437	+(C1+C2)/2	
4	0.6	0.6976763		
5	0.8	0.5272924		
6	1	0.3678794		
7	1.2	0.2369277		
8	1.4	0.1408584		
9	1.6	0.0773047		
10	1.8	0.0391638		
11	2	0.0183156		

The final spreadsheet then looks like this, with the left-hand sum in C1, the right-hand sum in C2, and the trapezoidal sum in C3.

	A	B	C	D
1	0	1	0.9800072	
2	0.2	0.9607894	0.7836703	
3	0.4	0.8521437	0.8818388	
4	0.6	0.6976763		
5	0.8	0.5272924		
6	1	0.3678794		
7	1.2	0.2369277		
8	1.4	0.1408584		
9	1.6	0.0773047		
10	1.8	0.0391638		
11	2	0.0183156		

SIMPSON'S RULE

Now we shall calculate the Simpson's rule approximation to

$$\int_0^2 e^{-x^2}\, dx$$

for $n = 10$. Since every other $f(x_k)$ is treated differently in Simpson's approximation, we enter the x_k's in two separate columns.

	A	B	C	D	E
1	0	+A1+2/10			
2	+A1+2*2/10				

We copy the formula in A2 into A3 through A6, and the formula in B1 into B2 through B5.

	A	B	C	D	E
1	0	0.2			
2	0.4	0.6			
3	0.8	1			
4	1.2	1.4			
5	1.6	1.8			
6	2				

We enter the formula for $f(x_k)$ in C1 and then copy it into C2 through C6 and D1 through D5.

	A	B	C	D	E
1	0	0.2	@EXP(−A1^2)		
2	0.4	0.6			
3	0.8	1			
4	1.2	1.4			
5	1.6	1.8			
6	2				

	A	B	C	D	E
1	0	0.2	1	0.9607894	
2	0.4	0.6	0.8521437	0.6976763	
3	0.8	1	0.5272924	0.3678794	
4	1.2	1.4	0.2369277	0.1408584	
5	1.6	1.8	0.0773047	0.0391638	
6	2		0.0183156		

Finally, in E1 we enter the formula for Simpson's rule, in the form

$$(@SUM(C1..C5)+@SUM(C2..C6)+4*@SUM(D1..D5))*2/(3*10)$$

(why does this calculate Simpson's rule?). The final spreadsheet is this, with the result in E1:

	A	B	C	D	E
1	0	0.2	1	0.9607894	0.8820748
2	0.4	0.6	0.8521437	0.6976763	
3	0.8	1	0.5272924	0.3678794	
4	1.2	1.4	0.2369277	0.1408584	
5	1.6	1.8	0.0773047	0.0391638	
6	2		0.0183156		

▶ ═══ **C.3** CALCULATING AVERAGES AND MOVING AVERAGES

The following table, from Example 5 in the Chapter on Applications of the Integral, shows Colossal Conglomerate Corp.'s closing stock prices for a 20-day period:

Day	1	2	3	4	5	6	7	8	9	10
Price	20	22	21	24	24	23	25	26	20	24
Day	11	12	13	14	15	16	17	18	19	20
Price	26	26	25	27	28	27	29	27	25	24

We shall demonstrate how to plot these data and the 5-day moving average using a computer spreadsheet.

We begin by entering the data in the spreadsheet, either as a row or a column. For this example, we shall enter the data in a column. (We show only the first 10 entries.)

	A	B
1	20	
2	22	
3	21	
4	24	
5	24	
6	23	
7	25	
8	26	
9	20	
10	24	
...	...	

We can now compute the moving average by first entering in cell B5 the formula @AVG(A1..A5).

	A	B
1	20	
2	22	
3	21	
4	24	
5	24	@AVG(A1. .A5)
6	23	
7	25	
8	26	
9	20	
10	24	
...	...	

When we copy this formula into the cells B6 through B20, we will have the 5-day moving average computed in those cells.

	A	B
1	20	
2	22	
3	21	
4	24	
5	24	22.2
6	23	22.8
7	25	23.4
8	26	24.4
9	20	23.6
10	24	23.6
...	...	...

Now we can use the computer to graph these data. If we select the A and B columns and then ask Lotus to draw a line graph, it will produce the graph in Figure 1 (we have also entered titles and other useful information in the graph).

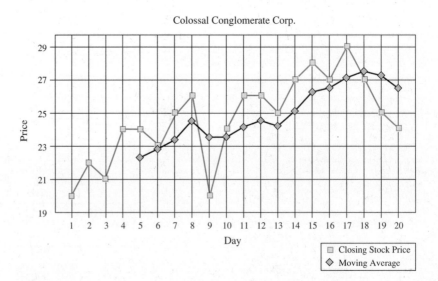

FIGURE 1

A Table of Integrals

Basic Integrals

1. $\displaystyle\int u^n\,du = \frac{u^{n+1}}{n+1} + C \qquad (n \neq -1)$

2. $\displaystyle\int u^{-1}\,du = \ln|u| + C$

3. $\displaystyle\int e^u\,du = e^u + C$

Integrals Containing $a + bu$

4. $\displaystyle\int (a + bu)^n\,du = \frac{(a + bu)^{n+1}}{b(n+1)} \qquad (n \neq -1)$

5. $\displaystyle\int (a + bu)^{-1}\,du = \frac{1}{b}\ln|a + bu| + C$

6. $\displaystyle\int \frac{u}{a + bu}\,du = \frac{1}{b^2}[bu - a\ln|a + bu|] + C$

7. $\displaystyle\int \frac{u}{(a + bu)^2}\,du = \frac{1}{b^2}\left[\frac{a}{a + bu} + \ln|a + bu|\right] + C$

8. $\displaystyle\int \frac{u}{(a + bu)^3}\,du = \frac{1}{b^2}\left[\frac{a}{2(a + bu)^3} - \frac{1}{a + bu}\right] + C$

9. $\displaystyle\int \frac{u^2}{a + bu}\,du = \frac{1}{b^2}\left[\frac{1}{2}(a + bu)^2 - 2a(a + bu) + a^2\ln|a + bu|\right] + C$

10. $\displaystyle\int \frac{u^2}{(a + bu)^2}\,du = \frac{1}{b^2}\left[bu - \frac{a^2}{a + bu} - 2a\ln|a + bu|\right] + C$

11. $\displaystyle\int \frac{du}{u(a + bu)} = \frac{1}{a}\ln\left|\frac{u}{a + bu}\right| + C$

12. $\displaystyle\int \frac{du}{u(a + bu)^2} = \frac{1}{a(a + bu)} + \frac{1}{a^2}\ln\left|\frac{u}{a + bu}\right| + C$

13. $\displaystyle\int \frac{du}{u^2(a + bu)} = -\frac{1}{au} + \frac{b}{a^2}\ln\left|\frac{a + bu}{u}\right| + C$

Integrals Containing $\sqrt{a + bu}$

14. $\displaystyle\int \sqrt{a + bu}\,du = \frac{2}{3b}(a + bu)^{3/2} + C$

15. $\displaystyle\int u\sqrt{a + bu}\,du = \frac{2}{15b^2}(3bu - 2a)(a + bu)^{3/2} + C$

16. $\displaystyle\int u^2\sqrt{a + bu}\,du = \frac{2}{105b^3}(15b^2u^2 - 12abu + 8a^2)(a + bu)^{3/2} + C$

17. $\displaystyle\int \frac{du}{\sqrt{a + bu}} = \frac{2}{b}\sqrt{a + bu} + C$

18. $\displaystyle\int \frac{u}{\sqrt{a + bu}}\,du = \frac{2}{3b^2}(bu - 2a)\sqrt{a + bu} + C$

19. $\displaystyle\int \frac{u^2}{\sqrt{a + bu}}\,du = \frac{2}{15b^3}(3b^2u^2 - 4abu + 8a^2)\sqrt{a + bu} + C$

20. $\displaystyle\int \frac{du}{u\sqrt{a + bu}} = \begin{cases} \dfrac{1}{\sqrt{a}}\ln\left|\dfrac{\sqrt{a + bu} - \sqrt{a}}{\sqrt{a + bu} + \sqrt{a}}\right| + C & \text{if } a > 0 \\[2ex] \dfrac{2}{\sqrt{-a}}\tan^{-1}\sqrt{\dfrac{a + bu}{-a}} + C & \text{if } a < 0 \end{cases}$

Integrals Containing $a^2 \pm u^2$ $(a > 0)$

21. $\displaystyle\int \frac{du}{a^2 + u^2} = \frac{1}{a}\tan^{-1}\frac{u}{a} + C$

22. $\displaystyle\int \frac{du}{a^2 - u^2} = \frac{1}{2a}\ln\left|\frac{a + u}{a - u}\right| + C$

Integrals Containing $\sqrt{u^2 \pm a^2}$ $(a > 0)$

23. $\displaystyle\int \sqrt{u^2 \pm a^2}\,du = \frac{u}{2}\sqrt{u^2 \pm a^2} \pm \frac{a^2}{2}\ln\left|u + \sqrt{u^2 \pm a^2}\right| + C$

24. $\displaystyle\int u\sqrt{u^2 \pm a^2}\,du = \frac{1}{3}(u^2 \pm a^2)^{3/2} + C$

25. $\displaystyle\int u^2\sqrt{u^2 \pm a^2}\,du = \frac{u}{8}(2u^2 \pm a^2)\sqrt{u^2 \pm a^2} - \frac{a^4}{8}\ln\left|u + \sqrt{u^2 \pm a^2}\right| + C$

26. $\displaystyle\int \frac{du}{\sqrt{u^2 \pm a^2}} = \ln\left|u + \sqrt{u^2 \pm a^2}\right| + C$

27. $\displaystyle\int \frac{u}{\sqrt{u^2 \pm a^2}}\,du = \sqrt{u^2 \pm a^2} + C$

28. $\displaystyle\int \frac{u^2}{\sqrt{u^2 \pm a^2}}\,du = \frac{u}{2}\sqrt{u^2 \pm a^2} \mp \frac{a^2}{2}\ln\left|u + \sqrt{u^2 \pm a^2}\right| + C$

29. $\displaystyle\int \frac{du}{u\sqrt{u^2 + a^2}} = -\frac{1}{a}\ln\left|\frac{a + \sqrt{u^2 + a^2}}{u}\right| + C$

30. $\displaystyle\int \frac{du}{u\sqrt{u^2 - a^2}} + \frac{1}{a}\sec^{-1}\frac{u}{a} + C$

31. $\displaystyle\int \frac{du}{u^2\sqrt{u^2 \pm a^2}} = \mp\frac{\sqrt{u^2 \pm a^2}}{a^2 u} + C$

32. $\displaystyle\int (u^2 \pm a^2)^{3/2}\, du = \frac{u}{8}(2u^2 \pm 5a^2)\sqrt{u^2 \pm a^2} + \frac{3a^4}{8}\ln\left|u + \sqrt{u^2 \pm a^2}\right| + C$

33. $\displaystyle\int \frac{du}{(u^2 \pm a^2)^{3/2}} = \pm\frac{u}{a^2\sqrt{u^2 \pm a^2}} + C$

Integrals Containing $\sqrt{a^2 - u^2}$

34. $\displaystyle\int \sqrt{a^2 - u^2}\, du = \frac{u}{2}\sqrt{a^2 - u^2} + \frac{a^2}{2}\sin^{-1}\frac{u}{a} + C$

35. $\displaystyle\int u\sqrt{a^2 - u^2}\, du = -\frac{1}{3}(a^2 - u^2)^{3/2} + C$

36. $\displaystyle\int u^2\sqrt{a^2 - u^2}\, du = \frac{u}{8}(2u^2 - a^2)\sqrt{a^2 - u^2} + \frac{a^4}{8}\sin^{-1}\frac{u}{a} + C$

37. $\displaystyle\int \frac{du}{\sqrt{a^2 - u^2}} = \sin^{-1}\frac{u}{a} + C$

38. $\displaystyle\int \frac{u}{\sqrt{a^2 - u^2}}\, du = -\sqrt{a^2 - u^2} + C$

39. $\displaystyle\int \frac{u^2}{\sqrt{a^2 - u^2}}\, du = -\frac{u}{2}\sqrt{a^2 - u^2} + \frac{a^2}{2}\sin^{-1}\frac{u}{a} + C$

40. $\displaystyle\int \frac{du}{u\sqrt{a^2 - u^2}} = -\frac{1}{a}\ln\left|\frac{a + \sqrt{a^2 - u^2}}{u}\right| + C$

41. $\displaystyle\int \frac{du}{u^2\sqrt{a^2 - u^2}} = -\frac{\sqrt{a^2 - u^2}}{a^2 u} + C$

42. $\displaystyle\int (a^2 - u^2)^{3/2}\, du = -\frac{u}{8}(2u^2 - 5a^2)\sqrt{a^2 - u^2} + \frac{3a^4}{8}\sin^{-1}\frac{u}{a} + C$

43. $\displaystyle\int \frac{du}{(a^2 - u^2)^{3/2}} = \frac{u}{a^2\sqrt{a^2 - u^2}} + C$

Answers to Odd-Numbered Exercises

Section 1.1 (page 14)

1. $P(-2, 4)$; $Q(1, 3)$; $R(-3, -4)$; $S(3, -3)$; $T(0, -2)$; $U(-\frac{9}{2}, 0)$; $V(\frac{5}{2}, \frac{3}{2})$; $W(\frac{5}{2}, 0)$

3.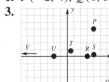

5.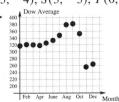

7.

Month	M	A	M	J	J	A	S	O	N	D	J	F	M
Deficit ($Billions)	-5.5	-7	-7.5	-6.5	-7.5	-8.5	-8.5	-7	-7.5	-6.5	-7.5	-8.0	-10.0

9. (a) 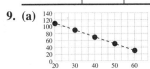 **(b)** A $10 increase in price has the effect of reducing sales by 20,000.

11.

Time	0	1	2	3	4
Height	500	480	440	350	250

13. (a) 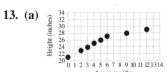 **(b)** 0.3 inches

15.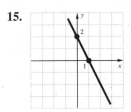

17.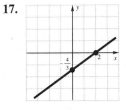

19.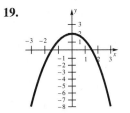

21.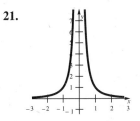

23. 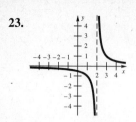   **25.** 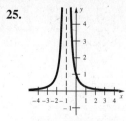   **27.** 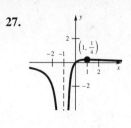   **29.**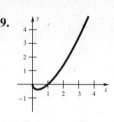

31. (a) 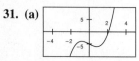   **31 (b)** Nothing is visible, since this viewing window is a square with sides of length 2 centered at the origin, and does not contain any points of the graph.

33. Replacing y with $(y - c)$ moves the curve up c units.   **35.** Replacing x with $(x - c)$ moves the curve to the right c units.

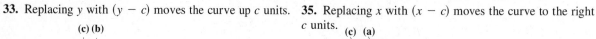

37. 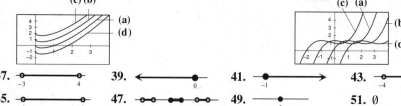   **39.**   **41.**   **43.**

45.   **47.**   **49.**   **51.** $\emptyset$

53. $[-1, 2]$ **55.** $(0, +\infty)$ **57.** $(-4, -2) \cup (-2, +\infty)$

Section 1.2 (page 32)

1. (a) $f(0) = 3$ **(b)** $f(1) = 6$ **(c)** $f(-1) = 2$ **(d)** $f(-3) = 6$ **(e)** $f(a) = a^2 + 2a + 3$
(f) $f(x + h) = (x + h)^2 + 2(x + h) + 3$ **(g)** $(f(x + h) - f(x))/h = 2x + h + 2$
3. (a) $g(1) = 2$ **(b)** $g(-1) = 0$ **(c)** $g(4) = 65/4$ **(d)** $g(x) = x^2 + \frac{1}{x}$ **(e)** $g(s + h) = (s + h)^2 + 1/(s + h)$
(f) $g(s + h) - g(s) = (s + h)^2 + 1/(s + h) - (s^2 + \frac{1}{s})$.
5. (a) $\phi(0) = \sqrt{3}$ **(b)** $\phi(-2) = \sqrt{7}$;   **(c)** $\phi(x + h) = \sqrt{(x + h)^2 + 3}$ **(d)** $\phi(x) + h = \sqrt{x^2 + 3} + h$
7. $-2x - h - 2$ **9.** $\dfrac{-2}{(x + h + 1)(x + 1)}$ **11.** $1 - \dfrac{1}{x(x + h)}$
13. $-\dfrac{2x + h}{x^2(x + h)^2}$ **15. (a)** yes; $f(4) = \dfrac{63}{16}$ **(b)** not defined   **(c)** not defined   **17. (a)** not defined
(b) not defined   **(c)** yes, $f(-10) = 0$ **19. (a)** yes, $f(0) = 1$ **(b)** not defined   **(c)** yes, $f(-3) = 2$
21. $(-\infty, +\infty)$ **23.** $[0, +\infty)$ **25.** $[1, +\infty)$ **27.** $(-\infty, 0) \cup (0, +\infty)$ **29.** $(-\infty, 0) \cup (0, +\infty)$
31. $(-\infty, 2) \cup (2, +\infty)$
33.   **(a)** $f(1) = 20$ **(b)** $f(2) = 30$ **(c)** $f(3) = 30$ **(d)** $f(5) = 22$ **(e)** $f(3) - f(2) = 0$
35. (a) $f(1) = 1.3$ **(b)** $f(-2) = 0$ **(c)** $f(0) = 2$ **(d)** $f(3) = 0$ **(e)** $f(3) - f(2) = -0.7$
37. (a) $f(-3) = -1$ **(b)** $f(0) = 1.25$ **(c)** $f(1) = 0$ **(d)** $f(2) = 1$ **(e)** $\dfrac{f(3) - f(2)}{3 - 2} = 0$
39. (a) $f(-3) = -0.5$ **(b)** $f(-2) = 0$ **(c)** $f(0) = 1$ **(d)** $f(2) = 0$ **(e)** $\dfrac{f(2) - f(0)}{2 - 0} = -\frac{1}{2}$
41. (a) (I)   **(b)** (IV)   **(c)** (V)   **(d)** (VI)   **(e)** (III)   **(f)** (II)   **43.** Equation: $y = x^3$; graph:

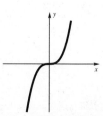

45. Equation: $y = x^4$; graph: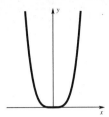

47. Equation: $y = \frac{1}{x^2}$; graph: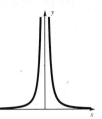

49.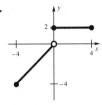

51.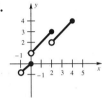

53.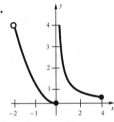

55.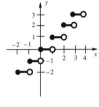

57.

59. Domain: $(-\infty, 1) \cup (1, 2) \cup (2, 3) \cup (3, 4) \cup (4, +\infty)$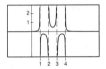

61.

63. Domain $[-5, 1) \cup (1, 2) \cup (2, 3) \cup (3, 5]$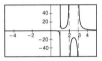

65. Domain: $[0, 1) \cup (1, 2) \cup (2, 5]$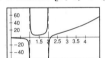

67. Lowest point approximately $(0.333, -0.385)$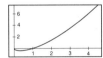

69. Domain: $[0, 1) \cup (1, +\infty)$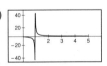
Never increasing

71. (a) 358,600 **(b)** 361,200 **(c)** $6.00 **73. (a)** $[0, 8]$ **(b)** $[0, +\infty)$ is not an appropriate domain, since it would predict investments in South Africa into the indefinite future with no basis. (It would also lead to preposterous results for large values of t.) **75. (a)** $12,000 **(b)** $N(q) = 2,000 + 100q^2 - 500q$; $N(20) = \$32,000$
77. (a) $C(x) = 8x + 80/x$, with domain $(0, +\infty)$ **(b)** Approximate value of x for lowest cost is 3.2 feet.

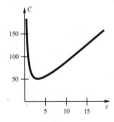

79. (a) $V(h) = \dfrac{1}{27}\pi h^3$, with domain $[0, +\infty)$ **(b)** $h(V) = (27V/\pi)^{1/3}$, with domain $[0, +\infty)$

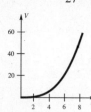

81. (a) $(0, +\infty)$ **(b)** $R(4000) = 30$ per hour   **(c)**

83. $G(Y_p) = 0.55Y_p$, with domain $(0, +\infty)$; $G(2) = 1.1$ trillion dollars. Thus inflation causes the actual GNP to fall below the projected GNP.   **85. (a)**   **(b)** 82%   **(c)** 13.7 months

87. (a) 31.22; the rocket ship appears to be 31.22 meters in length.   **(b)** $p = 0.8660$ warp, or 86.60% the speed of light

Section 1.3 (page 56)

1. 2   **3.** 2   **5.** -2 **7.** -1 **9.** 1/2   **11.** 0   **13.** $\sqrt{2}$;   **15.** infinite   **17.** $(d - b)/(c - a)$
19. parallel   **21.** neither   **23.** perpendicular   **25.** perpendicular
27. (a) (IV)   **(b)** (VII)   **(c)** (IX)   **(d)** (II)   **(e)** (I)   **(f)** (V)   **(g)** (VI)   **(h)** (III)   **(i)** (VIII)
29. $f(x) = 3x$ **31.** $f(x) = \frac{1}{4}x - 1$ **33.** $f(x) = -5x + 6$ **35.** $f(x) = -3x + \frac{9}{4}$
37. $f(x) = -x + 12$ **39.** $f(x) = 2x + 4$ **41.** $f(x) = x + 2$;   **43.** $x = 3$ (not a function)
45.   **47.**   **49.**   **51.**

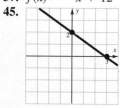   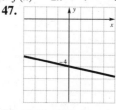   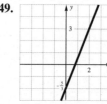   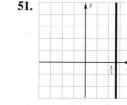

53.   **55.**   **57.** $x \approx 1.3$ **59.** $x \approx 1.0$

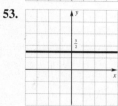   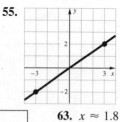   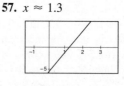   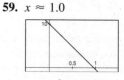

61. $x \approx -0.4$ **63.** $x \approx 1.8$ **65.** $x \approx -10.2$

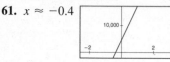      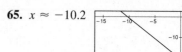

Section 1.4 (page 72)

1. $C(x) = 1500x + 1200$ **3.** $q = -40p + 2000$ **5. (a)** $R(t) = 8.50 + 0.95t$ **(b)** \$19.9 billion   **(c)** The model becomes unreasonable for large positive and negative values of t. For instance, it gives negative revenue for the year 1979 ($t = -9$) and predicts revenues that rise without bound in the future.   **7.** $f = 9c/5 + 32$; 86°F; 71.6°F; 14°F; 6.8°F   **9.** $C(x) = 88x + 20$ **(a)** \$196 **(b)** \$88 **(c)** \$88 **(d)** \$88 per tuxedo
11. $I(N) = 5N/100 + 50,000$; $N = \$1,000,000$; marginal income is $m = 5¢$ per dollar of net profit   **13.** Fixed cost = \$8,000, marginal cost = \$25 per bicycle   **15. (a)** $v(n) = 60,000 - 3,000n$ **(b)** 19.67 years **(c)** 20 years **(d)** after 20 years; the model predicts negative value   **17.** $t(s) = 16s$ (t = recovery time in hours and s = number of sets). When $t(15) = 240$ hr, or 10 days! This indicates that our linear model is reliable only for small numbers of sets. We need a nonlinear model to predict recovery time for arbitrary numbers of sets.   **19.** $P(x) = 100x - 5,132$, with domain [0, 405]. For profit, $x \geq 52$.
21. (a) $P(x) = 396 - 0.05x$ (millions of dollars), with domain [0, 1,100,000] **(b)** 47,920 homes damaged, or 4.36% of all the homes they insured   **23.** 5,000 units   **25.** $FC/(SP - VC)$
27. (a) $P(x) = 30x - 10,000$, with domain [0, $+\infty$) **(b)** Solve $P(1,000 + y) = 40y$ giving $y = 2,000$ new customers.   **29.** $b = -0.2365n + 55.2$; $b = 0$ when $n = 233.4$; about midway through the year 2053.
31. $p(n) = 3000n + 2500$; 29,500 gal   **33.** $L(n) = 12.2 - 0.28n$, with domain [0, 43.57]

35. $P(x) = 579.7x - 20,000$, with domain [0, $+\infty$); $x = 34.50$ g per day for breakeven
37. (a) $q(p) = -0.45p + 70.8$, with domain [0, $+\infty$) **(b)** 62 pounds per year **(c)** \$1.57 per pound; [0, 157]
(d) $R(p) = -0.45p^2 + 70.8p$, with domain [0,157]   **39. (a)** $n(t) = 245 - 3.583t$ (t = years since 1920)
(b) 1940   **(c)** The model becomes unreliable for values of t beyond 60, predicting a negative number of cases in 1990.   **41. (a)** $m = -1500$, $b = 6000$; $q = -1500p + 6000$. **(b)** 6,000 per week
(c) $R = p(-1500p + 6000) = -1500p^2 + 6000p$. **(d)** \$2 per hamburger; $R = \$6,000$

43. (a) $C(x) = 400x + 30,000$, with domain [0, 1,000] **(b)** $\overline{C}(x) = 400 + \dfrac{30,000}{x}$, with domain (0, 1,000] **(c)** 600 items per month   **45.** 60 units

Section 1.5 (page 87)

1. vertex: $(-\frac{3}{2}, -\frac{1}{4})$; y-intercept: 2; x-intercepts: $-2, -1$

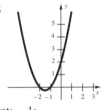

3. vertex: $(-\frac{1}{2}, -\frac{5}{4})$; y-intercept: -1; x-intercepts: $-1/2 \pm \sqrt{5}/2$

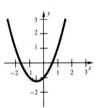

5. vertex: $(-2\sqrt{2}, -3)$; y-intercept: -1; x-intercepts: $2(-\sqrt{2} \pm \sqrt{3})$

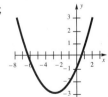

7. vertex: $(-1, 0)$; y-intercept: 1 x-intercept: -1

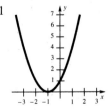

9. vertex: $(0,0)$; y-intercept: 0; x-intercept: 0

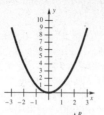

11. vertex $(0,1)$; y-intercept: 1; no x-intercepts

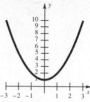

13. $R = -4p^2 + 100p$; maximum revenue when $p = \$12.50$.

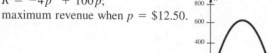

15. $R = -2p^2 + 400p$; maximum revenue when $p = \$100$

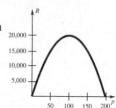

17. maximum revenue when $p = \$140$, $R = \$9,800$
19. maximum revenue with 70 houses, $R = \$9,800,000$
(b) $R = -0.5818p^2 + 36.245p$ **(c)** 31¢ per pound
20 mpg   **25.** \$10 per pound   **27. (a)** 4 seconds
Thus, doubling v_0 results in doubling t.

21. (a) $q = -0.5818p + 36.245$
23. maximum efficiency at 60 mph, efficiency =
(b) True; the time the ball is airborne is given by $t = v_0/16$.

Section 1.6 (page 94)

1. -3.45 ± 0.05, 1.45 ± 0.05 **3.** -2.55 ± 0.01 **5.** -1.886 ± 0.001, 0.503 ± 0.001, 1.622 ± 0.001
7. 1.17 ± 0.02 **9.** -0.47 ± 0.01, 0.54 ± 0.01, 3.94 ± 0.01 **11.** 1.34 ± 0.05 **13.** No solutions
15. 1.69 ± 0.05 **17.** They make a profit with 32 or more employees.   **19.** 6.95%

Chapter 1 Review Exercises (page 98)

1. $3x - y = 0$ **3.** $3x - y - 4 = 0$ **5.** $5x - 4y - 9 = 0$ **7.** $x + 3y + 5 = 0$ **9.** $x + 2y + 3 = 0$

11.
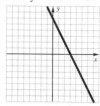

13.
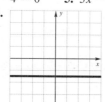

15.
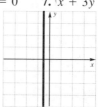

17.
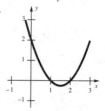

19.
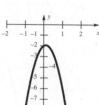

21.
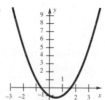

23.
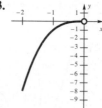

25.
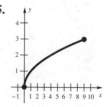

27. 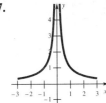 **29.** 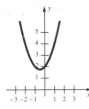 **31.** 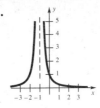 **33.**

35. 1.319 ± 0.005 **37.** $0, 2.924 \pm 0.005$ **39.** $0.68, 16.00 \pm 0.005$ **41. (a)** $\frac{1}{2}$ **(b)** $\frac{1}{3}$ **(c)** 1 **(d)** $\frac{1}{x}$

(e) $\dfrac{1}{x^2 + x + 2}$ **(f)** $x + \dfrac{1}{x^2 + 2}$ **43. (a)** 0 **(b)** 0 **(c)** $\sqrt{x + h - 1}$ **(d)** $\sqrt{x - 1} + h$

45. (a) 2 **(b)** $\dfrac{a^4 + 1}{a^2}$ **(c)** $\dfrac{(x + h)^2 + 1}{x + h} - h$ **(d)** $\dfrac{x + 1}{\sqrt{x}} + h$

47. $2x + 1 + h$ **49.** $\dfrac{-4}{[2(x + h) - 1][2x - 1]}$ **51.** $p = \dfrac{9}{2}n + 96$. At age 90, $p = \$388.50$

53. $s = 55t + 45{,}000$ **55.** $v = 32t$; acceleration $= 32$ ft/s/s **57.** 300,000 **59.** 12th year

61. $R = \$1{,}200$ when books were sold at \$10. Demand equation is $q = -6p + 180$; $R = -6p^2 + 180p$; $p = \$15$ maximizes revenue; total revenue at that price is \$1,350.

63. 50¢ **65. (a)** 21 **(b)** \$2 **67. (a)** $C(x) = 10{,}000 + 10x$ **(b)** $\overline{C}(x) = 10 + \dfrac{10{,}000}{x}$

(c) 4,000 fixtures **69. (a)** 43.59m **(b)** 0.9999, or 99.99% the speed of light **(c)** Its apparent length would be zero. **71. (a)** $I(x) = 100 + 2.75\sqrt{x}$ **(b)** $I(1) = 102.75, I(100) = 127.50$ **(c)** Approximately 1,322 sets per month. They will be lucky to have anyone work for them!

73. (a) $S(x) = 1.75x - 19$, with domain $[28, 48]$ **(b)** $S(x) = \begin{cases} 1.75x - 19 & \text{if } 28 \leq x \leq 48 \\ x + 17 & \text{if } 48 \leq x \leq 54 \\ 71 & \text{if } 54 \leq x \leq 58 \end{cases}$

75. $q = (500 - 100p)/p$; this gives the demand corresponding to the price p

▶ Chapter 2

Section 2.1 (page 115)

1. 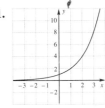 **3.** 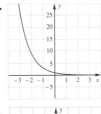 **5.** 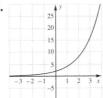 **7.**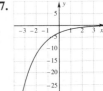

9. **11.**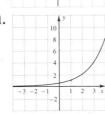

13. The graph of s is obtained by shifting the graph of f one unit to the right. **15.** The graph of r is obtained by inverting the graph of f in the y-direction, and then shifting it up one unit.

In answers 17 through 21, f_1 is solid and f_2 is dashed.

17. 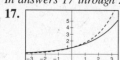   **19.** 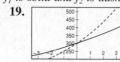   **21.**

23. (a) $A(t) = 10,000\,(1.03)^t$ **(b)** $13,439.16   **25. (a)** $A(t) = 10,000\,(1.00625)^{4t}$ **(b)** $11,327.08

27. (a) $A(t) = 10,000\left(1 + \frac{0.065}{365}\right)^{365t}$ **(b)** $19,154.30   **29. (a)** $A(t) = 10,000\,(1.002)^{12t}$ **(b)** $12,709.44

31. (a) $1,820.97   **(b)** $3,140.09   **33.** 4.8 years   **35.** $26.90   **37.** $2,109.60

39. $A(t) = 200,000(0.95)^{2t}$; $71,697.18

41. 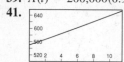   **43.**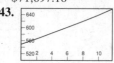

45. 31 years   **47.** 2.3 years   **49.** $A(r)$ is the amount that an item costing $1,000 now will cost in 10 years, given an annual rate of inflation r. We find $A(0.1) = 2,593.74$. Thus, at a rate of inflation of 10% per year, an item costing $1,000 now will cost $2,593.74 in 10 years. We find $A(1.15) = 2,110,496.32$. Thus, at a rate of inflation of 115% per year, an item costing $1,000 now will cost $2,110,496.32 in 10 years.

Section 2.2 (page 126)

1. 20.0855   **3.** 0.3679   **5.** 1.0305   **7.** 101.2578   **9.** 6,186.6127

11. 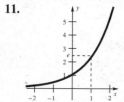   **13.** 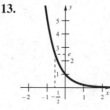   **15.**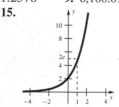

17. The graph of h is obtained from the graph of f by expanding it by a factor of 2 in the y-direction.

19. The graph of g is obtained from the graph of f by compressing it by a factor of 2 in the x-direction, and then reflecting it in the y-axis.   **21.** $16,487   **23.** $34,903   **25.** $13,331   **27.** $A(t) = 1,000e^{0.0235t}$; $A(5) = 1,124.68   **29.** $491.82   **31.** 4.08%   **33.** 22.14%   **35.** Ninth National has the lower effective rate.   **37.** $20,923.99   **39.** $A(t) = 100e^{-0.000121t}$; $A(10,000) \approx 29.82$ grams.   **41.** 200% per hour growth rate; 531,441,000 bugs   **43. (a)** $r(t) = 1.58 - 0.282t$;   **(b)** $\bar{r} = 1.58 - \dfrac{0.282t}{2}$ **(c)** $A(t) = Pe^{t\left(1.58 - \frac{0.282t}{2}\right)}$

(d) $5,820   **45.** $1083.29

47. 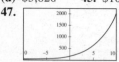   **49.**

51. 38,059 years old   **53.** 18 years; value = $17,280   **55.** 35 years   **57.** $548,812   **59.** 2002

Section 2.3 (page 141)

1. $\log_2 32 = 5$ **3.** $\log_3(\frac{1}{9}) = -2$ **5.** $\log 1{,}000 = 3$ **7.** $\ln y = x$ **9.** $\log_x y = -3$ **11.** $\ln 1 = 0$

13. $6^2 = 36$ **15.** $2^{-2} = \frac{1}{4}$ **17.** $10^8 = 100,000,000$ **19.** $e^{-1} = \frac{1}{e}$ **21.** $x^3 = y$ **23.** $e^y = -x$

25. $a + b$ **27.** $a - b$ **29.** $8a$ **31.** $b - (a + c)$ **33.** $-(a + 2c)$ **35.** $2b - 1$ **37.** 12   **39.** $x^2 y$

41. $\dfrac{x^2 y^4}{z^6}$ **43.** xy^{2x} **45.** 2   **47.** 0   **49.** $t = 0, -1$ **51.** 0.23105   **53.** 2.30259   **55.** 5.32193

57. 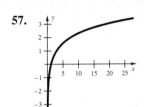   **59.** 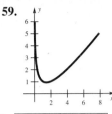   **61.**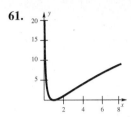

63.    **65.**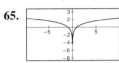

67. **(a)** about 5.012×10^{16} joules of energy   **(b)** about 2.24%   **(c)** proof   **(d)** 31.62
69. **(a)** 75 dB, 69 dB, 61 dB   **(b)** $D = 95.05 - 20 \log r$ **(c)** 56,559 feet   **71.** **(a)** $f(p) = 7.13 - 0.768p$
(b) $f(p) = 11.0p^{-0.750}$ **(c)** $f(p) = 11p^{-0.75}$ **(d)** $f(3.50) = 4.30$ servings per hour

Section 2.4 (page 152)

1. 8 years   **3.** 132 months   **5.** 5 years   **7.** 3 years   **9.** 13.43 years   **11.** 7.4 years   **13.** 9 days
15. 3.36 years   **17.** 11 years   **19.** 2.03 years   **21.** 55,259 years old   **23.** 3.8 months   **25.** 65,536,000
bacteria   **27.** 91,856 frogs   **29.** 311,000,000   **31.** 24 semesters   **33.** 2,360 million years   **35.** 3.89
days   **37.** $P = 23,200e^{-0.2261439t}$; 2,400 people in 1997   **39.** The least squares model gives a higher employment
figure for 1997.   **41.** 2002   **43.** 2.8 hours

Chapter 2 Review Exercises (page 158)

1. $\log_2 1024 = 10$ **3.** $\log 0.0001 = -4$ **5.** $\log_x y = e$ **7.** $3^4 = 81$ **9.** $10^3 = 1000$ **11.** $x^{-1} = y$
13. $\log_3 4 = \log 4 / \log 3 = 1.262$ **15.** 0 or -2 **17.** 0.5298   **19.** \$2,346.40   **21.** \$2,619.86
23. \$2,323.67   **25.** 10.2 years   **27.** 6 years   **29.** 10.8 years   **31.** \$11,956.00   **33.** \$25,633.04
35. 16.8 months   **37.** 0.301 seconds   **39.** Erewhon (effective yield = 27.12%)   **41.** 5.5 billion years
43. \$80   **45.** 18 minutes   **47.** \$0.83

Chapter 3

Section 3.1 (page 173)

In (1)–(9), d.q. stands for the difference quotient.

1. $d.q. = 2h$;

h	$d.q.$
1	2
0.1	0.2
0.01	0.02
0.001	0.002
0.0001	0.0002
-0.0001	-0.0002
-0.001	-0.002
-0.01	-0.02
-0.1	-0.2
-1	-2

3. $d.q. = \dfrac{\dfrac{1}{2+h} - \dfrac{1}{2}}{h} = -\dfrac{1}{2(2+h)}$

h	$d.q.$
1	-0.1667
0.1	-0.2381
0.01	-0.2488
0.001	-0.2499
0.0001	-0.24999
-0.0001	-0.25001
-0.001	-0.2501
-0.01	-0.2513
-0.1	-0.2632
-1	-0.5

5. $d.q. = \dfrac{(1 + h)^3 - 1}{h} = 3 + 3h + h^2$

h	$d.q.$
1	7
0.1	3.31
0.01	3.0301
0.001	3.0030
0.0001	3.0003
−0.0001	2.9997
−0.001	2.9970
−0.01	2.9701
−0.1	2.71
−1	1

7. $d.q. = 8 + h$

h	$d.q.$
1	9
0.1	8.1
0.01	8.01
0.001	8.001
0.0001	8.0001
−0.0001	7.9999
−0.001	7.999
−0.01	7.99
−0.1	7.9
−1	7

9. −2   **11.** 3   **13.** 0.03125   **15.** −0.17678   **17.** −5   **19.** −1.5

21.

h	1	0.1	0.01
v_{ave}	39	39.9	39.99

$v_{inst} = 40$ mph

23.

h	1	0.1	0.01
v_{ave}	140	66.2	60.602

$v_{inst} = 60$ mph

25.

h	1	0.1	0.01
v_{ave}	59.45	59.41	59.40

$v_{inst} = 59.4$ mph

27.

h	1	0.1	0.01
v_{ave}	9.443	9.938	9.994

$v_{inst} = 10$ mph

29.

h	50	10	1
C_{ave}	4.795	4.799	4.7999

$C'(1,000) = 4.8$

31.

h	50	10	1
C_{ave}	9.955	9.951	9.950

$C'(100) = 9.95$

33.

h	50	10	1
C_{ave}	99.93	99.91	99.90

$C'(100) = 99.9$

35. $q(100) = 50,000$, $q'(100) = -500$. A total of 50,000 pairs of sneakers can be sold at a price of $100, but the demand is decreasing at a rate of 500 pairs of sneakers per $1 increase in the price.

37. $P = \$257.07$ and $dP/dn = 5.07$. Your current profit is $257.07 per month, and is increasing at a rate of $5.07 per additional magazine in sales.   **39.** $E = 25.1$, $dE/dg = 2.51$. The professor's average class size is 25.1 students, and this is rising at a rate of 2.51 students per 1-point increase in grades awarded.   **41.** $A'(0.5) = 775$ points per day   **43. (a)** 60% of children can speak at the age of 10 months. Further, at the age of 10 months, this percentage is increasing by 18.2% per month. **(b)** As t increases, $p(t)$ approaches 100 (assuming all children eventually learn to speak) and $p'(t)$ approaches 0, since the percentage stops increasing.   **45. (a)** $R'(4) = -2.5$ thousand organisms per hour, per 1,000 new organisms. This means that the reproduction rate of organisms in a culture containing 4,000 organisms is declining at a rate of 2,500 organisms per hour for every 1,000 new organisms.
47. $l(.95) = 31.22$ meters and $l'(.95) = -304.24$ meters/warp. Thus, at a speed of warp 0.95, the rocket ship has an observed length of 31.22 meters and its length is decreasing at a rate of 304.24 meters per unit warp, or 3.0424 meters per one percent increase in the speed (measured in warp).   **49.** 1.000   **51.** 1.000   **53.** $S(5) = 109$, $S'(5) = 9.10$. After 5 weeks, sales are 109 pairs of sneakers per week, and sales are increasing at a rate of 9.1 pairs of sneakers per week each week.   **55.** $q(2) = 8,165$, $q'(2) = 599.2$. Thus, two months after the introduction, 8,165 video game units have been sold, and the demand is growing at a rate of 599.2 units per month.

Section 3.2 (page 185)

1. (a) R **(b)** P **3. (a)** P **(b)** R **5. (a)** Q **(b)** P **7. (a)** P **(b)** Q **9. (a)** Q **(b)** R **(c)** P
11. (a) R **(b)** Q **(c)** P **13. (a)** R **(b)** Q **(c)** P **15. (a)** $(1, 0)$ **(b)** none **(c)** $(-2, 1)$
17. (a) $(-2, 2)$, $(0, 1)$, $(2, 0)$ **(b)** none **(c)** none

19. The tangent to the graph of the function f at the general point where $x = a$ is the line passing through $(a, f(a))$ with slope $f'(a) = \lim\limits_{h \to 0} \dfrac{f(a + h) - f(a)}{h}$.   **21.** $y = f(a) + (x - a)f'(a)$.   **23.** (a) $2x + h$ (b) $2x$ (c) 4

25. (a) $-2x - h$ (b) $-2x$ (c) 2 **27.** (a) 3 (b) 3 (c) 3 **29.** (a) -2 (b) -2 (c) -2

31. (a) $6x + 3h$ (b) $6x$ (c) -6 **33.** (a) $1 - 2x - h$ (b) $1 - 2x$ (c) -3

35. (a) 3 (b) $y = 3x + 2$ **37.** (a) $\dfrac{3}{4}$ (b) $3x - 4y = -4$

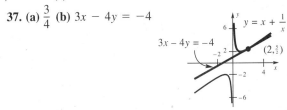

39. (a) $\dfrac{1}{4}$ (b) $x - 4y = -4$ **41.** (b)   **43.** (c)   **45.** (a)   **47.** -0.12 **49.** -0.15

51. -0.58 **53.** 0.26 **55.** 0.625 **57.** $1.425, 2.575$ **59.** $q(100) = 50,000$, $q'(100) = -500$. A total of 50,000 pairs of sneakers can be sold at a price of \$100, but the demand is decreasing at a rate of 500 pairs of sneakers per \$1 increase in the price.   **61.** $P = \$257.07$ and $dP/dn = 5.07$. Your current profit is \$257.07 per month, and this is increasing at a rate of \$5.07 per additional magazine in sales.   **63.** $E = 25.1$, $dE/dg = 2.51$. The professor's average class size is 25.1 students, and this is rising at a rate of 2.51 students per 1-point increase in grades awarded.   **65.** $A'(0.5) = 775$ points per day   **67.** (a) $R'(4) = -2.5$ thousand organisms per hour, per 1,000 new organisms. This means that the reproduction rate of organisms in a culture containing 4,000 organisms is declining at a rate of 2,500 organisms per hour for every 1,000 new organisms.   **69.** $l(.95) = 31.22$ meters and $l'(.95) = -304.24$ meters/warp. Thus, at a speed of warp 0.95, the rocket ship has an observed length of 31.22 meters and its length is decreasing at a rate of 304.24 meters per unit warp, or 3.0424 meters per one-percent increase in the speed (measured in warp).   **71.** $S(5) = 109$, $S'(5) = 9.10$. After 5 weeks, sales are 109 pairs of sneakers per week, and sales are increasing at a rate of 9.1 pairs of sneakers per week each week.   **73.** $q(2) = 8,165$, $q'(2) = 599.2$. Thus, two months after the introduction, 8,165 video game units have been sold, and the demand is growing at a rate of 599.2 units per month.

Section 3.3 (page 202)

1. 0 **3.** 4 **5.** (a) -2 (b) -1 **7.** (a) 1 (b) 1 **9.** (a) 0 (b) 2 (c) -1 (d) does not exist, since the left and right limits disagree (e) 2 **11.** (a) 1 (b) 1 (c) 2 (d) does not exist, since the left and right limits disagree (e) 1 **13.** (a) 1 (b) does not exist (c) does not exist (d) does not exist (e) not defined   **15.** (a) 1 (b) does not exist (c) -1 (d) does not exist (e) 2 (f) 0 **17.** (a) does not exist (b) 0 (c) 1.5 **19.** (a) 1 (b) 0 **21.** continuous on its domain   **23.** continuous on its domain   **25.** not continuous on its domain; discontinuous at $x = 0$ **27.** not continuous on its domain; discontinuous at $x = -1$ **29.** continuous on its domain   **31.** not continuous on its domain; discontinuous at $x = 0$ and $x = -2$ **33.** not continuous on its domain; discontinuous at $x = 0$ **35.** continuous on its domain   **37.** 1 **39.** 2 **41.** 0 **43.** 6 **45.** 4 **47.** 2 **49.** 0 **51.** 0 **53.** 12 **55.** 0 **57.** 2 **59.** -4 **61.** $2x - 2$ **63.** $-10x + 2$ **65.** $3t^2 + 1$ **67.** $4t^3 - 1$ **69.** $-\dfrac{6}{t^2}$ **71.** $1 - \dfrac{1}{x^2}$ **73.** $-\dfrac{1}{(x - 2)^2}$ **75.** $\dfrac{1}{2\sqrt{x} + 1}$ **77.** $-\dfrac{1}{2t\sqrt{t}}$

Section 3.4 (page 213)

1. $f'(x) = 3x^2$　　**3.** $f'(x) = -4x^{-3}$　　**5.** $f'(x) = -\frac{1}{4}x^{-3/4}$　　**7.** $f'(x) = 8x^3 + 9x^2$

9. $f'(x) = -1 - \frac{1}{x^2}$　　**11.** $f'(x) = \frac{1}{\sqrt{x}}$　　**13.** $f'(x) = x^{-2/3}$

15. $\frac{dy}{dx} = 10 \cdot 0 = 0$ by constant multiples and the power rule　　**17.** $\frac{d}{dx}(x^2 + x) = \frac{d}{dx}(x^2) + \frac{d}{dx}(x)$ (sum) =

$2x + 1$ (power rule)　　**19.** $\frac{d}{dx}(4x^3 + 2x - 1) = \frac{d}{dx}(4x^3) + \frac{d}{dx}(2x) - \frac{d}{dx}1$ (sum and difference) =

$4\frac{d}{dx}(x^3) + 2\frac{d}{dx}x - \frac{d}{dx}1$ (constant multiples) = $12x^2 + 2$ (power rule)　　**21.** $\frac{d}{dx}(x^{104} - 99x^2 + x) =$

$\frac{d}{dx}(x^{104}) - \frac{d}{dx}(99x^2) + \frac{d}{dx}x$ (sum and differences) = $104x^{103} - 99\frac{d}{dx}(x^2) + 1$ (constant multiples and power

rule) = $104x^{103} - 198x + 1$ (power rule)　　**23.** $s = t^{3/2} - t^{7/2} + t^{-3}$. Thus, $\frac{ds}{dt} = \frac{d}{dt}(t^{3/2} - t^{7/2} + t^{-3}) =$

$\frac{d}{dt}(t^{3/2}) - \frac{d}{dt}(t^{7/2}) + \frac{d}{dt}(t^{-3})$ (sums and differences) = $\frac{3}{2}t^{1/2} - \frac{7}{2}t^{5/2} - 3t^{-4}$ (power rule).

25. $\frac{d}{dr}\left(\frac{4\pi r^3}{3}\right) = \frac{4\pi}{3}\frac{d}{dr}(r^3)$ (constant multiple) = $\frac{4\pi}{3}(3r^2)$ (power rule) = $4\pi r^2$　　**27.** $2t + 20at^4$ (sum, constant

multiples, power rule)　　**29.** $\frac{1}{2}x^{-1/2} + \frac{3}{2}x^{1/2}$ (multiply out, then use sum and power rules)　　**31.** 3　　**33.** -2

35. $\frac{1}{32}$　　**37.** $-\frac{1}{4\sqrt{2}}$　　**39.** -5　　**41.** $-\frac{3}{2}$　　**43.** $y = 3x + 2$

45. $3x - 4y = -4$　　**47.** $x - 4y = -4$

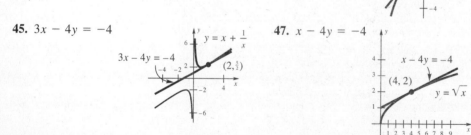

49. $f'(x) = 2x - 3$　　**51.** $f'(x) = 1 + \frac{1}{2\sqrt{x}}$　　**53.** $g'(x) = -\frac{2}{x^3} - \frac{6}{x^4}$　　**55.** $h'(x) = -\frac{1}{x^2} - \frac{2}{x^3} - \frac{3}{x^4}$

57. $r'(x) = \frac{1}{2\sqrt{x}} - \frac{1}{2x\sqrt{x}}$　　**59.** $f'(x) = 3x^2$　　**61.** $g'(x) = 1 - 4x$　　**63.** $1 - \frac{2}{x^3}$　　**65.** $2.6x^{0.3} + 1.2x^{-2.2}$

67. $3at^2 - 4a$　　**69.** $\frac{1}{2\sqrt{x}} + \frac{3\sqrt{x}}{2}$　　**71.** $5.15x^{9.3} - 99x^{-2}$　　**73.** $-\frac{2.31}{t^{2.1}} - \frac{0.3}{t^{0.4}}$　　**75.** $4\pi r^2$　　**77.** $-\frac{3}{4}$

79. no such values　　**81.** $x = 1, -1$　　**83.** $\frac{1}{4}$　　**85-89.** Proofs　　**91.** The rate of change of C_2 is twice the rate

of change of C_1.　　**93.** The rates of change of cost and revenue must be equal. This means that the next case costs as
much to make as it will bring in in revenue.　　**95.** $A'(0.5) = 775$ points per day　　**97.** $c'(15) \approx 1.389$,
$c'(30) \approx 0.011$. Thus, the hourly oxygen consumption is increasing at a slower rate when the chick hatches. (Notice
that the oxygen consumption is still increasing, however.)　　**99. (a)** 0, -32, -64, -96, -128 ft/s　**(b)** 2.5
seconds; downward at 80 ft/s　　**101.** $V'(10) = 4\pi(100) \approx 1{,}256.63$ cubic centimeters per centimeter increase in r.
When $r = 10$, the volume increases by approximately 1,256.63 cubic centimeters for every 1-centimeter increase in
the radius.

19. The tangent to the graph of the function f at the general point where $x = a$ is the line passing through $(a, f(a))$ with slope $f'(a) = \lim\limits_{h \to 0} \dfrac{f(a + h) - f(a)}{h}$.   **21.** $y = f(a) + (x - a)f'(a)$.   **23.** (a) $2x + h$ (b) $2x$ (c) 4

25. (a) $-2x - h$ (b) $-2x$ (c) 2   **27.** (a) 3 (b) 3 (c) 3   **29.** (a) -2 (b) -2 (c) -2

31. (a) $6x + 3h$ (b) $6x$ (c) -6 **33.** (a) $1 - 2x - h$ (b) $1 - 2x$ (c) -3

35. (a) 3 (b) $y = 3x + 2$

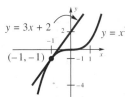

37. (a) $\dfrac{3}{4}$ (b) $3x - 4y = -4$

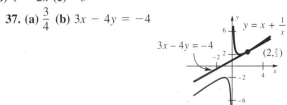

39. (a) $\dfrac{1}{4}$ (b) $x - 4y = -4$

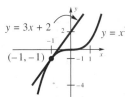

41. (b)   **43.** (c)   **45.** (a)   **47.** -0.12 **49.** -0.15

51. -0.58 **53.** 0.26   **55.** 0.625   **57.** 1.425, 2.575   **59.** $q(100) = 50{,}000$, $q'(100) = -500$. A total of 50,000 pairs of sneakers can be sold at a price of \$100, but the demand is decreasing at a rate of 500 pairs of sneakers per \$1 increase in the price.   **61.** $P = \$257.07$ and $dP/dn = 5.07$. Your current profit is \$257.07 per month, and this is increasing at a rate of \$5.07 per additional magazine in sales.   **63.** $E = 25.1$, $dE/dg = 2.51$. The professor's average class size is 25.1 students, and this is rising at a rate of 2.51 students per 1-point increase in grades awarded.   **65.** $A'(0.5) = 775$ points per day   **67.** (a) $R'(4) = -2.5$ thousand organisms per hour, per 1,000 new organisms. This means that the reproduction rate of organisms in a culture containing 4,000 organisms is declining at a rate of 2,500 organisms per hour for every 1,000 new organisms.   **69.** $l(.95) = 31.22$ meters and $l'(.95) = -304.24$ meters/warp. Thus, at a speed of warp 0.95, the rocket ship has an observed length of 31.22 meters and its length is decreasing at a rate of 304.24 meters per unit warp, or 3.0424 meters per one-percent increase in the speed (measured in warp).   **71.** $S(5) = 109$, $S'(5) = 9.10$. After 5 weeks, sales are 109 pairs of sneakers per week, and sales are increasing at a rate of 9.1 pairs of sneakers per week each week.   **73.** $q(2) = 8{,}165$, $q'(2) = 599.2$. Thus, two months after the introduction, 8,165 video game units have been sold, and the demand is growing at a rate of 599.2 units per month.

Section 3.3 (page 202)

1. 0   **3.** 4   **5.** (a) -2 (b) -1 **7.** (a) 1 (b) 1   **9.** (a) 0 (b) 2 (c) -1 (d) does not exist, since the left and right limits disagree (e) 2   **11.** (a) 1 (b) 1 (c) 2 (d) does not exist, since the left and right limits disagree (e) 1   **13.** (a) 1 (b) does not exist (c) does not exist (d) does not exist (e) not defined   **15.** (a) 1 (b) does not exist (c) -1 (d) does not exist (e) 2 (f) 0   **17.** (a) does not exist (b) 0 (c) 1.5   **19.** (a) 1 (b) 0   **21.** continuous on its domain   **23.** continuous on its domain   **25.** not continuous on its domain; discontinuous at $x = 0$ **27.** not continuous on its domain; discontinuous at $x = -1$ **29.** continuous on its domain   **31.** not continuous on its domain; discontinuous at $x = 0$ and $x = -2$ **33.** not continuous on its domain; discontinuous at $x = 0$ **35.** continuous on its domain   **37.** 1   **39.** 2   **41.** 0   **43.** 6   **45.** 4   **47.** 2   **49.** 0   **51.** 0   **53.** 12   **55.** 0   **57.** 2   **59.** -4 **61.** $2x - 2$ **63.** $-10x + 2$

65. $3t^2 + 1$ **67.** $4t^3 - 1$ **69.** $-\dfrac{6}{t^2}$ **71.** $1 - \dfrac{1}{x^2}$ **73.** $-\dfrac{1}{(x - 2)^2}$ **75.** $\dfrac{1}{2\sqrt{x + 1}}$ **77.** $-\dfrac{1}{2t\sqrt{t}}$

Section 3.4 (page 213)

1. $f'(x) = 3x^2$ **3.** $f'(x) = -4x^{-3}$ **5.** $f'(x) = -\dfrac{1}{4}x^{-3/4}$ **7.** $f'(x) = 8x^3 + 9x^2$

9. $f'(x) = -1 - \dfrac{1}{x^2}$ **11.** $f'(x) = \dfrac{1}{\sqrt{x}}$ **13.** $f'(x) = x^{-2/3}$

15. $\dfrac{dy}{dx} = 10 \cdot 0 = 0$ by constant multiples and the power rule   **17.** $\dfrac{d}{dx}(x^2 + x) = \dfrac{d}{dx}(x^2) + \dfrac{d}{dx}(x)$ (sum) $=$

$2x + 1$ (power rule)   **19.** $\dfrac{d}{dx}(4x^3 + 2x - 1) = \dfrac{d}{dx}(4x^3) + \dfrac{d}{dx}(2x) - \dfrac{d}{dx}1$ (sum and difference) $=$

$4\dfrac{d}{dx}(x^3) + 2\dfrac{d}{dx}x - \dfrac{d}{dx}1$ (constant multiples) $= 12x^2 + 2$ (power rule)   **21.** $\dfrac{d}{dx}(x^{104} - 99x^2 + x) =$

$\dfrac{d}{dx}(x^{104}) - \dfrac{d}{dx}(99x^2) + \dfrac{d}{dx}x$ (sum and differences) $= 104x^{103} - 99\dfrac{d}{dx}(x^2) + 1$ (constant multiples and power

rule) $= 104x^{103} - 198x + 1$ (power rule)   **23.** $s = t^{3/2} - t^{7/2} + t^{-3}$. Thus, $\dfrac{ds}{dt} = \dfrac{d}{dt}(t^{3/2} - t^{7/2} + t^{-3}) =$

$\dfrac{d}{dt}(t^{3/2}) - \dfrac{d}{dt}(t^{7/2}) + \dfrac{d}{dt}(t^{-3})$ (sums and differences) $= \dfrac{3}{2}t^{1/2} - \dfrac{7}{2}t^{5/2} - 3t^{-4}$ (power rule).

25. $\dfrac{d}{dr}\left(\dfrac{4\pi r^3}{3}\right) = \dfrac{4\pi}{3}\dfrac{d}{dr}(r^3)$ (constant multiple) $= \dfrac{4\pi}{3}(3r^2)$ (power rule) $= 4\pi r^2$ **27.** $2t + 20at^4$ (sum, constant

multiples, power rule)   **29.** $\frac{1}{2}x^{-1/2} + \frac{3}{2}x^{1/2}$ (multiply out, then use sum and power rules)   **31.** 3   **33.** -2

35. $\frac{1}{32}$ **37.** $-\dfrac{1}{4\sqrt{2}}$ **39.** -5 **41.** $-\frac{3}{2}$ **43.** $y = 3x + 2$

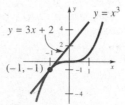

45. $3x - 4y = -4$

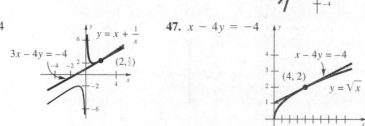

47. $x - 4y = -4$

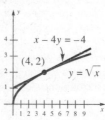

49. $f'(x) = 2x - 3$ **51.** $f'(x) = 1 + \dfrac{1}{2\sqrt{x}}$ **53.** $g'(x) = -\dfrac{2}{x^3} - \dfrac{6}{x^4}$ **55.** $h'(x) = -\dfrac{1}{x^2} - \dfrac{2}{x^3} - \dfrac{3}{x^4}$

57. $r'(x) = \dfrac{1}{2\sqrt{x}} - \dfrac{1}{2x\sqrt{x}}$ **59.** $f'(x) = 3x^2$ **61.** $g'(x) = 1 - 4x$ **63.** $1 - \dfrac{2}{x^3}$ **65.** $2.6x^{0.3} + 1.2x^{-2.2}$

67. $3at^2 - 4a$ **69.** $\dfrac{1}{2\sqrt{x}} + \dfrac{3\sqrt{x}}{2}$ **71.** $5.15x^{9.3} - 99x^{-2}$ **73.** $-\dfrac{2.31}{t^{2.1}} - \dfrac{0.3}{t^{0.4}}$ **75.** $4\pi r^2$ **77.** $-\frac{3}{4}$

79. no such values   **81.** $x = 1, -1$ **83.** $\dfrac{1}{4}$ **85-89.** Proofs   **91.** The rate of change of C_2 is twice the rate

of change of C_1.   **93.** The rates of change of cost and revenue must be equal. This means that the next case costs as
much to make as it will bring in in revenue.   **95.** $A'(0.5) = 775$ points per day   **97.** $c'(15) \approx 1.389$,
$c'(30) \approx 0.011$. Thus, the hourly oxygen consumption is increasing at a slower rate when the chick hatches. (Notice
that the oxygen consumption is still increasing, however.)   **99. (a)** 0, -32, -64, -96, -128 ft/s   **(b)** 2.5
seconds; downward at 80 ft/s   **101.** $V'(10) = 4\pi(100) \approx 1{,}256.63$ cubic centimeters per centimeter increase in r.
When $r = 10$, the volume increases by approximately 1,256.63 cubic centimeters for every 1-centimeter increase in
the radius.

Section 3.5 (page 226)

1.

h	50	10	1
C_{ave}	4.795	4.799	4.7999

$C'(1,000) = 4.8$

3.

h	50	10	1
C_{ave}	9.955	9.951	9.950

$C'(100) = 9.95$

5.

h	50	10	1
C_{ave}	99.93	99.91	99.90

$C'(100) = 100 - 1{,}000/100^2 = 99.9$

7. $C'(x) = 4$; $R'(x) = 8 - x/500$; $P'(x) = 4 - x/500$; $P'(x) = 0$ when $x = 2{,}000$. Thus, at a production level of 2,000, the profit is stationary (neither increasing nor decreasing) with respect to the production level. This may indicate a maximum profit at a production level of 2,000.   **9.** $C'(x) = 40 - .002x$. The cost is going up at a rate of $39.80 per teddy bear. The cost of producing the 101st teddy is $C(101) - C(100) = \$39.799$.   **11.** The profit on the sale of 1,000 videocassettes is $3,000, and is decreasing by approximately $3 per additional videocassette sold. **13.** $P = \$257.07$ and $dP/dn = 5.07$. Your current profit is $257.07 per month, and this would increase by approximately $5.07 for each increase by one magazine in sales.   **15.** $P'(50) = \$350$. This means that, at an employment level of 50 workers, the firm's daily profit will increase by approximately $350 per additional worker it hires.   **17. (a)** $4.47 per pound   **(b)** $R(q) = \dfrac{50000}{q^{0.5}}$ **(c)** $R(500) = \$2{,}236.07$. This is the monthly revenue that will result from setting the price at $4.47 per pound.   **(d)** $R'(q) = -\dfrac{25000}{q^{1.5}}$; $R'(500) = -\$2.24$ per additional pound of tuna demanded. Thus, at a demand level of 500 pounds of tuna, the revenue is decreasing by approximately $2.24 per additional pound.   **(e)** The fishery should raise the price in order to decrease demand, since the revenue drops when the price is lowered, even though the sales increase. It would be more economical to store the excess tuna.
19. (a) $R(q) = 152.33q - 2.22q^2$; $R'(q) = 152.33 - 4.44q$.   **(b)** $R'(50) = -69.67$¢ per pound of poultry. Thus if the price is raised, causing a decrease in demand of one pound of poultry, the revenue will increase by approximately 69.67¢.   **(c)** $P(q) = 142.33q - 2.22q^2$; $P'(q) = 142.33 - 4.44q$; $P'(50) = -79.67$¢ per pound. This means that, if a farmer lowers the price to increase the per capita demand by one pound, the annual profit will decrease by approximately 79.67¢.
21. (a) $R(p) = -\dfrac{4p^2}{3} + 80p$; $R'(p) = -\dfrac{8p}{3} + 80$ **(b)** (i) $R'(20) = \$26.67$ per $1 increase in price; (ii) $R'(30) = \$0$ per $1 increase in price; (ii) $R'(40) = -\$26.67$ per $1 increase in price   **(c)** If it charges $20 per ruby, it can increase its revenue by approximately $26.67 by raising the price $1. If it charges $30 per ruby, the revenue is not moving with the price. Thus, if they raise the price by $1, the revenue will not change significantly. If it charges $40 per ruby, the revenue would drop by approximately $26.67 if it were to raise the price another $1.
23. (a) $C'(q) = 200q$; $C'(10) = \$2{,}000$ per one-pound reduction in emissions.   **(b)** $S'(q) = 500$. Thus, $S'(q) = C'(q)$ when $500 = 200q$, or $q = \frac{5}{2}$ pounds per day reduction.   **(c)** $N(q) = C(q) - S(q) = 100q^2 - 500q + 4{,}000$. This is a parabola with lowest point (vertex) given by $q = -b/2a = 500/200 = \frac{5}{2}$. The net cost at this production level is $N(\frac{5}{2}) = \$3{,}375$ per day. The value of q is the same as that for part (b). The net cost to the firm is minimized at the reduction level for which the cost of controlling emissions begins to increase faster than the subsidy. This is why we get the answer by setting these two rates of increase equal to each other.   **25.** $M'(10) \approx 0.0002557$ mpg/mph. This means that, at a speed of 10 mph, the fuel economy is increasing at a rate of 0.0002557 miles per gallon for each 1-mph increase in speed. $M'(60) = 0$ mpg/mph. This means that, at a speed of 60 mph, the fuel economy is neither increasing nor decreasing with increasing speed. $M'(70) \approx -0.00001799$. This means that, at 70 mph, the fuel economy is decreasing by approximately 0.00001799 miles per gallon per 1-mph increase in speed. Thus, 60 mph is the most fuel-efficient speed for the car.   **27. (a)** $3.98 per 1,000 nautical miles   **(b)** The cost is dropping at a rate of 0.1¢ per extra ton.   **29.** (c)   **31.** (d)

Section 3.6 (page 239)

1. (a) diverges to $+\infty$ **(b)** 0   **3. (a)** diverges to $-\infty$ **(b)** 0   **5. (a)** diverges to $+\infty$ **(b)** diverges to $+\infty$ **(c)** diverges to $+\infty$ **7. (a)** -1 **(b)** diverges to $-\infty$ **(c)** does not exist, since the left and right limits disagree   **9. (a)** does not exist, since the left and right limits disagree   **(b)** diverges to $+\infty$ **(c)** -1 **11. (a)** does not exist, since the graph oscillates between -1 and 1   **(b)** does not exist for the reason given in (a)   **13.** diverges to $+\infty$

15. does not exist, since the left and right limits disagree **17.** $\frac{3}{2}$ **19.** $\frac{1}{2}$ **21.** does not exist **23.** 0
25. $\frac{3}{2}$ **27.** $\frac{1}{2}$ **29.** does not exist **31.** 0 **33.** 0 **35.** does not exist **37.** 1 **39.** 0 **41.** 0
43. jump discontinuity at $x = 0$ **45.** continuous everywhere **47.** removable discontinuity at $x = 0$
49. discontinuity at $x = 0$ **51.** not differentiable at $x = 0$ **53.** differentiable everywhere **55.** not
differentiable at $x = 1$ **57.** $\lim_{t \to +\infty} n(t) = 0$. Thus, if the trend were to continue indefinitely, the annual number of
DWI arrests in New Jersey will decrease to zero in the long term. **59.** $\lim_{t \to +\infty} p(t) = 100$, $\lim_{t \to +\infty} p'(t) = 0$. This tells us
that the percentage of children who learn to speak approaches 100% as their age increases, with the number of
additional children learning to speak approaching zero. **61.** $\lim_{t \to +\infty} q(t) = 10{,}000$, meaning that a total of 10,000 units
are sold over the life of the game

Chapter 3 Review Exercises (page 246)

1. (a) P **(b)** Q **(c)** R **3. (a)** Q, R **(b)** P **5. (a)** $(0, -1)$ **(b)** $(2, 1)$ **(c)** $(-2, 1)$
7. (a) $(0, 1)$ **(b)** $(-2, 0)$ **(c)** none **9.** $6x + 4x^{-2}$ **11.** x **13.** $1.2x^{0.2} + 2.3x^{1.3}$

15. $0.6s^{-0.8} - 44$ **17.** $3at^2 - 2bt$ **19.** $\dfrac{1}{\sqrt{x}} + \dfrac{1}{3\sqrt[3]{x^2}}$ **21.** $3 - \dfrac{3}{x^4}$ **23.** $\dfrac{1}{5} + \dfrac{5}{x^2}$ **25.** $\dfrac{3}{2\sqrt{x}}$

27. $6x^2 + 2x$ **29.** $-\dfrac{1}{2r\sqrt{r}} - \dfrac{2}{r^2}$ **31.** $y = -2x - 5$ **33.** $y = \dfrac{-4t}{5} + 1$ **35.** $y = \dfrac{3s}{4} + 1$

37. $y = -\dfrac{5t}{3} - 1$ **39.** $y = \dfrac{5}{4}t - 1$ **41.** $x = -\dfrac{3}{2}$ **43.** $x = \pm 1$ **45.** none

47. $x = -\sqrt[3]{2}$ **49.** -2 **51.** 0 **53.** 6 **55.** diverges to $+\infty$ **57.** 3 **59.** diverges to $+\infty$
61. diverges to $+\infty$ **63.** 0 **65. (a)** -1 **(b)** 1 **(c)** does not exist; left and right limits disagree **(d)** diverges
to $+\infty$ **(e)** 2 **(f)** 1 **67. (a)** -1 **(b)** diverges to $-\infty$ **(c)** 1 **69. (a)** diverges to $+\infty$ **(b)** diverges to
$-\infty$ **(c)** -1 **(d)** 1 **71. (a)** 1 **(b)** diverges to $-\infty$ **(c)** diverges to $+\infty$ **73.** discontinuous at -1 and 1
75. continuous **77.** discontinuous at -1 **79.** discontinuous at every integer **81.** 4 **83.** -2 **85.** -7
87. 0 **89.** 0 **91. (a)** 52, 66.4, 67.84, 67.984 ft/s **(b)** 68 ft/s **(c)** hits the ground after 6.25 seconds, and is
traveling 100 ft/s at the time **93.** $C'(x) = 60 - 0.002x$; $C'(50) = \$59.90$; $C(51) - C(50) = \$59.899$

95. $V'(1{,}000) = 300 \text{ cm}^3/\text{s}$ **97. (a)** $R(s) = s - 1{,}000 - \dfrac{s^{3/2}}{1{,}000}$ **(b)** $R'(100{,}000) = 0.526$; $R'(500{,}000) =$
$- 0.0607$ **(c)** $R'(s) > 0$ for $s < \$444{,}444.44$; $R'(s) < 0$ for $s > \$444{,}444.44$ **(d)** $R(444{,}444.44) = \$147{,}148.15$
99. (a) 60,912 frogs per year **(b)** 2006

▶ ## Chapter 4

Section 4.1 (page 262)

1. 3 **3.** $3x^2$ **5.** $2x + 3$ **7.** $\dfrac{2x}{3}$ **9.** $-\dfrac{2}{x^2}$ **11.** $(x^2 - 1) + 2x(x + 1) = (x + 1)(3x - 1)$

13. $(x^{-1/2} + 4)(x - x^{-1}) + (2x^{1/2} + 4x - 5)(1 + x^{-2})$ **15.** $8(2x^2 - 4x + 1)(x - 1)$

17. $\left(2x - \dfrac{1}{2\sqrt{x}}\right)\left(\sqrt{x} + \dfrac{1}{\sqrt{x}}\right) + (x^2 - \sqrt{x})\left(\dfrac{1}{2\sqrt{x}} - \dfrac{1}{2x^{3/2}}\right)$ **19.** $\dfrac{1}{2\sqrt{x}}\left(\sqrt{x} + \dfrac{1}{x^2}\right) + (\sqrt{x} + 1)\left(\dfrac{1}{2\sqrt{x}} - \dfrac{2}{x^3}\right)$

21. $-\dfrac{14}{(3x - 1)^2}$ **23.** $\dfrac{6x^2 - 4x - 7}{(3x - 1)^2}$ **25.** $\dfrac{5x^2 - 5}{(x^2 + x + 1)^2}$ **27.** $\dfrac{-1}{\sqrt{x}(\sqrt{x} - 1)^2}$ **29.** $-\dfrac{3}{x^4}$

31. $\dfrac{3x^2 - 2x - 13}{(3x - 1)^2}$

Section 3.5 (page 226)

1. h

h	50	10	1
C_{ave}	4.795	4.799	4.7999

$C'(1,000) = 4.8$

3. h

h	50	10	1
C_{ave}	9.955	9.951	9.950

$C'(100) = 9.95$

5. h

h	50	10	1
C_{ave}	99.93	99.91	99.90

$C'(100) = 100 - 1,000/100^2 = 99.9$

7. $C'(x) = 4$; $R'(x) = 8 - x/500$; $P'(x) = 4 - x/500$; $P'(x) = 0$ when $x = 2,000$. Thus, at a production level of 2,000, the profit is stationary (neither increasing nor decreasing) with respect to the production level. This may indicate a maximum profit at a production level of 2,000.   **9.** $C'(x) = 40 - .002x$. The cost is going up at a rate of $39.80 per teddy bear. The cost of producing the 101st teddy is $C(101) - C(100) = \$39.799$.   **11.** The profit on the sale of 1,000 videocassettes is $3,000, and is decreasing by approximately $3 per additional videocassette sold. **13.** $P = \$257.07$ and $dP/dn = 5.07$. Your current profit is $257.07 per month, and this would increase by approximately $5.07 for each increase by one magazine in sales.   **15.** $P'(50) = \$350$. This means that, at an employment level of 50 workers, the firm's daily profit will increase by approximately $350 per additional worker it hires.   **17. (a)** $4.47 per pound   **(b)** $R(q) = \dfrac{50000}{q^{0.5}}$ **(c)** $R(500) = \$2,236.07$. This is the monthly revenue that will result from setting the price at $4.47 per pound.   **(d)** $R'(q) = -\dfrac{25000}{q^{1.5}}$; $R'(500) = -\$2.24$ per additional pound of tuna demanded. Thus, at a demand level of 500 pounds of tuna, the revenue is decreasing by approximately $2.24 per additional pound.   **(e)** The fishery should raise the price in order to decrease demand, since the revenue drops when the price is lowered, even though the sales increase. It would be more economical to store the excess tuna.
19. (a) $R(q) = 152.33q - 2.22q^2$; $R'(q) = 152.33 - 4.44q$.   **(b)** $R'(50) = -69.67$¢ per pound of poultry. Thus if the price is raised, causing a decrease in demand of one pound of poultry, the revenue will increase by approximately 69.67¢.   **(c)** $P(q) = 142.33q - 2.22q^2$; $P'(q) = 142.33 - 4.44q$; $P'(50) = -79.67$¢ per pound. This means that, if a farmer lowers the price to increase the per capita demand by one pound, the annual profit will decrease by approximately 79.67¢.
21. (a) $R(p) = -\dfrac{4p^2}{3} + 80p$; $R'(p) = -\dfrac{8p}{3} + 80$ **(b)** (i) $R'(20) = \$26.67$ per $1 increase in price; (ii) $R'(30) = \$0$ per $1 increase in price; (ii) $R'(40) = -\$26.67$ per $1 increase in price   **(c)** If it charges $20 per ruby, it can increase its revenue by approximately $26.67 by raising the price $1. If it charges $30 per ruby, the revenue is not moving with the price. Thus, if they raise the price by $1, the revenue will not change significantly. If it charges $40 per ruby, the revenue would drop by approximately $26.67 if it were to raise the price another $1.
23. (a) $C'(q) = 200q$; $C'(10) = \$2,000$ per one-pound reduction in emissions.   **(b)** $S'(q) = 500$. Thus, $S'(q) = C'(q)$ when $500 = 200q$, or $q = \frac{5}{2}$ pounds per day reduction.   **(c)** $N(q) = C(q) - S(q) = 100q^2 - 500q + 4,000$. This is a parabola with lowest point (vertex) given by $q = -b/2a = 500/200 = \frac{5}{2}$. The net cost at this production level is $N(\frac{5}{2}) = \$3,375$ per day. The value of q is the same as that for part (b). The net cost to the firm is minimized at the reduction level for which the cost of controlling emissions begins to increase faster than the subsidy. This is why we get the answer by setting these two rates of increase equal to each other.   **25.** $M'(10) \approx 0.0002557$ mpg/mph. This means that, at a speed of 10 mph, the fuel economy is increasing at a rate of 0.0002557 miles per gallon for each 1-mph increase in speed. $M'(60) = 0$ mpg/mph. This means that, at a speed of 60 mph, the fuel economy is neither increasing nor decreasing with increasing speed. $M'(70) \approx -0.00001799$. This means that, at 70 mph, the fuel economy is decreasing by approximately 0.00001799 miles per gallon per 1-mph increase in speed. Thus, 60 mph is the most fuel-efficient speed for the car.   **27. (a)** $3.98 per 1,000 nautical miles   **(b)** The cost is dropping at a rate of 0.1¢ per extra ton.   **29.** (c)   **31.** (d)

Section 3.6 (page 239)

1. (a) diverges to $+\infty$ **(b)** 0   **3. (a)** diverges to $-\infty$ **(b)** 0   **5. (a)** diverges to $+\infty$ **(b)** diverges to $+\infty$ **(c)** diverges to $+\infty$ **7. (a)** -1 **(b)** diverges to $-\infty$ **(c)** does not exist, since the left and right limits disagree   **9. (a)** does not exist, since the left and right limits disagree   **(b)** diverges to $+\infty$ **(c)** -1 **11. (a)** does not exist, since the graph oscillates between -1 and 1   **(b)** does not exist for the reason given in (a)   **13.** diverges to $+\infty$

15. does not exist, since the left and right limits disagree   **17.** $\frac{3}{2}$ **19.** $\frac{1}{2}$ **21.** does not exist   **23.** 0
25. $\frac{3}{2}$ **27.** $\frac{1}{2}$ **29.** does not exist   **31.** 0   **33.** 0   **35.** does not exist   **37.** 1   **39.** 0   **41.** 0
43. jump discontinuity at $x = 0$ **45.** continuous everywhere   **47.** removable discontinuity at $x = 0$
49. discontinuity at $x = 0$ **51.** not differentiable at $x = 0$ **53.** differentiable everywhere   **55.** not
differentiable at $x = 1$ **57.** $\lim_{t \to +\infty} n(t) = 0$. Thus, if the trend were to continue indefinitely, the annual number of
DWI arrests in New Jersey will decrease to zero in the long term.   **59.** $\lim_{t \to +\infty} p(t) = 100$, $\lim_{t \to +\infty} p'(t) = 0$. This tells us
that the percentage of children who learn to speak approaches 100% as their age increases, with the number of
additional children learning to speak approaching zero.   **61.** $\lim_{t \to +\infty} q(t) = 10{,}000$, meaning that a total of 10,000 units
are sold over the life of the game

Chapter 3 Review Exercises (page 246)

1. (a) P **(b)** Q **(c)** R **3. (a)** Q, R **(b)** P **5. (a)** $(0, -1)$ **(b)** $(2, 1)$ **(c)** $(-2, 1)$
7. (a) $(0, 1)$ **(b)** $(-2, 0)$ **(c)** none   **9.** $6x + 4x^{-2}$ **11.** x **13.** $1.2x^{0.2} + 2.3x^{1.3}$

15. $0.6s^{-0.8} - 44$ **17.** $3at^2 - 2bt$ **19.** $\frac{1}{\sqrt{x}} + \frac{1}{3\sqrt[3]{x^2}}$ **21.** $3 - \frac{3}{x^4}$ **23.** $\frac{1}{5} + \frac{5}{x^2}$ **25.** $\frac{3}{2\sqrt{x}}$

27. $6x^2 + 2x$ **29.** $-\frac{1}{2r\sqrt{r}} - \frac{2}{r^2}$ **31.** $y = -2x - 5$ **33.** $y = \frac{-4t}{5} + 1$ **35.** $y = \frac{3s}{4} + 1$

37. $y = -\frac{5t}{3} - 1$ **39.** $y = \frac{5}{4}t - 1$ **41.** $x = -\frac{3}{2}$ **43.** $x = \pm 1$ **45.** none

47. $x = -\sqrt[3]{2}$ **49.** -2 **51.** 0   **53.** 6   **55.** diverges to $+\infty$ **57.** 3   **59.** diverges to $+\infty$
61. diverges to $+\infty$ **63.** 0   **65. (a)** -1 **(b)** 1 **(c)** does not exist; left and right limits disagree **(d)** diverges
to $+\infty$ **(e)** 2 **(f)** 1   **67. (a)** -1 **(b)** diverges to $-\infty$ **(c)** 1   **69. (a)** diverges to $+\infty$ **(b)** diverges to
$-\infty$ **(c)** -1 **(d)** 1   **71. (a)** 1 **(b)** diverges to $-\infty$ **(c)** diverges to $+\infty$ **73.** discontinuous at -1 and 1
75. continuous   **77.** discontinuous at -1 **79.** discontinuous at every integer   **81.** 4 **83.** -2 **85.** -7
87. 0   **89.** 0   **91. (a)** 52, 66.4, 67.84, 67.984 ft/s **(b)** 68 ft/s **(c)** hits the ground after 6.25 seconds, and is
traveling 100 ft/s at the time   **93.** $C'(x) = 60 - 0.002x$; $C'(50) = \$59.90$; $C(51) - C(50) = \$59.899$

95. $V'(1{,}000) = 300$ cm³/s   **97. (a)** $R(s) = s - 1{,}000 - \frac{s^{3/2}}{1{,}000}$ **(b)** $R'(100{,}000) = 0.526$; $R'(500{,}000) =$
-0.0607 **(c)** $R'(s) > 0$ for $s < \$444{,}444.44$; $R'(s) < 0$ for $s > \$444{,}444.44$ **(d)** $R(444{,}444.44) = \$147{,}148.15$
99. (a) 60,912 frogs per year **(b)** 2006

▶ Chapter 4

Section 4.1 (page 262)

1. 3   **3.** $3x^2$ **5.** $2x + 3$ **7.** $\frac{2x}{3}$ **9.** $-\frac{2}{x^2}$ **11.** $(x^2 - 1) + 2x(x + 1) = (x + 1)(3x - 1)$

13. $(x^{-1/2} + 4)(x - x^{-1}) + (2x^{1/2} + 4x - 5)(1 + x^{-2})$ **15.** $8(2x^2 - 4x + 1)(x - 1)$

17. $\left(2x - \frac{1}{2\sqrt{x}}\right)\left(\sqrt{x} + \frac{1}{\sqrt{x}}\right) + (x^2 - \sqrt{x})\left(\frac{1}{2\sqrt{x}} - \frac{1}{2x^{3/2}}\right)$ **19.** $\frac{1}{2\sqrt{x}}\left(\sqrt{x} + \frac{1}{x^2}\right) + (\sqrt{x} + 1)\left(\frac{1}{2\sqrt{x}} - \frac{2}{x^3}\right)$

21. $-\frac{14}{(3x - 1)^2}$ **23.** $\frac{6x^2 - 4x - 7}{(3x - 1)^2}$ **25.** $\frac{5x^2 - 5}{(x^2 + x + 1)^2}$ **27.** $\frac{-1}{\sqrt{x}(\sqrt{x} - 1)^2}$ **29.** $-\frac{3}{x^4}$

31. $\frac{3x^2 - 2x - 13}{(3x - 1)^2}$

33. $\dfrac{[(x+1)(x+2)+(x+3)(x+2)+(x+3)(x+1)](3x-1)-3(x+3)(x+1)(x+2)}{(3x-1)^2}$

35. $4x^3-2x$ **37.** 64 **39.** 3

41. $\dfrac{\left(2t-\dfrac{1}{2\sqrt{t}}\right)\left(\sqrt{t}+\dfrac{1}{\sqrt{t}}\right)-(t^2-\sqrt{t})\left(\dfrac{1}{2\sqrt{t}}-\dfrac{1}{2t\sqrt{t}}\right)}{\left(\sqrt{t}+\dfrac{1}{\sqrt{t}}\right)^2}$

43. $y=12x-8$ **45.** $y=\dfrac{x}{4}+\dfrac{1}{2}$ **47.** $y=-2$ **49.** $S'(5)=10$ (sales are increasing at a rate of 1,000 units per month); $p'(5)=-10$ (the price is dropping at a rate of \$10 per month); $R'(5)=900{,}000$ (revenue is increasing at a rate of \$900,000 per month)   **51.** decreasing at a rate of \$1 per day   **53.** decreasing at a rate of \$0.10 per month

55. $M'(x)=\dfrac{3600x^{-2}-1}{(3600x^{-1}+x)^2}$. $M'(10)\approx 0.0002557$ mpg/mph. This means that, at a speed of 10 mph, the fuel economy is increasing at a rate of 0.0002557 miles per gallon for each 1-mph increase in speed. $M'(60)=0$ mpg/mph. This means that at a speed of 60 mph, the fuel economy is neither increasing nor decreasing with increasing speed. $M'(70)\approx -0.00001799$. This means that, at 70 mph, the fuel economy is decreasing at a rate of 0.00001799 miles per gallon for each 1-mph increase in speed. Thus, 60 mph is the most fuel-efficient speed for the car.

57. \$111,870,000 per year   **59.** $R'(p)=-\dfrac{5.625}{(1+0.125p)^2}$; $R'(4)=-2.5$ thousand organisms per hour, per 1,000 new organisms. This means that the reproduction rate of organisms in a culture containing 4,000 organisms is declining at a rate of 2,500 organisms per hour for every 1,000 new organisms.

61. Oxygen consumption is decreasing at a rate of 1634 milliliters per day. This is due to the fact that the number of eggs is decreasing, since $C'(30)$ is positive.   **67.** (a)

Section 4.2 (page 277)

1. $4(2x+1)$ **3.** $-(x-1)^{-2}$ **5.** $2(2-x)^{-3}$ **7.** $\dfrac{1}{\sqrt{2x+1}}$ **9.** $-\dfrac{3}{(3x-1)^2}$

11. $4(x^2+2x)^3(2x+2)$

13. $-4x(2x^2-2)^{-2}$ **15.** $-5(2x-3)(x^2-3x-1)^{-6}$ **17.** $-\dfrac{6x}{(x^2+1)^4}$ **19.** $4\left(2t-\dfrac{1}{2\sqrt{t}}\right)(t^2-\sqrt{t})^3$

21. $-\dfrac{x}{\sqrt{1-x^2}}$ **23.** $\dfrac{\left(\dfrac{3}{2\sqrt{x}}+\dfrac{1}{2x\sqrt{x}}\right)}{2\sqrt{3\sqrt{x}-\dfrac{1}{\sqrt{x}}}}$ **25.** $-\dfrac{\left(\dfrac{1}{\sqrt{2x+1}}-2x\right)}{(\sqrt{2x+1}-x^2)^2}$ **27.** $\dfrac{-56(x+2)}{(3x-1)^3}$

29. $-\dfrac{1}{2}[(r+1)(r^2-1)]^{-3/2}(3r-1)(r+1)$ **31.** $-2(x^2-3x)^{-3}(2x-3)\sqrt{1-x^2}-\dfrac{x(x^2-3x)^{-2}}{\sqrt{1-x^2}}$

33. $(100x^{99}-99x^{-2})\dfrac{dx}{dt}$ **35.** $\left(-\dfrac{3}{r^4}+\dfrac{1}{2\sqrt{r}}\right)\dfrac{dr}{dt}$ **37.** $4\pi r^2 r'(t)$ **39.** $-47/4$

41. (a) $R(p)=-\dfrac{4p^2}{3}+80p$; $\left.\dfrac{dR}{dp}\right|_{q=60}=\$40$ per \$1 increase in price   (b) $-\$0.75$ per ruby   (c) $-\$30$ per ruby. Thus, at a demand level of 60 rubies per week, the weekly revenue is decreasing at a rate of \$30 per additional ruby demanded.   **43.** At an employment level of 10 engineers, it will increase its profit at a rate of \$146,454.90 per additional engineer hired.   **45.** $\dfrac{dM}{dB}\approx 0.000116$. This means that approximately 1.16 manatees are killed each year for every 10,000 registered boats.   **47.** 12π sq mi/hr

49. $200,000\pi$/week $= \$628,318.53$/week   **51. (a)** $q'(4) = 333.14$ units per month
(b) $R'(q) = 800$ **(c)** $R'(t) \approx \$266,512$ per month.   **53.** 3% per year   **55.** 8% per year

Section 4.3 (page 293)

1. $\dfrac{1}{x-1}$ **3.** $\dfrac{1}{x \ln 2}$ **5.** $\dfrac{2x}{x^2+3}$ **7.** e^{x+3} **9.** $-e^{-x}$ **11.** $4^x \ln 4$ **13.** $(2^{x^2-1})2x \ln 2$

15. $1 + \ln x$ **17.** $2x \ln x + \dfrac{x^2+1}{x}$ **19.** $10x(x^2+1)^4 \ln x + \dfrac{(x^2+1)^5}{x}$ **21.** $\dfrac{3}{3x-1}$ **23.** $\dfrac{4x}{2x^2+1}$

25. $\dfrac{\left(2x - \dfrac{1}{2\sqrt{x}}\right)}{x^2 - \sqrt{x}}$ **27.** $\dfrac{1}{(x+1)\ln 2}$ **29.** $2t \log_3\left(t + \dfrac{1}{t}\right) + \dfrac{(t^2+1)\left(1 - \dfrac{1}{t^2}\right)}{\left(t + \dfrac{1}{t}\right)\ln 3}$ **31.** $\dfrac{2 \ln|x|}{x}$

33. $\dfrac{2}{x} - \dfrac{2 \ln(x-1)}{x-1}$ **35.** $e^x(1+x)$ **37.** $\dfrac{1}{x+1} + 3e^x(x^3+3x^2)$ **39.** $e^x \ln|x| + \dfrac{e^x}{x}$ **41.** $2e^{2x+1}$

43. $(2x-1)e^{x^2-x+1}$ **45.** $2xe^{2x-1} + 2x^2 e^{2x-1}$ **47.** $4(e^{2x-1})^2$ **49.** $-\dfrac{4}{(e^x - e^{-x})^2}$

51. $\dfrac{-(2(\ln x)^{1/2} + (\ln x)^{-1/2})}{2x^2 \ln x}$ **53.** $\dfrac{1 - 2 \ln x}{2x^2(\ln x)^{1/2}}$ **55.** $\dfrac{1}{x \ln x}$ **57.** $\dfrac{1}{2x \ln x}$

59. $y = (e/\ln 2)(x-1) \approx 3.92(x-1)$ **61.** $y = x$ **63.** $y = -\dfrac{1}{2e}(x-1) + e$ **65.** $451.00 per year

67. $400,000 \ln 2 \approx 277,000$ people/yr   **69.** 0.000283 g/yr   **71.** $446.02 per year
73. $P'(22) \approx -1.3855$. This indicates that in ancient Rome, the percentage of people surviving was decreasing at a rate of 1.3855% per year at age 22.   **75.** 3,110,000 cases/month; 11,200,000 cases/month; 722,000 cases/month   **77.** Growing at a rate of 28.3309 billion packets per month each year.   **79. (a)** $p'(10) \approx 0.0931$, so that the percentage of firms using numeric control is increasing at a rate of 9.31% per year after 10 years. **(b)** 0.80. Thus, in the long term, 80% of all firms will be using numeric control. **(c)** same as (a) **(d)** 0. Thus, in the long run, the percentage of firms using numeric control will remain constant at 80%.   **81.** $n'(t)$ is a maximum when $t \approx 12.47$. ($n'(12.47) \approx 5.1673$). This means that the rate of growth of enrollment in HMOs reached a maximum of 5.167 million new enrollments per year at about June of 1987.

Section 4.4 (page 303)

1. $-\dfrac{2}{3}$ **3.** x **5.** $\dfrac{y-2}{3-x}$ **7.** $-y$ **9.** $-\dfrac{y}{x(1+\ln x)}$ **11.** $-\dfrac{x}{y}$ **13.** $-\dfrac{2xy}{x^2-2y}$ **15.** $\dfrac{6+9x^2y}{9x^3-x^2}$

17. $\dfrac{3y}{x}$ **19.** $\dfrac{p+10p^2q}{2p-q-10pq^2}$ **21.** $\dfrac{ye^x - e^y}{xe^y - e^x}$ **23.** $\dfrac{se^{st}}{2s - te^{st}}$ **25.** $\dfrac{ye^x}{2e^x + y^3 e^y}$ **27.** $\dfrac{y - y^2}{-1 + 3y - y^2}$

29. $-\dfrac{y}{x + 2y - xye^y - y^2 e^y}$ **31.** $(x^3+x)\sqrt{x^3+2}\left(\dfrac{3x^2+1}{x^3+x} + \dfrac{1}{2}\dfrac{3x^2}{x^3+2}\right)$ **33.** $x^x(1 + \ln x)$

35. 1   **37.** -2 **39.** -0.03314 **41.** 31.7295   **43.** -0.189783 **45.** 0   **47. (a)** 500 T-shirts

(b) $\dfrac{dq}{dp}\bigg|_{p=5} = -125$ T-shirts per dollar. Thus, when the price is set at $5, the demand is dropping at a rate of 125 T-shirts per $1 increase in price.

49. $\dfrac{dk}{de}\Big|_{e=15}$ = −0.307 carpenters per electrician. This means that, for a $200,000 house whose construction employs 15 electricians, adding one more electrician would cost as much as approximately 0.307 additional carpenters. In other words, one electrician is worth approximately 0.307 carpenters.   **51. (a)** 22.93 hours. (The other root is rejected, since it is larger than 30.)

(b) $\dfrac{dt}{dx}\Big|_{x=3.0}$ = −11.2132 hours per grade point. This means that, for a 3.0 student who scores 80 on the examination, 1 grade point is worth approximately 11.2132 hours.   **53.** $\dfrac{dr}{dy} = 2\dfrac{r}{y}$, so $\dfrac{dr}{dt} = 2\dfrac{r}{y}\dfrac{dy}{dt}$ by the chain rule   **55.** Let

$y = f(x)g(x)$. Then $\ln y = \ln f(x) + \ln g(x)$, and $\dfrac{1}{y}\dfrac{dy}{dx} = \dfrac{f'(x)}{f(x)} + \dfrac{g'(x)}{g(x)}$, so

$$\dfrac{dy}{dx} = y\left(\dfrac{f'(x)}{f(x)} + \dfrac{g'(x)}{g(x)}\right) = f(x)g(x)\left(\dfrac{f'(x)}{f(x)} + \dfrac{g'(x)}{g(x)}\right) = f'(x)g(x) + f(x)g'(x).$$

Section 4.5 (page 313)

1. $3x + 5$ **3.** $2 - 10x$ **5.** x **7.** $1 + x$ **9.** x **11.** $\dfrac{x}{2} + 1$ **13.** $1.3x - 0.3$

15. $\dfrac{x(e^2 - e^{-2}) + 3e^{-2} - e^2}{2}$ **17.** $0.5 - 0.05x$ **19.** $80\pi(x - 5)$ **21.** 4.0375 **23.** 6.98714

25. 4.0333 **27.** 1.3 **29.** -0.05 **31.** 36.67% **33.** -30% **35.** -20%
37. (a) $148x + 1{,}100$; $C(105) \approx \$16{,}640$. **(b)** 4.65% **39.** The average cost drops by approximately 0.13%.
41. (a) $N(q) = 120q^2 - 600q + 5{,}000$ **(b)** The daily net cost will increase by approximately 19.57%.
43. (a) $K \approx 21{,}930$; $q = 21{,}930p^{-1.040}$ **(b)** An approximate 20.8% drop in sales will result. **(c)** Profits will increase by approximately 59.2%.   **45.** $\pm 4\%$

Chapter 4 Review Exercises (page 318)

1. $3(x - 1)^2$ **3.** $-2(2x + 4)^{-2}$ **5.** $\dfrac{1}{2\sqrt{x + 1}}$ **7.** $-\dfrac{2}{(2x + 1)^2}$ **9.** $\dfrac{2x}{x^2 + 1}$ **11.** $\dfrac{2x + 1}{(x^2 + x)\ln 2}$

13. $-e^{-x}$ **15.** $e^x(1 + x)$ **17.** $9x^2 - 2x$ **19.** $4x\sqrt{x} + \dfrac{2x^2 - 1}{2\sqrt{x}}$ **21.** $\dfrac{-4}{(2x - 1)^2}$

23. $e^x(x^2 + 2x - 2)$ **25.** $4x \ln x + \dfrac{2x^2 - 1}{x}$ **27.** $\dfrac{(2x + 6x^{-4})(x + 1) - x^2 + 2x^{-3} - 1}{(x + 1)^2}$ **29.** $2e^{-2x}$

31. $-4(2x^2 - 2x + 1)^{-5}(4x - 2)$ **33.** $4(t^2 + e^{3t} + 2\sqrt{t})^3\left(2t + 3e^{3t} + \dfrac{1}{\sqrt{t}}\right)$ **35.** $\dfrac{e^x - 2x}{2\sqrt{e^x - x^2}}$

37. $-\dfrac{1}{2x\sqrt{1 - \ln x}}$ **39.** $\dfrac{1.3x^{0.3}(1 + x) - x^{1.3}}{(1 + x)^2}$ **41.** $0.1e^{0.1x}$ **43.** $-\dfrac{2e^x}{(1 + 2e^x)^2}$ **45.** $\dfrac{2b}{t}$

47. $\dfrac{2 \ln x}{x}$ **49.** $(2x - 3)e^{x^2 - 3x + 1}$ **51.** $y = 2$

53. $y = -2x + 1$ **55.** $y = -7t + 23$ **57.** $y = et - e$ **59.** $y = -x + 9/4$ **61.** $\dfrac{2x - 1}{2y}$

63. $-\dfrac{y}{x}$ **65.** $\dfrac{y - (x + y)^2}{x}$ **67.** $x^{x+1}\left(1 + \ln x + \dfrac{1}{x}\right)$ **69.** $x = \dfrac{1 - \ln 2}{2}$ **71.** no such values

73. $x = 0$ **75.** Yes: revenue is rising at a rate of $25 per dollar increase in price.   **77.** 9/20 gph/s   **79.** $500 per employee   **81.** $V'(1{,}000) = 600$ cm³/s   **83.** increasing at a rate of $924.18 per year   **85. (a)** 60,912 frogs per year **(b)** 2006   **87. (a)** growing at a rate of 1.168 trillion grams each year **(b)** 1941   **89. (a)** $9931.71 **(b)** save 0.00144 years per dollar   **91. (a)** Demand drops to about 967. **(b)** $dR/dp = q + p\,(dq/dp)$ is positive, so price should be raised.

Chapter 5

Section 5.1 (page 344)

1. Absolute minimum of -1 at $x = 3$ and $x = -3$, absolute maximum of 2 at $x = 1$; increasing on $[-3, 1]$, decreasing on $[1, 3]$　　**3.** Absolute minimum of 0 at $x = -3$ and $x = 1$, absolute maximum of 2 at $x = -1$ and $x = 3$; increasing on $[-3, -1]$ and $[1, 3]$, decreasing on $[-1, 1]$　　**5.** Local minimum of 1 at $x = -1$; increasing on $[-1, 0)$ and $(0, +\infty)$, decreasing on $(-\infty, -1]$　　**7.** Local maximum of 0 at $x = -3$, absolute minimum of -1 at $x = -2$, stationary point at $(1, 1)$; increasing on $[-2, +\infty)$, decreasing on $[-3, -2]$　　**9.** Local maximum at $(0.0, 2.0)$, absolute minimum at $(1.5, -0.3)$　　**11.** Local maximum at $(0.1, 0.8)$, absolute minimum at $(0.4, 0.7)$　　**13.** Local maximum at $(0.6, 0.3)$, local minimum at $(1.0, 0.0)$　　**15.** No maxima or minima　　**17.** Absolute maximum of 1 at 0; absolute minimum of -3 at 2; local maximum of -2 at 3　　**19.** Absolute minimum of -16 at -4; absolute maximum of 16 at -2; absolute minimum of -16 at 2; absolute maximum of 16 at 4　　**21.** Absolute minimum of -10 at -2; absolute maximum of 10 at 2　　**23.** Absolute minimum of -4 at -2; local maximum of 1 at -1; local minimum of 0 at 0　　**25.** Local maximum of 5 at -1; absolute minimum of -27 at 3　　**27.** Absolute minimum of 0 at 0　　**29.** Local minimum of $\frac{5}{3}$ at -2; local maximum of -1 at 0; local minimum of $\frac{5}{3}$ at 2　　**31.** Local maximum of 0 at $x = 0$; absolute minimum of $-2/(3\sqrt{3})$ at $x = 1/3$.　　**33.** Local maximum of 0 at $x = 0$; absolute minimum of -3 at $x = 1$　　**35.** No local extrema　　**37.** Absolute minimum of 1 at 1　　**39.** Local maximum of $1 + 1/e$ at -1; absolute minimum of 1 at 0; absolute maximum of $e - 1$ at 1.　　**41.** Local maximum at $(-6, -24)$, local minimum at $(-2, -8)$

43. Absolute maximum at $\left(\dfrac{1}{\sqrt{2}}, \sqrt{\dfrac{e}{2}}\right)$, absolute minimum at $\left(-\dfrac{1}{\sqrt{2}}, -\sqrt{\dfrac{e}{2}}\right)$

45. Stationary minimum at $x = -1$　　**47.** Stationary minima at $x = -3$ and $x = 1$, stationary maximum at $x = -1$　　**49.** Singular minimum at $x = 0$, stationary non-extreme point at $x = 1$　　**51.** Stationary maxima at $x = -3$ and $x = 1$, stationary minima at $x = -1$ and $x = 3$　　**53.** Local minima at $(0.15, -0.52)$ and $(2.45, 8.22)$, local maximum at $(1.40, 0.29)$　　**55.** Absolute maximum at $(-5, 700)$, local maxima at $(3.10, 28)$ and $(6, 40)$; absolute minimum at $(-2.10, -393)$; local minimum at $(5, 0)$

Section 5.2 (page 361)

1. $x = y = 5$; $P = 25$　　**3.** $x = y = 3$; $S = 6$　　**5.** $x = 2, y = 4$; $F = 20$　　**7.** $x = 20, y = 10, z = 20$; $P = 4,000$　　**9.** 5×5　　**11.** 5×10　　**13.** $p = \$10$　　**15.** $p = \$30$　　**17. (a)** \$1.41 per pound　**(b)** 5,000 pounds　**(c)** \$7,071.07 per month　　**19.** 34.5¢ per pound, for an annual (per capita) revenue of \$5.95.　　**21.** \$42.50 per ruby, for a weekly profit of \$408.33　　**23. (a)** 656 headsets, for a profit of \$28,120　**(b)** \$143 per headset　　**25.** 1600/27 cubic inches　　**27.** in 30 years' time　　**29.** 55 days　　**31.** Plant 25 additional trees.　　**33.** 38,730 CD players, giving an average cost of \$28 per CD player.　　**35.** It should remove 2.5 pounds of pollutant per day.　　**37.** $l = 30$ inches, $w = 15$ inches, $h = 30$ inches　　**39.** $l = 36$ inches, $w = h = 18$ inches, $V = 11,664$ cubic inches　　**41.** 1,600 copies. At this value of x, average profit equals marginal profit; beyond this the marginal profit is smaller than the average.　　**43.** decreasing most rapidly in 1963; increasing most rapidly in 1989　　**45.** Maximum when $t = 17$ days. This means that the embryo's oxygen consumption is increasing most rapidly 17 days after the egg is laid.　　**47.** $h = r = 11.7$ cm　　**49.** Absolute minimum of 5,137 at $n = 13.2$, absolute maximum of 34,040 at $n = 15$. Thus, the salary value per extra year of school is increasing most slowly (\$5,137 per year) at a level of 13.2 years of schooling, and most rapidly (\$34,040 per year) at a level of 15 years of schooling.　　**51.** You should sell them in 17 years' time, when they will be worth approximately \$3,960　　**53.** 71 employees　　**55. (d)**

Section 5.3 (page 379)

1. 6　　**3.** $\dfrac{4}{x^3}$　　**5.** $-0.96x^{-1.6}$　　**7.** $e^{-(x-1)}$　　**9.** $\dfrac{2}{x^3} + \dfrac{1}{x^2}$　　**11.** $a = -32$ ft/s^2; at $t = 2$, $a = -32$ ft/s^2　　**13.** $a = 2/t^3 + 6/t^4$ ft/s^2; at $t = 1$, $a = 8$ ft/s^2　　**15.** $a = -1/(4t^{3/2}) + 2$ ft/s^2; at $t = 4$, $a = 63/32$ ft/s^2.

17. $(1, 0)$ **19.** $(1, 0)$ **21.** none **23.** $(-1, 0), (1, 1)$ **25.** minimum at $(-1, 0)$, no points of inflection **27.** maximum at $(-2, 21)$; minimum at $(1, -6)$; point of inflection at $(-\frac{1}{2}, \frac{15}{2})$ **29.** minimum at $(-\frac{1}{2}, \frac{3}{4})$; maximum at $(0, 1)$; point of inflection at $(-\frac{1}{4}, \frac{7}{8})$. **31.** maximum at $(0, 0)$; minimum at $(1, -2)$; no inflection points **33.** minimum at $(1, 1)$; no points of inflection **35.** no local extrema; point of inflection at $(1, 1)$ and $(-1, 1)$. **37.** $(0.70, 2.32)$ **39.** $(-0.71, 0.61), (0.71, 0.61)$ **41.** $(0.79, -0.55), (0.21, 0.61)$ **43.** no points of inflection **45.** (a) 2 years into the epidemic (b) 2 years into the epidemic **47.** (a) 1992 (b) 1994 (c) 1990 **49.** Point of inflection at $(12.89, 193.8)$. The prison population was declining most rapidly at $t = 12.89$ (that is, in 1963), at which time the prison population was 193,800. **51.** (a) There are no points of inflection in the graph of S. (b) Since the graph is concave up, the derivative of S is increasing, and so the rate of *decrease* of SAT scores with increasing numbers of prisoners is diminishing. In other words, the apparent effect of more prisoners is diminishing. **53.** (a) $\left.\dfrac{d^2 n}{ds^2}\right|_{s=3} = -21.494$. Thus, for a firm with annual sales of \$3 million, the rate at which new patents are produced decreases with increasing firm size. This means that the returns (as measured in the number of new patents per increase of \$1 million in sales) are diminishing as the firm size increases. (b) $\left.\dfrac{d^2 n}{ds^2}\right|_{s=7} = 13.474$. Thus, for a firm with annual sales of \$7 million, the rate at which new patents are produced increases with increasing firm size by 13.474 new patents per \$1 million^2. (c) There is a point of inflection when $s \approx 5.4587$, so that in a firm with annual sales of \$5,458,700 per year, the number of new patents produced per additional \$1 million in sales is a minimum. **55.** $n = 10t - t^2$

57. (a) $a = 10, b = 19, k = 0.5; R = \dfrac{10}{1 + 19e^{-0.5t}}$ (b) $R''(0) = \$0.00106875$ million per month^2. This means that, when $t = 0$, the revenue is accelerating by \$0.00106875 million per month each month. **59.** (a) 10 years and 50 years (b) 9 years and 42 years (c) \$23 per year, after 9 years

Section 5.4 (page 398)

1.

x-intercept at $2 - \sqrt{3}$, y-intercept at 1
Absolute minimum at $(2, -3)$
Absolute maximum at $(0, 1)$
Local maximum at $(3, -2)$
No points of inflection

3.

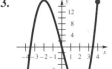

x-intercepts at 0 and $\pm 2\sqrt{3}$, y-intercept at 0
Absolute minima at $(-4, -16)$ and $(2, -16)$
Absolute maxima at $(-2, 16)$ and $(4, 16)$
Point of inflection at $(0, 0)$

5.

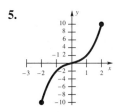

Intercept of 0 on both axes
Absolute minimum at $(-2, -10)$
Absolute maximum at $(2, 10)$
Point of inflection at $(0, 0)$

7.

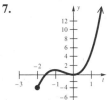

t-intercepts $-\frac{3}{2}$ and 0, y-intercept 0
Absolute minimum at $(-2, -4)$
Local minimum at $(0, 0)$
Local maximum at $(-1, 1)$
Point of inflection at $\left(-\frac{1}{2}, \frac{1}{2}\right)$

9.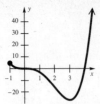

x-intercepts at 0 and 4, y-intercept at 0
Local maximum at $(-1, 5)$
Absolute minimum at $(3, -27)$
Points of inflection at $(0, 0)$ and $(2, -16)$

11.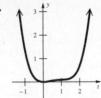

t-intercept 0, y-intercept, 0
Absolute minimum at $(0, 0)$
Points of inflection at $\left(\frac{1}{3}, \frac{11}{324}\right)$ and
$\left(1, \frac{1}{12}\right)$

13.

No t-intercept, y-intercept at -1
Local minima at $\left(-2, \frac{5}{3}\right)$ and $\left(2, \frac{5}{3}\right)$
Local maximum at $(0, -1)$
No points of inflection
Vertical asymptotes: $x = \pm 1$

15.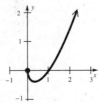

x-intercepts at 0 and 1, y-intercept at 0
Local maximum at $(0, 0)$
Absolute minimum at $\left(\dfrac{1}{3}, -\dfrac{2}{3\sqrt{3}}\right)$
No points of inflection

17.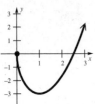

x-intercepts at 0 and $4^{2/3}$, y-intercept at 0
Local maximum at $(0, 0)$
Absolute minimum at $(1, -3)$
No points of inflection

19.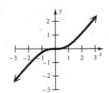

Intercept 0 on both axes
No maxima or minima
Point of inflection at $(0, 0)$

21.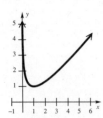

No intercepts
Absolute minimum at $(1, 1)$
No points of inflection
Vertical asymptote at $x = 0$

23.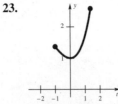

No t-intercepts, y-intercept at 1
Absolute minimum at $(0, 1)$
Absolute maximum at $(1, e - 1)$
Local maximum at $(-1, 1 + e^{-1})$
No points of inflection

25.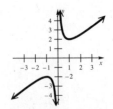

No intercepts
Local minimum at $(1, 2)$
Local maximum at $(-1, -2)$
No points of inflection
Vertical asymptote: $y = 0$

27.

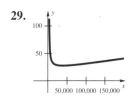

(a) Absolute maximum of 300,000 at $t = 0$. Local minimum of 242,612 at $t = 10$. Local maximum of 267,756 at $t = 30$. Points of inflection at (17.6393, 252,955) and (62.3607, 185,322). An appropriate domain is $[0, +\infty)$.
(b) Increasing most rapidly in 17.64 years, decreasing most rapidly in 0 years. (c) The model predicts that the collection of cars will eventually becomes worthless in terms of discounted value. This is saying that the value of the collection of classic cars, although increasing without bound, will eventually become worthless as a result of inflation. This is not reasonable if collectors continue to prize classic cars, so we may conclude that the given model is not a reliable one for long-term prediction.

29.

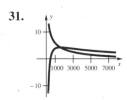

Absolute minimum at $(\sqrt{1,500,000,000}, 27.74597) \approx (38,730, 27.75)$ (a) This shows that the average cost per CD player is a minimum of $27.75 per player at a production level of 38,730 CD players. (b) $\lim_{x \to +\infty} \overline{C}(x) = +\infty$, showing that, when the production level is pushed higher and higher, the average cost per item rises without bound.
(c) $\lim_{x \to 0^+} \overline{C}(x) = +\infty$ showing that the average cost per item grows without bound as the number of items produced approaches zero

31.

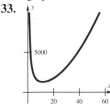

The graphs cross each other at (1,600, 4.25), which is the absolute maximum of the average profit function. When the average profit is maximized, the marginal profit and the average profit are equal.

33.

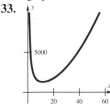

The limits at 0 and infinity are both infinite. Since the limit at zero is infinite, the amount of plastic needed to make very narrow buckets (small radius) would be very large. The reason for this is that the buckets would have to be extremely tall in order to hold the requisite volume. On the other hand, since the limit at infinity is also infinite, this says that a large amount of material would be needed to make very wide buckets (even though they would be very short).

35.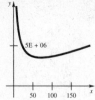

(a) Estimate the derivative at the point where $x = 50$. This is the slope of the tangent: $-\$15,858$ per year per employee.   (b) $\lim\limits_{x \to +\infty} C'(x) = +\infty$. This says that as the number of employees becomes large, the additional cost per employee becomes large without bound.

37.

Smallest rate of increase at $s = 5.46$

39.

(a) Absolute maximum at $x = 72.6$; points of inflection at $x = 67.4$ and $x = 77.8$
(b) 0.03%
(c) 72.6
(d) points of inflection

41.

Maxima at -4.7, 1.6, 7.9; minima at -7.9, -1.6, 4.7
Points of inflection at -9.4, -6.3, -3.1, 0, 3.1, 6.3, 9.4. The limit does not exist, because the graph oscillates between -1 and $+1$ as the x-coordinate approaches $+\infty$, or $-\infty$.

Section 5.5 (page 412)

1. $P = 10,000;\ \dfrac{dP}{dt} = 1,000$ **3.** Let R be the annual revenue of my company, and let q be annual sales. $R = 7,000$ and $\dfrac{dR}{dt} = 700$. Find $\dfrac{dq}{dt}$.   **5.** Let p be the price of a pair of shoes, and let q be the demand for shoes. $\dfrac{dp}{dt} = 5$. Find $\dfrac{dq}{dt}$.   **7.** Let T be the average global temperature, and let q be the number of pairs of Bermuda shorts sold per year. $T = 60$ and $\dfrac{dT}{dt} = 0.1$. Find $\dfrac{dq}{dt}$.   **9.** (a) $\dfrac{3}{5\pi} \approx 0.1910$ cm/s.   (b) $\dfrac{6}{7\sqrt{\pi}} \approx 0.4836$ cm/s.   **11.** $7\frac{1}{2}$ ft/s.

13. The price is decreasing at a rate of 31¢ per pound per month.   **15.** Monthly sales will drop at a rate of 40 T-shirts per month.   **17.** Raise the price by 3¢ per week.   **19.** The daily operating budget is dropping at a rate of $2.40 per year.

21. $\dfrac{3}{4\pi} \approx 0.2387$ ft/min   **23.** The y-coordinate is decreasing at a rate of 16 units per second.

25. $\dfrac{2300}{\sqrt{4100}} \approx 35.92$ mph   **27.** $1,814 per year   **29.** Their prior experience must increase at a rate of 0.9668 years every year.   **31.** $\dfrac{2500}{9\pi}\left(\dfrac{3}{5000}\right)^{2/3} \approx 0.6290$ m/s   **33.** $\dfrac{\sqrt{1 + 128\pi}}{4\pi} \approx 1.598$ cm/s   **35.** The average SAT score was 904.70 and decreasing at a rate of 0.11 per year.

Section 5.6 (page 422)

1. $E = 1.5$; the demand is going down 1.5% per 1% increase in price at that price level; revenue is maximized when $p = 25; weekly revenue at that price is $12,500.   **3. (a)** $E = \frac{6}{7}$; the demand is going down 6% per 7% increase in price at that price level; thus a price increase is in order.   **(b)** Revenue is maximized when $p = 100/3 \approx 33.33.   **(c)** Demand would be $(100 - 100/3)^2 = (200/3)^2 \approx 4,444$ cases per week.   **5. (a)** 1.71. Thus, the demand is elastic at the given tuition level, showing that an decrease in fees will result in an increase in revenue.   **(b)** The college should charge an average of $2,271.75 per student, and this will result in an enrollment of about 4,930 students, giving a revenue of $11,199,029.   **7. (a)** $E = -\dfrac{mp}{mp + b}$ **(b)** $p = -\dfrac{b}{2m}$ **9. (a)** $E = r$ **(b)** E is independent of p.   **(c)** If $r = 1$, then the revenue is not affected by the price. If $r > 1$, then the revenue is always elastic, while if $r < 1$, the revenue is always ineastic. This is an unrealistic model, since there should always be a price at which the revenue is a maximum.   **11. (a)** $E = 51$; the demand is going down 51% per 1% increase in price at that price level; thus, a large price decrease is advised.   **(b)** Revenue is maximized when $p = ¥0.50$.   **(c)** Demand would be $100e^{-3/4 + 1/2} \approx 77.88$ paint-by-number sets per month.   **13. (a)** $q = -1500p + 6000$ **(b)** $2 per hamburger, giving a total weekly revenue of $6,000   **15. (a)** $f(p) = 1,000e^{-0.3p}$ **(b)** at $p = $3, E = 0.9$; at $p = $4, E = 1.2$; at $p = $5, E = 1.5$ **(c)** $p = 3.33 **(d)** $p = 5.36. Selling at a lower price would increase demand, but you cannot sell more than 200 pounds anyway. You should charge as much as you can and still be able to sell all 200 pounds.   **17.** $\dfrac{Y}{Q} \cdot \dfrac{dQ}{dY} = \beta$. An increase in income of x% will result in an increase in demand of βx%. (Note that we should *not* take the negative here, as we expect an increase in income to produce an *increase* in demand.)

Chapter 5 Review Exercises (page 427)

1. Absolute minimum of $-3/2$ at $x = \frac{1}{2}$, local maximum of -1 at $x = 0$, absolute maximum of 11 at $x = 3$ **3.** Local maximum of 5 at $x = -1$, absolute minimum of -3 at $x = 1$ and $x = -2$ **5.** Absolute minimum of 0 at $t = -2$ and 0, local maximum of $\frac{1}{4}$ at $t = -1$ **7.** Absolute minimum of $-\frac{1}{9}$ at $t = -2$, absolute maximum of 3 at $t = 2$.   **9.** Absolute minimum of 0 at $t = 1$ **11.** Local maximum of 2 at $x = -1$, local minimum of -2 at $x = 1$ **13.** Absolute minimum of $\frac{1}{2}$ at $r = 1$ **15.** Absolute minimum of 2 at $t = 0$

17.   **19.**   **21.**   **23.**

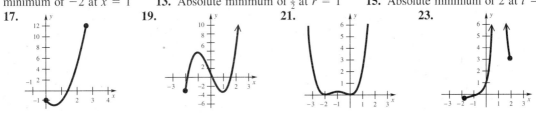

25. 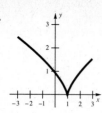   **27.** 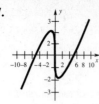   **29.** 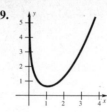   **31.**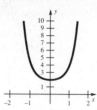

33. $x = 5$, $y = 5\sqrt{2}$, $P = 250$ **35.** $x = z = \frac{3}{2}$, $y = 6$, $S = 12$ **37.** width $1/\sqrt{3}$ and height $\frac{2}{3}$ **39.** 31.62 mph   **41.** $\sqrt{2}$ **43.** \$66.67   **45.** $q = 1{,}000$ books, \$40 per book   **47.** \$80 per room, giving 40 vacancies   **49.** Sales are increasing at a rate of 0.9 dozen roses per week.   **51.** R is decreasing at a rate of 34/49 ohm/s.   **53.** 2 ft/s   **55.** $E = \frac{1}{2}$ at a price of \$60   **57.** $p \approx \$35$

▶ Chapter 6

Section 6.1 (page 441)

1. $\frac{1}{6}x^6 + C$ **3.** $6x + C$ **5.** $\frac{x^2}{2} + C$ **7.** $\frac{x^3}{3} - \frac{x^2}{2} + C$ **9.** $x + \frac{x^2}{2} + C$ **11.** $-\frac{x^{-4}}{4} + C$

13. $\frac{u^3}{3} - \ln |u| + C$ **15.** $\frac{3x^5}{5} + 2x^{-1} - \frac{x^{-4}}{4} + 4x + C$ **17.** $2e^x + 5 \ln |x| + C$

19. $\ln |x| - \frac{2}{x} + \frac{1}{2x^2} + C$ **21.** $\frac{3x^{1.1}}{1.1} - \frac{x^{5.3}}{5.3} + C$ **23.** $\frac{x^{0.9}}{0.9} + \frac{40}{x^{0.1}} + C$ **25.** $-\frac{1}{x} - \frac{1}{x^2} + C$

27. $\frac{2x^{3/2}}{3} + C$ **29.** $\frac{3x^{4/3}}{2} - x^{1/2} + C$ **31.** $\ln |x| + \frac{4}{\sqrt{x}} + C$ **33.** $f(x) = \frac{x^2}{2} + 1$

35. $f(x) = e^x - x - 1$ **37.** $C(x) = 5x - x^2/20{,}000 + 20{,}000$ **39.** $C(x) = 5x + x^2 + \ln x + 994$
41. (a) $s = \frac{1}{3}t^3 + t + C$;   **(b)** $C = 1$; $s = \frac{1}{3}t^3 + t + 1$

Section 6.2 (page 451)

1. $\frac{1}{18}(3x + 1)^6 + C$ **3.** $\frac{1}{2}(-2x + 2)^{-1} + C$ **5.** $\frac{1}{24}(3x^2 + 3)^4 + C$ **7.** $\frac{2}{9}(3x^2 - 1)^{3/2} + C$
9. $-\frac{1}{2}e^{-x^2+1} + C$ **11.** $-\frac{1}{2}e^{-(x^2+2x)} + C$ **13.** $\frac{1}{2}(x^2 + x + 1)^{-2} + C$
15. $\frac{1}{3}(2x^3 + x^6 - 5)^{1/2} + C$ **17.** $4(\frac{1}{5}(x + 1)^{5/2} - \frac{1}{3}(x + 1)^{3/2}) + C$
19. $\frac{1}{7}(x - 2)^7 + \frac{1}{3}(x - 2)^6 + C$
21. $3e^{-1/x} + C$ **23.** $\frac{(x^2 + 1)^{2.3}}{4.6} + C$ **25.** $x + 3e^{3.1x-2} + C$ **27.** $20 \ln |1 - e^{-0.05x}| + C$
29. $\frac{1}{2}(e^{2x^2-2x} + e^{x^2}) + C$ **35.** $-e^{-x} + C$ **37.** $\frac{1}{2}e^{2x-1} + C$
39. $\frac{(2x + 4)^3}{6} + C$ **41.** $\frac{\ln |5x - 1|}{5} + C$ **43.** $\frac{(1.5x)^4}{6} + C$ **45.** $f(x) = \frac{1}{8}(x^2 + 1)^4 - \frac{1}{8}$
47. $f(x) = \frac{e^{x^2-1}}{2}$ **49. (a)** $s = \frac{1}{10}(t^2 + 1)^5 + \frac{1}{2}t^2 + C$ **(b)** $C = \frac{9}{10}$; $s = \frac{1}{10}(t^2 + 1)^5 + \frac{1}{2}t^2 + \frac{9}{10}$
51. $C(x) = 5x + \frac{2}{3}(x + 1)^{3/2} + 19{,}999.33$ **53.** $C(x) = 5x + \ln (x + 1) + 994.31$

Section 6.3 (page 464)

1. $C(x) = 10{,}000 + 100x + 0.005x^2$ **3.** $C(x) = 12{,}000 + 100x - 2{,}000e^{-0.01x}$
5. $C'(x) = 200 - 2x$, $C(x) = 200x - x^2$ **7.** $C(x) = 5{,}000(\ln (x + 10) - \ln 10) = 5{,}000 \ln \left(\frac{x}{10} + 1\right)$

9. $x = 1{,}000$, $\overline{C}(1{,}000) = 220$ **11.** $\overline{C}(x) = 20 - 400\,\dfrac{e^{-0.05x} - 1}{x}$ **13. (a)** $s(t) = 10t - \dfrac{0.2t^3}{3}$

(b) $s(7) \approx 47$ T-shirts   **15. (a)** (C) **(b)** $P(t) = 10t + 0.52t^3$ **(c)** $P(10) = \$620$ billion

17. (a) $E(t) = 460 \ln (1+e^{(t-4)}) - 8.349$; **(b)** $E(15) = \$5{,}052$ million

19. (a) $P(t) = 39.8107(6{,}000 + 2{,}000t)^{0.7}$ **(b)** 42,878 cellular phones

23. $\$940$ million   **25.** 3 years   **27.** 320 ft/s downwards.

31. $(1280)^{1/2} \approx 35.78$ ft/s   **33.** 113 feet below me   **35. (a)** 80 feet per second **(b)** 60 ft/s

(c) 1.25 seconds   **37.** $\sqrt{2} \approx 1.414$ times as fast   **39.** 50,724,000 feet   **41.** $88/3 \approx 29.33$ ft/s²

Section 6.4 (page 479)

1. $\dfrac{14}{3}$ **3.** 0   **5.** $\dfrac{40}{3}$ **7.** -0.9045 **9.** $2(e - 1)$ **11.** $\frac{1}{2}(e^1 - e^{-3})$ **13.** $50(e^{-1} - e^{-2}) \approx 11.627$

15. $\frac{2}{3}$ **17.** $e^{2.1} - e^{-0.1}$ **19.** 0   **21.** $\frac{5}{2}(e^3 - e^2)$ **23.** $\dfrac{3^{5/2}}{10} - \dfrac{3^{3/2}}{6} + \dfrac{1}{15}$ **25.** $[\ln 26 - \ln 7]/3$

27. $\frac{1}{2}$ **29.** $16/3$ **31.** $56/3$ **33.** $\frac{1}{2}$ **35.** 296 miles   **37.** $\$783$ **39.** $\$2.4$ billion

41. (a) $c(t) = 9.2 - 0.7636t$ **(b)** $\$55.0022$ billion   **43.** 60.7 milliliters   **45.** $\$5{,}052$ million   **47.** 42,878

cellular phones   **49.** 9 gallons   **51. (a)** $s(t) = 330e^{0.0872t}$ **(b)** $\$9{,}044$ billion (compared to $\$8{,}225$ billion)

(c) $\$7{,}011$ billion   **53.** 926 T-shirts

Section 6.5 (page 489)

1. Left sum: 4; right sum: 8; trap. sum: 6; error: 0   **3.** Left sum: 6; right sum: 6; trap. sum: 6; error: ± 1

5. Left sum: -0.4; right sum: 0.4; trap. sum: 0; error: 0   **7.** Left sum: 2.3129; right sum: 0.3130; trap. sum:

1.3130; error: ± 1 **9.** Left sum: 0.7456; right sum: 0.6456; trap. sum: 0.6956; error: ± 0.002

11. 3.0370   **13.** 0.8863   **15.** 0.2648   **17. (a)** Left sum: 3.1604; right sum: 3.1204; trap. sum: 3.1404

(b) Left sum: 3.1512; right sum: 3.1312; trap. sum: 3.1412   **19. (a)** Left sum: 0.0258; right sum: 0.0254; trap.

sum: 0.0256   **(b)** Left sum: 0.0257; right sum: 0.0255; trap. sum: 0.0256

21. (a) Left sum: 1.9998; right sum: 1.9998; trap. sum: 1.9998 **(b)** Left sum: 2.0000; right sum: 2.0000; trap.

sum: 2.0000   **23. (a)** $s(t) = -2.925t + 18.9625$ **(b)** The model gives total sales of $\$58.25$ billion. The actual

total sales were $\$66.4$ billion, much higher than the sales predicted by the model. One reason for this is that the 1992

sales were considerably higher than predicted by the linear model. (If you join the midpoint of the top of the first

column to that of the last, you will find that the 1992 sales "stick out" above this line.)

25. $\$1{,}799.01$ (same answer as given by the FTC)

27. Yes. Both estimates give an area of 420 square feet.   **29. (a)** 99.2% **(b)** $1.5 \times 10^{-14}\%$

Section 6.6 (page 497)

1. $A(x) = x$ **3.** $A(x) = \dfrac{(x - 2)^2}{2}$ **5.** $A(x) = c(x - a)$ **7.** $A(x) = \dfrac{x^2}{2}$ **9.** $A(x) = \dfrac{x^3}{3}$

11. $A(x) = e^x - 1$ **13.** $A(x) = \ln x$ **15.** $A(x) = x^2/2$ if $0 \le x \le 1$; $\frac{1}{2} + 2(x - 1)$ if $x > 1$

19. $f(0) = 0, f(0.5) = 0.46, f(1) = 0.75$ **21.** $f(-0.5) = -0.49, f(0.5) = 0.49$

Section 6.7 (page 509)

1. estimate $= 4$, error $= 0$ **3.** estimate $= 20.4525$, error ≤ 0.405, exact answer $= 20.25$

5. estimate $= 4.037$, error ≤ 0.053 **7.** 4, exact   **9.** 20.25, exact   **11.** 4.0470 ± 0.0034

13. all estimates $= 4$ **15. (a)** 20.252025 **(b)** 20.250506 **(c)** 20.250081

17. (a) 4.0470829 **(b)** 4.0471629 **(c)** 4.0471853   **19.** all estimates $= 4$ **21.** all

estimates $= 20.25$ **23. (a)** 4.0471895340 **(b)** 4.0471895604 **(c)** 4.0471895621   **25.** $n \ge 566$

27. $n \ge 15$ **29.** $n \ge 1{,}291$ **31.** $n \ge 69$ **33. (a)** 1.478940 **(b)** 1.478942857

35. (a) 2.003503 **(b)** 2.003497111

Chapter 6 Review Exercises (page 513)

1. $10x^{11}/11 + C$ **3.** $-1/x^3 + C$ **5.** $(2/3)x^{3/2} + \ln |x| + C$ **7.** $2e^x + 3x^2/2 - 4\ln |x| + C$
9. $(x + 2)^{11}/11 + C$ **11.** $(3/8)(x^2+1)^{4/3} + C$ **13.** $(1/2) \ln(x^2 + 1) + C$ **15.** $-(5/2)e^{-2x} + C$
17. $e^{x^2}/2 - (3/2)x^2 + C$ **19.** $\dfrac{4.7}{1.2} \ln |x^{1.2} - 4| + C$ **21.** $\dfrac{1}{0.6} \ln |1 + 2e^{0.3t}| + C$
23. $5/4$ **25.** $3/2$ **27.** $3/2 - 1/e$ **29.** $52/9$ **31.** $4 + \dfrac{2^5 - 1}{6}$ **33.** 0 **35.** $4/3$ **37.** $\ln 10$
39. $\dfrac{1 - e^{-25}}{2}$ **41.** $\dfrac{2\sqrt{2}}{3}$ **43.** $4/15$ **45.** $2/3$ **47.** Both sums are 1.46; trap. sum: 1.46
49. Left sum: -0.14, right sum: 0.55, trap. sum = 0.20
51. $n = 100$: left sum = 2.9257, right sum = 2.9257; $n = 200$: left sum = 2.9254, right sum = 2.9254
$n = 500$: left sum = 2.9253, right sum = 2.9253   **53.** $n = 100$: left sum = 0.3649, right sum = 0.3636;
$n = 200$: left sum = 0.3646, right sum = 0.3639; $n = 500$: left sum = 0.3644, right sum = 0.3641
55. $n = 50$: 1.49364830; $n = 100$: 1.49364827; $n = 500$: 1.49364827   **57.** $n = 50$: 0.23286788; $n = 100$:
0.23286795; $n = 500$: 0.23286795   **59.** $n = 50$: 2.925305; $n = 100$: 2.925304; $n = 500$: 2.925303
61. $n = 50$: 0.36423405; $n = 100$: 0.36423404; $n = 500$: 0.36423404   **63.** $n \geq 37$ **65.** $n \geq 34$
67. $n \geq 105$ **69.** $n \geq 13$ **71.** $n \geq 8$ **73.** $n \geq 8$ **75.** $n \geq 14$ **77.** $n \geq 6$
79. (a) $S(t) = -2.2t + 70$, $t =$ years since 1982   **(b)** actual = 704 million, estimated = 680 million
(c) 634 million   **81.** 943 ft   **83.** 6,560 ft   **85.** $49,850 + 50x + 40 \ln (x + 40)$
87. $75x + 25(1 - e^{-x})$ **89. (a)** $2 per watt of capacity   **(b)** $p = 2e^{-0.1597t}$ **(c)** 473.5 megawatts

Chapter 7

Section 7.1 (page 525)

1. $2e^x(x - 1) + C$ **3.** $-e^{-x}(2 + 3x) + C$ **5.** $e^{2x}(2x^2 - 2x - 1)/4 + C$
7. $-e^{-2x+4}(2x^2 + 2x + 3)/4 + C$ **9.** $-e^{-x}(x^2 + x + 1) + C$ **11.** $(x^4 \ln x)/4 - x^4/16 + C$
13. $(t^3/3 + t)\ln 2t - t^3/9 - t + C$ **15.** $\frac{3}{4}t^{4/3}(\ln t - \frac{3}{4}) + C$ **17.** e **19.** $38{,}229/286$
21. $(\frac{7}{2})\ln 2 - \frac{3}{4}$ **23.** $\frac{1}{4}$ **25.** $1 - 11e^{-10}$ **27.** $4 \ln 2 - \frac{7}{4}$ **29.** $28{,}800{,}000(1 - 2e^{-1})$ ft
31. $5001 + 10x - 1/(x + 1) - [\ln(x + 1)]/(x + 1)$ **33.** $33,598   **35.** $1,478 million

Section 7.2 (page 530)

1. $1 + x - \ln |1 + x| + C$ **3.** $\frac{1}{30}(6x - 2)(1 + 2x)^{3/2} + C$ **5.** $\dfrac{x}{2}\sqrt{x^2 + 4} + 2 \ln |x + \sqrt{x^2 + 4}| + C$

7. $\tan^{-1}x + C$ **9.** $\ln |x + \sqrt{3 + x^2}| + C$ **11.** $\frac{1}{4} \ln \left|\dfrac{2x + 1}{2x - 1}\right| + C$ **13.** $\frac{1}{3}(2x^2 - 1)^{3/2} + C$

15. $\dfrac{1}{2\sqrt{3}} \ln \left|\dfrac{\sqrt{3}x - 1}{\sqrt{3}x + 1}\right| + C$ **17.** $\tan^{-1}(x + 3) + C$

19. $\dfrac{2x - 1}{4}\sqrt{x^2 - x + 3} + \dfrac{11}{8} \ln \left|x - \dfrac{1}{2} + \sqrt{x^2 - x + 3}\right| + C$ **21.** $-\dfrac{1}{6x} + \dfrac{1}{3} \ln \left|\dfrac{2 + 4x}{x}\right| + C$

23. $\dfrac{1}{6} \ln \left|\dfrac{e^x + 3}{e^x - 3}\right| + C$ **25.** $\tan^{-1}(\ln x) + C$ **27.** $C(x) = 2[15x^2 - 12x + 8](1 + x)^{3/2} + 9{,}984$

29. $C(x) = 30 \ln |x + \sqrt{x^2 + 9}| + 10{,}000 - 30 \ln 3$ **31.** $5{,}000 \ln |x + \sqrt{x^2 + 1}|$

33. $\bar{C}(x) = \dfrac{10}{x}\left[\ln \left|\dfrac{x + \sqrt{4 + x^2}}{2}\right|\right] + \dfrac{10{,}000}{x}$ **35.** 119,004.17 ft

Section 7.3 (page 543)

1. $\frac{8}{3}$ **3.** 4   **5.** 1   **7.** $e - \frac{3}{2}$ **9.** $\frac{2}{3}$ **11.** $\frac{3}{10}$ **13.** $\frac{1}{20}$ **15.** $\frac{4}{15}$ **17.** $2 \ln 2 - 1$
19. $8 \ln 4 + 2e - 16$ **21.** 0.3222   **23.** 0.3222   **25.** $6.25   **27.** $512   **29.** $119.53   **31.** $900
33. $416.67   **35.** $326.27   **37.** $25   **39.** $0.50   **41.** $386.29   **43.** $225   **45.** $25.50
47. $12,684.60   **49.** $\bar{p} = \$3,655$, CS $= \$856,140$, PS $= \$3,715, 650$. The university would earn $3,715,650 less.

51. You save more.   **53.** CS $= \dfrac{1}{2m}(b - m\bar{p})^2$.   **55. (a)** $373.35 million   **(b)** This is the area of the region

between the graphs of $R(t)$ and $P(t)$ for $0 \le t \le 4$.   **(c)** Since the exponent for P is larger, this tells us that the ratio of profit to revenue was increasing; that is, costs accounted for a decreasing proportion of the revenues.
57. (b) $7,260 million   **(c)** The area between $y = 1,000,000(41.25q^2 - 697.5q + 3,210)$ and $y = 400,000,000$ for $0 \le q \le 12$.

Section 7.4 (page 554)

1. $14,403 million   **3.** $215.714 billion
5. Average $= 2$

7. Average $= 1$

9. Average $= (1 - e^{-2})/2$

11. Moving average: 2.6, 2.8, 3.0, 3.2, 3.4, 4.0, 4.6, 5.2, 5.8, 6.4, 7.0, 7.6, 7.8, 8.0, 8.2, 8.4

13. Moving average: $\bar{f}(x) = x^3 - (15/2)x^2 + 25x - 125/4$

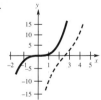

15. Moving average: $\bar{f}(x) = (3/25)[x^{5/3} - (x - 5)^{5/3}]$

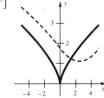

17. $\bar{f}(x) = \frac{2}{5}(e^{0.5x} - e^{0.5(x-5)})$

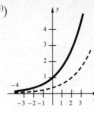

19. $\bar{f}(x) = \frac{2}{15}(x^{3/2} - (x-5)^{3/2})$

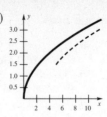

21. $10,410.88 **23.** $1500 **25.**

27.

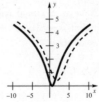

29. (a) $n(t) = 6.4 + 0.3t$, where t is the number of years since 1983 **(b)** $\bar{n}(t) = 5.8 + 0.3t$ **(c)** The slope of the moving average is the same as the slope of the original function. **(d)**

31. $\bar{f}(x) = mx + b - \dfrac{ma}{2}$

33. (a) $p(t) = -0.04t^2 + t + 12$ **(b)** 16.67% **(c)** 10.38% **35. (a)**

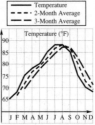

(b) The 24-month moving average is constant and equal to the year-long average of approximately 77°. **(c)** A quadratic model could not be used to predict temperatures beyond the given 12-month period, since temperature patterns are periodic, whereas parabolas are not.

37. (a) 78.10 **(b)** 7.48% Slightly less than the answer to Ex. 4. (This is expected because of compounding over eight years.) **39.**

41. (a) Average voltage over $[0, \frac{1}{6}]$ is zero; 60 cycles per second. **(b)**

(c) 116.673 volts.

Section 7.3 (page 543)

1. $\frac{8}{3}$ **3.** 4   **5.** 1   **7.** $e - \frac{3}{2}$ **9.** $\frac{2}{3}$ **11.** $\frac{3}{10}$ **13.** $\frac{1}{20}$ **15.** $\frac{4}{15}$ **17.** $2\ln 2 - 1$
19. $8\ln 4 + 2e - 16$ **21.** 0.3222   **23.** 0.3222   **25.** $6.25   **27.** $512   **29.** $119.53   **31.** $900
33. $416.67   **35.** $326.27   **37.** $25   **39.** $0.50   **41.** $386.29   **43.** $225   **45.** $25.50
47. $12,684.60   **49.** $\bar{p} = \$3,655$, CS $= \$856,140$, PS $= \$3,715, 650$. The university would earn \$3,715,650 less.

51. You save more.   **53.** CS $= \dfrac{1}{2m}(b - m\bar{p})^2$.   **55. (a)** \$373.35 million   **(b)** This is the area of the region
between the graphs of $R(t)$ and $P(t)$ for $0 \le t \le 4$.   **(c)** Since the exponent for P is larger, this tells us that the ratio
of profit to revenue was increasing; that is, costs accounted for a decreasing proportion of the revenues.
57. (b) \$7,260 million   **(c)** The area between $y = 1,000,000(41.25q^2 - 697.5q + 3,210)$ and $y = 400,000,000$ for
$0 \le q \le 12$.

Section 7.4 (page 554)

1. \$14,403 million   **3.** \$215.714 billion
5. Average $= 2$ **7.** Average $= 1$

9. Average $= (1 - e^{-2})/2$

11. Moving average: 2.6, 2.8, 3.0, 3.2, 3.4, 4.0, 4.6, 5.2, 5.8, 6.4, 7.0, 7.6, 7.8, 8.0, 8.2, 8.4

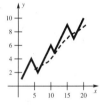

13. Moving average: $\bar{f}(x) = x^3 - (15/2)x^2 + 25x - 125/4$

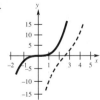

15. Moving average: $\bar{f}(x) = (3/25)[x^{5/3} - (x - 5)^{5/3}]$

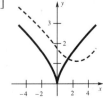

17. $\bar{f}(x) = \frac{2}{5}(e^{0.5x} - e^{0.5(x-5)})$

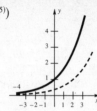

19. $\bar{f}(x) = \frac{2}{15}(x^{3/2} - (x-5)^{3/2})$

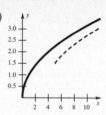

21. $10,410.88   **23.** $1500   **25.**

27.

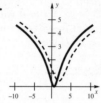

29. (a) $n(t) = 6.4 + 0.3t$, where t is the number of years since 1983   **(b)** $\bar{n}(t) = 5.8 + 0.3t$ **(c)** The slope of the moving average is the same as the slope of the original function.   **(d)**

31. $\bar{f}(x) = mx + b - \dfrac{ma}{2}$

33. (a) $p(t) = -0.04t^2 + t + 12$ **(b)** 16.67%   **(c)** 10.38%   **35. (a)**

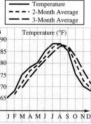

(b) The 24-month moving average is constant and equal to the year-long average of approximately 77°.   **(c)** A quadratic model could not be used to predict temperatures beyond the given 12-month period, since temperature patterns are periodic, whereas parabolas are not.

37. (a) 78.10   **(b)** 7.48% Slightly less than the answer to Ex. 4. (This is expected because of compounding over eight years.)   **39.**

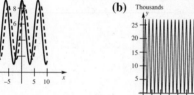

41. (a) Average voltage over $[0, \frac{1}{6}]$ is zero; 60 cycles per second.   **(b)**   **(c)** 116.673 volts.

Section 7.5 (page 569)

1. diverges   **3.** converges to $2e$ **5.** converges to e^2 **7.** converges to $\frac{1}{2}$ **9.** converges to $\frac{1}{108}$
11. converges to $3 \cdot 5^{2/3}$ **13.** diverges   **15.** diverges   **17.** converges to $\frac{5}{4}(3^{4/5} - 1)$ **19.** diverges
21. converges to 0 **23.** diverges   **25.** diverges   **27.** No; You will not sell more than 2,000 of them.
29. \$2,596.85 million   **31.** \$70,833   **33.** \$25,160 billion   **35.** $\int_0^{+\infty} q(t)\,dt$ diverges, indicating that there is no bound to the expected future exports of pork. $\int_{-\infty}^{0} q(t)\,dt$ converges to approximately 8.3490, indicating that total exports of pork prior to 1985 amounted to approximately \$8.3490 million.
37. (a) 2.467 meteors   **(b)** The integral diverges. We can interpret this as saying that the number of impacts by meteors smaller than 1 megaton is very large. (This makes sense because, for example, this number includes meteors no larger than a grain of dust.)   **39.**   **(a)** $\Gamma(1) = 1; \Gamma(2) = 1$
41. 1   **43.** 0.1586

Section 7.6 (page 579)

In the answers to Exercises 1–9, A and C denote arbitrary constants.

1. $y = \dfrac{x^3}{3} + \dfrac{2x^{3/2}}{3} + C$ **3.** $\dfrac{y^2}{2} = \dfrac{x^2}{2} + C$ **5.** $y = Ae^{x^2/2}$ **7.** $y = -\dfrac{2}{(x+1)^2} + C$ **9.** $y = \sqrt{(\ln|x|)^2 + D}$

11. $y = \dfrac{x^4}{4} - x^2 + 1$ **13.** $y = (x^3 + 8)^{1/3}$ **15.** $y = 2x$ **17.** $y = e^{x^2/2} - 1$ **19.** $y = -\dfrac{2}{\ln(x^2 + 1) + 2}$

21. With $s(t)$ = monthly sales after t months, $\dfrac{ds}{dt} = -0.05s$; $s = 1{,}000$ when $t = 0$. Solution: $s = 1{,}000e^{-0.05t}$.

23. With $S(t)$ = total sales after t months, $\dfrac{dS}{dt} = -0.1(100{,}000 - S)$; $S(0) = 0$. Solution: $S = 100{,}000(1 - e^{-0.1t})$.

27. $S = \dfrac{2/1999}{e^{-0.5t} + 1/1999}$ **29. (a)** $y = be^{Ae^{-at}}$, $A = $ constant   **(b)** $y = 10e^{-0.69315e^{-t}}$

Graph:

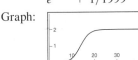

Graph:

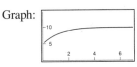

It will take about 27 months to saturate the market.

Chapter 7 Review Exercises (page 585)

1. $(x^2 - 2x + 4)e^x + C$ **3.** $(1/3)x^3 \ln 2x - x^3/9 + C$ **5.** $e - 2$ **7.** $(2e^3 + 1)/9$

9. $\dfrac{1}{6}\tan^{-1}\left(\dfrac{2x}{3}\right) + C$ **11.** $\dfrac{1}{2}\ln|2x + \sqrt{9 + 4x^2}| + C$

13. $\dfrac{x+1}{2}\sqrt{x^2 + 2x + 2} + \dfrac{1}{2}\ln|x + 1 + \sqrt{x^2 + 2x + 2}| + C$ **15.** $\frac{1}{2}\tan^{-1}(e^{2x}) + C$ **17.** 0.41817

19. 1.14779   **21.** $\frac{1}{4}$ **23.** $2\sqrt{e}$ **25.** diverges   **27.** diverges   **29.** 2   **31.** $\dfrac{1}{2}\left[\dfrac{3}{2^{1/3}} - 1\right]$

33. $3 - 4\sqrt{2} + e^2 + \dfrac{2}{e^2}$ **35.** $\dfrac{2\sqrt{2}}{3}$ **37.** 1   **39.** 1.01295   **41.** $e - 2$ **43.** 2.66229

45. 0.418171   **47.** 0.66799

49. $\bar{f}(x) = \dfrac{3}{10}\left[x^{5/3} - (x-2)^{5/3}\right]$

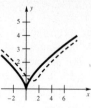

51. $\bar{f}(x) = \frac{1}{2}\left[x \ln x - (x-2)\ln(x-2) - 2\right]$

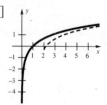

53. $y = -\dfrac{3}{x^3} + C$ **55.** $y = Ae^{(x+1)^2/2} - 1$

57. $\dfrac{y^2}{2} = x + C$ **59.** $y = \sqrt{2 \ln |x| + 1}$ **61.** $y = 2\sqrt{x^2 + 1}$ **63. (a)** Approximately 920. There were approximately 920,000 deaths resulting from motor vehicles in the period 1973–1992.   **(b)** Approximately 280. There were approximately 280,000 more deaths resulting from motor vehicles than guns in the period 1973–1992.
65. \$47,259.56   **67.** Equilibrium price = \$5, CS = \$82.11, PS = \$41.67   **69. (a)** 1,600. This is the total distance the cannonball has traveled.   **(b)** T = 20 secs. Area = 3,200   **71. (a)** Average = 322 pedestrians per year.
(b) 2-year moving average: **73.** \$9,705.91   **75.** $y = (100 + 0.05t)^2$

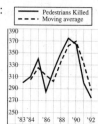

Chapter 8

Section 8.1 (page 597)

1. $f(0, 0) = 1; f(1, 0) = 1; f(0, -1) = 2; f(a, 2) = a^2 - a + 5; f(y, x) = y^2 + x^2 - y + 1;$
$f(x + h, y + k) = (x + h)^2 + (y + k)^2 - (x + h) + 1$ **3.** $f(0, 0) = \sqrt{5}; f(1, 0) = 2; f(0, -1) = \sqrt{10};$
$f(a, 2) = |a - 1|; f(y, x) = \sqrt{(y-1)^2 + (x-2)^2}; f(x + h, y + k) = \sqrt{(x + h - 1)^2 + (y + k - 2)^2}$
5. $g(0, 0, 0) = 1; g(1, 0, 0) = g(0, 1, 0) = e; g(z, x, y) = e^{x+y+z}; g(x + h, y + k, z + l) = e^{x+h+y+k+z+l}$
7. $g(0, 0, 0)$ does not exist; $g(1, 0, 0) = g(0, 1, 0) = 0; g(z, x, y) = xyz/(x^2 + y^2 + z^2);$
$g(x + h, y + k, z + l) = (x + h)(y + k)(z + l)/[(x + h)^2 + (y + k)^2 + (z + l)^2]$ **9.** $\sqrt{2}$ **11.** $\sqrt{a^2 + b^2}$
13. $\frac{1}{2}$ **15.** circle with center $(2, -1)$ and radius 3   **17.** circle with center $(-3, -2)$ and radius $\sqrt{6}$
19. 7,071,068 (to the nearest item)
21. (a) $100 = K(1,000)^a(1,000,000)^{1-a}; 10 = K(1,000)^a(10,000)^{1-a}$ **(b)** $\log K - 3a = -4; \log K - a = -3$
(c) $a = 0.5, K \approx 0.003162$ **(d)** $P = 71$ pianos (to the nearest piano)
23. $s(i, t) = 0.33i + 0.02t + 2.32$
25. (a) The demand function increases with increasing values of r.
(b) $Q(2 \times 10^8, 0.5, 500) = 90,680$. This means that if the total real income in Great Britain is 2×10^8 units of currency, and if the average retail price of beer is 0.5 units of currency per unit of beer, and if the average retail price of all other commodities is 500 units of currency, then 90,680 units of beer will be sold per year.

27. $U(11, 10) - U(10, 10) \approx 5.75$. This means that, if your company now has 10 copies of Macro Publish and 10 copies of Turbo Publish, then the purchase of one additional copy of Macro Publish will result in a productivity increase of approximately 5.75 pages per day.

29. (a) $(a, b, c) = (3, \frac{1}{4}, \frac{1}{\pi})$; $(a, b, c) = (\frac{1}{\pi}, 3, \frac{1}{4})$.

(b) $a = (3/4\pi)^{1/3}$. The resulting ellipsiod is a sphere with radius a.

31.

x	y	$f(x, y) = x^2\sqrt{1 + xy}$
-1	-1	$\sqrt{2}$
1	12	$\sqrt{13}$
0.3	0.5	0.096514
41	42	$1{,}681\sqrt{1{,}723}$

33.

x	y	$f(x, y) = x \ln(x^2 + y^2)$
3	1	$3 \ln 10$
1.4	-1	1.5193
e	0	$2e$
0	e	0

35. (a) 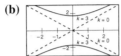 **(b)**  **(c)**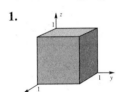

37. **(a)** 4.118×10^{-3} gram per square meter **(b)** the total weight of sulfates in the earth's atmosphere

39. **(a)** The value of N would be doubled. **(b)** $N(R, f_p, n_e, f_l, f_i, L) = R f_p n_e f_l f_i L$, where here L is the average lifetime of an intelligent civilization. **(c)** The function is not linear, since it involves a product of variables.
(d) By taking the logarithm of both sides, since this would yield the linear function
$\ln(N) = \ln(R) + \ln(f_p) + \ln(n_e) + \ln(f_l) + \ln(f_i) + \ln(L)$.

41. (a) The model predicts 8.06%. (The actual figure was 8.2%, showing the accuracy of the model.) **(b)** Foreign manufacturers, since each 1% gain of the market by foreign manufacturers decreases Chrysler's share by 0.8%—the largest of the three.

Section 8.2 (page 611)

1. 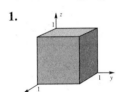   **3.** 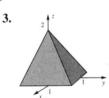   **5.** 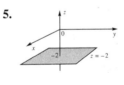   **7.**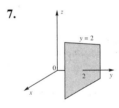

9. 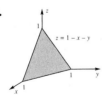   **11.** (H)   **13.** (B)   **15.** (F)   **17.** (C)

19. 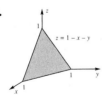   **21.** 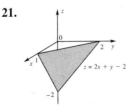   **23.** 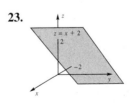   **25.**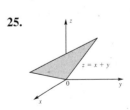

27.

29.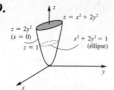

31.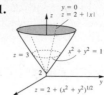

33.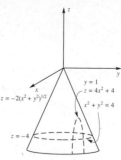

35.

37.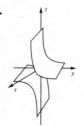

39.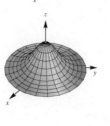

41.

The level curve at $z = 3$ has the form $3 = x^{0.5}y^{0.5}$, or $y = 9/x$, and shows the relationship between the number of workers and the operating budget at a production level of 3 units.

43.

The level curve at $z = 0$ consists of the nonnegative y-axis ($x = 0$) and tells us that zero utility corresponds to zero copies of Macro Publish, regardless of the number of copies of Turbo Publish. (Zero copies of Turbo Publish does not necessarily result in zero utility, according to the formula.)

45. Both slices are quarter-ellipses. (We only see the portion in the first quadrant because $e \geq 0$ and $k \geq 0$.) The level curve $C = 30,000$ represents the relationship between the number of electricians and the number of carpenters used in building a home that costs \$30,000. A similar relationship corresponds to for the level curve $C = 40,000$.

47. The following figure shows several level curves together with several lines of the form $h + w = c$.

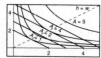

From the figure, thinking of the curves as contours on a map, we see that the largest value of A anywhere along any of the lines $h + w = c$ occurs midway along the line, when $h = w$. Thus, the largest-area rectangle with a fixed perimeter occurs when $h = w$ (that is, when the rectangle is a square).

Section 8.3 (page 621)

1. $f_x(x, y) = 6x + 1; f_x(1, -1) = 7$
$f_y(x, y) = -3y^2; f_y(1, -1) = -3$
$f_{xx}(x, y) = 6; f_{xx}(1, -1) = 6$
$f_{yy}(x, y) = -6y; f_{yy}(1, -1) = 6$
$f_{xy}(x, y) = f_{yx}(x, y) = 0; f_{xy}(1, -1) = f_{yx}(1, -1) = 0$

3. $f_x(x, y) = 6xy; f_x(1, -1) = -6$
$f_y(x, y) = 3x^2; f_y(1, -1) = 3$
$f_{xx}(x, y) = 6y; f_{xx}(1, -1) = -6$
$f_{yy}(x, y) = 0; f_{yy}(1, -1) = 0$
$f_{xy}(x, y) = f_{yx}(x, y) = 6x; f_{xy}(1, -1) = f_{yx}(1, -1) = 6$
5. $f_x(x, y) = 2xy^3 - 3x^2y^2 - y; f_x(1, -1) = -4$
$f_y(x, y) = 3x^2y^2 - 2x^3y - x; f_y(1, -1) = 4$
$f_{xx}(x, y) = 2y^3 - 6xy^2; f_{xx}(1, -1) = -8$
$f_{yy}(x, y) = 6x^2y - 2x^3; f_{yy}(1, -1) = -8$
$f_{xy}(x, y) = f_{yx}(x, y) = 6xy^2 - 6x^2y - 1; f_{xy}(1, -1) = f_{yx}(1, -1) = 11$
7. $f_x(x, y) = 6y(2xy + 1)^2; f_x(1, -1) = -6$
$f_y(x, y) = 6x(2xy + 1)^2; f_y(1, -1) = 6$
$f_{xx}(x, y) = 24y^2(2xy + 1); f_{xx}(1, -1) = -24$
$f_{yy}(x, y) = 24x^2(2xy + 1); f_{yy}(1, -1) = -24$
$f_{xy}(x, y) = f_{yx}(x, y) = 6(2xy + 1)^2 + 24xy(2xy + 1) = 6(6xy + 1)(2xy + 1);$
$f_{xy}(1, -1) = f_{yx}(1, -1) = 30$
9. $f_x(x, y) = e^{x+y}; f_x(1, -1) = 1$
$f_y(x, y) = e^{x+y}; f_y(1, -1) = 1$
$f_{xx}(x, y) = e^{x+y}; f_{xx}(1, -1) = 1$
$f_{yy}(x, y) = e^{x+y}; f_{yy}(1, -1) = 1$
$f_{xy}(x, y) = f_{yx}(x, y) = e^{x+y}; f_{xy}(1, -1) = f_{yx}(1, -1) = 1$
11. $f_x(x, y) = 3x^{-0.4}y^{0.4}; f_x(1, -1)$ not defined
$f_y(x, y) = 2x^{0.6}y^{-0.6}; f_y(1, -1)$ not defined
$f_{xx}(x, y) = -1.2x^{-1.4}y^{0.4}; f_{xx}(1, -1)$ not defined
$f_{yy}(x, y) = -1.2x^{0.6}y^{-1.6}; f_{yy}(1, -1)$ not defined
$f_{xy}(x, y) = f_{yx}(x, y) = 1.2x^{-0.4}y^{-0.6}; f_{xy}(1, -1)$ and $f_{yx}(1, -1)$ not defined
13. $f_x(x, y) = 4.92x^{0.2}e^{-0.2y}; f_x(1, -1) = 4.92e^{0.2}$
$f_y(x, y) = -0.82x^{1.2}e^{-0.2y}; f_y(1, -1) = -0.82e^{0.2}$
$f_{xx}(x, y) = -0.984x^{-0.8}e^{-0.2y}; f_{xx}(1, -1) = -0.984e^{0.2}$
$f_{yy}(x, y) = 0.164x^{1.2}e^{-0.2y}; f_{yy}(1, -1) = 0.164e^{0.2}$
$f_{xy}(x, y) = f_{yx}(x, y) = -0.984x^{0.2}e^{-0.2y}; f_{xy}(1, -1) = f_{yx}(1, -1) = -0.984e^{0.2}$
15. $f_x(x, y) = 0.2ye^{0.2xy}; f_x(1, -1) = -0.2e^{-0.2}$
$f_y(x, y) = 0.2xe^{0.2xy}; f_y(1, -1) = 0.2e^{-0.2}$
$f_{xx}(x, y) = 0.04y^2e^{0.2xy}; f_{xx}(1, -1) = 0.04e^{-0.2}$
$f_{yy}(x, y) = 0.04x^2e^{0.2xy}; f_{yy}(1, -1) = 0.04e^{-0.2}$
$f_{xy}(x, y) = f_{yx}(x, y) = 0.2(1 + 0.2xy)e^{0.2xy}; f_{xy}(1, -1) = f_{yx}(1, -1) = 0.16e^{-0.2}$
17. $f_x(x, y, z) = yz; f_x(0, -1, 1) = -1$
$f_y(x, y, z) = xz; f_y(0, -1, 1) = 0$
$f_z(x, y, z) = xy; f_z(0, -1, 1) = 0$
19. $f_x(x, y, z) = 4/(x + y + z^2)^2; f_x(0, -1, 1)$ not defined
$f_y(x, y, z) = 4/(x + y + z^2)^2; f_y(0, -1, 1)$ not defined
$f_z(x, y, z) = 8z/(x + y + z^2)^2; f_z(0, -1, 1)$ not defined
21. $f_x(x, y, z) = e^{yz} + yze^{xz}; f_x(0, -1, 1) = e^{-1} - 1$
$f_y(x, y, z) = xze^{yz} + e^{xz}; f_y(0, -1, 1) = 1$
$f_z(x, y, z) = xy(e^{yz} + e^{xz}); f_z(0, -1, 1) = 0$
23. $f_x(x, y, z) = 0.1\dfrac{y^{0.4}z^{0.5}}{x^{0.9}}; f_y(x, y, z) = 0.4\dfrac{x^{0.1}z^{0.5}}{y^{0.6}}; f_z(x, y, z) = 0.5\dfrac{x^{0.1}y^{0.4}}{z^{0.5}}; f_x(0, -1, 1)$ is not defined; $f_y(0, -1, 1)$
is not defined, $f_z(0, -1, 1)$ is not defined.
25. $f_x(x, y, z) = yze^{xyz}; f_y(x, y, z) = xze^{xyz}; f_z(x, y, z) = xye^{xyz};$
$f_x(0, -1, 1) = -1, f_y(0, -1, 1) = f_z(0, -1, 1) = 0$

27. $f_x(x, y, z) = 0$; $f_y(x, y, z) = -\dfrac{600z}{y^{0.7}(1 + y^{0.3})^2}$; $f_z(x, y, z) = \dfrac{2,000}{1 + y^{0.3}}$;
$f_x(0, -1, 1) = 0$; $f_y(0, -1, 1)$ is not defined; $f_z(0, -1, 1)$ is not defined.

29. $\partial c/\partial x = -0.8$, showing that Chrysler's percentage of the market decreases by 0.8% for every 1% rise in foreign manufacturers' share. $\partial c/\partial y = -0.2$, showing that Chrysler's percentage of the market decreases by 0.2% for every 1% rise in Ford's share. $\partial c/\partial z = -0.7$, showing that Chrysler's percentage of the market decreases by 0.7% for every 1% rise in G.M.'s share.

31. The marginal cost of cars is $6,000 + 1,000e^{-0.01(x+y)}$ per car. The marginal cost of trucks is $4,000 + 1,000e^{-0.01(x+y)}$ per truck. Both marginal costs decrease as production rises.

33. $\overline{C}(x, y) = \dfrac{200,000 + 6,000x + 4,000y - 100,000e^{-0.01(x+y)}}{x + y}$

$\overline{C}_x(50, 50) = -\$2.64$ per car. This means that at a production level of 50 cars and 50 trucks per week, the average cost per car is decreasing by $2.64 for each additional car manufactured. $\overline{C}_y(50, 50) = -\$22.64$ per truck. This means that at a production level of 50 cars and 50 trucks per week, the average cost per truck is decreasing by $22.64 for each additional truck manufactured.

35. No; your marginal revenue from the sale of cars is $15,000 - 2,500/\sqrt{x + y}$ per car and $10,000 - 2,500/\sqrt{x + y}$ per truck from the sale of trucks. These increase with increasing x and y. In other words, you will earn more revenue per vehicle with increasing sales, and so the rental company will pay more for each additional vehicle it buys.

37. $P_z(10, 100,000, 1,000,000) \approx 0.0001010$ papers/$

39. **(a)** $U_x(10, 5) = 5.18$, $U_y(10, 5) = 2.09$. This means that if 10 copies of Macro Publish and 5 copies of Turbo Publish are purchased, the company's daily productivity is increasing at a rate of 5.18 pages per day for each additional copy of Macro purchased and by 2.09 pages per day for each additional copy of Turbo purchased. **(b)** $U_x(10, 5)/U_y(10, 5) \approx 2.48$ is the ratio of the usefulness of one additional copy of Macro to one of Turbo. Thus, with 10 copies of Macro and 5 copies of Turbo, the company can expect approximately 2.48 times the productivity per additional copy of Macro than Turbo.

41. 6×10^9 n/s

43. **(a)** $A_P(100, 0.1, 10) = 2.59$; $A_r(100, 0.1, 10) = 2,357.95$; $A_t(100, 0.1, 10) = 24.72$. Thus, for a $100 investment at 10% interest invested for 10 years, the accumulated amount is increasing at a rate of $2.59 per $1 of principal, at a rate of $2,357.95 per increase of 1 in r (note that this would correspond to an increase in the interest rate of 100%), and at a rate of $24.72 per year. **(b)** $A_P(100, 0.1, t)$ tells you the rate at which the accumulated amount in an account bearing 10% interest with a principal of $100 is growing per $1 increase in the principal, t years after the investment.

45. **(a)** $P_x = Ka(y/x)^b$ and $P_y = Kb(x/y)^a$. They are equal precisely when $a/b = (x/y)^b(x/y)^a$. Substituting $b = 1 - a$ now gives $a/b = x/y$. **(b)** The given information implies that $P_x(100, 200) = P_y(100, 200)$. By part (a), this occurs precisely when $a/b = x/y = \frac{100}{200} = \frac{1}{2}$. But $b = 1 - a$, so $a/(1 - a) = \frac{1}{2}$, giving $a = \frac{1}{3}$ and $b = \frac{2}{3}$.

47. Decreasing at 0.007457 cc/s

Section 8.4 (page 632)

1. local maximum at P and R, saddle point at Q **3.** critical point (neither saddle point nor local extremum) at Q, non-critical point at P and R **5.** non-critical point at P and Q, saddle point at R

7.

9. minimum of 1 at $(0, 0)$ **11.** maximum of $\frac{3}{2}$ at $(-\frac{1}{2}, \frac{1}{2})$
13. maximum of 0 at $(0, 0)$, saddle points at $(\pm 4, 2, -16)$
15. minimum of 1 at $(0, 0)$

17. minimum of -16 at $(-2, \pm 2)$; $(0, 0)$ a critical point which is not a local extremum
19. saddle point at $(0, 0, -1)$ **21.** maximum of e at $(-1, 0)$ **23.** minimum of $3(2^{2/3})$ at $(2^{1/3}, 2^{1/3})$
25. minimum of 4 at both $(1, 1)$ and $(-1, -1)$ **27.** 400 5-speeds and 2,000 10-speeds

29. $\overline{C}_x = \partial/\partial x[C/(x + y)] = [(x + y)C_x - C]/(x + y)^2$. This is zero when $(x + y)C_x = C$, or $C_x = \dfrac{C}{x + y} = \overline{C}$.
Similarly, $\overline{C}_y = 0$ when $C_y = \overline{C}$. This is reasonable because if the average cost is decreasing with increasing x, then the average cost is greater than the marginal cost C_x. Similarly, if the average cost is increasing with increasing x, the average cost is less than the marginal cost C_x. Thus, if the average cost is stationary with increasing x, then the average cost equals the marginal cost C_x. (The situation is similar for the case of increasing y.)
31. They should charge \$580.81 for the Ultra Mini and \$808.08 for the Big Stack.

Section 8.5 (page 643)

1. maximum value of 8 at $(2, 2)$, minimum value of 0 at $(0, 0)$ **3.** maximum value of 9 at $(-2, 0)$, minimum value of 0 at $(1, 0)$ **5.** maximum value of e^4 at $(0, \pm 2)$, minimum value of 1 at $(0, 0)$ **7.** maximum value of e^4 at $(\pm 1, 0)$, minimum value of 1 at $(0, 0)$ **9.** maximum value of 5 at $(\frac{1}{2}, \frac{1}{2})$, minimum value of 3 at $(1, 1)$
11. maximum value of $\frac{161}{9}$ at $(1, 9)$ and $(9, 1)$, minimum value of 17 at $(1, 1)$ **13.** maximum value of 8 at $(2, 0)$, minimum value of -1 at $(-1, 0)$ **15.** maximum value of 8 at $(2, 0)$, minimum value of 0 at $(0, 0)$
17. $(1/\sqrt{3}, 1/\sqrt{3}, 1/\sqrt{3})$, $(-1/\sqrt{3}, -1/\sqrt{3}, 1/\sqrt{3})$, $(1/\sqrt{3}, -1/\sqrt{3}, -1/\sqrt{3})$, $(-1/\sqrt{3}, 1/\sqrt{3}-1/\sqrt{3})$
19. For minimum cost of \$16,600, make 100 5-speeds and 80 10-speeds. For maximum cost of \$17,400, make 100 5-speeds and 120 10-speeds.   **21.** For a maximum profit of \$4,500, sell 50 copies of Walls and 150 copies of Doors. For a minimum profit of \$0, sell nothing.   **23.** hottest point: $(1, 1)$; coldest point: $(0,0)$
25. hottest points: $(-1/2, \pm\sqrt{3}/2)$; coldest point: $(1/2, 0)$ **27.** $(-\frac{5}{9}, \frac{5}{9}, \frac{25}{9})$ **29.** $(0, \frac{1}{2}, -\frac{1}{2})$
31. $1 \times 1 \times 2$ **33.** $(2l/h)^{1/3} \times (2l/h)^{1/3} \times 2^{1/3}(h/l)^{2/3}$, where l = cost of lightweight cardboard, and h = cost of heavy-duty cardboard per square foot   **35.** 11,664 cubic inches (18 inches $\times$ 18 inches $\times$ 36 inches)
37. $1 \times 1 \times 1/2$ **39.** Produce 150 quarts of vanilla and 100 quarts of mocha, for a profit of \$325.   **41.** Offer 100 sections of Finite Math and no sections of Calculus.   **43.** Use 7.280 servings of Mixed Cereal and 1.374 servings of Tropical Fruit Dessert.   **45.** Buy 100 of each.

Section 8.6 (page 656)

1. $y = 3x/2 - 2/3$ 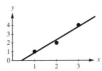   **3.** $y = 0.4118x + 0.9706$ 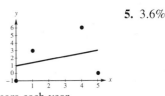   **5.** 3.6%

7. 21.3 billion barrels   **9.** Life expectancy was increasing by 0.291 years each year.
11. $P = 101.08t - 330.33$; $P(13) = \$984$ million.   **13.** \$6755 billion   **15.** $P = 5,835.53e^{0.020769t}$ (t = time since 1790); $P(2,000) = 457,353,000$.   **17.** Linear function; $y = 217.69t + 718.36$ (t = time in years since 1980). High in 2000: 5,072.   **19.** $A = 40.65e^{0.0605t}$ (t = time in years since 1970). Index in the year $2000 = A(30) = 249.63$.   **21.** $q = 3803p^{-2.058}$

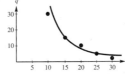

23. Taking the natural log of both sides gives $\ln P = \ln K + a \ln x + (1 - a) \ln y$. This leads to the linear relationship $\ln (P/y) = \ln K + a \ln (x/y)$. Thus, given a number of data points (x_i, y_i, P_i), we can use them to calculate the best-fit linear relationship between $\ln (P/y)$ and $\ln (x/y)$. The slope is then a and the intercept is $\ln K$.

Section 8.7 (page 670)

1. $-\frac{1}{2}$ **3.** $e^2/2 - 7/2$ **5.** $(e^3 - 1)(e^2 - 1)$ **7.** $\frac{7}{6}$ **9.** $\frac{1}{2}[e^3 - e - e^{-1} + e^{-3}]$ **11.** $\frac{1}{2}$
13. $\frac{1}{2}(e - 1)$ **15.** $\frac{45}{2}$ **17.** $\frac{8}{3}$ **19.** $\frac{4}{3}$ **21.** 0   **23.** $\frac{2}{3}$ **25.** 2   **27.** 1   **29.** 0   **31.** $\frac{1}{3}$
33. $\int_0^1 \int_0^{1-x} f(x, y)\, dy\, dx$

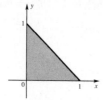

35. $\int_0^1 \int_{x^2-1}^1 f(x, y)\, dy\, dx$

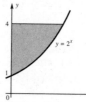

37. $\int_1^4 \int_1^{2/\sqrt{y}} f(x, y)\, dx\, dy$

39. $\int_1^4 \int_0^{\ln_2 y} f(x, y)\, dx\, dy$

41. $\frac{4}{3}$ **43.** $\frac{1}{6}$ **45.** 161,781 gadgets   **47.** Maximum revenue is $375,500. Minimum revenue is $256,000. Average revenue is $312,750.   **49.** Maximum revenue is $20,000. Minimum revenue is $15,000. Average revenue is $17,500.   **51.** 8,216   **53.** 1 degree

Chapter 8 Review Exercises (page 678)

1. $g(0, 0, 0) = 0$, $g(1, 0, 0) = 1$, $g(0, 1, 0) = 0$, $g(x, x, x) = x^3 + x^2$, $g(x, y + k, z) = x(y + k)(x + y + k - z) + x^2$ **3.** $g(0, 0, 0) = 0$, $g(1, 0, 0) = 1$, $g(0, 1, 0) = 0$, $g(x, x, x) = xe^{x^2+x}$ $g(x, y + k, z) = xe^{x(y+k)+z}$ **5.** $2\sqrt{2}$
7. $|b - c|$ **9.** 2,511,886   **11.** $f_x = 2x + yz$, $f_y = xz$, $f_z = xy$, $f_{xy} = z$, $f_{xz} = y$, $f_{zz} = 0$
13. $f_x = (-x^2 + y^2 + z^2)/(x^2 + y^2 + z^2)^2$, $f_y = -2xy/(x^2 + y^2 + z^2)^2$, $f_z = -2xz/(x^2 + y^2 + z^2)^2$, $f_x(0, 1, 0) = 1$ **15.** 0   **17.** decreasing by 1.02266×10^{-12} newtons per second   **19.** absolute minimum at $(1, \frac{3}{2})$ **21.** saddle point at $(1, 0)$ **23.** saddle point at $(0, 0)$ **25.** critical point at $(0, 0)$ (neither a local extremum nor a saddle point)   **27.** absolute maximum at each point on the circle $x^2 + y^2 = 1$ **29.** $(0, 2, \sqrt{2})$
31. $(1, 0, 1)$ **33.** coldest point: $(1, 0)$; hottest point: $(3, 0)$ **35.** Producing 1,000 pencils and 1,200 pens costs the least. Producing 900 or 1,100 pencils and 1,500 pens costs the most.   **37.** $y = -x/2 + (5/6)$
39. $y = -0.1471x + 1.6176$ **41.** $P = 0.16037e^{0.0013240t}$. The fit is a poor one. $P(2,000) \approx 2.27$ billion, $P(2050) \approx 2.42$ billion.   **43.** $\frac{4}{15}(5^{5/2} - 3^{5/2} - 32 + 2^{5/2})$ **45.** $\frac{1}{2}\ln 5$ **47.** 2
49. $\frac{2}{15}(5^{5/2} - 3^{5/2} - 32 + 2^{5/2})$ **51.** $\frac{1}{4}\ln 5$ **53.** $\frac{2}{\pi}$ **55.** $\frac{40}{3}$ **57.** $230.50

▶ ## Appendix A

Section A.1 (page A-5)

1. -48 **3.** -1 **5.** 1   **7.** 33   **9.** $\frac{5}{18}$ **11.** 6   **13.** $\frac{43}{16}$ **15.** $3*(2-5)$ **17.** $3/(2-5)$
19. $(3-1)/(8+6)$ **21.** $3-(4+7)/8$ **23.** $(2/3)/5$ **25.** $3^\wedge(4-5)*6$ **27.** $3*(1+4/100)^\wedge(-3)$
29. $3(1-(-1/2)^\wedge 2)^\wedge 2+1$

Section A.2 (page A-10)

1. 27   **3.** -36 **5.** $\frac{4}{9}$ **7.** $-\frac{1}{8}$ **9.** 16   **11.** 2   **13.** 32   **15.** 2   **17.** x^5 **19.** $-\dfrac{y}{x}$ **21.** $\dfrac{1}{x}$

23. $x^3 y$ **25.** $\dfrac{z^4}{y^3}$ **27.** $\dfrac{x^6}{y^6}$ **29.** $\dfrac{x^4 y^6}{z^4}$

Section A.3 (page A-15)

1. 2   **3.** $\frac{1}{2}$ **5.** $\frac{4}{3}$ **7.** $\frac{2}{5}$ **9.** 3.606   **11.** $\frac{3}{2}$ **13.** 2   **15.** 2   **17.** ab **19.** $x + 9$ **21.** $x\sqrt[3]{a^3 + b^3}$

23. $\dfrac{2y}{\sqrt{x}}$ **25.** $3^{1/2}$ **27.** $x^{3/2}$ **29.** $(xy^2)^{1/3}$ **31.** $\dfrac{x^2}{x^{1/2}}$ **33.** $\sqrt[3]{2^2}$ **35.** $\sqrt[3]{x^4}$ **37.** $\sqrt[5]{\sqrt{x}\sqrt[3]{y}}$ **39.** 64

41. $\sqrt{3}$ **43.** $\dfrac{1}{x}$ **45.** xy **47.** $\left(\dfrac{y}{x}\right)^{1/3}$ **49.** ± 4 **51.** $\pm \frac{2}{3}$ **53.** $-1, -\frac{1}{3}$ **55.** -2

Section A.4 (page A-18)

1. $4x^2 + 6x$ **3.** $2xy - y^2$ **5.** $x^2 - 2x - 3$ **7.** $2y^2 + 13y + 15$ **9.** $4x^2 - 12x + 9$ **11.** $x^2 + 2 + \dfrac{1}{x^2}$

13. $4x^2 - 9$ **15.** $y^2 - \dfrac{1}{y^2}$ **17.** $2x^3 + 6x^2 + 2x - 4$ **19.** $x^4 - 4x^3 + 6x^2 - 4x + 1$ **21.** $y^5 + 4y^4 + 4y^3 - y$

Section A.5 (page A-22)

1. $x(2 + 3x)$ **3.** $2x^2(3x - 1)$ **5.** $(x - 1)(x - 7)$ **7.** $(x - 3)(x + 4)$ **9.** $(2x + 1)(x - 2)$
11. $(2x + 3)(3x + 2)$ **13.** $(3x - 2)(4x + 3)$ **15.** $(x + 2y)^2$ **17.** $(x^2 - 1)(x^2 - 4)$
19. $x = 0, -\frac{2}{3}$ **21.** $x = 0, \frac{1}{3}$ **23.** $x = 1, 7$ **25.** $x = 3, -4$ **27.** $x = -\frac{1}{2}, 2$ **29.** $x = -\frac{3}{2}, -\frac{2}{3}$
31. $x = \frac{2}{3}, -\frac{3}{4}$ **33.** $x = -2y$ **35.** $x = \pm 1, \pm 2$ **37.** $(x + 1)(2x + 5)$
39. $(x^2 + 1)^5(x + 3)^3(x^2 + x + 4)$ **41.** $-x^3(x^3 + 1)\sqrt{x + 1}$ **43.** $(x + 2)\sqrt{(x + 1)^3}$

Section A.6 (page A-25)

1. $\dfrac{2x^2 - 7x - 4}{x^2 - 1}$ **3.** $\dfrac{3x^2 - 2x + 5}{x^2 - 1}$ **5.** $\dfrac{x^2 - x + 1}{x + 1}$ **7.** $\dfrac{x^2 - 1}{x}$ **9.** $\dfrac{2x - 3}{x^2 y}$ **11.** $\dfrac{(x + 1)^2}{(x + 2)^4}$

13. $\dfrac{-1}{\sqrt{(x^2 + 1)^3}}$ **15.** $\dfrac{-(2x + y)}{x^2(x + y)^2}$

Section A.7 (page A-29)

1. -1 **3.** $-\frac{3}{2}$ **5.** -1 **7.** 1   **9.** $-\dfrac{b}{2}$ **11.** $-\dfrac{b}{a}$ **13.** ± 1 **15.** 0   **17.** ± 2 **19.** 1

21. $-1, -6$ **23.** no solutions   **25.** no solutions

Section A.8 (page A-40)

1. $x = 5$ **3.** $x = -\dfrac{4}{3}$ **5.** $x = \dfrac{(c - b)}{a}$ **7.** $x = -4, \frac{1}{2}$ **9.** no solutions   **11.** $x = \pm\sqrt{\dfrac{5}{2}}$

13. $x = -1$ **15.** $x = -1, 3$ **17.** $x = \dfrac{1 + \sqrt{5}}{2}, \dfrac{1 - \sqrt{5}}{2}$ **19.** $x = 1$ **21.** $x = \pm 1, \pm 3$

23. $x = \pm\sqrt{\dfrac{-1 + \sqrt{5}}{2}}$ **25.** $x = -1, -2, -3$ **27.** $x = -3$ **29.** $x = 1$ **31.** $y = -2$

33. $x = 1, \pm\sqrt{5}$ **35.** $x = \pm 1, \pm\dfrac{1}{\sqrt{2}}$ **37.** $x = -2, -1, 2, 3$

Section A.9 (page A-46)

1. $x = 0, 3$ **3.** $x = \pm\sqrt{2}$ **5.** $x = -1, -\frac{5}{2}$ **7.** $x = -3$ **9.** $x = 0, -1$ **11.** $x = -1$ ($x = -2$ is not a solution.)   **13.** $x = -2, -\frac{3}{2}, -1$ **15.** $x = -1$ **17.** $x = \pm\sqrt[4]{2}$ **19.** $x = \pm 1$ **21.** $x = \pm 3$ **23.** $x = \frac{2}{3}$ **25.** $x = -4, -\frac{1}{4}$

Subject Index

Company Index